CONCEPTS

Solution of a differential equation is a function that renders it an identity when substituted for the unknown (the *dependent* variable). A solution of an initial-value problem satisfies the differential equation and initial condition(s). Solutions of most initial-value problems *exist* and are *unique*; i.e., solution graphs do not cross.

A *homogeneous* differential equation has the trivial (identically zero) solution.

Linear differential equation: Unknown appears only in "line-like" terms, as in $a_2(x)y'' + a_1(x)y' + a_0(x)y = f(x)$.

Superposition: Linear combinations of linear differential equations are solved by linear combinations of their solutions.

Linearly independent functions: e.g., y_1, y_2 are if $C_1 y_1(x) + C_2 y_2(x) = 0$ throughout an interval only if $C_1 = C_2 = 0$.

General solution of linear differential equation: A particular solution of nonhomogeneous equation plus linear combination of linearly independent solutions of homogeneous equation; e.g., for second-order differential equation, $y_g = y_p + C_1 y_1 + C_2 y_2$. Use initial or boundary conditions to determine arbitrary constants.

Steady state or *equilibrium*: A constant solution.

Stable steady state: Small perturbations of steady state decay to zero as $t \to \infty$.

Concepts: page references

	1st-order DE	2nd-order DE	1st-order system
Solution of			
DE, IVP	10, 134	233	366
BVP	—	438	—
IBVP		Partial DE, p. 463	
Existence and			
uniqueness	183–194	245*	425*
Homogen. DE	135	234	369
Linear DE	136, 138	234	369
Superposition	139	242	379*
Linear indep.	—	237	371
Lin. ind. test	—	253	372
General soln.	143	238–239	373–374
Stability	45, 118	245	414–415

Abbreviations: Differential equation (DE), initial-value problem (IVP), boundary-value problem (BVP), initial-boundary value problem (IBVP). * See exercise, project, or remark.

ANALYTIC TOOLS

Separation of variables: For $dy/dt = Y(y)T(t)$, integrate $dy/Y(y) = dt/T(t)$ with $T, Y \neq 0$; e.g., example 12, p. 17. For linear equations:

- Find a pa[rticular solution formula] [variati]on of pa-*rameters*: [...] [assu]med from homogen[...] [...] varying"; e.g., for f[...] [...]; then substitute, so[...] [...]42, p. 180.

For linear, constant-coe[...] [...]

- Find homogeneous solution via *characteristic equation*: Assume $y_h = e^{rt}$ (scalar equation) or $\mathbf{y}_h = \mathbf{p}e^{rt}$ (system). Substitute, solve characteristic equation for r. Either exponentials with distinct r values or real and imaginary parts of complex-valued exponentials form linearly independent solutions; e.g., example 24, p. 270, or example 15, p. 394.

- Find particular solution via *undetermined coefficients*: Particular solution has form similar to polynomial, exponential, or sine/cosine forcing term; e.g., assume $y_p = A \cos pt + B \sin pt$ if forcing term includes $\cos pt$ and/or $\sin pt$. Substitute, equate coefficients of like terms to find undetermined coefficients; e.g., example 40, p. 300.

Solution behavior observed from:

- *Characteristic equation* includes time constant for growth or decay (p. 119), half-life (exercise 9, p. 34); undamped (p. 272), overdamped (p. 269), underdamped (p. 271), and critically damped (p. 270) oscillations; natural frequency (p. 278).

- *Undetermined coefficients* include resonance, resonant frequency (p. 309, 313); beating (p. 312).

Convert a higher-order equation to an *equivalent system* by introducing one new variable for each derivative; e.g., example 1, p. 367.

Linearized stability analysis: $y(t) = y_{ss} + p(t)$. Assume perturbation p is small initially. Then substitute y, expand nonlinear terms in Taylor series, approximate by discarding powers of p, and analyze the (approximate) linear equation for growth or decay of p; e.g., example 47, p. 323.

Power series about ordinary point x_0: Assume $y = \sum_n c_n(x - x_0)^n$, substitute, and equate powers of $x - x_0$ to find c_n; e.g., example 11, p. 590.

Method of Frobenius about regular singular point x_0: Assume $y = (x - x_0)^r \sum_n c_n(x - x_0)^n$, substitute, and equate powers of $x - x_0$ to find c_n after requiring $c_0 \neq 0$ to find r; e.g., example 17, p. 606.

(Continued on back endpaper)

Differential Equations
Modeling with MATLAB

Differential Equations
Modeling with MATLAB

Paul W. Davis
Worcester Polytechnic Institute

PRENTICE HALL
Upper Saddle River, NJ 07458

Library of Congress Cataloging-in-Publication Data

DAVIS, PAUL W.
   Differential equations: modeling with MATLAB / Paul W. Davis.
   p. cm.
   Includes bibliographic references and index.
   ISBN 0-13-736539-X
   1. Differential equations—Data processing. 2. MATLAB. I. Title.
   QA371.5.D37 D38 1999
   515'.35'0285--dc21   99-18101
                                                                 CIP

Acquisitions editor: George Lobell
Editor in chief: Jerome Grant
Senior managing editor: Linda Mihatov Behrens
Executive managing editor: Kathleen Schiaparelli
Assistant Vice President of production and manufacturing: David W. Riccardi
Director of marketing: Melody Marcus
Marketing assistant: Amy Lysik
Manufacturing manager: Trudy Pisciotti
Manufacturing buyer: Alan Fischer
Editorial assistants: Gale Epps, Nancy Bauer
Director of creative services: Paula Maylahn
Associate creative director: Amy Rosen
Art manager: Gus Vibal
Art editor: Grace Hazeldine
Art director: Maureen Eide
Assistant to art director: John Christiana
Interior/cover designer: Joseph Sengotta

Printed in the United States of America

10 9 8 7 6 5 4 3 2

ISBN 0-13-736539-X

Prentice-Hall International (UK) Limited, *London*
Prentice-Hall of Australia Pty. Limited, *Sydney*
Prentice-Hall Canada, Inc., *Toronto*
Prentice-Hall Hispanoamericana, S.A., *Mexico*
Prentice-Hall of India Private Limited, *New Delhi*
Prentice-Hall of Japan, Inc., *Tokyo*
Simon & Schuster Asia Pte. Ltd., *Singapore*
Editora Prentice-Hall do Brasil, Ltda., *Rio de Janeiro*

To Sharon

Contents

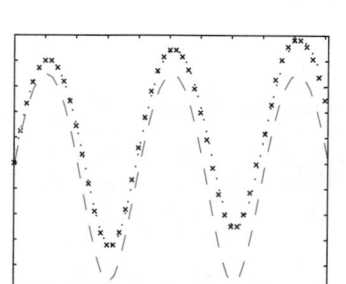

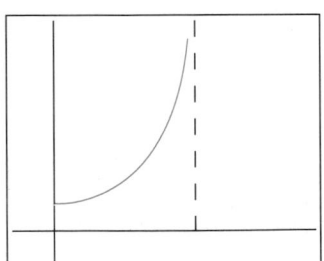

5 ■ TWO-DIMENSIONAL MODELS: OSCILLATING SYSTEMS   203

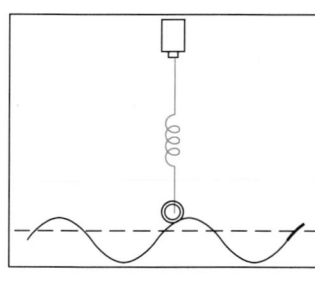

6 ■ ANALYTIC TOOLS FOR TWO DIMENSIONS   233

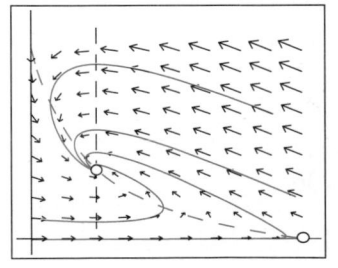

7 ■ GRAPHICAL TOOLS FOR TWO DIMENSIONS   334

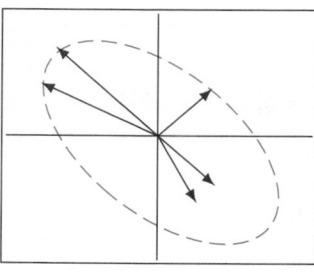

8 ▪ ANALYTIC TOOLS FOR HIGHER DIMENSIONS   365

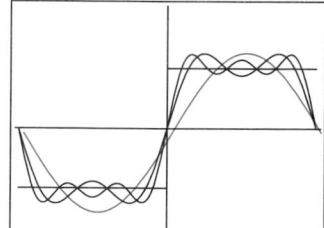

9 ▪ DIFFUSION MODELS AND BOUNDARY-VALUE PROBLEMS   428

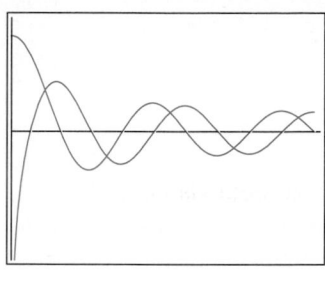

APPENDICES   621

BIBLIOGRAPHY   642

SOLUTIONS TO SELECTED EXERCISES   644

INDEX   679

Preface

This text introduces differential equations from four perspectives, physical, analytical, numerical, and graphical. Differential equations are derived to model physical phenomena. Information about those phenomena is extracted from the differential equations with analytical, numerical, and graphical tools. The cycle is completed by using the results of the analysis to better understand the phenomena being modeled.

Those explorations are supported by DELAB, a graphical user interface that provides convenient access to MATLAB's numerical and graphical power as well as to its symbolic toolbox. (MATLAB is a powerful, high-level language for scientific computing and visualization that includes a special student edition.) Users can choose the tools—say, an analytic solution formula or a set of numerical solution values or a phase plane graph—that best answer the questions at hand. The DELAB software is available from **www.wpi.edu/~pwdavis/DELab**.

The overriding goal of this text is developing the ability to use mathematical ideas to provide answers to the difficult open-ended questions that are part of professional practice in science and engineering. In pursuit of that goal, this text reinforces many of the central ideas of calculus, and it introduces the fundamental concepts of differential equations, all the while challenging the reader to think in four different frames, the physical, analytical, numerical, and graphical.

TO THE READER

This text is filled with guideposts to your reading. Important terms are often introduced intuitively, first identified in *italics*, then shown in **bold face** when they are defined more carefully. Labels in the margins also highlight important concepts. Use the explanatory material that accompanies these definitions to form your own intuitive framework for each concept.

This text, like most books in mathematics and science, requires several readings for full understanding. On the first reading, look for the overall structure and for the key points that are included in summaries. Take details for granted.

On your second or third reading, begin answering (with pencil and paper as needed) the questions and challenges labeled Stop and Think These test your understanding and improve your mastery of the material.

The computational and visual power of MATLAB is an important component of learning about differential equations. Suggestions and instructions for its use, primarily through the package DELAB, are scattered throughout the text. You will see that DELAB is organized around three mathematical perspectives, analytical, numerical, and graphical. Work near a computer when you can so that you can explore new ideas from each perspective.

Be alert for patterns. For example, …

- Models are derived by coupling a physical law (perhaps Newton's law $F = ma$ or a conservation law) with experimental observations, perhaps a constitutive law like Hooke's law $F = kz$.

- Definitions of basic terms like *solution*, *homogeneous*, *linear*, etc. are presented successively for first-order equations, second-order equations, and systems. Find the similarities among each of these cases to see the pattern of generalization.

- Most solution methods are introduced in a simple first-order setting, then generalized to higher-order equations and systems of first-order equations.

- Ideas are explored from as many as four perspectives, the physical, analytical, numerical, and graphical. Explore each perspective, being conscious of which is in view at any given time.

Marginal notes highlight the common patterns of analysis.

The exercises for each section contain an exercise guide, a table that lists those exercises that use particular ideas developed in that section. Use it to select exercises to practice the topics that trouble you most. Review the topics listed in the exercise guide to check your mastery of the concepts in that section.

Summaries appear at key points throughout the text. Read them carefully to consolidate your understanding of the material. Consult them when you need help separating the forest from the trees.

Notes in smaller type are scattered throughout the text to provide additional information. Skip them on your first reading, but return to them later.

Figures, tables, examples, and referenced equations are numbered consecutively within each chapter. For example, equation (5.1) is the first numbered equation of chapter 5. Citations to the bibliography at the end of the text are enclosed in square brackets; e.g., [9].

TO THE INSTRUCTOR

Suggestions and support for using this text, including several "road maps" for possible course organization as well as additional exercises and test questions, are available from `www.wpi.edu/~pwdavis/ModelingWithMatlab`. Comments, questions, and suggestions are welcome; please send them to `pwdavis@wpi.edu`.

ACKNOWLEDGMENTS

Many friends, colleagues, and students—happily, there is much overlap among those sets—have contributed to this book. I offer them all my grateful thanks. At the risk of omitting some, they include: from WPI, Joseph Fehribach, Arthur Heinricher, Roger Lui, Peter Schultz, Dalin Tang, Brian Ball, Nathan

Gibson, Jae Lee, and Casey Richardson; Douglas Borden (US Coast Guard Academy), James Lang (Valencia Community College), P.J. McKenna (University of Connecticut), Cleve Moler (The MathWorks, Inc.), Marie Planchard (Massachusetts Bay Community College), Paul Porch (Mt. Hood College); the reviewers of this manuscript, including Mohammad Tavakoli (Chaffey College), Michael Kirby (Colorado State University), Joan Remski (University of Michigan – Dearborn), Thomas T. Read (Western Washington University), Hendrik J. Kuiper (Arizona State University), Lee Johnson (Virginia Polytechnic Institute and State University), Mikhail Shvartsman (University of St. Thomas).

George Lobell of Prentice Hall is a wise and far-sighted editor. Nick Romanelli, Dennis Kletzing, and their colleagues are responsible for the creative and exacting production of this text. Working with all of them has been a pleasure.

Above all, Sharon deserves more than I can say.

Paul Davis

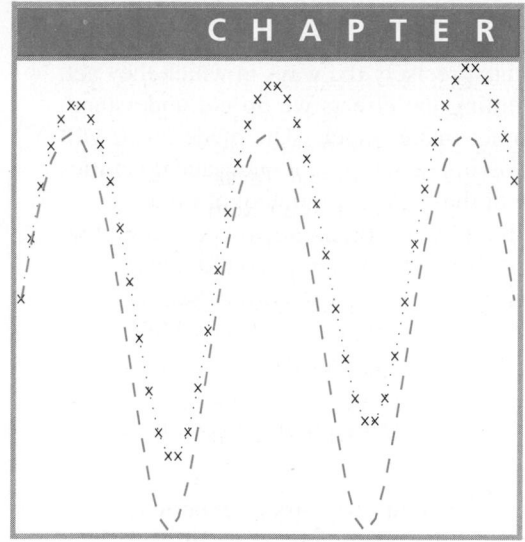

1

Prologue

1.1 ▪ GOALS

Differential equations constitute a language that scientists and engineers use to make careful mathematical statements about the problems they confront. This text is an introduction to that language, which emphasizes understanding rather than the formalities of grammar. It discusses three major issues:

1. *Formulating a model*, using differential equations;
2. *Analyzing the model*, both by solving the differential equation and by extracting qualitative information about the solution from the differential equation;
3. *Interpreting the analysis* in light of the physical setting modeled in step 1.

It aims to develop skill with solution techniques as well as judgment and sensitivity in using differential equations to understand complex physical phenomena.

1.2 ▪ A MODELING EXAMPLE

What factors affect the acceleration, velocity, and displacement of a rock thrown vertically upward?

This simple question may not have a simple answer. Does the weight of the rock matter? Gravity is obviously important, but what about wind and air resistance? Does the shape of the rock matter? The speed with which the rock leaves our hand ought to affect how high it goes. Could a long wind-up or a sharp whack with a bat force it higher, even though its starting speed is the same?

A mathematical **model** allows us to make a list of just which effects will be considered, and it permits us to state precisely the ways in which they act. We can simplify the problem by omitting the effects we do not understand, can not describe clearly, or think do not matter much. The model itself will be a set of mathematical relations involving quantities we know and quantities we do not know, such as the velocity of the rock as a function of time.

Intuition suggests that gravity should be the predominant factor affecting the rock's motion, so we will try to include it and omit everything else, such as wind, the shape of the rock, etc. If we are wrong, the error should appear in one of two ways: The model will not be complete without adding more information, or the model will predict something that counters our physical experience.

To derive a model in this setting, we have two tools at our disposal: a physical law and an experimental fact.

Newton's law.   The total force F_t on a body of mass m experiencing acceleration a is the product of mass and acceleration: $F_t = ma$.

Experimental fact.   The force of gravity on a body is proportional to its mass.

Both *law* and *fact* are misused. A *law* is usually a postulate that has repeatedly led to mathematical models predicting physically reasonable behavior. A *fact* is a summary of physical observations that are sufficiently accurate for the purpose at hand. Such a "fact" is inevitably a simplification whose validity depends upon the circumstances. For example, the force of gravity actually varies with location on the earth's surface and with distance above the surface of the earth, complications we ignore in a first attempt at modeling.

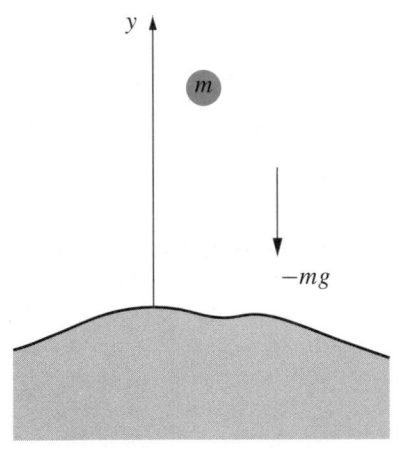

FIGURE 1.1   A coordinate system for the rock model. This orientation means positive velocities are upwards, negative are downwards.

Newton's law is given in an unambiguous mathematical statement, but the experimental fact is not. To restate the latter in mathematical terms, we need notation and agreement on a coordinate system to determine the signs of forces and the like. Borrowing the notation of Newton's law, taking "up" as the positive direction, and letting g be the (positive) constant of proportionality, the experimental fact can be written as

$$F_g = -mg,$$

where F_g is the force due to gravity acting on a body of mass m. The minus sign enters because of the up-is-positive orientation chosen in figure 1.1.

Neglecting air resistance and other troublesome effects means that the force of gravity is the only force acting on the rock. Hence, the total force on the rock is just the force of gravity; i.e., $F_t = F_g$ or, when m is canceled,

$$
\begin{aligned}
F_t &= F_g, \\
ma &= -mg, \\
a &= -g.
\end{aligned}
\tag{1.1}
$$

MODEL 1

Equation (1.1) is a model for the motion of a rock thrown vertically upwards, subject only to the influence of gravity.

ANALYSIS

The mass m of the rock appears nowhere in (1.1). Therefore, our model predicts that the motion of the rock is independent of its mass.

INTERPRETATION

Does our physical experience suggest that the motion of the rock is independent of its mass? Can't a marble be thrown higher than a bowling ball? Is it the mass of the rock that matters, then, or the speed with which you release it? Or is it the size of the rock, a factor that the model (1.1) has neglected?

That wave of questions certainly raises doubts about the validity of the model, but there is a clue that mass may not matter in Galileo's legendary experiments in Pisa: Mass did not affect the speed of an object's fall. So it may not be unreasonable to suggest that the motion of the rock, at least in its fall back to earth, is independent of its mass.

Galileo actually rolled brass spheres down an inclined plane, timing them with a water clock.

Now recast this model in search of more information about the motion of the rock. Since acceleration is the derivative of velocity with respect to time, the equation $a = -g$ can also be written

MODEL 2

$$\frac{dv}{dt} = -g, \tag{1.2}$$

a model that now incorporates velocity. Equation (1.2) is a **differential equation** for the unknown *function $v(t)$*. A **solution** of (1.2) is a function that reduces (1.2) to an equality.

The *analysis* of a differential equation model might involve

- finding a formula for a solution function,
- sketching graphs of solutions, or
- finding numerical values of the solution at specific times.

That is, the information contained in a differential equation can be seen from three perspectives: analytic (e.g., a solution formula), graphical, or numeric. A fourth perspective is physical, as when we ask, "What does all this mean?" in the context of the original physical problem.

ANALYSIS

Begin the analysis of (1.2) from the graphical perspective. Since g is a positive number, the differential equation $dv/dt = -g$ asserts that $dv/dt < 0$; i.e., the slope of the velocity-versus-time curve is always negative. The velocity of the rock is a decreasing function of time.

Figure 1.2, a **direction field** diagram, displays a series of tick marks on a v versus t graph; each tick is aimed in the direction given by the slope relation $dv/dt = -g$. To draw a graph of the solution function $v(t)$, pick a starting point, then "go with the flow" of the direction field. The lines in figure 1.2 are the result. That a graph of velocity versus time should be a straight line is consistent with the model: $dv/dt = -g$ says that the slope dv/dt is constant.

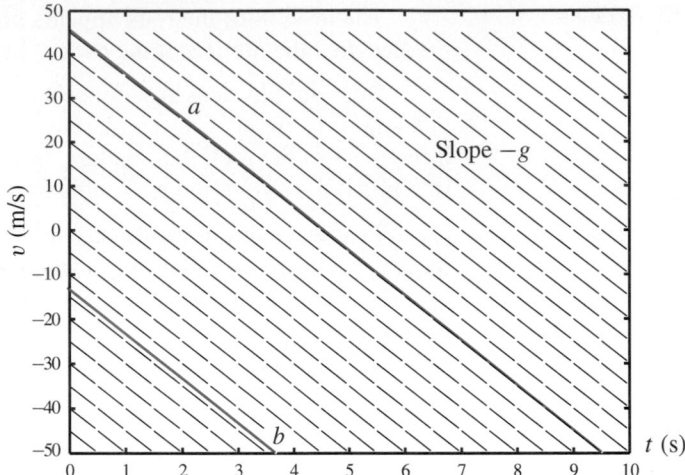

FIGURE 1.2 The direction field (slopes of v vs. t) predicted by model 2, $dv/dt = -g$.

**MATLAB**

MATLAB can draw a direction field diagram to produce your own version of figure 1.2. Start MATLAB, then type `delab` in the MATLAB command window to start DELAB. For guidance, select `Help, Textbook`, then go to chapter 1, figure 1.2. Experiment with different choices of the range of variables plotted on each axis.

INTERPRETATION

If velocity is decreasing with time, then it must eventually become negative, as the graphs of figure 1.2 illustrate. Recalling the agreement to take the upwards direction as positive, we conclude that negative velocity corresponds to the rock falling.

Knowing that the rock eventually will fall hardly earns gasps of admiration, but it does point out a flaw in the model. Both lines in figure 1.2 (and any other lines with negative slopes you care to draw) become more and more negative as time increases. That is, the rock continues to fall faster and faster, forever.

The flaw, of course, is simple. This model omitted the ground, which stops the fall of the rock. However, our interpretation has identified a clear limitation: The model $dv/dt = -g$ is not valid for all time.

MORE ANALYSIS

Some additional thought indicates that line b in figure 1.2 cannot be a reasonable graph of velocity versus time if the rock is thrown upwards initially because velocity is never positive there. We need a graph like line a, which includes a period of positive velocity corresponding to throwing the rock up into the air.

Stop and Think

1.1 Describe the physical situation represented by line b in figure 1.2.

But we cannot draw the graph more precisely without more information. We know that the line has slope $-g$. If we had the intercept with either the t-axis or the v-axis, then we could draw the unique line that is the graph of velocity versus time.

MORE INTERPRETATION

These intercepts are physically significant. The t-intercept (horizontal axis intercept) is a point where velocity is momentarily zero, as the rock's velocity changes from positive to negative. Plainly, this transition occurs at the highest point of the rock's trajectory, and the t-intercept must be the time from release to peak. On the other hand, the v-intercept (vertical axis intercept) is the velocity at time $t = 0$, the instant the rock was released.

The velocity at $t = 0$, written $v(0)$, is a quantity we can control directly as we release the rock, whereas the time from release to peak is a bit more remote. It appears then that we ought to specify $v(0)$, the *initial velocity*, as part of the model if we want to obtain an unambiguous graph of velocity versus time.

Prompted by this argument, we propose a new model,

MODEL 3

$$\frac{dv}{dt} = -g, \tag{1.3}$$

$$v(0) = v_i, \tag{1.4}$$

where the constant v_i is the specified initial velocity. The extra data given by (1.4), a specification of the value of the unknown function (or one of its derivatives) at the initial time, is called an **initial condition**.

ANALYSIS

Using the reasoning discussed before, we can draw a unique graph of velocity versus time as shown in figure 1.3. Throwing the rock upward corresponds to $v_i > 0$.

FIGURE 1.3 The particular velocity-versus-time curve obtained from model 3. The v-intercept is specified by equation (1.4), $v(0) = v_i$.

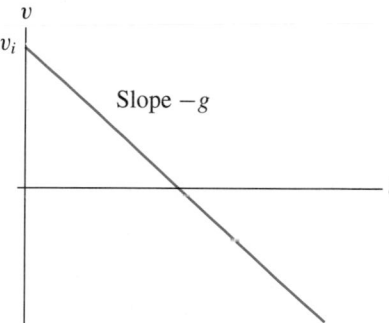

INTERPRETATION

Our ability to draw a unique velocity-versus-time curve suggests that we have a firm grasp on the motion of the rock, at least within the bounds set by the assumptions underlying the model. The model (1.3–1.4) still does not involve the mass of the rock, but it does indicate that we need an additional piece of data, the initial velocity. Perhaps the motion of a massive rock differs from that of a smaller one because we cannot give the same initial velocity to both. If there is only one velocity-versus-time curve corresponding to the model (1.3–1.4), then the motion of a marble and a bowling ball should be identical if we can impart the *same initial velocity* to both.

What about the height y of the rock above the surface of the earth? Since $y' = dy/dt = v$ by definition, a model for height *and* velocity would involve two differential equations

MODEL 4

$$y' = v,$$
$$v' = -g, \tag{1.5}$$

known as a **system** of differential equations, subject to two initial conditions, one for each unknown function,

$$y(0) = y_i, \quad v(0) = v_i. \tag{1.6}$$

1.2.1 Examples

■ **EXAMPLE 1** *The Guiness Book of World Records claims that Nolan Ryan pitched a ball at 45.1 m/s (100.9 mi/h) while playing for the Houston Astros in 1974. Put Mr. Ryan on his back and implement the model (1.3–1.4) in this case.*

His world's record pitch corresponds to an initial velocity of $v_i = 45.1$ m/s. Using $g = 9.8$ m/s^2, the governing model is

$$\frac{dv}{dt} = -9.8$$

$$v(0) = 45.1.$$

According to this model, any object that can be given an initial velocity of 45.1 m/s will reach the same height as Nolan Ryan's baseball. Obviously, a bowling ball must be thrown by an arm stronger even than Ryan's to achieve the same initial velocity. ■

■ **EXAMPLE 2** *For Ryan's record-setting vertical pitch, construct a table of values of velocity versus time at intervals of 0.5 s. Use the table to estimate the time to reach maximum height.*

The first entry in the table is easy: at $t = 0$, $v = 45.1$ m/s.
To find the second entry, ask, "How much does v change over the first half second?" Answer: multiply rate of change of v by elapsed time (as in "rate times time"). The *change* in v at the end of the first half second is

$$(dv/dt) \cdot 0.5 = -9.8 \cdot 0.5 = -4.9 \text{ m/s.}$$

Add this change to the initial velocity to find v at the end of the first half-second step:

$$v_1 = v_i + (dv/dt) \cdot 0.5 = 45.1 - 4.9 = 40.2 \text{ m/s.}$$

Repeat this process to find v at the end of the second half-second step (giving v at $t = 1$ s):

$$v_2 = v_1 + (dv/dt) \cdot 0.5 = 40.2 - 4.9 = 35.3 \text{ m/s.}$$

TABLE 1.1   Values of v in steps of 0.5 s for Nolan Ryan's vertical pitch

t (s)	v (m/s)
0	45.1
0.5	40.2
1.0	35.3
1.5	30.4
2.0	25.5
2.5	20.6
3.0	15.7
3.5	10.8
4.0	5.9
4.5	1.0
5.0	−3.9
5.5	−8.8
6.0	−13.7
6.5	−18.6
7.0	−23.5
7.5	−28.4
8.0	−33.3
8.5	−38.2
9.0	−43.1
9.5	−48.0

In general,

$$v_{n+1} = v_n + (dv/dt) \cdot 0.5 = v_n - g \cdot 0.5, \quad n = 0, 1, \ldots, 13.$$

(Note that the subscript gives the step number.) Table 1.1 contains the complete results.

From the graphical perspective, this process is just following the slope lines in the direction field diagram with steps of 0.5 s along the t axis. The first few of these steps are illustrated in figure 1.4.

When does the ball reach maximum height? Answer: when its velocity is zero. Table 1.1 suggests that time-to-peak is between 4.5 and 5 s. ■

This table of solution values was constructed using the **Euler method**, the simplest process for finding numerical values of solutions of initial-value problems. The recipe for stepping through the direction field diagram can be written

new value = old value + (slope) × (time step).

This process is exact *only if* the slope is constant, as it is in this example: $dv/dt = -9.8$.

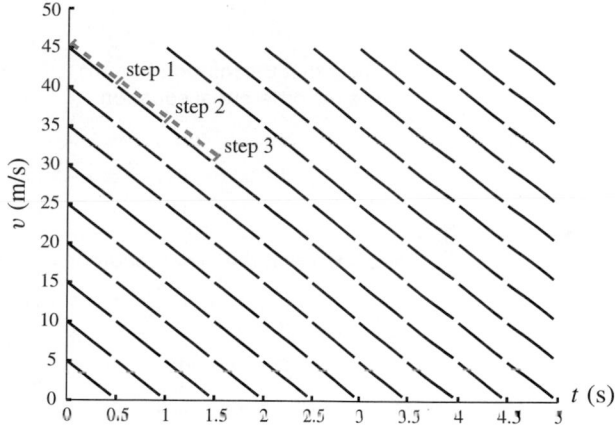

FIGURE 1.4   Constructing a table of values of the solution of $v' = -9.8$, $v(0) = 45.1$ m/s, in steps of 0.5 s is the same as following the slope lines in the direction field diagram starting from (0, 45.1).

MATLAB

DELAB can implement the Euler method. Type `delab` in the MATLAB command window. For guidance, select `Help`, `Textbook`, then go to chapter 1, example 2.

Stop and Think

1.2 Why is it reasonable to stop table 1.1 at $t = 9.5$ s?

1.3 Use the `Euler` numerical tool in DELAB with smaller steps to refine the estimate of the time when Ryan's vertical pitch reaches its maximum altitude.

■ **EXAMPLE 3**   *Find a formula for the velocity of Ryan's vertical pitch as a function of time. Use it to find the exact time required to reach maximum height.*

An equivalent request is, "Find the solution of the initial-value problem $dv/dt = -9.8$, $v(0) = 45.1$ m/s."

Integrate the differential equation,

$$v(t) = \int -9.8 \, dt = -9.8t + C;$$

then use the initial condition $v(0) = 45.1$ to find C:

$$v(0) = -9.8 \cdot 0 + C = 45.1 \implies C = 45.1.$$

The solution is $v(t) = -9.8t + 45.1$.

The time-to-peak t_p satisfies $v(t_p) = 0$. Solving $v(t_p) = -9.8t_p + 45.1 = 0$ yields

$$t_p = \frac{-45.1}{-9.8} = 4.602 \text{ s. } ■$$

To check that the solution formula $v = -9.8t + 45.1$ just found is correct, substitute it into the differential equation

$$
\begin{aligned}
v' &= -9.8 \\
(-9.8t + 45.1)' \; ? &= ? \; -9.8 \\
-9.8 &= -9.8 \; \checkmark
\end{aligned}
$$

and verify that the initial condition is satisfied,

$$
\begin{aligned}
v(0) &= 45.1 \\
(-9.8t + 45.1)|_{t=0} \; ? &= ? \; 45.1 \\
45.1 &= 45.1 \; \checkmark
\end{aligned}
$$

■ **EXAMPLE 4**   *How does time-to-peak for Ryan's vertical pitch change if gravity changes, say by moving him to the top of a mountain? To be specific, if the acceleration of gravity is reduced by 2%, how much does time-to-peak change? Does it increase or decrease?*

To find a formula for time-to-peak t_p in terms of the acceleration of gravity g, repeat the analysis of the previous example with g in place of 9.8:

$$v(t) = \int -g \, dt = -gt + C.$$

From $v(0) = 45.1$, deduce $C = 45.1$. Hence, $v(t) = -gt + 45.1$. Since $v(t_p) = 0$, $gt_p = 45.1$ or

$$t_p = 45.1/g.$$

Evidently, decreasing g *increases* t_p. To estimate the change in t_p due to a change $\Delta g = -0.02g$ in g (0.02 for 2%, $-$ for the *decrease* in g), use the derivative dt_p/dg computed from $t_p = 45.1/g$:

$$\Delta t_p \approx \frac{dt_p}{dg}\Delta g = -\frac{45.1}{g^2}(-0.02g) = \frac{0.02 \cdot 45.1}{g}.$$

To find the percent change in t_p, divide by $t_p = 45.1/g$:

$$\frac{\Delta t_p}{t_p} = 0.02.$$

Hence, a 2% *decrease* in the acceleration of gravity will *increase* time-to-peak by 2%. ■

1.3 ■ DIFFERENTIAL EQUATIONS AND SOLUTIONS

1.3.1 Some Terminology

Both models 2 and 3 involved a *differential equation*, an equation for an unknown function $v(t)$ that includes the derivative of the function. Model 4 involves two differential equations for two unknown functions, $y(t)$ and $v(t)$. Other modeling processes can lead to differential equations involving higher derivatives of the unknown (see exercise 5) or both the unknown and its derivative (see equation (2.1)).

   A **differential equation** is a relation involving one or more derivatives of an unknown function and perhaps the function itself. The unknown function, e.g., $v(t)$ in (1.3),

$$\frac{dv}{dt} = -g,$$

is called the **dependent variable**, and its argument, e.g., t in the last equation, is called the **independent variable**. The order of the highest derivative of the dependent variable appearing in the equation is called the **order** of the differential equation; e.g., $v' = -g$ is a *first*-order equation, and $y' = v$, $v' = -g$ is a *first*-order system.

■ **EXAMPLE 5**   *Identify the dependent variable, the independent variable, and the order of the differential equation*

$$y'' + 2y' + 16\sin y = -3\cos \pi t.$$

   Since y is differentiated, it is the dependent variable. The order of the differential equation is two because the highest derivative that appears is the second derivative y''.

   The independent variable is perhaps less obvious; y'' could denote d^2y/dx^2 or d^2y/ds^2 or d^2y/dr^2 or $\ldots$. But the variable that appears on the right side is t. We conclude that the independent variable must be t. ■

The term g in the differential equation $v' = -g$ is called a **parameter**. It represents a constant in a given setting (say, 9.8 m/s² on the surface of the earth), but it would change if we used different units or moved our rock tossing experiment to another planet.

The initial velocity v_i is also a parameter in the model (1.3–1.4),

$$\frac{dv}{dt} = -g, \quad v(0) = v_i.$$

It is a constant for a given trial, but we could reasonably ask how the motion of the rock changes as we vary v_i, perhaps by asking someone other than Nolan Ryan to pitch.

Since dv/dt, $v'(t)$, and v' are all symbols for the derivative of v with respect to t, the following are equivalent forms of the same differential equation:

$$\frac{dv}{dt} = -g,$$

$$v'(t) = -g,$$

$$v' = -g.$$

Of course, other letters can represent the same variables. The differential equation

$$y'(x) = k,$$

where k is a constant, is a mathematical problem that is equivalent to those listed in the preceding paragraph. We can identify y with v, x with t, and k with $-g$. (The context will say that v' means dv/dt while y' means dy/dx.)

A **solution of a differential equation** is a sufficiently differentiable function that renders the relation defining the equation an identity when substituted for the unknown. For example, some solutions of $v' = -g$ are

$$v(t) = -gt + 2,$$
$$v(t) = -gt - 6.3,$$
$$v(t) = -gt + C, \quad C \text{ any constant,}$$

as you can verify by differentiating each expression for $v(t)$ and substituting in the equation $v' = -g$.

When can we determine the arbitrary constant C in a solution such as $v = -gt + C$? When we are given additional information such as the initial velocity condition $v(0) = v_i$. Setting $t = 0$ in the solution $v(t) = -gt + C$ yields

$$v(0) = C,$$

while the initial condition requires

$$v(0) = v_i.$$

Hence, $C = v_i$, and a solution of the model $v' = -g$, $v(0) = v_i$, is

$$v(t) = -gt + v_i. \tag{1.7}$$

This solution is just the equation of the straight line velocity-versus-time graph shown in figure 1.3, a graph that was drawn without actually solving the initial-value problem $v' = -g$, $v(0) = v_i$. Using $v(0) = v_i$ to find C is the same as using that initial condition to determine which of the lines in figure 1.2 is the actual graph of the rock's velocity.

> Note the pattern: A solution technique (say, integration) gives a solution formula containing C; the extra data $v(0) = v_i$ determines C.

Extra information such as $v(0) = v_i$ is called an **initial condition**. A problem such as

$$v' = -g, \quad v(0) = v_i,$$

that couples the auxiliary data $v(0) = v_i$ with a differential equation, is called an **initial-value problem** if all of the data is given at one value of the independent variable. The reason for the name *initial value* is obvious in this model, where the data are specified at $t = 0$.

Initial conditions specify the value of the dependent variable and perhaps one or more of its derivatives at a fixed value of the independent variable. These values are called **initial values**. The number of initial values is equal to the order of the differential equation, and the highest derivative whose initial value is specified is one less than the order of the differential equation.

More generally, the *generic first-order differential equation* is often written

FIRST-ORDER DIFFERENTIAL EQUATION

$$y' = f(x, y);$$

e.g., we could write the rock tossing equation as $v' = f(t, v)$ where f is the constant function $f(t, v) = -g$. In the same spirit, the *generic first-order initial-value problem* is often written

FIRST-ORDER INITIAL-VALUE PROBLEM

$$y' = f(x, y), \quad y(x_0) = y_i.$$

> More general forms are possible; e.g., $f(x, y, y') = 0$ includes the equation $(y')^2 + y^2 = 0$ considered in exercise 26.

Definition 1. *The function u is a **solution** on the interval $x_0 \leq x \leq x_1$ of the first-order differential equation $dy/dx = f(x, y)$ if u is continuously differentiable and if*

$$\frac{du}{dx} = f(x, u) \quad \text{for } x_0 \leq x \leq x_1.$$

*Further, u is a **solution of the initial-value problem** $dy/dx = f(x, y)$, $y(x_0) = y_i$, on the interval $x_0 \leq x \leq x_1$ if u is continuously differentiable,*

$$\frac{du}{dx} = f(x, u) \quad \text{for } x_0 \leq x \leq x_1, \text{ and } u(x_0) = y_i.$$

Recall that a function u is continuously differentiable on an interval if its derivative is continuous there; that is, if $u'(a)$ and $\lim_{x \to a} u'(x)$ are defined and $\lim_{x \to a} u'(x) = u(a)$ for each point a in the interval.

This definition of *solution* can be relaxed slightly to permit piecewise continuously differentiable solutions, functions whose derivatives are discontinuous at only a finite number of points.

Parallel definitions for second-order equations appear in section 6.1 and for systems of first-order equations in section 8.1.

For example, $dv/dt = -g$ has $f(t, v) = -g$. A solution of this differential equation on the interval $0 \le t < \infty$ is the function defined by $v(t) = -gt + C$, C an arbitrary constant: v satisfies the differential equation $v' = f(t, v)$ (Substitute!), and $v' = -g$ is continuous on $0 \le t < \infty$. Similarly, the function defined by $v(t) = -gt + v_i$ is a solution on $0 \le t < \infty$ of the initial-value problem $dv/dt = -g$, $v(0) = v_i$ because v is a solution of the differential equation on $0 \le t < \infty$ and $v(0) = v_i$. Even more concretely, example 3 illustrates the solution $v(t) = -9.8t + 45.1$ of the initial-value problem $dv/dt = -9.8$, $v(0) = 45.1$ m/s, that models Nolan Ryan's vertical pitch.

Verify a proposed solution formula by substitution in the differential equation and in the initial condition, if appropriate, then verify its continuity properties, unless they are obvious.

A first-order differential equation $y' = f(x, y)$ is *linear* if the "right-hand side" function $f(x, y)$ is a linear (straight line-like) function of the dependent variable y. That is, $y' = f(x, y)$ is linear if it can be written in the form

LINEAR FIRST-ORDER EQUATION

$$y' = g(x)y + h(x)$$

for some functions $g(x)$, $h(x)$ depending on the *independent variable only*. An equation is *nonlinear* if it is not linear. Generally, linear equations are easier to analyze than nonlinear equations; the significance of linearity is explored more fully in subsequent chapters.

Stop and Think   **1.4** Explain why the expression $g(x)y + h(x)$ is "straight line-like" in y.

■ **EXAMPLE 6**   *Classify the equations $v' = -g$, $P' = kP$, $P' = aP - sP^2$, $u' = (1 - t)u + \cos t$, and $u' = (1 - t)u + \cos u$ as linear or nonlinear.*

Compare each of the given first-order equations with the standard linear form $y' = g(x)y + h(x)$.

The rock model equation $v' = -g = 0 \cdot v - g$ is linear; make the identifications $y \leftarrow v$, $x \leftarrow t$, $g(x) \leftarrow 0$, $h(x) \leftarrow -g$.

The equation $P' = kP = kP + 0$ is linear; make the identifications $y \leftarrow P$, $x \leftarrow t$, $g(x) \leftarrow k$, $h(x) \leftarrow 0$. (This equation arises in a population model; see section 2.1.)

The equation $P' = aP - sP^2$ is *non*linear; make the identifications $y \leftarrow P$, $x \leftarrow t$, $g(x) \leftarrow a$, $h(x) \leftarrow 0$. But what about sP^2? It can not be identified with $h(x)$ because sP^2 depends upon the *dependent* variable; h can be a function only of the *independent* variable $x \leftarrow t$. (This equation arises in another population model; see section 2.2.)

The equation $u' = (1 - t)u + \cos t$ is linear; make the identifications $y \leftarrow u$, $x \leftarrow t$, $g(x) \leftarrow (1 - t)$, $h(x) \leftarrow \cos t$. But $u' = (1 - t)u + \cos u$ is *non*linear; the identification $h(x) \leftarrow \cos u$ fails because u is the *dependent* variable and h can be a function only of the *independent* variable $x \leftarrow t$. ■

"Finding a solution" of a differential equation usually means finding a formula for the unknown function(s) defined by the differential equation, but solution formulas are neither the end of the story nor always a good beginning. The analysis and interpretation stages of the modeling process demand an understanding of the qualitative behavior of a solution (if a solution even exists!). There may be no closed-form solution formula, or such a formula may be too complicated to offer much understanding, or the formula may not include all possible solution functions. Perhaps much of the desired qualitative behavior can be discovered without even knowing a solution formula. Consequently, finding a solution formula is only one part of using the language of differential equations productively.

1.3.2 More Examples

■ **EXAMPLE 7** *Solve the initial-value problem*

$$\frac{dy}{dx} = xe^{x^2}, \quad y(0) = 2.$$

Since the unknown function does not appear on the right side of the differential equation, use antidifferentiation to obtain

$$y(x) = \int xe^{x^2}\,dx + C = \tfrac{1}{2}e^{x^2} + C.$$

To evaluate the arbitrary constant C, apply the initial condition:

$$y(0) - \tfrac{1}{2} + C - 2, \quad C - \tfrac{3}{2}.$$

Hence, the solution of the initial-value problem is

$$y(x) = \tfrac{1}{2}e^{x^2} + \tfrac{3}{2}. \quad ■$$

MATLAB

The analytic tools in DELAB can find solution formulas for many differential equations and initial-value problems. Start DELAB and choose **Help, Textbook**, then go to chapter 1, example 7.

■ **EXAMPLE 8** *Find the solution of the second-order differential equation*

$$\frac{d^2z}{ds^2} = \sin s.$$

As in the preceding example, use antidifferentiation, but two steps are required now because of the second derivative:

$$\frac{dz}{ds} = \int \sin s \, ds + C_1 = -\cos s + C_1,$$

$$z(s) = \int (-\cos s + C_1) \, ds + C_2 = -\sin s + C_1 s + C_2.$$

Since no initial conditions are given, we can not determine the arbitrary constants C_1, C_2. ■

> The problem statement asks for *the* solution. But this differential equation has many solutions, one for each value of C_1 and C_2. We can not speak of *the* solution.

■ **EXAMPLE 9**   *Verify that*

$$z(s) = -\sin s + C_1 s + C_2 \tag{1.8}$$

is indeed a solution of the differential equation

$$\frac{d^2 z}{ds^2} = \sin s. \tag{1.9}$$

Substituting the solution (1.8) into the differential equation (1.9) requires computing $z''(s)$. From the solution formula (1.8), calculate

$$z'(s) = -\cos s + C_1$$
$$z''(s) = \sin s.$$

Hence, (1.8) is a solution of the differential equation (1.9). ■

■ **EXAMPLE 10**   *Verify that for any value of the constant C,*

$$y(t) = Ce^{4t}$$

is a solution for all t of the differential equation

$$\frac{dy}{dt} = 4y.$$

Substitute the proposed solution $y(t) = Ce^{4t}$ into the differential equation $y' = 4y$:

$$
\begin{aligned}
y &= Ce^{4t}, \\
y' &= 4y, \\
(Ce^{4t})' \; &?=? \; 4(Ce^{4t}), \\
C4e^{4t} &= 4(Ce^{4t}). \; \checkmark
\end{aligned}
$$

For any value of C, the expression $y = Ce^{4t}$ is a solution for all t of the differential equation $dy/dt = 4y$. ■

The differential equation $dy/dt = 4y$ also has a less interesting solution, the function $y(t) \equiv 0$. Obviously, $d0/dt = 4 \cdot 0$. The identically zero solution is called the **trivial solution**.

Trivial solution

1.3.3 Separation of Variables: A Quick Look

How did we find the solution to an equation such as $dy/dt = 4y$? Simple antidifferentiation fails:

$$y(t) = \int \frac{dy}{dt}\, dt = \int 4y\, dt.$$

We can not evaluate the antiderivative $\int 4y\, dt$ because we do not know the formula for y as a function of t. (If we did, we would already know the solution!)

A technique that works for this particular equation and for many other first-order equations is **separation of variables**. To illustrate it quickly here, notice that $dy/dt = 4y$ can be written with all the y variables on one side of the equation and all the t variables on the other,

Separate variables.

$$\frac{dy}{y} = 4\, dt,$$

providing $y \neq 0$. This *separation of variables* is accomplished by thinking of the derivative dy/dt as a quotient of the differentials dy and dt.

We will verify $y \neq 0$ once we have a formula for y.

Proceeding formally (that is, without careful mathematical justification), find the antiderivative of each side of the separated equation,

Integrate.

$$\int \frac{dy}{y} = \ln|y| = \int 4\, dt = 4t + c.$$

The two different arbitrary constants that arose from the two antiderivatives were collected on the right side of the equality and their difference labeled c.

Exponentiate both sides of $\ln|y| = 4t + c$ to invert the natural logarithm and obtain

Solve for y.

$$|y| = e^{4t+c} = e^c e^{4t} = |C| e^{4t},$$

where $|C| = e^c$.

Since $|y| = \pm y$, according to whether y is positive or negative, $y = \pm C e^{4t}$. The exponential is always positive, so the sign of $C e^{4t}$ is determined by the sign of C; that is, the solution is

$$y = C e^{4t},$$

with C chosen to be either positive or negative, as appropriate. If $C \neq 0$, then $y \neq 0$, as required. To verify that this solution is correct, turn back to example 10.

In general, separation of variables can find a formula for the solution of differential equations $y' = f(t, y)$ that can be written in the *separable form*

Separable differential equation

$$\frac{dy}{dt} = T(t)Y(y).$$

The formal manipulation of the derivative as a quotient of differentials yields

$$dy = T(t)Y(y)\,dt,$$

$$\frac{dy}{Y(y)} = T(t)\,dt,$$

$$\int \frac{dy}{Y(y)} = \int T(t)\,dt + c.$$

If possible, carry out the indicated integration and solve for y. For example, in the equation $dy/dt = 4y$, we chose $T(t) = 1$ and $Y(y) = 4y$, integrated $dy/4y = dt$, and solved for y using the exponential function.

> **Separation of variables.**   To solve $dy/dt = T(t)Y(y)$, integrate $dy/Y(y) = T(t)dt$ and solve for y, if possible.

The formal manipulations of separation of variables can be justified using the chain rule and the definition of antiderivative; see exercise 27 of section 4.2.

Stop and Think

1.5 The equation $dy/dt = 4y$ can also be separated by choosing $T(t) = 4$ and $Y(y) = y$. Does that choice also lead to the solution formula $y = Ce^{4t}$?

1.6 Comment on the following assertion: Even if the result of integrating $dy/Y(y) = dt/T(t)$ can not be solved easily for y, that formula still defines y as an implicit function of t.

■ **EXAMPLE 11**   *Use separation of variables to find a solution of the initial-value problem*

$$\frac{dy}{dt} = 4y - 6, \quad y(0) = -3.$$

First use separation of variables to find a function that satisfies the differential equation; then use the initial condition to determine the arbitrary constant in the solution formula.

Choose $T(t) = 1$, $Y(y) = 4y - 6$ and separate variables to obtain

Separate variables.

$$dy = (4y - 6)\,dt$$

$$\frac{dy}{4y - 6} = dt,$$

providing $4y - 6 \neq 0$.

Antidifferentiation follows:

Integrate.

$$\int \frac{dy}{4y - 6} = \tfrac{1}{4}\ln|4y - 6| = \int dt = t + c.$$

Multiplying by 4, using the exponential to invert the natural logarithm, writing $|C| = e^{4c}$, and disposing of the absolute value signs as above gives a solution of the differential equation,

Solve for y.

$$y(t) = \tfrac{3}{2} + Ce^{4t}.$$

Use the initial condition $y(0) = -3$ to determine C:

Find C.

$$y(0) = \tfrac{3}{2} + C = -3, \quad C = -\tfrac{9}{2}.$$

The solution of the initial-value problem is obtained by substituting this value of C into the solution formula of the preceding paragraph. (Check $4y - 6 \neq 0$ for yourself.)

$$y(t) = \tfrac{3}{2} - \tfrac{9}{2}e^{4t}. \quad ■$$

■ **EXAMPLE 12**   *Use separation of variables to find a solution of*

$$\frac{du}{dt} = u^2 \cos \pi t, \quad u(0) = -1/2.$$

For what values of t is the solution you find valid?

Choose $T(t) = \cos \pi t$ and $U(u) = u^2$ to write $du/dt = u^2 \cos \pi t$ as $du/dt = T(t)U(u)$, then separate variables:

Separate variables.

$$\frac{du}{u^2} = \cos \pi t \, dt,$$

providing $u \neq 0$. Then antidifferentiate:

Integrate.

$$\int \frac{du}{u^2} = \int \cos \pi t \, dt,$$

$$\frac{1}{u} = \frac{1}{\pi} \sin \pi t + C.$$

Solving for the dependent variable u yields

Solve for u.

$$u(t) = -\frac{\pi}{\sin \pi t + C}. \tag{1.10}$$

To determine C, set $t = 0$ and solve:

Find C.

$$u(0) = -\pi/C = -1/2 \implies C = 2\pi.$$

The solution of the initial-value problem is

$$u(t) = -\frac{\pi}{\sin \pi t + 2\pi}.$$

This formula is valid for any value of t for which $\sin \pi t \neq -2\pi$, i.e., for all t.

There was some algebraic sleight of hand. Inverting the equation before (1.10), multiplying through by -1, and multiplying the right side by π/π actually yields

$$u(t) = \frac{-\pi}{\sin \pi t + \pi C}.$$

Since C is an arbitrary constant, πC can be written as C for simplicity. ■

Stop and Think   **1.7** Suppose the initial condition in the previous example were $u(0) = -2$. Would the solution formula be valid for all t?

> **MATLAB**
>
> DELAB can find this analytic solution, too. Choose `Help, Textbook` in DELAB, then go to chapter 1, example 12.

1.3.4   The Euler Method

If an analytic formula for the solution of an initial-value problem is too complicated or not available, an alternative is a table of (approximate) values of the solution. The simplest way to compute such a table is the technique illustrated in example 2: take discrete time steps to follow the slopes given by the direction field. In words, this process is

> **Euler method**
>
> new value = old value + slope × time step.

See figure 1.4, p. 7.

For the initial-value problem $v' = -9.8$, $v(0) = 45.1$ m/s of example 2, slope is always $v' = -9.8$. The first "old value" is the initial value $v(0) = 45.1$ m/s. If v_0 denotes $v(0)$ and Δt denotes the time step, the recipe for the **Euler method** applied to this initial-value problem is

$$v_{n+1} = v_n + v'\Delta t = v_n - 9.8\Delta t, \quad n = 0, 1, 2, \ldots$$
$$v_0 = 45.1$$

Example 2 illustrates the Euler method with $\Delta t = 0.5$.

■ **EXAMPLE 13**   *Estimate the maximum height of Nolan Ryan's vertical pitch by applying the Euler method with $\Delta t = 0.5$ to the system*

$$y' = v,$$
$$v' = -9.8$$

that models both vertical height y and velocity v.

The Euler method requires conditions to give the "old values" for the first step. If Ryan does indeed release the pitch straight up, it would leave his hand at a height of about 2.5 m (roughly 8 ft). Hence, choose $y(0) = 2.5$ m and $v(0) = 45.1$ m/s, the speed of the world-record pitch. Example 2 found that the ball reaches maximum height between 4 and 5 s; to play it safe, take 12 steps of 0.5 s each to reach $t = 6$ s.

The "new = old + slope × time step" recipe is applied twice, once for y and once for v, beginning with $y_0 = 2.5$ and $v_0 = 45.1$. The slope y' depends on v; use the old value since that is the only one available:

$$y_1 = y_0 + y_0' \Delta t = 2.5 + v_0 \cdot 0.5 = 2.5 + 45.1 \cdot 0.5 = 25.05,$$
$$v_1 = v + 0 + v_0' \Delta t = 45.1 - 9.8 \cdot 0.5 = 40.2.$$

The values $y_1 = 25.05$, $v_1 = 40.2$ approximate y, v at $t = 1\Delta t = 0.5$ s.

The general recipe is

$$y_{n+1} = y_n + v_n \Delta t, \qquad y_0 = 2.5 \text{ s}$$
$$v_{n+1} = v_n - 9.8 \Delta t, \qquad v_0 = 45.1 \text{ m/s}$$
$$n = 0, 1, 2, \ldots, 11.$$

Table 1.2 shows the resulting values.

Table 1.2 shows that the maximum height is approximately 117.75 m. Why *approximately*? Because the solution has been sampled at only discrete values of t and, more importantly, because "*slope × time step*" *is exact only if the slope is constant*. The slope of y vs. t is not constant because $y' = v$ and v varies with time. ■

TABLE 1.2 **Values of y and v in steps of 0.5 s for Nolan Ryan's vertical pitch computed from $dy/dt = v$, $dv/dt = -9.8$, $y(0) = 2.5$ m, $v(0) = 45.1$ m/s using the Euler method.**

t (s)	y (m)	v (m/s)
0	2.5	45.1
0.50	25.05	40.20
1.00	45.15	35.30
1.50	62.80	30.40
2.00	78.00	25.50
2.50	90.75	20.60
3.00	101.05	15.70
3.50	108.90	10.80
4.00	114.30	5.90
4.50	117.25	1.00
5.00	117.75	-3.90
5.50	115.80	-8.80
6.00	111.40	-13.70

MATLAB

To confirm that one or the other of these graphs—y or v vs. t—does indeed have varying slope, use DELAB to plot solutions of $y' = v$, $v' = -9.8$. Which has the varying slope? For guidance, select **Help, Textbook**, then go to chapter 1, example 13.

■ **EXAMPLE 14** *Use the Euler method to plot an approximate solution of the initial-value problem of example 12, $u' = u^2 \cos \pi t$, $u(0) = \pi/2$, $0 \le t \le 10$.*

Since $u' = u^2 \cos \pi t$, the Euler recipe "new = old + slope × step size" is

$$u_{n+1} = u_n + u' \Delta t = u_n + (u_n^2 \cos \pi t_n) \Delta t,$$
$$u_0 = -1/2, \quad n = 0, 1, 2, \ldots.$$

Here u_n is the approximation to $u(t_n)$, $t_n = n\Delta t$.

What is a good choice of Δt? That question has no easy answer. For starters, note that $\cos \pi t$ has period 2; Δt should be chosen to capture some of the variation in this part of the slope term. Dividing the period of 2 into 20 parts suggests $\Delta t = 0.1$.

Beginning with $t_0 = 0$ and $u_0 = -1/2$, the first few steps are

$$u_1 = u_0 + (u_0^2 \cos \pi t_0) \Delta t$$
$$= 0.5 + ((-0.5)^2 \cos 0) 0.1 = -0.4750,$$
$$u_2 = u_1 + (u_1^2 \cos \pi t_1) \Delta t$$
$$= 0.5 + ((-0.5)^2 \cos 0.25\pi) 0.1 = -0.4535,$$
$$u_3 = u_2 + (u_2^2 \cos \pi t_2) \Delta t$$
$$= 0.5 + ((-0.5)^2 \cos 0.50\pi) 0.1 = -0.4369,$$

$$\vdots$$

A graph of these values appears in figure 1.5.

That figure also includes a plot of the exact solution found in example 12. Evidently, Euler with $\Delta t = 0.1$ has done a poor job of approximation; a smaller value is called for. Further numerical experiments show that $\Delta t = 0.01$ is a reasonable choice for a visually accurate plot. ■

MATLAB

Use the Euler method in DELAB to experiment with the effect of various step sizes on the approximate solution of this initial-value problem. For guidance, go to chapter 1, example 14 in **Help, Textbook**.

Stop and Think

1.8 Look carefully at the first few steps of the Euler method shown in figure 1.5. Do you think Euler is overestimating or underestimating u? Is u decreasing or increasing over those first few steps?

FIGURE 1.5 A plot of the Euler approximation with $\Delta t = 0.1$ and the exact solution of $u' = u^2 \cos \pi t$, $u(0) = -1/2$.

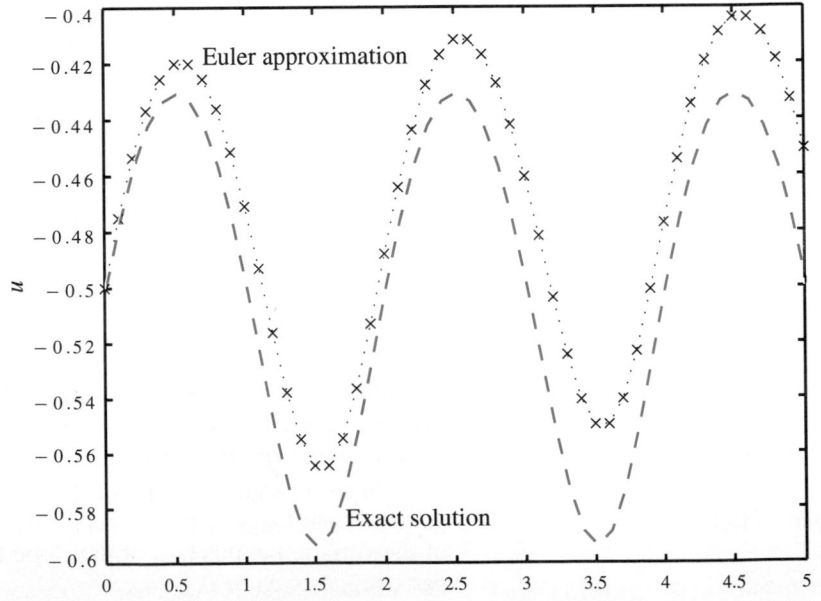

1.3.5 Summary

Initial-value problems can be models of physical phenomena. The behavior of the solution of an initial-value problem provides guidance about the behavior of the physical phenomenon it models.

An understanding of solution behavior can come from several sources, including an analytic formula for the solution, a graph of the solution sketched from slope information in the differential equation, or a table of approximate numerical values of the solution.

The ensuing chapters will develop additional models and expand the range of tools for understanding solution behavior.

1.4 ■ CHAPTER 1 SUMMARY

Model

In our context, a *model* involves a differential equation whose unknown function is a quantity of interest, such as the velocity of the thrown rock. The differential equation in the model is usually derived from a *physical law* such as Newton's law using an *experimental fact* to relate the quantities of interest. For example, we used the experimental fact that the force of gravity is proportional to mass, $F_g = -mg$.

Differential equation

Solution

Order

A *differential equation* is a relation involving one or more derivatives of an unknown function and perhaps the function itself. A *solution of a differential equation* is a sufficiently smooth function which renders the differential equation an identity when it is substituted for the unknown function. The *order of a differential equation* is the order of the highest derivative appearing in the equation.

Dependent variable

Independent variable
Parameter

The unknown function is called the *dependent variable*. The argument of the unknown function, the variable with respect to which it is differentiated, is the *independent variable*. A *parameter* is a term that is constant for any particular occurrence of the differential equation.

Initial-value problem

Initial value, initial condition

An *initial-value problem* is a differential equation coupled with a specification of the value of the dependent variable and perhaps one or more its derivatives at a fixed value of the independent variable. These values are called *initial values*, and such specifications are called *initial conditions*.

If the differential equation simply gives an expression for a derivative of the unknown in terms of the independent variable, it can be solved by *antidifferentiation*. If all the dependent variables in a first-order differential equation can be collected on one side of the equation and all the independent variables on the other, then the equation can be solved by *separation of variables*.

Separation of variables

Direction field

Euler method

A first-order differential equation gives the slope of its solution. Those slopes can be represented graphically in *direction field diagrams*. They are also the basis for building a table of solution values using the "new value = old value + slope × step size" process of the *Euler method*.

1.5 ■ CHAPTER EXERCISES

EXERCISE GUIDE

To gain experience ...	Try exercises
Deriving models	5(a), 8–9(a–b), 17–18
Analyzing and interpreting models	1, 4, 5(b–d), 8–9(c–e), 10–12
With the basic ideas of these models	2–3
Solving differential equations and initial-value problems	6–7, 13, 16, 19(a)
With differential equations concepts	14–15, 19

1. Use the solution $v(t) = -gt + v_i$ of the rock model $v' = -g$, $v(0) = v_i$, to determine the time t_p required for the rock to reach the peak of its trajectory. (*Hint*: What is its velocity at the peak?) Locate t_p on the graph of figure 1.3.

2. Suppose we had decided to formulate model 3,

$$\frac{dv}{dt} = -g, \quad v(0) = v_i,$$

by specifying the t-intercept t_p of the velocity-versus-time curve (see figure 1.3) instead of the v-intercept v_i. Write down the model in this form. Determine the constant C in the solution formula $v(t) = -gt + C$ in terms of t_p.

3. *A model with too much auxiliary data*: Suppose we try to specify both initial velocity v_i and time-to-peak t_p in the rock tossing model; that is, suppose we pose the model

$$v'(t) = -g, \tag{1.11}$$

$$v(0) = v_i, \quad v(t_p) = 0. \tag{1.12}$$

(The second condition takes advantage of the observation that velocity is zero at the peak of the rock's trajectory.) Show that the two conditions in (1.12) contradict one another unless $t_p = v_i/g$. In other words, show that the model (1.11–1.12) has no solution unless the auxiliary data in (1.12) satisfy a special relation. (A problem of this sort is called *overdetermined* because only part of the data is truly necessary to specify the solution.)

Interpret the observations of this exercise geometrically in terms of figure 1.3. Does figure 1.3 show the relation that must hold between the two pieces of auxiliary data in (1.12)?

4. You have been asked to determine the velocity with which a sling shot releases a stone. Propose an experiment based on the model $v' = -g$, $v(0) = v_i$, to accomplish this task. (*Hint*: See exercises 1 and 10.)

5. (a) Model 2, $v' = -g$, was obtained from $a = -g$ by writing acceleration as the derivative of velocity. Continue this logic by writing velocity as the derivative of height y: $v = dy/dt$. (Figure 1.1 will need a y-axis. Be sure to mark its origin.)

   Derive the following model involving height for the motion of the rock:

$$\frac{d^2y}{dt^2} = -g. \tag{1.13}$$

   Solve this equation by antidifferentiation and argue that two pieces of initial data must be prescribed to determine all arbitrary constants. Suggest reasonable choices for this data, and state the resulting initial-value problem.

   (b) Provide some analysis and interpretation of this initial-value problem model. For example, what is the time required to fall from the peak of the trajectory back to the point of release? What is the rock's velocity at the instant it returns to the point of release?

   What type of curve has a constant second derivative, as in (1.13)? How much data is needed to specify such a curve? Sketch such a curve and show the data you propose to use on the sketch.

   (c) What is the maximum height of the rock? How does the maximum height depend upon the parameters in the model? (Interpret; do more than recite equations.)

   (d) Solve the initial-value problem from part (a), unless you have already. Compare your analysis with this solution.

6. Use antidifferentiation to find solutions involving the specified number of arbitrary constants for the following differential equations. Confirm your results using the analytic tools in DELAB.

   (a) $dy/dt = \cos t$, one constant

   (b) $d^2w/dt^2 = \sqrt{t}$, two constants

(c) $dy/dx = 4e^x$, one constant

7. Solve the following initial-value problems. Confirm your results using the analytic tools in DELAB.

(a) $dy/dt = \cos t$, $y(0) = 0$

(b) $d^2w/dt^2 = \sqrt{t}$, $w(0) = 2$, $w'(0) = -2$

(c) $dy/dx = 4e^x$, $y(2) = \pi$

8. *A model of the rock's motion with air resistance*: To include the effects of air resistance, we can appeal to an additional experimental fact:

> The force due to air resistance is proportional to the velocity, and it acts in the direction opposite to velocity.

(a) *Formulate a mathematical statement of the experimental fact*: Argue that a correct mathematical expression of this experimental fact is $F_a = -kv$, where k is a positive constant of proportionality. (We expect k to depend upon the shape of the rock but not its mass.)

(b) *Derive a model*: The total force acting on the rock in this model is $F_g + F_a$. Couple this observation with Newton's law to derive a more sophisticated model of the rock's motion,

$$\frac{dv}{dt} = -g - \frac{k}{m}v, \quad v(0) = v_i.$$

(c) *Analyze the model*: How does the v-versus-t curve of this model compare with that of figure 1.3, the corresponding curve for model 3? Do the two curves begin at the same point? How do their slopes compare at $t = 0$? How do their slopes compare when $v = 0$? How does the slope of the graph of this model change as the velocity decreases from v_i to zero? Use DELAB as appropriate.

(d) *Interpret*: Does air resistance appear to modify the rock's motion in a manner consistent with physical intuition? Which term in the differential equation accounts for air resistance? Are the comments in the text about the effects of the rock's mass on its motion still valid?

(e) *Interpret the model in a different setting*: Suppose the rock were dropped from a balloon. Would this model still apply? Would the velocity of the falling rock increase indefinitely?

9. Shotgun pellets and other small spherical metal particles are formed by dropping a bit of molten metal down a long hollow tube, called a *shot tower*. As the mass falls, air resistance retards its motion with a force proportional to the *square* of its velocity.

Let m denote the mass of the particle and v its velocity. Let r be the proportionality constant for the force of air resistance. Take *up* as the *positive* direction.

(a) *Formulate a mathematical statement of the experimental facts*: Write an expression for the force of gravity on the pellet. Write an expression for the force of air resistance on the pellet. (Be careful with signs!)

(b) *Derive a model*: Use Newton's law ($F_{total} = ma$) to derive a model for the velocity of the pellet.

(c) *Analyze the model*: How does the v-versus-t curve for this model compare with that of figure 1.3, the corresponding curve for model 3?

(d) *Interpret*: Will the pellet fall faster and faster or will it attain a constant terminal velocity? If the latter occurs, find the value of the terminal velocity.

10. How long would Nolan Ryan's vertical pitch require to return to its point of release? (See example 1.)

11. Use the slope and intercept information in the model for Nolan Ryan's vertical pitch, $v' = -9.8$, $v(0) = 45.1$, to draw a careful graph of the solution of this initial-value problem. (See example 1.) Use a piece of graph paper so that the slope and the intercept with the vertical axis can be plotted exactly. Find a solution formula and confirm that its graph is the one you drew.

12. Use the model of exercise 5 to determine the maximum height Nolan Ryan's vertical pitch could reach. (See example 3.)

13. Use separation of variables to solve each of the following differential equations. Confirm your results using the analytic tools in DELAB. State the range of values of the independent variable for which the solution is valid. Classify each equation as linear or nonlinear.

(a) $dy/dt = -8y$

(b) $dP/dt = 0.0102P$

(c) $dP/dt = kP$, where k is a constant

(d) $dP/dt = 0.015P - 0.209$

(e) $y' = y^3 - y$

(f) $dT/dt = -0.0002(T - 5)$

(g) $dT/dt = -(Ak/cm)(T - T_{out})$, where A, k, c, m, T_{out} are all constants

(h) $dy/dx = e^{2x}$

(i) $dy/dx = e^{2y}$

(j) $du/dy = y^2 - 1$

14. (a) Which of the differential equations in the preceding exercise have the same generic form as the differential equation in part (a); that is, which equations have the form $dy/dx = ay$ for some constant a? In each case, identify the variables that correspond to y and to x and to the parameter a.

   (b) Which of the differential equations in the preceding exercise have the same generic form as the differential equation in part (d); that is, which equations have the form $dy/dx = ay+b$ for some constants a, b? In each case, identify the variables that correspond to y and to x and to the parameters a and b.

   (c) Which of the differential equations in the preceding exercise have a generic form that is completely distinct from all others in that exercise? In each case, write the form of the equation using y for the dependent variable, x for the independent variable, and $a, b, c, \ldots$ for constants.

15. Which of the differential equations listed in exercise 13 possess the trivial solution? Which do not? Verify the claim in each case by substituting the identically zero function into the equation and showing that equality does or does not hold.

16. Give an example of a differential equation (other than the one in the text) involving an unknown function and only its first derivative that can not be solved by separation of variables.

17. Follow the steps given below to derive a model describing the height above the surface of the earth of a rocket launched vertically upwards. Use Newton's law ($F = ma$) and the following experimental facts:

   Fact 1.   The force of gravity on the rocket is proportional to its mass and *inversely* proportional to the square of its distance from the *center* of the earth.

   Fact 2.   A known function $T(t)$ gives the thrust force of the rocket at time t measured from the instant of ignition.

   (a) Introduce a coordinate system and write a formula for the force F_g of gravity given by the first experimental fact in terms of the height $y(t)$ of the rocket above the *surface* of the earth at time t.

   (b) Use Newton's law to obtain a differential equation relating the height $y(t)$, one or more of its derivatives, the gravity expression from part (a), and the thrust $T(t)$ of the rocket to one another.

   (c) Complete the model by adding appropriate initial conditions. (Is the rocket moving at the instant its motor is ignited?)

18. In his *Dialogues Concerning Two New Sciences* (see [22, p. 344]), Galileo asserts the proposition

   The spaces described by a body falling from rest with a uniformly accelerated motion are to each other as the squares of the time intervals employed in traversing these distances.

   (A freer translation of Galileo's Latin might replace *spaces described* with *distances traveled*.)

   (a) Derive a model that relates the displacement of the body to a uniform acceleration a. (*Hint*: Can you use something like (1.13)?)

   (b) Use the solution of the model to prove Galileo's assertion that distance traveled is proportional to the square of the elapsed time.

19. Differential equations of the form $dy/dt = ky$ are often called *equations of growth and decay*. Explain this name as follows.

   (a) Solve the equation and argue that y is increasing for all t if $k > 0$ and $y(0) > 0$.

   (b) Solve the equation and argue that y is decreasing for all t if $k < 0$ and $y(0) > 0$.

   (c) Argue directly from the differential equation that y is increasing if $k > 0$ and $y(0) > 0$, because $y' > 0$.

   (d) Argue directly from the differential equation that y is decreasing if $k < 0$ and $y(0) > 0$, because $y' < 0$.

   We encounter both forms of this equation in the next chapter.

20. Describe the physical situation represented by line b in figure 1.2.

21. For each of the following equations,
   (i) Sketch a direction field diagram by hand.
   (ii) Confirm your sketch using the graphical tools in DELAB.
   (iii) Identify the range of initial values (if any) for which the solution is increasing.

   (a) $dy/dt = -8y$

   (b) $dP/dt = 0.0102P$

   (c) $dP/dt = 0.015P - 0.209$

   (d) $y' = y^3 - y$

   (e) $dT/dt = -0.0002(T - 5)$

   (f) $dy/dx = e^{2x}$

   (g) $dy/dx = e^{2y}$

   (h) $du/dy = u^2 - 1$

22. Sketch by hand a direction field diagram for $y' = ky$, $k > 0$ and another one for $k < 0$. Confirm your sketch using the graphical tools in DELAB.

23. For each of the following initial-value problems,
    - (i) Use two steps of the Euler method and a calculator to approximate the solution at $t = 1$.
    - (ii) Confirm your answer using the **Euler** method from the numerical tools in DELAB.
    - (iii) Does the solution appear to increase or decrease during the first step? Is this behavior consistent with the sign of the slope given by the differential equation?

    (a) $dy/dt = -8y$, $y(0) = 2$

    (b) $dP/dt = 0.015P - 0.209 \cos \pi t$, $P(0) = 8$

    (c) $T' = -0.02(T - 5)$, $T(0) = 4$

    (d) $du/dy = u^2 - 1$, $u(0) = 0.1$

24. Solve $u' = u^2 \cos \pi t$, $u(0) = -2\pi$ via separation of variables. Show that the resulting solution formula is not valid for all t. Confirm your solution formula using the analytic tools in DELAB. Does DELAB give any hint that this solution formula is not valid for all t?

25. Show by substitution that the initial-value problem $y' = \sqrt{1 - y^2}$, $y(0) = 1$, has *two* solutions,
$$y_1(t) = \sin(t + \pi/2)$$
$$y_2(t) = 1.$$

Are these solutions continuously differentiable? Which solution would you find using separation of variables? Which solution is provided by the analytic tools in DELAB? Are there hints in either approach of the existence of a second solution? Can this initial-value problem be written in the generic form $y' = f(x, y)$, $y(0) = y_i$?

26. Argue that the only solution of the first-order differential equation $(y')^2 + y^2 = 0$ is $y \equiv 0$. That is, show that there is a first-order equation whose solution contains *no arbitrary constants*. Can this equation be written in the generic form $y' = f(x, y)$? (This example is from [23].)

27. In order to solve $dy/dt = T(t)Y(y)$, separation of variables formally manipulates the derivative dy/dt as if it were a quotient of differentials. Justify the result of that abuse of notation as follows. Assuming the quantities in question are all defined, let
$$g(y) = \int \frac{dy}{Y(y)}, \quad h(t) = \int T(t)\, dt.$$
Suppose y is a function of t that is defined implicitly by
$$g(y) = h(t) + c.$$
Compute the implicit derivative dy/dt using
$$\frac{dg}{dy} \frac{dy}{dt} = \cdots$$
and show that it satisfies $dy/dt = T(t)Y(y)$. *Hint:* To make this general process more concrete, first carry out these calculations for $dy/dt = 4y$ with $T = 1$, $Y = 4y$.

1.6 ■ CHAPTER PROJECTS

1. **A more sophisticated model of the rock's motion.** To include the effects of air resistance in a slightly more accurate manner than in exercise 7, page 23, appeal to an experimental fact that better represents the variation of air resistance with velocity over a large range of velocities:

    The force due to air resistance is proportional to the *square* of the velocity, and it acts in the direction opposite to velocity.

    (a) Argue that a correct mathematical expression of this experimental fact is $F_a = -kv|v|$, where k is a positive constant of proportionality. Couple this observation with Newton's law to derive a more sophisticated model of the rock's motion,
    $$\frac{dv}{dt} = -g - \frac{k}{m}v|v|, \quad v(0) = v_i.$$

    (b) How does the v-versus-t curve of this model compare with that of figure 1.3, the corresponding curve for model 3? Do the two curves begin at the same point? How do their slopes compare at $t = 0$? How do their slopes compare when $v = 0$? How does the slope of the graph of this model change as the velocity decreases from v_i to zero?

(c) Does air resistance appear to modify the rock's motion in a manner consistent with physical intuition? Which term in the differential equation accounts for air resistance? How does the rock's mass affect its motion? Does the velocity of the falling rock increase indefinitely? Could one speak of a *terminal velocity*?

2. **Parachute design.** In some circumstances, rescue supplies, radio beacons, remote sensors, and similar equipment are dropped from airplanes flying at relatively high altitudes. To minimize drift due to wind, the parachute on the package is not deployed until the package has reached a lower, predetermined altitude. Given the size of the package and the altitude at which the parachute will open, how should the parachute be designed to prevent shattering the payload? Or given the parachute and the payload, when should the parachute open?

The parachute can be characterized by its drag coefficient K: the drag force exerted by the parachute is proportional with constant K to its velocity and acts in the opposite direction. The payload will be safe if it hits the ground with speed less than some value S. To be concrete, suppose that the package will be dropped from an altitude of 3,000 m and that the parachute will open at 1,000 m.

(a) Argue that a reasonable model for the altitude x and velocity v of the falling package is

$$x' = v,$$
$$mv' = -g - k(x)v,$$
$$x(0) = 3{,}000, \quad v(0) = 0,$$

where

$$k(x) = \begin{cases} 0, & x > 1{,}000 \\ K, & x \le 1{,}000. \end{cases}$$

Explain the assumptions underlying this model.

(b) Analyze the model: Find a range of values K that guarantee the package will land with speed less than S.

(c) Generalize this analysis by replacing the fixed altitudes 3,000 m and 1,000 m with parameters, say x_d for the drop altitude and x_p for the altitude at which the parachute opens. Determine a safe range of values for the parachute drag coefficient in terms of the maximum impact speed and the altitude parameters. Given the parachute drag coefficient, determine the minimum parachute deployment altitude.

(d) Provide some general recommendations regarding trade-offs among parachute size, deployment altitude, and impact speed.

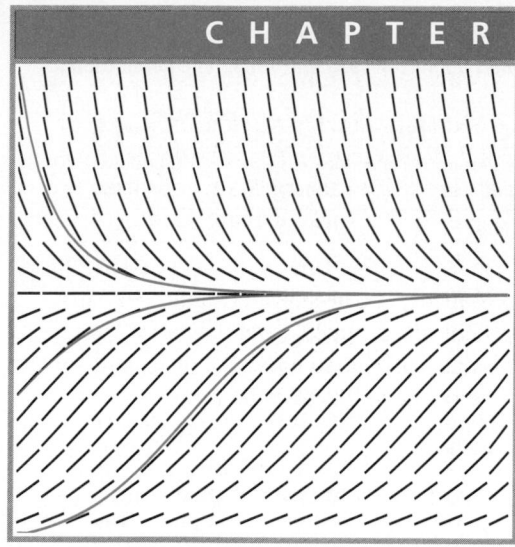

2

Models from Conservation Laws

2.1 ■ SIMPLE POPULATION MODELS

Populations of animals, insects, chemicals, and even money provide a rich source of models with a common dominant feature, growth. Later in this chapter, heat-flow models offer another alternative, the potential for decay.

The common feature in these models is their dependence upon a *conservation law*, a principle as simple as the one that governs a bank account: change in balance equals deposits minus withdrawals. The derivation of a differential equation model begins by accounting for all additions and subtractions over a short period of time.

2.1.1 Data

Population surveys reveal an apparent innocuous fact: Most countries in the world have populations that increase between 1% and 3% per year; that is, each succeeding year's population is 0.01 to 0.03 greater than the preceding year's. Some typical annual fractional increases are listed in table 2.1. These numbers suggest that the populations of industrialized nations increase at about a one percent annual rate, while some third-world nations run closer to a two or three percent annual rate.

Do these modest differences in annual rates of population increase have any great significance? Do they foreshadow future population imbalances? What population levels might we anticipate? How does varying the annual rate of increase affect population levels a generation or two hence? How will limited resources, immigration, disease, and other external factors influence future population levels?

This section and the next will develop, analyze, and interpret mathematical models that will provide some answers to these questions. Few of these questions have a single, simple answer; don't let fancy mathematics and

TABLE 2.1   **Populations and annual growth rates of selected countries and regions for 1978 and 1985. The population data are from [5, pp. 147, 166–169].**

| Country | 1978 | | 1985 | | 1997 |
	Population (millions)	Growth Rate	Population (millions)	Growth Rate	Population (millions)
World Total	4,258	0.0183	4,837	0.0165	5,840
Argentina	26.39	0.0129	30.56	0.0154	34.67
Brazil	115.40	0.0282	135.56	0.0216	162.66
Denmark	5.10	0.0039	5.11	0.0020	5.21
German Federal Republic	61.31	0.0005	61.02	0.0005	83.57
German Democratic Republic	16.76	−0.0007	16.64	−0.0012	(unified)
Great Britian	49.12	0.0012	49.92	0.0032	
India	638.39	0.0197	750.86	0.0203	952.11
Indonesia	145.10	0.0232	163.39	0.0217	206.61
Japan	114.90	0.0084	120.75	0.0061	125.45
P.R.China	933.03	0.0123	1,059.52	0.0120	1,210.00
Papau, New Guinea	2.99	0.0301	3.33	0.0300	4.39
United States	218.72	0.0085	239.28	0.0097	266.48

Note:

$$1978 \text{ annual growth rate} = \frac{(1979 \text{ population}) - (1978 \text{ population})}{1978 \text{ population}}$$

The 1985 annual growth rate is defined similarly.

strings of numbers persuade you that any question about issues as complex as human population has a unique answer.

Students of science and engineering who will never encounter population models might well ask, "Why waste time on something I don't need? Let's get to the differential equations I'm going to use in circuit theory of thermodynamics or . . . ."

Population models are a good setting for studying modeling because little specialized knowledge is required at an introductory level. Learning will not be complicated by excess technical baggage, and the skills of modeling, analysis, and interpretation you develop in this setting can easily be transferred to your specialty. For example, predicting the mass of a radioactive isotope involves populations of atoms, and determining the charge on a capacitor involves populations of electrons.

World population is also an important social concern. Predictions of population trends and proposals for solving population problems often are based on mathematical models. A critical understanding of modeling can make you a more effective world citizen.

The modest spread of annual rates of increase shown in table 2.1 could presage dramatic differences in future populations. For example, pushing a few buttons on a calculator (see exercise 15) reveals the following:

If the initial population is 1,000,000, then the population 10 years later is

- 1,104,622 at an annual rate of increase of 1% and
- 1,343,916 at an annual rate of increase of 3%.

If these population differences do not seem significant, look ahead 50 years!

Aren't these numbers enough? No! We need models that will illuminate the effects of parameters like the annual growth rate. Numerical calculations can show trends, but they probably can not identify the mechanisms by which different parameter values affect population growth.

2.1.2 A Law and Some Facts

To begin modeling, we need governing laws and experimental facts, such as Newton's law and "the force of gravity is proportional to mass" fact used in the rock motion models of chapter 1. The basic law is so obvious in this setting that it hardly seems worth stating:

Law of conservation of population. The net change in a population over a given period of time is the number of individuals added less the number of individuals removed.

Conservation: change = number added − number lost

This conservation law is the $F = ma$ of population models.

The conservation law is used here in a consistent pattern: First write mathematical expressions for the change in population, for the number added, and for the number lost over some arbitrary interval of time. Then let that interval become arbitrarily small, leading to a differential equation that relates *rates* of change.

The resulting balance of rates often seems painfully obvious, hardly worth the trouble of the derivation, but starting with a conservation law is a fruitful approach to modeling more complicated phenomena that arise throughout science and engineering. For example, you may encounter this same approach in other disciplines (e.g., conservation of energy in physics and thermodynamics or conservation of momentum in dynamics), although final expression may be left in terms of amounts, not rates.

The experimental facts needed depend upon the causes of population change we choose to consider. For the moment, restrict attention to births and "natural" deaths only; neglect migration, wars, disasters, plagues, etc.

The population of the United States over the last few decades corresponds to these conditions. The data of table 2.2 suggest that the net annual growth rate has hovered around 0.01 over that period. An easy simplification is to suppose that the birth rate and the death rate are each constant. This idea leads us to accept the following.

Experimental fact 1. The fraction of individuals in a population who reproduce in a unit time period is constant.

Experimental fact 2. The fraction of individuals in a population who die in a unit time period is constant.

These facts play the same role for this simple population model that the "force of gravity is proportional to mass" does for the rock model of chapter 1.

The constants mentioned in these experimental facts are loosely called the **birth rate** and the **death rate**, respectively. More precisely, they are the number of deaths and births per unit time per individual. They are illustrated

TABLE 2.2 **United States population data for 1970–91 taken from [4, 15]. The arithmetic mean of the growth rates is 0.0102.**

Year	Population (millions)	Annual Growth Rate
1970	205.05	0.0127
1971	207.66	0.0108
1972	209.90	0.0096
1973	211.91	0.0092
1974	213.85	0.0099
1975	215.97	0.0096
1976	218.04	0.0101
1977	220.24	0.0107
1978	222.59	0.0111
1979	225.06	0.0120
1980	227.76	0.0104
1981	230.14	0.0103
1982	232.52	0.0098
1983	234.80	0.0094
1984	237.00	0.0096
1985	239.28	0.0097
1986	241.61	0.0096
1987	243.92	0.0090
1988	246.11	0.0090
1989	247.34	
1990	249.92	
1991	253.69	

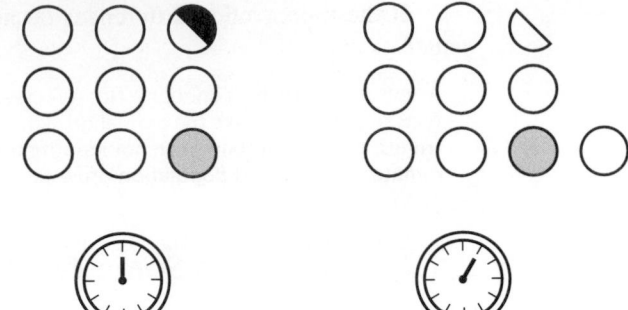

FIGURE 2.1   A fixed fraction of this population reproduces (light shading) and a fixed fraction (dark shading) dies during a unit time. In this example, $k_b = 1/9$ and $k_d = 1/18$.

in figure 2.1. If time is measured in years, then these constants would have units of per year per person.

2.1.3   A Model

The conservation law relates population change to the number added or lost over a given period of time. Those additions and loses are computed as follows.

Let k_b denote the birth rate constant and k_d the death rate constant mentioned in the experimental facts. These numbers are the *fraction* of the population giving birth or dying *per unit time*.

If the size of the population at some time t is $P(t)$, then the number of people being added to the population *per unit time* at that instant is $k_b P(t)$. That is,

Rate per unit time

- $k_b P(t)$ is the *rate per unit time* at which individuals are being added to the population due to births at time t, and

- $k_d P(t)$ is the *rate per unit time* at which individuals are being removed from the population due to deaths at time t.

To convert these *rates* into *amounts* over the period of time from t to $t + \Delta t$, multiply rate by elapsed time Δt:

Number added

- The number of individuals added *due to births* from t to $t + \Delta t$ is approximately $k_b P(t) \Delta t$.

Number lost

- The number of individuals lost *due to deaths* from t to $t + \Delta t$ is approximately $k_d P(t) \Delta t$.

"Rate times time" is only approximate because the rates $k_b P(t)$ and $k_d P(t)$ are not constant.

The main entries in the law of conservation of population are now at hand:

$$\text{change in population from } t \text{ to } t + \Delta t = P(t + \Delta t) - P(t)$$

$$\text{number added from } t \text{ to } t + \Delta t = k_b P(t) \Delta t$$

$$\text{number removed from } t \text{ to } t + \Delta t = k_d P(t) \Delta t.$$

Recall that change is "final − initial": $P(t + \Delta t) - P(t)$.

The model now follows from using the conservation law to connect the experimental facts, just as the rock model came from using Newton's law to relate total force to that of gravity:

Conservation law

$$\text{change} \quad = \text{ number added} \ - \text{ number lost}$$
$$P(t + \Delta t) = \quad k_b P(t) \Delta t \quad - \quad k_d P(t) \Delta t. \tag{2.1}$$

To obtain a differential equation, divide this last expression by Δt and let Δt approach zero:

$$\frac{P(t + \Delta t) - P(t)}{\Delta t} = (k_b - k_d) P(t),$$

$$\lim_{\Delta t \to 0} \frac{P(t + \Delta t) - P(t)}{\Delta t} = \lim_{\Delta t \to 0} (k_b - k_d) P(t),$$

$$\frac{dP}{dt} = (k_b - k_d) P.$$

Technically, P is a function that counts people; it takes only integer values. Since it is not even continuous, it can never have a derivative.

Here is the usual way around this problem: If the population is large, then the addition of one person to the population produces only a small relative change in the value of P. So smooth out those integer jumps in P, pretend P is continuous, even differentiable, and ignore fractions that occur because of this smoothing.

Intuition and our experience with the velocity model in chapter 1, which also involved a first-order equation, both suggest that we need an initial condition, the size of the starting population. Let P_i denote the population at the beginning of the first year of study. Label the first year as t_0. Then the initial condition is $P(t_0) = P_i$.

The complete (simple) population model is

$$\frac{dP}{dt} = kP, \tag{2.2}$$

SIMPLE POPULATION MODEL

$$P(t_0) = P_i, \tag{2.3}$$

where $k = k_b - k_d$. The constant $k = k_b - k_d$ is called the **net growth rate** or the **intrinsic rate of population growth** for the population. This growth rate is "net" because it includes both births and deaths. It is positive if the fraction of the population that reproduces is greater than the fraction of the population that dies per unit time.

In some populations of radioactive isotopes, k is negative because atoms "die" as they decay, but none are "born."

The net **annual growth rate** is the change in population over the given year divided by the population at the start of the year. (See the note in table 2.1.) The growth rates illustrated in the tables are *net* growth rates. We can estimate intrinsic growth rates from annual growth rates, but we are making the classic approximation, *equating an average rate of change over a finite time interval to the instantaneous rate*: $dP/dt = kP$.

The differential equation $dP/dt = kP$ appearing in this model is a consequence of the postulate that birth and death rates are constant. That equation says that the rate of change of population dP/dt is proportional to the current population P. The constant of proportionality is k, the population growth rate parameter.

2.1.4   Analysis

Separation of variables can find a formula for the function $P(t)$ that satisfies the population model initial-value problem (2.2–2.3). But first let us learn what we can about the behavior of the population function directly from the differential equation, beginning with a specific case.

ANALYSIS—SPECIFIC CASE

■ **EXAMPLE 1**   *Implement the population model $dP/dt = kP$, $P(t_0) = P_i$, for the United States beginning with 1970. Comment on the sign of the slope of the graph of P versus t. Draw a direction field diagram.*

The mean of the growth rates shown in table 2.2, page 29, is 0.0102. Hence, choose $k = 0.0102$ yr^{-1}. Using this value of k, $t_0 = 1970$, and $P_i = 205.05$, the model (2.2–2.3) becomes

$$\frac{dP}{dt} = 0.0102P,$$

$$P(1970) = 205.05 \quad (P \text{ in millions}).$$

The slope of the population curve at $t = 1970$ is positive:

$$\frac{dP}{dt} = (0.0102)(205.05) = 2.09 \text{ million people per year.}$$

Thus, the population grows initially, and it continues to grow faster and faster because the rate of growth (the slope) increases with the population: $P' = 0.0102P$. ■

MATLAB

The graphical tools in DELAB can draw the direction field for $P' = 0.0102\,P$. For guidance, start DELAB, select **Help, Textbook**, then go to chapter 2, figure 2.2.

Experiment with the range of values displayed. Click on an initial point to draw a solution curve through the direction field. Compare your results with figure 2.2.

Stop and Think

2.1 How do the values of the population predicted in figure 2.2 compare with the actual values in table 2.2?

2.2 The slopes in figure 2.2 become steeper with higher populations. Is that behavior consistent with the differential equation $P' = 0.0102P$? How does doubling the population affect the slope?

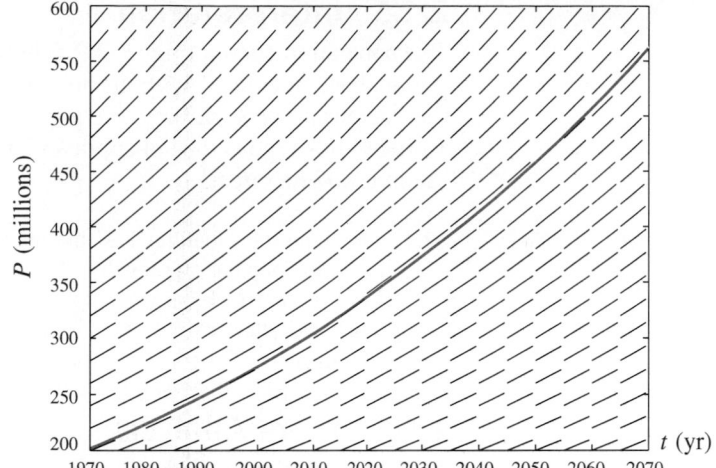

FIGURE 2.2 The direction field for the U.S. population model $P' = 0.0102P$ and a solution curve starting at $P(1970) = 205.05$.

2.3 How would figure 2.2 change if it were redrawn for a population with a *higher* growth rate, say for Brazil? How would the figure change if it were redrawn for a population with a *lower* growth rate, say Denmark? How would it change if the growth rate were *negative*?

To see that these same ideas apply in general, analyze the model $dP/dt = kP$, $P(t_0) = P_i$, for arbitrary (positive) values of the growth rate parameter k.

If you become confused in a general analysis, replace the letters with specific numbers. Here, take $k = 0.0102$, $t_0 = 1970$, and $P_i = 205.05$.

ANALYSIS—GENERAL CASE

If the initial population P_i is positive (negative populations are unlikely!), then at $t = t_0$ the slope of the P-versus-t curve is positive, for the differential equation $dP/dt = kP$ combines with the initial condition $P(t_0) = P_i$ to yield

$$\frac{dP}{dt} = kP_i > 0 \quad \text{at } t = t_0.$$

Hence, the population increases initially.

Indeed, the equation $dP/dt = kP$ forces the P-versus-t curve to have a positive slope as long as P remains positive. Since P begins with a positive value, the graph of P versus t always has a positive slope.

As P increases, the right-hand side of $dP/dt = kP$ becomes larger. Since P increases with time, the slope of the population curve increases with time, just like the curve in figure 2.2. Put differently, the curve in figure 2.2 is *concave up* because $P' = kP$ yields

$$\frac{d^2P}{dt^2} = \frac{dP'}{dt} = \frac{d(kP)}{dt} = kP' = k^2P > 0.$$

Population curves predicted by $P' = kP$ are concave up because $P'' > 0$ (if $k > 0$).

2.1.5 Interpretation

INTERPRETATION

The population curve of figure 2.2, whose shape was suggested by the preceding analysis, increases without bound at an ever more rapid rate. It appears that there is no upper limit to the populations increase at more rapid rates. If a model of this type is indeed applicable to human populations, then overpopulation is no idle threat!

Stop and Think

2.4 What important features has this model omitted that might limit such ever-increasing population growth? Might the death rate increase as larger populations competed for a fixed supply of food? Such possibilities are considered in the next section.

2.1.6 More Analysis

To obtain an alternate view of the predictions of this model, return to the specific case of the United States population beginning with 1970. The next example finds a formula for population as a function of time by solving the initial-value problem that constitutes the model.

■ **EXAMPLE 2**   *Use separation of variables to find the solution of the model of the population of the United States beginning in 1970,*

$$\frac{dP}{dt} = 0.0102P,$$

$$P(1970) = 205.05.$$

(See example 1.) Does the solution formula exhibit the same unbounded increase suggested by the preceding analysis of the general equation $dP/dt = kP$?

To solve the differential equation, follow *precisely* the steps used to solve $dy/dt = 4y$ in section 1.3.3, replacing y with P, 4 with k, and so on.

Collect terms involving P on the left (assuming $P \neq 0$), t on the right:

Separate variables.

$$\frac{dP}{P} = 0.0102 \, dt.$$

Integrate both sides:

Integrate.

$$\ln |P| = 0.0102t + c.$$

Exponentiate to remove the natural logarithm:

Solve for P.

$$|P| = e^{0.0102t+c}.$$

Finally, write $|C| = e^c$ and let the sign of C determine the sign of P:

$$P(t) = Ce^{0.0102t}.$$

This function solves $dP/dt = 0.0102P$ for *any* value of the arbitrary constant C. Clearly, $P \neq 0$ if $C \neq 0$.

Now choose C to satisfy the initial condition $P(1970) = 205.05$. Substituting $t = 1970$ and $P = 205.05$ into $P(t) = Ce^{0.0102t}$ leads to

Find C.

$$205.05 = Ce^{0.0102 \cdot 1970}.$$

$$C = 205.05e^{-0.0102 \cdot 1970}.$$

Substituting this expression for C into the solution formula $P(t) = Ce^{0.0102t}$ yields the solution of the initial-value problem

$$\frac{dP}{dt} = 0.0102P, \quad P(1970) = 205.05,$$

$$P(t) = 205.05e^{0.0102(t-1970)}. \tag{2.4}$$

To check this solution, verify $dP/dt = 0.0102P$ and $P(1970) = 205.05$.

Rather than evaluate $C = 205.05e^{-0.0102 \cdot 1970}$, it was left in this form to emphasize both the starting time 1970 and the initial population 205.05 in the solution formula (2.4).

Does this solution formula exhibit the sort of unbounded growth shown in figure 2.2? Yes, the population is growing exponentially! ■

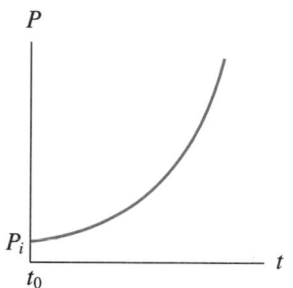

FIGURE 2.3 Exponential growth in the simple population model.

Is there a similar formula for the solution of the general population model $dP/dt = kP$, $P(t_0) = P_i$? Certainly. Just mimic the steps in the preceding example to obtain $P(t) = P_i e^{k(t-t_0)}$. See part (f) of exercise 2. This solution is plotted in figure 2.3.

MATLAB

The analytic tools in DELAB can solve the initial-value problem $P' = kP$, $P(t_0) = P_i$. For guidance, start DELAB, select **Help, Textbook**, then go to chapter 2, example 2. Does DELAB obtain $P(t) = P_i e^{k(t-t_0)}$?

SUMMARY

Compare the two analyses of the simple population model. One used slope information obtained directly from the differential equation. The other found a formula for the solution of the initial-value problem. Both concluded that population is unbounded.

The next two examples illustrate other ways in which a formula for the solution of an initial-value problem can be useful.

■ **EXAMPLE 3** *Use the model for the U.S. population beginning in 1970 to predict the time required for the population to double.*

The governing model is $P' = 0.0102P$, $P(1970) = 205.05$. The question asks for the time t_2 when $P(t_2) = 2 \cdot 205.05$. Use the solution formula

(2.4) to find it:

$$P(t) = 205.05e^{0.0102(t-1970)},$$
$$P(t_2) = 2 \cdot 205.05,$$
$$2 \cdot 205.05 = 205.05e^{0.0102(t_2-1970)},$$
$$2 = e^{0.0102(t_2-1970)},$$
$$\ln 2 = 0.0102(t_2 - 1970),$$
$$t_2 = (\ln 2)/0.0102 + 1970,$$
$$t_2 \approx 67.96 + 1970 \approx 2037.$$

The population of the U.S. doubles in about 68 years, *independent of the starting population.* ■

■ **EXAMPLE 4**   *A 1997 newspaper article [3] claims, "... since 1982, the number of stockbrokers has exploded by 123 percent." Suppose the population of stockbrokers is governed by the simple population model $P' = kP$. Find the "birth rate" k and compare it with that of typical populations. Is the verb "exploded" justified?*

The newspaper statistics assert that

$$P(1997) = (1 + 1.23)P(1982) = 2.23P(1982).$$

(The *increase* in population is $1.23P(1982)$.) The governing initial-value problem is $P' = kP$, $P(1982) = P_i$; 1982 is the natural starting date in this setting. Our problem is to find k using what we know about the behavior of the solution of this initial-value problem.

Separation of variables provides a solution formula:

$$P(t) = P_i e^{k(t-1982)}.$$

> MATLAB
> Compare this formula with that obtained using DELAB. For guidance, select **Help, Textbook**, then go to chapter 2, example 4. Are the two results the same?

Now use $P(1997) = 2.23P(1982) = 2.23P_i$ to find k:

$$P(1997) = P_i e^{k(1997-1982)},$$
$$2.23P_i = P_i e^{k \cdot 15},$$
$$\ln 2.23 = 15k,$$
$$k = (\ln 2.23)/15 \approx 0.0535 \text{ yr}^{-1}.$$

At better than 5% per year, the population of brokers is increasing much faster than the U.S. population, whose rate is about 1%. ■

> MATLAB
>
> To visualize the difference between the two growth rates, plot solutions of $P' = kP$, $k = 0.01$ (roughly, the U.S. population) and $k = 0.05$ (the brokers), on the same figure. Use the same initial value in both cases, say $P_i = 100$, and for convenience, $t_0 = 0$. For guidance, select **Help, Textbook**, then go to chapter 2, example 4. What differences do you observe? How does your perception of the differences in growth rate change with the elapsed time that is plotted?

Thomas Malthus (1766–1834) is credited with the first observation of the hypothesis underlying the model (2.2–2.3): The rate of increase of a population is proportional to its size. This and other population models are discussed in Smith's article [24]. The sobering predictions described in that article may pique your interest in population modeling.

Frauenthal's monograph [9] gives a good view of some of the scientific challenges of demography. Murray [18] provides a comprehensive account of many applications of mathematics to biology. Other excellent discussions of applications of mathematics to biology include [8] and [12].

For more information about the role of conservation laws in applied mathematics, see Lin and Segel's pioneering text [17].

2.1.7 Exercises

Exercise Guide	
To gain experience . . .	**Try exercises**
Deriving models	8, 10
Analyzing models from differential equations	4, 6, 9(a), 11(a), 14, 18(a), 19–20
Analyzing models from a solution	3, 5, 7, 9(b–c), 11(b–c), 13(b–c)
Interpreting models	1(b), 4–5, 7, 14
With concrete examples of models	1–5, 7, 12–13
Solving problems like these models	1–3, 5, 7, 9(b), 11(b)–13(b)
With the basic ideas of these models	4, 6, 12, 15–17
With the mathematical subtleties of model derivation	21–22

1. Example 2 used separation of variables to find the solution
$$P(t) = 205.05e^{0.0102(t-1970)}$$
of the initial-value problem
$$\frac{dP}{dt} = 0.0102P, \quad P(1970) = 205.05$$
modeling the population of the United States during the 1970s. (Population is measured in millions.)

(a) Verify that this function actually solves the initial-value problem by showing that it satisfies both the differential equation and the initial condition.

(b) Use this solution formula to compute $P(1978)$ and $P(1988)$. How accurately has this model predicted the actual populations shown in table 2.2?

2. Use separation of variables to solve each of the given differential equations. Verify that you have actually obtained

a solution in each case by substituting your proposed solution into the original differential equation and verifying that it reduces to an identity. If a differential equation is accompanied by an initial condition, use it to determine the arbitrary constant C that appears in your solution. Show that your analytic work is consistent with that of DELAB.

(a) The differential equation given in example 1 as part of the model of the population of the United States in the decade of the 1970s,

$$\frac{dP}{dt} = 0.0102P.$$

(b) The model for the population of the United States in the decade of the 1970s,

$$\frac{dP}{dt} = 0.0102P, \quad P(1970) = 205.05.$$

(c) A model for the population of Brazil beginning in 1978,

$$\frac{dP}{dt} = 0.0282, \quad P(1978) = 115.40.$$

(d) The differential equation from a model for the population of Brazil beginning in 1978,

$$\frac{dP}{dt} = 0.0282P.$$

(e) The differential equation for the general population model derived in this section,

$$\frac{dP}{dt} = kP.$$

(*Hint*: In section 1.3.3, we first applied separation of variables to an equation that looks very much like this one.)

(f) The general population model derived in this section,

$$\frac{dP}{dt} = kP, \quad P(t_0) = P_i.$$

3. For each of the indicated regions or countries,

(i) Write a population model in the form $dP/dt = kP$, $P(t_0) = P_i$, using 1978 as the initial year. Obtain growth rates and initial population data from table 2.1.

(ii) Use separation of variables or DELAB to find the solution of the initial-value problem.

(iii) Evaluate your solution at $t = 1985$, 1997, and compare the population predicted by your model with that given in table 2.1. What might account for the differences you observe?

(a) The world

(b) United States

(c) Denmark

(d) People's Republic of China (P.R.C.)

(e) German Democratic Republic

(f) Papau, New Guinea

4. Table 2.1 shows that the population of the German Democratic Republic has a negative growth rate in both 1978 and 1985. Is its population growing or decaying? What can you say about the relative values of k_b and k_d for the German Democratic Republic?

5. Use the data of table 2.1 for 1985 to write a model for the population of the German Democratic Republic. Solve the initial-value problem, and use the solution of your model to determine when the population will have declined to half of its 1985 value. This period is called the *half life* of this declining population. Repeat the calculation using the data for 1978. Does the half life appear to be sensitive to the value of the growth rate parameter?

6. What does the model $P' = (k_b - k_d)P$, $P(0) = P_i$, predict if $k_b < k_d$? Is that behavior consistent with the physical significance of the inequality $k_b < k_d$?

7. One way of gauging the speed with which a population is growing is to determine the time required for its size to double. You can use the data of table 2.1 and the solution of the appropriate model to make this calculation.

(a) Using the data for 1985, determine the time required for the population of the world to double.

(b) Using the data for 1985, determine the time required for the population of Indonesia to double.

(c) Using the general population equation $dP/dt = kP$, $k > 0$, find a formula in terms of the growth rate k for the time required for a population to double. Does the initial population matter? Find the time needed to double the slowest and fastest growing 1985 populations in table 2.1. Comment on the difference.

8. Consider a "population" of atoms of a radioactive isotope. No new isotope is created, but a fixed fraction s of the atoms decays to a more stable state ("dies") in unit time. State a form of the law of conservation of population that applies to isotopes, state the appropriate experimental fact(s), and derive the model

$$\frac{dN}{dt} = -sN, \quad N(0) = N_i,$$

for the number of atoms $N(t)$ of the isotope at time t. Explain the physical significance of N_i. (Actually, physicists and chemists do not use units that directly count the number of isotopes, as we have suggested here. Instead, they use *moles*, multiples of Avogadro's number— 6.023×10^{23}—of isotopes.)

9. (a) Argue directly from the differential equation in the model of the preceding exercise that the number of isotopes N is decreasing with time.

   (b) Use separation of variables or DELAB to find an analytic solution of the initial-value problem of the preceding exercise. Argue that your solution formula shows that N is in fact decaying to zero.

   (c) For a given value of the decay constant s, find the time required for the amount of isotope present to be reduced to half of its initial value. This time is known as the *half life* of the isotope. Show that the half life is independent of the initial amount of isotope.

10. Consider a "population" of dollars in a bank account. No money is being removed from the account. A fixed fraction i of the amount in the account is added to it each day as the account earns daily interest. (The daily interest rate i corresponds to a daily birth rate k_b for a population.) State a form of the law of conservation of population that applies to money in the bank, state the appropriate experimental fact(s), and derive the model

$$\frac{dN}{dt} = iN, \quad N(0) = N_i,$$

for the number of dollars $N(t)$ in the account at time t. Explain the physical significance of N_i.

11. (a) Argue directly from the differential equation in the model of the preceding exercise that the amount of money N in the account is increasing with time.

   (b) Use separation of variables or DELAB to find an analytic solution of the initial-value problem of the preceding exercise. Argue that your solution formula shows that N is in fact increasing without bound.

   (c) For a value of the daily interest rate i of your choice, use the solution formula to determine the time required for the initial sum in the account to double. Show that the time required to double the amount in the account is independent of the initial deposit.

12. The data in table 2.3 give the population of the United States at 10-year intervals.

   (a) Has the annual growth rate been constant? Estimate values of k, the growth rate parameter in the governing equation $P' = kP$, for the decades 1790–1800, 1850–1860, 1900–1910, and 1960–1970.

   (b) Use the annual growth rate estimated for the decade 1790–1800 to predict the 1970 population using the

approach of exercise 3. (Write an initial-value problem model using the data in table 2.3, solve the initial-value problem, and evaluate the solution at 1970.) How does your result compare with the actual population? Repeat, using the 1850 annual growth rate.

TABLE 2.3 United States population from 1790–1990 (from [4, p. 7] and [15, p. 6]).

Year	Population (millions)	Year	Population (millions)
1790	3.93	1890	63.06
1800*	5.30	1900	76.09
1810*	7.22	1910	92.41
1820	9.62	1920	106.46
1830	12.90	1930	123.19
1840*	17.12	1940	132.12
1850	23.26	1950	151.68
1860	31.51	1960*	180.67
1870	39.91	1970	204.88
1880	50.26	1980	227.72
		1990	249.92

*Indicates decades in which the land area included was increased.

13. The data in table 2.3 give the population of the United Sates at 10-year intervals. It can be used to estimate the net annual growth rate for any of the decades shown there.

   (a) Use the data of table 2.3 to estimate the annual growth rate of the United States for the decade 1790–1800. Using that data, state an initial-value problem of the form $P' = kP$, $P(t_0) = P_i$, which models the population of the United States beginning in 1790. (Give appropriate numerical values for the parameters k, t_0, P_i.)

   (b) Use separation of variables or DELAB to find a solution of the initial-value problem you found in part (a). Compare the value given by your solution formula with the population value in table 2.3 in 1970. What factors might account for any discrepancies you observe?

   (c) Repeat the preceding two steps beginning with 1850.

14. The analysis of the population model

$$\frac{dP}{dt} = kP, \quad P(t_0) = P_i,$$

showed that the slope of the population curve is always positive when P is positive. Further, it suggested that the slope increases as P increases. Confirm this suggestion by showing that the population curve is concave up because $P''(t) > 0$. (Calculate an expression for P'' directly from $P' = kP$.) Is this behavior consistent with the sketch of figure 2.2?

Determine $P''(1970)$ for the United States population model of examples 1 and 2. According to this model, how fast is population growth accelerating?

Can the solution of this initial-value problem exhibit local maxima or minima?

15. Work out the details of the back-of-the-envelope calculation mentioned near the end of section 2.1.1. That is, show that if the initial population is 1,000,000, and the annual growth rate is 1%, then the population after 10 years is 1,104,622, while at a growth rate of 3% the population rises to 1,343,916. (If the annual growth rate is 3%, then $P_1 = 1.03 P_0$, $P_2 = 1.03 P_1 = (1.03)^2 P_0$, and so forth, where P_n is the population at the end of year n.) *Hint*: You can use DELAB's **Numerical tools** if you can establish a connection between this exercise and finding an approximate solution of a certain initial-value problem using the Euler method.

16. The calculations in the preceding exercise suggest a population model of the form $P(t) = (1 + k)^t P(0)$, where k is the annual growth rate. Derive such a model, and explore its qualitative features. Are there limits to growth in such a model? Is your answer to that question consistent with the hypotheses upon which the model is based?

17. The analysis of the model $P' = kP$, $P(t_0) = P_i$, suggested that the population curve predicted by this model would have the general shape of figure 2.2.

(a) Starting with the population of Brazil in 1978 (see table 2.1), approximate the curve of figure 2.2, by drawing a straight line through the point $(1978, P(1978))$ with slope dP/dt given by $dP/dt = kP$. Use this line to estimate $P(1979)$. Then repeat the line-drawing process beginning at $(1979, P(1979))$. (The differential equation $dP/dt = kP$ will give you the slope of this new line segment using your value for $P(1979)$.) Continue for 10 years. What is the connection between this process and the Euler method? *Hint*: Once you establish the connection, you can use DELAB's **Numerical tools, Visual demos** to display this process.

(b) Establish a connection between this approximation technique and the model of exercise 15.

18. Consider the U.S. population model $P' = 0.0102P$, $P(1970) = 205.05$. As appropriate, use DELAB.

(a) Apply the Euler method to estimate the year the population of the U.S. will reach one billion. Why is this value only an estimate?

(b) Use the exact solution of this initial-value problem to determine the year the population of the U.S. will reach one billion. What confidence do you have in this value?

19. Choose a country of your choice from table 2.1. Write a version of the simple population model that is appropriate for it, and use the Euler method to estimate its population in the year 2100. As appropriate, use DELAB.

20. Use birth rate data from table 2.1 to write population equations for Denmark and the P.R.C. What differences would you expect between the direction fields for the two equations? After entering the appropriate equation, select DELAB's **Numerical tools, Direction field** menu to draw the direction fields and confirm those differences. Are there additional differences you didn't expect?

21. Example 3, page 35, used the solution of the initial-value problem $P' = 0.0102P$, $P(1970) = 205.05$, a model for the population of the U.S., to determine that the population doubles every $(\ln 2)/0.0102 \approx 69$ yr. Use one step of the Euler method to estimate the time required for the population of the U.S. to double. Compare the two values and comment on the differences.

22. A 1997 newspaper article [3] asserts that "the U.S. population has increased 14% since 1982...." Use this information to estimate the growth rate constant k for the U.S. population. Compare the value you obtain with those in table 2.2, page 29. Comment on any differences you observe.

23. Find a formula for the time required for a population governed by $P' = kP$ to double. Show that the doubling time depends only on k and not on the initial population.

24. Use one step of the Euler method to find a formula for the time required for a population governed by $P' = kP$ to double. Compare this result with the exact expression obtained in the previous exercise and comment on the relationship.

25. Suppose a population is governed by $P' = kP$ and that you know two values of the population P_0, P_1 at times t_0, t_1. Find the value of the growth rate constant k.

26. Suppose a population is governed by $P' = kP$ and that you know two values of the population P_0, P_1 at times t_0, t_1. Use one step of Euler's method with $\Delta t = t_1 - t_0$ to find an approximate formula for the growth rate constant k. Compare this result with the exact expression obtained in the previous exercise and comment on the relationship.

27. In deriving the population models in this section, we have repeatedly used logic that amounts to the following:

The net growth rate at time t is $kP(t)$. If Δt is not too large, then the increase in population is $kP(t)\Delta t$, the product of rate with time.

But no matter how small Δt is, the rate $kP(t)$ will not be constant. Could the approximation we are making here lead to a basic flaw in our population models?

This exercise asks you to carry out a more precise derivation which shows that the approximation we described in the preceding paragraph is valid; our models are not flawed on this account.

To be definite, limit your attention to a population which has a constant net growth rate and which is free of immigration or emigration.

(a) Argue that the increase in population from time t to $t + \Delta t$ is

$$\int_t^{t+\Delta t} kP(s)\,ds.$$

(*Hint*: The instantaneous rate of change—the population "velocity"— is $kP(t)$. How do you obtain distance traveled from the velocity function? Note that the variable of integration is s to avoid confusion with the particular time t at which the interval of interest begins.)

(b) Use the preceding result and the law of conservation of population to show that

$$P(t + \Delta t) - P(t) = \int_t^{t+\Delta t} kP(s)\,ds.$$

What is the corresponding equation in the derivation in the text?

(c) The next steps in the text's derivation are division by Δt and finding the limit as Δt approaches zero. The left-hand term in the preceding equation obviously reduces to dP/dt. Argue that the right-hand term can be written

$$\lim_{\Delta t \to 0} \frac{f(t + \Delta t) - f(t)}{\Delta t} = f'(t),$$

where $f(t) = \int_0^t kP(s)\,ds$.

(d) Use the fundamental theorem of calculus (see the Appendix) to show that $f'(t) = kP(t)$.

(e) Conclude that this more careful derivation leads to the same model: $dP/dt = kP$, $P(t_0) = P_i$.

28. The preceding exercise used the fundamental theorem of calculus in a careful derivation of the population model equation $dP/dt = kP$. Instead, use the mean-value theorem for integrals to show that

$$\int_t^{t+\Delta t} kP(s)\,ds = kP(\tilde{t})\Delta t$$

for some $\tilde{t}$, $t < \tilde{t} < t + \Delta t$. Then carry out a careful derivation of $dP/dt = kP$ similar to that in the preceding exercise. What becomes of $\tilde{t}$ as Δt approaches zero? (You can find a statement of the mean-value theorem for integrals in your calculus text.)

2.2 ■ EMIGRATION AND COMPETITION

Incorporating two new effects—emigration and competition—in the conservation law leads to two new population models. The derivations and analyses are similar: Account for gains and losses in the conservation law to derive a model, then use graphical tools (primarily) to predict population behavior.

2.2.1 Emigration

In 1994, about one million people, most of them Hutus, fled Rwanda to escape retribution after months of Hutu attacks against rival Tutsis. In the fall of 1989, before Germany was reunited, the German Democratic Republic briefly permitted as many as $12,000$ people per day, or 4.39×10^6 people per year to leave. In the spring of 1980, the Cuban government allowed about $5,000$ people per day (1.83×10^6 per year) to exit. In the middle of the last century, hundreds of thousand of people fled Ireland during the potato famine.

How does such emigration affect population levels predicted by the simple population model of the previous section?

With apologies to those who know their prefixes, *emigration* is people leaving.

A Model

To apply the law of conservation of population, we must account for all ways of increasing the population and all ways of decreasing the population over an arbitrary time interval Δt. Assume that birth and death continue to operate as in the simple population model. The only new effect is emigration.

Suppose individuals are leaving the population at a rate $E(t)$; appropriate units are millions of people per year. Then the usual "rate times time" argument shows that the number lost from t to $t + \Delta t$ due to emigration is $E(t)\Delta t$.

Now compute the population change terms for which the conservation law accounts:

- The number of individuals added from t to $t + \Delta t$ *due to births* is approximately $k_b P(t)\Delta t$.
- The number of individuals lost from t to $t + \Delta t$ *due to deaths* is approximately $k_d P(t)\Delta t$.
- The number of individuals lost from t to $t + \Delta t$ *due to emigration* is approximately $E(t)\Delta t$.

Recall that "rate times time" is only approximate because the rates $k_b P(t)$, $k_d P(t)$, and $E(t)$ are not constant.

The main entries in the law of conservation of population are:

$$\text{change in population from } t \text{ to } t + \Delta t = P(t + \Delta t) - P(t),$$

$$\text{number added by birth from } t \text{ to } t + \Delta t = k_b P(t)\Delta t,$$

$$\text{number removed by death from } t \text{ to } t + \Delta t = k_d P(t)\Delta t,$$

$$\text{number removed by emigration from } t \text{ to } t + \Delta t = E(t)\Delta t.$$

As before, the model follows from the conservation law:

$$
\begin{array}{ccccc}
\text{change} & = & \text{number added} & - & \text{number lost} \\
P(t + \Delta t) - P(t) & = & k_b P(t)\Delta t & - & (k_d P(t)\Delta t + E(t)\Delta t).
\end{array}
\tag{2.5}
$$

Stop and Think

2.5 Compare the conservation law expression (2.5) with its analog (2.1) from the derivation of the simple population model. Which terms are different—number added or number lost? Explain.

To obtain a differential equation, divide by Δt and let Δt approach zero:

$$\frac{P(t + \Delta t) - P(t)}{\Delta t} = (k_b - k_d)P(t) - E(t),$$

$$\lim_{\Delta t \to 0} \frac{P(t + \Delta t) - P(t)}{\Delta t} = \lim_{\Delta t \to 0} (k_b - k_d)P(t) - E(t),$$

$$\frac{dP}{dt} = (k_b - k_d)P.$$

(margin notes)

Emigration rate

Number emigrating

Conservation law

Write $k = k_b - k_d$ and add the usual initial condition to complete the model:

EMIGRATION MODEL

$$\frac{dP}{dt} = kP - E, \qquad (2.6)$$

$$P(t_0) = P_i. \qquad (2.7)$$

Analysis

Rather than seek a formula for the population function $P(t)$ by solving the initial-value problem (2.6–2.7), we will learn as much as we can about the behavior of this population *directly* from the differential equation $dP/dt = kP - E$.

To begin an analysis of this model, note that dP/dt might be positive or negative at $t = t_0$. What is the critical initial population level that guarantees $dP/dt > 0$ at $t = t_0$? Can you argue that dP/dt remains positive if it is positive at $t = t_0$? That it remains negative if it is negative at $t = t_0$?

Stop and Think   **2.6** Answer the questions posed in the previous paragraph.

Interpretation

If these questions can be answered *yes*, then it appears that a sufficiently small initial population could be completely eliminated by emigration while a sufficiently large population could survive. Alternatively, the value of the initial population could be used to determine the maximum rate of emigration a population could sustain.

■ **EXAMPLE 5**   *Using the following data, implement the emigration model (2.6–2.7) for Ireland during the period 1847–1850, the worst of the horrors of the potato famine.*

An average of about 209,000 people per year left Ireland during this period. To simplify the situation, take a constant emigration rate equal to this average value:

$$E = 0.209 \text{ million people per year.}$$

One might estimate a net annual rate of population increase of 1.5%, or $k = 0.015$. The population of Ireland in 1847 was about 8 million. The model (2.6–2.7) then has the form

POTATO FAMINE MODEL

$$\frac{dP}{dt} = 0.015P - 0.209, \qquad (2.8)$$

$$P(1847) = 8. \qquad (2.9)$$

The slope of the population curve at $t = 1847$ is

$$\frac{dP}{dt} = 0.015P(1847) - 0.209$$

$$= 0.015 \cdot 8 - 0.209 = -0.089.$$

The population is decreasing! Convince yourself that the slope of this model's population curve will be forever negative. Had emigration continued unchecked, Ireland would have become deserted. ■

A direction field diagram for the potato famine model (2.8–2.9) appears in figure 2.4. The initial population $P(1847) = 8$ is in a region of decreasing slopes, confirming the depressing analysis of the example: the population of Ireland was decreasing because the emigration rate 0.209 million/yr exceeded the net birth rate in 1847, $0.015 \cdot 8 = 0.120$ million/yr.

FIGURE 2.4   Direction field for $P' = 0.015P - 0.209$.

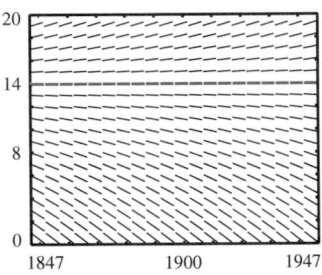

Stop and Think   **2.7** Identify the initial population point $P(1847) = 8$ on the direction field of figure 2.4. Trace along the direction field to estimate the year in which Ireland would have been deserted had emigration continued unabated.

> **MATLAB**
>
> Use the graphical tools in DELAB to draw your own version of the direction field in figure 2.4. For guidance, select **Help, Textbook**, then go to chapter 2, figure 2.4.

Could we obtain a formula for the solution of the initial-value problem that models Ireland during the potato famine? Certainly. See either exercise 1 or 2(b). Could we obtain a formula for the solution of the general emigration model $dP/dt = kP - E$, $P(t_0) = P_i$? Again, yes, at least if E is constant. See exercise 2(g).

■ **EXAMPLE 6**   *Could the population of Ireland have been maintained at a constant level during the potato famine? Obtain an answer using the differential equation $P' = 0.015P - 0.209$ derived in the previous example.*

STEADY-STATE SOLUTION

Such a constant population, if one exists, is called a **steady state**. A **steady-state solution** of a differential equation is a constant solution.

Assume there is such a constant solution; call it P_{ss}. To find its value, substitute it into $P' = 0.015P - 0.209$. Remember that $P'_{ss} = 0$ because P_{ss} is a constant:

$$0 = 0.015P_{ss} - 0.209,$$
$$P_{ss} = 0.209/0.015 \approx 14.$$

Hence, there is a steady-state population of about 14 million people. ■

The steady-state solution found in the previous example appears in figure 2.4: it is the horizontal line, corresponding to zero slope, at $P \approx 14$ on the vertical axis.

Notice that a slight increase in population above the steady-state value would move above the horizontal line into a region of population increase; the population would continue to grow. A slight decrease in population would have the opposite effect, leading to a continuing decrease in population. This steady state is called **unstable** because it disappears if it is disturbed. It would never be seen in reality for very long since an increase or decrease of just one person would cause the population to grow or decay away from the steady state.

STABLE STEADY STATE

A steady state y_{ss} of a first-order differential equation is called **asymptotically stable** or simply **stable** if solutions whose initial values are sufficiently close to y_{ss} approach that value in the limit: $\lim_{t \to \infty} y(t) = y_{ss}$. See example 9 for an example of a stable steady state. Steady states and stability are discussed in more detail beginning in section 3.3.

MATLAB

To demonstrate the instability of the steady-state solution $P_{ss} \approx 14$ of (2.8), draw its direction field using DELAB. Click near $P_{ss} \approx 14$ and watch the solutions diverge from it, carried away by the flow of the direction field.

■ **EXAMPLE 7** *Draw a direction field diagram for the general emigration equation $P' = kP - E$, E a constant.*

Mimic the observations for the specific equation $P' = 0.015P - 0.209$: the slope $P' = kP - E$ can be positive or negative, depending on whether P is large or small. The steady-state population, the one where $P' = 0$, separates the two cases.

The (constant) steady-state solution P_{ss} satisfies

Find P_{ss}.

$$P'_{ss} = 0 = kP_{ss} - E.$$

Hence, $P_{ss} = E/k$. Mark this point on the P axis in figure 2.5, then find the sign of P' for values of P above and below it.

If $P > E/k = P_{ss}$, then

$P > P_{ss} \Longrightarrow P' > 0$

$$P' = kP - E > kP_{ss} - E = 0$$

and the population is increasing. Similarly, if $P < E/k = P_{ss}$, then

$P < P_{ss} \Longrightarrow P' < 0$

$$P' = kP - E < kP_{ss} - E = 0$$

and the population is decreasing. That is, above $P = E/k$ the direction arrows point up, and below $P = E/k$ they point down.

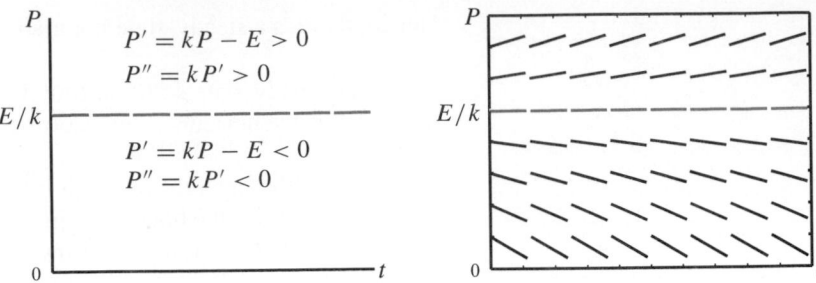

FIGURE 2.5   Direction field and supporting analysis for $P' = kP - E$, E constant.

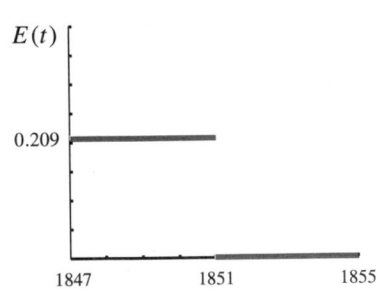

FIGURE 2.6   Suppose emigration ends abruptly in 1851 as the potato famine wanes; the rate is millions/year.

The second derivative tells how the angle of the direction arrows changes with P:

$$P'' = (kP - E)' = kP'.$$

Hence, P'' has the same sign as P'. When the direction field arrows are sloping up, they trace paths that are concave up; i.e., the slope of upward pointing arrows increases with P. The situation is reversed when $P' < 0$.

Figure 2.5 summarizes this analysis and shows some typical direction field arrows.   ■

The previous discussion of the potato famine model assumes emigration continues forever, i.e., that $E \equiv 0.209$. In fact, the worst of the famine ended about 1851. Emigration should have tapered off after that. A crude representation of that change is the emigration rate expression displayed in figure 2.6,

$$E(t) = \begin{cases} 0.209, & 1847 \leq t < 1851 \\ 0 & 1851 \leq t \end{cases} \tag{2.10}$$

■   **EXAMPLE 8**   *The emigration model (2.8–2.9) with constant emigration rate predicts perpetually decreasing population. Does the same behavior occur with the modified emigration rate (2.10)?*

For $1847 \leq t < 1851$, the governing equation is

$$P' = 0.015P - 0.209.$$

At $t = 1847$, $P = 8$ and $P'(1847) = 0.015 \cdot 8 - 0.209 < 0$. The population is decreasing initially, just as in the model (2.8–2.9).

But for $1851 \leq t$, $P' = 0.015P > 0$ if $P > 0$. So unless the population has vanished, it will begin to increase after 1851, unlike the steady decrease predicted by (2.8–2.9). The population begins to recover after 1851, as the direction field diagram of figures 2.7 and 2.8 illustrate.   ■

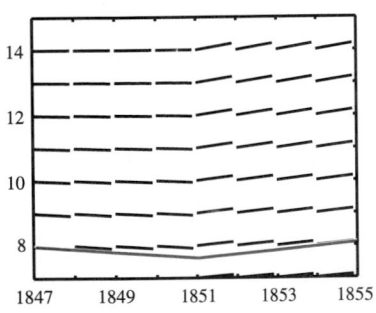

FIGURE 2.7   If emigration ends in 1851, the population of Ireland begins to recover.

Stop and Think   **2.8** Can the differential equation $P' = 0.015P - E(t)$, where $E(t)$ is given by (2.10), be solved easily by separation of variables? Is the difficulty with this particular choice of $E(t)$ or with E varying with t?

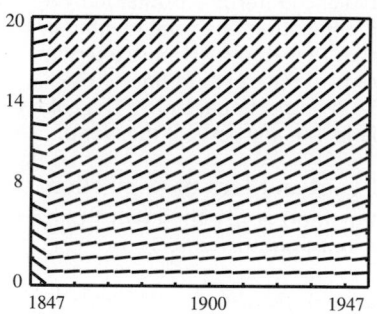

FIGURE 2.8 The direction field of $P' = 0.015P - E(t)$, $E(t)$ the variable emigration rate defined by (2.10).

MATLAB

Asking MATLAB to draw a direction field for $P' = 0.015P - E(t)$ with $E(t)$ defined by (2.10) requires a "one line" formula for $E(t)$. The MATLAB function `sign` provides the tool; `sign(x)` is $+1$, -1, or 0 according to whether `x` is positive, negative, or zero. (Type `help sign` at the MATLAB prompt for more information.)

The function pictured in figure 2.6 is "on" (with a value of 0.209) for $1847 \leq t < 1851$ and "off" otherwise. The MATLAB function `(1 - sign(x))/2` is "on" (with a value of 1) for `x < 0` and "off" for `x > 0`.

To write $E(t)$ in terms of `(1 - sign(x))/2`, replace `x` with `t - 1851` to make the switch occur at `1851`. Multiply by `0.209` to get the correct "on" value. The resulting MATLAB representation for (2.10) is `0.209*sign(t - 1851)/2`.

2.2.2 Competition

The models examined so far have omitted an important constraint, competition. As population grows, food and space can become scarce, causing death by starvation and disease. To model a population whose growth is limited by competition for resources, suppose

- the birth rate is constant,
- the death rate *rises* with population to reflect competition.

> If data were available to support these assumptions, they could be labeled *experimental facts*. Without supporting data, they are just the simplest way of tentatively introducing the effects of competition.

To distinguish this case from the simple population model (2.2–2.3), let a (rather than k_b) denote the constant fraction of the population that reproduces per unit time. Then

Rate added

- $aP(t)$ is the *rate per unit time* at which individuals are being added to the population at time t due to births.

The simplest way to model a death rate that increases with population is to replace the constant death rate k_d of the simple population model (2.2–2.3) with a term proportional to P, say, sP for some (positive) constant s. That is, the fraction of the population which dies per unit time is sP. Just as $k_d \cdot P$ is the rate per unit time at which individuals are being removed in the simple population model,

Rate lost

- $sP(t) \cdot P(t) = s(P(t))^2$ is the *rate per unit time* at which individuals are being removed from the population at time t due to deaths.

> Another route to the same death rate expression is to suppose that the death rate varies in a way that reflects the number of encounters between individuals. It is those encounters in the search for food and shelter that can limit population growth.

A single individual in a population of size P could encounter $P-1$ other individuals. Since there are P single individuals in the population, the total number of encounters among all individuals is proportional to $P(P-1) \approx P^2$. Since a certain fraction of encounters between competitors leads to the death of one of them, the number of individuals who die per unit time is proportional to P^2. This cold logic should not conceal the reality of this scenario in many parts of the world.

We now have expressions for rates of gain and loss of population. It remains only to find the corresponding amounts and apply the law of conservation of population. "Rate times time" gives amounts:

Number added

- The number of individuals added from t to $t + \Delta t$ *due to net births* is approximately $k_b P(t) \Delta t$.

Number lost

- The number of individuals lost from t to $t + \Delta t$ *due to competition* is approximately $s(P(t))^2 \Delta t$.

The main entries in the law of conservation of population are:

$$\text{change in population from } t \text{ to } t + \Delta t = P(t + \Delta t) - P(t),$$
$$\text{number added by birth from } t \text{ to } t + \Delta t = aP(t)\Delta t,$$
$$\text{number removed by competition from } t \text{ to } t + \Delta t = s(P(t))^2 \Delta t.$$

As before, the model follows from the conservation law:

Conservation law

$$
\begin{array}{ccccc}
\text{change} & = & \text{number added} & - & \text{number lost} \\
P(t + \Delta t) - P(t) & = & aP(t)\Delta t & - & s(P(t))^2 \Delta t.
\end{array}
\tag{2.11}
$$

Stop and Think   **2.9** Compare the conservation law expression (2.11) with its analogs (2.1) and (2.5) from the derivations of the simple population model and the emigration model. Which terms are different? Explain.

As usual, derive the differential equation by dividing by Δt and letting Δt approach zero:

$$\frac{P(t + \Delta t) - P(t)}{\Delta t} = aP(t) - sP(t)^2,$$

$$\lim_{\Delta t \to 0} \frac{P(t + \Delta t) - P(t)}{\Delta t} = \lim_{\Delta t \to 0} aP(t) - sP(t)^2,$$

$$\frac{dP}{dt} = aP - sP^2.$$

Add an initial condition to complete the model:

LOGISTIC MODEL

$$\frac{dP}{dt} = aP - sP^2, \tag{2.12}$$

$$P(t_0) = P_i. \tag{2.13}$$

The differential equation $P' = aP - sP^2$ is often called the **logistic equation**. (*Logistic* refers to the science of numbering things.) The Belgian mathematician P. F. Verhulst (1804–1849) derived it in 1838.

Because of the P^2 term, the logistic equation is *nonlinear*, distinguishing it from the simple population equation $P' = kP$ and the emigration equation $P' = kP - E$, both of which are linear.

■ **EXAMPLE 9**   *Using the parameters given below for a population of algae, determine if competition places an upper limit on population.*

One study [5] concluded that for the alga *genera fosiella* one could choose $a = 0.12$/day and $s = 0.024$/mm^2-day. Thus, the governing equation is $P' = 0.12P - 0.024P^2$. (The raw data is in table 2.4, page 55.)

Population, or *biomass*, of algae is measured by the area covered, not by counting individuals.

Since $aP - sP^2$ must have units of rate of population change—mm^2/day in this case—the dimensions of a and s must be such that aP and sP^2 each have these same units, mm^2/day.

MATLAB

MATLAB can draw the direction field. See the right half of figure 2.9, which also includes solution trajectories starting at $P(0) = 0.25$, 3.0, and 9.5 mm^2. Confirm the analysis shown in figure 2.9 using DELAB's graphical tools. For guidance, select **Help, Textbook**, then go to chapter 2, figure 2.9.

The direction field in figure 2.9 suggests that small populations grow (e.g., those starting at 0.25 and 3.0 mm^2) while large populations (e.g., $P(0) = 9.5$ mm^2) decay. All solutions seem to be converging on a (stable) steady state $P_{ss} \approx 5$ mm^2.

It seems that *competition prevents unbounded growth* and that there is a natural population level of 5 mm^2. This steady state is called the **carrying capacity** of the alga population. The carrying capacity is a *stable* steady state because all nearby solutions converge to it as time passes. ■

Stop and Think   **2.10** Using the data of the previous example, estimate the fraction of encounters between two algae that lead to the death of one of the competitors.

■ **EXAMPLE 10**   *Confirm analytically the direction field behavior shown in the right half of figure 2.9 for the alga competition model* $P' = 0.12P - 0.024P^2$.

Steady states solve

$$P'_{ss} = 0 = 0.12P_{ss} - 0.024P_{ss}^2.$$

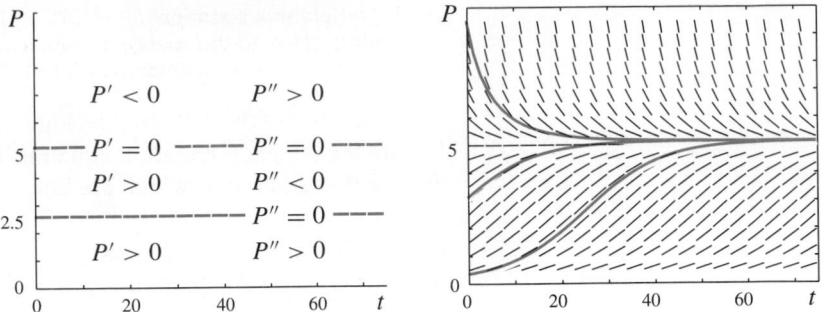

FIGURE 2.9   Left: analysis of the direction field for the alga competition equation $P' = 0.12P - 0.024P^2$. Right: direction field with solution curves starting at $P(0) = 0.25, 3.0, 9.5$ mm^2.

The quadratic $0.12P_{ss} - 0.024P_{ss}^2 = P_{ss}(0.12 - 0.024P_{ss})$ has two roots,

$$P_{ss} = 0, \quad P_{ss} = 0.12/0.024 = 5.$$

Mark these values on the P axis in the left half of figure 2.9.

Now find the sign of $P' = 0.12P - 0.024P^2$ on the intervals between the steady states, $0 < P < 5$ and $5 < P$. Since P' can have only one sign between its zero values, it suffices to test the sign of P' at just one point within each interval. Say, use $P = 2$ to discover that $P' > 0$ for $0 < P < 5$, and use $P = 6$ to show that $P' < 0$ for $5 < P$. Add this information to the left half of figure 2.9.

Use the *chain rule* to compute P'' from $P' = 0.12P - 0.024P^2$:

$$P'' = \frac{dP'}{dt} = \frac{d}{dt}(0.12P - 0.024P^2)$$

$$= 0.12\frac{dP}{dt} - 0.024\frac{d(P^2)}{dt}$$

$$= 0.12\frac{dP}{dt} - 0.024 \cdot 2P\frac{dP}{dt}$$

$$= (0.12 - 0.048P)P'.$$

The chain rule says that $d(P^2)/dt = 2PdP/dt$ for the same reason it says that $d(\cos^2 \pi t)/dt = 2(\cos \pi t)d(\cos \pi t)/dt$.

First find the zeros of P''; then find its sign between the zeros, just as with P'. Since $P'' = (0.12 - 0.048P)P'$,

$$P'' = 0 \quad \text{if } 0.12 - 0.048P = 0 \text{ or if } P' = 0.$$

Hence, $P'' = 0$ if $P = 0.12/0.048 = 2.5$ or if $P = 0$ or $P = 5$. Mark these values on the P axis in figure 2.9.

Test the sign of P'' at a point within each of the intervals $0 < P < 2.5$, $2.5 < P < 5, 5 < P$. The results appear in figure 2.9.

Conclusions: the graph of P is increasing for $0 < P < 5$, is concave up for $0 < P < 2.5$, is concave down for $2.5 < P < 5$, and has an inflection point at $P = 2.5$, half way to the carrying capacity of $P_{ss} = 5$. The graph of P is decreasing and concave up for $5 < P$.

Compare with example 7, page 45, a similar analysis of an emigration equation. ■

Stop and Think   **2.11** Does figure 2.9 suggest that the steady state $P_{ss} = 0$ of $P' = 0.12P - 0.024P^2$ is stable or unstable?

2.2.3 Summary of Single-Species Population Models

CONSERVATION OF POPULATION

Population models are derived by coupling the law of *conservation of population* with experimental facts about birth, death, emigration, or competition. The conservation law replaced Newton's law that was used in the rock model. Conservation of population requires that the change in population over a period of time equal the number added less the number lost.

The models are all initial-value problems. The differential equation for the simplest growth model is

SIMPLE GROWTH

$$\frac{dP}{dt} = kP.$$

The growth rate constant is $k = k_b - k_d$, the difference of the birth rate and the death rate. The differential equation for a model of the same population subject to emigration at a rate E is

EMIGRATION

$$\frac{dP}{dt} = kP - E.$$

The logistic equation

LOGISTIC

$$\frac{dP}{dt} = aP - sP^2$$

arises in a model that supposes that the death rate increases with population. Each of the models is completed by imposing the initial condition

$$P(t_0) = P_i.$$

Direction field diagrams guide a graphical analysis of each model. The direction information comes from the slope expression on the right side of the differential equation $dP/dt = \cdots$.

2.2.4 Exercises

1. This exercise leads you through the process of finding a solution of the initial-value problem

$$\frac{dP}{dt} = 0.015P - 0.209, \quad P(1847) = 8,$$

which models the Irish potato famine, and it asks you to use the solution formula you obtain to corroborate the analysis of the text.

(a) Separate variables in the differential equation so that you are left with the antidifferentiation problem

$$\int \frac{dP}{0.015P - 0.209} = \int dt.$$

Evaluate each of these antiderivatives.

(b) Exponentiate, bring a constant out of the exponential, and obtain the formula

$$P(t) = Ce^{0.015t} + \frac{0.209}{0.015},$$

where C is an arbitrary constant.

(c) Using the initial condition $P(1847) = 8$ to evaluate the constant C, obtain the solution

$$P(t) = \left(8 - \frac{0.209}{0.015}\right) e^{0.015(t-1847)} + \frac{0.209}{0.015}$$

of the complete initial-value problem.

(d) Verify by substituting into the differential equation and the initial condition that you do indeed have a function which satisfies both. Do the analytic tools in DELAB obtain the same solution?

(e) From an analysis of the differential equation itself, the text concludes that this model predicts the population of Ireland will decline. Show that the solution formula you have just obtained predicts precisely the same effect. When would the population of Ireland have been depleted? Could you have obtained that information without the solution formula?

2. Use separation of variables to solve each of the differential equations listed below. Verify that you have actually obtained a solution in each case by substituting your proposed solution into the original differential equation. If a differential equation is accompanied by an initial condition, use it to determine the arbitrary constant C that appears in your solution. Compare your solution formulas with those obtained by DELAB.

(a) The differential equation given in example 5 as part of the model of Ireland during the potato famine,

$$\frac{dP}{dt} = 0.015P - 0.209.$$

(b) The model given in example 5, of Ireland during the potato famine,

$$\frac{dP}{dt} = 0.015P - 0.209, \quad P(1847) = 8.$$

(c) The differential equation from a possible model for the German Democratic Republic during the fall of 1989, assuming emigration continued unchecked,

$$\frac{dP}{dt} = -0.0012P - 4.39.$$

(Population is in millions, time in years. From where in the text did we get these numbers?)

(d) A possible model for the German Democratic Republic during the fall of 1989, assuming emigration continued unchecked,

$$\frac{dP}{dt} = -0.0012P - 4.39, \quad P(1989) = P_i.$$

(The growth rate value is that for 1985. The initial population appears just as the parameter P_i since we do not have its exact value. How might you use the data of table 2.1 to make a reasonably accurate estimate of P_i?)

(e) The differential equation from a model of the population of Cuba during the spring of 1980, assuming emigration continued unchecked,

$$\frac{dP}{dt} = kP - 1.825.$$

(Population is in millions, time in years. The growth rate appears just as the parameter k since we do not have its value. Where might you find a value for k?)

(f) The differential equation of the population model with constant emigration rate E,

$$\frac{dP}{dt} = kP - E.$$

(What difficulties might you encounter if E were a function of time rather than a constant?)

(g) The population model with constant emigration rate E,

$$\frac{dP}{dt} = kP - E, \quad P(t_0) = P_i.$$

(h) The differential equation of the logistic population model,

$$\frac{dP}{dt} = aP - sP^2.$$

(i) The logistic population model,

$$\frac{dP}{dt} = aP - sP^2, \quad P(t_0) = P_i.$$

3. Suppose the government had decided to limit Irish emigration in 1847 to a level that would keep the population constant. Use the data of example 5 to determine the number of people per year who could have emigrated then.

4. Follow the first four steps of exercise 1 to obtain a formula for the solution of the general emigration model

$$\frac{dP}{dt} = kP - E, \quad P(0) = P_i.$$

Derive from your formula an expression for the critical initial population level that is required to prevent depletion of

the population by a given emigration rate E. Find the time when the population reaches zero if the initial population is less than this critical value.

5. (a) Adapt the data from table 2.1 and the law of conservation of population to derive an emigration model for the German Democratic Republic during the fall of 1989, when approximately 12,000 people per day were emigrating.

(b) Find the solution of this model. If this rate of emigration had continued indefinitely, when would the country have been stripped of its population?

(c) What is the maximum emigration rate the government could have permitted without causing the population to decline to zero? Is emigration the sole cause of population decline in this model?

6. A population of yeast spores increases due to reproduction by a fixed proportion R (i.e., by $100R\%$) each day. The spores are harvested at a rate of H spores per day.

(a) Write an expression for the *rate* per day at which the number of yeast spores is increasing due to reproduction alone.

(b) Write an expression for the *rate* per day at which the number of yeast spores is decreasing due to harvesting alone.

(c) State a law of conservation of population and use it to derive a model for the number of yeast spores in this population.

7. For most of the past two centuries, the number of immigrants entering the United States has exceeded the number of emigrants departing. Derive an *immigration model* for such a situation. Carefully state the growth rate assumptions you use, and precisely define any terms you introduce to characterize immigration. Derive your model from the law of conservation of population.

8. The preceding exercise requests a careful derivation of a model for a population subject to *immigration*. The text derives a model for a population with constant net growth rate subject to *emigration*; the governing differential equation is the emigration equation

$$P'(t) = kP(t) - E(t).$$

(a) Call the rate of immigration $I(t)$. Comparing this situation with the emigration model suggests that an immigration model equation should have the form $P'(t) = kP \pm I(t)$. Which sign do you think is correct? Explain your reasoning.

(b) Derive the model as requested in the preceding exercise. Did you guess the form of the differential equation correctly? If not, what did you do wrong?

9. A population of amoeba in a particular region of a petri dish is under observation. The birth rate of the amoeba is proportional to the population. The rate at which amoeba die is proportional to the *square* of the population (to account for crowding effects).

   In addition, amoeba move into or out of this region in response to an attractant released by other amoeba. The rate of movement of amoeba is proportional to the difference between the population in the region and the population A_{out} outside the region. The amoeba move *toward the higher population* level.

   Let $A(t)$ denote the population of amoeba in the region at time t.

   (a) Letting b denote the proportionality constant, write an expression for the *number* (not rate) of amoeba added to the population in the region by births over a time period of length Δt.

   (b) Letting d denote the proportionality constant, write an expression for the *number* (not rate) of amoeba lost from the population in the region by deaths over a time period of length Δt.

   (c) Letting r denote the proportionality constant, write an expression for the *number* (not rate) of amoeba added to the population in the region by the effect of the attractant over a time period of length Δt. (Note that population is added if $A > A_{out}$ and lost if $A < A_{out}$.)

   (d) State an appropriate conservation law for the amoeba population in the region. Derive a model for the population function $A(t)$ from this law.

10. A population of yeast spores increases by a fixed proportion k each day. Spores are harvested at a rate of E spores per day. Argue that this spore population is governed by the emigration model

$$P' = kP - E, \quad P(t_0) = P_i.$$

   Derive this model from the law of conservation of population.

11. An amount of money m is earning interest at a daily rate of I (or $100I$ percent). Regular deposits are made at a rate of $D(t)$ dollars per day. The account was opened with a starting balance of S dollars. Determine which three of the following initial-value problems could *not* possibly be a reasonable model of this situation. *Justify* rejecting each unacceptable choice.

   (a) $m' + Im = D, m(0) = S$

   (b) $m' - Im = -D, m(0) = S$

   (c) $m' - Im = D, m(0) = S$

   (d) $m' + Im = -D, m(0) = S$

   (Interest is being compounded continuously on the funds in this account.)

12. Show that the one initial-value problem you did not reject in the previous exercise is the correct model by deriving it from the "law of conservation of money": The net change in the balance in an account over an interval of time equals the amount added less the amount removed.

13. Assuming that the rate of deposit D is constant, solve the one initial-value problem you did not reject in exercise 11. Show that the solution function $m(t)$ you obtain exhibits behavior that is consistent with the situation being modeled there, a bank account that is earning interest and receiving regular deposits.

14. One of the initial-value problems listed in exercise 11 has exactly the same form as the emigration model

$$\frac{dP}{dt} = kP - E, \quad P(t_0) = P_i.$$

   Identify it. (Recall that exercise 11 considers a model of the balance in a bank account earning interest and receiving regular deposits.) By comparing the physical situations being modeled (population with emigration, bank account with regular deposits), argue that this initial-value problem can not possibly model the bank account.

15. The analysis of the model

$$P' = kP - E, \quad P(t_0) = P_i,$$

   raises several questions. Answer them and pursue the directions suggested by the interpretation of that model:

   What is the critical initial population level that guarantees $dP/dt > 0$ at $t = t_0$? Can you argue that dP/dt remains positive if it is positive at $t = t_0$? That it remains negative if it is negative at $t = t_0$?

   It appears that a sufficiently small initial population could be completely eliminated by emigration while a sufficiently large population could survive. Instead of determining a critical initial population level, use the initial population to determine the maximum rate of emigration a population could sustain.

16. Use the model $P' = kP - E, P(t_0) = P_i,$ to determine the maximum emigration rate five countries of your choice from table 2.1, page 28, could sustain. (*Hint:* Use the answer to the challenge posed by the last sentence of exercise 15.)

17. The analysis of the emigration model

$$\frac{dP}{dt} = kP - E, \quad P(t_0) = P_i,$$

showed that the slope of the population curve at t_0 could be either positive or negative, depending on the size of the initial population P_i. If the emigration rate E is constant, argue that a population curve that starts with a positive (negative) slope always has a positive (negative) slope. Further, argue that the second derivative of the solution of this initial-value problem has the same sign as the first derivative, so that an increasing (decreasing) population curve is concave up (down). (*Hint:* Calculate P'' directly from $P' = kP - E$.)

Can population ever become negative in this model? Can the population curve exhibit local maxima or minima?

18. The logistic model $dP/dt = aP - sP^2$, $P(0) = P_i$, was derived in the text by assuming that the number of individuals dying per unit time is proportional to the square of the population. This exercise outlines a different approach.

The competition effect is difficult to assess quantitatively because it could affect both births and deaths, but it should act to decrease the net growth rate k as population increases. A linear, decreasing relationship between k and P may be the easiest way to achieve this variation. Mathematically, we can write such a variable growth rate as

$$g(P) = a - sP$$

for some (positive) constants a, s. The variable growth rate g replaces the constant growth rate k used earlier.

The last equation is our new experimental "fact," such as it is. If s is a small positive constant, then $g(P)$ has the qualitative behavior sketched in figure 2.10. The constants a and s must be chosen to fit specific population data. Derive the logistic equation model by invoking conservation of population with population additions due only to the *net* growth term $g(P)P$ and with no removal term.

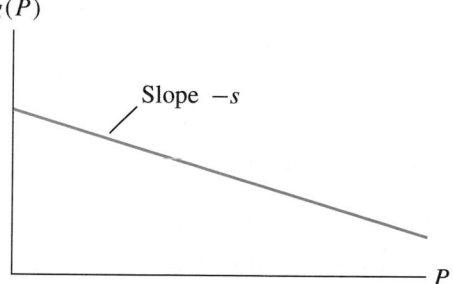

$g(P)$

Slope $-s$

P

FIGURE 2.10 A growth rate that reflects competition by decreasing with increasing population.

19. In deriving the death rate term in the competition model (2.12–2.13), the text says

> The simplest way to model a death rate that increases with population is to replace the constant death rate ... with a term proportional to P, say, sP for some (positive) constant s.

But the death rate expression sP suggests that the death rate is practically zero for small populations when the minimum death rate probably ought to be some nonzero constant.

To remedy this apparent defect, replace sP with $r + sP$, where r and s are positive constants. That is, suppose that

$$(r + sP(t)) \cdot P(t) = rP(t) + s(P(t))^2$$

is the rate per unit time at which individuals are being removed from the population due to deaths at time t.

Derive the governing differential equation. Show that it has the same form as the logistic equation (2.12). Argue that the term analogous to aP in (2.12) represents a *net* birth rate.

20. Table 2.4 shows population data for the alga *genera fosiella*. Using the logistic parameters given in example 9, solve (analytically or numerically, using DELAB as appropriate) the competition model for this population. Comment on the correspondence between the data of table 2.4 and the solution you obtain. Was the population near its maximum sustainable size?

TABLE 2.4 **Population data for the alga** *genera fosiella*. **From [5].**

Time (days)	Biomass (mm²)	Time (days)	Biomass (mm²)
11	0.00476	39	1.7
15	0.0105	44	2.45
18	0.0207	54	3.5
23	0.0619	64	4.5
26	0.337	74	5.09
31	0.74		

21. Argue that the logistic equation $P' = aP - sP^2$ has the steady state $P_{ss} = a/s$. Generalize the direction field argument of figure 2.9, page 50, to show that all populations, small or large, tend toward $P_{ss} = a/s$ as time passes.

Explain why a/s is called the *carrying capacity* of a population governed by the logistic equation. Define $K = a/s$ and show that the logistic equation can be written $P' = aP(1 - P/K)$, thereby emphasizing that the carrying capacity K is a steady state.

22. Repeat exercise 27 of section 2.1 for a population with a constant net growth rate k that is losing individuals to emigration at a rate of $E(t)$ individuals per unit time.

23. Repeat exercise 27 of section 2.1 for a population with a variable net growth rate $g(P) = a - sP$. Assume there is neither emigration nor immigration.

2.3 ■ A HEAT-FLOW MODEL

A CPU in a critical control system nearly overheats. Will it cool sufficiently between duty cycles? The rate of cooling of molten metal in a casting determines the microstructure and ultimate strength of the part. How does the thickness of the walls of the mold affect the rate of cooling?

We could ask similar questions about heat loss in a chemical reactor or about a temperature-sensitive experiment aboard the space shuttle. A parallel set of questions could inquire how rapidly devices warm up when they are exposed to hotter surroundings.

The mathematical models that provide answers to these questions all have similar form, and they arise from a conservation law, just like population models. To introduce such models and to minimize technical jargon, we will seek answers to a question about a very common object:

How does insulation affect the way the temperature drops in an unheated home?

We need two experimental facts.

Experimental fact 1.   The heat energy Q of a body is proportional to its mass and to its absolute temperature T.

Heat energy is exhibited in the vigor of the random motion of the molecules of the body. The constant of proportionality is called the **specific heat** of the body; a given weight of a material with a high specific heat, such as water, holds more heat energy per degree of temperature than an equal mass of material of lower specific heat, such as copper or air. The relationship is

$$Q = cmT, \tag{2.14}$$

where c denotes the specific heat.

TABLE 2.5 **Specific heat (cal/g·K) and thermal conductivity (cal/s·K·cm^2) of selected materials. The thermal conductivity values are for a sample 1 cm thick.**

Material	Specific heat, c	Thermal conduc., k
Water	1.0	—
Glass	0.20	0.0025
Marble	0.21	0.0071
Copper	0.09	0.908
Wood	0.42	0.0003
Air	0.24	—
Red brick	—	0.0015
Flannel	—	0.00023

If Q is measured in calories, m in grams, and T in degrees Kelvin (degrees Celsius plus $273°$), then specific heat c has units of calories per gram per degree Kelvin (cal/g·K). Table 2.5 lists specific heats of some common materials.

The Kelvin temperature scale is the Celsius scale with its zero shifted $273°$ from the freezing point of water down to "absolute zero." For example, a comfortable room temperature of $20°$C is $293°$K.

Although scientists usually use the calorie as a unit of heat energy, engineers in the United States often use the British thermal unit (Btu). One Btu is about 252 cal.

A calorie is the energy required to raise the temperature of one gram of water from $15°$C to $16°$C. Consequently, the specific heat of water is $c = 1$ cal/g·K.

We will also use the following.

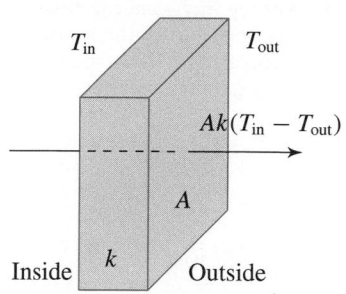

FIGURE 2.11 Rate of heat flow through a surface of area A. The direction from the inside toward the outside is positive.

change = amount added − amount removed

Introduce notation.

Rate times elapsed time Δt

Experimental fact 2. The rate of flow of heat-energy through a unit area of a surface (such as the wall of a house) is proportional to the temperature difference on either side of that surface. The heat flows in the direction of decreasing temperature, from hot to cold.

The proportionality constant, denoted by k, is known as the **thermal conductivity**. It depends on both the material and the thickness of the wall. If we take the direction inside-towards-outside as positive, the *rate of heat flow per unit area* is

$$k(T_{in} - T_{out}),$$

where T_{in} is the temperature of the inside surface and T_{out} is the temperature of the outside surface. See figure 2.11.

Thermal conductivity k for a material of given thickness has units of calories per second (rate of heat-energy flow) per degree Kelvin (the temperature difference) per square centimeter (surface area). Table 2.5 lists thermal conductivities of some common materials.

As with population models, the governing law is one of conservation:

Law of conservation of heat energy. The net change in the heat energy of a body over a given period of time is the amount of heat energy added less the amount of heat energy removed.

Heat energy is redundant. Heat *is* energy. But the redundancy provides useful reinforcement.

The net change of heat energy within a house is reflected in the change in the temperature of its interior as air, floors, walls, furniture, stereos, etc. gain or lose heat. Causes of heat-energy change include loss or gain by conduction through the walls and roof, heat supplied by external sources such as radiant heating by the sun, heat loss or gain from internal sinks or sources such as air conditioning or a furnace, and so on.

For the moment, suppose that this model home is unheated, without air conditioning, and situated in the shade; that is, it can gain or lose heat only by conduction through its walls and roof.

To derive a model for the temperature of the house, introduce notation for the quantities mentioned in the experimental facts, then write expressions for the terms in the law of conservation of heat energy.

Let $T(t)$ denote the temperature of the interior of the house at time t, and let T_{out} denote the outside temperature. Let A represent the total area of walls and roof combined, and let k be the thermal conductivity of these surfaces, assuming roof and walls are equally well insulated. Finally, let $Q(t)$ be the total heat energy of the house at time t.

The conservation law requires an accounting of the change in heat energy of the house over time. Heat will flow through the walls and roof, and the second experimental fact says that through a typical section of wall or roof

$$\text{\textit{rate} of heat flow out \textit{per unit area}} = k(T(t) - T_{out}),$$

$$\text{\textit{rate} of heat flow out through}$$
$$\text{\textit{walls and roof combined}} = A \cdot k(T(t) - T_{out}),$$

$$\text{\textit{total} heat flow out over time } \Delta t = Ak(T(t) - T_{out}) \cdot \Delta t.$$

The expression $Ak(T(t) - T_{out})\Delta t$ is indeed the total energy *lost*—energy flow *out*—because it is positive when $T(t) > T_{out}$. Heat energy flows out of the house when it is warmer than its surroundings. If the house is cooler than its surroundings, then $Ak(T(t) - T_{out})\Delta t < 0$; that is, the house is actually gaining heat because heat loss is negative.

Now apply the conservation law:

Conservation law

$$\begin{array}{ccccc}
\text{change in energy} & = & \text{amount added} & - & \text{amount lost} \\
Q(t + \Delta t) - Q(t) & = & 0 & - & Ak(T(t) - T_{out})\Delta t.
\end{array} \quad (2.15)$$

Stop and Think **2.12** Suppose the house contained a furnace that was releasing heat at a specified rate, thereby adding heat to the house. Where would that term appear in the conservation law?

2.13 Compare this conservation law expression with those used in the derivation of the population models in sections 2.1 and 2.2. Identify as many similarities as you can.

Obtain a differential equation with the usual manipulations (divide by Δt and let $\Delta t \to 0$):

$$\frac{dQ}{dt} = -Ak(T(t) - T_{out}).$$

The two unknown functions, the heat energy Q and temperature T, are related by the first experimental fact: $Q = cmT$. Hence,

$$\frac{dQ}{dt} = cm\frac{dT}{dt}.$$

Here m is the mass of the entire house and its contents, and c is the (average) specific heat of the material in the house.

Combining the last two displayed equations yields a differential equation for a single unknown, the temperature T of the interior of the house. Adding the starting temperature of the house T_i as an initial condition completes the model for the temperature of an unheated house,

HEAT-FLOW MODEL

$$\frac{dT}{dt} = -\frac{Ak}{cm}(T(t) - T_{out}), \quad (2.16)$$

$$T(0) = T_i. \quad (2.17)$$

Stop and Think **2.14** Is the heat-loss equation (2.16) linear or nonlinear?

ANALYSIS

If the house is initially warmer than the outside, that is, if $T_i > T_{out}$, then the model equation (2.16) reveals

$$T' = -\frac{Ak}{cm}(T - T_{out}) < 0;$$

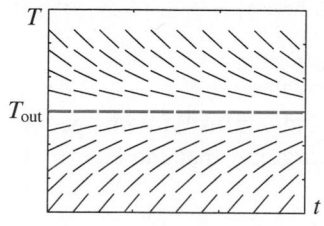

FIGURE 2.12 Direction field for the heat-flow equation $T' = -(Ak/cm)(T - T_{out})$, T_{out} constant

the temperature T of the house will decrease. As long as $T > T_{out}$, this decrease will continue. If $T = T_{out}$, then $T' = 0$, suggesting that the temperature of the house would remain at this constant value or *steady state* if it ever reached it.

Furthermore, if T_{out} is constant, then

$$T''(t) = \frac{dT'}{dt} = \frac{d\left(-(Ak/cm)(T - T_{out})\right)}{dt} = -\frac{Ak}{cm}T',$$

and the sign of T'' is opposite that of T'; that is, the graph of temperature versus time is concave down when it is increasing and concave up when it is decreasing. The direction field for the heat-flow equation (2.16) should look something like figure 2.12.

Stop and Think

2.15 Examine figure 2.12. Do you think the steady-state $T = T_{out}$ is stable or unstable?

INTERPRETATION

The differential equation $T' = -(Ak/cm)(T - T_{out})$ reveals that the rate of change of temperature T' is proportional to the thermal conductivity k of the walls and roof of the house. Everything else being equal, the rate of temperature drop can be reduced by a factor of two if the thermal conductivity of the walls and roof are reduced by a factor of two through the addition of more insulation.

> Home insulation is rated by an R factor; R denotes *resistance*. It is the reciprocal of conductivity k. Doubling R corresponds to reducing k by a factor of two. A common attic insulation is rated R 25 for an 8 in. thickness.
>
> The building industry's R factor has units hr·F·ft^2/Btu. The formula for conversion from thermal conductivity k in cal/s·K·cm^2 is $R = 3.45 \times 10^{-4}/k$, where k is measured for a sample 1 cm thick and R is the thermal resistance per inch of material.

■ **EXAMPLE 11** *For a typical ranch house, the quotient Ak/cm might have the value 0.012 min^{-1}. The interior of the house is at a comfortable $20°C$ when the furnace quits. How cold will the house be one hour later?*

Since the outside temperature is not given, T_{out} must be carried as a parameter. An analytic solution of (2.16–2.17) probably will provide the most convenient answer to the question.

In this case, use $Ak/cm = 0.012$ and $T_i = 20 + 273 = 293°$K. Time zero is the instant the furnace stopped. Hence, the heat-flow model (2.16–2.17) takes the form

$$\frac{dT}{dt} = -0.012(T(t) - T_{out}), \quad T(0) = 293. \tag{2.18}$$

Assume T_{out} is constant so that separation of variables can be applied.

If $T - T_{out} \neq 0$, then the separated differential equation is

Separate.

$$\frac{dT}{T - T_{out}} = -0.012 \, dt.$$

Integrate to obtain

Integrate.

$$\ln |T - T_{out}| = -0.012t + b,$$

where b denotes the constant of integration to avoid confusion with specific heat c. Exponentiation, writing $|C| = e^b$, and arguing that the sign of C determines the sign of $T - T_{out}$ yield

Solve for T.

$$T(t) = T_{out} + Ce^{-0.012t}.$$

The initial condition $T(0) = 293$ requires

Find C.

$$293 = T_{out} + C \quad \text{or} \quad C = 293 - T_{out}.$$

The temperature of the house in degrees Kelvin t minutes after the failure of the furnace is

$$T(t) = T_{out} + (293 - T_{out})e^{-0.012t}. \tag{2.19}$$

Since $e^{-0.012 \cdot 60} \approx 0.49$, the temperature one hour after the furnace fails is $T \approx 293 \cdot 0.49 + T_{out} \cdot 0.51 = 143 + 0.51 T_{out}$. ■

Stop and Think   **2.16** Suppose $T_{out} = 5°C$. How long until the temperature of the house reaches $10°C$?

2.17 Explain the following alternate form of the conclusion of the previous example: In one hour, the difference between the temperature of the house and the temperature outside drops by nearly 50%.

2.18 Use the solution formula (2.19) to verify that $T - T_{out} \neq 0$, as required in the first step of separation of variables.

2.19 Check that (2.19) actually solves the initial-value problem (2.18) in two steps:

(a) Verify by substitution that (2.19) solves the differential equation

$$T' = -0.012(T(t) - T_{out}).$$

(b) Verify that (2.19) satisfies the initial condition $T(0) = 293$.

Using degrees C

Degrees Kelvin are a nuisance for discussions of everyday phenomena like the temperature of a house. What does the heat-flow model (2.16–2.17) look like if temperature is expressed in degrees Celsius instead?

Let S denote temperature in degrees Celsius: $T = S + 273$. Then $T' = S'$, $T - T_{out} = S - S_{out}$, and $S' = -(Ak/cm)(S - S_{out})$. The model is unchanged!

For the rest of this section, T will take the more convenient units of $°C$.

■ **EXAMPLE 12**   *Suppose the house contains a furnace with a heat output rate of F cal/min. Could*

$$T' = -\frac{Ak}{cm}(T - T_{out}) + \frac{F}{cm} \tag{2.20}$$

be a correct governing equation?

Since F is (presumably) a positive number, the effect of the term F/cm is to increase T', an effect consistent with the action of a furnace. Furthermore, the furnace represents a source of heat, and it should contribute a term involving F to the conservation expression (2.15). A careful derivation would be required to verify that this equation is indeed correct, not just plausible. (See exercise 12.) ■

■ **EXAMPLE 13**   *Suppose (2.20) is indeed a reasonable governing equation for a house with a furnace. The output rate of the furnace should depend on the temperature of the house, zero if $T > T_{set}$, where T_{set} is the temperature at which the thermostat is set, and positive if $T < T_{set}$. Explore variations in the temperature of the house using this sort of model.*

Write the temperature-dependent output rate $F(T)$ of the furnace as

$$F(T) = \begin{cases} R, & T < T_{set} \\ 0, & T \geq T_{set} \end{cases} \tag{2.21}$$

where R is some positive constant. Accept the argument of the previous example that a plausible governing model is $T' = -(Ak/cm)(T - T_{out}) + F(T)/cm$, $T(0) = T_i$.

Suppose it is a cold day so that $T_{set} > T_{out}$.

If $T > T_{set}$, then $F(T) = 0$ and

$$T' = -\frac{Ak}{cm}(T - T_{out}) < -\frac{Ak}{cm}(T_{set} - T_{out}) < 0.$$

Hence, temperature will decrease until T is just below T_{set} and the furnace turns on.

Once the furnace is running because $T < T_{set}$, the governing equation becomes

$$T' = -\frac{Ak}{cm}(T - T_{out}) + \frac{R}{cm}.$$

Since $T < T_{set}$,

$$T' > -\frac{Ak}{cm}(T_{set} - T_{out}) + \frac{R}{cm}.$$

If the furnace is powerful enough—if $R/cm > -(Ak/cm)(T_{set} - T_{out})$ — then $T' > 0$ and the temperature will increase.

Stop and Think   **2.20** Work through the details of the linchpin argument of the last paragraph:

$$T < T_{set} \implies T' > -\frac{Ak}{cm}(T_{set} - T_{out}) + \frac{R}{cm}.$$

But it could be too cold outside! If $-(Ak/cm)(T_{set} - T_{out}) + R/cm < 0$, then the furnace will run continuously while the temperature of the house continues to drop. ■

MATLAB

To pursue a numerical view of the previous example, choose $Ak/cm = 0.012 \, \text{min}^{-1}$, as in example 11, and $T_i = 15°C$. Suppose $T_{out} = 5°C$ and $T_{set} = 20°C$.

From the discussion on page 47 of the on-off emigration rate, a MATLAB representation for the furnace output rate $F(T)$ in (2.21) is

```
(R*(1 + sign(20 - T)))/2
```

If T $> 20 = T_{set}$, then `sign(20 - T)` $= -1$ and the complete expression is zero. Similarly, it has value R if T $< 20 = T_{set}$.

Finally, in this setting, a reasonable value for R/cm is $0.40°C\cdot\text{min}$.

Use these parameters and the numerical tools in DELAB to find an approximate solution of the initial-value problem

$$T' = -0.012(T - 5) + F(T), \quad T(0) = 15°C,$$

using $\Delta t = 5$ min. For guidance, select **Help, Textbook**, then go to chapter 2, example 13.

A plot of this numerical approximation appears in figure 2.13. If the numerical approximation can be believed, the house appears to need about 20 min to warm up.

Is the overshoot of the set point temperature a numerical artifact or an accurate representation of the behavior predicted by the model? How about the saw tooth behavior after that? Will the house be maintained at a steady temperature? How often will the furnace cycle? Further analysis is certainly needed; see project 5, page 82.

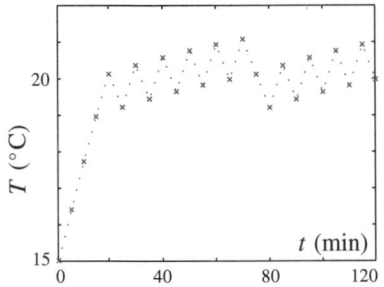

FIGURE 2.13 An Euler approximation using $\Delta t = 5$ min to the solution of a furnace model.

2.3.1 Exercises

Exercise Guide	
To gain experience ...	**Try exercises**
Deriving models	4, 5, 11–12(a), 13
Analyzing models from differential equations	4(b), 8, 11(b–c), 12(c), 16–17
Analyzing models from a solution	4(c), 5(b–c), 10, 15, 18(b)
Interpreting models	4(b), 8, 10–11(b), 12(d)
With concrete examples of models	4(a), 5(a), 9
Solving problems like these models	1, 4(c), 5(b), 10(a), 11(d), 14(b), 18(a)
With the basic ideas of these models	5(a), 7–9, 14
With the mathematical form of these models	6, 12(b)

1. Solve each of the following heat-loss model initial-value problems. What is the relation between the direction field for each of these equations and that of figure 2.12, page 59? Does DELAB produce the same analytic solution?

   (a) A model for the temperature of a typical ranch house at 20°C when its furnace breaks,

   $$\frac{dT}{dt} = -0.012(T - T_{\text{out}}), \quad T(0) = 293.$$

   (Time is in minutes and temperature in degrees Kelvin. Add the details omitted from the example in the text.)

   (b) The general heat-loss model,

   $$\frac{dT}{dt} = -\frac{Ak}{cm}(T - T_{\text{out}}), \quad T(0) = T_i.$$

   (c) A model for the temperature of a house whose roof and walls have different thermal conductivities,

   $$\frac{dT}{dt} = -\frac{A_r k_r + A_w k_w}{cm}(T - T_{\text{out}}), \quad T(0) = T_i.$$

2. A power transistor of mass m and surface area A is immersed in an oil-cooling bath whose temperature is T_{out}. The thermal conductivity through the wall of the transistor housing to the oil bath is k. The specific heat of silicon is c. Derive a model for the temperature T of the transistor between duty cycles, when it is generating no heat of its own. Comment on any similarities with the house heat-loss model

   $$T' = -\frac{Ak}{cm}(T - T_{\text{out}}), \quad T(0) = T_i.$$

3. Suppose a diligent homeowner has insulated her attic so that its thermal conductivity is less than that of the house walls. Derive a model similar to

   $$T' = -\frac{Ak}{cm}(T - T_{\text{out}}), \quad T(0) = T_i,$$

   for her house.

4. An individual of mass 70 kg will be traveling in winter temperatures of 5°C. He can choose to wear a suit of 1 cm thickness of red flannel or 1 cm of copper. (In one case he will be disguised as Santa Claus and in the other, as a droid.)

   (a) Using appropriate parameter values from table 2.5, write a form of the heat-loss model

   $$T' = -\frac{Ak}{cm}(T - T_{\text{out}}), \quad T(0) = T_i,$$

   for the body temperature of this intrepid explorer while wearing each of these costumes. Neglect the heat generated by his body. (Since the human body is largely

water, you can approximate this individual's specific heat as 1 cal/g·K. What should you use for an initial temperature? Can you estimate the area of a human body? Take care with units.)

   (b) Find the ratio of the rate of temperature decrease wearing copper to that wearing flannel. Which costume would you recommend for warmth? (*Neither* would be the answer if style were the main consideration.)

   (c) Find the temperature of his body after a 30-minute exposure in each of these costumes, again neglecting the heat his body generates.

   Should you think of his clothing as a reservoir of heat or as insulation? How should you regard his body?

5. The example in the text considers a house for which Ak/cm has the value 0.012. As appropriate, use DELAB for the following.

   (a) What would that value be if the insulation in the walls and roof were doubled, thereby halving the thermal conductivity of those surfaces?

   (b) If the house is initially at 20°C, what is its temperature after one hour without heat?

   (c) For two specific values of outside temperature of interest to you, determine how much warmer this better insulated house is than the house in the example after an hour without heat.

6. (a) Is the heat-loss model

   $$T' = -\frac{Ak}{cm}(T - T_{\text{out}}), \quad T(0) = T_i,$$

   similar in mathematical form to any of the models derived in sections 2.1 or 2.2? Identify any such similarities precisely.

   (b) If you find a population model that is similar to this heat-loss model, indicate which variables and parameters in the two models are analogous.

   (c) Write one differential equation that includes both models as special cases, depending on whether the unknown is interpreted as population or temperature.

7. The equation

   $$T' = -\frac{Ak}{cm}(T - T_{\text{out}})$$

   contains the parameter m, the mass of the body losing heat. How might you reasonably compute m when modeling heat loss in a house?

8. How does changing any one of the parameters A, k, c, or m affect the rate of temperature drop in the model

$$T' = -\frac{Ak}{cm}(T - T_{\text{out}}), \quad T(0) = T_i?$$

If the surface area were doubled, what could you do to keep the rate of temperature drop the same?

9. The temperature of a house dropped from $25°C$ to $19°C$ in 45 minutes when the outside temperature was $-2°C$. Estimate Ak/cm for this house.

10. One way of quantifying the relative efficiency of the insulation of a house is determining the time required for the temperature of an unheated house to drop half the way from its initial value of T_i to its final value of T_{out}.

   (a) Find a formula for the time required for this temperature drop.

   (b) If the total mass of the house could be doubled, perhaps by adding stone flooring or an indoor swimming pool, how would this decay time change? To increase this decay time, should you add materials with a high specific heat or a low specific heat?

11. The rate of radiation of heat energy from a so-called black body depends on the *fourth* power of the temperatures: The rate of heat energy radiated from a mass m at temperature T with surface area A and radiation constant κ is $A\kappa(T_{\text{in}}^4 - T_{\text{out}}^4)$.

   (a) Using this experimental fact, derive the model

   $$\frac{dT}{dt} = -\frac{A\kappa}{cm}(T^4 - T_{\text{out}}^4), \quad T(0) = T_i.$$

   (b) Does this model exhibit a temperature decrease when the body is warmer than its surroundings? Is a steady state temperature of T_{out} still possible?

   (c) How do rates of temperature decrease compare between this model and the familiar model

   $$\frac{dT}{dt} = -\frac{Ak}{cm}(T - T_{\text{out}}), \quad T(0) = T_i.$$

   (d) What antiderivative must you evaluate to solve this new model by separation of variables? List a set of steps that might evaluate it. Would you anticipate finding a simple solution formula? What do the analytic tools in DELAB offer?

   (e) Using $Ak/cm = 0.01$ min^{-1}, $\kappa = 10^{-5}k$, $T_i = 294°\text{K}$, $T_{\text{out}} = 293°\text{K}$, a value of T_{out} of your choice, and the Euler method, compare temperature values predicted by these models over $0 < t < 1$.

12. (a) The model

   $$T' = -\frac{Ak}{cm}(T - T_{\text{out}}), \quad T(0) = T_i,$$

   assumes the house contains no heat source. Derive a similar model for the temperature of a house that contains a furnace by incorporating into the conservation law a term that adds heat energy. Specifically identify the parameter(s) that characterizes the furnace. (*Hint:* You could follow the lead of heating contractors and specify the heat output rate of the furnace; they usually use units of Btu/h. What term would such an energy source add to the conservation law expression (2.15), page 58?)

   (b) How does the mathematical form of your new model equation compare with $T' = -(Ak/cm)(T - T_{\text{out}})$?

   (c) Use your model to determine the rate at which the furnace must release heat to maintain the house at a specified constant temperature.

   (d) A heating contractor would like a simple formula for calculating the size (heat release rate) of the furnace to be installed in a house. One possibility is to choose the output rate of the furnace so that it can keep the house at a constant comfortable temperature on the coldest day of the winter if it runs continuously. What can you suggest?

13. A power transistor of mass m and surface area A is immersed in an oil cooling bath whose temperature is T_{out}. The thermal conductivity through the wall of the transistor housing to the oil bath is k. The specific heat of silicon is c. When it is operating, it is generating heat internally at a rate of R cal/s. By allowing for this extra source of heat in the conservation law, derive a model for the temperature T of the transistor during a duty cycle. Comment on any similarities with the house heat-loss model

   $$T' = -\frac{Ak}{cm}(T - T_{\text{out}}), \quad T(0) = T_i.$$

   Do you see any similarities with the preceding problem, which seeks a model for a house with a furnace?

14. (a) Repeat exercise 17(a) of section 2.1 for the heat-loss model

   $$T' = -\frac{Ak}{cm}(T - T_{\text{out}}), \quad T(0) = T_i.$$

   The quantity Ak/cm might be about 0.012 min^{-1} for a typical ranch house. Try 5-minute time steps, and estimate the temperature change in one hour for conditions that interest you. What is the connection between this process and the Euler method? If you see the connection, use DELAB.

(b) Compare your estimate with the temperature given by the exact solution of the model for these conditions.

15. Example 11 found a relation between the outside temperature and the temperature of a house that was at 20°C when its furnace failed: $T(t) = T_{out} + (293 - T_{out})e^{-0.012t}$. How sensitive is the temperature of the house to changes in the outside temperature? (That is, find an expression for the rate of change of T with respect to T_{out}.) Given a change ΔT_{out} in the outside temperature, estimate the corresponding change $\Delta T(t)$ in the temperature of the house at time t.

   This is a question about *parametric dependence*. How does behavior predicted by a model depend on the parameters in the model?

16. Use DELAB to draw a direction field diagram for the heat-flow equation of example 11, $T' = -0.012(T - T_{out})$, with $T_{out} = 5°C$. Compare with figure 2.12, page 59, and comment on similarities.

17. Discussing a house heated by a furnace, example 13, page 61, concludes, "But it could be too cold outside! If

$$-\frac{Ak}{cm}(T_{set} - T_{out}) + \frac{R}{cm} < 0,$$

then the furnace will run continuously while the temperature of the house continues to drop."

   How cold is too cold? Given the thermostat setting T_{set} and the characteristics of the house (A, k, c, m), find the critical outside temperature below which the furnace will run continuously while the temperature of the house drops.

18. The temperature of an unheated house is modeled by (2.16–2.17), page 58. Write an expression for T_{out} that simulates periodic swings between a daytime temperature of 15°C and a nighttime temperature of 5°C.

   (a) Can the resulting model be solved by separation of variables?

   (b) Using $Ak/cm = 0.012$ min^{-1} and an initial temperature of your choice, use DELAB to conduct a numerical study of the temperature variations of the house. Does its temperature eventually follow the variations of the outside temperature? If so, how long before the two temperatures fall into the same rhythm? What is the lag between the outside temperature and the house temperature?

2.4 ■ MULTIPLE SPECIES

Previous sections in this chapter applied conservation laws to derive models of a single species or models of a single quantity (heat energy). Each of those models involved just one differential equation.

Applying conservation laws to several species at once leads to models involving several coupled differential equations, *systems* of differential equations. For example, populations of foxes and rabbits or owls and field mice are interconnected by a predator–prey relationship. When prey is plentiful, the predators flourish and the prey is consumed. Then the predator population dwindles because food is scarce, and the population of prey recovers.

Other multiple species situations include models of several populations competing for the same resource and models of epidemics that track those who are infective and those who are merely susceptible to infection as separate species.

2.4.1 Predator and Prey

A classic model of an interconnected system is that of predator and prey, such as foxes and rabbits. Foxes consume rabbits as their food, and the fox population grows when rabbits are plentiful. But a larger fox population decimates the rabbits, and the fox population declines from lack of food. Then the rabbit population recovers in the face of smaller numbers of foxes.

To model this situation easily, simplify it drastically. Rather than confidently advance experimental facts, we instead propose some tentative ideas that will account for gains and losses in separate conservation expressions for foxes and for rabbits.

In the absence of foxes, suppose that the growth rate of the rabbits is constant:

Tentative Idea 1. The fraction of the rabbit population that reproduce in a unit time period is constant.

If b_R denotes the constant of proportionality and $R(t)$ is the rabbit population at time t, then approximately

Rate $b_R R$ times time Δt

$$b_R R(t) \Delta t$$

rabbits are added to the population by reproduction during the time interval from t to $t + \Delta t$.

If foxes are the only predators of rabbits and if we ignore natural causes of death, then the death rate of rabbits should increase in proportion to the number of foxes. Hence, we propose:

Tentative Idea 2. The fraction of the rabbit population that die in a unit time period is proportional to the fox population.

If the constant of proportionality is β and $F(t)$ is the fox population, then

- the *fraction* of the rabbits that die per unit time is $\beta F(t)$,
- the *number* of rabbits that die *per unit time* is $\beta F(t) R(t)$, and
- the *number* that die *from t to $t + \Delta t$* is approximately

Rate $\beta F R$ times time Δt

$$\beta F(t) R(t) \Delta t. \qquad (2.22)$$

This last expression can also be interpreted as saying that a certain fraction of the encounters between rabbits and foxes will cause the demise of a rabbit.

To further simplify the situation, suppose that the fraction of foxes that die per unit time is constant:

Tentative Idea 3. The fraction of foxes that die in a unit time period is constant.

If the constant of proportionality is d_F, then the number of foxes lost from t to $t + \Delta t$ is approximately

Rate $d_F F$ times time Δt

$$d_F F(t) \Delta t.$$

If foxes require rabbits for survival, their birth rate should depend on the number of rabbits available.

Tentative Idea 4. The fraction of the fox population that reproduce in a unit time period is proportional to the *rabbit* population.

If the constant of proportionality is α, then

- the *fraction* of foxes that reproduce per unit time is $\alpha R(t)$,
- the *number* of foxes born *per unit time* is $\alpha R(t) F(t)$, and
- the *number* of foxes born *from t to* $t + \Delta t$ is approximately

Rate $\alpha R F$ times time Δt

$$\alpha R(t) F(t) \Delta t. \tag{2.23}$$

This last expression can be interpreted as saying that a certain proportion of the encounters between a rabbit and a fox will provide enough nutrition to permit the birth of a new fox.

A model of the fox and rabbit populations is derived by combining these tentative ideas using a familiar law.

Law of conservation of population. The net change in a population over a given period of time is the number of individuals added less the number of individuals removed.

Conservation is applied twice, once for each of the populations. Conservation of rabbits requires

CONSERVATION OF RABBITS

$$\begin{array}{rccc} \text{change in rabbits} & = & \text{number added} & - & \text{number lost} \\ R(t + \Delta t) - R(t) & = & b_R R(t) \Delta t & - & \beta F(t) R(t) \Delta t. \end{array}$$

Conservation of foxes requires

CONSERVATION OF FOXES

$$\begin{array}{rccc} \text{change in foxes} & = & \text{number added} & - & \text{number lost} \\ F(t + \Delta t) - F(t) & = & \alpha R(t) F(t) \Delta t & - & d_F F(t) \Delta t. \end{array}$$

Divide both equations by Δt and let Δt approach zero to obtain a pair of first-order equations,

PREDATOR–PREY EQUATIONS

$$F' = -(d_F - \alpha R) F \tag{2.24}$$

$$R' = (b_R - \beta F) R. \tag{2.25}$$

The model is completed by imposing two initial conditions, the starting size of each population,

$$F(0) = F_i, \quad R(0) = R_i.$$

The differential equations (2.24–2.25) form a *nonlinear* system; both equations are nonlinear because of the product terms RF and FR, each vaguely reminiscent of the P^2 term in the single-species logistic equation.

These equations are often called the *Lotka-Volterra equations*. V. Volterra proposed them in 1926 to explain variations in fish populations. A. J. Lotka had developed the same model in 1920 to explain periodic chemical reactions.

Murray [18, p. 68] provides a sobering discussion of the limitations of this model when it is applied to field data.

Stop and Think
2.21 Compare the net growth rate terms — the right-hand sides — of equations (2.24) and (2.25) with that of the logistic (competition) equation $P' = aP - sP^2$. What similarities do you see?

2.22 Could you add a logistic growth term to (2.25) to limit the growth of rabbits?

ANALYSIS

How will the two populations fare? Is there an equilibrium or steady state in which the two populations co-exist? Can either be driven to extinction?

Since $F' = R' = 0$ at a steady state (if one exists), (2.24–2.25) immediately yield

$$-(d_F - \alpha R_{ss})F_{ss} = 0$$
$$(b_R - \beta F_{ss})R_{ss} = 0. \tag{2.26}$$

This pair of equations produces two possible steady states:

$$F_{ss} = b_R/\beta, \quad R_{ss} = d_F/\alpha$$
$$F_{ss} = 0, \quad\quad R_{ss} = 0.$$

Either both species survive or both species die.

Stop and Think

2.23 Evidently, these steady states were obtained by solving the pair of simultaneous equations (2.26); in each equation, either the term in parentheses is zero or the other factor is zero. With two values from the first equation and two from the second, it seems as if there should be four pairs of steady-state values. Why are there only two?

■ **EXAMPLE 14**   *What happens if these steady states are perturbed? Could one of the species die out? Consider a concrete case in which $d_F = 0.04$, $\alpha = 0.0011$, $b_R = 0.06$, $\beta = 0.0009$.*

This example is asking a question about the stability of the steady states, a topic discussed in more detail in chapters 3 and 7.

With these parameter values, the system of predator-prey equations is

$$F' = -(0.04 - 0.0011R)F, \tag{2.27}$$
$$R' = (0.06 - 0.0009F)R, \tag{2.28}$$

and the possible steady states are

$$F_{ss} \approx 67, \quad R_{ss} \approx 36$$
$$F_{ss} = 0, \quad\quad R_{ss} = 0.$$

Near $F_{ss} \approx 67$, $R_{ss} \approx 36$

To study behavior near the nonzero steady state, apply the Euler method to approximate the solution of (2.27–2.28) with initial conditions $F(0) = 60$, $R(0) = 40$, say. The result appears in the upper graph of figure 2.14. The fox and rabbit populations appear to oscillate around their steady-state values, but more analysis is needed to be certain. (Do the populations really vary periodically? Is the growth in the population peaks numerical error or a prediction of the model?)

Notice that the rabbit population rises ahead of the fox population. When more food is available, the fox population grows. Eventually, foxes overwhelm rabbits and drive the rabbit population down. Facing a shortage of food, the fox population then declines as well.

Suppose the zero steady state is perturbed by adding one fox and two rabbits: $F(0) = 1$, $R(0) = 2$. With these initial conditions, Euler produces the approximate solution of (2.27–2.28) shown in the lower part of figure 2.14.

The fox and rabbit populations exhibit similar tandem behavior, but if there are oscillations, the period is longer than in the upper graph. The range of population values is also an order of magnitude greater. More analysis—and application of the Euler method over a longer time interval—is required. ■

Near $F_{ss} = 0$, $R_{ss} = 0$

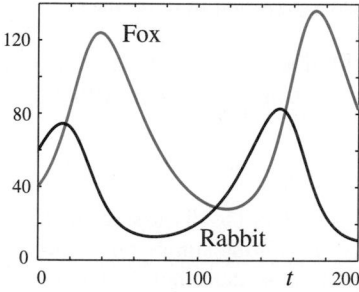

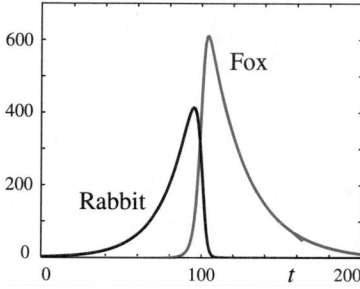

FIGURE 2.14   Response of the fox and rabbit populations to perturbations of their two steady states, $F_{ss} \approx 67$, $R_{ss} \approx 36$ (upper graph) and $F_{ss} = R_{ss} = 0$ (lower graph). The graphs were computed using the Euler method with $\Delta t = 1$.

MATLAB

To produce figure 2.14, use DELAB's numerical tools to find an Euler approximation to the solution of an initial-value problem for a system of two differential equations. Choose $\Delta t = 1$. Consider initial conditions such as $F(0) = 60$, $R(0) = 40$ and $F(0) = 1$, $R(0) = 1$ to start near each of the two steady states. For guidance, select `Help, Textbook`, then go to chapter 2, figure 2.14.

2.4.2 Epidemics

The spread of disease in a population can be modeled by dividing the population into three subgroups:

Susceptibles, those susceptible to infection but not yet infected,

Infectives, those actually infected with the disease and capable of spreading it, and

Recovereds, those who have recovered from the disease (or, more morbidly, have been removed from the population) and are no longer prey to the disease.

At the simplest level, childhood diseases like chicken pox and measles follow this pattern.

The following set of tentative ideas or hypotheses is sufficient to derive a simple model of the spread of such diseases. Let $S(t)$, $I(t)$, and $R(t)$ denote the number of susceptibles, infectives, and recovered at time t.

Tentative Idea 1.   The fraction of the susceptibles who become infected per unit time is proportional to the number of infectives.

If b is the constant of proportionality, then

- the *fraction* of susceptibles infected per unit time is bI,
- the *number* of susceptibles infected *per unit time* is $(bI)S$, and
- the *number* infected *from t to $t + \Delta t$* is approximately

Rate bIS times time Δt $bIS\Delta t$.

This last expression can also be interpreted as saying that a certain fraction of the encounters between infective and susceptible individuals results in the infection of a susceptible.

Stop and Think   **2.24** Compare this idea with tentative ideas 2 and 4 in the discussion of the predator–prey model in section 2.4.1. Are all three describing interactions in essentially the same way?

Tentative Idea 2.   A fixed fraction of the infectives recovers per unit time.

If r denotes the constant fraction of infectives who recover, then the recovery rate is rI and the number who recover in time Δt is approximately

Rate rI times time Δt
$$rI\Delta t.$$

Stop and Think   **2.25** Compare this relation with tentative ideas 1 and 3 in the discussion of the predator–prey model in section 2.4.1. Compare with the simple population model derived in section 2.1. Are these "birth rates" or "death rates"? Or does the choice depend on which population is being considered?

Now apply conservation three times, once for each of the three subpopulations, susceptibles, infectives, and recovereds. Note that susceptibles are lost to infectives and infectives are lost to recovereds.

Conservation of susceptibles requires

CONSERVATION OF
SUSCEPTIBLES
$$\text{change in susceptibles} = \text{number added} - \text{number lost}$$
$$S(t + \Delta t) - S(t) = 0 - bI(t)S(t)\Delta t.$$

Conservation of infectives requires

CONSERVATION OF
INFECTIVES
$$\text{change in infectives} = \text{number added} - \text{number lost}$$
$$I(t + \Delta t) - I(t) = bI(t)S(t)\Delta t - rI(t)\Delta t.$$

Conservation of recovereds requires

CONSERVATION OF
RECOVEREDS
$$\text{change in recovereds} = \text{number added} - \text{number lost}$$
$$R(t + \Delta t) - R(t) = rI(t)\Delta t - 0.$$

The usual division by Δt and the limiting process converts the three conservation laws into a system of three differential equations,

$$S' = -bIS,$$
$$I' = bIS - rI,$$
$$R' = rI.$$

Observe that R appears only in the third equation; the first two are not coupled to the third. There are really just two governing equations.

More generally, conservation permits eliminating one equation. Since births, deaths, and other changes in the total population have been neglected, $P = S + I + R = \text{constant}.$

Stop and Think **2.26** Sum the three differential equations to show that $S + I + R =$ constant because $(S + I + R)' = 0$.

Hence, one of the dependent variables, say $R = P - S - I$, can be eliminated. Adding initial conditions for the two remaining variables gives the so-called *SIR model* for spread of an epidemic:

SIR MODEL

$$S' = -bIS,$$
$$I' = bIS - rI, \tag{2.29}$$
$$S(0) = S_i, \quad I(0) = I_i. \tag{2.30}$$

ANALYSIS

What behavior does such a model predict? To find steady states, solve

$$S'_{ss} = 0 = -bI_{ss}S_{ss} \implies I_{ss} = 0 \text{ or } S_{ss} = 0$$
$$I'_{ss} = 0 = I_{ss}(bS_{ss} - r) \implies I_{ss} = 0 \text{ or } S_{ss} = r/b.$$

Evidently, only $I_{ss} = 0$ satisfies both equations. If $I_{ss} = 0$, then S_{ss} is arbitrary.

What happens if a few infectives are introduced into the population, i.e., if $I_i > 0, 0 < S_i \leq P$? Regardless of the (nonzero) values of I and S, the number of susceptibles will decrease:

$$S' = -bIS < 0.$$

How will the illness progress? The sign of I', the rate of change of the number of infectives, depends on the size of S. If $I > 0$, then

$$I' = I(bS - r) \begin{cases} > 0, & S > r/b \\ \leq 0, & 0 \leq S \leq r/b. \end{cases}$$

Evidently, infectives can grow in number if there are enough susceptibles (if $S_i > r/b$). But eventually, susceptibles will have declined to the point that infectives must also decrease until the disease has run its course.

The course of two outbreaks of disease is shown in figure 2.15 in a **phase plane** diagram, a plot of I versus S. Two initial points (S_i, I_i) are shown. The initial level of infectives is the same, but the lower curve begins with $S_i < r/b$ and the upper curve begins with $S_i > r/b$.

The arrows show the direction of increasing time. Both outbreaks tend toward the horizontal axis, $I = 0$, meaning both die out in time, but the one with the larger number of susceptibles is a real epidemic: the number of infectives rises dramatically before the disease finally runs its course.

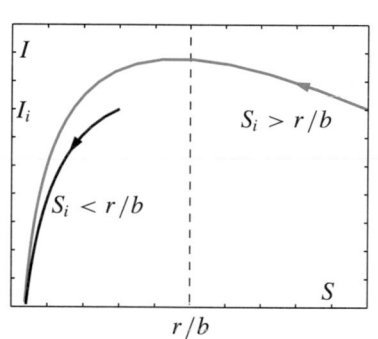

FIGURE 2.15 A phase plane diagram of the course of two outbreaks of disease, one with $S_i < r/b$ (lower curve) in which infectives steadily decline, one with $S_i > r/b$ (upper curve) in which infectives peak before declining.

Some engineers and scientists would refer to the point $(I(t), S(t))$ in a phase plane diagram as a **state vector**. This terminology envisions a vector pointing from the origin of figure 2.15 to a point on the trajectory, tracing the motion of the point toward the $I = 0$ axis as time passes.

INTERPRETATION

The steady state is disease-free ($I = 0$) with any number of susceptibles (S arbitrary). Over time the infection will die out, but if there are enough susceptibles, it will peak before it does. That is, a true epidemic seems to require a sufficient number of susceptibles, independent of the initial number of infectives.

Stop and Think **2.27** In figure 2.15 mark the peak of the number of infectives for each of the two out-breaks shown there. Verify that one outbreak never exceeds its initial number of infectives while the other peaks before dying out.

Modeling Reinfection

In some cases, those who have recovered may again become susceptible, per-haps because their initial infection was not sufficiently severe. To model such a situation, suppose that some fraction g of those who have recovered become susceptible again. Then the governing equations are

$$S' = -bIS + gR,$$
$$I' = bIS - rI,$$
$$R' = rI - gR.$$

Stop and Think **2.28** What "tentative idea" would account for the term gR? What is its effect on the number of susceptibles and recovereds?

2.29 Sum these equations and argue that the overall population $P = S + I + R$ is constant.

Since the overall population $P = S + I + R$ is constant, one variable can again be eliminated, leading to the so-called *SIR model*

SIRS MODEL

$$S' = -bIS + g(P - S - I),$$
$$I' = bIS - rI, \tag{2.31}$$

$$S(0) = S_i, \quad I(0) = I_i. \tag{2.32}$$

Stop and Think **2.30** Do you think the SIRS model equations are linear or nonlinear? What about the SIR model? (*Hint*: Compare them with the predator–prey equations (2.24–2.25)).

2.4.3   Competition

Two species can compete with one another for a common resource, say, rabbits and moose that compete to feed on the grass in the same meadow. The growth rate of each species will be affected by its birth rate behavior, which could be limited as in the single-species logistic equation (2.12), $P' = aP - sP^2$, and by the presence of the other species, which is consuming their common food supply. The effects of those interactions might be proportional to the product of the two populations, like the interaction terms (2.22) and (2.23) in the predator–prey model (2.24–2.25).

Tentative ideas or experimental facts could define precisely each of these effects. Then we could write expressions for the number of rabbits or moose added or removed over time Δt by birth and by competition. Finally, we could invoke two conservation laws, one for each population, to derive a pair of differential equations. With the two populations labeled y and z and with appropriate units, the result would be a pair of *competition equations*,

COMPETITION EQUATIONS

$$y' = y(1 - y - ez), \tag{2.33}$$
$$z' = rz(1 - z - fy). \tag{2.34}$$

A complete model requires the addition of initial conditions, $y(0) = y_i$, $z(0) = z_i$.

Are these equations plausible representations of competing species? The first equation provides logistic-like, limited growth for y in the first two slope terms, $y' = y - y^2 - \cdots$. The last term in (2.33), $y' = \cdots - eyz$, acts to decrease y' because of the negative sign (assuming that $e > 0$ and that y, z remain positive), and the magnitude of this effect increases with the size z of the other population.

Stop and Think

2.31 Use similar arguments to demonstrate that (2.34) is also plausible.

2.32 One could say that the competition equations are an amalgam of terms from the logistic equation $P' = aP - sP^2$ and from the predator–prey equations (2.24–2.25). Identify each of the terms in (2.33–2.34) as coming from one or the other.

These competition equations *were not derived carefully*! (See exercises 5 and 6.) But they are sufficiently plausible to warrant analysis here and in chapter 7.

Exercise 5 requests a careful derivation using conservation laws. Then exercise 6 asks for the introduction of *dimensionless variables*, new time and population gauges that lack standard dimensions but measure those variables against scales that occur naturally in the problem.

In these dimensionless variables, the populations are measured against their carrying capacities, that is, against the long-term steady state they would attain if logistic growth alone governed. So one would expect $y \to 1$ and $z \to 1$ if logistic growth controlled. Rather than counting individuals, y and z can be thought of as representing a fraction of the logistic carrying capacity. In (2.33–2.34), one unit of t is the time required for the y population to increase by a factor $e \approx 2.718$ in the absence of all other influences.

Using these variables has the advantages of minimizing the number of parameters and of making the analysis apply to a range of situations, not just a single pair of species.

■ **EXAMPLE 15**   *Consider the competition equations*

$$y' = y(1 - y - 2z), \tag{2.35}$$
$$z' = z(1 - z - 2y). \tag{2.36}$$

Show that they have a steady state at $y_{ss} = 1/3$, $z_{ss} = 1/3$. Conduct some numerical experiments to determine the stability of this steady state.

The steady-state claim is easy to verify: just show that both equations are satisfied when the given values of y_{ss} and z_{ss} are substituted.

To explore the stability of this steady state, consider a nearby initial point, say $y(0) = 0.32$, $z(0) = 0.30$. Now apply the Euler method from DELAB to (2.35–2.36) using these initial conditions. A plot of the resulting approximate solution appears in figure 2.16.

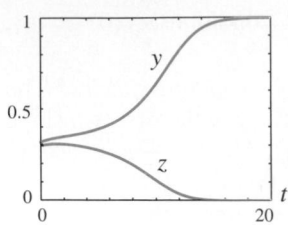

FIGURE 2.16   The y population survives, the z population dies out from $y(0) = 0.32$, $z(0) = 0.30$.

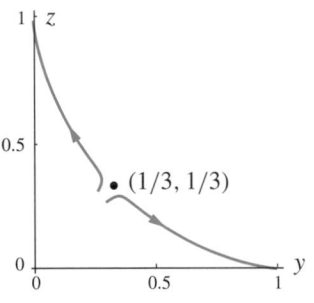

FIGURE 2.17   A phase plane plot of the history of two y-z population pairs. Both started near the steady state $y_{ss} = 1/3$, $z_{ss} = 1/3$. The z species survives in the upper left curve, y in the lower right. The arrows indicate increasing time.

PREDATOR–PREY

SIR

COMPETITION

> **MATLAB**
>
> To generate figure 2.16, follow *exactly* the same steps as in example 14, page 68.

Figure 2.16 illustrates the *principle of competitive exclusion*. The y population grows toward $y = 1$, its carrying capacity limit, while the z population dies out. One species has driven the other to extinction!

The situation is symmetric: starting with $y(0) = 0.30$, $z(0) = 0.32$ leads to extinction of the y population.

Figure 2.17 shows these two instances of competitive exclusion on a plot of z versus y, a *phase plane* diagram for (2.35–2.36). Time is invisible so we can't see the rate of growth or decay, but the curves in figure 2.17 clearly trace the fate of the two populations from initial state to survival or extinction, as the case may be.

The lower right curve, or **trajectory**, is tending toward the point $y = 1$, $z = 0$, which represents survival of y and extinction of z. The time behavior of this trajectory is plotted in figure 2.16. The upper left curve predicts extinction of y and survival of z. The steady state $y_{ss} = 1/3$, $z_{ss} = 1/3$ is just the point $(1/3, 1/3)$ on this plot.

Since neither of the trajectories starting near $(1/3, 1/3)$ return to it, we conjecture that this steady state is unstable. ■

2.4.4   Summary

Conservation laws are the common thread for these models.

The *predator–prey* equations can be appropriate when one species eats the other (say, foxes eat rabbits):

$$F' = -(d_F - \alpha R)F,$$
$$R' = (b_R - \beta F)R.$$

Epidemics involve three species within one population—susceptibles, infectives, and recovereds—but the *SIR* equations can be written with just two of them, susceptibles and infectives, because the sum of the three is constant:

$$S' = -bIS,$$
$$I' = bIS - rI.$$

Two species (say, moose and rabbits) may compete for a common food supply. With appropriate time units and scales for the two competing populations, the *competition* equations assume an especially simple form,

$$y' = y(1 - y - ez),$$
$$z' = rz(1 - z - fy).$$

All of these models involve systems of nonlinear differential equations.

2.4.5 Exercises

1. "Improve" the predator–prey equations

$$F' = -(d_F - \alpha R)F,$$
$$R' = (b_R - \beta F)R.$$

derived in the text by adding one more effect,

> The fraction of rabbits who die due to natural causes in a unit time period is constant.

Call the constant of proportionality d_R and derive the governing equations. Show that the resulting equations have the same mathematical form with b_R replaced by another expression. What is the physical interpretation of the coefficient that replaces b_R?

2. The rabbit population equation from the predator–prey model can be written

$$R' = (b_R - \beta F)R.$$

The logistic equation can be written

$$P' = (a - sP)P.$$

Argue that both of these equations model populations with growth rates that decline as a certain parameter increases. Explain the significance of that parameter in each case.

3. Use the fox and rabbit population versus time curves in the upper part of figure 2.14 to guide a sketch of a phase plane diagram (R versus F). Mark the steady state $F_{ss} \approx 67$, $R_{ss} \approx 36$. Where is it relative to the trajectory you have traced? If the fox and rabbit populations vary periodically, should you obtain a closed curve in the phase plane?

4. Referring to the competition equation (2.33), page 72, the text claims "the first two slope terms, $y' = y - y^2 - \dots$ provide logistic-like, limited growth for y." Verify this assertion by drawing a direction field diagram for $y' = y - y^2$. What is the maximum value of y this model can sustain? Confirm your analysis using DELAB.

5. **Derivation of competition equations.** Consider two species, say moose and rabbits, that are competing for a common resource. Denote their populations by M and R.

   (a) Suppose that the fraction of moose that give birth per unit time is a_M, that the fraction of moose that die per unit time due to competition with other moose is $s_M M$, and that the number of moose that die per unit time due to competition with rabbits is $s_{MR} R$. Apply a conservation law to derive the moose population equation

   $$M' = M(a_M - s_M M - s_{MR} R). \qquad (2.37)$$

   (b) State a parallel set of assumptions that involve the parameters a_R, s_R, s_{RM} for the rabbit population. Use a conservation law to derive the rabbit population equation

   $$R' = R(a_R - s_R R - s_{MR} M). \qquad (2.38)$$

   (c) Add initial conditions to complete a moose–rabbit competition model.

6. The preceding exercise derived a pair of competition equations. This exercise asks you to reduce them to the competition equations (2.33–2.34) presented in the text by measuring time and populations in appropriate units. (This process in known as introducing *dimensionless variables*.)

   (a) Define new time and population scales:

   $$\tau = a_M t, \quad y = \frac{M}{a_M/s_M}, \quad z = \frac{R}{a_R/s_R}.$$

   Substitute for M, M', R, and R' in (2.37) and (2.38) in terms of y, $dy/d\tau$, z, and $dz/d\tau$. Note that the chain rule requires

   $$\frac{dM}{dt} = \frac{d(a_M/s_M)y}{d\tau} \cdot \frac{d\tau}{dt} = \left(\frac{a_M}{s_M}\right)\frac{dy}{d\tau}a_M$$

with a similar expression for dR/dt.

Simplify and show that (2.33–2.34) result when y' and z' are understood to mean $dy/d\tau$ and $dz/d\tau$. Write r, e, and f in terms of a_M, s_M, ...

(b) Explain the relation between the old and new population and time scales. (See the note on page 73.) (*Hint* for the population scales: if the moose population were governed by the logistic equation $M' = a_M M - s_M M^2$ (no rabbits present), it would tend to the steady-state *carrying capacity* a_M/s_M. See exercise 21, page 55, of section 2.2.)

7. Exercise 5 derived the equation

$$M' = M(a_M - s_M M - s_{MR} R)$$

to govern competition between moose (M) and rabbits (R).

(a) What is the (nonzero) steady-state moose population in the absence of rabbits? *Equivalent question*: What is the maximum sustainable moose population (or moose population *carrying capacity*) in the absence of rabbits?

(b) Use the appropriate equation from exercise 5 to provide the same information for the rabbit population.

8. Explore the situation symmetric to that considered in example 15. Use the Euler method from DELAB to solve $y' = y(1 - y - 2z)$, $z' = z(1 - z - 2y)$, $y(0) = 0.30$, $z(0) = 0.32$. Show that the y population becomes extinct.

9. The previous problem used a numerical method to show that reversing the sizes of the initial populations in the competition model produced the symmetric response: the y population became extinct instead of the z population, as in example 15. Argue that the form of the differential equations makes this outcome inevitable. Because the equations are symmetric in the dependent variables, the two initial-value problems are actually identical. (Would this argument be valid if one initial-value problem could have *two* different solutions?)

10. Show that the three SIRS equations,

$$S' = -bIS + gR,$$
$$I' = bIS - rI,$$
$$R' = rI - gR,$$

force $P = S + I + R$ to be constant.

11. State the necessary additional tentative idea(s) and derive the SIRS model (2.31–2.32).

12. Figure 2.15, page 71, is a phase plane plot of S versus I for two sets of initial conditions in the SIR model (2.29–2.30). Use the information in that figure to sketch graphs of S and I versus t

13. Find the steady state(s) of the SIRS system (2.31),

$$S' = -bIS + g(P - S - I),$$
$$I' = bIS - rI.$$

Provide a physical interpretation of each.

14. Choose $b = 2$, $r = 1$ in the SIR equations (2.29). Determine two sets of initial conditions that you would expect to produce the two curves seen in figure 2.15, page 71, an epidemic (I increases before dying out) and simple extinction of the disease. Use DELAB and the Euler method to approximate S and I. Do they exhibit the expected behavior in each case?

15. The phase plane of figure 2.17, page 74, shows two competition trajectories, one leading to extinction of y, the other to extinction of z. Figure 2.16, page 74, shows the time history of the extinction of z. Use the information in figure 2.17 to sketch the time history of the extinction of y.

16. Explain the role of each of the terms appearing in the competition equations (2.33–2.34). What effect does each represent? Does it tend to increase or decrease the population in question?

2.5 ■ CHAPTER EXERCISES

Exercise Guide	
To gain experience ...	**Try exercises**
Deriving models	1(a), 2–4, 6, 8(a), 9, 11–12(a), 19–22
Analyzing models	1(b, d–e), 5, 7, 10, 12, 18, 23–25
Solving initial-value problems	1(c), 8(b–c), 11(b–c)

1. You are given the following:

   Experimental fact. A radioactive isotope decays at a rate proportional to its mass.

   Let $k > 0$ be the constant of proportionality and let $y(t)$ be the mass of the isotope present at time t.

   (a) Derive the governing differential equation

   $$\frac{dy}{dt} = -ky,$$

   and complete the model by imposing appropriate initial conditions.

   (b) Analyze this model, and interpret your analysis in the light of your intuitive ideas about radioactive decay.

   (c) Solve your model initial-value problem and interpret the behavior of the solution in the light of your intuitive ideas about radioactive decay.

   (d) Find the time required for the mass of the isotope to decay to one half its original value. Show that this **half life** is independent of the original mass.

   (e) Suggest appropriate units for y and k.

2. Repeat parts (a)–(c) of the preceding problem in the case in which the isotope is bombarded by a neutron beam which creates a new unstable isotope at a rate of Q g/s. (Your model will include the differential equation $dy/dt + ky = Q$.)

   Conceivably, the processes of radioactive decay and creation of a new isotope by the neutron beam could reach an equilibrium in which losses balance gains and the amount of isotope remains constant. Is this possible in your model? If so, what is the steady-state amount of isotope predicted by the model? Do both the differential equation and the solution of the initial-value problem predict the same steady-state amount of isotope?

3. Use the following information to derive a model for the charge q on the capacitor in the RC circuit shown in figure 2.18:

   Kirchhoff's voltage law. The sum of the voltage changes around a closed loop is zero.

   Experimental fact 1. The voltage drop v_R across a resistor is proportional to dq/dt with proportionality constant R; R is the *resistance*.

   Experimental fact 2. The voltage drop v_C across a capacitor is proportional to $q(t)$ with proportionality constant $1/C$; C is the *capacitance*.

   The following steps will lead you to the desired model:

   (a) Write a mathematical statement of each experimental fact.

   (b) Write Kirchhoff's voltage law in terms of v_R, v_C.

   (c) Combine the results of the preceding two steps to derive the governing differential equation

   $$R\frac{dq}{dt} + \frac{1}{C}q = 0.$$

   Complete the model for a series RC circuit by imposing an appropriate initial condition.

   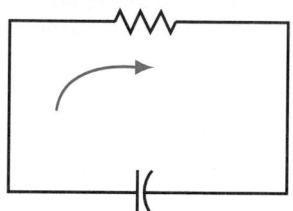

   FIGURE 2.18 A source-free series RC circuit.

4. Suppose the RC circuit of the preceding problem now includes a voltage source (imposed emf) E. Derive the following model for this circuit:

   $$R\frac{dq}{dt} + \frac{1}{C}q = E, \quad q(0) = q_i.$$

   See figure 2.19. Note that the source E increases voltage when the circuit is traversed in the direction of the arrow, while the resistor R and the capacitor C decrease voltage.

   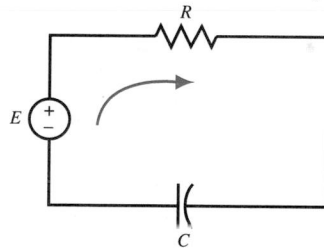

   FIGURE 2.19 A series RC circuit with a voltage source E.

5. Analyze the models of the RC circuits in the two preceding problems by answering the following questions for each:

   (a) By studying the sign of dq/dt in the differential equation, determine whether a given initial charge q_i on the capacitor will grow or decay. (Both R and C are positive constants.)

   (b) By studying the solution of each initial-value problem, determine whether a given initial charge q_i on the capacitor will grow or decay.

   (c) If charge stops flowing in either circuit, what is the steady-state charge on the capacitor?

6. The preceding problems deal with *RC* circuits. Figure 2.20 shows an unusual *RC* circuit, Ben Franklin's kite flying in a lightning storm. The lightning arises from a voltage difference between the atmosphere and the kite. Current can flow down the wet kite string to the Leyden jar, a kind of capacitor, on the ground. In principle, you could calculate the capacitance of the Leyden jar and the resistance of the kite string. Derive a model that could help determine the potential difference that caused the lightning bolt. Describe how you would use the model to estimate this potential difference.

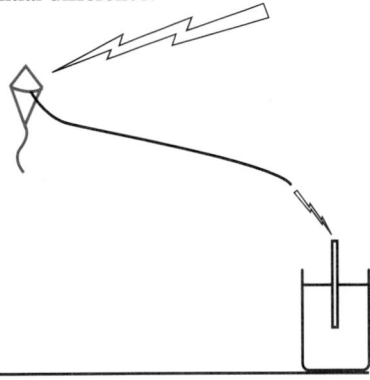

FIGURE 2.20 A schematic diagram of Ben Franklin's kite flying in a lightning storm; charge accumulates in the Leyden jar at the lower right.

7. Move the setting of the heat-loss model (2.16–2.17) from New England to the Southwest: A house whose air conditioner has failed will begin to heat up because the air around it is so hot. Derive a model for this setting, in which $T_{out} > T_i$. Show that you obtain *precisely* the same model:

$$T' = -\frac{Ak}{cm}(T - T_{out}), \quad T(0) = T_i.$$

Perhaps this model would be better called a *heat-flow model*, since it can represent either heat loss or heat gain.

8. A particular chemical reaction produces a chemical at a rate proportional to the concentration of that chemical; the proportionality constant, called the reaction rate coefficient, is k. Diluting the mixture reduces the concentration of the chemical at a rate of d gm/cm³·s; d is positive. Which initial-value problem is an appropriate model of this situation? Justify your rejection of each unacceptable choice.

(a) $c'(t) + kc(t) = d, c(0) = c_i$

(b) $c'(t) + kc(t) = -d, c(0) = c_i$

(c) $c'(t) - kc(t) = -d, c(0) = c_i$

(d) $c'(t) - kc(t) = d, c(0) = c_i$

9. Consider the situation described in the previous problem.

(a) Extract from the problem statement the necessary experimental facts, state an appropriate conservation law, and derive a model. If the model you derive is not the same as the initial-value problem you chose in the previous problem, find your error and explain it.

(b) Solve the initial-value problem you accepted in the previous problem. Show that the behavior of this solution is consistent with the physical situation being modeled.

(c) Solve each of the three initial-value problems you rejected in the previous problem. Show that each solution exhibits some physically unacceptable behavior.

10. A thin-walled mold containing molten plastic will be plunged into a cold-water bath to solidify the plastic rapidly. Introduce the necessary notation, and derive a model for the temperature of the plastic in the mold. Does this model look like any of the models studied in this chapter?

11. A radioactive isotope is decaying at a rate proportional to its mass; the proportionality constant is $k > 0$. A neutron beam bombarding the sample creates new isotope at a rate of $i(t)$ g/s. If $c(t)$ represents the amount of isotope at time t, which initial-value problem is a correct model of this situation? Justify your rejection of each unacceptable choice.

(a) $c'(t) + kc(t) = i(t), c(0) = c_i$

(b) $c'(t) + kc(t) = -i(t), c(0) = c_i$

(c) $c'(t) - kc(t) = i(t), c(0) = c_i$

(d) $c'(t) - kc(t) = -i(t), c(0) = c_i$

12. Consider the situation described in the previous problem.

(a) Extract from the problem statement the necessary experimental facts, state an appropriate conservation law, and derive a model. If the model you derive is not the same as the initial-value problem you chose in the previous problem, find your error and explain it.

(b) Solve the initial-value problem you accepted in the previous problem. Show that the behavior of this solution is consistent with the physical situation being modeled.

(c) Solve each of the three initial-value problems you rejected in the previous problem. Show that each solution exhibits some physically unacceptable behavior.

13. Waste water dumped into waterways often contains organic matter whose decay consumes oxygen in the water, depriving fish and other aquatic life of the oxygen they

need to survive. Environmentalists term such pollution *biological oxygen demand* (BOD); it is measured in units of amount of oxygen per unit volume of water.

(a) Experiments suggest that BOD decays at a rate proportional to the amount present. Derive a model for the BOD level in a closed lake into which BOD is dumped at a given rate.

(b) Either by studying the differential equation in your model or by solving the initial-value problem, determine the fate of the lake. Will it be overwhelmed by BOD and die, or can the natural decay of BOD balance the steady influx of pollutants?

(c) Could the lake cleanse itself of BOD if the pollution were halted? Characterize how quickly the water quality would improve in terms of the parameters you have introduced in your model.

14. An entrepreneur has established a line of credit at a bank. She is charged a daily interest rate of 0.0493% on the total amount she owes the bank at the end of each business day. She borrows at a rate of $b(t)$ dollars per day, and her initial loan was in the amount a_i. If $a(t)$ denotes the amount she *owes* t days after the initial loan, which of the following would be a reasonable model for the amount she owes? Justify your rejection of each unacceptable choice.

(a) $a'(t) + 0.000493a = b(t)$, $a(0) = a_i$

(b) $a'(t) - 0.000493a = b(t)$, $a(0) = a_i$

(c) $a'(t) + 0.000493a = -b(t)$, $a(0) = a_i$

(d) $a'(t) - 0.000493a = -b(t)$, $a(0) = a_i$

15. An insect population (denoted by y) is being destroyed by an insecticide at a constant rate d. It is increasing at a rate proportional to the current population; the constant of proportionality is g. Which of the following initial-value problems might be a reasonable model of this situation? Justify your rejection of each unacceptable choice.

(a) $y' + gy = d$, $y(0) = y_i$

(b) $y' - gy = d$, $y(0) = y_i$

(c) $y' + gy = -d$, $y(0) = y_i$

(d) $y' - gy = -d$, $y(0) = y_i$

16. A possible model for the temperature $T(t)$ of a house heated by a wood stove is

$$T' = -\frac{Ak}{cm}(T - T_{\text{out}}) \pm \frac{S}{cm}, \quad T(0) = T_i,$$

where S is a (positive) parameter representing the constant heat release rate of the stove. Should the last term in the differential equation be $+S/cm$ or $-S/cm$? Justify your choice.

17. Consider the equation $y' = ky + f(t)$.

(a) Examine each of the models derived in this and the previous chapter and in the exercises you have been assigned. Which can be written in this general form? In each case, identify the unknown $y(t)$, the constant k, and the forcing term $f(t)$.

(b) When $f(t) \equiv 0$, equations of this form are sometimes called *equations of growth or decay*. Why? How does the sign of k determine whether growth or decay occurs?

(c) Why could $f(t)$ reasonably be labeled a "source term"? What must be its sign to make it a genuine source? What is its sign when it is a so-called sink, or loss, term?

18. How might you use the data of table 2.1, page 28, for the G.D.R. and the G.F.R. to estimate the growth rate of a united Germany?

19. A trout farm raises fish in a closed pond. Young fish are added to the pond at a fixed rate A, and mature fish are harvested at a rate R. Derive a model for the population F of the fish in the pond, assuming that a fixed fraction of the trout reproduce per unit time.

20. Table 2.6, page 80, shows two population models and the mechanisms for adding or removing population which each considers. The entry for the simple population model is complete.

(a) What additional steps are required to derive the simple population model $P' = kP$, $P(0) = P_i$ from the expression "net change (in the population) over $\Delta t = \cdots$"? What law(s), if any, must you invoke?

(b) Complete the table entry for the emigration model and derive the model $P' = kP - E$, $P(0) = P_i$. Answer the previous questions for this model. Are your answers any different?

(c) Develop a table entry for the logistic model $P' = aP - sP^2$, $P(0) = P_i$.

(d) Develop a table entry for the "population" of isotopes described in exercise 1.

(e) Develop a table entry for the "population" of isotopes described in exercise 2.

(f) Develop a table entry for the "population" of chemical reactant described in exercise 8.

(g) Develop a table entry for the "population" of biological oxygen demand (BOD) described in exercise 13.

(h) Develop a table entry for the "population" of dollars borrowed on the line of credit described in exercise 14.

TABLE 2.6 Quantities added to and removed from populations by various effects.

Model Name	Mechanisms Considered by the Model	Number Added over Δt	Number Lost over Δt
Simple	Natural birth at rate $k_b P$	$k_b P \Delta t$	
	Natural death at rate $k_d P$		$k_d P \Delta t$
	Net change over $\Delta t = k_b P \Delta t - k_d P \Delta t$		
Emigration	Natural birth at rate $k_b P$	$k_b P \Delta t$	
	Natural death at rate $k_d P$		$k_d P \Delta t$
	Emigration at rate E		$E \Delta t$
	Net change over $\Delta t = \cdots$		

(i) Develop a table entry for the population of insects described in exercise 15.

21. Construct a table analogous to table 2.6 for each of the following heat-flow models:

    (a) the house in exercise 7,

    (b) the plastic mold described in exercise 10,

    (c) the house heated by a wood stove described in exercise 16.

22. The preceding two problems consider population models (in various guises) and heat-flow models. Comment on the similarities between the tables derived in those two exercises. Can you draw a parallel between the mechanisms for change considered in each case? Can you draw a parallel between the underlying conservation laws?

23. Compare the form of the predator–prey equations (2.24–2.25) and the competition equations (2.33–2.34). Which terms do they have in common? Which terms are different? How are those similarities and differences reflected in the phenomena being modeled?

24. Find the steady-state solution of $y' = ky$, k a constant. Sketch two direction field diagrams, one for $k > 0$, one for $k < 0$, and from them formulate a conjecture about the range of values of k for which this steady state is stable. Use an analytic solution to confirm (or correct) your conjecture. Which model(s) involve special cases of this equation?

25. Repeat the preceding exercise for $y' = k(y - a)$, k and a constants.

26. Show that the competition system (2.33–2.34), page 72, has the steady-state solution $y = 0$, $z = 0$. Use DELAB's phase plane graphical tool for systems of two equations to formulate a conjecture about the stability of this steady state. (Use an intuitive notion of stability. A deeper study awaits in chapter 8.)

27. Introducing *dimensionless* variables can reduce the clutter of parameters with which some models are afflicted.

    (a) Argue that the constant k in the growth model $y' = ky$ has units of inverse time (e.g., s^{-1}).

    (b) Define the dimensionless time $s = kt$ and the new variable Y, $Y(s) \equiv y(s/k)$. Carefully express dY/ds in terms of dy/dt and show that $dy/dt = ky$ becomes simply $dY/ds = Y$.

    (c) Apply these techniques to the emigration equation $P' = kP - E$ and to the heat-loss equation $T' = -(Ak/cm)(T - T_{out})$.

2.6 ■ CHAPTER PROJECTS

1. **Fitting population data.** The simplest population equation is $P' = kP$. Argue that a plot of $\log P$ versus t should be a straight line for populations that are accurately modeled by this equation.

    (a) Examine the United States population data in table 2.2, page 29, and find some periods of time over which this model is not unreasonable.

What is the significance of the slope and the intercept of a line drawn through a plot of the *logarithm* of the population values versus time?

(b) Draw a line by eye through this log data plot. Use that line to estimate the growth rate k. Repeat for every time period for which the model $P' = kP$ seems reasonable.

(c) Using MATLAB or a spreadsheet that will fit a line to data using least squares, estimate k for each of the cases considered in part (b). How do these estimated values compare with the annual growth rates listed in table 2.2?

2. **Analyzing the logistic model.** The curve-sketching tools developed in calculus can provide a qualitative interpretation of the logistic model

$$\frac{dP}{dt} = aP - sP^2, \quad P(t_0) = P_i.$$

The idea is to learn as much as possible about the sign of the first and second derivatives of P to aid in sketching the population curve. (Plotting points precisely requires a formula for the solution of this initial-value problem, but plotting points does not give much more information than this project.)

(a) Write the differential equation as $P' = (a - sP)P$ and argue that the sign of $P'(t_0)$ depends upon the relative values of P_i and the ratio a/s.

(b) Compute P'' directly from $P' = aP - sP^2$ and use information about the sign of P' to find the sign of P'' in the intervals $0 < P < a/2s$, $a/2s < P < a/s$, $a/s < P$. Does the population curve exhibit any inflection points?

(c) Note that $P' = 0$ when $P = a/s$. Argue that populations that are initially larger than a/s decay to the level a/s while those initially smaller than a/s grow to a/s. In either case, the population curve does not actually cross the horizontal line $P = a/s$. (Take this last assertion on faith for the moment; it is a consequence of the uniqueness of the solution of initial-value problems. See section 4.6.) What is the growth rate of this population when $P = a/s$?

(d) Can the population curve exhibit any local maxima or minima?

(e) Combine this information into sketches of the population curves for initial populations in each of the three intervals listed in part (b).

(f) The ratio a/s is sometimes called the *carrying capacity* of the population. Explain this term.

(g) What happens to the growth rate of the population at $P = a/2s$?

3. **Growth rate in the logistic model.** By writing the logistic equation (2.12), page 48, in the form $dP/dt = (a - sP)P$, we can think of it as arising from the simple population equation $dP/dt = kP$ when the constant growth rate k is replaced by a declining growth rate $g(P) = a - sP$.

Does the declining growth rate function $g(P)$ fit the growth rates obtained for the United States from table 2.3, page 39, over any periods of time? Attempt to estimate the parameters *a* and *s* (using regression software in a spread sheet program if you wish). Use the results of the analysis requested in project 2 (specifically, that all populations tend toward the value a/s) to predict a maximum United States population.

4. **How hot is the amplifier?** A stereo system amplifier is enclosed in a steel cabinet. Much of the electric power it consumes is released as heat by the components inside the cabinet. If the amplifier were not ventilated, what temperature might it reach?

   A typical amplifier enclosure is about $45 \times 15 \times 30$ cm. Its mass is about 9 kg. The thermal conductivity of the steel cabinet is about 0.05 cal/s·K·cm^2. The specific heat of its components is about 0.08 cal/g·K. If its power consumption in watts is P, then its components are releasing heat at about $4P$ cal/s. (1 watt $= 4.186$ cal/s; assume the other 0.186 cal/s goes to audible power.)

   (a) Write an expression for the *rate* at which heat energy is lost through the walls of the amplifier cabinet. Write an expression for the *rate* at which heat energy is being released by the components inside the cabinet. Use these expressions and the law of conservation of heat energy to derive a model for the temperature of the amplifier.

   (b) Obtain from your model an expression for the constant operating temperature of the unvented amplifier. What is the operating temperature of a 20-watt amplifier? A 50-watt amplifier? Would you recommend adding ventilation slots to amplifiers to lower their operating temperatures?

5. **A furnace model.** Example 13, page 61, and the subsequent discussion introduced a model of the temperature of a house in which the output of its furnace depends on the temperature. An Euler approximation of the solution of a specific case ($T' = -0.012(T - 5) + F(T)$, $T(0) = 15°C$) of this model appears in figure 2.13, page 62.

   Is the overshoot of the set point temperature shown in figure 2.13 an artifact of the Euler method approximate or is it an accurate representation of the behavior predicted by the model? How about the saw tooth behavior after that? Will the house be maintained at a steady temperature? How often will the furnace cycle? Might the model of the furnace be improved by having separate on and off temperatures a few degrees apart, like real thermostatic controls?

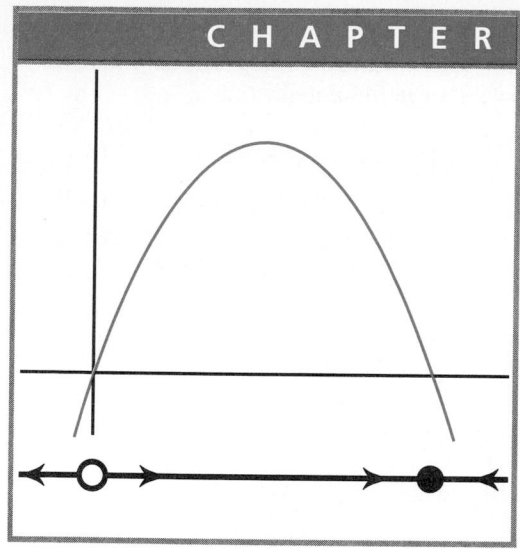

Numerical and Graphical Tools

This chapter develops numerical techniques for approximating solutions of initial-value problems and graphical techniques for visualizing the behavior of such solutions. The focus is on problems with one unknown, like the logistic population model or the heat-loss model. The numerical tools extend painlessly to models in higher dimensions such as predator-prey and epidemics, and the graphical tools have two-dimensional analogs.

3.1 ■ NUMERICAL METHODS

We have one tool for the numerical approximation of the solution of an initial-value problem, the Euler method. Its simplicity is appealing,

$$\text{new value} = \text{old value} + \text{slope} \times \text{time step},$$

but there are many better methods available. Interpreting the Euler method from a variety of perspectives sets the stage for developing some of those better numerical methods.

3.1.1 Interpretations of the Euler Method

To illustrate and interpret the Euler method, we will use a model for the temperature of an unheated house when the outside temperature swings periodically between a daytime temperature of 15°C and a nighttime temperature of 5°C (figure 3.1).

Writing the model requires a formula for T_{out} as a function of time. If temperature is measured in °C, time t in min, and if $t = 0$ is noon, the time of peak in temperature, then one function with the required periodic variation is

$$T_{out} = 10 + 5\cos(2\pi t/1{,}440).$$

(Note that $\cos(2\pi t/1{,}440)$ has period 1,440 min = 24 h. See exercise 18, page 65.)

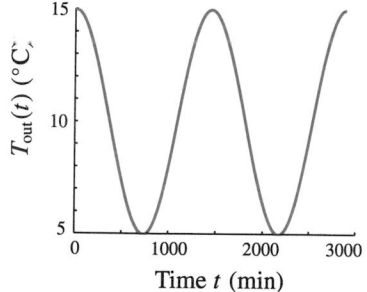

FIGURE 3.1 Periodically varying outside temperature $T_{out}(t)$.

If the house is at 20° at noon ($t = 0$), the corresponding heat-loss model is $T' = -0.012(T - T_{out}(t))$, $T(0) = 20$, or in full detail,

Heat-loss model with time-dependent T_{out}

$$T' = -0.012[T - (10 + 5\cos(2\pi t/1{,}440))], \tag{3.1}$$

$$T(0) = 20. \tag{3.2}$$

Stop and Think   **3.1** Verify that $T_{out}(t)$ does indeed vary periodically over the course of a day from a high of 15°C to a low of 5°C. Does it reach its peak at noon?

If T_n denotes the "old" value of T and T_{n+1} the "new" value, then the Euler method recipe for approximating the solution of (3.1–3.2) is

$$\text{new} = \text{old} + \qquad\qquad \text{slope} \times \text{step size},$$

$$T_{n+1} = T_n + [-0.012(T_n - (10 + 5\cos(2\pi t_n/1{,}440)))]\,\Delta t, \tag{3.3}$$

$$T_0 = 20, \quad n = 0, 1, 2, \dots.$$

Stop and Think   **3.2** The pattern of the Euler method may not be apparent in the detail of the expression for T_{out}. To see the pattern clearly, write the Euler recipe when the heat-loss model is expressed in the form

$$T' = -0.012(T - T_{out}(t)), \quad T(0) = 20.$$

For the generic initial-value problem

$$y' = f(t, y), \quad y(0) = y_i,$$

a general statement of the Euler method is

Euler method

$$\text{new} = \text{old} + \text{slope} \times \text{step size},$$

$$y_{n+1} = y_n + \qquad f(t_n, y_n)\Delta t, \tag{3.4}$$

$$y_0 = y_i, \quad n = 0, 1, 2, \dots.$$

Stop and Think   **3.3** Translate the heat-loss model (3.1–3.2) into y-t notation, then write the formula for $f(t, y)$. Use the general statement of the Euler method just given to write the Euler method for (3.1–3.2). Verify that you obtain the right method.

Suggestion: Carry out these steps whenever you encounter a general statement: write out the general case for a specific example. Use the specific example to give meaning to the form of the general statement whenever you encounter it.

MATLAB

The Euler method is among the numerical tools in DELAB that are available after entering or choosing an equation. For more information, start DELAB, select **Help, Textbook**, then go to chapter 3, Euler method.

Interpretation 1: Following the Direction Field

The Euler method takes steps of finite length Δt through a direction field diagram.

Each Euler step is in the direction whose slope is determined by the differential equation. The first step begins at the point given by the initial condition. Subsequent steps begin where their predecessors ended.

For example, the first step of an Euler method approximate solution of (3.1–3.2) using $\Delta t = 50$ min begins at the "old" point $t_0 = 0$, $T_0 = 20$ and moves in the direction given by the slope

$$T_0' = -0.012[20 - (10 + 5\cos(2\pi \cdot 0/1{,}440))] = -0.06$$

at the old point to arrive at the new point

$$t_1 = 0 + 50 = 50,$$
$$T_1 = 20 + [-0.012(20 - (10 + 5\cos(2\pi \cdot 0/1{,}440)))]\,50 = 17.00.$$

Because slope T_0' is negative, temperature is decreasing.

Figure 3.2 illustrates this and several subsequent steps. Each step follows the slope of the direction field at its starting point.

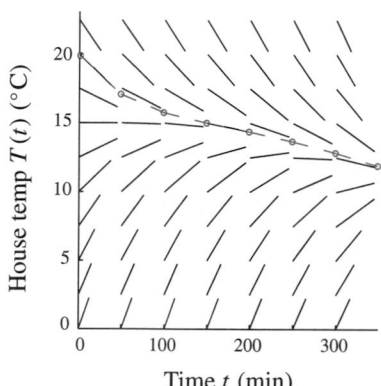

FIGURE 3.2 The first seven steps of the Euler approximation to the solution of the heat-loss model (3.1–3.2) computed using (3.3) with $\Delta t = 50$ min.

(House temp $T(t)$ (°C) vs. Time t (min))

MATLAB

To draw your own version of figure 3.2 or to watch the Euler method follow a direction field, use the Euler and visual demo tools in DELAB. For guidance, select **Help, Textbook**, then go to chapter 3, figure 3.2.

Stop and Think

3.4 Does the first Euler step in figure 3.2 begin where it should? Is its direction consistent with that given by the direction field lines?

3.5 Describe the behavior of the Euler approximation beginning at $T(0) = 0$. Does it increase or decrease initially? Might it eventually follow the same path as the one in the figure beginning at $T(0) = 20$? Verify your conjecture using the Euler method in DELAB.

3.6 Why do the direction field lines in figure 3.2 at $t = 0$ just above and below $T = 15$ point toward one another? Why is the direction line at $t = 0$, $T = 15$ horizontal?

Think of the exact solution of an initial-value problem as following a direction field diagram using infinitesimally small steps. It is able to sense instantaneously every change of direction. In contrast, the Euler method can only approximate the solution of an initial-value problem because it takes a finite step in a fixed direction, missing the instantaneous direction changes.

Stop and Think

3.7 Suppose the slope of the direction field never changes, as in the equation $y' = -9.8$ for example. Can Euler find *exact* solutions of the initial-value problems involving this differential equation?

Interpretation 2: Tangent to the Solution Curve

The Euler approximation is the line tangent to the exact solution curve.

The slope of a line is the ratio of the change in y to the change in x,

$$\text{slope } m = \frac{\Delta y}{\Delta x}.$$

Since Δy is just the new value of y minus the old, another version of the straight-line relation is

$$\text{new } y = \text{old } y + \text{slope} \times x \text{ step.}$$

This word equation looks suspiciously like the Euler method. How does the notation compare with the first step of Euler for the heat-loss model (3.1–3.2)? To make that comparison, write y_0, T_0 for the old y, T values:

$$\text{Euler:} \quad T = T_0 + [-0.012(T_0 - (5\cos(2\pi t_0/1{,}440)))]\,\Delta t,$$
$$\text{line:} \quad y = y_0 + \hspace{3.5cm} m\,\Delta x.$$

The Euler value comes from a line with slope

$$m = [-0.012(T_0 - (10 + 5\cos(2\pi t_0/1{,}440)))],$$

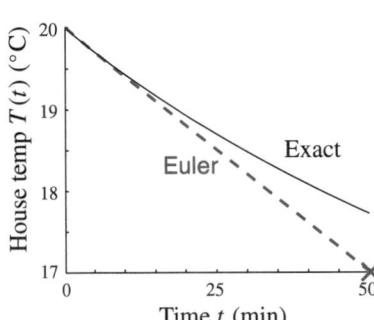

FIGURE 3.3 At the beginning of a step, the Euler method draws a line tangent to the exact solution curve through that point; this figure illustrates the first step of (3.3).

the *slope term from the differential equation* (3.1) evaluated at the starting time t_0 and the old temperature T_0. Euler is drawing a line with the same slope as the exact solution. In other words, *Euler is drawing a line tangent to the exact solution* at the starting point of the Euler step.

Figure 3.3 shows the first step of the Euler method approximation computed from (3.3); it is a line tangent to the exact solution. (Figure 3.3 is a close up of figure 3.2.)

This approximation process is called **local linearization**. Clearly, the accuracy of an Euler approximation is limited by the wiggles in the exact solution graph; the less the exact solution behaves like a straight line, the less accurate is Euler's straight-line approximation.

Stop and Think   **3.8** Explain *local* and *linearization* in the phrase *local linearization*.

■ **EXAMPLE 1**   *Estimate the time required for the temperature of the house to drop from* 20°C *to* 18°C.

Use linear approximation; that is, apply one step of Euler with $T_0 = 20$, $T_1 = 18$ and solve to find Δt:

$$18 = 20 + [-0.012(20 - (10 + 5\cos(2\pi \cdot 0/1{,}440)))]\,\Delta t$$
$$= -0.06\Delta t.$$

Hence, the house temperature drops from 20°C to 18°C in about

$$\Delta t = \frac{18 - 20}{-0.06} \approx 33 \text{ min.} \quad ■$$

Interpretation 3: Approximating the Derivative

The Euler method is equivalent to stopping short of the limit in the definition of the first derivative.

Since the derivative is defined by the limit

$$y'(t) = \lim_{\Delta t \to 0} \frac{y(t + \Delta t) - y(t)}{\Delta t},$$

an obvious approximation is

$$y'(t) \approx \frac{y(t + \Delta t) - y(t)}{\Delta t}. \tag{3.5}$$

Use this expression to approximate the derivative in the generic differential equation $y' = f(t, y(t))$:

$$\frac{y(t + \Delta t) - y(t)}{\Delta t} \approx f(t, y(t)).$$

Solve for the "new" value $y(t + \Delta t)$ in terms of both the "old" value $y(t)$ and f evaluated at that old value of y. The familiar Euler method results:

$$y(t + \Delta t) \approx y(t) + f(t, y(t))\Delta t.$$

Stop and Think   **3.9** Apply this "approximate the derivative" argument to the heat-loss equation (3.1) to derive the Euler method formulation (3.3).

The Euler method can't be completely accurate because it depends on stopping the limiting process in the definition of the limit; instead of an instantaneous rate of change over an infinitesimally small period of time, it uses an average rate of change over a finite interval of time.

Interpretation 4: Taylor Polynomial

Using the Euler method is equivalent to approximating the exact solution using the first two terms of its Taylor series.

Before reading this section, you may wish to review Taylor polynomials; see appendix section 3.

The mission of the heat-loss model (3.1–3.2) is, "Given the starting temperature of the house, find its temperature at any later time." The most naive answer is, "If the house temperature starts at 20°C, it will always be about 20°C." That uninspired answer—treating the time varying function $T(t)$ as if it were a constant—ignores the information in the differential equation about the rate of change of temperature.

How does one improve the constant approximation $T(t) \approx 20$ if some information about $T'(t)$ is available? Taylor's theorem (see appendix section 3) provides some guidance:

$$T(t_0 + \Delta t) = T(t_0) + T'(t_0)\Delta t + \tfrac{1}{2}T''(t_0)(\Delta t)^2 + \cdots. \tag{3.6}$$

The first term alone gives the naive constant approximation

$$T(t_0 + \Delta t) \approx T(t_0) = 20°C.$$

Using *two* terms improves this approximation by adding some rate of change information,

$$T(t_0 + \Delta t) \approx T(t_0) + T'(t_0)\Delta t. \tag{3.7}$$

Euler is this two-term Taylor approximation; just define $t_1 = t_0 + \Delta t$ to recover the case $n = 1$ in (3.3).

Stop and Think

3.10 Describe precisely the rate of change information used by the two-term Taylor approximation (3.7).

3.11 Repeat this analysis for the generic equation $y' = f(t, y)$. Write out the corresponding Taylor polynomial approximations and derive the general Euler method formula (3.4).

How could the accuracy of the Euler method be improved? One approach is to keep the next term in the Taylor polynomial (3.6), the one involving T''. One version of this improvement leads to the *Heun method*, which is described later.

Stop and Think

3.12 The previous discussion of Euler as a tangent approximation suggested that the Euler method should perform better with problems whose solution curves don't wiggle too much. Argue that wiggling, or large changes in slope, are signaled by large values of $|T''|$. How does that observation relate to the Taylor polynomial view of the Euler method?

3.1.2 Accuracy Examples

Suppose we wanted to follow the temperature of the house using the Euler method over a period of, say, two days (or 2,880 min). If we didn't want to work too hard, we might take large steps; 19 steps with $\Delta t = 150$ min would just about cover the two days ($19 \cdot 150 = 2,850$).

The result of applying the Euler method as in (3.3) with $\Delta t = 150$ is shown in figure 3.4. The results shown there are troubling because intuition suggests that the temperature of the house should oscillate regularly and that the transition around $t = 0$ ought to be smooth.

Is the problem with our intuition or with the error in the Euler method? Would a smaller value of Δt improve the picture? What is the connection between the size of Δt and the error in the approximate solution?

A later discussion will answer that last question in some detail. For the moment, we will explore the error in the Euler method approximation to the solution of an even simpler initial-value problem with an easy exact solution.

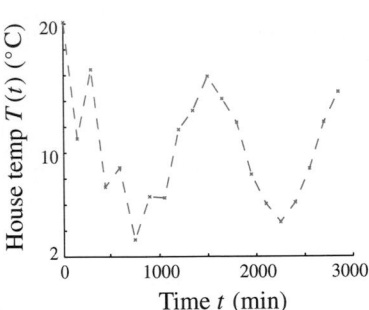

FIGURE 3.4 The (not very good) Euler approximation with $\Delta t = 150$ min to the solution of the heat-loss model (3.1–3.2).

■ **EXAMPLE 2** *Use the Euler method to approximate the solution of the initial-value problem*

$$\frac{dP}{dt} = 0.0282P, \quad P(1978) = 115.40, \tag{3.8}$$

a model for the population of Brazil using the data from table 2.1, page 28. *Estimate population in the year* 2008. (*Time t is measured in years and population P in millions of people.*)

For simplicity, choose $\Delta t = 10$. (Then there are just three steps from 1978 to 2008.) Beginning with the initial value $P_0 = 115.40$ and the initial time $t_0 = 1978$ in (3.4), calculate

$$P_1 = P_0 + 0.0282 P_0 \cdot 10$$
$$= 115.40 + 0.0282 \cdot 115.40 \cdot 10 = 147.94.$$

Hence, $P(1988) = P(1978 + \Delta t) \approx P_1 = 147.94$.

Repeat with another step of length $\Delta t = 10$ from $t_1 = t_0 + \Delta t = 1988$ and $P_1 = 147.94$:

$$P_2 = P_1 + 0.0282 P_1 \cdot 10$$
$$= 147.94 + 0.0282 \cdot 147.94 \cdot 10 = 189.66,$$

to find $P(1998) = P(1978 + 2\Delta t) \approx P_2 = 189.66$. Yet another step yields

$$P_3 = P_2 + 0.0282 P_2 \cdot 10$$
$$= 189.66 + 0.0282 \cdot 189.66 \cdot 10 = 243.15,$$

the estimate we sought,

$$P(2008) = P(1978 + 3\Delta t) \approx P_3 = 243.15 \text{ million people.} \blacksquare$$

Distinguish parentheses from subscripts in the notation: e.g., $P(1998)$ denotes the value at $t = 1998$ of the function $P(t)$ that exactly solves the initial-value problem (3.8) while P_2 denotes the value after two steps of the Euler approximation to the solution of (3.8). If $\Delta t = 10$, then two steps of Euler is approximating $P(t)$ at $t = 1978 + 2\Delta t = 1998$. Hence, P_2 is the Euler approximation to $P(1998)$.

Stop and Think

3.13 If $\Delta t = 3$, what value of $P(t)$ does P_2 approximate? Fill in the missing information: $P(t_0 + n\Delta t) \approx P_?$.

TABLE 3.1 **Values of the approximate solution of** $P' = 0.0282P$, $P(1978) =$ 115.40, **computed using the Euler method with** $\Delta t = 10$.

n	t_n	P_n
0	1978	115.40
1	1988	147.94
2	1998	189.66
3	2008	243.15

■ **EXAMPLE 3** *Determine the error in the Euler method approximation using* $\Delta t = 10$ *to the solution of* $P' = 0.0282P$, $P(1978) = 115.40$, *at* $t = 2008$.

From table 3.1 the Euler method approximation is $P(2008) = P(t_3) \approx P_3 = 243.15$.

Separation of variables gives the exact solution of the initial-value problem,

$$P(t) = 115.40 e^{0.0282(t-1978)}.$$

The value of the exact solution at $t = 2008$ is

$$P(2008) = 115.40 e^{0.0282 \cdot 30} = 268.92.$$

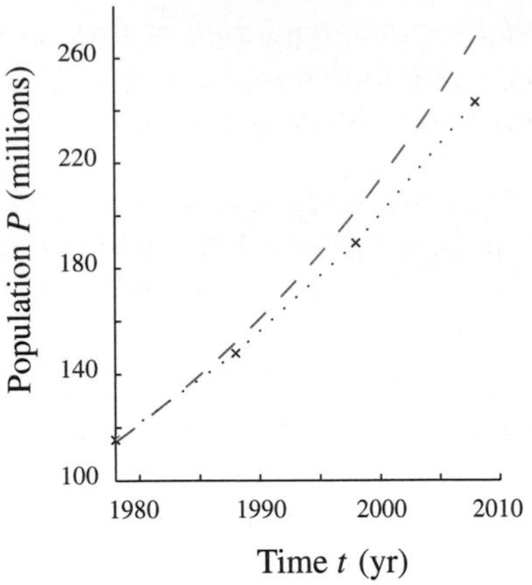

FIGURE 3.5 The Euler method approximate solution (lower curve) of the population model in example 3 and its exact solution (upper curve) from table 3.1.

The error in the approximation at $n = 3$ is

$$e_3 = |P(2008) - P_3| = |268.92 - 243.15| = 25.77,$$

an error of almost 10%. Figure 3.5 shows this approximate solution. ■

TABLE 3.2 Global error $e_n = |P(t_N) - P_N|$ **at** $t_N = 2008$ **in the Euler method approximate solutions of** $P' = 0.0282P$, $P(1978) = 115.40$ **using various values of** Δt.

Δt	N	Error e_N	Percent error
10.	3	25.76	9.58%
5.	6	14.28	5.31%
2.5	12	7.55	2.80%
1.25	24	3.88	1.44%
0.625	48	1.97	0.73%
0.3125	96	0.99	0.36%

As the last lines of the preceding example suggest, a useful measure of error is its size relative to the exact solution, the **relative error**,

$$\text{relative error} = \frac{|\text{actual error}|}{|\text{exact value}|}.$$

Relative error is often expressed as a per cent. For example, the relative error at the end of step 3 in the last example is

$$\text{relative error} = \frac{|P(t_3) - P_3|}{|P(t_3)|} = \frac{25.77}{268.92} = 9.6\%.$$

Using smaller values of Δt improves the situation somewhat, as table 3.2 shows; halving Δt (and doubling the computational work by doubling the number of steps) halves the error. (These values were generated using DELAB.)

```
MATLAB
To generate a table like 3.2, in DELAB select Help, Textbook, then go
to chapter 3, table 3.2.
```

The important problem of estimating the accuracy of an approximate solution when we can not check it against the exact solution is a major task of the discipline of *numerical analysis*. The Catch-22 of scientific computing is that we never know the exact answer; if we did, we would not be doing the computing. Later parts of this section introduce some of the fundamental ideas of error analysis.

3.1.3 A Better Method: Heun

The error in the Euler method seems to occur because we take a step whose direction is fixed to be the slope of the direction field at the beginning of the

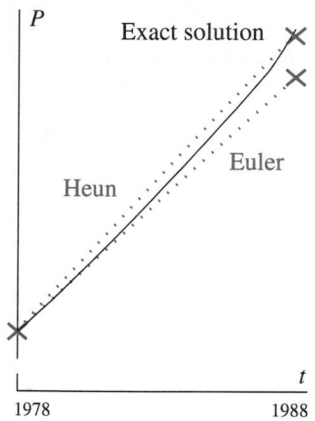

FIGURE 3.6 The tangent line Euler approximation and the average slope Heun approximation for the population equation $P'(t) = 0.0282P(t)$.

step; see figure 3.5, for example. If we had an idea of the slope at the point at the *end* of the step, perhaps we could modify the direction of our step to use the knowledge of the direction field both at the *beginning* of the step and at the *end* of the step.

Consider, for example, the situation shown in figure 3.5. From the differential equation $P'(t) = 0.0282P(t)$, the slope at the beginning of the first step is $P_0' = 0.0282P_0 = 3.254$. The slope at the end of that step is $P_1' = 0.0282P_1 = 4.172$.

Now back up, remembering the slope information we just obtained. Take a single step from 1978 to 1988 using the *average* of these two slopes, the one from the beginning of the step and the one from the end of the conventional Euler step. This technique is called the **modified Euler method** or the **Heun method**. See figure 3.6.

To restate the Heun method in the notation of the population model, begin at the point (t_0, P_0). First compute the Euler method approximation, denoted by an overbar,

$$\bar{P}_1 = P_0 + 0.0282P_0\Delta t.$$

The slope of the direction field at $(t_1, \bar{P}_1)$ is $0.0282\bar{P}_1$. Now take a step from the initial point in the direction whose slope is the *average* of the initial and final slopes. That is, compute

$$P_1 = P_0 + \tfrac{1}{2}(0.0282P_0 + 0.0282\bar{P}_1)\Delta t$$

to obtain the approximate solution at time t_1.

For the generic initial-value problem

$$y' = f(t, y), \quad y(0) = y_i,$$

a general statement of the **modified Euler method** or **Heun method** is

$$
\begin{aligned}
&\text{new} = \text{old} + \text{average slope} \times \text{step size,} \\
&\bar{y}_{n+1} = y_n + f(t_n, y_n)\Delta t, \\
&y_{n+1} = y_n + \tfrac{1}{2}[f(t_n, y_n) + f(t_{n+1}, \bar{y}_{n+1})]\Delta t, \\
&y_0 = y_i, \quad n = 0, 1, 2, \ldots .
\end{aligned}
\tag{3.9}
$$

The preceding statement of the Heun method emphasizes its geometric interpretation, but it wastes effort by evaluating $f(t_n, y_n)$ twice. A more efficient form is

Heun method

$$
\begin{aligned}
k_1 &= f(t_n, y_n)\Delta t, \\
k_2 &= f(t_{n+1}, y_n + k_1)\Delta t, \\
y_{n+1} &= y_n + (k_1 + k_2)/2, \\
y_0 &= y_i, \quad n = 0, 1, 2, \ldots .
\end{aligned}
\tag{3.10}
$$

> **MATLAB**
>
> The Heun method is available from the numerical tools menu of DELAB.

Stop and Think **3.14** Show that (3.10) and (3.9), the two forms of the Heun method, are indeed equivalent.

■ **EXAMPLE 4**   *Repeat example 3 using the Heun method in place of the Euler method.*

Begin with $P(1978) = 115.40$. Using $\Delta t = 10$ and (for geometric emphasis) the inefficient form (3.9), the pattern of the calculations is:

$$\bar{P}_1 = 115.40 + 0.0282 \cdot 115.40 \cdot 10 = 147.94,$$

$$P_1 = 115.40 + \tfrac{1}{2}0.0282(115.40 + 147.94)10 = 152.53,$$

$$\bar{P}_2 = 152.53 + 0.0282 \cdot 152.53 \cdot 10 = 195.54,$$

$$P_2 = 152.53 + \tfrac{1}{2}0.0282(152.53 + 195.54)10 = 201.61,$$

$$\vdots$$

We find ultimately that $P_3 = 266.48$, the Heun method approximation to $P(2008)$. These values appear in table 3.3.

Borrowing the exact solution from example 3 yields the error in the Heun method approximation,

$$P(2008) - P_3 = 268.92 - 266.48 = 2.44,$$

an error of less than 1%, a significant improvement over the Euler method. ■

A graph of the Heun method approximation obtained in the last example appears in figure 3.7. The improved accuracy is evident in comparison with figure 3.5, the corresponding plot for the Euler method.

In addition to the geometric argument of averaging slopes, the Taylor polynomial interpretation of Euler also explains why Heun is more accurate. Euler keeps only the first two terms of the Taylor polynomial (3.6),

$$y(t_0 + \Delta t) \approx y(t_0) + y'(t_0)\Delta t.$$

The next term in the Taylor expansion is $y''(t_0)(\Delta t)^2/2$. Incorporating it in some form would improve upon Euler.

Since y'' is the derivative of $y' = f(t, y)$, approximate y'' just as in (3.5), page 87, using $y'' \approx \Delta y'/\Delta t$ in place of $y' \approx \Delta y/\Delta t$,

$$y''(t_0) \approx \frac{y'(t_0 + \Delta t) - y'(t_0)}{\Delta t} = \frac{f(t_0 + \Delta t, y(t_0 + \Delta t)) - f(t_0, y(t_0))}{\Delta t}.$$

Now replace $y(t_0 + \Delta t)$ by its Euler approximation $\bar{y}_1 = y(t_0) + y'(t_0)\Delta t$ to obtain

$$y''(t_0) \approx \frac{f(t_0 + \Delta t, \bar{y}_1) - f(t_0, y(t_0))}{\Delta t}.$$

TABLE 3.3   Values of the approximate solution of $P' = 0.0282P$, $P(1978) = 115.40$ **computed using the Heun method with** $\Delta t = 10$.

n	t_n	P_n
0	1978	115.40
1	1988	152.53
2	1998	201.61
3	2008	266.48

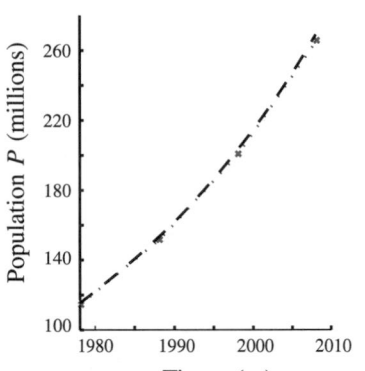

FIGURE 3.7   The Heun method approximate solution of the population model in example 4 and its exact solution; see table 3.3.

EULER APPROXIMATION

Using this last expression to approximate $y''(t_0)$, the *three-term* Taylor polynomial yields

$$y(t_0 + \Delta t) \approx y(t_0) + y'(t_0)\Delta t + \tfrac{1}{2}y''(t_0)(\Delta t)^2$$

$$\approx y(t_0) + f(t_0, y(t_0))\Delta t + \frac{f(t_0 + \Delta t, \bar{y}_1) - f(t_0, y(t_0))}{2\Delta t}(\Delta t)^2$$

$$= y(t_0) + \tfrac{1}{2}[f(t_0, y(t_0)) + f(t_0 + \Delta t, \bar{y}_1)]\Delta t,$$

the Heun method approximation. Hence, Heun can be said to improve upon Euler because it keeps an approximation to the third term in the Taylor polynomial expansions of the solution; that is, it preserves some information about y'', the rate of change of the slope of the solution, information that Euler ignores.

Stop and Think   **3.15** Mimic the preceding analysis for the population equation $P' = 0.0282P$.

3.16 Mimic the preceding analysis for the time-varying heat-loss equation, $T' = -0.012(T - T_{out}(t))$.

3.1.4   Global and Local Error

A comparison of figures 3.5 and 3.7 shows that the Heun method is more accurate than the Euler method, at least for the population model solved there. What precisely is the accuracy advantage of Heun over Euler, if indeed there is one in general? What improvement in accuracy can we expect with either method if we reduce the step size? How might we construct methods that are better than either of these? This section develops answers to these questions.

Our primary interest is in the **global error** after N steps with a given method,

GLOBAL ERROR

$$e_N = |y(t_N) - y_N|.$$

Here $y(t)$ is the *exact* solution of the initial-value problem $y' - f(t, y)$, $y(t_0) = y_i$, and y_N is the approximate solution at some time t_N of interest; y_N is the last computed value in the sequence $y_0, y_1, \ldots, y_N$ generated with one of these methods. Examples 3 and 4 computed the global error e_3 at $t_3 = 2008$ for Euler and Heun applied to a population model.

We can not expect to find a computable formula for global error; if we had such a formula, then we could find the error and add it to the approximate solution y_N to obtain the exact solution. But we can determine the proportionate change in global error due to a change in Δt.

A key contributor to global error is the **local truncation error**, the error at the end of *one* step of a method, *beginning with an exact solution value*,

LOCAL TRUNCATION ERROR

$$E_L = |y(t_1) - y_1|,$$

where $y(t_0) = y_0$. This difference is called *truncation* error because it arises from the approximation $y'(t) \approx (y(t + \Delta t) - y(t))/\Delta t$, which truncates the

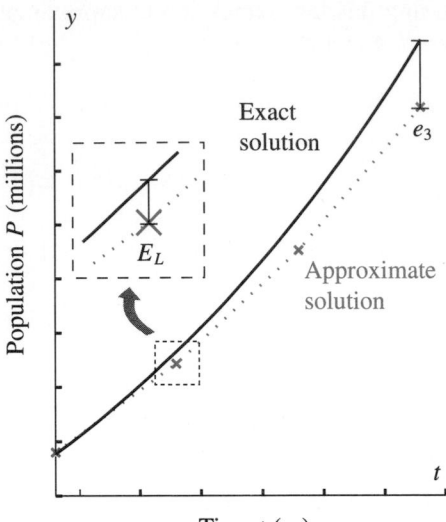

FIGURE 3.8   Local truncation error E_L and global error e_N at $N = 3$ in a numerical solution method.

limit in the definition of derivative. Figure 3.8 illustrates these two forms of error.

What is the relation between local truncation error and global error? To approximate the solution at some fixed time T using a step size Δt requires $N = T/\Delta t$ steps. The error accumulates as we step from point to point, in part from the local truncation error that is introduced even when the step begins with an exact solution value and in part from the effect of beginning each step after the first with an incorrect solution value. If the latter effect is not too pronounced, intuition suggests that the global error accumulates in the same way as the sum of the local truncation errors. Since there are $N = T/\Delta t$ steps, global error might behave something like the sum of N local error values—that is, like

$$NE_L = T\frac{E_L}{\Delta t}.$$

Indeed, if the solution of the initial-value problem has sufficiently smooth derivatives, then it can be proven that

Global error $e_N = |y(t_N) - y_N|$ at time T is proportional to $E_L/\Delta t$:

$$e_N \sim \frac{E_L}{\Delta t}.$$

Global error vs. local truncation error

Error in the Euler Method

To find an expression for the global error in the Euler method, we must evaluate the local truncation error E_L. Since $y' = f(t, y)$, the Euler method formula $y_1 = y_0 + f(t_0, y_0)\Delta t$ can also be written

One step of the Euler method

$$y_1 = y_0 + y_0'\Delta t,$$

where $y_0' = f(t_0, y_0)$. This expression is dangerously close to a Taylor expansion for the exact solution $y(t_1)$ about the initial point t_0,

Taylor expansion for exact solution

$$y(t_1) = y(t_0) + y'(t_0)(t_1 - t_0) + \frac{y''(t_0)(t_1 - t_0)^2}{2}$$
$$+ \cdots + \frac{y^{(n-1)}(t_0)(t_1 - t_0)^{n-1}}{(n-1)!} + R_n, \qquad (3.11)$$

where the remainder term is

$$R_n = \frac{y^{(n)}(\xi)(t_1 - t_0)^n}{n!}$$

for some ξ, $t_0 \le \xi \le t_1$. (See Taylor's theorem, appendix section 3.) This observation is the key to analyzing local truncation error.

To make a closer connection between the Euler method

Euler

$$y_1 = y_0 + y_0'\Delta t$$

and the Taylor expansion, write $t_1 - t_0 = \Delta t$ in (3.11), assume y has two continuous derivatives, and stop the expansion at $n = 2$:

Exact

$$y(t_1) = y(t_0) + y'(t_0)\Delta t + R_2.$$

Then the local truncation error is

$$E_L = y(t_1) - y_1$$
$$= (y(t_0) + y'(t_0)\Delta t + R_2) - (y_0 + y_0'\Delta t)$$
$$= y(t_0) - y_0 + (y'(t_0) - y_0')\Delta t + R_2.$$

The definition of local truncation error requires that the Euler method step begin with the value of the exact solution; i.e., $y_0 = y(t_0)$. As a consequence, the approximate and exact initial derivative values are identical as well,

$$y_0' = f(t_0, y_0) = f(t_0, y(t_0)) = y'(t_0).$$

Hence, the local truncation error expression for the Euler method reduces to

Local error for the Euler method

$$E_L = R_2 = \frac{y''(\xi)}{2}\Delta t^2,$$

where $t_0 \le \xi \le t_1$. (Note that Δt^2 means $(\Delta t)^2$.) This analysis has proved the following theorem.

Theorem 1 (Local truncation error for the Euler method). *Let the solution $y(t)$ of the initial-value problem $y' = f(t, y)$, $y(t_0) = y_i$, have two continuous derivatives in the interval $[t_0, t_1]$. Then within that interval, the local truncation error in the Euler method applied to this problem is $E_L = y''(\xi)\Delta t^2/2$ for some ξ in $[t_0, t_1]$.*

The content of this formal statement is simple: Reducing Δt by 2 will decrease the local truncation error in the Euler method by 4, and so on.

Since

Local truncation error for the Euler method is proportional to Δt^2

and since global error is proportional to $E_L/\Delta t$, we conclude that

Global error for the Euler method is proportional to Δt

Global error for the Euler method

$$e_N \sim \Delta t$$

if the exact solution has two continuous derivatives. The direct proportionality between global error and Δt is illustrated clearly in table 3.2, page 90.

Stop and Think   **3.17** Suppose you plotted the error vs. Δt data in table 3.2, then drew a straight line through the points. What slope would that line have? What does the slope of that line tell you about the relation between error and step size?

Halving Δt will halve the global error (and halving Δt will require twice as much computation to reach the same ending time T). Because its global error is proportional to Δt to the *first* power, the Euler method is called a **first-order method**.

Finding Better Methods

If global error for any method is always proportional to $E_L/\Delta t$, then a method with improved global error must exhibit local truncation error E_L proportional to a power of Δt higher than the Δt^2 factor in the local error expression for the Euler method. To improve local truncation error, an improved method needs to match more than the first two terms in the Taylor expansion (3.11) of the exact solution.

Computational experience shows that one improvement on Euler is the Heun method,

Heun method

$$k_1 = f(t_n, y_n)\Delta t,$$
$$k_2 = f(t_{n+1}, y_n + k_1)\Delta t,$$
$$y_{n+1} = y_n + (k_1 + k_2)/2.$$

It averages two slope values to step from y_n to y_{n+1}. Generalizing this idea, seek a method of the form

General method

$$k_1 = f(t_n, y_n)\Delta t,$$
$$k_2 = f(t_n + \alpha\Delta t, y_n + \beta k_1)\Delta t, \qquad (3.12)$$
$$y_{n+1} = y_n + ak_1 + bk_2,$$

where the constants a, b, α, β are to be determined to match as many terms as possible in the Taylor expansion (3.11).

Obviously, the Heun method corresponds to choosing $\alpha = \beta = 1$ and $a = b = 1/2$.

To parallel the analysis of local truncation error in the Euler method, we need to write the first step of this method as if it were a Taylor expansion.

Letting $y_0' = f(t_0, y_0)$, we have

$$k_1 = y_0' \Delta t,$$
$$k_2 = f(t_0 + \alpha \Delta t, y_0 + \beta k_1) \Delta t,$$
$$y_1 = y_0 + a y_0' \Delta t + b k_2.$$

It remains to write k_2 using a two-variable Taylor expansion about (t_0, y_0) (see section 3 of the appendix):

$$k_2 = f(t_0 + \alpha \Delta t, y_0 + \beta k_1) \Delta t$$

$$= f(t_0, y_0) \Delta t + \frac{\partial f(t_0, y_0)}{\partial t} (t_0 + \alpha \Delta t - t_0) \Delta t$$

$$+ \frac{\partial f(t_0, y_0)}{\partial y} (y_0 + \beta k_1 - y_0) \Delta t + \mathcal{O}(\Delta t^3)$$

$$= y_0' \Delta t + \frac{\partial f(t_0, y_0)}{\partial t} \alpha \Delta t^2 + \frac{\partial f(t_0, y_0)}{\partial y} \beta y_0' \Delta t^2 + \mathcal{O}(\Delta t^3).$$

We have used $y_0' = f(t_0, y_0)$ and $k_1 = y_0' \Delta t$. The term $\mathcal{O}(\Delta t^3)$ ($\mathcal{O}$ for *order of*) denotes terms multiplied by Δt^n, $n \geq 3$; those terms decay to zero at least as fast as Δt^3.

The *big O* notation $\mathcal{O}(\Delta t^3)$ is a convenient way of lumping together terms that are less significant when Δt is small than those with factors of Δt or Δt^2. More precisely, we write $g(\Delta t) = \mathcal{O}(\Delta t^3)$ for a given function g if

$$\lim_{\Delta t \to 0} \frac{g(\Delta t)}{\Delta t^3} = c$$

for some constant c.

From this expression for k_2, the first step of the general method is

y_1, approximate solution value

$$y_1 = y_0 + a y_0' \Delta t + b k_2$$

$$= y_0 + a y_0' \Delta t$$

$$+ b \left(y_0' \Delta t + \frac{\partial f(t_0, y_0)}{\partial t} \alpha \Delta t^2 + \frac{\partial f(t_0, y_0)}{\partial y} \beta y_0' \Delta t^2 + \mathcal{O}(\Delta t^3) \right)$$

$$= y_0 + (a + b) y_0' \Delta t + \left(\alpha b \frac{\partial f(t_0, y_0)}{\partial t} + \beta b \frac{\partial f(t_0, y_0)}{\partial y} y_0' \right) \Delta t^2$$

$$+ \mathcal{O}(\Delta t^3). \tag{3.13}$$

We wish to compare this expression with the Taylor expansion about t_0 of the exact solution $y(t_1)$,

$y(t_1)$, exact solution value

$$y(t_1) = y(t_0) + y'(t_0) \Delta t + \tfrac{1}{2} y''(t_0) \Delta t^2 + R_3$$

$$= y_0 + y_0' \Delta t + \tfrac{1}{2} y''(t_0) \Delta t^2 + R_3.$$

The terms involving Δt^2 in the Taylor expansion multiply y'', while the corresponding terms in the method multiply various partial derivatives of f. To reconcile those terms, we use $y' = f(t, y)$ and the chain rule to write

$$y''(t) = \frac{\partial f(t, y(t))}{\partial t} + \frac{\partial f(t, y(t))}{\partial y} y'(t). \tag{3.14}$$

Stop and Think   **3.18** Verify this relation for $T' = -0.012(T - T_{\text{out}}(t))$.

Evaluated at $t = t_0$ with $y(t_0) = y_0$ and $y'(t_0) = y_0'$, (3.14) becomes

$$y''(t_0) = \frac{\partial f(t_0, y_0)}{\partial t} + \frac{\partial f(t_0, y_0)}{\partial y} y_0'.$$

Hence, the Taylor expansion for the exact solution is

Exact solution value

$$y(t_1) = y_0 + y_0'\Delta t + \left(\frac{1}{2}\frac{\partial f(t_0, y_0)}{\partial t} + \frac{1}{2}\frac{\partial f(t_0, y_0)}{\partial y}y_0' \right)\Delta t^2 + R_3. \tag{3.15}$$

Now we can match terms in the method (3.13) with those in the Taylor expansion (3.15) for the exact solution in an attempt to make the order of the local truncation error E_L as high as possible. Comparing coefficients of like powers of Δt in (3.13) and (3.15) we find

$$\begin{aligned}
\Delta t: \quad & a + b = 1, \\
\Delta t^2: \quad & \alpha b = \tfrac{1}{2}, \\
& \beta b = \tfrac{1}{2}.
\end{aligned} \tag{3.16}$$

(A more detailed analysis would show that we can not match the coefficients Δt^3.) For *any* choice of a, b, α, β satisfying these three equations, the method (3.12) will have local truncation error

$$E_L = y(t_1) - y_1 = R_3 - \mathcal{O}(\Delta t^3).$$

But the Taylor remainder term R_3 is proportional to Δt^3, providing the exact solution $y(t)$ has three continuous derivatives. Consequently, the local truncation error in this method is proportional to Δt^3 and *the global error is proportional to Δt^2* if a, b, α, β satisfy (3.16). We have proved:

Theorem 2. *When the general method (3.12) is applied to an initial-value problem whose solution has three continuous derivatives, the global error is proportional to Δt^2 if the coefficients in the method satisfy (3.16).*

Corollary 3. *Global error in the Heun method is proportional to Δt^2 when it is applied to initial-value problems whose solutions have three continuous derivatives.*

■ **EXAMPLE 5**   *Show that the Heun method, one version of the general second-order method (3.12), has global error at $t = 2008$ that is proportional to Δt^2 when it is applied to $P' = 0.0282P$, $P(1978) = 115.40$.*

The exact solution is available in example 3, page 89; $P(2008) = 268.92$.

Use the Heun method in DELAB to approximate $P(2008)$. Choose various values of N. For each N, let

$$\Delta t = \frac{2008 - 1978}{N},$$

then compute P_N and the global error $e_N = |P(2008) - P_N|$. If global error is indeed proportional to Δt^2, then $e_N / \Delta t^2$ should be approximately constant, at least for smaller values of Δt. Table 3.4 illustrates exactly this behavior. ■

TABLE 3.4 **Global error** $e_N = |P(t_N) - P_N|$ **at** $t_N = 2008$ **in the Heun method approximate solutions of** $P' = 0.0282P$, $P(1978) = 115.40$ **using various values of** Δt.

Δt	N	Error e_N	$e_N / \Delta t^2$
10.	3	2.4368	0.0244
5.	6	0.6777	0.0271
2.5	12	0.1787	0.0286
1.25	24	0.0459	0.0294
0.625	48	0.0116	0.0298
0.3125	96	0.0029	0.0300
0.1562	192	0.0007	0.0301
0.0781	384	0.0002	0.0301

MATLAB

For guidance in constructing a table like 3.4, select **Help, Textbook,** then go to chapter 3, table 3.4.

Stop and Think

3.19 Deduce from table 3.4 an estimate of the proportionality constant c in $e_N \approx c(\Delta t)^2$, the global error relation for the Heun method.

Any method of the form (3.12) with a, b, α, β satisfying the conditions (3.16) is called a **second-order Runge-Kutta** method. It is of *second* order because its global error is proportional to Δt to the *second* power. Halving Δt decreases the global error by a factor of four while only doubling the amount of work. Clearly, these Runge-Kutta methods are more efficient computationally than the Euler method.

Of course, one member of this family is the familiar Heun method. As we observed above, it corresponds to the choices $\alpha = \beta = 1$ and $a = b = 1/2$, which satisfy the conditions (3.16).

3.1.5 An Even Better Method: Fourth-Order Runge-Kutta

Precisely these same ideas can be used to develop a family of *fourth-order Runge-Kutta methods*. Beginning with a general formulation similar to (3.12), but one that averages four slope values rather than two, we can find conditions similar to (3.16) on a set of constants to ensure that the local truncation error is proportional to Δt^5. Hence, these methods are globally fourth order. For the initial-value problem $y' = f(t, y)$, $y(t_0) = y_i$, one choice of those constants leads to:

Fourth-order Runge-Kutta method

$$k_1 = f(t_n, y_n)\Delta t,$$
$$k_2 = f(t_n + \Delta t/2, y_n + k_1/2)\Delta t,$$
$$k_3 = f(t_n + \Delta t/2, y_n + k_2/2)\Delta t,$$
$$k_4 = f(t_n + \Delta t, y_n + k_3)\Delta t,$$
$$y_{n+1} = y_n + (k_1 + 2k_2 + 2k_3 + k_4)/6,$$
$$y_0 = y_i, \quad n = 0, 1, 2, \ldots.$$

> **MATLAB**
> This fourth-order Runge-Kutta method is among the numerical tools in DELAB that are available after entering or choosing an equation. For more information, start DELAB, select `Help, Textbook`, then go to chapter 3, Runge-Kutta method.

This method, or one of its fourth-order relatives, is commonly called the **Runge-Kutta method**. The second-order analogs we derived previously are known by other names such as Heun or modified Euler or, for another choice of a, b, α, β, *midpoint* (see exercise 45). There are Runge-Kutta methods of order greater than four, too.

Ortega and Poole [19] discuss in more detail the analysis of errors in numerical methods for initial-value problems.

3.1.6 Numerical Stability

The size of the step Δt used in a numerical method is determined by the accuracy desired in the approximate solution to the initial-value problem being studied. But there can be another influence on Δt as well, introduced by the subtle requirements of *numerical stability*. The numerical method can introduce artifacts that are not part of the solution of the original initial-value problem. Sometimes these components of the numerical approximation blow up like perturbations of an unstable steady state, even though the exact solution of the initial-value problem should be decaying.

Numerical stability is tested by applying the method in question to a special test problem, $y' = ky$ with $k < 0$. Of course, this problem has the steady state $y_{ss} = 0$, to which all of its solutions decay: $\lim_{t \to \infty} y(t) = 0$ for every solution $y(t)$. *The essence of numerical stability is preserving the stability of $y_{ss} = 0$.*

NUMERICALLY STABLE METHOD

Suppose a numerical method using a given value of Δt is applied to $y' = ky$, $k < 0$, to generate a sequence of approximate solutions $y_0, y_1, \ldots$. (The initial value of y doesn't matter.) Then the method is said to be **numerically stable** for that value of Δt if $y_n \to 0$ as $n \to \infty$. That is, *the method is numerically stable if its solutions decay to zero as do the exact solutions.*

■ **EXAMPLE 6**   *Find the range of values of Δt for which the Euler method is stable.*

Applying Euler to $y' = ky$ with some arbitrary starting value y_0 leads to the sequence of values

$$y_1 = y_0 + (ky_0)\Delta t = (1 + k\Delta t)y_0,$$

$$y_2 = y_1 + (ky_1)\Delta t = (1 + k\Delta t)y_1 = (1 + k\Delta t)^2 y_0,$$

$$\vdots$$

$$y_n = y_{n-1} + (ky_{n-1})\Delta t = (1 + k\Delta t)y_{n-1} = \cdots = (1 + k\Delta t)^n y_0.$$

Stop and Think   **3.20** Verify this calculation.

Ignore the trivial case $y_0 = 0$. As $n \to \infty$, $y_n = (1 + k\Delta t)^n y_0 \to 0$ only if $|1 + k\Delta t| < 1$. That condition is equivalent to

$$-1 < 1 + k\Delta t < 1,$$
$$-2 < k\Delta t < 0.$$

Since $k < 0$, Δt must satisfy the *numerical stability condition for the Euler method*,

Stability condition for the Euler method

$$\Delta t < -\frac{2}{k}. \tag{3.17}$$

Stop and Think   **3.21** Verify this string of inequalities, particularly the stability condition.

The Euler method is numerically stable if $\Delta t < -2/k$. ■

Figure 3.9 shows the consequences of numerical instability. Euler with $\Delta t = 0.05$ is used to solve $y' = ky$, $y(0) = 1$. In the upper graph, Δt satisfies the stability condition (3.14) for $k = -30$:

$$\Delta t = 0.05 < -\frac{2}{-30} \approx 0.067.$$

The numerical solution is not particularly accurate, but it does decay toward zero, as does the exact solution $y = e^{-30t}$. In the lower graph, Δt does *not* satisfy the stability condition for $k = -40$:

$$\Delta t = 0.05 \not< -\frac{2}{-40} = 0.05.$$

The numerical solution oscillates between ± 1 instead of decaying. An initial-value problem involving $y' = -40y$ can not be solved reliably using the Euler method unless $\Delta t < -2/(-40) = 0.05$.

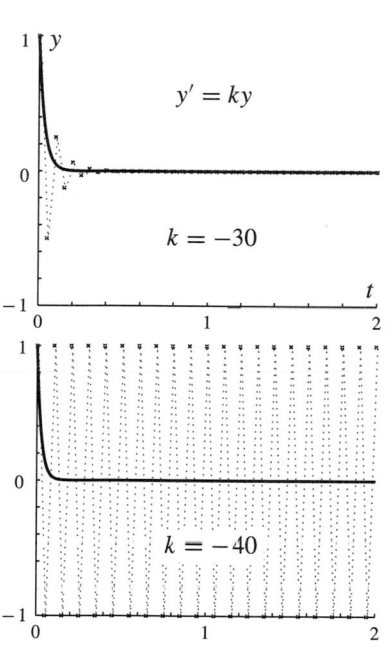

FIGURE 3.9 The Euler method with $\Delta t = 0.05$ applied to $y' = ky$. It is stable for $k = -30$ (upper graph, colored) but unstable for $k = -40$ (lower graph, colored). The exact solutions are solid curves.

> **MATLAB**
>
> Using DELAB, apply the Euler method to $y' = -30y$, $y(0) = 1$, with $\Delta t = 0.06$ and $\Delta t = 0.07$. What do you expect will happen? Are your expectations met?

If a method has a relatively tight stability restriction, as does Euler, it can force the use of a correspondingly small time step, even though accuracy requirements might be met with a much larger step. For example, the step size of $\Delta t = 0.05$ seems unnecessarily small to follow the tail of the decaying exponentials in figure 3.9—they are practically flat—yet the lower graph in figure 3.9 shows clearly that even $\Delta t = 0.05$ is too large. The next example introduces an *implicit method*. Methods of this sort are attractive because they can reduce or avoid such stability restrictions on step size.

■ **EXAMPLE 7**   *A variant of Euler is the* implicit Euler method,

IMPLICIT EULER METHOD

$$y_{n+1} = y_n + f(t_{n+1}, y_{n+1})\Delta t.$$

(*See exercises* 46 *and* 52 *for geometric motivation and an explanation of the adjective* implicit.) *Find the range of values of* Δt *for which it is numerically stable.*

To test numerical stability, apply the method to $y' = ky$, $k < 0$:

$$y_{n+1} = y_n + (ky_{n+1})\Delta t.$$

Solve for the "new" value y_{n+1}:

$$y_{n+1} = \frac{y_n}{1 - k\Delta t} = \frac{y_{n-1}}{(1 - k\Delta t)^2} = \cdots = \frac{y_0}{(1 - k\Delta t)^{n+1}}.$$

Since $k < 0$, $1 < 1 - k\Delta t$ and

$$y_{n+1} = \frac{y_0}{(1 - k\Delta t)^{n+1}} \to 0$$

for *all* $\Delta t > 0$. The implicit Euler method is *never* unstable numerically. ■

Stop and Think   **3.22** Why not use implicit Euler all the time? Try applying it to a nonlinear equation, say, $y' = y - y^3$.

3.1.7   Summary: Numerical Methods

The (absolute value of the) difference between the exact solution and an approximate solution after one step of a numerical method is *local truncation error*. After more than one step, the accumulated error is the *global error*. Providing the exact solution of the given initial-value problem is sufficiently smooth, global error is proportional to a power of the step size. That power is the *order of the method*.

Local truncation error

Global error

Order of a method

To approximate the solution of $y' = f(t, y)$, $y(t_0) = y_i$, all methods choose $y_0 = y_i$ and $t_n = t_0 + n\Delta t$.

The Euler method

Euler

$$y_{n+1} = y_n + f(t_n, y_n)$$

is of order one; global error is proportional to Δt.

The Heun method,

$$k_1 = f(t_n, y_n)\Delta t,$$
$$k_2 = f(t_{n+1}, y_n + k_1)\Delta t,$$
$$y_{n+1} = y_n + (k_1 + k_2)/2,$$

Heun

like its other second-order Runge-Kutta relatives, has global error proportional to Δt^2.

Fourth-order Runge-Kutta,

Fourth-order Runge-Kutta

$$k_1 = f(t_n, y_n)\Delta t,$$
$$k_2 = f(t_n + \Delta t/2, y_n + k_1/2)\Delta t,$$
$$k_3 = f(t_n + \Delta t/2, y_n + k_2/2)\Delta t,$$
$$k_4 = f(t_n + \Delta t, y_n + k_3)\Delta t,$$
$$y_{n+1} = y_n + (k_1 + 2k_2 + 2k_3 + k_4)/6,$$

exhibits global error proportional to Δt^4.

To apply any method to a system of equations, apply it to each of the equations in the system; e.g., example 13, page 18.

3.1.8 Exercises

EXERCISE GUIDE	
To gain experience . . .	**Try exercises**
Applying the Euler method	1(a), 3(a), 5, 7(a), 9(a), 11(a), 14, 16(a), 26(a), 27–29(a)
Applying the Heun method	2(a), 4(a), 6, 8(a), 10(a), 12(a), 13, 15, 16(b), 25(a), 33
With error in the Euler method	1(b), 3(b), 7(b), 9(b), 11(b), 14, 16(c), 24, 26(b–c), 27–29(b), 38, 41
With error in the Heun method	2(b), 4(b), 8(b), 10(b), 12(b), 15, 16(c), 25(b–c), 39, 42
With error in the Runge-Kutta method	40
Interpreting the Euler method geometrically	30, 32(b)
Interpreting the Heun method geometrically	31–32
Interpreting other methods geometrically	48, 49, 53(a)
Studying solution behavior numerically	17–23, 27–29(c)
Examining computational efficiency	43, 44
With variants of familiar methods	45, 46, 52
Analyzing local truncation error	47, 50, 51
With numerical stability	53, 54

1. (a) Using five steps of the Euler method, find an approximation to the solution of $y' + 2y = \cos t$, $y(0) = -4$, at $t = 2$.

(b) Find the exact solution and determine the error in your numerical approximation.

2. (a) Using five steps of the Heun method, find an approximation to the solution of $y' + 2y = \cos t$, $y(0) = -4$, at $t = 2$.

(b) Find the exact solution and determine the error in your numerical approximation.

3. (a) Choosing $\Delta t = 0.5$, use the Euler method to find an approximation to the solution of $u' = 1 - u$, $u(0) = 4$, at $t = 2$.

(b) Find the exact solution and determine the error in your numerical approximation.

4. (a) Choosing $\Delta t = 0.5$, use the Heun method to find an approximation to the solution of $u' = 1 - u$, $u(0) = 4$, at $t = 2$.

(b) Find the exact solution and determine the error in your numerical approximation.

5. Use four steps of the Euler method to approximate the solution of $u' = u^2 - 1$, $u(0) = -2$, at $t = 1$.

6. Use four steps of the Heun method to approximate the solution of $u' = u^2 - 1$, $u(0) = -2$, at $t = 1$.

7. (a) Use *one step* of the Euler method to estimate the time required for the population governed by $P' = kP$, $P(0) = P_i$, to double.

(b) Find an exact formula for the time at which the population has doubled and check the accuracy of your answer.

8. Repeat exercise 7 using the Heun method.

9. (a) Use *one step* of the Euler method to estimate the time required for the difference between the initial temperature and the outside temperature to decrease to one-half its initial size for the heat-flow model

$$T' = -\frac{Ak}{cm}(T - T_{\text{out}}), \quad T(0) = T_i.$$

(b) Find an exact formula for the time at which the temperature difference is at one-half its initial value and check the accuracy of your answer.

10. Repeat exercise 9 using the Heun method.

11. (a) Table 3.1 contains the first few steps of the approximate solution by the Euler method of the initial-value problem $P' = 0.0282P$, $P(1978) = 115.40$. Extend that table to the year 2048. What is the corresponding value of n?

(b) Find the exact solution of this initial-value problem and then use it to add a third column to table 3.1, the error $P(t_n) - P_n$. How does the accuracy of the approximate solution vary with n?

12. Repeat the preceding problem for table 3.3, page 92, which applies the Heun method to the same initial-value problem. Which method is more accurate for this problem, Euler or Heun?

13. Example 4 approximates the solution to a model for the population of Brazil $P' = 0.0282P$, $P(1978) = 115.40$, using the inefficient form (3.9) of the Heun method with $\Delta t = 10$.

(a) Repeat those calculations using the efficient form. Verify that you obtain the same values of P_n.

(b) How many multiplication operations do you save at each time step? How many would you save if $0.0282P$ were replaced by a function which required 100 multiplications for its evaluation?

14. For each of the following initial-value problems:

    (i) Use 10 steps of the Euler method on a calculator or in a program of your own to find an approximate value of the solution at the indicated value of the independent variable.

    (ii) Use DELAB to confirm the values you just computed.

    (iii) Using an analytic solution method of your choice, find the exact value of the solution at the given value of the independent variable.

    (iv) Calculate the error in the solution.

    (v) Using DELAB, experimentally determine the number of steps (or equivalently, the step size) the Euler method requires to reduce the error you found in part (iii) by a factor of 10. Is the result consistent with the global error relation $e_N \sim \Delta t$?

(a) $P' = 0.0282P$, $P(1978) = 115.40$; approximate $P(2010)$. (This initial-value problem is the population model for Brazil mentioned in the text.)

(b) $T' = -0.72(T - 10)$, $T(0) = 28$; approximate $T(10)$. (This initial-value problem is the heat-loss model (2.6.–2.7) with time measured in hours and temperature in degrees Celsius.)

(c) $P'(t) = 0.015P - 0.209$, $P(1847) = 8$; approximate $P(1850)$. (This initial-value problem is a model for the population of Ireland during the potato famine; see example 5, page 43.)

15. Complete the preceding problem using the Heun (modified Euler) method.

16. Choose a country from the population data of table 2.1 of chapter 2. Determine the values of the parameters k and P_i for an initial-value problem of the form $P' = kP$, $P(1978) = P_i$.

(a) Use the Euler method to approximate the population in 1997. Compare with the appropriate entry in table 2.1 and comment on the differences you observe.

(b) Repeat part (a) using the Heun method.

(c) Compare your numerical results, the data from the table, and the exact solution of the model. Comment on the differences you observe.

17. Using DELAB and the approximation method of your choice, verify the accuracy of the solution curve sketches you obtained in project 2 of chapter 2 for the population model with competition

$$P'(t) = aP - sP^2, \quad P(t_0) = P_0.$$

Choose appropriate values of the parameters a, s, P_0, and t_0 to duplicate the conditions you considered in that earlier project.

18. Using DELAB and the approximation method of your choice, estimate the year in which the population of Ireland would have dropped to zero according to the model developed in example 5, page 43,

$$P'(t) = 0.015P - 0.209, \quad P(1847) = 8.$$

(Recall that population is measured in millions and time in years.)

19. Using DELAB and the approximation method of your choice, graph representative solutions of $y' = y^3 - y$. Comment on the behavior associated with different ranges of initial conditions.

20. Using DELAB and the approximation method of your choice, graph representative solutions of $y' = y - y^3$. Comment on the behavior associated with different ranges of initial conditions.

21. Using DELAB and the approximation method of your choice, select the one graph from among those accompanying exercise 7, page 117, which most accurately represents the solution of the initial-value problem $y' = 1 - y^2$, $y(0) = 0$. Justify your choice.

22. Using DELAB and the approximation method of your choice, select the one graph from among those accompanying exercise 8, page 117, which most accurately represents the solution of the initial-value problem $y' = (4 - y)^3$, $y(0) = 0$. Justify your choice.

23. Using DELAB and the approximation method of your choice, select the one graph from among those accompanying exercise 9, page 117, which most accurately represents the solution of the initial-value problem $y' = y^2$, $y(0) = 1$. Justify your choice.

24. Derive the exact solution formula used in example 3 and verify the exact value of $P(2008)$ exhibited there.

25. (a) Using a calculator or DELAB, verify the Heun method approximation P_3 exhibited in example 4.

(b) Repeat the calculation with $\Delta t = 5$. For what value of n is P_n an approximation to $P(2008)$? By what factor has the error changed? Is that factor consistent with corollary 3, page 98.

(c) Compare this change in error with that obtained for the Euler method in the corresponding part of the following problem. Which method gives the greater improvement in accuracy in return for doubling the computational effort by halving Δt?

26. (a) Using a calculator, DELAB, or the formula in the following problem, verify the Euler method approximation P_3 exhibited in example 3.

(b) Repeat the calculation with $\Delta t = 5$. For what value of n is P_n an approximation to $P(2008)$? By what factor has the error in the approximation to $P(2008)$ changed? Is that factor consistent with the global error relation $e_N \sim \Delta t$?

(c) Compare this change in error with that obtained for the Heun method in the corresponding part of exercise 25. Which method gives the greater improvement in accuracy in return for doubling the computational effort by halving Δt?

27. (a) Argue that the solution $P(t)$ of the population model (2.1–2.2),

$$P'(t) = kP(t), \quad P(t_0) = P_0,$$

is approximated using the Euler method at time $t_{n+1} = t_0 + (n + 1)\Delta t$ by

$$P_{n+1} = (1 + k\Delta t)P_n = (1 + k\Delta t)^2 P_{n-1}$$
$$= \cdots = (1 + k\Delta t)^{n+1} P_0.$$

(b) How does this expression compare with the exact solution $P(t_{n+1})$?

(c) Provide a physical interpretation for the approximation formula $P_{n+1} = (1 + k\Delta t)^{n+1} P_0$ obtained in part (a). Is your interpretation approximately consistent with the assumptions used in deriving the differential equation model?

28. Repeat the previous exercise for the heat-loss model

$$T'(t) = -\frac{Ak}{cm}(T(t) - T_{out}), \quad T(0) = T_i.$$

29. Try to repeat the preceding two exercises for the population model with competition (2.5),

$$P'(t) = aP - sP^2, \quad P(t_0) = P_0.$$

What property of this differential equation apparently prevents your writing a closed-form expression analogous to $P_{n+1} = (1 + k)^{n+1} P_0$ for the Euler method approximation to the solution of this problem?

30. Draw a *careful* graph of the exact solution both of the initial-value problem $P' = 0.0282P$, $P(1978) = 115.40$, considered in example 3 and of the first few steps of the Euler method approximation using $\Delta t = 20$. (That is, draw an analog of figure 3.5 for this example.) Explain geometrically why the error in the Euler method, the difference between the exact solution and the approximation, grows with each step.

31. Carry out the preceding problem using the Heun method rather than the Euler method.

32. When the Heun method is applied to the initial-value problem $y' = f(t, y)$, $y(t_0) = y_0$, the second step computes

$$y_{n+1} = y_n + \tfrac{1}{2}(f(t_n, y_n) + f(t_{n+1}, \bar{y}_{n+1}))\Delta t.$$

    (a) Which of the terms in this expression correspond to the slope of the differential equation's direction field at the *beginning* of the step?

    (b) Which of the terms in this expression correspond to the slope of the direction field at the *end* of an Euler method step?

    (c) What term tells you that you are averaging these two slopes, not weighting one more heavily than another?

33. Write a single expression for the Heun method approximation P_{n+1} to the solution of the simple population model equation $P' = kP$. (Eliminate the $\bar{P}_{n+1}$ term.)

34. There have been a number of suggestions in our earlier analytic studies of the population model with competition,

$$P'(t) = aP - sP^2, \quad P(t_0) = P_0,$$

    that the added term sP^2 acts to provide an upper bound on population, in contrast to the simple population model that predicts continued exponential growth.

    Suppose we apply this model to the population of Brazil beginning in 1978. A comparison with the simple model considered in the text,

$$\frac{dP}{dt} = 0.0282P, \quad P(1978) = 115.40,$$

    suggests that we might take $a = 0.0282$ with the obvious choices for t_0, P_0.

    (a) Using DELAB and the approximation method of your choice, conduct experiments to determine a value of s that gives a stable population level in about the year 2050. What is the value of the limiting population?

    (b) Confirm this result analytically.

35. Use DELAB to plot an Euler approximation to the solution of the heat-loss model (3.1), page 84, over two full days.

What choice of Δt appears to produce plausible solution curves? When do the maxima and minima of temperature occur? Are those times consistent with your intuition about this problem?

36. Repeat the previous exercise with Heun. Does this method appear to need larger or smaller values of Δt than Euler to produce reasonable solution curves? Explain the differences you observe.

37. Repeat exercise 35 with fourth-order Runge-Kutta. Does this method appear to need larger or smaller values of Δt than Euler to produce reasonable solution curves? Larger or smaller values of Δt than Heun? Explain the differences you observe.

38. Using the Euler method and DELAB, approximate the solution of $y' + xy = 4x$, $y(0) = 2$, at $x = 4$ with 20, 40, 60, and 80 steps. Determine the global error in your approximations to $y(4)$. Plot error versus Δt to verify that the global error in the Euler method is proportional to Δt.

39. Repeat exercise 38 for the Heun method, plotting global error against the appropriate power of Δt.

40. Repeat exercise 38 for the fourth-order Runge-Kutta method, plotting global error against the appropriate power of Δt.

41. The text claims that the global error in the Euler method is proportional to Δt. Reducing Δt by one-half should reduce the error by roughly the same amount. Test this assertion using DELAB and an initial-value problem of your choice. Compute an approximation to the solution at a fixed value of the independent variable using the Euler method with 20, 40, and 80 steps. Compare the three approximate solution values you obtain with the exact solution. Do you believe that the error in the Euler method is proportional to Δt?

42. Since the Heun method is a second-order Runge-Kutta method, its global error is proportional to Δt^2. Reducing Δt by one-half should reduce the error by about $1/4$. Test this assertion using DELAB and an initial-value problem of your choice. Compute an approximation to the solution at a fixed value of the independent variable using the Heun method with 20, 40, and 80 steps. Compare the three approximate solution values you obtain with the exact solution. Do you believe that the error in the Heun method is proportional to Δt^2?

43. The computational cost of an algorithm is sometimes measured by the number of times it must evaluate a complicated function that appears in it. An appropriate measure for the Euler and Heun algorithms is the number of times each has to evaluate the function f appearing in the differ-

ential equation $y' = f(t, y)$. The trick is balancing computational cost against accuracy.

Using the global error results in this section, determine how Δt must be changed for each method to reduce the global error at a given time by one-fourth. How will the total computational cost of using each algorithm change? What recommendations would you make regarding the trade-off between accuracy and computational cost?

44. Referring to the second-order Runge-Kutta methods (3.12), the book says

> Halving Δt decreases the global error by a factor of four while only doubling the amount of work. Clearly, these Runge-Kutta methods are more efficient than the Euler method.

Justify these statements.

45. Another acceptable choice of constants for the second-order Runge-Kutta method (3.12) is $a = 0$, $b = 1$, $\alpha = \beta = 1/2$, leading to the *midpoint method*.

(a) Use a sketch or other geometric argument to explain the name *midpoint*.

(b) Verify directly that its local truncation error E_L is proportional to Δt^3 and, hence, that its global error is of second order.

(c) Using DELAB or a program of your own, verify as follows that the global error in this method is indeed of second order: Approximate the solution of the initial-value problem $y' = -y$, $y(0) = 1$, at $t = 1$ using this method with 10, 20, 40, and 80 steps; compare the approximate and exact solutions in each case; show that the error is proportional to Δt^2.

46. A variant of the Euler method, known as *backward Euler* or *implicit Euler*, is

$$y_{n+1} = y_n + f(t_{n+1}, y_{n+1})\Delta t.$$

(See the note following exercise 52 for an explanation of the adjective *implicit*.)

(a) Use a sketch or other geometric argument to explain the name *backward*.

(b) Verify directly that its local error E_L is proportional to Δt^2 and, hence, that its global error is of first order.

(c) Using DELAB, a programmable calculator, or a program of your own, verify as follows that the global error in this method is indeed of first order: Approximate the solution of the initial-value problem $y' = -y$, $y(0) = 1$, at $t = 2$ using this method with 10, 20, and 40 steps; compare the approximate and exact solutions in each case; show that the error is proportional to Δt.

47. Is there a choice of the constants a, b, α, β that reduces the general method (3.12) to the Euler method? Does that choice of constants satisfy the second-order error conditions (3.16)?

48. Explain the geometric significance of the terms $k_1/\Delta t$ and $k_2/\Delta t$ that appear in the general second-order Runge-Kutta method (3.12).

49. The general second-order Runge-Kutta method (3.12) computes $y_{n+1} = y_n + ak_1 + bk_2$. We found that a, b must satisfy $a + b = 1$; see (3.16). Can you interpret this method as stepping from y_n to y_{n+1} by averaging two slopes, as we did with the Heun method?

50. By writing out the terms in (3.13) that involve Δt^3, find the exact expression for the local truncation error in the Heun method. For the Heun method, formulate the analog of the Euler local error result, theorem 1. Must you demand more smoothness of the exact solution than with the Euler method?

51. By writing out the terms in (3.13) that involve Δt^3, find the exact expression for the local truncation error in the general second-order Runge-Kutta method (3.12). What choices of a, b, α, β make the coefficient of this error term a minimum?

52. Suppose you tried to improve the Heun method by eliminating the first Euler step. Then you would be using an approximation scheme that has the form

$$y_{n+1} = y_n + \tfrac{1}{2}(f(t_n, y_n) + f(t_{n+1}, y_{n+1}))\Delta t.$$

(a) Give a verbal description of the slopes you would be averaging in this process.

(b) Calculate a few iterations for the population model $P' = 0.0282P$, $P(1978) - 115.40$.

(c) Try to calculate a few iterations for the population model with competition

$$P'(t) = aP - sP^2, \quad P(t_0) = P_0.$$

Choose values of the parameters a, s, t_0, and P_0 that suit you. What happens?

(d) What is the difference between the two differential equations? How does that difference affect your ability to implement this method?

This method is known as an *implicit method*. The adjective *implicit* is used because, in general, we can not solve this iteration scheme to give an explicit expression for the approximation y_{n+1} at the end of the step.

53. By examining the formula for the Euler method applied to $y' = ky$, verify the Euler method results for $k = -30, -40$ displayed in figure 8. How would the lower graph in figure 8 change with $k = -41$?

54. Produce a version of figure 8 which demonstrates the stability and instability of the Euler method using $\Delta t = 0.05$ for $y' = ky$ with $k = -30, -40$, and using the implicit Euler method $y_{n+1} = y_n + f(t_{n+1}, y_{n+1})\Delta t$.

3.2 ■ GRAPHS, DIRECTION FIELDS, AND PHASE LINES

3.2.1 Solution Graphs

The basic shape of the graph of a solution of a differential equation tells a great deal about the behavior of the phenomena that the equation models. That qualitative behavior, a "feel" for the solution, is often more important than the precise details of a point-by-point plot of an exact solution. Much of this qualitative information comes directly from the differential equation itself.

The tools for sketching solution graphs are familiar from plotting curves in calculus. There you plotted a few points and examined the first derivative to find regions where the function was increasing or decreasing. The first derivative also determined the location of critical points, those points that were candidates for local maxima or minima because the first derivative failed to exist or was zero there. The second derivative showed concavity and located inflection points, where the concavity of the curve changed direction.

In summary, that information about the graph of $y(t)$ is as follows:

- $y'(t) > 0 \Rightarrow$ the graph of y is increasing at t; $y'(t) < 0 \Rightarrow$ the graph of y is decreasing at t;

- $y'(t) = 0 \Rightarrow$ possible local maximum (peak) or minimum (valley) at t; ditto if $y'(t)$ fails to exist;

- $y''(t) > 0 \Rightarrow$ the graph of y is concave up at t; $y''(t) < 0 \Rightarrow$ the graph of y is concave down at t;

- $y''(t) = 0 \Rightarrow$ possible inflection point (transition from concave up to concave down or vice versa) at t.

Stop and Think   **3.23** If $y''(t) > 0$, then $y'(t)$ is increasing because its derivative $y'' = dy'/dt$ is positive. Are curves with increasing slope concave up or concave down? Use that reasoning to explain "$y''(t) > 0 \Rightarrow$ the graph of y is concave up."

Those same tools still work, even without an explicit formula for the solution function. An initial condition gives one point on the solution curve. A first-order differential equation provides an expression for the first derivative of the unknown, although that expression may well involve the unknown itself. An expression for the second derivative can be calculated from the differential equation, sometimes more easily than from the solution formula itself (assuming we can even find such a formula).

■ **EXAMPLE 8**   *Sketch graphs of solutions of the rock-tossing model*

$$v'(t) = -g, \quad v(0) = v_i,$$

for various values of initial velocity v_i; $g > 0$ is the acceleration of gravity.

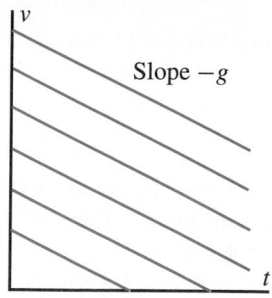

FIGURE 3.10 A family of solution curves for the initial-value problem $v' = -g$, $v(0) = v_i$, the rock-tossing model, for various values of v_i.

The differential equation $v' = -g$ says that the v curve must have constant, negative slope. The initial condition $v(0) = v_i$ determines a point $(0, v_i)$ of this downward-sloping straight line.

Nonbelievers can compute $v''(t)$ from the differential equation $v' = -g$:

$$v'' = \frac{dv'}{dt} = \frac{d(-g)}{dt} = 0.$$

Since $v'' = 0$, the graph of v must indeed be a line. ■

A family of solution curves of $v' = -g$, $v(0) = v_i$, for different choices of the initial velocity v_i is shown in figure 3.10. Do all portions of such solution curves necessarily represent physically valid behavior?

■ **EXAMPLE 9** *Sketch a family of solution curves of the simple population model*

$$P' = kP, \quad P(0) = P_i,$$

for various values of the initial population P_i.

Using logic that has been employed before, note that $P' = kP$ and $k > 0$ together force P' to have the same sign as P. If P_i is positive, then $P'(0)$ will be positive, causing P to increase beyond its initial value and thereby remain positive. Hence, P' will always be positive if P_i is positive. Likewise, P' will always be negative if P_i is negative.

The derivative $P' = kP$ is zero only when $P = 0$, and it is defined for all finite P. Therefore, the only critical points occur when $P = 0$. But since $P' = 0$ when $P = 0$, the population remains constant at $P = 0$; a population that starts with nothing never grows.

To examine concavity, compute P'' from $P' = kP$,

$$P'' = \frac{dP'}{dt} = \frac{d(kP)}{dt} = kP' = k^2 P.$$

The sign of the second derivative is the same as the sign of P. The graph of a positive population is concave up; that of a negative population, concave down.

Which of these cases occurs—an increasing, concave up population graph or the converse—is determined solely by the initial value P_i. Population curves that begin with a positive value are increasing and concave up. Those that begin with a negative value are decreasing and concave down, as figure 3.11 illustrates. (Of course, negative populations have no direct physical meaning. Only the first quadrant of figure 3.11 can be assigned any physical significance.) ■

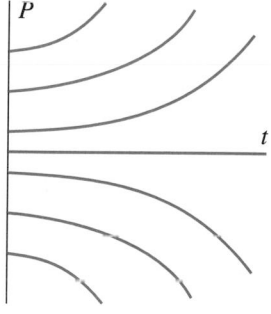

FIGURE 3.11 Graphs of the solution of the simple population model $P' = kP$, $P(0) = P_i$, for various values of P_i. Solution curves are shown for $P < 0$, but these normally have no physical significance.

Compare the conclusions of this example with the exponential solution $P = Ce^{kt}$ of $P' = kP$.

■ **EXAMPLE 10** *Sketch a family of solution curves of the population model with emigration,*

$$P' - kP = -E, \quad P(0) = P_i,$$

for various values of the initial population P_i when the emigration rate E is constant.

Mimic the techniques of the previous example.

Since $P' = kP - E$, P' is positive when $kP - E$ is positive; i.e., $P' > 0$ if $P > E/k$. Likewise, $P' < 0$ if $P < E/k$. If the initial population P_i exceeds E/k, then $P'(0)$ is positive, and $P(t)$ will remain greater than E/k for all t. If $P_i < E/k$, then $P'(0) < 0$, and $P(t) < E/k$ for all t.

If $P_i = E/k$, the $P'(0) = 0$. In this case, $P' = kP - E$ has the constant solution $P(t) = E/k$, as direct substitution confirms.

Since E is constant, use the differential equation to compute

$$P'' = \frac{d(kP - E)}{dt} = kP'.$$

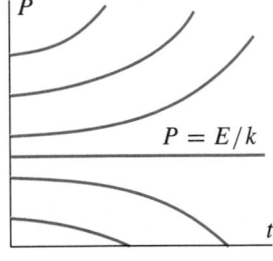

FIGURE 3.12 Graphs of the solution of the emigration model $P' = kP - E$, $P(0) = P_i$, for various values of P_i. Note that $P_i = E/k$ leads to the constant solution $P(t) = E/k$.

Hence, P'' and P' have the same sign; the population curve is concave up when it is increasing, and it is concave down when it is decreasing.

Evidently, the relative sizes of P_i and E/k determine whether the solution increases or decreases. Initial populations greater than E/k lead to increasing, concave up population curves. Those below E/k lead to decreasing, concave down population curves. If $P_i = E/k$, then the population remains at this constant value. This behavior is illustrated in figure 3.12. ■

The initial population level E/k separates populations that grow at an increasingly more rapid rate from those that decay more and more rapidly. Compare the conclusions of this example with a formula for the solution of $P' = kP - E$, $P(0) = P_i$.

■ **EXAMPLE 11**   *Describe the slope and concavity characteristics of the graph of the solution of the logistic model*

$$P' = aP - sP^2, \quad P(0) = P_i,$$

for various positive values of P_i.

From the differential equation $P' = aP - sP^2$, we have

$$P' = P(a - sP) > 0$$

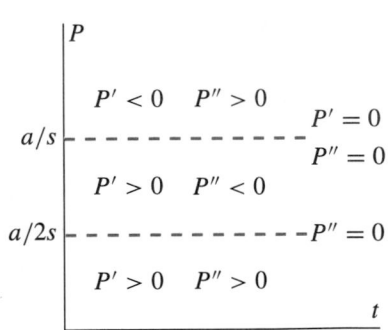

FIGURE 3.13 A summary of the slope and concavity behavior of the logistic equation $P' = aP - sP^2$.

The chain rule yields $(P^2)' = 2PP'$.

when both $P > 0$ and $a - sP > 0$. Since we are considering only $P > 0$, we conclude that $P' > 0$ for $a - sP > 0$ or, equivalently, for $P < a/s$. Conversely, $P' < 0$ for $P > a/s$.

Again compute P'' directly from the differential equation:

$$P'' = \frac{d(aP - sP^2)}{dt} = aP' - 2sPP' = P'(a - 2sP).$$

The sign of P'' is determined by the sign of P' and by the sign of $a - 2sP$. The latter is positive when $P < a/2s$ and negative when $P > a/2s$. In addition, $P'' = 0$ when either $P' = 0$ or $P = a/2s$.

This information is summarized in figure 3.13. ■

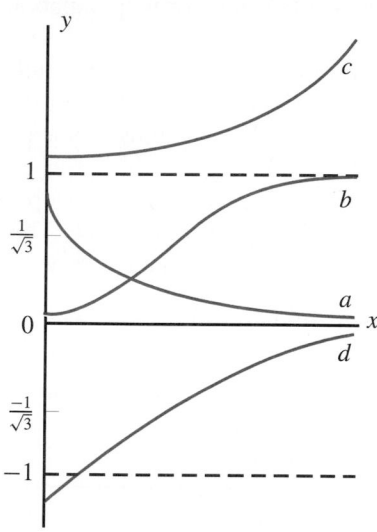

FIGURE 3.14 Graphs of four proposed solutions of $y' = y^3 - y$.

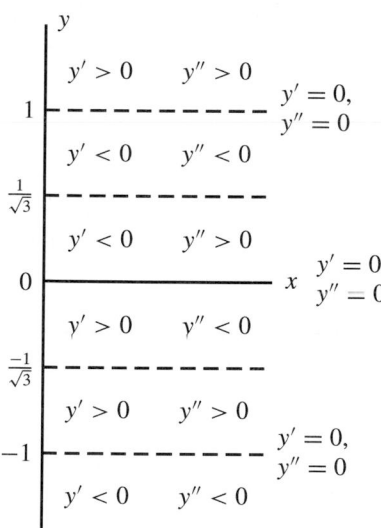

FIGURE 3.15 A summary of the slope and concavity behavior of $y' = y^3 - y$.

■ **EXAMPLE 12** *Which of the graphs in figure 3.14 could not be a solution of the differential equation*

$$y' = y^3 - y?$$

Writing the differential equation as $y' = y(y^2 - 1)$ reveals that $y' = 0$ when $y = -1, 0, 1$. The sign of y' is determined on the intervals $y < -1$, $-1 < y < 0$, $0 < y < 1$, and $1 < y$ by examining the signs of the two factors y and $y^2 - 1$.

Using the chain rule, as in the preceding example, reveals

$$y'' = \frac{dy'}{dx} = 3y^2 y' - y' = y'(3y^2 - 1).$$

Hence, $y'' = 0$ when $y' = 0$ or when $y = \pm\sqrt{1/3} = \pm 0.577\ldots$. The signs of these two factors determine the sign of y''.

The results of this analysis are summarized in figure 3.15.

Curve a of figure 3.14 is decreasing and concave up over most of the interval $0 < y < 1$. Its negative slope is consistent with the slope behavior summarized in figure 3.15. However, figure 3.15 indicates that a solution of $y' = y^3 - y$ should have an inflection point at $y = \sqrt{1/3}$, a condition violated by curve a in figure 3.14. Hence, curve a of figure 3.14 is not a graph of a solution of $y' = y^3 - y$.

Curve b of figure 3.14 can be rejected immediately because it is increasing on $0 < y < 1$, where solutions of $y' = y^3 - y$ must be decreasing.

Curve c of figure 3.14 is increasing and concave up for $y > 1$. Both types of behavior are consistent with the properties of figure 3.15. It is a reasonable candidate for a graph of a solution of $y' = y^3 - y$.

A portion of curve d of figure 3.14 is increasing for $y < -1$ while figure 3.15 requires that solutions be decreasing for $y < -1$.

Only curve c of figure 3.15 can possibly represent a graph of a solution of $y' = y^3 - y$. ■

If this equation arose from a model, you could now make some judgment of the model's suitability for the problem at hand. Could you obtain this picture of a solution of $y' = y^3 - y$ as easily from a solution formula? What would a solution formula for this equation look like?

MATLAB

DELAB can sketch graphs of solutions of equations like $y' = y^3 - y$ for initial conditions of your choice. For guidance, select **Help, Textbook**, then go to chapter 3, example 12.

3.2.2 Direction Fields

The preceding examples captured the qualitative behavior of solutions by sketching their graphs. A principal tool was the equation itself, for it determined the slope of its solution curve at any point. For example, the simple population model $P' = kP$, $P(0) = P_i$, says that the slope of its population curve at $t = 0$ is $P'(0) = kP(0) = kP_i$.

The differential equation is a kind of compass that guides us as we shuffle along the path of a solution curve. The equation specifies the slope of the curve at each point, heading us in the proper direction. How do those compass headings change as we move around the plane?

One way to capture this compass-like information is to plot the slopes determined by the differential equation at a variety of points of the plane. This plane full of slopes is called the **direction field** of the differential equation.

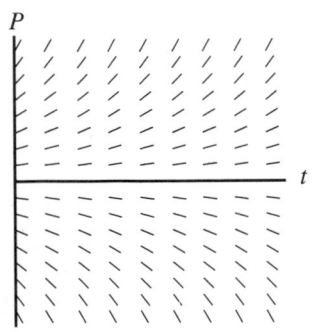

FIGURE 3.16 A direction field for the simple population equation $P' = kP$; compare with figure 3.11.

> **MATLAB**
>
> DELAB automates the task of plotting a direction field diagram. For guidance, select **Help**, **Textbook**, then go to chapter 3, example 13. Interpretation is your job!

■ **EXAMPLE 13**   *Plot the direction field of the simple population equation*

$$P' = kP.$$

Since the right side of this equation is not a function of the independent variable t, the slope of a solution curve depends only on P. In other words, all solution curves have the same slope $P' = kP$ at the same value of P. The line segments in figure 3.16 indicating the directions dictated by the differential equation do not change along horizontal lines.

Since $P' = kP$ and $k > 0$, the slopes must increase as P increases. This behavior is illustrated in figure 3.16 as well.

Do the solution curves for this equation, which we sketched in figure 3.11, follow the directions prescribed in figure 3.16? Is figure 3.16 indeed a set of compass readings that would guide us along the solution curves shown in figure 3.11? ■

Stop and Think   **3.24** Describe the connections between tracing a solution graph on a direction field diagram and computing an approximate solution of an initial-value problem using the Euler method.

■ **EXAMPLE 14**   *Figure 3.17 shows the direction field for the emigration equation $P' = kP - E$, E constant. What does that figure reveal about the significance of the population level $P = E/k$?*

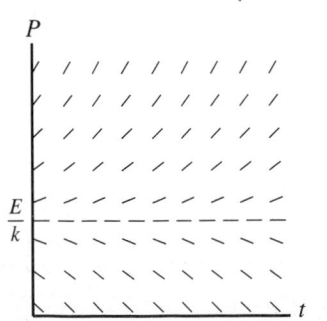

FIGURE 3.17 A direction field for the emigration equation $P' = kP - E$, E constant; compare with figure 3.12.

When $P = E/k$, the direction lines are horizontal. A population that begins at E/k stays at that equilibrium value; such a solution is called a *steady state*. Since the lines point downward for $P < E/k$, populations initially below this level will decrease. Furthermore, the rate of decrease will become more rapid because the slope of the lines is steeper as P decreases. The situation reverses when $P > E/k$.

Evidently, $P = E/k$ is a critical initial population level, separating growing populations from decaying populations. Populations that are too small will die out. Larger ones can overcome the loss due to emigration and grow. ■

The solution $P = E/k + (P_i - E/k)e^{kt}$ of the emigration model $P' = kP - E$, $P(0) = P_i$, contains this same information, though in a less obvious form. But a family of solution curves such as figure 3.12 or a direction field such as figure 3.17 makes immediately evident the presence of the critical population level E/k.

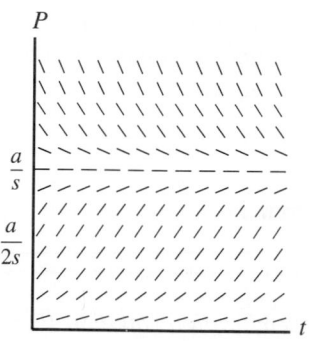

FIGURE 3.18 A direction field for the logistic equation $P' = aP - sP^2$; compare with figure 3.13.

■ **EXAMPLE 15** *Plot a direction field of the logistic equation*

$$P' = aP - sP^2.$$

Like the simple population model $P' = kP$, the value of the slope P' is independent of t.

Figure 3.13 summarizes the study of this equation from an earlier example. The sign of the derivative is reflected in the slope of the direction lines in figure 3.18. The concavity information tells whether slopes will increase or decrease as we move vertically through the direction field toward larger values of P.

To complete the study of this direction field, note that the direction lines are horizontal at $P = a/s$, where $P' = 0$. Given an initial point, could you use the directions prescribed in figure 3.18 to trace a solution curve? Would its slope and concavity behavior be consistent with figure 3.13? ■

Stop and Think **3.25** Study figure 3.18 carefully. As a population curve grows toward $P = a/s$, what becomes of its slope? Can the curve cross $P = a/s$? How do populations larger than a/s behave? Does the logistic equation $P' = aP - sP^2$ predict bounded or unbounded populations?

3.2.3 Phase Lines

First-order equations whose right-hand side does *not* depend on the independent variable are called **autonomous**. That is, autonomous equations are those of the form $y' = f(y)$; f does not depend on the independent variable t. The emigration model $P' = kP - E$ is autonomous so long as the emigration rate E is a constant. The heat-loss equation (3.1) with varying outside temperature

$$T' = -0.012\,[T - (10 + 5\cos(2\pi t/1{,}440))]$$

is *nonautonomous*.

Stop and Think **3.26** If T_{out} is constant, is the heat-loss equation $T' = -(Ak/cm)(T - T_{\text{out}})$ autonomous or nonautonomous? How about the logistic equation $P' = aP - sP^2$? The emigration equation $P' = kP - E(t)$ with varying emigration rate $E(t)$?

The growth and decay behavior and the steady-state solutions of autonomous differential equations can be summarized in a simple way on a **phase line**, in essence the vertical axis of a direction field diagram. For the equation $y' = f(y)$, a phase line records the values of y for which

- $y' = 0$, corresponding to constant or steady-state solutions (horizontal lines on a direction field diagram);
- $y' > 0$, corresponding to increasing solutions (upward line in a direction field); and
- $y' < 0$, corresponding to decreasing solutions (downward line in a direction field).

Since the slope $y' = f(y)$ of the solution of an autonomous equation does not depend upon t, these three pieces of information are almost all there is to know about the direction field of an autonomous equation.

MATLAB

Phase lines can be added to direction field diagrams in DELAB by selecting the appropriate option.

■ **EXAMPLE 16**   *Sketch the phase line for the emigration equation $P' = kP - E$, E constant.*

To sketch a direction field for this equation, example 14, page 112, argued that

- $P' = 0$ (constant or steady-state solution) if $P = E/k$,
- $P' > 0$ (increasing solution) if $P > E/k$, and
- $P' < 0$ (decreasing solution) if $P < E/k$.

To represent this information on a single axis, use a circle to mark constant solutions and arrows to mark increasing (arrow to the right) or decreasing (arrow to the left) solutions.

Figure 3.19 illustrates the result. Note that the arrows are pointing away from the constant solution $P = E/k$. Such a steady state is called *unstable* because nearby solutions move away from it. The phase line is essentially the vertical axis in a direction field diagram like figure 3.17 laid on its side. ■

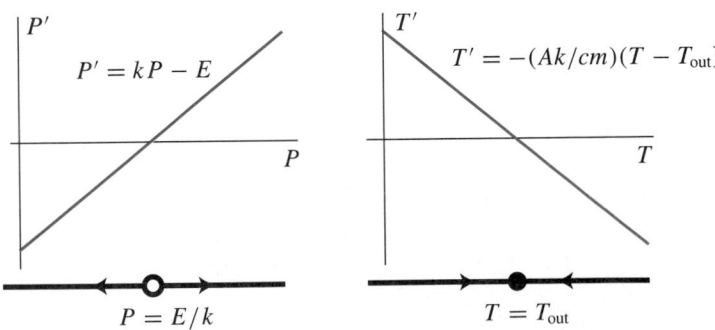

FIGURE 3.19   The phase line for the emigration equation $P' = kP - E$ (left) and the phase line for the heat-loss equation $T' = -(Ak/cm)(T - T_{out})$ (right) and graphs of their forcing terms.

■ **EXAMPLE 17**   *Sketch the phase line for the heat-loss equation*

$$T' = -\frac{Ak}{cm}(T - T_{out}),$$

T_{out} *constant.*

A familiar argument yields

- $T' = 0$ (constant or steady-state solution) if $T = T_{out}$,
- $T' > 0$ (increasing solution) if $T < T_{out}$, and
- $T' < 0$ (decreasing solution) if $T > T_{out}$.

Figure 3.19 shows the corresponding phase line. Note that the arrows are pointing toward the constant solution $T = T_{out}$. Such a steady state is called *stable* (more precisely, *asymptotically stable*) because nearby solutions move toward it. ■

Stop and Think   **3.27** Stand the emigration phase line in figure 3.19 on its end and compare with the vertical axis in the emigration direction field of figure 3.17, page 112. Do the two figures indeed contain the same information? Repeat the comparison for the heat-loss equation of the previous example.

Since the sign of $f(y)$ determines the sign of $y' = f(y)$, there is a close geometric connection between the graph of f and the phase line for $y' = f(y)$. When $f(y)$ is above the axis, y' is positive; when $f(y)$ crosses the horizontal axis, y has a steady state; and so on. Compare the phase line diagrams in figure 3.19 with the graphs of the right-hand side functions shown there. Figure 3.20 illustrates this connection for the logistic equation.

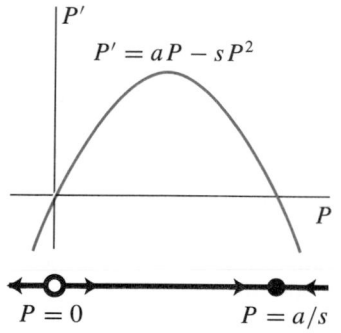

FIGURE 3.20   The phase line for the logistic equation and the graph of its forcing term $f(P) = aP - sP^2$.

3.2.4 Exercises

EXERCISE GUIDE	
To gain experience . . .	**Try exercises**
Drawing direction fields and phase lines	1–3(a)
Interpreting direction fields and phase lines	1(b), 2(b, c)
Sketching solution graphs	2–3(b), 4–5, 6(a), 7–9, 11
Solving differential equations	2(c), 4(c), 6(b)
Analyzing and interpreting solution behavior	3(c), 5, 6(c), 10, 12

1. (a) Sketch a direction field and a phase line for the rock-tossing equation $v' = -g$. Mark your axes carefully, and use the units you prefer. (On the earth, $g = 9.8$ m/s² or $g = 32$ ft/s².) Confirm your sketch using DELAB.

   (b) How would your picture of the direction field change if your model were moved to the moon, which has a gravitational constant smaller than the earth's?

   (c) Answer these same questions for the phase line.

2. (a) Sketch a direction field and a phase line for the differential equation

   $$P' = 0.015P - 0.209$$

   from the model for the population of Ireland during the potato famine of the last century. (The units for P are millions of people and those for t are years.) Confirm your sketch using DELAB.

   (b) Using the differential equation, sketch graphs of the solution for various initial populations. What initial population level separates growing from dwindling populations? Confirm your sketch and your conclusions using DELAB. Illustrate this same information on a phase line diagram.

   (c) Find a solution of this equation and use it to confirm your qualitative analysis.

3. (a) Figure 3.13 summarizes the slope and concavity behavior of the solution of the logistic equation

   $$P' = aP - sP^2.$$

   Supply the details needed to complete the construction of this figure from the start given in the text. Use the information in that figure to draw a direction field diagram and a phase line diagram.

   (b) Sketch graphs of the solutions of this equation for various initial populations. Consider initial values in the intervals $(0, a/2s]$ and $(a/2s, a/s]$ as well as values larger than a/s.

   (c) The quantity a/s is sometimes called the **carrying capacity** of the population in this model. Use the curves you drew in part (b) to justify this name.

4. (a) Figure 3.15 summarizes the slope and concavity behavior of the solutions of the differential equation

   $$y' = y^3 - y.$$

   Supply the details needed to complete the derivation of figure 3.15 from the start given in the text. Sketch a phase line for this equation as well.

   (b) Sketch a solution of this equation satisfying the initial condition $y(0) = y_i$ for y_i in each of the indicated intervals. Confirm each sketch using DELAB.

   (i) $\sqrt{1/3} < y_i < 1$

   (ii) $0 < y_i < \sqrt{1/3}$

   (iii) $-\sqrt{1/3} < y_i < 0$

   (iv) $-1 < y_i < -\sqrt{1/3}$

   (v) $y_i < -1$.

   (c) Sketch the phase line diagram for this equation.

   (d) Find a solution of the initial-value problem for a value of y_i within each of the intervals in part (b). Is the qualitative behavior you sketched evident from solution formulas? Do solution formulas provide any information that is not evident in your graphs?

5. Differential equations of the form

   $$u'(t) = u^2 - u$$

   arise in the study of materials whose elastic behavior changes dramatically with temperature. The dependent variable u represents the magnitude of a disturbance in the material and t is time. Large disturbances tend to grow, while small ones tend to die out in time.

   Show that this equation exhibits such behavior by sketching its solutions for initial values between 0 and 1 and for initial values greater than 1. What happens when $u(0) = 1$? Illustrate your conclusions on a phase line diagram.

   Confirm your analysis using DELAB.

6. (a) Sketch the best graph you can of the solution of the initial-value problem

   $$y' = y - y^3, \quad y(0) = 0.1.$$

   Carefully mark regions where the graph is increasing, decreasing, concave up, or concave down. Show inflection points. Summarize this information on a phase line diagram. Confirm your sketch using DELAB.

   (b) Find a solution of this initial-value problem.

   (c) Which approach would give the information in your graph more easily, analysis of the differential equation as in part (a) or analysis of the solution derived in part (b)? Does the solution formula provide some information that isn't apparent from the graph?

7. Which of the graphs in figure 3.21 might be a graph of a solution of $y' = 1 - y^2$, $y(0) = 0$? Justify rejecting each unacceptable choice and keeping each acceptable choice. Confirm your choice using DELAB.

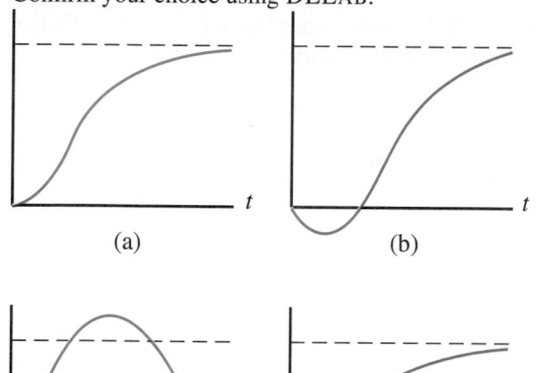

(a) (b)

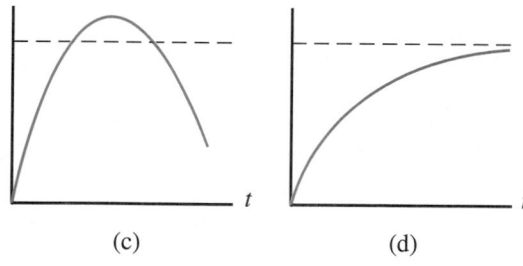

(c) (d)

FIGURE 3.21 Possible solution graphs for the initial-value problem $y' = 1 - y^2$, $y(0) = 0$.

8. Which of the graphs in figure 3.22 might be a graph of a solution of $y' = (4 - y)^3$, $y(0) = 0$? Justify rejecting each unacceptable choice and keeping each acceptable choice. Confirm your choice using DELAB.

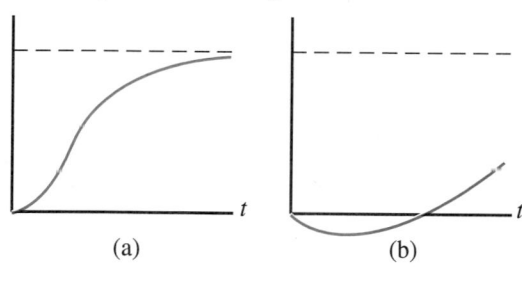

(a) (b)

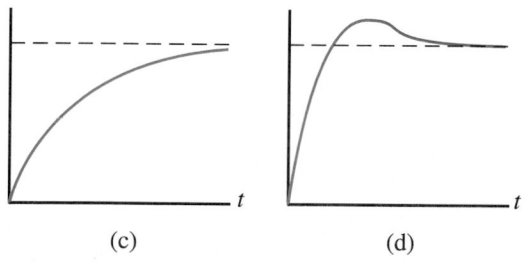

(c) (d)

FIGURE 3.22 Possible solution graphs for the initial-value problem $y' = (4 - y)^3$, $y(0) = 0$.

9. Which of the graphs in figure 3.23 might be a graph of a solution of $y' = y^2$, $y(0) = 1$? Justify rejecting each unacceptable choice and keeping each acceptable choice. Confirm your choice using DELAB.

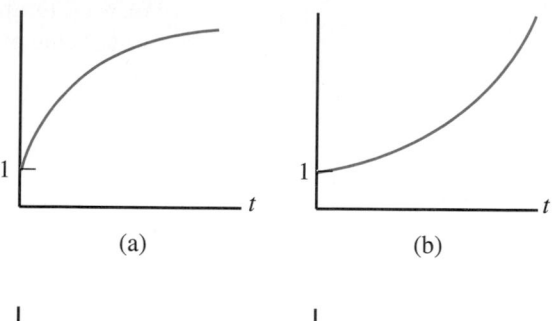

(a) (b)

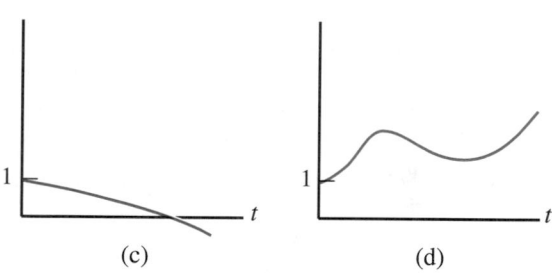

(c) (d)

FIGURE 3.23 Possible solution graphs for the initial-value problem $y' = y^2$, $y(0) = 1$.

10. What range of values of the constant k will ensure that the solution of the initial-value problem $y' + ky = 0$, $y(0) = 1$, decays? Could this problem reasonably model a population of some sort? The mass of a decaying radioactive isotope?

11. Sketch a direction field defined by each of the following differential equations. Confirm your sketch using DELAB. Sketch the corresponding phase line for each equation.

(a) $u' = u^2 - u$

(b) $y' = y - y^3$

(c) $y' = 1 - y^2$

(d) $y' = (4 - y)^3$

(e) $y' = y^2$

12. Study figure 3.18, page 113, carefully. As a population curve grows toward $P = a/s$, what becomes of its slope? Can the curve cross $P = a/s$? How do populations larger than a/s behave? Does the logistic equation $P' = aP - sP^2$ predict bounded or unbounded populations? Is this behavior obvious from the solution formula

$$P(t) = \frac{Cae^{at}}{1 + Cse^{at}},$$

which can be obtained by separation of variables?

3.3 ■ STEADY STATES, STABILITY, AND LINEARIZATION

3.3.1 Overview

We want to understand more fully the long-time behavior of the solution of initial-value problems such as the heat-loss model

$$\frac{dT}{dt} = -\frac{Ak}{cm}(T - T_{\text{out}}), \quad T(0) = T_i.$$

There are two basic questions:

- Are there steady states?
- If there are steady states, are they stable—do nearby solutions approach them as time passes?

In this section, the notion of steady-state solution is reviewed and new ideas are added (transient solution, time constant for decay), the intuitive notion of a stable steady state is given a more careful definition, and the powerful technique of linearization is introduced for analyzing the stability of steady states.

3.3.2 Steady States and Stability

Recall that a *steady-state* or *equilibrium solution* of a differential equation is a solution that is a constant function.

Intuitively, a steady state y_{ss} is *stable* if all solutions whose initial values are near y_{ss} approach y_{ss} in the limit. A formal definition makes *near* more precise:

Definition 1. *The steady state* y_{ss} *is* **asymptotically stable** *if there is some* $\delta > 0$ *such that every solution having* $y_{\text{ss}} - \delta \leq y(0) \leq y_{\text{ss}} + \delta$ *satisfies* $\lim_{t \to \infty} y(t) = y_{\text{ss}}$. *Otherwise, the steady state is unstable.*

> There are other variations of the stability/instability dichotomy, some of which will be explored for systems of two first-order equations in chapter 8; e.g., solutions that start near the steady state may not approach it, but they may not get farther away either, so-called *neutral stability*.

By definition, a **steady-state** or **equilibrium solution** of a differential equation is a constant solution, if such a solution exists. The **transient** component of a solution is that part (if any) whose limit is zero for large values of the independent variable. To find a steady-state solution y_{ss} of an autonomous equation $y' = f(y)$, simply seek a constant solution; that is, solve $f(y_{\text{ss}}) = 0$. The transient part of a solution is then $y_{\text{trans}}(t) = y(t) - y_{\text{ss}}$, provided $\lim_{t \to \infty} y_{\text{trans}}(t)$ exists and is zero.

To find a steady state, solve $f(y_{\text{ss}}) = 0$.

Stop and Think   **3.28** Show that the rock model equation $v' = -g$ has no steady-state solutions. What would such a steady-state solution represent physically? Is the lack of a steady-state solution consistent with the assumptions underlying the model?

■ **EXAMPLE 18**   *Find a steady-state solution T_{ss} of*

$$T' = -\frac{Ak}{cm}(T - T_{out}),$$

T_{out} *constant, and identify the transient part of solutions that approach this steady state.*

Assume there is a constant solution T_{ss} and substitute it into the differential equation:

Seek T_{ss} = constant.

$$T'_{ss} = 0 = -\frac{Ak}{cm}(T_{ss} - T_{out}).$$

Solving yields $T_{ss} = T_{out}$. See figure 3.25.

The transient is easy to identify as well. If T_{out} is constant, a solution of the initial-value problem $T' = -(Ak/cm)(T - T_{out})$, $T(0) = T_i$, is

$$T(t) = T_{out} + (T_i - T_{out})e^{-Akt/cm}. \tag{3.18}$$

Stop and Think   **3.29** Derive the solution formula (3.18) using separation of variables: integrate $dT/(T - T_{out}) = -(Ak/cm)\,dt$, etc. Do the analytic tools in DELAB produce the same solution?

Then the transient is

Transient part of solution

$$\begin{aligned} T_{trans} &= T(t) - T_{ss} \\ &= T_{out} + (T_i - T_{out})e^{-Akt/cm} - T_{out} \\ &= (T_i - T_{out})e^{-Akt/cm} \to 0 \text{ as } t \to \infty. \blacksquare \end{aligned} \tag{3.19}$$

Figure 3.24 illustrates the transient and steady-state parts of the solution of the heat-loss model. The steady state T_{out} is evidently stable.

Time constant

When decay of a transient is governed by an exponential term of the form e^{-rt}, $1/r$ is called the **time constant**. The time constant characterizes the rate of decay of the transient and, hence, the speed of approach to the underlying steady state. When $t = 1/r$, then $e^{-rt} = e^{-1} \approx 0.37$; the transient has decayed to roughly one-third its original size in time $1/r$. The time constant in the heat-loss model is $1/(Ak/cm) = cm/Ak$.

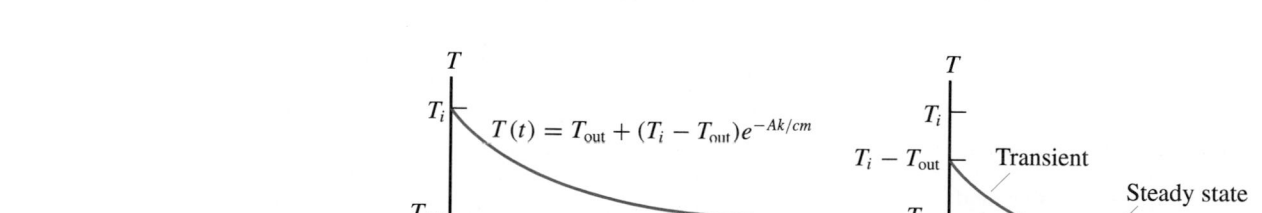

FIGURE 3.24 The transient and steady-state parts of the solution of the heat-loss model $T' + (Ak/cm)T = (Ak/cm)T_{out}$, $T(0) = T_i$.

Stop and Think   **3.30** One of the purposes of mathematical modeling is identifying a parameter group such as the time constant, whose value succinctly characterizes the behavior of the system being studied. For example, the time constant of your house tells you whether it is well or poorly insulated. Does added insulation increase or decrease the time constant? Does the heat-loss time constant cm/Ak actually have units of time?

3.31 Explain why $T' = -(Ak/cm)(T - T_{out})$ has no steady-state solution if T_{out} is not constant. For example, in equation (3.1),

$$T' = -0.012\,[T - (10 + 5\cos(2\pi t/1,440))]\,,$$

is T_{out} constant or variable? Does this equation have a steady-state solution?

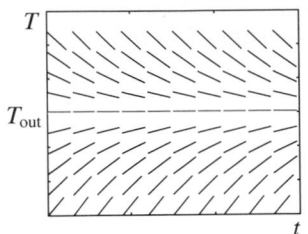

FIGURE 3.25   A direction field diagram for the heat-loss model $T' = -(Ak/cm)(T - T_{out})$, T_{out} constant, illustrating the *stability* of the steady state $T_{ss} = T_{out}$.

Note that we can select a steady-state solution by starting on it, by choosing $T(0) = T_{ss}$. Furthermore, in the heat-loss model, *every* initial state T_i eventually finds its way to this equilibrium,

$$T_{ss} = \lim_{t \to \infty} T(t) = T_{out},$$

as illustrated in the direction field diagram of figure 3.25 and in the behavior of the transient (3.19).

The steady state T_{out} of the heat-loss equation is *stable*; examine the solution formula (3.18) or the direction field diagram in figure 3.25. Both show $\lim_{t\to\infty} T(t) = T_{out}$.

On the other hand, the steady state E/k of the emigration equation $P' = kP - E$ is *unstable*. That instability is evident from the direction field diagram of figure 3.26. The flow shown by the lines is always away from the equilibrium E/k; even a small departure from this state can never return.

Stop and Think   **3.32** Explain why the stability properties of the steady states of the heat-loss and emigration equations are also obvious from the phase line diagrams of figure 3.19, page 114.

3.33 The logistic equation $P' = aP - sP^2$ has equilibria at $P = 0, a/s$. Study the direction field diagram of figure 3.18, page 113, or the phase line diagram of figure 3.20, page 115. What do you judge the stability of these steady states to be?

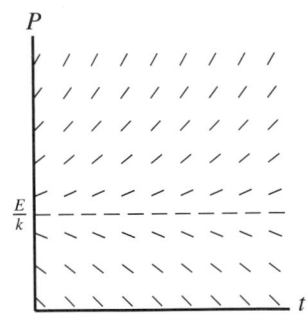

FIGURE 3.26   A direction field diagram for the emigration equation $P' = kP - E$, E constant, illustrating the *instability* of the steady state $P_{ss} = E/k$.

Not all solutions exhibit steady states or transients, and not all steady states are attained as a transient dies away. For example, the emigration model

$$P' = kP - E, \quad P(0) = P_i,$$

with E constant, has the solution

$$P(t) = \frac{E}{k} + \left(P_i - \frac{E}{k}\right)e^{kt}.$$

But this solution has no transient because $\lim_{t\to\infty} P(t)$ does not exist. On the other hand, we attain the equilibrium $P_{ss} = E/k$ if (and only if) $P_i = E/k$, because this equilibrium is unstable.

Stop and Think **3.34** Does the simple population equation $P' = kP$ have a steady state? A *stable* steady state? Does its exponential solution $P = P_i e^{kt}$ exhibit a transient? Does the situation change if $k < 0$?

Perturbing a steady state corresponds to starting the system at an initial value near the equilibrium. If the perturbed problem has a decaying transient, then the equilibrium is stable. For example, if the house in the heat-loss model is at its equilibrium temperature T_{out}, then a perturbation of that steady state is modeled by the heat-loss initial-value problem with $T(0) \approx T_{out}$. Since that problem always has a decaying transient, the equilibrium T_{out} is stable. If the perturbation grows, then the equilibrium is unstable, as is the population level E/k in the emigration model.

> An apple resting in the bottom of a bowl is in a stable equilibrium. A small perturbation, pushing the apple slightly away from that steady state, starts a motion that dies away as the apple settles back to the bottom of the bowl.
>
> An apple balanced on the edge of a bowl is in an unstable equilibrium. The smallest perturbation, giving it an initial position only slightly different from the steady state, will lead to a large, permanent departure from that tenuous equilibrium on the edge of the bowl. The apple will roll either to the bottom of the bowl or onto the table.

■ **EXAMPLE 19**   *Use definition 1 to verify that the equilibrium solution $T_{ss} = T_{out}$ of the heat-loss equation $T' = -(Ak/cm)(T - T_{out})$ is stable.*

We must find $\delta > 0$ with the property that $\lim_{t\to\infty} T(t) = T_{ss}$ when $T(t)$ solves

$$\frac{dT}{dt} = -\frac{Ak}{cm}(T - T_{out}), \quad T(0) = T_i,$$

for $T_{out} - \delta < T_i < T_{out} + \delta$; that is, we must specify precisely how near to T_{out} a solution must start in order to return to it in the limit of large time.

But we have already seen that the solution

$$T(t) = T_{out} + (T_i - T_{out})e^{-Akt/cm}$$

of this initial-value problem approaches T_{out} in the limit,

$$\lim_{t\to\infty}\left(T_{out} + (T_i - T_{out})e^{-Akt/cm}\right) = T_{out},$$

for *every* value of T_i. Hence, T_{out} is an equilibrium of the heat-loss equation that is stable to *arbitrarily large* perturbations; that is, δ in definition 1 can be chosen to be arbitrarily large. ■

The direction field of figure 3.25 and the phase line of figure 3.27 both illustrate this stability: everything is headed toward $T_{ss} = T_{out}$.

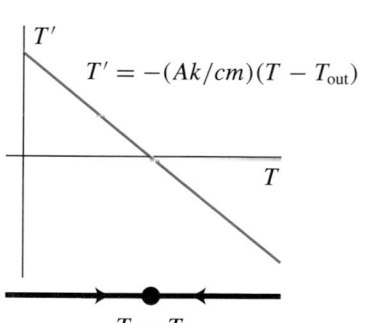

FIGURE 3.27 A phase line diagram for $T' = -(Ak/cm)(T - T_{out})$ illustrating the stability of the steady state $T_{ss} = T_{out}$.

3.3.3 Linearization

The stability of the steady state $T_{ss} = T_{out}$ is easy to analyze because the governing linear equation has a simple solution formula. Nonlinear equations

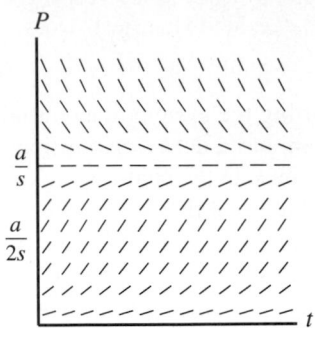

FIGURE 3.28   A direction field for the logistic equation $P' = aP - sP^2$.

like the logistic equation can be *linearized* near the steady state. The resulting analysis is not exact, but the burden of stability falls back onto a simple linear differential equation.

■ **EXAMPLE 20**   *Show that the logistic equation $P' = aP - sP^2$ has the steady state $P_{ss} \equiv a/s$ and that this equilibrium is stable.*

Constant solutions of the logistic equation must satisfy

$$0 = aP_{ss} - sP_{ss}^2 = P_{ss}(a - sP_{ss}).$$

Hence, $P_{ss} = 0$ and $P_{ss} = a/s$ are equilibrium solutions.

The direction field diagram of figure 3.28 suggests that $P_{ss} = a/s$ is a stable equilibrium because the flow of the direction field is toward this constant solution.

Stop and Think   **3.35** Argue from figure 3.28 that in definition 1, we could choose δ in the range $0 < \delta < a/s$.

To prove stability, we must show that solutions of

$$P' = aP - sP^2, \quad P(0) = P_i,$$

satisfy

$$\lim_{t \to \infty} P(t) = a/s$$

when P_i is near a/s. That is, we must show that small perturbations of the steady state a/s die out with time.

Solution = steady state + perturbation

To study such a perturbation $q(t)$ explicitly, define it by

$$P(t) = \frac{a}{s} + q(t),$$

where P is the solution of $P' = aP - sP^2$, $P(0) = P_i$. Note that the initial value for q is $q(0) = P_i - a/s$; "starting near $P_{ss} = a/s$" means "choose $P_i \approx a/s$" or "choose $q(0)$ to be small."

Now substitute this proposed solution into $P' = aP - sP^2$ and simplify:

Substitute to find equation for perturbation.

$$\left(\frac{a}{s} + q\right)' = a\left(\frac{a}{s} + q\right) - s\left(\frac{a}{s} + q\right)^2$$

$$q' = aq - 2s\frac{a}{s}q - sq^2 + a\frac{a}{s} - s\left(\frac{a}{s}\right)^2$$

$$q' = -aq - sq^2$$

Stop and Think   **3.36** Which points in the preceding analysis used $P_{ss}' = 0$? Which used $aP_{ss} - sP_{ss}^2 = 0$? Explain the origin of these two relations.

Unfortunately, the perturbation is defined by a nonlinear initial-value problem that is no easier to solve than the original logistic equation *unless we use the assumption that the perturbation q is small to linearize*, that is, unless we discard nonlinear terms that are smaller than the remaining linear terms.

Our decision to examine solutions that start near the steady state a/s means that the perturbation is small initially; $q(0) = P_i - a/s$ and $P_i \approx a/s$. If q is small, then q^2 is even smaller. The approximation $q^2 \approx 0$ reduces $q' = -aq - sq^2$ to the *linear* equation

Linearized stability equation

$$q' = -aq.$$

This *linearized equation* determines the fate of the perturbation q of the steady state a/s so long as the perturbation is small.

> The linearized equation is an approximation, and the approximation fails if the neglected term sq^2 ever becomes large compared with the remaining term aq.

Since $q' = -aq$, $a > 0$, has exponentially decaying solutions, the perturbation dies away to zero and $\lim_{t \to \infty} P(t) = a/s$. The steady state a/s is stable to perturbations that are near enough a/s to permit the linearizing approximation $q^2 \approx 0$. ■

Stop and Think

3.37 To give "near enough a/s" a more concrete expression, return to the nonlinear equation defining the perturbation q, $q' = -aq - sq^2$. What is the range of initial values $q(0) = P_i - a/s$ that guarantees $q'(0) < 0$? Does $\lim_{t \to \infty} q(t) = 0$ follow from $q'(0) < 0$? What value of δ in definition 1 is determined by this analysis? Is that value consistent with the direction field diagram in figure 3.28?

The technique we have just illustrated, finding an equation for the perturbation and linearizing that equation by neglecting small nonlinear terms, is called a *linear stability analysis*.

Linear stability analysis

1. Define the perturbed solution $y(t) = y_{ss} + q(t)$.
2. Substitute y and simplify.
3. If necessary, linearize by *assuming* that the perturbation q remains small for all time and by eliminating terms involving powers of q greater than one.
4. Examine the resulting (possibly approximate) linear differential equation for $q(t)$ to determine if it grows (y_{ss} is unstable) or decays (y_{ss} is stable).

The essence of a linear stability analysis is linear approximation.

Stop and Think

3.38 Identify in example 20 each of the steps just listed.

> The process of linear stability analysis seems reasonable, but it hinges on the assumption that the behavior of the linearized equation reflects the behavior of the full nonlinear equation. That assumption cries out for a careful proof, but we will not provide one here.

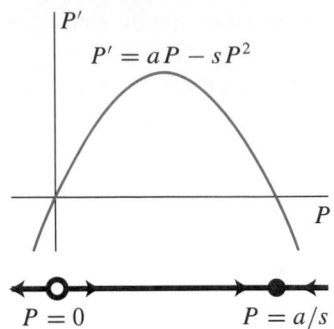

FIGURE 3.29 A phase line diagram for $P' = aP - sP^2$. The horizontal arrows on either side of the equilibrium points $P = 0, a/s$ show how the sign of P' near those points determines their stability.

Compare these last two examples. The heat-loss equation is *linear*. Its equilibrium solution is stable for *every* initial condition; starting near the equilibrium is not important. The logistic equation is *nonlinear*. We had to start near enough the equilibrium to be able to neglect the nonlinear term sq^2 to conduct a linear stability analysis. We can conclude only that the steady state is stable to *small* perturbations.

A phase line diagram, derived from a plot of P' versus P as shown in figure 3.29, provides another view of the stability of the steady states $P_{ss} = 0$, a/s. The parabolic curve in the figure is the graph of $P' = aP - sP^2$ versus P. When the curve is above the P axis, $P' > 0$ and vice versa. It crosses the horizontal axis at the equilibrium points $P = 0, a/s$ because $P' = 0$ at equilibria.

Perturbations of these steady states correspond to moving to the left or right along the P axis away from $P = 0$ or $P = a/s$. For example, a perturbation to the right from $P = a/s$ moves to a point on the P axis where $P' < 0$. But $P' < 0$ forces the perturbed P to decrease back towards its equilibrium at a/s, as the left-pointing arrow just to the right of $P = a/s$ shows. Similarly, perturbation of the steady state $P = 0$ moves into regions where the sign of P' pulls the perturbation farther from the unstable state $P = 0$.

Stop and Think   **3.39** What value of the parameter δ from definition 1 can be read from the phase line diagram in figure 3.29? Is that consistent with the value of δ predicted by the direction field diagram of figure 3.28? Should the phase line and direction field diagrams be consistent in such matters?

3.3.4 Exercises

EXERCISE GUIDE	
To gain experience . . .	**Try exercises**
Finding transients and steady states	4–6(a), 9–11(a), 14, 16(a, d), 17–18, 19–20
Determining stability of equilibria	4(b), 5(b), 6(b), 8, 9–11(b), 12(a–b), 13(a), 16(b–c), 18
Using direction field or phase line diagrams	4–5(b), 9–11(d), 13(b), 18, 21
Finding time constants, etc.	2(a–b), 3, 4(c), 17
Analyzing and interpreting solution behavior	1, 2(c–d), 3, 4(e), 6–8, 9–11(c), 12(c), 13(b)

1. Move figure 3.25 from New England to the Southwest. Redraw it for the case $T_i < T_{out}$. (If necessary, convince yourself that the model

$$T' = -\frac{Ak}{cm}(T - T_{out}), \quad T(0) = T_i,$$

remains valid in this case.) Is the description "heat-loss model" still accurate?

2. For a typical ranch house, the parameter group Ak/cm in the heat-loss model has the value 2×10^{-4} s^{-1}.

(a) What is the corresponding time constant?

(b) How long is required for the temperature difference

$T - T_{out}$ to drop to half its initial value?

(c) How could you use such information to select a heating (or cooling) plant for the house? Is the information you need for this design decision contained in the transient or the steady-state portion of the solution?

(d) Would these considerations be fundamentally different if your concern were the temperature of a power transistor or the cargo bay of the space shuttle instead of a house?

3. The text asks whether added insulation increases or decreases the time constant for heat loss from a house. What do you think? Verify that the time constant cm/Ak has units of time.

4. A mass m falling through air experiences a resisting force proportional to its velocity. The proportionality constant k varies with shape. A model for the velocity v of the mass is

$$v' = -g - \frac{k}{m}v, \quad v(0) = v_i.$$

(See exercise 8 of the chapter 1 exercises.)

(a) Find and sketch the graphs of the transient and steady-state parts of the solution of this problem. Explain their physical significance. (*Hint*: Sky divers speak of a *terminal velocity*.)

(b) Determine the stability of the steady state. Explain the physical significance of your stability conclusion. Illustrate with a direction field or phase line diagram (drawn by DELAB if you wish), and explain the connection between the diagram and your stability conclusion.

(c) What is the time constant for the transient? What does it mean?

5. A more sophisticated model of the situation in the preceding problem supposes that air resistance is proportional to the square of the velocity of the mass. The resulting model is

$$v' = -g - \frac{k}{m}v|v|, \quad v(0) = v_i.$$

(See project 1 of chapter 1.)

(a) Find the equilibrium solution(s) of this model.

(b) Determine their stability and interpret your results. Illustrate with a direction field or phase line diagram (drawn by DELAB if you wish), and explain the connection between the diagram and your stability conclusion.

6. A population of yeast spores increases due to reproduction by a fixed proportion R each day. The spores are harvested at a rate of H spores per day. A model for the spore population y is the initial-value problem

$$y' - Ry = -H, \quad y(0) = y_i.$$

(See exercise 6, section 2.2.4.)

(a) Is there a steady-state yeast population? If so, what is its value? Explain the physical significance of this equilibrium.

(b) Is this equilibrium stable? (A direction field diagram might help you decide.) What is the physical significance of your stability conclusions?

7. The text finds that perturbations q of the steady state a/s of the logistic equation $P' = aP - sP^2$ are governed by the linearized equation $q' = -aq$. It states, "Since $q' = -aq$ has exponentially decaying solutions . . . ." Verify this claim.

8. The text claims that the steady state $P \equiv 0$ of the simple population equation $P' = kP$ is unstable. Verify this claim and explain the physical significance of this instability.

9. A particular chemical reaction produces a chemical at a rate proportional to the concentration of that chemical; the proportionality constant, called the reaction rate, is k. Diluting the mixture reduces the concentration of the chemical at a constant rate of d g/cm^3·s. A model for this situation is the initial-value problem

$$c'(t) - kc(t) = -d, \quad c(0) = c_i.$$

(See exercise 8 of the chapter 2 exercises.)

(a) Find the steady state of this equation.

(b) Show that this steady state is stable.

(c) Explain the physical significance of this equilibrium.

(d) Illustrate with a direction field or phase line diagram (drawn by DELAB if you wish), and explain the connection between the diagram and your stability conclusion.

10. A radioactive isotope is decaying at a rate proportional to its mass; the proportionality constant is $k > 0$. A neutron beam bombarding the sample creates new isotopes at a constant rate of i g/s. A model for this situation is the initial-value problem

$$c'(t) + kc(t) = i, \quad c(0) = c_i.$$

(See exercise 11 of the chapter 2 exercises.)

(a) Find the steady state of this equation.

(b) Show that this steady state is unstable.

(c) Explain the physical significance of this equilibrium.

(d) Illustrate with a direction field or phase line diagram (drawn by DELAB if you wish), and explain the connection between the diagram and your stability conclusion.

11. A capacitor, a resistor, and a constant voltage E are connected in series. A model for the charge q on the capacitor in this RC circuit is the initial-value problem

$$Rq' + \frac{1}{C}q = E, \quad q(0) = q_i.$$

(See exercise 4 of the chapter 2 exercises.)

(a) Find the steady state of this equation.

(b) Show that this steady state is stable.

(c) Explain the physical significance of this equilibrium.

(d) Illustrate with a direction field or phase line diagram (drawn by DELAB if you wish), and explain the connection between the diagram and your stability conclusion.

12. The text observes that the logistic equation $P' = aP - sP^2$ has the equilibria 0 and a/s.

(a) Verify this claim.

(b) Use the direction field diagram of figure 3.18 to determine the stability of each equilibrium.

(c) Explain the physical significance of your observations.

13. The text shows that the steady state a/s of the logistic equation $P' = aP - sP^2$ is stable.

(a) Use a linear stability analysis to determine the stability of the other equilibrium, $P \equiv 0$.

(b) Explain the physical significance of these stability observations. Illustrate with a direction field or phase line diagram (drawn by DELAB if you wish), and explain the connection between the diagram and your stability conclusion.

14. State some conditions which guarantee that the linear equation $y' + g(x)y = f(x)$ has a steady-state solution.

15. State conditions which guarantee that the constant-coefficient linear equation $y' + ky = f(x)$ has a steady-state solution.

16. Consider the constant-coefficient equation $b_1 y' + b_0 y = h$ subject to the constant forcing term h.

(a) Find the steady state(s), if any, of this equation.

(b) Find conditions which guarantee that this equation has a decaying transient.

(c) Find conditions which guarantee that the steady state you found in part (a) is stable.

(d) Show that the following model equations assume the form $b_1 y' + b_0 y = h$ for appropriate choices of b_1, b_0, h. Verify that their specific steady-state and transient properties match your general conclusions.

(i) The heat-loss equation $T' = -(Ak/cm)(T - T_{out})$ with constant outside temperature T_{out}.

(ii) The emigration equation $P' = kP - E$ with constant emigration rate E.

(iii) The RC circuit equation $Rq' + q/C = E$ with constant imposed voltage E. (See exercise 4 of the chapter 2 exercises.)

(iv) The rock-tossing model with air resistance $v' = -g - kv/m$. (See exercise 8 of the chapter 1 exercises.)

17. Argue that *any* RC circuit modeled by

$$R\frac{dq}{dt} + \frac{1}{C}q = E(t), \quad q(0) = q_i,$$

where $E(t)$ is an imposed voltage, has a transient with time constant RC. Does this circuit necessarily have a steady-state charge level q_{ss}? (See exercise 4 of the chapter 2 exercises.)

18. For each of the following equations,

(i) Find all equilibria.

(ii) Find a linear equation governing a perturbation of each steady state.

(iii) Indicate whether the perturbation equation is exact or approximate.

(iv) Determine the stability of each equilibrium.

(v) Sketch a direction field to confirm your stability analysis. Explain the connection between the diagram and the stability conclusion.

(vi) Sketch a phase line diagram similar to figure 3.29 to confirm your stability analysis. Explain the connection between the diagram and the stability conclusion.

(vii) Compare your diagram(s) with one(s) drawn by DELAB. Comment on differences and similarities.

(a) $y' = 1 - y$

(b) $y' = 1 - y^2$

(c) $y' = y - y^3$

(d) $w' = (2 - w)(4 + w)$

(e) $P' = aP - sP^3$

(f) $v' = -g - kv^2/m$

(g) $v' = -g + kv^2/m$

(h) $y' = 1 - e^{y-2}$

19. What is the maximum number of steady states that the linear equation $y' + g(x)y = f(x)$ can have?

20. The nonlinear equation $y' = y^3 - y$ is one example of a general class of nonlinear, first-order equations of the form $y' + n(y) = 0$; here $n(y) = y - y^3$.

   (a) Show that the steady states of $y' + n(y) = 0$ are the solutions of $n(y_{ss}) = 0$. Illustrate this general result for $y' = y^3 - y$.

   (b) If $n(y)$ is a quadratic, what is the maximum number of steady states that $y' + n(y) = 0$ can have? Must it have any equilibrium points at all?

   (c) Give an example of a nonlinear first-order equation having the equilibrium points ± 2.

   (d) What can you say about the possible number of equilibrium points if $n(y)$ is a fourth-order polynomial?

21. *Linear stability analysis can be misleading.* Consider the simple equation $y' = y^3$. Show that a linear stability analysis of a perturbation $p(t)$ of the steady state $y_{ss} = 0$ predicts $p' = 0$. Apparently, perturbations will neither grow nor decay, a condition known as *neutral stability*. Using a phase line diagram, a phase plane diagram, or other arguments, show that $y_{ss} = 0$ is actually unstable: solutions that start near $y = 0$ grow.

   Repeat this analysis for $y' = -y^3$ but show that $y_{ss} = 0$ is stable.

3.4 ■ CHAPTER EXERCISES

EXERCISE GUIDE	
To gain experience ...	**Try exercises**
Finding transients and steady states, determining stability	1, 3–7, 8(c), 10, 11(c), 12(d), 13
Using numerical approximations	9(b), 12(d), 13
Sketching and interpreting graphs, direction fields, and phase lines	2, 8(a, d), 10, 11(a, d), 12(a, c, e), 13
Analyzing and interpreting models and solution behavior	2, 3(b, c), 4–5, 7, 8(b), 9(a), 11(b)

1. Find the transient and steady state of the solution of $y' + 2xy = x$, $y(0) = 2$.

2. The differential equation $dy/dt = ky$, k a constant, is sometimes called an *equation of growth and decay*. Explain why this is a good name. What must you know to determine whether this differential equation is describing growth or decay? How does the sign of k affect the stability of the steady state? Compare direction field and phase line diagrams for the cases $k > 0$ and $k < 0$.

3. A tank is full of saltwater. Brine in the tank is drained from the bottom so that 10% of the mixture leaves every minute. Water enters the top at an equal rate so that the tank is always full. In addition, salt is stirred into the tank at the rate of 5 lb/min.

   Let $s(t)$ denote the total weight of salt in the tank at time t and s_i the total weight of salt in the tank at $t = 0$. Then a model for this situation is the initial-value problem

   $$s' + 0.1s = 5, \quad s(0) = s_i.$$

   (a) Solve this initial-value problem.

   (b) Find the transient and steady state of the solution of this initial-value problem. Explain the physical significance of each.

   (c) Argue that the terms in the differential equation do indeed provide a reasonable model for the situation described.

4. The initial-value problem

   $$P'(t) - 0.02P(t) = -1,000, \quad P(0) = P_i,$$

   is a model of a population of microbes in which 2% of the current population reproduces per unit time. The term $-1,000$ represents a constant death rate due to the presence of a toxic substance in the microbes' environment.

   Argue that a very small initial population will die out while a large initial population will grow. Find the threshold initial population that separates *small* from *large*. That is, find a population P_t with the property that P is increasing if $P_i > P_t$, but P is decreasing if $P_i < P_t$. What happens if $P_i = P_t$?

5. Section 2.3 considered a heat-loss model that included a term $F(t)$ representing the heat-energy output rate of a furnace,

$$T'(t) = -\frac{Ak}{cm}(T - T_{\text{out}}) + \frac{F(t)}{cm}, \quad T(0) = T_i.$$

In one example, F was regarded as a constant; in another, it varied with temperature T. Here we wish to consider solutions of this initial-value problem when $F(t)$ is allowed to vary with time.

If $F(t)$ varies with time, we might expect it to range from a minimum of zero, when the furnace is turned off, to a maximum of H say. Two possible choices are
(i) $F(t) = H(1 + \sin \omega t)/2$,
(ii) $F(t) = H \sin^2 \omega t$,
where ω is constant in each case. Find the solution of the initial-value problem for each choice of $F(t)$. Does either exhibit a steady state? Does either approach a periodic function in the limit $t \to \infty$? Is that limiting function, if it exists, a solution of the differential equation? Speculate about a definition of *stable periodic solution*. How might *transient solution* be defined in such a context?

6. Consider the differential equation $y' + ky = f(t)$, k a constant. This equation arises in a model in which y is the concentration of a quantity. For $t > 0$, $f(t) < 0$. Is the function f modeling a source or a sink (loss) of y? What conditions must you impose on f to ensure that a steady state exists? That the steady state is stable?

7. Describe a physical situation of which the initial-value problem

$$T'(t) = -\frac{Ak}{cm}(T - T_{\text{out}}) - \frac{F}{cm}, \quad T(0) = T_i,$$

might be a model; here $F > 0$ is a constant. (Note the sign in front of F, which is opposite that of the usual furnace equation considered previously.) Find the steady state. How does changing the sign of F affect its stability? The behavior of the transient? The steady state? Are these changes consistent with the physical interpretation you have given this initial-value problem?

8. Exercise 14 in the chapter 2 exercises considers the following situation.

An entrepreneur has established a line of credit at a bank. She is charged a daily interest rate of 0.0493% on the total amount she owes the bank at the end of each business day. She borrows at a rate of $b(t)$ dollars per day, and her initial loan was in the amount a_i. The amount she *owes* t days after the initial loan is $a(t)$.

That problem offers four initial-value problems as possible models of this situation.
(i) $a'(t) + 0.000493a = b(t)$, $a(0) = a_i$
(ii) $a'(t) - 0.000493a = b(t)$, $a(0) = a_i$
(iii) $a'(t) + 0.000493a = -b(t)$, $a(0) = a_i$
(iv) $a'(t) - 0.000493a = -b(t)$, $a(0) = a_i$

(a) Assume that the function $b(t)$ is a positive constant. Without actually solving these equations, sketch graphs of the solutions of each of these four initial-value problems. Whenever appropriate, comment upon differences in solution behavior caused by different choices of the initial value a_i.

(b) How does your intuition about the situation just described suggest the solution of a correct model should behave? Which initial-value problem gives solution graphs closest to the behavior you expect in a good model?

(c) Which of the initial-value problems possess a steady state? When appropriate, sketch the steady state and transient. Which of these equilibria are stable?

(d) Sketch phase line or direction field diagrams for the four equations. Use those diagrams to select the most reasonable model.

9. The initial-value problem $m' = Im$, $m(0) = m_i$, can be regarded as a model of the amount of money m in a bank account that pays continuously compounded interest at the rate I; m_i is the initial deposit in the account.

(a) Derive this model.

(b) Approximate the solution of this initial-value problem using the Euler method with $\Delta t = 1$. Argue that $m_0, m_1, \ldots$, then represent the balance at the end of day 0, 1, ... in a bank account that pays interest that is compounded daily (assuming t is measured in days).

10. Each of the sketches of figure 3.30 represents the solution of some initial-value problem. Sketch the transient and steady-state parts of each solution that possess a steady state.

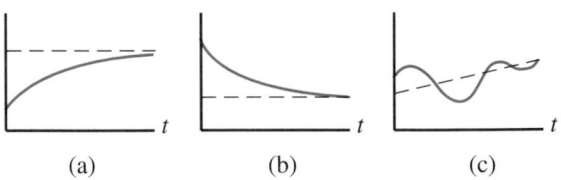

(a)   (b)   (c)

FIGURE 3.30 Exercise 10.

11. Exercise 15 of the chapter 2 exercises considers the following situation.

An insect population (denoted by y) is being destroyed by an insecticide at a constant rate d. It is increasing at a rate proportional to the current population; the constant of proportionality is g.

That problem offers four initial-value problems as possible models of this situation.

    (i) $y' + gy = d$, $y(0) = y_i$

    (ii) $y' - gy = d$, $y(0) = y_i$

    (iii) $y' + gy = -d$, $y(0) = y_i$

    (iv) $y' - gy = -d$, $y(0) = y_i$

(a) Assume that the parameters g and d are positive constants. Without actually solving these equations, sketch graphs of the solutions of each of these four initial-value problems. Whenever appropriate, comment upon differences in solution behavior caused by different choices of the initial value y_i.

(b) How does your intuition about the situation described here suggest the solution of a correct model should behave? Which initial-value problem gives solution graphs closest to the behavior you expect in a good model?

(c) Which of the initial-value problems possess a steady state? When appropriate, sketch the steady state and transient. Which of these equilibria are stable?

(d) Sketch phase line or direction field diagrams for the four equations. Use those diagrams to select the most reasonable model.

12. (a) Without actually solving the equation, sketch graphs of the solution of $w' = (1 - w)(1 + w)$ for a variety of initial values, positive, negative, and zero.

(b) Solve this equation and confirm that your sketches are correct.

(c) Which is easier, sketching these solutions directly from the equation or sketching them from a solution formula?

(d) Find the steady states of this equation and determine their stability using a linear stability analysis.

(e) Confirm the stability analysis using a direction field or a phase line diagram.

13. For each of the following:

    (i) Identify the steady-state solution(s).

    (ii) Construct a direction field or a phase line diagram and use it to formulate a conjecture about the stability of the steady state(s).

    (iii) Use the numerical solution tools in DELAB to test your conjecture for several initial values close to each steady state.

    (iv) Confirm your results through a linear stability analysis.

(a) $y' = 4 - y^2$

(b) $y' = y^2 - y$

(c) $y' = 3y - y^2$

(d) $y' = 1 + \sin y$ (Test only one steady state.)

3.5 ■ CHAPTER PROJECTS

1. **Numerical methods via integration.** Formally integrating the generic initial-value problem

$$y' = f(t, y), \quad y(t_i) = y_i$$

leads to

$$y(t) = y_i + \int_{t_i}^{t} f(s, y(s)) \, ds.$$

Think of the integrand as a function of s alone, say $g(s) = f(s, y(s))$. Then the integral $\int_{t_i}^{t} g(s) \, ds$ can be approximated in many different ways, including the rectangle rule

$$\int_{t_i}^{t} g(s) \, ds \approx g(t_i) \Delta t$$

and the trapezoid rule

$$\int_{t_i}^{t} g(s)\,ds \approx [g(t_i) + g(t)]\frac{\Delta t}{2},$$

where $\Delta t = t - t_i$. If $t = t_{n+1}$ and $t_i = t_n$, then each approximation to the integral gives a different numerical method for approximating the solution of the initial-value problem.

Identify the two methods given here. What integral approximation corresponds to the Heun method? Describe the *midpoint rule*, the method that results if the rectangle rule is used with g evaluated at the midpoint of the interval rather than at the end. What happens if the evaluation point is the *right* end of the interval?

Determine the local truncation error in the midpoint rule. Conduct some numerical experiments in MATLAB to compare it with Euler and Heun.

2. **Multipoint methods.** Push the ideas of the previous project a step further. Approximate the integral with Simpson's rule, and choose $t = t_{n+2}$, $t_i = t_n$, and t_{n+1} as the midpoint of $[t_n, t_{n+2}]$. State the resulting *multipoint* (or *Adams*) method that gives y_{n+2} in terms of y_{n+1} and y_n.

Conduct some numerical experiments with this method. Use one step of the Heun method to approximate $y(t_1)$, given $y(t_0)$, then use the new method to approximate $y(t_2)$ and subsequent values. Estimate the order of its global error.

3. **The logistic map and chaos.** Applying the Euler method to the logistic equation $P' = aP - sP^2$ leads to

$$P_{n+1} = (1 + a\Delta t)P_n - s\Delta t P_n^2, \quad n = 0, 1, 2, \ldots.$$

By defining $b = 1 + a\Delta t$ and $y_n = s\Delta t P_n/b^2$, those Euler iterations can be written

$$y_{n+1} = b(y_n - y_n^2), \quad n = 0, 1, 2, \ldots.$$

This relation is the *logistic map*. For initial values y_0 in $0 < y_0 < 1$ and b in $0 < b < 4$, it exhibits behavior much richer than anything we might anticipate from the logistic population model.

For some values of b (say, $b = 3.2$), the iterations quickly settle into cycles of period 2, alternating between two fixed values with each iteration, regardless of the initial value. As b increases, say to 3.5, the period suddenly doubles to 4; every fourth iterate is the same. Then the period doubles again to 8 just before $b = 3.6$. This *period doubling* continues as b increases. Interspersed among the periodic iterations are ranges of values of b for which the iterates seem randomly to cover almost every point in an interval, regions of *chaotic behavior*. Furthermore, changing y_0 in such a region leads to a completely different pattern of iterates, almost as if the iterates were chosen at random.

Write a simple MATLAB program to iterate the logistic map using given values of b and y_0. Skip the first 10–20 iterates to allow transient effects to disappear. Experiment with b in the neighborhood of 2.8, 3.2, 3.45, 3.5, 3.55, 3.6, and 3.8. If you are more ambitious, provide some graphical output by marking the iterates on the unit interval. (A period-two mapping will appear as two points, period-four as four points, and so on. Chaotic behavior will begin to fill an interval.)

> The chaotic behavior of the logistic map, its apparent randomness for certain values of b, arises not from some external uncertainty but from its nonlinearity. The change in behavior, the period doubling, as b is varied is called *bifurcation*.

> Nonlinear dynamics, of which the logistic map is a simple example, offer new insights into such phenomena as variations in the weather and turbulent fluid flow. For an account of many of the people and ideas involved in these discoveries, see Gleick's popular book [10]. The mathematical concepts of nonlinear dynamics are introduced in Baker and Gollub [2].

4. **Heat loss in humans.** Think of a human being as a house with a furnace. Heat is lost through the skin to the surroundings, clothing acts as insulation, and metabolism generates heat as if it were a furnace.

   (a) Use these ideas to derive a model for the temperature of a human. Take care to define the meaning of each parameter as it is introduced.

   (b) Suppose an individual weighing 70 kg can choose to wear a 0.5-cm layer of either flannel or copper while venturing into 5°C weather. Use the data of table 2.5, page 56, to determine the heat output rate required to maintain the usual body temperature of 37°C in each of these outfits.

   (c) Show that the normal body temperature is a stable equilibrium of this model. What would be the consequences if this temperature were an *unstable* equilibrium?

5. **General stability analysis.** Provide the framework for a general stability analysis of the first-order equation $y' = f(y)$.

   Suppose y_0 is an equilibrium. What can be said about $f(y_0)$? Find the linearized equation that governs a small perturbation of the steady state $y \equiv y_0$. State conditions on f that guarantee stability and instability of this steady state. Are there any undecided situations? Can those be resolved? State your results carefully in the form of a theorem, and organize your arguments into a proof.

6. **Spruce budworm population model.** The spruce budworm rapidly defoliates the Canadian balsam fir, destroying entire forests in a few years. Birds are its principal natural enemy.

   (a) Suppose the budworm population $P(t)$ has a population-dependent intrinsic growth rate $k = a - sP$, as in the logistic population model. Further, suppose that predation by birds removes budworms from the population at a rate $R(P)$. Derive the budworm population model

$$\frac{dP(t)}{dt} = aP - sP^2 - R(P), \quad P(0) = P_i.$$

$R(P)$

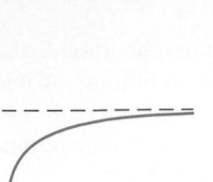

FIGURE 3.31   The predation rate $R(P)$ rises suddenly when the budworm population reaches a critical level.

(b) The predation function $R(P)$ exhibits threshold behavior, as illustrated in figure 3.31: once the budworm population reaches a critical level, birds begin consuming them at a nearly constant rate. (Might spraying insecticides by humans be modeled similarly?) One functional form that models this behavior is

$$R(P) = \frac{\beta P^2}{\alpha^2 + P^2},$$

where α, β are parameters. For example, when α is small, the threshold population is low, but the increase in predation rate is dramatic. The parameter β is the maximum predation rate.

(c) The budworm equation now involves four parameters, a, s, α, and β. To reduce the number of parameters, define the dimensionless variables and parameters

$$u(\tau) = \frac{P}{\alpha}, \quad A = \frac{\alpha a}{\beta}, \quad S = \frac{\alpha^2 s}{\beta}, \quad \tau = \frac{\beta t}{\alpha}.$$

Use the chain rule,

$$\frac{dP(t)}{dt} = \frac{du(\tau)}{d\tau}\frac{d\tau}{dt} = \cdots,$$

to write the budworm model as

$$\frac{du(\tau)}{d\tau} = Au - Su^2 - \frac{u^2}{1+u^2}, \quad u(0) = u_i,$$

where $u_i = P_i/\alpha$. The differential equation now contains only two parameters, A and S.

(d) Argue that steady states of this dimensionless model occur when

$$A - Su = \frac{u}{1+u^2}.$$

Plot the curve on the right and the line on the left for various (positive) values of A, S. Show that this equation can have one, two, or three steady states, depending on how A, S affect the intersection between the line and curve.

(e) For the cases of one or three steady states, use a phase line diagram like that of figure 3.29, page 124, to ascertain the stability of the steady state(s). Interpret these results physically.

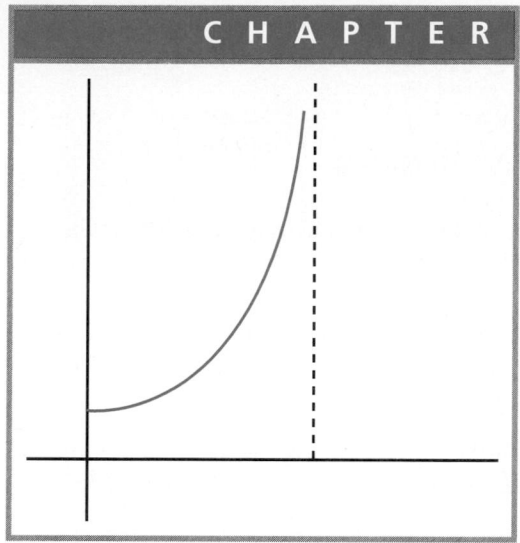

Analytic Tools for One Dimension

This chapter develops analytic solution techniques—methods for deriving solution formulas—for certain first-order differential equations. It begins with definitions and terminology that classify differential equations in order to determine the structure of solutions and to guide the choice of solution methods. These analytic solution tools complement the numerical and graphical tools of the previous chapter.

4.1 ■ BASIC DEFINITIONS

4.1.1 Equations and Solutions

We recall the terminology and definitions introduced in section 1.3 as well as extending and making more precise a number of important concepts.

The population equation

Example of a first-order equation

$$\frac{dP}{dt} = kP$$

is of **first order**. The unknown function or dependent variable is P and the independent variable is t. In the heat-loss equation

$$\frac{dT}{dt} = -\frac{Ak}{cm}(T - T_{\text{out}}),$$

the dependent variable is T and t is again the independent variable. More generally, in the generic first-order differential equation

$$\frac{dy}{dx} = f(x, y),$$

y is the **dependent variable** and x is the **independent variable**.

The generic first-order **initial-value problem** seeks a function satisfying the two requirements,

Generic first-order initial-value problem

$$\frac{dy}{dx} = f(x, y), \quad y(x_0) = y_i.$$

The **initial condition** is $y(x_0) = y_i$. A familiar initial-value problem is the simple population model

$$\frac{dP}{dt} = kP, \quad P(t_0) = P_i.$$

Recall the formal definition given in section 1.3:

Definition 1. *The function u is a solution on the interval $x_0 \leq x \leq x_1$ of the first-order differential equation $dy/dx = f(x, y)$ if u is continuously differentiable and if*

$$\frac{du}{dx} = f(x, u) \quad \text{for } x_0 \leq x \leq x_1.$$

*Further, u is a **solution of the initial-value problem** $dy/dx = f(x, y)$, $y(x_0) = y_i$, on the interval $x_0 \leq x \leq x_1$ if u is continuously differentiable,*

$$\frac{du}{dx} = f(x, u) \quad \text{for } x_0 \leq x \leq x_1 \quad \text{and} \quad u(x_0) = y_i.$$

Equivalently, u is a solution of an initial-value problem if u is a solution of the differential equation and if u has the correct initial value, $u(x_0) = y_i$. The interval of existence of a solution can be unbounded in either direction.

The same symbol is commonly used for both the unknown in a differential equation and for a specific solution function, rather than distinguishing them by y and u as in the preceding definition. Example 1 is an illustration.

■ **EXAMPLE 1**   *Show that $P' = kP$ has the solutions $P(t) \equiv 0$ and $P(t) = e^{kt}$ for all values of t.*

The differential equation is of first order, and both the functions 0 and e^{kt} have continuous first derivatives for all t.

Substituting $P \equiv 0$ into $P' = kP$ obviously yields an identity,

Substitute to check a solution

$$\frac{d0}{dt} = 0 = k0,$$

as does substituting $P = e^{kt}$,

$$\frac{dP}{dt} = \frac{de^{kt}}{dt} = ke^{kt} = kP.$$

Both functions are solutions of $P' = kP$. ■

■ **EXAMPLE 2**   *Show that $P(t) = 6e^{kt}$ is a solution of the initial-value problem*

$$P' = kP, \quad P(0) = 6.$$

The proposed solution $6e^{kt}$ obviously solves the differential equation:

Check the differential equation.

$$P' = (6e^{kt})' = 6ke^{kt} = k(6e^{kt}) = kP.$$

To check the initial condition, set $t = 0$ in the proposed solution:

Check the initial condition.

$$P(0) = 6e^{k0} = 6.$$

Hence, the function $P(t) = 6e^{kt}$ solves this initial-value problem because it satisfies both the differential equation and the initial condition. ■

Trivial solution

An identically zero solution such as $P \equiv 0$ in example 1 is called the **trivial solution**. *Trivial solution* does not mean *no solution*. The identically zero function is a legitimate, if boring, function.

Homogeneous differential equation

A differential equation is **homogeneous** if it possesses a trivial solution. Otherwise, the equation is **nonhomogeneous**. Example 1 shows that $P' = kP$ is homogeneous by showing that $P \equiv 0$ is a solution. The heat-loss equation $T' = -(Ak/cm)(T - T_{\text{out}})$ is *non*homogeneous because $T \equiv 0$ is *not* a solution:

$$\frac{d0}{dt} = 0 \neq -\frac{Ak}{cm}(0 - T_{\text{out}}).$$

(Is this equation homogeneous for some value of T_{out}? For some value of k?)

■ **EXAMPLE 3** *For what values of A is the equation*

$$P' - kP = A \sin \omega t$$

homogeneous?

To test homogeneity, substitute the proposed solution $P \equiv 0$:

$$0' - k0 = 0 = A \sin \omega t.$$

Hence, $P \equiv 0$ is a solution and the equation is homogeneous only if $A = 0$. (Although $A \sin \omega t$ is zero for isolated values of t such as $t = 0, \pi, 2\pi, \ldots$, this relation must hold for an interval of t values if $P \equiv 0$ is to be a solution.) ■

Like differential equations, initial conditions are called **homogeneous** if they are satisfied by the trivial solution and **nonhomogeneous** otherwise. The initial condition $P(0) = P_i$ is nonhomogeneous unless $P_i = 0$.

4.1.2 Linear Equations

Linear differential equations, a loosely defined concept heretofore, are important because new solutions can be formed from sums and constant multiples of old solutions.

Definition 2. *A **linear, first-order ordinary differential equation** is a first-order differential equation that can be written in the form*

LINEAR DIFFERENTIAL
EQUATION

$$a_1(x)y' + a_0(x)y = r(x). \tag{4.1}$$

Note that the "right-hand side" or **forcing function** $r(x)$ depends *only* upon the independent variable x and *not* upon the unknown y. Each term on the left-hand side of the prototype linear equation (4.1) involves but one occurrence of the unknown or its derivative. Neither y nor y' appears as an argument of a function. They are not raised to powers other than one, and y and y' do not multiply one another. That is, the unknown appears only in linear factors.

A first-order equation that can not be written in the form (4.1) is called **nonlinear**.

Stop and Think   **4.1** Preceding this formal definition, section 1.3 had stated, "$y' = f(x, y)$ is linear if it can be written in the form $y' = g(x)y + h(x)$." Is this form included in the formal definition (4.1)? Are the two forms completely equivalent?

■ **EXAMPLE 4**   *Show that the population equation $P' = kP$ is linear.*

This equation is linear because it can be written in the form (4.1)

$$1P' - kP = 0.$$

(The coefficient 1 was added in front of P' for emphasis.) We make the identification $a_1 = 1$, $a_0 = -k$, $r = 0$, $y = P$, and $x = t$. ■

■ **EXAMPLE 5**   *Show that the logistic equation $P' = aP - sP^2$ is nonlinear.*

Attempting to put the logistic equation in the form (4.1) forces $r = -sP^2$:

$$1P' - kP = -sP^2.$$

The definition of a linear first-order equation requires that the term r depend only on the independent variable, whereas this choice of r is a function of the unknown P.

We can not transpose the term $-sP^2$ to the left-hand side of the equation because the general form (4.1) permits on left only terms involving the unknown raised to the first power. Trying to write sP^2 in the form $(sP)P$ to obtain

$$P' + (-k + sP)P = 0,$$

where $a_1 = 1$ and $a_0 = -k + sP$ does not help either, because the coefficient a_0 then depends on the unknown. ■

When an equation is written in the linear form (4.1) with all unknowns collected on the left, linearity is, roughly speaking, a property only of the left-hand side. Think of "input" to the differential equation as substituting an expression in the left-hand side, and think of "output" as the resulting expression on the right side of the equality. Then linear equations are important because output is proportional to input.

> A rubber band appears to be linear. If you pull it (input) twice as hard, it stretches (output) twice as far. But if you pull it too hard, it becomes nonlinear. First it refuses to stretch farther. Then it breaks.

■ **EXAMPLE 6**   *Is the equation*

$$P' - kP = A \sin \omega t$$

linear?

This equation is linear for the same reason that $P' - kP = 0$ is linear: We can still make the identification $a_1 = 1$, $a_0 = -k$, $y = P$, and $x = t$, as we did before. The forcing term on the right just changes to $r = A \sin \omega t$. ■

The Operator Perspective

Definition 2 is useful for identifying linear differential equations, but it reveals nothing about their important properties. One route to understanding the importance of linear equations is through a quick, informal study of the concept of an operator. Loosely, an *operator* is a function whose input and output are functions: an operator is a rule that unambiguously transforms one function into another. A *first-order differential operator* is an operator that involves no derivative other than the first.

For example, differentiation defines the *derivative operator*, whose rules everyone knows and loves,

Derivative operator

$$\mathcal{D}y \equiv y'$$

for each differentiable function y; e.g.,

$$\mathcal{D}x^2 - 2x,$$
$$\mathcal{D} \cos \pi x = -\pi \sin \pi x.$$

(Of course, $\mathcal{D}$ is not defined for non-differentiable functions like $|x|$.) The rock model equation $v' = -g$ can be written in terms of this operator as $\mathcal{D}v = -g$.

As with functions, new operators can be defined in terms of old ones. For example, we could define the "heat-loss operator" in terms of $\mathcal{D}$:

$$\mathcal{M}_h T \equiv \mathcal{D}T + (Ak/cm)T = T' + (Ak/cm)T,$$

or the "logistic operator,"

$$\mathcal{M}_\ell P \equiv \mathcal{D}P - aP + sP^2 = P' - aP + sP^2.$$

Then the familiar heat-loss equation $T' = -(Ak/cm)(T - T_{\text{out}})$ can be written compactly as

$$\mathcal{M}_h = (Ak/cm)T_{\text{out}}$$

and the logistic equation $P' = aP - sP^2$ as

$$\mathcal{M}_\ell P = 0.$$

In fact, any first-order equation can be written in the form $\mathcal{M}y = r(x)$ for an appropriate form of the operator $\mathcal{M}$. For example, by defining the *first-order linear differential operator*

LINEAR DIFFERENTIAL OPERATOR

$$\mathcal{L}_1 y \equiv a_1(x)y' + a_0(x)y,$$

then the equation (4.1) in the definition of a linear equation can be written $\mathcal{L}_1 y = r(x)$.

Stop and Think

4.2 Verify that $\mathcal{M}_h T = (Ak/cm)T_{\text{out}}$ is indeed equivalent to $T' = -(Ak/cm)(T - T_{\text{out}})$. Repeat for $\mathcal{M}_\ell P = 0$ and the logistic equation.

4.3 What is the operator $\mathcal{M}$ that permits writing the simple population equation $P' = kP$ as $\mathcal{M}P = 0$? Use that operator to rewrite the emigration equation $P' = kP - E$.

Definition 3. *An operator $\mathcal{L}$ is **linear** if for every pair of functions y, z and constants C_1, C_2,*

$$\mathcal{L}(C_1 y + C_2 z) = C_1 \mathcal{L}y + C_2 \mathcal{L}z, \tag{4.2}$$

*providing $\mathcal{L}y$ and $\mathcal{L}z$ are defined. An operator is **nonlinear** if it is not linear.*

∎ **EXAMPLE 7**   *Show that the heat-loss operator $\mathcal{M}_h T = T' + (Ak/cm)T$ is linear and that the logistic operator $\mathcal{M}_\ell P = P' - aP + sP^2$ is nonlinear.*

Testing linearity

Suppose y and z are differentiable functions. To assess linearity, evaluate each side of (4.2) for each of the operators in question.
For the heat-loss operator, calculate

$$\begin{aligned}
\mathcal{M}_h(C_1 y + C_2 z) &= (C_1 y + C_2 z)' + (Ak/cm)(C_1 y + C_2 z) \\
&= C_1(y' + (Ak/cm)y) + C_2(z' + (Ak/cm)z) \\
&= C_1 \mathcal{M}_h y + C_2 \mathcal{M}_h z.
\end{aligned}$$

Hence, $\mathcal{M}_h$ is a linear operator.
For the logistic operator, calculate

$$\begin{aligned}
\mathcal{M}_\ell(C_1 y + C_2 z) &= (C_1 y + C_2 z)' - a(C_1 y + C_2 z) + s(C_1 y + C_2 z)^2 \\
&= C_1(y' - ay + sy^2) + C_2(z' - az + sz^2) + 2sC_1 C_2 yz \\
&= C_1 \mathcal{M}_\ell y + C_2 \mathcal{M}_\ell z + 2sC_1 C_2 yz \\
&\neq C_1 \mathcal{M}_\ell y + C_2 \mathcal{M}_\ell z.
\end{aligned}$$

Hence, $\mathcal{M}_\ell$ is *not* a linear operator. Of course, the problem with $\mathcal{M}_\ell$ is the nonlinear term sP^2. ∎

Stop and Think

4.4 Show that the simple population operator $\mathcal{M}P \equiv P' - kP$ is linear.

4.5 Verify that the first-order linear differential operator $\mathcal{L}_1 y \equiv a_1(x)y' + a_0(x)y$ is indeed a linear operator in the sense of definition 3.

Definition 4 (*Alternate definition of first-order linear differential equation*). *Suppose $\mathcal{M}$ is a first-order differential operator. Then $\mathcal{M}y = r(x)$ is a **linear differential equation** if and only if $\mathcal{M}$ is a first-order linear differential operator.*

■ **EXAMPLE 8** *Use definition 4 to decide the linearity of the equations*

$$P' = kP, \quad P' - kP = A \sin \omega t, \quad P' = aP - sP^2.$$

Define $\mathcal{M}P \equiv P' - kP$. Then $P' = kP$ and $P' - kP = A \sin \omega t$ are equivalent to $\mathcal{M}P = 0$ and $\mathcal{M}P = A \sin \omega t$, respectively. Leaving it to the reader to verify that $\mathcal{M}$ is indeed linear, we claim that these two equations are linear; compare with examples 4 and 6.

Example 7 showed that the logistic operator $\mathcal{M}_\ell P = P' - aP + sP^2$ is nonlinear. Hence, the logistic equation $\mathcal{M}_\ell P = 0$ or, in more familiar form, $P' = aP - sP^2$, is nonlinear, consistent with example 5. ■

Stop and Think

4.6 Is the differential equation $a_1(x)y' + a_0(x)y = r(x)$ appearing in definition 2 indeed linear in the sense of the alternate definition 3?

Are the two definitions 2 and 3 of linear differential equation equivalent? This is a central mathematical question. Using the word *linear* in each could be just an accident of language.

An affirmative answer to the preceding Stop and Think question settles half of that equivalence: *linear* in the sense of definition 2 is *linear* in the sense of definition 4. The other half depends on showing that a first-order linear differential operator $\mathcal{L}$ has the form $\mathcal{L}y = a_1(x)y' + a_0(x)y$.

Stop and Think

4.7 Suppose you are given a first-order linear differential operator $\mathcal{L}$. As a first step toward showing it has the form $\mathcal{L}y = a_1(x)y' + a_0(x)y$, might it be reasonable to choose $a_0(x) = \mathcal{L}1$ and $a_1(x) = \mathcal{L}x - (\mathcal{L}1)x$?

4.1.3 Superposition

Linear differential equations are important because they obey the principle of superposition.

Theorem 1 (**Principle of superposition**). *Let $y(x)$ be a solution of the general linear first-order equation with forcing term $r(x)$,*

$$a_1(x)y' + a_0(x)y = r(x).$$

Let $z(x)$ be a solution of the same differential expression with forcing term $s(x)$,

$$a_1(x)z' + a_0(x)z = s(x).$$

Then the sum $u(x) = C_1 y(x) + C_2 z(x)$ *is a solution of the equation with forcing term* $C_1 r(x) + C_2 s(x)$,

$$a_1(x)u' + a_0(x)u = C_1 r(x) + C_2 s(x).$$

Proof. The sum u must have the same differentiability properties as the given solutions y and z. We simply need to verify by substitution that it satisfies the appropriate differential equation:

$$a_1(x)u' + a_0(x)u = a_1[C_1 y + C_2 z]' + a_0[C_1 y + C_2 z]$$
$$= C_1[a_1 y' + a_0 y] + C_2[a_1 z' + a_0 z]$$
$$= C_1 r(x) + C_2 s(x).$$

These calculations freely interchange the derivative operation, addition, and multiplication by a constant; i.e., they make full use of the *linearity* of the first derivative, a property you have used often but may not have called by that name. ■

The principle of superposition is self-evident from the alternate definition 3 of linear differential equation.

A schematic representation of the principle of superposition is

Solution of equation with forcing term $C_1 r(x) + C_2 s(x)$	$= C_1$	Solution of equation with forcing term $r(x)$	$+ C_2$	Solution of equation with forcing term $s(x)$

The principle of superposition says that adding two inputs to a linear equation gives as output the sum of their individual outputs, doubling the input to a linear equation doubles the output, and so on.

From the principle of superposition, we immediately deduce that

- If y_h is a solution of a linear, homogeneous equation, then $C y_h$ is again a solution for any constant C.

- If y_p is any particular solution of a linear, *non*homogeneous equation and if y_h is any solution of the corresponding homogeneous equation, then $y_p + C y_h$ is again a solution of the *non*homogeneous equation.

Exercise 50 asks you to prove these assertions.

The following examples illustrate these two points.

■ **EXAMPLE 9**   *The equation* $P' = kP$ *is homogeneous and linear. It was shown above to have the solution* $P_h = e^{kt}$. *Show that any multiple of* P_h *is again a solution.*

Let C be an arbitrary constant. To show that $C e^{kt}$ is a solution of $P' = kP$, substitute into the equation we hope it satisfies:

$$\frac{d(Ce^{kt})}{dt} = C\frac{d(e^{kt})}{dt} = C(ke^{kt}) = k(Ce^{kt}).$$

Since $(Ce^{kt})' = k(Ce^{kt})$, the expression Ce^{kt} is indeed a solution of $P' = kP$ for *any* value of C. ■

Use $g = 9.8$ m/s^2.

■ **EXAMPLE 10** *The rock-tossing equation $v' = -9.8$ is linear and nonhomogeneous. One particular solution is $v_p = -9.8t$. Find additional solutions of $v' = -9.8$ by adding to $v_p = -9.8t$ solutions of the corresponding homogeneous equation $v' = 0$.*

Obviously, one solution of the homogeneous equation $v' = 0$ is $v_h = 1$. If C is an arbitrary constant, are expressions of the form $v = -9.8t + Cv_h = -9.8t + C$ solutions of $v' = -9.8$?

Of course, the solutions $v = -9.8t + C$ are just what we obtained by integrating $v' = -9.8$. ■

■ **EXAMPLE 11** *The heat-loss equation*

$$\frac{dT}{dt} = -\frac{Ak}{cm}(T - T_{\text{out}})$$

is linear and nonhomogeneous. When T_{out} is constant, one particular solution is $T_p = T_{\text{out}}$. Construct new solutions of this equation by adding to T_p constant multiples of a solution of the related homogeneous equation.

Rewritten in the standard linear form, this equation is

$$\frac{dT}{dt} + \frac{Ak}{cm}T = \frac{Ak}{cm}T_{\text{out}}. \tag{4.3}$$

Since the derivative of a constant is zero, $T_p = T_{\text{out}}$ is obviously a solution, as claimed. (To derive this solution, see exercise 43.)

To obtain the homogeneous form of this equation, replace the forcing term $(Ak/cm)T_{\text{out}}$ by zero:

$$\frac{dT}{dt} + \frac{\Lambda k}{cm}T = 0.$$

A solution (obtained by separation of variables) is $T_h = e^{-Akt/cm}$.

Now we must verify that for any constant C,

$$T = T_p + CT_h = T_{\text{out}} + Ce^{-Akt/cm}$$

is a solution of the nonhomogeneous heat-loss equation (4.3). Direct substitution of the formulas for T_p and T_h would suffice. Instead, we will use our knowledge of which equation is satisfied by each of the terms in the proposed solution $T_p + CT_h$.

Since T_p solves the nonhomogeneous equation (4.3), we can write

$$\frac{dT_p}{dt} + \frac{Ak}{cm}T_p = \frac{Ak}{cm}T_{\text{out}}.$$

Likewise, since T_h solves the homogeneous equation $T' + (Ak/cm)T = 0$, we have

$$\frac{dT_h}{dt} + \frac{Ak}{cm}T_h = 0.$$

Now substitute the proposed solution $T = T_p + CT_h$ into the left side of the heat-loss equation (4.3). From the preceding two equations satisfied by T_p and T_h, we obtain

$$\frac{d(T_p + CT_h)}{dt} + \frac{Ak}{cm}(T_p + CT_h)$$

$$= \frac{dT_p}{dt} + \frac{Ak}{cm}T_p + C\left(\frac{dT_h}{dt} + \frac{Ak}{cm}T_h\right)$$

$$= \frac{Ak}{cm}T_{\text{out}} + C \cdot 0$$

$$= \frac{Ak}{cm}T_{\text{out}}.$$

Thus, $T_p + CT_h$ is again a solution of the nonhomogeneous heat-loss equation. ■

In this example, we used our knowledge of which equations T_p and T_h satisfied, not the formulas for these functions. When T_p was the input to the left-hand side,

$$\frac{dT}{dt} + \frac{Ak}{cm}T,$$

the output was the desired right-hand side,

$$\frac{Ak}{cm}T_{\text{out}}.$$

When CT_h was the input to this same left-hand side, the output was zero,

$$C \cdot 0 = 0,$$

because T_h solves the homogeneous equation. Hence, input of the combination $T_p + CT_h$ gave the same output as input of T_p alone.

To state this observation differently, $T_p + CT_h$ and T_p are solutions of the same differential equation because that equation is linear and because T_h solves the homogeneous version of that equation. Adding a multiple of T_h to the solution contributes nothing to the right-hand side of the equation.

What advantage does the solution $T_p + CT_h$ have over the solution T_p? It contains an arbitrary constant C whose value we can choose to satisfy an initial condition. This property is the basis of a key concept, the *general solution* of a linear equation.

4.1.4 General Solution

Particular solution

First we formalize some terminology. A **particular solution** of a differential equation is just that, any specific solution of the equation, regardless of its initial value—it just has to satisfy the differential equation. The homogeneous equation corresponding to a given first-order linear equation is that equation obtained by replacing the forcing term $r(x)$ by zero in the general linear form (4.1); the resulting equation does indeed have the solution $y = 0$. A **homogeneous solution** of a first-order linear equation is any *nontrivial* solution of the corresponding homogeneous equation.

Homogeneous solution

> *Homogeneous solution* is shorter than the more accurate *nontrivial solution of the corresponding homogeneous equation*.

Definition 5. *Let y_p be any particular solution of the first-order linear equation (4.1). Let y_h be a homogeneous solution. Let C be an arbitrary constant. Then the combination*

GENERAL SOLUTION

$$y_g = y_p + C y_h$$

*is a **general solution** of* (4.1).

The subscript g in y_g denotes *general*. A general solution contains a whole family of solutions, one for each value of C, the constant of integration. A general solution has two building blocks, a particular solution of the original equation and a nontrivial solution of the corresponding homogeneous equation.

The principle of superposition guarantees that a general solution of a linear equation is *always* a solution. We will see later that *every* solution of a linear differential equation can be obtained from a general solution by choosing C appropriately. Many solution techniques are tailored to finding either a particular solution of the nonhomogeneous equation or a solution of the homogeneous equation.

■ **EXAMPLE 12**   *Explain why*

$$v(t) = -9.8t + C$$

is a general solution of the rock-tossing equation $v' = -9.8$.

The differential equation $v' = -9.8$ is certainly first order and linear; *general solution* is defined. This proposed general solution is just the result of integrating $v' = -9.8$. Clearly, we can identify $v_p = -9.8t$ as a particular solution because $v_p' = (-9.8t)' = -9.8$. Any nonzero constant is a nontrivial solution of the corresponding homogeneous equation $v' = 0$; we could choose the homogeneous solution $v_h = 1$. Then $v = -9.8t + C$ is a general solution because it has the form $v = v_p + C v_h$, where v_p is a particular solution of the nonhomogeneous equation and v_h is a nontrivial solution of the homogeneous equation. ■

■ **EXAMPLE 13**   *Write a general solution of the heat-loss equation*

$$\frac{dT}{dt} = -\frac{Ak}{cm}(T - T_{\text{out}})$$

for T_{out} constant.

Since this is a first-order linear equation, we can sensibly seek a general solution. From the work in example 11, we have a particular solution $T_p = T_{\text{out}}$ and a homogeneous solution $T_h = e^{-Akt/cm}$. (If you are in doubt, verify by substitution that T_p is a solution of the given equation and that T_h is a solution of the corresponding homogeneous equation $T' + (Ak/cm)T = 0$.)

Then a general solution is

$$T_g = T_p + CT_h = T_{\text{out}} + Ce^{-Akt/cm}.$$

Now put that constant C in the general solution to use. ■

■ **EXAMPLE 14**   *Find a solution of the initial-value problem modeling heat loss from a house,*

$$\frac{dT}{dt} = -\frac{Ak}{cm}(T - T_{\text{out}}), \quad T(0) = T_i,$$

when T_{out} is constant.

Set $t = 0$ in the general solution $T_g = T_{\text{out}} + Ce^{-Akt/cm}$ of the previous example and choose C to satisfy $T(0) = T_i$:

$$T_g(0) = T_{\text{out}} + C = T_i$$

To check, substitute the solution in the equation and verify $T(0) = T_i$.

or $C = T_i - T_{\text{out}}$. The solution of the initial-value problem is

$$T(t) = T_{\text{out}} + (T_i - T_{\text{out}})e^{-Akt/cm}. \ ■$$

By definition, a homogeneous equation always has the trivial solution. Using the trivial solution as a particular solution, we see that a *general solution of a homogeneous equation* is just a constant multiple of any nontrivial solution.

■ **EXAMPLE 15**   *Find a general solution of the homogeneous population equation $P' = kP$.*

The obvious particular solution is the trivial solution, $P_p = 0$. A nontrivial solution is $P_h = e^{kt}$. A general solution is $P_g = P_p + CP_h = 0 + Ce^{kt} = Ce^{kt}$, just the solution we obtained earlier by separation of variables.

A particular solution is *any* function that solves the equation.

We also could have chosen $P_p = e^{kt}$. Then we would have written $P_g = P_p + CP_h = (1 + C)e^{kt}$. But $1 + C$ is a single arbitrary constant that we could relabel C. We would find ourselves back at $P_g = Ce^{kt}$.

The homogeneous equation $P' = kP$ has the trivial solution, too. Could we choose $P_h = 0$? No! The homogeneous solution must be nontrivial. ■

Observing that the general solution of a homogeneous equation has the form Cy_h leads to an alternate statement of the definition of general solution:

A *general solution* of a *non*homogeneous equation is the sum of a particular solution of the *non*homogeneous equation and a general solution of the *homogeneous* equation.

Schematically, that relationship is

| General solution of *non*homogeneous equation | = | Particular solution of *non*homogeneous equation | + | General solution of *homogeneous* equation |

■ **EXAMPLE 16** *When the emigration rate E is constant, a solution of the nonhomogeneous emigration equation*

$$P' - kP = -E$$

is E/k. Find a general solution of this equation and verify that it satisfies the equation.

We are given a particular solution, $P_p = -E/k$. (You can verify this solution by substitution. To derive it, see exercise 44.)

The corresponding homogeneous equation is $P' - kP = 0$. From example 9, its general solution is Ce^{kt}. Hence, a general solution of the nonhomogeneous emigration equation is $P_g = -E/k + Ce^{kt}$.

To verify that P_g is indeed a solution, substitute $P_p(t) + CP_h(t)$ into $P' - kP = -E$:

$$(P_h + P_p)' - k(P_h + P_p) = (P'_p - kP_p) + (P'_h - kP_h)$$
$$= -E + 0 = -E. ■$$

Did we need to know the actual formulas for $P_h(t)$ and $P_p(t)$ in the last example? Why not?

To check a solution, substitute.

4.1.5 Summary

Homogeneous differential equations have the **trivial** (identically zero) solution. **Nonhomogeneous** equations do not.

Linear first-order equations are those that can be written in the special form

LINEAR EQUATION

$$a_1(x)y' + a_0(x)y = r(x).$$

Every linear equation has a **general solution** of the form

GENERAL SOLUTION

$$y_g = y_p + Cy_h,$$

where y_p is any **particular** solution of the given equation and y_h is a **homogeneous** solution, a nontrivial solution of the corresponding homogeneous equation obtained by setting the original forcing function r to zero. The general solution $y_p + Cy_h$ solves the same equation as does the particular solution

y_p because of the **principle of superposition:** Roughly, the sum of the inputs gives the sum of the individual outputs.

Given a general solution of a differential equation, an initial-value problem is easy to solve: Just choose the constant C to satisfy the initial condition.

MATLAB

The analytic solution tools in DELAB can find analytic solutions of first-order equations. For guidance, select **Help, Textbook**, then go to chapter 4, section 4.1. You will need to decide for yourself whether the solution provided by DELAB is the one you need.

4.1.6 Exercises

EXERCISE GUIDE	
To gain experience . . .	**Try exercises**
With linear versus nonlinear	1–26, 29(a–b), 38–39, 42, 48
With homogeneous versus nonhomogeneous	1–28, 29(c–d), 32–33
Checking solutions of equations	30, 41, 43–44(b, d)
With particular solutions	29(e), 30, 38–39, 40–41(a), 43–44(a)
With homogeneous solutions	29(f), 31, 38–39, 40–41(b), 43–44(c), 49
Finding and using general solutions	34–39, 40–41(c), 42, 43–44(e), 45, 47
Solving initial-value problems	34–37(b), 46
With the principle of superposition	38–42, 46, 50

For the equations in exercises 1–24:

(a) Classify each as linear or nonlinear and as homogeneous or nonhomogeneous, justifying each classification.

(b) Identify the variables corresponding to x and y in the generic form $y' = f(x, y)$ and find the form of f.

(c) Define an operator $\mathcal{M}$ and a function r so that each equation can be written in the form $\mathcal{M}y = r$ (or $\mathcal{M}v = \cdots$ or $\mathcal{M}u = \cdots$ or $\ldots$).

1. $v' = -9.8$

2. $v' = -g$

3. $u' = u^2 \cos \pi t$

4. $dy/dt = \cos t$

5. $dy/dt = 4e^t$

6. $v' = -g - (k/m)v$

7. $v' = -g - (k/m)v|v|$

8. $P' = 0.0102P$

9. $dT/dt = -0.0002(T - 5)$

10. $dP/dt = 0.015P - E$

11. $du/dy = y^2 - 1$

12. $P' = kP - E$

13. $P' = kP$

14. $P' = aP - sP^2$

15. $T' = -(Ak/cm)(T - T_{\text{out}})$

16. $y' + xy = 0$

17. $y' + 3y = -3x$

18. $y' + xy = -3x^2$

19. $y' + xy = -3y$

20. $y' + xy = -3y^2$

21. $u'(t) - u(t)^2 + u(t) = 0$

22. $u'(t) - t^2 + u(t) = 0$

23. $\sin \pi x = x^3 y'(x)$

24. $\sin \pi y = x^3 y'(x)$

25. Write the heat-loss equation

$$\frac{dT}{dt} = -\frac{Ak}{cm}(T - T_{out})$$

in the standard linear form

$$a_1(x)y' + a_0(x)y = r(x),$$

and verify that the homogeneous form of this equation is

$$\frac{dT}{dt} = -\frac{Ak}{cm}T.$$

26. Write the emigration equation

$$\frac{dP}{dt} = kP - E$$

in the standard linear form

$$a_1(x)y' + a_0(x)y = r(x),$$

and verify that the homogeneous form of this equation is

$$\frac{dP}{dt} = kP.$$

27. For what value(s) of T_{out} is the heat-loss equation

$$T' = -\frac{Ak}{cm}(T - T_{out})$$

homogeneous? For what value(s) of k? A? c? m?

28. For what value(s) of E is the emigration equation $P' = kP - E$ homogeneous? For what value(s) of k?

29. Either explicitly or implicitly, example 11 makes the following claims about the heat-loss equation

$$\frac{dT}{dt} = -\frac{Ak}{cm}(T - T_{out}) \qquad (4.4)$$

and its corresponding homogeneous equation

$$\frac{dT}{dt} = -\frac{Ak}{cm}T. \qquad (4.5)$$

Verify each.

(a) Equation (4.4) is linear. (Explicitly identify a term appearing in (4.4) with each of the symbols a_1, a_0, r, y, and x appearing in the definition (4.1) of linear equation.

(b) Equation (4.5) is linear.

(c) Equation (4.4) is nonhomogeneous. (Are there any values of the parameters in (4.4) for which it is homogeneous? If so, what are the parameters and what are those values?)

(d) Equation (4.5) is homogeneous. (Does (4.4) reduce to the homogeneous equation (4.5) for the parameter values you found in part (c)?)

(e) The function $T_p = T_{out}$, T_{out} a constant, is a solution of (4.4).

(f) The function $T_h = e^{-Akt/cm}$ is a solution of (4.5).

30. You are reading the text and encounter the statement, "Let $y_p(x)$ denote a particular solution of the general linear, first-order equation $a_1(x)y' + a_0(x)y = r(x)$." Use that statement to complete the equation

$$a_1(x)y'_p + a_0(x)y_p = ?.$$

31. You are reading the text and encounter the statement, "Let y_h denote any solution of the homogeneous equation $a_1(x)y' + a_0(x)y = 0$." Use that statement to complete the equation

$$a_1(x)y'_h + a_0(x)y_h = ?.$$

32. Verify each of the following statements about linear, first-order differential equations.

(a) If a particular solution is the identically zero function, then the equation is homogeneous.

(b) If the equation is homogeneous, then a particular solution is the identically zero function.

(c) Only homogeneous equations have trivial solutions. (Is this statement related to either of the preceding?)

33. Consider the statement, "If the equation is linear, then it is homogeneous if the forcing term is zero, and it is nonhomogeneous otherwise."

(a) Give examples of a homogeneous linear equation and a nonhomogeneous linear equation which illustrate this statement.

(b) Prove this statement for the general first-order linear equation

$$a_1(x)y' + a_0(x)y = r(x).$$

34. (a) Solve the heat-loss equation

$$T' = -\frac{Ak}{cm}(T - T_{out})$$

by separation of variables, assuming T_{out} is constant. Show that you obtain a general solution. Does DE-LAB provide a general solution?

(b) Use this general solution to solve the initial-value problem $T' = -(Ak/cm)(T - T_{out})$, $T(0) = T_i$. Does DELAB find the same solution?

35. (a) Solve the population equation $P' = kP$ by separation of variables. Show that you obtain a general solution. Does DELAB provide a general solution?

(b) Use this general solution to solve the initial-value problem $P' = kP$, $P(t_0) = P_i$. Does DELAB find the same solution?

36. (a) Solve the emigration equation $P' = kP - E$ by separation of variables, assuming E is constant. Show that you obtain a general solution. Does DELAB provide a general solution?

(b) Use this general solution to solve the initial-value problem $P' = kP - E$, $P(t_0) = P_i$. Does DELAB find the same solution?

37. (a) Solve the rock-tossing equation $v' = -g$ by antidifferentiation. Show that you obtain a general solution. Does DELAB provide a general solution?

(b) Use this general solution to solve the initial-value problem $v' = -g$, $v(0) = v_i$. Does DELAB find the same solution?

38. You are given the following information: $T_p(t)$ is a function that solves the equation

$$\frac{dT}{dt} = -\frac{Ak}{cm}(T - T_{out}),$$

and $T_h(t)$ solves

$$\frac{dT}{dt} = -\frac{Ak}{cm}T.$$

Show that for any value of the constant C, $T_p + CT_h$ solves the first equation. (You could find formulas for T_p and T_h. Do not. Use only the information given here. There is an example in the text that will be of great help.)

39. You are given the following information: $P_p(t)$ is a function that solves the equation

$$\frac{dP}{dt} = kP - E,$$

and $P_h(t)$ solves

$$\frac{dP}{dt} = kP.$$

Show that for any value of the constant C, $P_p + CP_h$ solves the first equation. (You could find formulas for P_p and P_h. Do not. Use only the information given here.)

40. Example 11 shows $T_p + CT_h$ is a solution of the heat-loss equation

$$\frac{dT}{dt} = -\frac{Ak}{cm}(T - T_{out})$$

if T_p is a particular solution of this equation and T_h is a solution of the related homogeneous equation. That example never uses the formulas for either of those solutions,

$$T_p = T_{out}, \qquad T_h = e^{-Akt/cm}.$$

(The expression for T_p assumes T_{out} is constant.) *Using these formulas:*

(a) Verify that T_p is a particular solution of

$$T' = -\frac{Ak}{cm}(T - T_{out}).$$

(b) Verify that T_h is a solution of the corresponding homogeneous equation. State that equation explicitly.

(c) Verify that $T_p + CT_h$ is a solution of

$$T' = -\frac{Ak}{cm}(T - T_{out})$$

for any value of the constant C.

41. According to the principle of superposition, $P_p + CP_h$ should be a solution of the emigration equation

$$\frac{dP}{dt} = kP - E$$

if P_p is a particular solution of this equation and P_h is a solution of the related homogeneous equation. Assuming E is constant, formulas for those solutions are

$$P_p = E/k, \qquad P_h = e^{kt}.$$

Using these formulas:

(a) Verify that P_p is a particular solution of $P' = kP - E$.

(b) Verify that P_h is a solution of the corresponding homogeneous equation. State that equation explicitly.

(c) Verify that $P_p + CP_h$ is a solution of $P' = kP - E$ for any value of the constant C.

42. Example 11 argues that $T_p + CT_h$ is a solution of the heat-loss model equation stated there. Mimic those steps to show that $y_p + Cy_h$ is a solution of the general linear equation $a_1(x)y' + a_0(x)y = r(x)$. Instead of a direct verification, could you have appealed to the principle of superposition? Recall that y_p is a particular solution of $a_1(x)y' + a_0(x)y = r(x)$ and that y_h is a solution of the corresponding homogeneous equation.

43. Example 11 claims that a particular solution of the heat-loss model equation

$$\frac{dT}{dt} = -\frac{Ak}{cm}(T - T_{out})$$

is $T_p = T_{out}$ (assuming T_{out} is constant) and that a corresponding homogeneous solution is $T_h = e^{-Akt/cm}$.

(a) Obtain this particular solution yourself by

    (i) Assuming the solution has the form $T_p = L$, where L is some constant.

    (ii) Substituting this guess into the equation and solving for L.

    (Do not worry about making an assumption about a solution you do not yet know. If the assumption is wrong, you will encounter a contradiction of some sort. If it is correct, you will find the solution. Nothing ventured, nothing gained!)

(b) Verify by substitution that $T_p = T_{out}$ is indeed a solution of the equation.

(c) Obtain the homogeneous solution using separation of variables.

(d) Verify by substitution that the homogeneous solution is correct.

(e) Write a general solution of this differential equation. Does DELAB provide a general solution, too?

44. Example 16 claims that a particular solution of the population model equation with emigration

$$P' - kP = -E$$

is $P_p = E/k$ (assuming E is constant) and that a corresponding homogeneous solution is $P_h = e^{kt}$.

(a) Obtain this particular solution yourself by

    (i) Assuming the solution has the form $P_p = c$, where c is some constant.

    (ii) Substituting this guess into the equation and solving for c.

(b) Verify by substitution that $P_p = E/k$ is indeed a solution of the equation.

(c) Obtain the homogeneous solution using separation of variables.

(d) Verify by substitution that the homogeneous solution is correct.

(e) Write a general solution of this differential equation. Does DELAB provide a general solution, too?

45. Use the information given in each part to write an expression that is a general solution of the differential equation described there. If the information is insufficient and you can not find what you need or if a general solution is not appropriate, explain.

(a) A nonhomogeneous differential equation has the solution $\sqrt{x}$. The homogeneous version of this equation has the solution e^x.

(b) A linear, first-order, nonhomogeneous differential equation has the solution $\sqrt{x}$. The homogeneous version of this equation has the solution e^x.

(c) The equation $y' + m(x)y = 4x + 1$ has the solution $p(x)$. The equation $y' + m(x)y = 0$ has the solution $h(x)$.

(d) The equation $y' - y = e^x$ has the solution xe^x.

46. If possible, use the given information to find a solution of the indicated initial-value problem. If you can not solve the given initial-value problem, explain why.

(a) Given: y is a solution of $u' - 4u = 3\tan 2x$, $u(1) = 4$. Solve $u' - 4u = 3\tan 2x$, $u(1) = 12$.

(b) Given: y is a solution of $u' - 4u^2 = 3\tan 2x$, $u(1) = 4$. Solve $u' - 4u^2 = 3\tan 2x$, $u(1) = 12$.

(c) Given: $P' = \cos \pi x - 4xP$, $H' = -4xH$, $P(0) = H(0) = -2$. Solve $y' + 4xy = \cos \pi x$.

(d) Given: $P' = \cos \pi x - 4xP$, $H' = -4xH$, $P(0) = -2$, $H(0) = 0$. Solve $y' + 4xy = \cos \pi x$.

(e) Given: $P' = \cos \pi x - 4xP$, $H' = -4xH$, $P(0) = H(0) = -2$. Solve $y' - 24xy = -6\cos \pi x$.

47. Verify that

$$y(x) = C \exp\left(-\int_0^x (a_0(s)/a_1(s))\,ds\right)$$

is a general solution of $a_1(x)y' + a_0(x)y = 0$ if $a_1 \neq 0$.

48. Recall the often-repeated calculus slogan, "The derivative of a sum is the sum of the derivatives." Explain why the principle of superposition is a consequence of this statement. (The principle of superposition follows from the linearity of the derivative operation.)

49. *Prove*: The differential equation $y' = f(x, y)$ is homogeneous if and only if $f(x, 0) = 0$ for all x.

50. By making appropriate choices of r, s, C_1, and C_2 in the principle of superposition, verify two claims made in the text:

(a) If y_h is a solution of a linear, homogeneous equation, then Cy_h is again a solution for any constant C.

(b) If y_p is any particular solution of a linear, nonhomogeneous equation and if y_h is any solution of the corresponding homogeneous equation, then $y_p + Cy_h$ is again a solution of the nonhomogeneous equation.

4.2 ■ SEPARATION OF VARIABLES

We review the **separation of variables** solution technique introduced in chapter 1.

Linear, first-order, homogeneous differential equations, among others, are **separable**: All terms involving the unknown (dependent) variable can be placed on one side of the equality, and all terms involving the independent variable can be placed on the other.

Apply separation of variables to a general linear, first-order, homogeneous differential equation

$$a_1(x)\frac{dy}{dx} + a_0(x)y = 0.$$

Replacing dy/dx by a quotient of the differentials dy and dx and collecting all terms involving y on the left and all terms involving x on the right lead to

Separate variables.

$$\frac{dy}{y} = -\frac{a_0}{a_1}dx, \tag{4.6}$$

as long as $a_1 \neq 0$ and $y \neq 0$. An antiderivative yields

$$\ln|y(x)| = -\int_0^x \frac{a_0(s)}{a_1(s)}\,ds + c. \tag{4.7}$$

Invert the logarithm with the exponential to obtain

$$|y| = \exp\left(-\int_0^x (a_0/a_1)\,ds + c\right) = |C|\exp\left(-\int_0^x (a_0/a_1)\,ds\right),$$

where $|C| = e^c$. But the absolute value signs around y can be removed because the expression

$$y = Ce^{-\int_0^x (a_0/a_1)\,ds} \tag{4.8}$$

has constant sign and satisfies the differential equation whether C is positive or negative. Furthermore, $y \neq 0$ unless $C = 0$, in which case y is the trivial solution. We have a formula (4.8) for a solution of the linear homogeneous equation $a_1 y' + a_0 y = 0$.

Do not memorize formula (4.8)! It is the first of many, too many to memorize. Understand the process instead. Test your understanding by deriving this formula with the book closed.

Stop and Think　**4.8** Test your understanding of separation of variables for linear equations by repeating it for $P' - kP = 0$. (*Hint*: Look ahead to the next example.)

The manipulations involving differentials that lead to (4.6) are purely formal. How can we be sure that the expression in (4.8) is really a solution of $a_1 y' + a_0 y = 0$? One approach is to substitute the solution expression (4.8)

Check solutions by substitution!

directly into the differential equation, as exercise 23 requests.

A second approach is to replace the potentially dubious manipulations leading to (4.6) with obviously correct procedures. To that end, write $a_1 y' + a_0 y = 0$ as

$$\frac{1}{y}\frac{dy}{dx} + \frac{a_0(x)}{a_1(x)} = 0,$$

where we assume as before that $a_1 \neq 0$ and $y \neq 0$. Knowing the result we seek, we recognize that the left-hand side is just a derivative,

$$\frac{1}{y}\frac{dy}{dx} + \frac{a_0(x)}{a_1(x)} = \frac{d}{dx}\left(\ln|y| + \int \frac{a_0(x)}{a_1(x)}\, dx \right),$$

providing a_0/a_1 is integrable. Hence,

$$\ln|y| + \int \frac{a_0(x)}{a_1(x)}\, dx = c.$$

We have found our way by more certain means to (4.7), and the derivation of the final formula for y can now continue as before.

Not only has separation of variables given us a solution, it has produced a *general solution* of this first-order, linear, homogeneous equation. The expression (4.8) has the structure of an arbitrary constant multiplying a nontrivial solution of the homogeneous equation. We will see later that *every* solution of the linear equation $a_1 y' + a_0 y = r$ can be obtained from the general solution by choosing an appropriate value for C (providing $a_1 \neq 0$).

■ **EXAMPLE 17**   *Solve the population model initial-value problem*

$$\frac{dP}{dt} = 0.02P, \quad P(1960) = 1.2.$$

We begin solving any linear initial-value problem by finding a general solution, which separation of variables can provide in this case:

Separate variables.

$$\int \frac{1}{P}\, dP = \int 0.02\, dt,$$

$$\ln|P| = 0.02t + c,$$

$$P(t) = Ce^{0.02t}.$$

To reach the last step, replace e^c by $|C|$.

Use the initial condition to find C.   Now use the initial condition $P(1960) = 1.2$ to evaluate the arbitrary constant C. We find $P(1960) = 1.2 = Ce^{(0.02)(1960)} = Ce^{39.2}$ or $C = 1.2e^{-39.2}$. Hence, the solution of this initial-value problem is

Solution of initial-value problem

$$P(t) = 1.2e^{-39.2}e^{0.02t}.$$

Use $39.2 = 0.02 \cdot 1960$ to write this solution more naturally as

$$P(t) = 1.2e^{0.02(t-1960)}.$$

Now the growth rate parameter 0.02, the starting time 1960, and the initial population level 1.2 are all readily apparent. ■

> MATLAB
>
> DELAB can find solutions of first-order initial-value problems. For guidance, select `Help, Textbook`, then go to chapter 4, example 17.

Of course, separation of variables is not limited to linear equations.

■ **EXAMPLE 18**   *Find a solution of the logistic equation*

The solution formula (4.8) does not apply to this nonlinear equation.

$$P' = aP - sP^2.$$

The appropriate separation is

Separate variables, assuming $aP - sP^2 \neq 0$.

$$\int \frac{dP}{aP - sP^2} = \int dt.$$

The antiderivative on the left can be evaluated using partial fractions:

$$\frac{1}{a} \ln \left| \frac{P}{a - sP} \right| = t + c.$$

Multiplying by a, inverting the logarithm with the exponential function, and solving the resulting expression for $P(t)$ finally yields

$$P(t) = \frac{Cae^{at}}{1 + Cse^{at}}, \tag{4.9}$$

where $|C| = e^{ac}$. These manipulations assume P, $a - sP \neq 0$, and they consider separately the cases $a - sP > 0$ and $a - sP < 0$. ■

This solution formula (4.9) does not have the form of a general solution; it is not a constant multiplying a homogeneous solution. Of course, the logistic equation is nonlinear, and *general solution* is defined only for linear equations.

■ **EXAMPLE 19**   *Find the solution of the logistic population model*

$$P' = aP - sP^2, \quad P(0) = \frac{a}{s}.$$

To evaluate the constant C in the solution formula (4.9), we set $t = 0$:

$$P(0) = \frac{Ca}{1 + Cs} = \frac{a}{s}.$$

Simplification yields

$$Ca = \frac{a}{s} + Ca,$$

a relation that holds for *no finite value* of C. Is there no solution of this initial-value problem?

There is a solution,

$$P(t) \equiv \frac{a}{s}.$$

Direct substitution shows $P = a/s$ solves $P' = aP - sP^2$. Obviously, $P(0) = a/s$. We simply could not obtain this solution from the separation of variables formula (4.9). ■

Stop and Think

4.9 What condition does the initial value a/s violate?

4.10 Does DELAB solve this initial-value problem successfully?

General solution is defined only for linear equations, and every solution of a linear initial-value problem can be obtained from the general solution by an appropriate choice of the arbitrary constant. But not every solution of a nonlinear equation can be obtained from a solution expression that contains an arbitrary constant.

> Some texts define *general solution* to be any solution that contains an arbitrary constant, whether the differential equation is linear or nonlinear. We have just seen that a "general solution" of a nonlinear equation is not general enough to include all solutions. Hence, we do not define *general solution* for nonlinear equations.

Having developed this solution method, a natural question is as follows: To which first-order equations can separation of variables be applied? Formula (4.8) shows it can be applied to any homogeneous, first-order linear equation.

Can separation of variables be applied to an arbitrary *non*homogeneous linear first-order equation? In general, the answer is *no*, unless, for example, the coefficients and the forcing term are all constant.

■ **EXAMPLE 20**   *Attempt to solve the differential equation*

$$\frac{dP}{dt} = kP - \alpha e^{-t},$$

which could arise in an emigration model with a decaying emigration term $E = \alpha e^{-t}$.

The forcing term $-\alpha e^{-t}$ gets in the way. Attempting to put P on the left and t on the right always leads to expressions like

$$dP = (kP - \alpha e^{-t})\, dt.$$

We can not find the antiderivative of the expression on the right; we can not compute $\int P(t)\, dt$ if we do not know the form for P as a function of t. Attempting to divide through by P leads to a similar quandry. ■

Note that the same equation with a *constant* emigration rate E can be solved by separation of variables.

Stop and Think   **4.11** Verify the claim in the previous sentence by outlining the steps needed to solve $P' = kP - E$, E constant, by separation of variables.

4.12 Can DELAB solve $P' = kP - \alpha e^{-t}$? If so, it must know something more than separation of variables!

Exercise 26 asks you to demonstrate that any differential equation of the form

$$\frac{dy}{dx} = Y(y)X(x)$$

can be solved by separation of variables as long as $Y \neq 0$. An example follows.

■ **EXAMPLE 21**   *Solve*

$$\frac{dy}{dx} = y^4 \sin x, \quad y(\pi/2) = \frac{1}{2}.$$

Separating variables leads to

$$\frac{dy}{y^4} = \sin x \, dx.$$

An antiderivative and a little algebra give

$$y(x) = \frac{1}{(3\cos x + C)^{1/3}}.$$

The initial condition forces $1/C^{1/3} = 1/2$ or $C = 8$. Hence, a solution of this initial-value problem is

$$y(x) = \frac{1}{(3\cos x + 8)^{1/3}}. \quad ■$$

Stop and Think   **4.13** The differential equation here is an example of the general form $y' = Y(y)X(x)$ with $Y(y) = y^4$ and $X(x) = \sin x$. Does the solution we just obtained fulfill the condition $Y(y) \neq 0$? What happens if the initial condition is changed to $y(\pi/2) = 2$?

4.14 Does DELAB find the same solution as the last example?

4.2.1 Exercises

EXERCISE GUIDE	
To gain experience . . .	**Try exercises**
Checking solutions	13, 17(a), 23(a)
Using separation of variables to solve equations	1–12, 15–16, 18–19, 21–22(a), 24, 25(b)
With general solutions	3, 5–10, 12, 15, 21–22(a), 23(b)

EXERCISE GUIDE (Continued)	
To gain experience . . .	Try exercises
Solving initial-value problems	2–8, 16, 17(c), 18, 20(a), 21–22(b), 23(c)
With conditions for validity of separation of variables	17(b–c), 18, 25(a, c), 26
With linear, homogeneous, etc.	14
Analyzing models using solutions	4, 20(b–d), 21(b–c), 22(c)

Solve 1–12. State the conditions under which your solution is valid. If separation of variables can not be applied, explain why. If separation of variables can be applied, compare your solution with that given by DELAB.

1. $y' = y^2 \cos x$

2. $y' = y^2 \cos x$, $y(\pi) = 1/2$

3. $P' = kP$, $P(t_0) = P_i$

4. $P' = kP^2$, $P(0) = P_i$ (If $k > 0$ and this were a population model, what disaster would it predict?)

5. $T' = -(Ak/cm)(T - T_{out})$, $T(0) = T_i$, T_{out} constant

6. $P' = kP - E$, $P(t_0) = P_i$, E constant

7. $P' = 0.015P - 0.209$, $P(1847) = 8$

8. $P' = 0.015P - 0.209(1 - \sin \pi t)$, $P(1847) = 8$

9. $y' = 2y + 1$

10. $y' = 2y + x$

11. $y' = 2y^2 + 1$

12. $y' = 2x(y + 1)$

13. Check each of the solutions you obtained in exercises 1–12 by substituting into the differential equation and by verifying the initial value, if appropriate.

14. Which of the differential equations given in exercises 1–12 are linear? Homogeneous?

15. Solve each of the following separable differential equations for μ. (These equations arise in the course of using an *integrating factor*, another tool for solving first-order equations. The integrating factor μ is defined by the condition

$$(ye^{\mu(x)})' = e^{\mu(x)}(y' + f(x)y).$$

Multiplying a differential equation by an integrating factor reduces the differential expression to the derivative of a product. See chapter project 1.)

(a) $\mu' = x\mu$

(b) $\mu' = \mu \cos x$

(c) $\mu' = (2/x)\mu$, $x \geq 1$

(d) $\mu'(t) = (Ak/cm)\mu(t)$

(e) $\mu' = g(x)\mu$, g continuous for all x.

16. Example 21 solves the initial-value problem $y' = y^4 \sin x$, $y(\pi/2) = 1/2$, but it omits most of the details. Fill them in.

17. Example 21 uses separation of variables to solve the initial-value problem $y' = y^4 \sin x$, $y(\pi/2) = 1/2$. It obtains the solution

$$y(x) = \frac{1}{(3 \cos x + 8)^{1/3}}.$$

(a) Verify that this expression is a solution of the initial-value problem.

(b) Verify that the separation of variables process never involves dividing by zero.

(c) What happens if the initial condition is replaced by $y(\pi/2) = 2$?

18. Example 19 showed that the initial-value problem

$$\frac{dP}{dt} = aP - sP^2, \quad P(0) = \frac{a}{s}$$

for the logistic equation could not be solved using the solution formula (4.9) obtained by separation of variables,

$$P(t) = \frac{Cae^{at}}{1 + Cse^{at}}.$$

Study the steps in the derivation of this solution formula, and determine what condition is violated by a solution that assumes the initial value $P(0) = a/s$.

19. Supply the details of the derivation of the solution (4.9) of the logistic equation. In particular, evaluate the antiderivative

$$\int \frac{dP}{aP - sP^2},$$

and carefully solve

$$\frac{1}{a} \ln \left| \frac{P}{a - sP} \right| = t + c, \quad \text{for } P.$$

20. This exercise asks you to analyze the behavior of the logistic population model

$$P' = aP - sP^2, \quad P(0) = P_i,$$

using the solution (4.9) of the differential equation obtained from separation of variables,

$$P(t) = \frac{Cae^{at}}{1 + Cse^{at}}.$$

As you will see, having a solution formula is only a beginning to understanding the behavior predicted by the model.

(a) Use formula (4.9) for the solution of the logistic equation to write an expression for the solution of the initial-value problem. Must you exclude any values of P_i? (*Hint:* See example 19.)

(b) Show that $\lim_{t \to \infty} P(t) = P_\infty$ exists and find its value. How does this behavior for large time differ from that of the simple population model $P' = kP$, $P(0) = P_i$?

(c) Show that if $0 < P_i < P_\infty$, then population is increasing for all time. What happens if $P_\infty < P_i$? If $P_i = P_\infty$?

(d) Based on this analysis. provide an interpretation of the population behavior predicted by the logistic model.

21. Consider the emigration equation

$$\frac{dP}{dt} = kP - E, \quad \text{where } E \text{ is constant.}$$

(a) Find the general solution.

(b) Find the solution of this equation subject to the initial condition $P(t_0) = P_i$.

(c) Verify that there is a critical initial population level in the sense that P will decay if P_i is too small while it will grow if P_i is sufficiently large.

22. Consider the heat-loss equation

$$\frac{dT}{dt} = -\frac{Ak}{cm}(T - T_{out}), \quad \text{where } T_{out} \text{ is constant.}$$

(a) Find the general solution of this equation.

(b) Find the solution of this equation subject to the initial condition $T(0) = T_i$.

(c) How do the relative values of T_i and T_{out} influence the behavior of the solution? Is this behavior intuitively reasonable?

23. The text used separation of variables to solve the general homogeneous linear equation

$$a_1(x)y' + a_0(x)y = 0.$$

For $a_1 \neq 0$, it obtained the solution

$$y = C \exp \left(-\int_0^x (a_0/a_1) \, ds \right).$$

(a) Verify by substitution that this expression is indeed a solution.

(b) Verify that this expression is a *general* solution.

(c) Write the solution of this differential equation subject to the initial condition $y(0) = y_i$.

24. Use separation of variables to derive the solution

$$y(x) = y_i \exp \left(-\int_{x_0}^x (a_0/a_1) \, ds \right)$$

of the initial-value problem

$$a_1(x)y' + a_0(x)y = 0, \quad y(x_0) = y_i.$$

Assume $a_1 \neq 0$ and a_1, a_0 continuous.

25. The text contains the following paragraph:

> Can separation of variables be applied to an arbitrary *non*homogeneous linear first-order equation? In general, the answer is no, unless, for example, the coefficients and the forcing term are all constant.

(a) Show that if the coefficients a_1, a_0 and the forcing term f in $a_1 y' + a_0 y = f$ are independent of x, then the equation is separable.

(b) Obtain a formula for the general solution of this equation.

(c) Can you find an example of the equation $a_1 y' + a_0 y = f$ that is separable without the coefficients and the forcing term being constant?

26. Show that any equation of the form

$$\frac{dy}{dx} = Y(y)X(x)$$

can be solved by separation of variables on an interval in which $Y \neq 0$. Write an expression that defines solutions of this equation. What conditions must be satisfied by X and Y to guarantee that the antiderivatives in this solution actually exist? Is the logistic equation $P' = aP - sP^2$ in this form?

4.3 ■ CHARACTERISTIC EQUATIONS

A **constant-coefficient**, first-order, linear differential equation is one of the form

$$b_1 y' + b_0 y = f(x),$$

where b_0, b_1 are constants. This section and the next develop two methods for solving these special equations, one for homogeneous and the other for nonhomogeneous equations. Although more general methods can always be applied to constant-coefficient equations, these special methods are simple enough to deserve separate study.

Begin with the homogeneous equation $b_1 y' + b_0 y = 0$. A simple example is the population equation

$$P' - kP = 0.$$

Solving this equation amounts to finding a function whose derivative is a constant multiple of itself: $P' = kP$.

What function is its own derivative, give or take a constant multiple? The exponential, of course.

To find out precisely which exponential function satisfies $P' - kP = 0$, let this last bit of intuition guide us. *Assume* the solution is an exponential of the form

$$P(t) = e^{rt}$$

for some yet-to-be-determined constant r, and substitute $P = e^{rt}$,

$$re^{rt} - ke^{rt} = 0.$$

Since $e^{rt} \neq 0$ for any finite r and t, this last equality holds only if

$$r - k = 0.$$

We conclude that $r = k$, a constant value consistent with our initial assumption. Hence, a nontrivial solution of $P' = kP$ is $P(t) = e^{kt}$, and a general solution is

$$P_g(t) = Ce^{kt},$$

where C is an arbitrary constant. This is the same general solution as obtained by separation of variables.

The key equation

Characteristic equation

$$r - k = 0$$

that determined r is called the **characteristic equation** of the differential equation $P' = kP$. It is *characteristic* because it has captured the character of the behavior of the solution by determining the growth or decay factor r.

For the first-order, linear, constant-coefficient, homogeneous differential equation

$$b_1 y' + b_0 y = 0$$

the steps in this method are:

> **Characteristic equation method (first order)**
>
> 1. Substitute $y = e^{rx}$, r constant, and simplify to obtain the characteristic equation
> $$b_1 r + b_0 = 0.$$
>
> 2. Solve the characteristic equation for $r = -b_0/b_1$ and write the general solution
> $$y_g = C e^{-b_0 x/b_1}.$$

Do not memorize this formula! Understand the process.

Figure 4.1 illustrates the bottom line of this method: solutions of the constant-coefficient, homogeneous equation $b_1 y' + b_0 y = 0$ are either growing in absolute value or decaying exponentially, depending on the sign of b_0/b_1.

Stop and Think

4.15 There is one solution of $b_1 y' + b_0 y = 0$ that is neither growing nor decaying. Where should it appear in figure 4.1?

4.16 Which solutions in figure 4.1 are increasing with t? Which are decreasing? Is the sign of b_0/b_1 the sole determinant of this behavior?

This method seems to be based on guessing. How can we *assume* anything about an unknown solution of a differential equation? Guessing has no place in mathematics.

Not so. Guessing, or following intuition, is an important part of creative mathematics.

We made an educated guess about the form of the solution. If our guess were wrong, we would encounter some inconsistency; e.g., k would not be constant. We can always verify the final solution by substituting it into the differential equation. The process is foolproof.

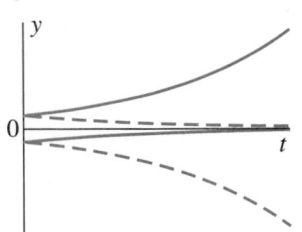

FIGURE 4.1   Solutions of $b_1 y' + b_0 y = 0$ are either growing in absolute value (solid curve) or decaying (dashed).

■ **EXAMPLE 22**   *The data of table* 2.1, *page* 28, *suggest that a model for the growth of the world's population is*

$$P' = 0.0183P, \quad P(1978) = 4{,}258,$$

where time is measured in years and population in millions of people. Solve this initial-value problem by the characteristic equation method.

Since this differential equation is homogeneous and has constant coefficients, the characteristic equation method is applicable. Instead of using the general solution $P_g(t) = C e^{kt}$ with $k = 0.0183$, we attack this problem directly.

Assume a solution of the form $P(t) = e^{rt}$, substitute it into $P' = 0.0183P$ to obtain

$$r e^{rt} - 0.0183 e^{rt} = 0,$$

and divide by $e^{rt} \neq 0$ to find the characteristic equation

Characteristic equation

$$r - 0.0183 = 0.$$

Since $r = 0.0183$, a general solution of $P' = 0.0183P$ is

General solution

$$P(t) = Ce^{0.0183t}.$$

Use the initial condition $P(1978) = 4{,}258$ to find

Find C.

$$C = 4{,}258e^{-(0.0183)(1978)}.$$

The solution of the initial-value problem is

Solution of initial-value problem

$$P(t) = 4{,}258e^{0.0183(t-1978)}. \quad ■$$

Could we obtain a characteristic equation for a *non*homogeneous equation like $P' - kP = \alpha e^{-t}$? What would happen when we attempted to divide out the exponential factor?

What would happen if an equation were homogeneous but lacked constant coefficients? Would the resulting "characteristic equation" necessarily give a constant value for r?

These questions are explored in the next three examples.

■ **EXAMPLE 23** *Try to solve the nonhomogeneous equation*

$$P' - kP = \alpha e^{-t}$$

by the characteristic equation method.

Assuming $P = e^{rt}$, where r is a constant to be determined, substituting, and dividing by e^{rt} yields

$$r = k - \alpha\frac{e^{-t}}{e^{rt}} = k - \alpha e^{-(1+r)t}.$$

This "characteristic equation" certainly does not give a constant value for r, contradicting the initial assumption. Hence, this differential equation does not have a solution of the form $P = e^{rt}$, r constant. The characteristic equation method has failed. ■

■ **EXAMPLE 24** *Try to solve the variable-coefficient equation*

$$P' = t^2 P$$

by the characteristic equation method.

Assuming $P = e^{rt}$, substituting, and dividing by e^{rt} yields

$$r = t^2,$$

another "characteristic equation" that fails to have a constant solution consistent with the initial assumption. ■

■ **EXAMPLE 25**   *Try to solve the nonlinear logistic equation*

$$P' = aP - sP^2$$

by the characteristic equation method.

Assuming $P = e^{rt}$, substituting, and dividing by e^{rt} yields

$$r = a - se^{rt},$$

yet another "characteristic equation" that fails to have a constant solution. The fundamental assumption that r is constant has been violated again. ■

To put the moral of these three examples in positive terms, the characteristic equation method finds a general solution of *constant-coefficient*, *linear*, *homogeneous* differential equations. Check the form of the equation before applying the method.

MATLAB

The analytic solution option in DELAB automatically checks the form of the equation and chooses an appropriate analytic solution method, if one is available.

So why worry about deciding whether or not the characteristic equation method is appropriate? Because it describes the structure of half of the general solution of first-order, constant-coefficient equations: the homogeneous solution is an exponential, growing or decaying depending on the sign of b_0/b_1.

4.3.1 Exercises

EXERCISE GUIDE	
To gain experience . . .	**Try exercises**
Using the characteristic equation method	1–15, 18
With the limitations of this method	19–21
Finding homogeneous solutions	18
With linear, homogeneous, constant-coefficient, etc.	12, 19–21
Analyzing models using solutions	16–17
Comparing with other solution methods	14–15, 20

In exercises 1–11, find the characteristic equation and use it to find either a general solution or a solution of the initial-value problem. Compare your solution with that provided by DE-LAB. If the characteristic equation method is not applicable, explain why.

1. $y' - 6y = 0$

2. $xy' - 6y = 0$

3. $2u' + 7u = 0, u(-2) = 6$

4. $dw/dx = x$

5. $dw/dx = w$

6. $P' - kP = 0$

7. $dT/dt + (Ak/cm)(T - T_{out}) = 0, T(0) = T_i$

8. $y' - 3y^3 = 0, y(1) = 2$

9. $y' - 3y = 0, y(2) = 4e^6$

10. $4u' + 2u = 0$

11. $v' = -g$

12. Classify each of the equations in exercises 1–11 as

 (a) Linear or nonlinear.

 (b) Homogeneous or nonhomogeneous.

 (c) Constant-coefficient or variable-coefficient.

13. Verify, as follows, the two steps of the characteristic equation method given in the text.

 (a) Show that the characteristic equation for the general first-order, linear, homogeneous, constant-coefficient differential equation $b_1 y' + b_0 y = 0$ is

 $$b_1 r + b_0 = 0.$$

 Begin by assuming a solution of the form $y(x) = e^{rx}$.

 (b) Show that a general solution of $b_1 y' + b_0 y = 0$ is

 $$y(x) = Ce^{-b_0 x/b_1}.$$

14. Solve the general constant-coefficient, linear, first-order, homogeneous differential equation $b_1 y' + b_0 y = 0$ by *separation of variables*. Show that you obtain the same general solution as given by the characteristic equation method.

15. (a) Verify that $b_1 y' + b_0 y = 0$ is a linear equation in the sense of definition 1, page 134.

 (b) Verify that the operator $\mathcal{L}$ defined by $\mathcal{L}y \equiv b_1 y' + b_0 y$ is a linear operator in the sense of definition 2, page 136.

16. Recall the expression e^{rt}, the form we assumed for the solution of $P' - kP = 0$. The quotient $1/r$ is sometimes called the **time constant** for the solution. Explore the meaning of this term by proving the following:

 (a) If $r > 0$, then e^{rt} doubles in a time interval proportional to $1/r$.

 (b) If $r < 0$, then e^{rt} is halved in a time interval proportional to $-1/r$.

 (c) The time constant has units of time.

17. See exercise 16 for the definition of *time constant*.

 (a) What are the time constants for the populations of three countries of your choice from table 2.1, page 28? If these populations obey the simple population equation $P' = kP$, how much time would each need to double in size? Comment on these results.

 (b) What is the time constant for the heat-loss equation

 $$\frac{dT}{dt} = -\frac{Ak}{cm}(T - T_{out})?$$

 Comment on the relation between the parameters in this exercise and the size of the time constant.

18. Find *homogeneous solutions* of the following equations using the methods of this section.

 (a) The emigration equation $P' = kP - E$.

 (b) The equation for emigration with decay, $P' = kP - \alpha e^{-t}$.

 (c) The heat-loss equation $T' = -(Ak/cm)(T - T_{out})$.

 (d) The equation for Ireland during 1847–1850, the height of the potato famine, $P' = 0.015P - 0.209$.

 (e) The rock-tossing equation $v' = -g$.

19. Attempt to apply the characteristic equation approach to the constant-coefficient, *non*homogeneous differential equation

 $$y' - 2y = x^3.$$

 Comment on the results.

20. Attempt to apply the characteristic equation approach to the homogeneous, *non*constant-coefficient differential equation

 $$y' - xy = 0.$$

 Comment on the results. How might you find a general solution of this equation?

21. Attempt to apply the characteristic equation approach to the constant-coefficient, homogeneous, *non*linear differential equation

 $$y' + 4\sin y = 0.$$

 Comment on the results. Verify that this equation is indeed homogeneous.

4.4 ■ UNDETERMINED COEFFICIENTS

The method of characteristic equations easily finds solutions of constant-coefficient linear homogeneous equations. Another method, known as *undetermined coefficients*, complements by solving a large class of constant-coefficient *non*homogeneous equations. Together, the two methods produce a general solution; the method of characteristic equations provides the homogeneous solution and the method of undetermined coefficients provides a particular solution.

The next few pages provide motivation and examples. To cut directly to the chase, turn to table 4.1, page 168.

4.4.1 The Main Ideas

Consider the problem of finding a particular solution of a *constant-coefficient*, nonhomogeneous equation such as the emigration equation

$$P'(t) - kP(t) = -E$$

when the emigration rate E is constant.

The physical situation suggests an educated guess of the functional form of a particular solution. Under the right conditions, might not growth exactly balance emigration and the population remain constant? That steady-state constant population value is a simple particular solution.

To pursue that idea, assume that $P' - kP = -E$ has a particular solution P_p which is constant, say

$$P_p \equiv A,$$

for some yet-to-be-determined constant A. (The subscript p denotes *particular solution*.) To find A, substitute this proposed solution into the differential equation. Since $A' = 0$,

$$P'_p - kP_p = A' - kA = -kA = -E.$$

Hence, $A = E/k$, and a particular solution of the emigration equation $P' - kP = -E$ is

$$P_p = E/k.$$

Of course, this is exactly the process used to find steady states: assume there is a constant solution, substitute, and solve for the constant.

Try this educated guessing approach on another problem,

$$y' - 2y = 18e^{-4x}.$$

We must guess a solution of a form that can be combined on the left side of this equation with its derivative to yield the given forcing function e^{-4x} on the right. *Since derivatives of exponentials are again exponentials*, we guess

Guess a particular solution.

$$y_p(x) = Ae^{-4x},$$

where A is an undetermined constant. We guessed an exponential of the form e^{-4x} because that exponential must appear as the forcing term on the right when the guess is substituted into the left side of the equation $y' - 2y = 18e^{-4x}$.

Substituting $y_p(x) = Ae^{-4x}$ into $y' - 2y = 18e^{-4x}$ yields

Substitute.

$$-4Ae^{-4x} - 2Ae^{-4x} = 18e^{-4x}.$$

Canceling the common factor e^{-4x} gives an equation for the unknown constant A,

Find the undetermined coefficient A.

$$-6A = 18$$

or $A = -3$. Hence, a particular solution of $y' - 2y = 18e^{-4x}$ is

Write the particular solution.

$$y_p(x) = -3e^{-4x}.$$

You have just seen two simple examples of the **method of undetermined coefficients**, a grand name for what might better be called educated guessing. In each of these examples, the *undetermined coefficient* is the unknown constant A.

These examples suggest the following *preliminary rules* to guide our guessing of the form of a particular solution of constant-coefficient, linear equations:

1. If the forcing function is a constant, guess that a particular solution is a constant.

2. If the forcing function is an exponential, guess that a particular solution is a constant multiple of that exponential.

■ **EXAMPLE 26** *Find a particular solution of*

$$P'(t) - 0.015P(t) = -0.209,$$

equation 2.8, page 43, from the Irish potato famine model.

Guess.

Rule 1 above suggests that the guess $P_p(t) = A$, where A is the undetermined coefficient. Substituting our guess into the differential equation immediately yields

Substitute.

$$-0.015A = -0.209,$$

Find the coefficient.

or $A = 0.209/0.015 = 13.9$. Hence, a particular solution of

$$P'(t) - 0.015P(t) = -0.209$$

Write the solution.

is $P_p = 13.9$. ■

This example is no more than a special case of the opening discussion of solving $P' = kP - E$ to find $P_p = E/k$.

■ **EXAMPLE 27**   *Find a particular solution of*

$$P'(t) - kP(t) = -\alpha e^{-t},$$

a population equation with a decaying emigration rate.

Rule 2 above suggests the guess

Guess.

$$P_p(t) = Ae^{-t}.$$

Here A is the undetermined coefficient. (Of course, in the differential equation k and α are fixed parameters.) Substituting $P_p = Ae^{-t}$ into $P' - kP = -\alpha e^{-t}$ yields

Substitute.

$$-Ae^{-t} - kAe^{-t} = -\alpha e^{-t},$$

or

$$A(1 + k) = \alpha.$$

Find the coefficient.

If $k \neq -1$, the last expression yields $A = \alpha/(1 + k)$. Hence, a particular solution of $P' - kP = -\alpha e^{-t}$ is

Write the solution.

$$P_p(t) = \frac{\alpha}{1 + k}e^{-t}, \quad k \neq -1. \ ■$$

Of course, the differential equation loses its meaning as a population model when the growth rate k becomes negative.

4.4.2   The Homogeneous Solution Problem

Why did the restriction $k \neq -1$ arise in the last example? Certainly, we needed to avoid dividing by zero in $A(1 + k) = \alpha$. To see if there are any deeper reasons, reconsider this differential equation in the special case $k = -1$,

$$P' + P = -\alpha e^{-t}.$$

Since the forcing term involves e^{-t}, we could again try to guess $P_p = Ae^{-t}$. Substituting that proposed particular solution into $P' + P = -\alpha e^{-t}$ yields

$$P'_p + P_p = -Ae^{-t} + Ae^{-t} = -\alpha e^{-t}$$

or

$$0 = -\alpha e^{-t},$$

an equation that holds only if $\alpha = 0$. What went wrong?

When we substituted the proposed particular solution into the left side of the differential equation, it reduced to zero. Why? Because the proposed particular solution Ae^{-t} is actually a solution of the *homogeneous* equation $P' + P = 0$! The constant A can never be evaluated because it disappears when the proposed solution is substituted into the differential equation.

The same phenomenon appears in the next example as well. It will guide us toward a solution to this dilemma.

■ **EXAMPLE 28** *Find a particular solution of the rock-tossing equation,*

$$v'(t) = -g,$$

where g is the (constant) acceleration of gravity.

Since the forcing term in this equation is a constant, rule 1 suggests a particular solution of the form

$$v_p(t) = A,$$

where A is an undetermined constant. Substituting into $v' = -g$ yields

$$0 = -g.$$

The guess for the particular solution has again been reduced to zero when substituted into the differential equation. The form assumed for the particular solution actually satisfies the *homogeneous* equation. The undetermined constant disappears; it can never be found. ■

The guessing rules must be modified when the proposed particular solution contains a solution of the corresponding homogeneous equation. Reexamine the last example to learn what that modification should be.

Directly integrating $v' = -g$ yields

$$v(t) = -gt + C,$$

a general solution (*general* because the arbitrary constant C is a general solution of the homogeneous equation $v'(t) = 0$ and $-gt$ is a particular solution of the nonhomogeneous equation).

The first guess for a particular solution was $v_p = A$. The particular solution just obtained is $v_p(t) = -gt$, which has the form of a constant multiplied by t. That is, the actual particular solution has the form t *times the initial guess*. Try using this idea again.

■ **EXAMPLE 29** *Find a particular solution of*

$$v'(t) = g$$

by assuming a solution of the form "t multiplying the first guess," which was $v_p = A$.

The particular solution is to have the form

Guess

$$v_p(t) = At,$$

where A is a constant to be determined. Substituting in the differential equation yields

Substitute

$$v_p'(t) = (At)' = A = -g.$$

Find the coefficient

Write the solution

The undetermined constant A has been found: $A = -g$. A particular solution of the given differential equation is $v_p(t) = -gt$. ■

Now the tentative rule for solving the nonhomogeneous equation seems to be:

If the first guess for the particular solution satisfies the homogeneous differential equation, multiply that guess by the independent variable.

Try that idea on another problem.

■ **EXAMPLE 30**   *Using the "multiply the first guess by t" idea, find a particular solution of*

$$P' + P = \alpha e^{-t}.$$

The first guess, which turned out to solve the homogeneous equation $P' + P = 0$, was $P_p = Ae^{-t}$. The idea just developed suggests we should try instead

Guess.
$$P_p = Ate^{-t}.$$

Substituting this expression into $P' + P = \alpha e^{-t}$ yields

Substitute.
$$Ae^{-t} + At(-e^{-t} + e^{-t}) = -\alpha e^{-t}$$

or

Find the coefficient.
$$A = -\alpha.$$

Write the solution.   We have found the particular solution $P_p = -\alpha t e^{-t}$ of $P' + P = \alpha e^{-t}$. ■

We conclude that rules 1 and 2 must be supplemented by adding a third rule:

3. If the proposed particular solution contains a solution of the homogeneous equation, multiply that proposed solution by the independent variable.

■ **EXAMPLE 31**   *Find a particular solution of*

$$y' - 3y = 6e^{3x}.$$

Make a first guess.   Rule 2 first suggests the particular solution $y_p(x) = Ae^{3x}$. To check whether this proposed solution could be a solution of the corresponding homogeneous equation

$$y' - 3y = 0,$$

we could substitute directly into that equation. Alternatively, we could find the general solution of this homogeneous equation and determine by inspection whether the proposed particular solution solves the homogeneous equation.

To follow the latter tack, use the characteristic equation method. Substitute the assumed homogeneous solution $y_h(x) = e^{rx}$ into $y' - 3y = 0$ to obtain the characteristic equation $r - 3 = 0$. Since $r = 3$, the general solution of the homogeneous equation is

Check for homogeneous solution in first guess.
$$y_h(x) = Ce^{3x}.$$

Hence, the proposed particular solution $y_p(x) = Ae^{3x}$ is a solution of the homogeneous equation. Rule 3 now requires multiplying the first guess by x. The new candidate for the particular solution is

Make a better guess.

$$y_p(x) = Axe^{3x}.$$

Substituting this expression into the nonhomogeneous differential equation $y' - 3y = 6e^{3x}$ yields

$$(Ae^{3x} + 3Axe^{3x}) - 3Axe^{3x} = 6e^{3x},$$

Find the coefficient.

from which we obtain $A = 6$. Hence, a particular solution of $y' - 3y = 6e^{3x}$ is

Write the particular solution.

$$y_p(x) = 6xe^{3x}.$$

With no additional effort, we can combine this particular solution with $y_h(x) = Ce^{3x}$, the general solution of the homogeneous equation found earlier, to obtain a general solution of the nonhomogeneous equation $y' - 3y = 6e^{3x}$,

Combine particular and homogeneous solutions to write the general solution.

$$y_g(x) = Ce^{3x} + 6xe^{3x}.$$

The effort of finding the homogeneous solution was not wasted; it led directly to a general solution of this equation. ■

4.4.3 The Whole Story

The basic steps in the **method of undetermined coefficients** are:

1. Assume a particular solution of the appropriate form containing one or more undetermined coefficients.

2. Substitute the assumed solution form into the differential equation and solve for coefficients by equating similar terms.

In the preceding examples, the particular solution is assumed to be the same type of function as the forcing term. That approach is also successful with forcing functions that are cosines, sines, or polynomials. The appropriate educated guess for the form of the particular solution in each case is shown in table 4.1, which summarizes and extends the educated guessing rules developed in the previous examples. That table is the guide to step 1 of the undetermined coefficient solution method.

Recall that the *order* or *degree* of a polynomial is the highest power appearing in it; e.g., $1 - x^4$ is a fourth-order polynomial, and the constant 6 is a polynomial of order zero (because $6 = 6x^0$).

The contents of table 4.1 can be restated as follows:

1. If the forcing function is
   (a) An exponential, the trial particular solution is an exponential.

TABLE 4.1   **Particular solutions via undetermined coefficients for the constant-coefficient, first-order linear equation** $b_1 y' + b_0 y = f(x)$.

$b_1 y' + b_0 y = f(x)$	
Forcing Term $f(x)$	**Trial Particular Solution** $y_p(x)$
$a_n x^n + \cdots + a_1 x + a_0$	$A_n x^n + \cdots + A_1 x + A_0$
$a e^{qx}$	$A e^{qx}$
$a \cos px + b \sin px$	$A \cos px + B \sin px$

If the assumed form of the particular solution solves the corresponding homogeneous equation, multiply the assumed form by x.

Lowercase letters a, a_i, b are constants given in the forcing function. The coefficients to be determined are denoted by uppercase letters A, A_i, B.

    (b) A polynomial (including a constant), the trial particular solution is a polynomial of the same order.

    (c) A sine and/or a cosine, the trial particular solution includes both a sine *and* a cosine.

2. If the trial particular solution solves the corresponding homogeneous equation, multiply the trial solution by the independent variable.

These rules also cover combinations of the functions listed in table 4.1. For example, if the forcing term is a polynomial multiplying an exponential, say $x^3 e^{-2x}$, then the first guess for a particular solution is $(A_3 x^3 + \cdots + A_1 x + A_0)e^{-2x}$. We will neglect this additional complexity until we study higher-order equations.

Note that forcing functions involving *either* a sine or a cosine give rise to particular solution forms involving *both* a sine and a cosine. The following example illustrates this situation.

■   **EXAMPLE 32**   *Find a particular solution of*

$$y' - 2y = 13 \cos 3x.$$

Table 4.1 recommends the form $y_p(x) = A \cos 3x + B \sin 3x$. In anticipation of substituting into the left side of the differential equation, compute

Guess.

$$y_p'(x) = -3A \sin 3x + 3B \cos 3x.$$

Note that the $\cos 3x$ term gave rise to $\sin 3x$ in the derivative. If the initial guess had included only the cosine, substituting into the left side of the equation would have given an equation of the form

$$-3A \sin 3x - 2A \cos 3x = \cos 3x,$$

which can never be satisfied for any choice of A so long as x is free to vary.

Now, substitute the candidate particular solution and its derivative into $y' - 2y = 13 \cos 3x$ in hopes of determining the constants A and B. Collecting coefficients of like functions yields

Substitute.

$$(-3A - 2B)\sin 3x + (3B - 2A)\cos 3x = 13\cos 3x.$$

Since no $\sin 3x$ term appears on the right, the coefficient of $\sin 3x$ on the left must be zero:

Find the coefficients.

$$-3A - 2B = 0.$$

Similarly, the coefficient of $\cos 3x$ on the left must equal the coefficient of $\cos 3x$ on the right:

$$3B - 2A = 13.$$

Solving this pair of simultaneous linear equations for the unknowns A and B yields $A = -2$, $B = 3$. A particular solution of $y' - 2y = 13\cos 3x$ is

Write the solution.

$$y_p(x) = -2\cos 3x + 3\sin 3x. \quad ■$$

To solve the simultaneous equations from the last example,

$$-3A - 2B = 0,$$
$$-2A + 3B = 13,$$

combine the two equations to eliminate one of the unknowns. For example, multiply the first equation by 2, the second by -3, and add to eliminate A, leaving $-13B = -39$, or $B = 3$. Then either equation yields $A = -2$.

> ### MATLAB
>
> If tools like MATLAB can solve equations like these, either analytically or numerically, why bother to learn the method? One answer is that the method says clearly what types of behavior can be expected from solutions of a given equation. The next Stop and Think explores this idea.

Stop and Think   **4.17** What can undetermined coefficients predict about the possible behavior of the solution to a differential equation? To be more precise, suppose a first-order, linear, constant-coefficient differential equation has one of the following as its forcing function:

(a) a constant,

(b) an increasing exponential,

(c) a decreasing exponential, or

(d) a sine or a cosine.

Will some (or all or at least one ...) solutions of the equation exhibit the same behavior? Be as precise as you can in each answer.

4.4.4 More Examples

The particular solutions in these examples are constructed using undetermined coefficients. Homogeneous solutions are found using characteristic equations, and these are coupled with the particular solutions to form general solutions and to solve initial-value problems.

■ **EXAMPLE 33**   *Find a particular solution of*

$$2y' + 5y = 15x^3 - x.$$

Since the forcing term here is a polynomial of order 3, table 4.1 recommends assuming a particular solution that is also a polynomial of order 3,

Guess.

$$y_p(x) = A_3 x^3 + A_2 x^2 + A_1 x + A_0.$$

Substituting this form of y_p and collecting coefficients of like powers of x yields

Substitute.

$$5A_3 x^3 + (6A_3 + 5A_2)x^2 + (4A_2 + 5A_1)x + (2A_1 + 5A_0) = 15x^3 - x.$$

Equate coefficients of corresponding powers of x on either side of this equation:

Find the coefficients.

$$x^3 : \quad 5A_3 = 15,$$
$$x^2 : \quad 6A_3 + 5A_2 = 0,$$
$$x^1 : \quad 4A_2 + 5A_1 = -1,$$
$$x^0 : \quad 2A_1 + A_0 = 0.$$

Beginning with the first equation and solving successively gives the values of the undetermined coefficients:

$$A_3 = 3, \quad A_2 = \frac{-18}{5}, \quad A_1 = \frac{67}{25}, \quad A_0 = \frac{-134}{25}.$$

Hence, a particular solution of $2y' + 5y = 15x^3 - x$ is

Write the solution.

$$y_p(x) = 3x^3 - \frac{18x^2}{5} + \frac{67x}{25} - \frac{134}{25}. \quad ■$$

Stop and Think   **4.18** There is a small shortcut in the last example, not checking explicitly whether any part of the proposed particular solution is a solution of the corresponding homogeneous equation. Why are polynomials not solutions of this constant-coefficient, first-order linear homogeneous equation? (*Hint:* What does the characteristic equation tell you?) Why was the constant term not a candidate solution of the homogeneous equation?

■ **EXAMPLE 34**   *Use table 4.1 to determine the* form *of the particular solution of*

$$y' = 15x^3 - x.$$

Do not evaluate the undetermined coefficients in the proposed particular solution.

The forcing term is the same as that of the last example, but the differential expression on the left is different. Since the forcing terms are identical, the proposed particular solution should be the same as that in the previous example, providing no part of the guess is a solution of the homogeneous version of this equation, $y' = 0$. Hence, we tentatively propose a particular solution of the form

First guess

$$y_p(x) = A_3x^3 + A_2x^2 + A_1x + A_0.$$

Check for homogeneous solution.

But $y' = 0$ obviously has the general solution $y_h = C$. Since a constant A_0 also appears in the assumed form of the particular solution, table 4.1 dictates multiplying the entire guess by x. Hence, the correct form for the proposed particular solution of this equation is

Better guess

$$y_p(x) = x(A_3x^3 + A_2x^2 + A_1x + A_0). \quad ■$$

To appreciate the need for a particular solution of this form, simply solve the differential equation $y' = 15x^3 - x$ by antidifferentiation:

$$y(x) = \frac{15x^4}{4} - \frac{x^2}{2} + C.$$

The term of highest order in this solution is x^4, as is the term $x(A_3x^3 + \cdots)$ in the particular solution form of the last example. But the highest-order term would have been only x^3 if the first form of the particular solution had not been multiplied by x to compensate for the presence of a homogeneous solution.

■ **EXAMPLE 35**　　*Use table 4.1 to determine the* form *of the particular solution of*

$$3y' + 6y = 1 + 2\cos 4x + \sin 4x - 3\sin \pi x - 6e^{-2x} + x^3.$$

Do not evaluate the undetermined coefficients in the proposed particular solution.

The power of linearity!

This forcing term is a combination of the various forcing terms appearing in table 4.1. Since the differential equation is linear, we can break this problem into a series of smaller tasks. In each subproblem, seek a particular solution of an equation whose forcing term is just one of the entries in table 4.1. The sum of these individual particular solutions will be a particular solution of the original equation. (The *principle of superposition* justifies this decomposition.)

To identify these subproblems, collect the terms on the right side of the differential equation into groups of similar type:

Make a set of guesses.

TABLE 4.2

Forcing Term Group	Proposed Particular Solution
polynomial: $1 + x^3$	$A_3x^3 + A_2x^2 + A_1x + A_0$
sine, cosine: $2\cos 4x + \sin 4x$	$B_c \cos 4x + B_s \sin 4x$
sine, cosine: $-3\sin \pi x$	$D_c \cos \pi x + D_s \sin \pi x$
exponential: $6e^{-2x}$	Ee^{-2x}

Two groups of sines and cosines appear, one with argument $4x$, the other with argument πx.

Check for homogeneous solution.

To determine if any of these assumed forms should be multiplied by x, find a general solution of the homogeneous equation $3y' + 6y = 0$. Its characteristic equation is $3r + 6 = 0$, which yields $r = -2$ and

$$y_h(x) = Ce^{-2x}.$$

Since the exponential group appearing last in this list contains a term that is a constant multiple of e^{-2x}, that particular solution term must be multiplied by x to yield the revised form of a particular solution,

Improve one guess.

$$Exe^{-2x}.$$

None of the other proposed solution forms contains e^{-2x}, so no other is changed.

The constants in each group can be determined individually; e.g., E can be chosen so that $y_p(x) = Exe^{-2x}$ is a particular solution of

Substitute in separate equations.

$$3y' + 6y = -6e^{-2x},$$

and B_c and B_s can be chosen so that $y_p(x) = B_c \cos 4x + B_s \sin 4x$ is a particular solution of

$$3y' + 6y = 2\cos 4x + \sin 4x.$$

In accord with the principle of superposition, the sum of the particular solutions for all these individual forcing term groups will be a particular solution of the original equation

$$3y' + 6y = 1 + 2\cos 4x + \sin 4x - 3\sin \pi x - 6e^{-2x} + x^3. \quad \blacksquare$$

Recall that a general solution of a first-order, linear, nonhomogeneous differential equation is a sum of a general solution of the corresponding homogeneous equation and a particular solution of the original nonhomogeneous equation, $y_p = Cy_h + y_p$.

We now have the tools to find these two component parts for constant-coefficient, linear, first-order differential equations. The characteristic equation method finds the general solution of the homogeneous equation, and the method of undetermined coefficients finds particular solutions of equations whose forcing terms are exponentials, polynomials, or sines and cosines.

■ **EXAMPLE 36**   *Find a general solution of*

$$y' - 2y = 13\cos 3x.$$

We need two components, a particular solution of the given nonhomogeneous equation and a general solution of the corresponding homogeneous equation. The particular solution component is available from example 32,

Particular solution

$$y_p(x) = -2\cos 3x + 3\sin 3x.$$

The corresponding homogeneous equation is

$$y' - 2y = 0.$$

Its characteristic equation is $r - 2 = 0$. Hence, a general solution of the homogeneous equation is

Homogeneous solution

$$y_h(x) = Ce^{2x},$$

and a general solution of the nonhomogeneous equation $y' - 2y = 13\cos 3x$ is

General solution

$$y(x) = Ce^{2x} - 2\cos 3x + 3\sin 3x. \quad \blacksquare$$

■ **EXAMPLE 37**   *Find a general solution of*

$$2y' + 5y = 15x^3 - x.$$

Again, we need both a particular solution, such as the one provided by example 33,

Particular solution

$$y_p(x) = 3x^3 - \frac{18x^2}{5} + \frac{67x}{25} - \frac{134}{25},$$

and a general solution of the corresponding homogeneous equation

Homogeneous solution

$$2y' + 5y = 0.$$

Since the characteristic equation for this homogeneous differential equation is $2r + 5 = 0$ (yielding $r = -5/2$), its general solution is $y_h(x) = Ce^{-5x/2}$. Thus, a general solution of $2y' + 5y = 15x^3 - x$ is

General solution

$$y(x) = Ce^{-5x/2} + 3x^3 - \frac{18x^2}{5} + \frac{67x}{25} - \frac{134}{25}. \quad \blacksquare$$

To see that a general solution is the appropriate starting point for solving a linear initial-value problem, think of all the unknown terms in the differential equation grouped on the left and the forcing terms on the right. Substitute the general solution in the left side.

Then the particular solution ensures that the correct forcing term will appear on the right. Although the homogeneous solution contributes zero to the forcing term, it adds an arbitrary constant in the solution itself. It is that constant that is chosen to satisfy the initial condition.

Stop and Think   **4.19** Illustrate the claims of the previous two paragraphs about the role of the homogeneous and particular solutions using the emigration equation $P' - kP = -E$ using $P_h = Ce^{kt}$ and $P_p = E/k$.

The next example illustrates the contributions of each component of the general solution to the forcing term, and the succeeding examples use general solutions to solve initial-value problems.

■ **EXAMPLE 38**   *Example 36 showed that a general solution of*

$$y' - 2y = 13\cos 3x$$

is

$$y(x) = Ce^{2x} - 2\cos 3x + 3\sin 3x.$$

What does each part of this general solution contribute to the right-hand side of $y' - 2y = 13\cos 3x$?

The differential equation is written with all its unknown terms on the left. If we substitute the general solution in the left side, we find

$$
\begin{aligned}
y' - 2y &= (Ce^{2x})' - 2Ce^{2x} \\
&\quad + (-2\cos 3x + 3\sin 3x)' - 2(-2\cos 3x + 3\sin 3x) \\
&= 2Ce^{2x} - 2Ce^{2x} + 6\sin 3x + 9\cos 3x + 4\cos 3x - 6\sin 3x \\
&= 0 + 13\cos 3x = 13\cos 3x.
\end{aligned}
$$

The term Ce^{2x}, which is the general solution of the homogeneous differential equation, contributes zero to the right-hand side while the particular solution $-2\cos 3x + 3\sin 3x$ has provided the correct forcing term $13\cos 3x$ on the right.

This result is another example of the *principle of superposition*. ■

To solve a linear initial-value problem, find a general solution of the given differential equation and choose the arbitrary constant in it to satisfy the initial condition. The next examples illustrate the process.

■ **EXAMPLE 39**   *Find the solution of the initial-value problem*

$$y' - 2y = 13\cos 3x, \quad y(\pi/2) = 5.$$

Example 36 provides a general solution,

General solution

$$y(x) = Ce^{2x} - 2\cos 3x + 3\sin 3x.$$

Although the homogeneous solution term Ce^{2x} adds nothing to this expression's ability to satisfy the given nonhomogeneous differential equation, it does provide the arbitrary constant we need to satisfy the initial condition $y(\pi/2) = 5$.

Applying the initial condition to this general solution yields

$$
\begin{aligned}
y(\pi/2) &= Ce^{\pi} - 2\cos(3\pi/2) + 3\sin(3\pi/2) \\
&= Ce^{\pi} - 3.
\end{aligned}
$$

To satisfy $y(\pi/2) = 5$, we choose C so that $Ce^\pi - 3 = 5$ or

Find C.

$$C = 8e^{-\pi}.$$

Using this value of C in the general solution $y(x) = Ce^{2x} - 2\cos 3x + 3\sin 3x$ gives the solution of the initial-value problem,

Solution of initial-value problem

$$y(x) = 8e^{x-\pi} - 2\cos 3x + 3\sin 3x.$$

To verify that we have correctly solved the initial-value problem, substitute this solution into both the differential equation $y' - 2y = 13\cos 3x$ and the initial condition $y(\pi/2) = 5$. ■

MATLAB

Another source for a general solution of a differential equation is the analytic tool menu in DELAB. But if the goal is solving an initial-value problem, the arbitrary constant must still be found as in the previous example, regardless of the source of the general solution formula.

■ **EXAMPLE 40**   *Find the solution of the initial-value problem*

$$2y' + 5y = 15x^3 - x, \quad y(0) = -7.$$

The pattern is precisely that of the preceding example. The general solution found in example 37 is

General solution

$$y(x) = Ce^{-5x/2} + 3x^3 - \frac{18x^2}{5} + \frac{67x}{25} - \frac{134}{25}.$$

Requiring $y(0) = -7$ yields

$$y(0) = C - \frac{134}{25} = -7,$$

or

Find C.

$$C = -\frac{41}{25}.$$

The solution of the initial-value problem is

Solution of initial-value problem

$$y(x) = \frac{-41e^{-5x/2}}{25} + 3x^3 - \frac{18x^2}{5} + \frac{67x}{25} - \frac{134}{25}.$$

Again, check this solution by showing that it satisfies both the differential equation $2y' + 5y = 15x^3 - x$ and the initial condition $y(0) = -7$. ■

4.4.5 Exercises

EXERCISE GUIDE	
To gain experience . . .	**Try exercises**
Using undetermined coefficients	1–18, 21
With particular solutions	1–18, 22(a)
With homogeneous solutions	19(a), 23–24
With general solutions	19(b), 22(b)
Solving initial-value problems	19(c), 22(c)
With the foundations of undetermined coefficients	20–21, 25–27

In exercises 1–18, use undetermined coefficients to find a particular solution of the given equation. If undetermined coefficients are not applicable, explain why.

1. $u' - 4u = 4e^{\pi x}$

2. $u' - 4u = 4\cos \pi x$

3. $u' - 4u = 4\tan \pi x$

4. $u' - 4u = \pi e^{4x}$

5. $u' - 4u = 4x$

6. $u' - 4u = 4x + \pi e^{4x}$

7. $u' - 4xu = \pi e^{-4x}$

8. $u' - 4u = 4u$

9. $u' - 4u = 0$

10. $dy/dx = (6 - x)(6 + x) - 2y$

11. $y' - 3x = 2y$

12. $y' - 3x^2 = 2y$

13. $y' - 3x^2 = 2y^2$

14. $T' + 4T = (2 - t)^2$

15. $y' + \pi y^2 = \cos x$

16. $3w' + 2w = t + 2$

17. $dx/dt = x + 4 - t$

18. $2u' = u - 4(\sin 2t + 5t)$

19. In those of exercises 1–18 in which you found a particular solution,

   (a) Find a nontrivial solution of the corresponding homogeneous equation.

   (b) Write a general solution of the nonhomogeneous equation.

   (c) Find the solution of the equation that has the value 1 when the independent variable is zero.

   (d) Compare these results with the corresponding solutions from DELAB.

20. The method of undetermined coefficients finds particular solutions of constant-coefficient, linear, nonhomogeneous equations. To which of the equations in exercises 1–18 is this method *not* applicable because the equation

   (a) has variable coefficients?

   (b) is nonlinear?

   (c) is homogeneous?

21. In each of the following, find the *form* of a particular solution. *Do not evaluate* the coefficients.

   (a) $y' + 3y = 1 + x^4 - e^{3x} + 2\cos 3x$

   (b) $y' + 3y = 1 + x^4 - e^{-3x} + 2\cos 3x$

   (c) $y' + 3x = 1 + x^4 - e^{-3x} + 2\sin 3x$

   (d) $y' + 3y = 2 - \sin 2x + \pi \cos(\pi x/4) + (5 - 3x)^2$

   (e) $y' + 3y = (x + 2)^2 - (x - 3)^3 + 6(2 + \pi e^{-3x})$

   (f) $P' + kP = Ae^{kt} - \sin kt$, k, A constant

22. Assuming that E is constant,

   (a) Verify that E/k is a particular solution of

$$P' - kP = -E.$$

   (b) Obtain a general solution of this equation.

   (c) Solve the initial-value problem

$$P' - kP = -E, \quad P(t_0) = P_0,$$

the population model with emigration.

23. The text claims that the general linear, constant-coefficient, homogeneous equation

$$b_1 y' - b_0 y = 0$$

has a general solution that is either an exponential or a constant (if $b_0 = 0$). Use the characteristic equation method to verify this assertion.

24. Table 4.1 contains the statement, "If the assumed form of the particular solution solves the corresponding homogeneous equation, multiply the assumed form by x." Does this warning apply to sine or cosine forcing functions? Why? Can a sine or a cosine be a solution of a constant-coefficient, linear, first-order, homogeneous equation?

25. To see that the method of undetermined coefficients is essentially limited to constant-coefficient equations, attempt to solve the variable coefficient equation

$$xy' + y = e^{4x}$$

by assuming a particular solution of the form

$$y_p(x) = Ae^{4x},$$

A an undetermined constant. What goes wrong?

26. Here is an idea for solving the logistic equation

$$P' - aP = sP^2.$$

When the logistic equation is written in this form, the left side is linear with constant coefficients. The right side is a polynomial, so table 4.1 suggests a particular solution that is a polynomial of the same order, $P_p = A_2 P^2 + A_1 P + A_0$. All I have to do is substitute to find the undetermined constants A_2, A_1, A_0.

What is wrong with this idea? What is a reasonable method for solving this equation?

27. The text suggests that table 4.1 could be extended to provide particular solution forms for forcing terms that are combinations of the functions shown there; e.g., a forcing term that is a product of a polynomial and an exponential should lead to a particular solution that is a product of a polynomial of the same order with a similar exponential. With that hint,

   (i) Find the *form* of a particular solution of each of the following.

   (ii) Find a particular solution of each of the following.

   (iii) Compare your results with those of DELAB.

   (a) $y' + y = xe^x$

   (b) $y' + y = xe^{-x}$

   (c) $y' + y = x \cos 2x - \sin x$

   (d) $y' + y = e^x \cos 2x$

4.5 ■ VARIATION OF PARAMETERS

Separation of variables can quickly solve the homogeneous equation $y' + g(x)y = 0$ if $g(x)$ is continuous, but separation often fails for the nonhomogeneous equation $y' + g(x)y = f(x)$. The technique of *variation of parameters* provides the nonhomogeneous solution, the missing piece of a general solution.

4.5.1 Separation of Variables and the Homogeneous Solution

Recall how separation of variables solves $y' + g(x)y = 0$. From

$$\frac{dy}{y} = -g(x)\, dx,$$

we obtain

$$\ln|y(x)| = -\int_a^x g(s)\, ds + c.$$

To avoid confusion with the independent variable x, the variable of integration is denoted by s.

Exponentiation and setting $C = \pm e^c$ yield the *general solution of the homogeneous equation* $y' + g(x)y = 0$,

General solution of
$y' + g(x)y = 0$

$$y_h(x) = C \exp\left(-\int_a^x g(s)\, ds\right).$$ (4.10)

The lower limit of integration a is arbitrary. It is usually chosen to be the value of x at which an initial condition is given.

■ **EXAMPLE 41**   *Solve $y' + 6x^2 y = 0$, $y(-1) = 2$.*

Separation, integration, and exponentiation yield a general solution:

$$\frac{dy}{y} = -6x^2\, dx,$$

$$\ln|y(x)| = -6\int_{-1}^x s^2\, ds,$$

General solution

$$y_g(x) = C \exp\left(-6\int_{-1}^x s^2\, ds\right).$$

The initial condition $y(-1) = 2$ is easy to satisfy because the initial point $x = -1$ has been chosen as the lower limit of integration:

$$y_g(-1) = Ce^0 = C = 2.$$

Hence, the solution of the initial-value problem is

$$y(x) = C \exp\left(-6\int_{-1}^x s^2\, ds\right).$$

Of course, the integral in the exponent can be evaluated to give

Solution of initial-value problem

$$y(x) = 2e^{-2(x^3+1)}. \quad ■$$

4.5.2   Variation of Parameters and the Nonhomogeneous Equation

To solve the nonhomogeneous equation

$$y' + g(x)y = f(x),$$

we use a technique known as *variation of parameters*. Although it may seem quite arbitrary at first, closer inspection shows it to be a consequence of linearity.

Suppose we know a nontrivial solution y_h of the homogeneous equation

$$y_h' + g(x)y_h = 0,$$

for example from (4.10). **Variation of parameters** begins by assuming that a particular solution of the *non*homogeneous equation can be written in the form

Assumed form of particular solution

$$y_p(x) = u(x)y_h(x),$$

where u is to be determined.

> This method is also called *variation of constants* because the expression $u(x)y_h(x)$ looks like the homogeneous solution $Cy_h(x)$ with the constant C allowed to vary.

Substituting $y_p = uy_h$ in the nonhomogeneous equation

$$y' + g(x)y = f(x)$$

yields

$$(uy_h)' + g(x)uy_h = f(x)$$
$$u(y_h' + g(x)y_h) + u'y_h = f(x)$$
$$u'(x) = \frac{f(x)}{y_h(x)}.$$

Stop and Think **4.20** What happened to the term $y_h' + g(x)y_h$ in parentheses in the second equation?

4.21 How can you be sure that $y_h \neq 0$? Why does it matter? (Exercise 14 also asks you to answer these questions.)

The last equation yields

$$u(x) = \int_a^x \frac{f(s)}{y_h(s)} \, ds + C.$$

A particular solution of $y' + g(x)y = f(x)$ is

Particular solution of
$y' + g(x)y = f(x)$

$$y_p(x) = y_h(x) \int_a^x \frac{f(s)}{y_h(s)} \, ds + Cy_h(x). \tag{4.11}$$

Closer inspection reveals that we have a *general* solution of $y' + g(x)y = f(x)$; the first term is a particular solution of the nonhomogeneous equation, and the second is a constant multiple of a nontrivial solution of the corresponding homogeneous equation.

> The lower limit a on the integral $\int_a^x \frac{f(s)}{y_h(s)} \, ds$ can be any constant. The most convenient choice is usually the point at which an initial value is given.

From (4.10) with $C = 1$, say, we can choose $y_h = \exp\left(-\int_a^x g(s)\,ds\right)$. Then a general solution of the nonhomogeneous equation is

General solution of
$y' + g(x)y = f(x)$

$$y_g(x) = \int_a^x f(s) \exp\left(-\int_s^x g(r)\,dr\right) ds + C \exp\left(-\int_a^x g(s)\,ds\right). \quad (4.12)$$

Don't memorize this formula. Understand the process of variation of parameters. The key is $y_p(x) = u(x)y_h(x)$.

Variation of parameters. To find a particular solution y_p of $y' + g(x)y = f(x)$:

1. Find a nontrivial solution y_h of the corresponding homogeneous equation $y' + g(x)y = 0$.
2. Assume $y_p(x) = u(x)y_h(x)$.
3. Substitute $y_p(x)$ and solve for u'.
4. Integrate u' to find u. Set $y_p = uy_h$.

Stop and Think   **4.22** Argue that the calculations in steps 2 and 3 do *not* require a formula for y_h. In fact, the steps in variation of parameters could be carried out in the order 2-3-1-4: the actual formula for y_h is not needed until the final step of integrating u'.

■ **EXAMPLE 42**   *Solve $y' + 6x^2y = \cos \pi x$, $y(-1) = -8$.*

From the previous example, we know a solution of the homogeneous equation $y' + 6x^2y = 0$,

Homogeneous solution

$$y_h(x) = 2e^{-2(x^3+1)}.$$

To use variation of parameters to solve the nonhomogeneous equation, assume a particular solution has the form $y_p(x) = u(x)y_h(x)$ and substitute. *Without using the formula for y_h*, we find

Assume $y_p(x) = u(x)y_h(x)$.

$$(uy_h)' + 6x^2uy_h = \cos \pi x,$$

$$u(y_h' + 6x^2y_h) + u'y_h = \cos \pi x,$$

$$u'(x) = \frac{\cos \pi x}{y_h(x)}.$$

Hence,

A particular solution that is also a
general solution

$$y_p(x) = y_h(x) \int_{-1}^x \frac{\cos \pi s}{y_h(s)}\,ds + Cy_h(x).$$

Since the first term in the last expression solves the nonhomogeneous equation and the second is a general solution of the homogeneous equation, this expression is, in fact, a general solution of $y' + 6x^2y = \cos \pi x$.

Choosing the lower limit of integration to be the initial point -1 makes evaluation of C easy:

Find C from the initial condition.

$$y_p(-1) = 0 + Cy_h(-1) = C \cdot 2 = -8.$$

Hence, $C = -4$.

Using $y_h(x) = 2e^{-2(x^3+1)}$, we can write the solution of the initial-value problem as

Solution of initial-value problem

$$y(x) = 2e^{-2(x^3+1)} \int_{-1}^{x} \frac{\cos \pi s}{2e^{-2(s^3+1)}} \, ds - 8e^{-2(x^3+1)}$$

$$= \int_{-1}^{x} e^{-2(x^3-s^3)} \cos \pi s \, ds - 8e^{-2(x^3+1)}. \quad ■$$

In the grand scheme of things, the impact of the variation of parameters idea is not as an explicit solution formula such as the one we just obtained. Rather, it is the idea that a solution of the *non*homogeneous equation can be written as the product uy_h, where y_h is a nontrivial solution of the homogeneous equation and u is the solution of a *very* simple differential equation. Much of our general understanding of differential equations is built upon this representation of a solution.

Stop and Think   **4.23** Suppose you have available computer algebra software such as MATLAB's Symbolic Toolbox that can calculate definite integrals of arbitrary functions. How might you couple that capability with the variation of parameters solution formula (4.12) to find analytic solutions of linear, first-order differential equations?

4.5.3 Exercises

EXERCISE GUIDE	
To gain experience . . .	**Try exercises**
Solving homogeneous equations via separation of variables	1–10, 16(a)
Solving nonhomogeneous equations via variation of parameters	1–10, 16(b)
Solving initial-value problems	1–10, 16
Verifying solutions and general solutions	11, 12, 15(b)
Studying the behavior of solutions	17, 18

In exercises 1–10,

(i) Find a general solution of the homogeneous form of the equation by separation of variables.

(ii) Find a particular solution of the nonhomogeneous equation using variation of parameters.

(iii) Combine the preceding results to write a general solution of the given differential equation.

(iv) Write a solution of the given initial-value problem.

(v) Compare your results with those of DELAB.

1. $y' - 2xy = x^2$, $y(0) = 3$

2. $T' = -(Ak/cm)(T - T_{out})$, $T(0) = T_i$, T_{out} constant

3. $v' = -g$, $v(0) = v_i$, g constant

4. $dy/dt + ky = Q$, $y(0) = y_i$, k, Q constant

5. $x' - x \sin \pi t = 4 \sin \pi t$, $x(\pi/4) = 3$

6. $4y' + 12x^3 y = x^3 + 1$, $y(1) = 1$

7. $y - y' = \sqrt{x}$, $y(1) = 4$

8. $R \, dq/dt + q/C = E$, $q(0) = q_i$, R, C, E constant

9. $P' - kP = A \cos 2\pi t$, $P(0) = P_i$, k, A constant

10. $P' - kP = -E$, $P(t_0) = P_i$, k, E constant

11. Using the fundamental theorem of calculus, verify by direct substitution that

$$y_h(x) = C \exp\left(-\int_a^x g(s) \, ds\right)$$

is a solution of $y' + g(x)y = 0$.

12. Using the fundamental theorem of calculus and the fact that y_h satisfies $y_h' + g(x)y_h = 0$, verify by direct substitution that

$$y(x) = y_h(x) \int_a^x \frac{f(s)}{y_h(s)} \, ds + C y_h(x)$$

is a solution of $y' + g(x)y = f(x)$.

13. Example 42 uses variation of parameters to solve $y' + 6x^2 = \cos \pi x$. Without using the actual formula

$$y_h(x) = 2e^{-2(x^3+1)}$$

for a solution of $y' + g(x)y = 0$, it substitutes $y_p = u y_h$ into $y' + 6x^2 = \cos \pi x$ and obtains

$$(u y_h)' + 6x^2 u y_h = \cos \pi x$$

$$u(y_h' + 6x^2 y_h) + u' y_h = \cos \pi x$$

$$u'(x) = \frac{\cos \pi x}{y_h(x)}.$$

(a) Using the formula for y_h, verify these steps.

(b) Explain why you do not really need a formula for y_h. Why is it enough to know that y_h is a nontrivial solution of $y' + g(x)y = 0$?

14. To apply variation of parameters to the equation

$$y' + g(x)y = f(x),$$

the text substitutes $y(x) = u(x)y_h(x)$ and obtains

$$(u y_h)' + g(x)u y_h = f(x)$$

$$u\boxed{(y_h' + g(x)y_h)} + u' y_h = f(x)$$

$$u'(x) = \frac{f(x)}{y_h(x)}.$$

(a) What happened to the term $y_h' + g(x)y_h$ in parentheses in the second equation? (It is boxed here for emphasis.) Would this term arise if the equation were nonlinear?

(b) How can we be sure that $y_h \neq 0$?

15. The text uses $y_h = \exp\left(-\int_a^x g(s) \, ds\right)$ to derive the general solution

$$y(x) = \int_a^x f(s) \exp\left(-\int_s^x g(r) \, dr\right) ds$$

$$+ C \exp\left(-\int_a^x g(s) \, ds\right)$$

of $y' + g(x)y = f(x)$ from the variation of parameters formula

$$y_p(x) = y_h(x) \int_a^x \frac{f(s)}{y_h(s)} \, ds + C y_h(x).$$

(a) Verify by substitution that the first expression is a solution of $y' + g(x)y = f(x)$.

(b) Verify that it is a *general* solution.

(c) Carry out the details of deriving the first expression from the second.

16. Use the results of this section to find formulas for the solution of each of the following initial-value problems. Can DELAB solve such general problems?

(a) $y' + g(x)y = 0$, $y(a) = y_i$

(b) $y' + g(x)y = f(x)$, $y(a) = y_i$

17. Use the representation $y_h = C \exp\left(-\int_a^x g(s) \, ds\right)$, equation (4.10), for the solution of

$$y' + g(x)y = 0$$

obtained in this section to prove the following. Assume g is continuous for all x.

(a) The solution of the initial-value problem $y' + g(x)y = 0$, $y(0) = y_i$, is either identically zero or never zero. (*Hint*: What value must y_i have to give this initial-value problem the trivial solution?)

(b) If $g(x) \geq \delta$ for all x and for some $\delta > 0$, then the solution y of $y' + g(x)y = 0$, $y(0) = y_i$, satisfies

$$\lim_{x \to \infty} y(x) = 0.$$

18. Use the representation

$$y_g(x) = \int_a^x f(s) \exp\left(-\int_s^x g(r)\,dr\right) ds$$

$$+ \, C \exp\left(-\int_a^x g(s)\,ds\right),$$

equation (4.12), for a general solution of

$$y' + g(x)y = f(x)$$

to find a set of conditions on g, f which guarantee that every solution will decay to zero as x grows without bound.

19. Why do we require that the coefficient $g(x)$ in the equation $y' + g(x)y = 0$ be continuous in order to derive the solution formula

$$y_h(x) = C \exp\left(-\int_a^x g(s)\,ds\right)?$$

4.6 ■ UNIQUENESS AND EXISTENCE

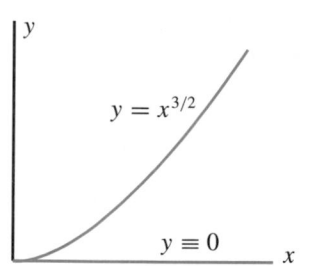

$y = x^{3/2}$

$y \equiv 0$

FIGURE 4.2 Two distinct solutions of the initial-value problem $y' = 3y^{1/3}/2$, $y(0) = 0$: $y = x^{3/2}$ and $y \equiv 0$.

Separation of variables shows that the initial-value problem $y' = 3y^{1/3}/2$, $y(0) = 0$, has the solution $y(x) = x^{3/2}$. But the Euler method approximation to the solution of this initial-value problem is $y_n = 0$, $n = 0, 1, \ldots$, suggesting that the exact solution is $y(x) \equiv 0$. Both appear in figure 4.2. Which is right, separation of variables or Euler? Or both? Or neither?

Does an initial-value problem always have *exactly one* solution? Does an initial-value problem always have at least one solution?

The answer to both questions is *no*. For instance, example 44, page 186, shows that $y' = 3y^{1/3}/2$, $y(0) = 0$, is an initial-value problem that actually has *three* distinct solutions.

Under what conditions does an initial-value problem have at least one solution? Under what conditions does an initial-value problem have exactly one solution? These are the questions of *existence* and *uniqueness*.

Stop and Think

4.24 Verify that $y(x) = x^{3/2}$ is a solution of $y' = 3y^{1/3}/2$, $y(0) = 0$.

4.25 Verify that the Euler approximation to the solution of $y' = 3y^{1/3}/2$, $y(0) = 0$, is $y_n = 0$, $n = 0, 1, \ldots$, regardless of the value of Δt.

4.6.1 Linear Equations: Existence and Uniqueness

Here we prove that if f and g are continuous, then the first-order linear equation

$$y' + g(x)y = f(x) \tag{4.13}$$

has a solution defined on the entire interval in which f, g are continuous, a result known as an *existence theorem*. We also prove that under the same conditions, the initial-value problem

$$y' + g(x)y = f(x), \quad y(x_0) = y_i,$$

has exactly one solution, a *uniqueness theorem*.

The existence theorem tells us that linear, first-order models such as the emigration model or the heat-loss model are not vacuous; e.g., the model equations $P' = kP - E$ and $T' = -(Ak/cm)(T - T_{out})$ always have solutions. In fact, these particular equations have solutions that are defined for all time. (The physical significance of those solutions is another question.)

The uniqueness theorem says that a model with a prescribed initial condition can describe only one specific response. For example, the emigration model with a specified initial population can describe only one pattern of population variation with time. Changing that initial population will change the resulting population curve. Furthermore, two different population curves starting from different initial values can never cross. Otherwise, their common point would be an initial population value with two distinct solutions emanating from it, in violation of the uniqueness theorem.

Finally, the combination of the existence and uniqueness theorems allows us to assert that *every solution* of the linear equation $y' + g(x)y = f(x)$ can be represented as a *general solution* with an appropriate choice of the arbitrary constant.

EXISTENCE THEOREM

Theorem 2 (Existence for linear equations). *Let f, g be continuous on $x_1 \leq x \leq x_2$. Then there exists a continuously differentiable solution of*

$$y' + g(x)y = f(x)$$

defined by

$$y(x) = \int_a^x \exp\left(-\int_s^x g(r)\,dr\right) f(s)\,ds$$

$$+ C \exp\left(-\int_a^x g(r)\,dr\right), \tag{4.14}$$

$$x_1 \leq x \leq x_2,$$

where C is an arbitrary constant and a is a constant in $x_1 \leq x \leq x_2$.

Proof. Solve $y' + g(x)y = f(x)$ using variation of parameters. The solution formula (4.14) results. (Exercise 12 asks you to supply the details.)

Expression (4.14) certainly defines a function y that is continuous and satisfies $y' + g(x)y = f(x)$ for $x_1 \leq x \leq x_2$. Since f, g, and y are all continuous on that interval, $y' = f(x) - g(x)y$ is continuous there as well. ■

The constant coefficients and forcing terms in equations such as $P' = kP - E$ and $T' = -(Ak/cm)(T - T_{out})$ are certainly continuous for all t. Hence, their solutions are defined for all t. But the existence of a solution over long intervals of the independent variable can fail if the equation is not linear.

■ **EXAMPLE 43**   *The first-order* nonlinear *equation $y' - y^2 = 0$ has constant (hence, continuous for all x) coefficients, but a solution starting at $x = 0$ with initial value $y(0) = 1/2$ is undefined at $x = 2$.*

Use separation of variables to solve $y' - y^2 = 0$:

$$\frac{dy}{y^2} = dx, \quad \frac{-1}{y} = x + C, \quad y = \frac{-1}{x + C}.$$

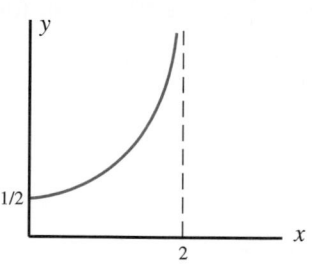

FIGURE 4.3 The solution of $y' - y^2 = 0$, $y(0) = 1/2$, does not exist at $x = 2$ even though the (constant) coefficients of the equation are continuous for all x.

UNIQUENESS THEOREM

The initial condition $y(0) = 1/2$ requires $C = -2$. Hence, a solution of $y' - y^2 = 0$, $y(0) = 1/2$, is

$$y = \frac{1}{2 - x}.$$

But this solution is obviously undefined at $x = 2$. See figure 4.3.

Solutions of *nonlinear* equations may not exist throughout the full interval where their coefficients are continuous. ■

The proof of the existence theorem is constructive; we have actually exhibited the solution in equation (4.14). Exercise 11 asks you to show that this expression is a *general solution*.

The presence of the arbitrary constant C in the solution expression (4.14) confirms what we already knew: The equation $y' + g(x)y = f(x)$ has an infinite number of solutions, one for each value of C. We can not speak of uniqueness, of there being only one solution, unless we add an initial condition that fixes the value of the constant C.

Theorem 3 (Uniqueness for linear initial-value problems). *Let f, g be continuous on $x_1 \leq x \leq x_2$. Let $x_1 \leq x_0 \leq x_2$. Then the linear initial-value problem*

$$y' + g(x)y = f(x), \quad y(x_0) = y_i,$$

has at most one continuously differentiable solution defined for $x_1 \leq x \leq x_2$.

Proof. Suppose there are two continuously differentiable functions u, v that are solutions of this initial-value problem for $x_1 \leq x \leq x_2$:

$$u' + g(x)u = f(x), \quad u(x_0) = y_i,$$
$$v' + g(x)v = f(x), \quad v(x_0) = y_i.$$

Let $z = u - v$. Then from the principle of superposition, theorem 1, page 139, z satisfies the differential equation with forcing term $f(x) - f(x)$; i.e.,

$$z' + g(x)z = 0.$$

Now multiply this equation by the function

$$\mu = \exp\left(\int_{x_0}^{x} g(s)\,ds\right).$$

Observe that the left side of the equation after multiplication can be written as the derivative of a product (i.e., $\mu(z' + g(x)z) = (z\mu)'$), then integrated:

$$\exp\left(\int_{x_0}^{x} g(s)\,ds\right)[z' + g(x)z] = 0,$$

$$\left[z\exp\left(\int_{x_0}^{x} g(s)\,ds\right)\right]' = 0,$$

$$z\exp\left(\int_{x_0}^{x} g(s)\,ds\right) = C.$$

Here C is an arbitrary constant and $x_1 \leq x \leq x_2$. To evaluate C, set $x = x_0$. Then $z(x_0) = u(x_0) - v(x_0) = 0$, and we find $C = 0 \cdot e^0 = 0$. Hence, for $x_1 \leq x \leq x_2$,

$$z(x) \exp \left(\int_{x_0}^{x} g(s)\, ds \right) = 0.$$

Since the exponential is never zero, we conclude that

$$z(x) \equiv 0, \quad x_1 \leq x \leq x_2.$$

We have shown that any two solutions of this initial-value problem must be identical. Hence, there is at most one solution. ■

The proof of this theorem is equivalent to showing that the homogeneous initial-value problem

$$y' + g(x)y = 0, \quad y(0) = y_i,$$

has only the trivial solution.

The function μ used in the proof is known as an *integrating factor*. See chapter project 1, page 68, for its use as a solution technique.

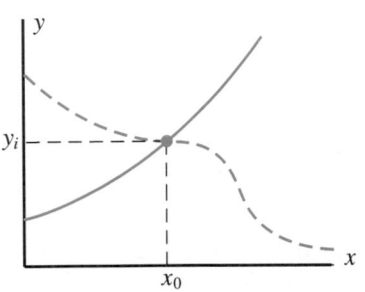

FIGURE 4.4   Two solution curves of a linear initial-value problem can not cross. Otherwise, there would be two distinct solutions satisfying the one initial condition $y(x_0) = y_i$.

An important consequence of the uniqueness theorem is the observation that no two solution curves of a linear differential equation can intersect at a single point. If they did, we could choose that crossing point as an initial condition. But two distinct solutions can not originate from the same initial value without violating theorem 3. Figure 4.4 shows this impossible situation. *Two solutions of a linear differential equation that agree at a single point must be identical* throughout their interval of definition.

Stop and Think   **4.26** Why do the intersecting solution curves in figure 4.2 not contradict the linear uniqueness theorem 3?

But this uniqueness result can fail if the differential equation is nonlinear.

■ **EXAMPLE 44**   *Although its (constant) coefficients are continuous, the* non*linear initial-value problem*

$$y' = \tfrac{3}{2}y^{1/3}, \quad y(0) = 0,$$

has three *distinct solutions.*

This initial-value problem clearly has the trivial solution $y \equiv 0$. Separation of variables reveals two others:

$$\frac{2}{3}\frac{dy}{y^{1/3}} = dx, \quad y^{2/3} = x + C.$$

The initial condition $y(0) = 0$ requires $C = 0$. Two other solutions appear to be $y = \pm x^{3/2}$.

Since $y = \pm x^{3/2}$ are zero at $x = 0$, the separation of variables derivation of these solutions is flawed by division by zero. However, direct substitution, which exercise 19 requests, verifies that $\pm x^{3/2}$ are indeed solutions. Hence, the nonlinear equation $y' = 3y^{1/3}/2$ has *three* solutions satisfying the initial condition $y(0) = 0$; they are $y = 0, \pm x^{3/2}$. Two of them are illustrated in figure 4.2. ■

We can combine theorems 2 and 3 into a single existence and uniqueness theorem for the linear initial-value problem. Exercise 21 asks you for a proof.

Corollary 4 (Existence and uniqueness). *Let f, g be continuous on $x_1 \le x \le x_2$. Let $x_1 \le x_0 \le x_2$. Then the linear initial-value problem*

$$y' + g(x)y = f(x), \quad y(x_0) = y_i,$$

has exactly one *continuous differentiable solution defined on $x_1 \le x \le x_2$. That solution is*

$$y(x) = \int_{x_0}^{x} \exp\left(-\int_{s}^{x} g(r)\,dr\right) f(s)\,ds + y_i \exp\left(-\int_{x_0}^{x} g(r)\,dr\right), \quad (4.15)$$

the general solution (4.14) with $C = y_i$.

These results establish an important fact about a general solution.

> A **general solution** includes *every* possible solution of a linear equation

such as (4.15), providing it has continuous forcing term and coefficients.

To establish this claim in one direction, note that a general solution, if one exists, is always a solution by its very construction. Theorem 2 proves the existence of a general solution (4.14).

In the other direction, to show that every solution of a linear initial-value problem is just the general solution (4.14) with an appropriate choice of C, suppose that $u(x)$ is a solution of $y' + g(x)y = f(x)$ defined on $x_1 \le x \le x_2$. From corollary 4, there is exactly one solution of the initial-value problem

$$y' + g(x)y = f(x), \quad y(x_0) = u(x_0).$$

That solution must be $u(x)$, and it can be written using (4.15) with $y_i = u(x_0)$. But that expression is then the general solution (4.14) with $C = u(x_0)$.

These arguments establish the *power of the general solution*: There are no solutions of the linear differential equation $y' + g(x)y = f(x)$ with f and g continuous that can not be obtained from a general solution.

4.6.2 Uniqueness for Nonlinear Equations

Under what conditions can we guarantee that the general nonlinear initial-value problem

$$y' = f(x, y), \quad y(0) = y_i, \quad (4.16)$$

has no more than one solution, i.e., that its solution is *unique*? One answer is to limit consideration to equations of the form $y' = -g(x)y + f(x)$ with f and g continuous, since the uniqueness theorem 3, page 185, for linear equations guarantees then that there is no more than one solution. But we want to treat more than just linear equations.

To state a uniqueness theorem for the general nonlinear equation, we ask, in effect, that $f(x, y)$ be able to be linearized. We can then prove uniqueness just as in the proof of the linear uniqueness theorem.

For the nonlinear initial-value problem as for the linear initial-value problem, we find the following.

> **Consequence of uniqueness:** Two solutions of an initial-value problem that agree at one point must be identical throughout their common interval of existence.

In other words, *if uniqueness holds*, two solution graphs can not cross at an isolated point. Either the two graphs are identical because they are superimposed or they never cross. Figure 4.2, page 183, is impossible for initial-value problems that satisfy the hypotheses of a uniqueness theorem.

UNIQUENESS THEOREM

Theorem 5 (Uniqueness).   *Suppose that*

  i. *$u(x)$, $v(x)$ are both (continuously differentiable) solutions of the initial-value problem (4.16) for $x_0 \leq x \leq x_1$ for some x_1, and*

  ii. *$f(x, y)$ and $\partial f(x, y)/\partial y$ are continuous in x and y for $x_0 \leq x \leq x_1$ and for*

$$\min\{u(x), v(x) \mid x_0 \leq x \leq x_1\}$$
$$\leq y \leq \max\{u(x), v(x) \mid x_0 \leq x \leq x_1\}. \qquad (4.17)$$

Then

Statement of uniqueness

$$u(x) \equiv v(x), \quad x_0 \leq x \leq x_1;$$

i.e., the initial-value problem (4.16) *has at most one solution.*

Proof. We must show $u(x) \equiv v(x)$, $x_0 \leq x \leq x_1$. We prove the theorem by contradiction. Assume there is some interval $a < x < b$, $x_0 \leq a < b \leq x_1$, where $u \neq v$ while $u(a) = v(a)$. (One candidate for a is x_0. If there were no such interval $a < x < b$, then u and v would be identical at every point of $x_0 \leq x \leq x_1$.)

Let $w(x) = u(x) - v(x)$. Since u, v both satisfy the given differential equation, we have $w' = f(x, u) - f(x, v)$. We can apply Taylor's theorem, section A.3, to the second argument of f to obtain

$$w' = f(x, u) - f(x, v) = \frac{\partial f(x, \xi)}{\partial y} w(x),$$

where ξ lies between $u(x)$ and $v(x)$.   But the continuous function $|\partial f(x, \xi)/\partial y|$ is bounded, say by L, for $x_0 \leq x \leq x_1$, and ξ in the interval (4.17). Hence,

$$w' \leq L|w|, \quad x_0 \leq x \leq x_1.$$

Without loss of generality, we can suppose that

$$w(x) = u(x) - v(x) > 0, \quad a < x < b,$$

so that $|w(x)| = w(x)$ for $a < x < b$ and

$$w' \le Lw, \quad a \le x < b.$$

Now multiply this differential *inequality* by the (positive) function $\mu(x) = e^{-Lx}$. Since

$$e^{-Lx}(w' - Lw) = (we^{-Lx})',$$

the differential inequality $w' - Lw \le 0$ becomes

$$(we^{-Lx})' \le 0.$$

Integrating from $x = a$ yields

$$w(x)e^{-Lx} - w(a)e^{-La} \le 0, \quad a \le x < b.$$

By hypothesis, $w(a) = u(a) - v(a) = 0$, hence,

$$w(x)e^{-Lx} \le 0,$$

forcing $w(x) \le 0$, $a < x < b$, in violation of the hypothesis $w(x) > 0$, $a < x < b$.

We conclude that $w(x) = u(x) - v(x) \equiv 0$, $x_0 \le x \le x_1$. ■

The function μ is an integrating factor, just like the function of the same name used in the proof of theorem 2.

■ **EXAMPLE 45**   *Recall example 44, page* 186: *the initial-value problem*

$$y' = \tfrac{3}{2}y^{1/3}, \quad y(0) = 0,$$

has three *distinct solutions,* $y = 0$, $\pm x^{3/2}$. *Has the uniqueness theorem 5 failed?*

No. The second hypothesis of the uniqueness theorem is violated, for with

$$f(x, y) = \tfrac{3}{2}y^{1/3},$$

the partial derivative

$$\frac{\partial f}{\partial y} = \frac{1}{2y^{2/3}} \tag{4.18}$$

is hardly continuous at $y = 0$.

But solutions of

$$y' = \tfrac{3}{2}y^{1/3}, \quad y(0) = 1,$$

are unique because the partial derivative (4.18) is certainly continuous in a neighborhood of $y = 1$. (Since $y' > 0$ for $y > 0$, solutions of this initial-value problem increase so that $y(x) \geq 1$.) ■

■ **EXAMPLE 46**   *Figure 4.5 is a direction field diagram of the logistic equation $P' = aP - sP^2$. Several solution trajectories are shown, including the stable steady state $P_{ss} = a/s$. Do solutions that are approaching P_{ss} reach it in finite time?*

If the uniqueness theorem applies, then the answer is *no*; the only solution of $P' = aP - sP^2$ that satisfies $P(t) = a/s$ for t finite is the steady state itself, for which $P(t) \equiv P_{ss}$ for all t. If two continuously differentiable solutions cross at one point, they must be equal throughout their interval of existence.

To test uniqueness, examine the behavior of $f(t, P) = aP - sP^2$. Since both f and $\partial f/\partial P = a - 2sP$ are continuous for all t and P, the uniqueness theorem applies. ■

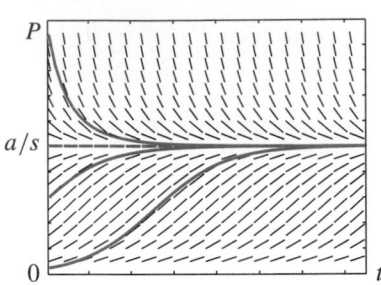

FIGURE 4.5   A direction field diagram and several solution trajectories for the logistic equation $P' = aP - sP^2$.

4.6.3 Existence for Nonlinear Equations

Under what conditions does the initial-value problem (4.16), $y' = f(x, y)$, $y(0) = y_i$, have a solution? How large is the x interval upon which solutions are defined?

The answer to the first question is provided by an *existence theorem*. The second question asks about the size of the *maximum interval of existence* of the solution of an initial-value problem.

The existence theorem hinges on two key ideas. One is the equivalence between the initial-value problem

$$y' = f(x, y), \quad y(0) = y_i,$$

and the **integral equation**

$$y(x) = y_i + \int_0^x f(s, y(s))\, ds. \tag{4.19}$$

If the function y satisfies the integral equation, then obviously it satisfies the initial condition $y(0) = y_i$. Differentiating with respect to x and using the fundamental theorem of calculus recovers the differential equation itself. To go from the initial-value problem to the integral equation, integrate the differential equation and use $y(0) = -y_i$.

The second important idea is *successive approximation*, a technique that leads from the integral equation to a sequence whose limit is a solution of the integral equation and, hence, of the initial-value problem. Successive approximation begins with some guess $y_0(x)$ for the solution of the initial-value problem and computes an improved (we hope) guess $y_1(x)$ using one side of the integral equation (4.19),

$$y_1(x) = y_i + \int_0^x f(s, y_0(s))\, ds.$$

The process repeats endlessly:

$$y_{n+1}(x) = y_i + \int_0^x f(s, y_n(s)) \, ds, \quad n = 0, 1, 2, \ldots . \tag{4.20}$$

The following example should build confidence in successive approximation.

■ **EXAMPLE 47** *Use successive approximation* (4.20) *to find the solution of*

$$y' = ky, \quad y(0) = 1.$$

Lacking a better idea, we choose the initial value $y_i = 1$ as the starting approximation: $y_0(x) = 1$. Then with $f(x, y) = ky$, the next approximation is

$$y_1(x) = 1 + \int_0^x ky_0(s) \, ds = 1 + \int_0^x k \, ds = 1 + kx.$$

A second iteration yields

$$y_2(x) = 1 + \int_0^x k(1 + ks) \, ds = 1 + kx + \frac{(kx)^2}{2},$$

and a third

$$y_3(x) = 1 + \int_0^x k \left(1 + ks + \frac{(ks)^2}{2} \right) ds = 1 + kx + \frac{(kx)^2}{2} + \frac{(kx)^3}{3!}.$$

Studying the pattern of the iterates reveals that

$$y_n(x) = 1 + kx + \frac{(kx)^2}{2} + \cdots + \frac{(kx)^n}{n!},$$

the Taylor polynomial approximation to the exact solution e^{kx}. Clearly, the limit of the sequence $\{y_n(x)\}$ is the exact solution of this initial-value problem. ■

EXISTENCE THEOREM

Theorem 6 (Existence). *Let $f(x, y)$ and $\partial f(x, y)/\partial y$ be defined and continuous in a domain D that includes the initial point $(0, y_i)$. Then the initial-value problem*

$$y' = f(x, y), \quad y(0) = y_i,$$

has a solution $y(x)$ defined in a neighborhood of $x = 0$.

Of course, $f(x, y)$ must be defined in a neighborhood of the initial point $(0, y_i)$. The hypothesis that f be continuous allows a limit of the sequence of successive approximations to be brought inside the function. The requirement that $\partial f/\partial y$ be continuous can be relaxed somewhat, but we will not pursue that additional generality here.

Proof. For some $a, b > 0$, the domain D within which f is defined includes the rectangle

$$R = \{(x, y) : |x| \leq a, |y - y_i| \leq b\}.$$

Let M denote the maximum value of $|f(x, y)|$ on R and L the maximum value of $\partial f(x, y)/\partial y$ on R.

Mimicking the previous example, we construct a sequence of successive approximations $\{y_n(x)\}$ from (4.20) beginning with $y_0(x) = y_i$. Since f is defined in a neighborhood of $(0, y_i)$, $f(s, y_i)$ is defined providing $|s| \leq a$. Hence,

$$y_1(x) = y_i + \int_0^x f(s, y_i)\, ds$$

is defined for $|x| \leq a$.

Calculating the next iterate requires that $f(s, y_1(s))$ be defined. Hence, we must guarantee $|s| \leq a$ and $|y_1(s) - y_i| \leq b$ to remain in R. Since $|f| \leq M$ on R, (4.20) yields

$$|y_1(x) - y_i| = \left| \int_0^x f(x, y_i)\, ds \right| \leq \int_0^x |f(x, y_i)|\, ds \leq Mx.$$

To keep $(x, y_1(x))$ in R, we must further restrict x so that $Mx \leq b$; that is, we now demand $x \leq \min(a, b/M)$.

This argument applies to each of the iterates: $y_n(x)$ is defined so long as $|x| \leq \min(a, b/M)$.

To examine the convergence of the iterates, restrict x still further by requiring $|x| \leq \min(a, b/M, \alpha/L)$ for some number $\alpha < 1$. In addition, if $y(x), z(x)$ are any two functions defined for such values of x, let

$$\|y - z\| = \max\{|y(x) - z(x)| : |x| \leq \min(a, b/M, \alpha/L)\}.$$

We will use $\| \cdots \|$ to compare the maximum difference between iterates of (4.20).

To compare $y_m(x)$ and $y_n(x)$ for arbitrary integers m, n, compute

$$|y_m(x) - y_n(x)| = \left| \int_0^x [f(s, y_{m-1}(s)) - f(s, y_{n-1}(s))]\, ds \right|$$

$$\leq \int_0^x |f(s, y_{m-1}(s)) - f(s, y_{n-1}(s))|\, ds$$

$$\leq \int_0^x \left| \frac{\partial f(s, \xi(s))}{\partial y} \right| |y_{m-1}(s) - y_{n-1}(s)|\, ds$$

$$\leq L \int_0^x |y_{m-1}(s) - y_{n-1}(s)|\, ds$$

$$\leq Lx \|y_{m-1} - y_{n-1}\|.$$

Since $Lx \leq \alpha$, this string of inequalities yields

$$\|y_m - y_n\| \leq \alpha \|y_{m-1} - y_{n-1}\|. \tag{4.21}$$

Since $\alpha < 1$, we have just shown that the rule or *mapping* (4.20) that takes y_{m-1} into y_m is a *contraction mapping*. The distance between iterates as measured by $\| \cdots \|$ contracts with each iteration.

If $m > n$, then

$$\|y_m - y_n\| \leq \alpha \|y_{m-1} - y_{n-1}\| \leq \alpha^2 \|y_{m-2} - y_{n-2}\|$$
$$\leq \cdots \leq \alpha^n \|y_{m-n} - y_0\|.$$

But the triangle inequality applies to the maximum expression $\| \cdots \|$,

$$\|y + z\| \leq \|y\| + \|z\|.$$

Using the triangle inequality and the contraction relation (4.21), we obtain

$$\|y_{m-n} - y_0\| = \|y_{m-n} - y_{m-n-1} + \cdots + y_2 + y_1 - y_0\|$$
$$\leq \|y_{m-n} - y_{m-n-1}\| + \cdots + \|y_2 - y_1\| + \|y_1 - y_0\|$$
$$\leq (\alpha^{m-n-1} + \cdots + \alpha + 1)\|y_1 - y_0\|$$
$$\leq \frac{b}{1 - \alpha}.$$

This last inequality used $1 + \alpha + \cdots + \alpha^{m-n-1} \leq 1/(1-\alpha)$ from the sum of a geometric series as well as $\|y_1 - y_0\| = \|y_1 - y_i\| \leq b$.

Combining the results of these two blocks of inequalities, we have

$$\|y_m - y_n\| \leq \frac{\alpha^n b}{1 - \alpha}, \quad m \geq n.$$

That is, since $\alpha < 1$, the maximum difference between arbitrary elements of the sequence $\{y_n(x)\}$ can be made as small as desired by choosing n (and, hence, m) sufficiently large. Such a sequence is known as a *Cauchy sequence*, and it can be shown to *converge uniformly* to a continuous function $y(x)$ for $|x| \leq \min(a, b/M, \alpha/L)$.

We can take limits on both sides of the successive approximation expression (4.20) and use uniform convergence and the continuity of f to bring the limit on the right side under the integral and inside f. We find that the limiting function $y(x)$ satisfies (4.19), the integral equation equivalent to the initial-value problem (4.16). Hence, $y(x) = \lim_{t \to \infty} y_n(x)$ is a solution of the initial-value problem. ■

Since the requirements for f in the existence theorem are the same as those of the uniqueness theorem, the solution whose existence we have just demonstrated is the *only* solution of the initial-value problem.

Corollary 10 (Existence and uniqueness). *If f and $\partial f/\partial y$ are defined and continuous on some domain that includes the initial point $(0, y_i)$, then the initial-value problem (4.16) has exactly* one *solution defined for x in some neighborhood of $x = 0$.*

In the proof of the existence theorem, x was restricted to a neighborhood of $x = 0$ by the requirement $x \leq \min(a, b/M, \alpha/L)$. Consequently, this result is a *local existence theorem*.

Could x be allowed to grow to larger values? The primary limitation seems to be keeping the solution point $(x, y(x))$ within the domain D upon which $f(x, y)$ is defined. Roughly speaking, if the solution is *not* defined for all x, then we can show that as x approaches some finite value,

- Either the solution point $(x, y(x))$ leaves the domain D because x is nearing the limit for which $f(x, \cdot)$ is defined
- Or y is growing without bound.

Because any two solutions that agree within a neighborhood of $x = 0$ must agree throughout their common interval of definition, we can always choose the solution with the larger interval of existence. Consequently, we can speak of a **maximum interval of existence** of an initial-value problem. That interval may be infinite or finite, and the breakdown at a finite endpoint can occur for either of the reasons listed here. The following example exhibits the second phenomena, a blowup at finite x.

■ **EXAMPLE 48**   *Show that the initial-value problem*

$$y' = y^2, \quad y(0) = 1,$$

has a finite maximum interval of existence.

Solving by separation of variables, we find

$$y(x) = \frac{1}{1 - x}.$$

Clearly, the interval of existence of this solution is $-\infty < x < 1$. Since $f(x, y) = y^2$ is defined for all y, the interval of existence is limited because the solution of the initial-value problem grew without bound, not because f became undefined at a finite value of x. ■

Existence and uniqueness are widely discussed. Accessible references include the texts of Hurewicz [13], Sánchez (treating systems) [20], and Simmons [23]. A classic advanced treatment is that of Coddington andLevinson [6].

4.6.4 Exercises

EXERCISE GUIDE	
To gain experience . . .	**Try exercises**
Applying existence and uniqueness criteria	1–10, 26–30, 38
With the proof of the linear existence theorem	11, 21
With the proof of the linear uniqueness theorem	12, 13, 17, 18, 21

EXERCISE GUIDE (Continued)	
To gain experience . . .	**Try exercises**
With the proof of the nonlinear uniqueness theorem	31
With the proof of the nonlinear existence theorem	34–36, 39
Limitations of the linear existence theorem	14–16
Limitations of the linear uniqueness theorem	19, 20
With general solutions	11, 22
With the behavior of solutions	23, 24
With successive approximations	34–36
With integral equations	37
With interval of existence	32–33

Exercises 1–10 list initial-value problems that model a variety of situations considered earlier in the text or in exercises. For which of these models do the results of this section guarantee the existence of a unique solution? For what range of values of the independent variable? If the theorems in this section do not apply, explain why.

1. Rock model: $v' = -g$, $v(0) = v_i$.

2. Rock model with air resistance: $v' = -g - kv$, $v(0) = v_i$.

3. Improved rock model with air resistance: $v' = -g - kv|v|$, $v(0) = v_i$.

4. Simple population model: $P' = kP$, $P(t_0) = P_i$.

5. Emigration model: $P' = kP - E$, $P(t_0) = P_i$.

6. Logistic population model: $P' = aP - sP^2$, $P(t_0) = P_i$.

7. Potato famine model: $P' = 0.015P - 0.209$, $P(1847) = 8$.

8. Potato famine model with time-dependent emigration rate: $P' - 0.015P = -E(t)$, $P(1847) = 8$, where

$$E(t) = \begin{cases} 0.209, & 1847 \le t \le 1851 \\ 0, & 1851 \le t. \end{cases}$$

(This modification abruptly ends emigration in 1851, when the effects of the famine had begun to wane.)

9. Heat-loss model: $T' = -(Ak/cm)(T - T_{out})$, $T(0) = T_i$.

10. Heat-loss model with furnace: $T' = -(Ak/cm)(T - T_{out}) + F/cm$, $T(0) = T_i$.

11. Show that the solution formula

$$y(x) = \int_a^x \exp\left(-\int_s^x g(r)\,dr\right) f(s)\,ds$$
$$+ C \exp\left(-\int_a^x g(r)\,dr\right)$$

obtained in the existence theorem 2 is, in fact, a *general solution* of $y' + g(x)y = f(x)$.

12. Supply the details of the proof of theorem 2. In particular, carry out the variation of parameters construction mentioned there.

13. The solution formula

$$y(x) = \int_a^x \exp\left(-\int_s^x g(r)\,dr\right) f(s)\,ds$$
$$+ C \exp\left(-\int_a^x g(r)\,dr\right)$$

given in the existence theorem 2 is the same as equation (4.12), page 180, which was obtained using separation of variables and variation of parameters. Prove theorem 2 directly using separation of variables and variation of parameters.

14. Find the interval of existence of a solution of the nonlinear differential equation $y' - y^2 = 0$ satisfying the initial condition $y(0) = y_i$. Express the interval of existence in terms of y_i, considering y_i positive, negative, and zero. Does the solution exist for all $x \ge 0$ for any y_i?

15. Example 43 exhibits a nonlinear equation having a solution that does not exist throughout the full interval in which the

(constant) coefficients of the differential equation are continuous. Construct a similar example using $y' - y^3 = 0$.

16. Example 43 exhibits a nonlinear equation having a solution that does not exist throughout the full interval in which the (constant) coefficients of the differential equation are continuous. Using equations of the form $y' - y^n = 0$, $n > 2$, construct a family of examples of this phenomenon.

17. To prove the uniqueness theorem 3 the text supposes that there are two functions u, v satisfying the same initial-value problem,

$$u' + g(x)u = f(x), \quad u(x_0) = y_i,$$
$$v' + g(x)v = f(x), \quad v(x_0) = y_i.$$

It then appeals to the principle of superposition to show that $z = u - v$ is a solution of the homogeneous differential equation $z' + g(x)z = 0$. Instead, directly verify by substitution that z solves this equation.

18. Verify that $\mu = \exp\left(\int_{x_0}^{x} g(s)\, ds\right)$ is indeed an *integrating factor* for the homogeneous equation $z' + g(x)z = 0$ considered in the proof of the uniqueness theorem 3. That is, show that $\mu(z' + g(x)z) = (\mu z)'$.

19. Example 44 exhibits three solutions of the nonlinear initial-value problem

$$y' = \tfrac{3}{2}y^{1/3}, \quad y(0) = 0.$$

They are $y = 0$, $\pm x^{3/2}$. Verify by substitution that each of these functions is indeed a solution of this initial-value problem.

20. Example 44 shows that the nonlinear initial-value problem $y' = 3y^{1/3}/2$, $y(0) = 0$, does not have a unique solution. Construct a family of examples of this phenomenon using $y' = Ay^\alpha$, $y(0) = 0$, with $0 < \alpha < 1$. Choose A to be a convenient value.

21. Prove corollary 4, existence and uniqueness for the linear initial-value problem.

22. Corollary 4 displays the solution (4.15)

$$y(x) = \int_{x_0}^{x} \exp\left(-\int_{s}^{x} g(r)\, dr\right) f(s)\, ds$$
$$+ y_i \exp\left(-\int_{x_0}^{x} g(r)\, dr\right)$$

of the initial-value problem $y' + g(x)y = f(x)$, $y(x_0) = y_i$. Argue that if y_i is interpreted as an arbitrary constant, then this expression is a general solution of $y' + g(x)y = f(x)$.

23. Consider the equation $y' + g(x)y = 0$ where g is continuous and $g(x) \geq \delta > 0$ for $0 \leq x < \infty$.

(a) Show that the general solution of this equation decays to zero as x grows.

(b) Use the ideas of this section to argue that *all* nontrivial solutions of this equation decay as x grows.

24. Formulate and solve a problem similar to the preceding one in the case when g is *negative* and continuous for $0 \leq x < \infty$.

25. Write each of the following differential equations in the form $y' = f(x, y)$, compute $\partial f/\partial y$, and give the values of y for which this derivative is continuous for all x.

(a) $y' - ky + y^3 = 0$

(b) $y' - \sin(ty) = t^2 - 1$

(c) $P' = aP - sP^2$ (logistic population equation)

(d) $P' = kP - E(t)$ (emigration equation)

(e) $v' = -g - kv|v|/m$ (improved air resistance model)

26. For what values of y_i does the uniqueness theorem 5 guarantee unique solutions of the initial-value problem

$$y' = \tfrac{3}{2}y^{1/3}, \quad y(0) = y_i?$$

27. For what values of the parameter α does the uniqueness theorem 5 guarantee that the initial-value problem

$$y' = y^\alpha, \quad y(0) = 0,$$

has only the trivial solution?

28. Suppose $f(x, y)$ and $\partial f/\partial y$ are both continuous in x and y and that $f(x, 0) = 0$ for all x. Argue that the initial-value problem

$$y' = f(x, y), \quad y(x_0) = 0,$$

has only the trivial solution.

29. While the logic is ultimately circular, use the uniqueness theorem 5 to prove that the linear initial-value problem

$$y' + g(x)y = 0, \quad y(0) = y_i,$$

has at most one solution if g is continuous. Explicitly display the partial derivative $\partial f/\partial y$ for this equation.

30. Let $p(x, y)$ be a polynomial in y with coefficients that are continuous functions of x:

$$p(x, y) = a_0(x) + a_1(x)y + a_2(x)y^2 + \cdots + a_n(x)y^n.$$

Use the uniqueness theorem 5 to argue that the initial-value problem

$$y' = p(x, y), \quad y(0) = y_i,$$

has at most one solution any value of y_i. Use the existence theorem 6 to argue that it has at least one solution. Which of the models that we have studied can be cast in this form?

31. Why in the proof of theorem 5 was it important that the differential inequality $w' \le Lw$ be multiplied by a *positive* function $\mu = e^{Lt}$?

32. How does the maximum interval of existence of $y' = y^2$, $y(0) = y_i$, depend upon y_i?

33. Show that $y' = y^{\alpha+1}$, $y(0) = 1$, has a finite interval of existence for $\alpha > 0$.

34. Show that the successive approximation scheme (4.20) leads to the correct solution of the emigration model $P' = kP - E$, $P(0) = P_i$, when E is constant.

35. Show that the successive approximation scheme (4.20) leads to the correct solution of $y' = -2y$, $y(0) = y_i$.

36. Use successive approximation (4.20) to approximate the solution of $y' = y$, $y(0) = y_i$. Show that the sequence you obtain is converging to the correct solution.

37. Write the integral equation (4.19) that is equivalent to the initial-value problem $y' = x^2$, $y(0) = 2$. Carry out the integration and show that you obtain the correct solution of the initial-value problem. Why are you able to carry out the integration?

38. What hypotheses must you impose on $r(x)$, $g(x)$ to ensure that the linear initial-value problem $y' + g(x)y = r(x)$, $y(0) = y_i$, has a unique solution? Are those hypotheses consistent with corollary 4, page 187, the existence and uniqueness result for linear problems?

39. How does the proof of the existence theorem 6 change if the initial condition is $y(x_i) = y_i$ rather than $y(0) = y_i$?

4.7 ■ CHAPTER EXERCISES

EXERCISE GUIDE	
To gain experience ...	**Try exercises**
Applying basic definitions	1, 24, 26–28
Finding homogeneous solutions	26–27, 39
Finding particular solutions	35–38(a), 39
Finding general solutions	2–25, 26–27(a), 35–38(b)
Solving initial-value problems	2–9, 14, 17, 26–27(b), 28, 29–30(a), 35–38(c)
Using an integrating factor	34
Analyzing solution behavior	29–30(b), 31–33, 35–38(d), 39

1. Write an example of a first order differential equation

    (a) That is homogeneous but can not be solved by separation of variables.

    (b) That is linear and homogeneous but can not be solved by separation of variables.

    (c) That is linear and has constant coefficients but can not be solved by undetermined coefficients.

    (d) That is linear and has constant coefficients but can not be solved by the characteristic equation method.

    If you believe that any of these requests is impossible, explain why.

In exercises 2–25, find either a general solution of the given differential equation or the solution of the given initial-value problem, as appropriate. Compare your results with those of DELAB. If the analytic solution methods available to you fail or if a general solution is not appropriate, explain.

2. $y' - e^x y = 0$, $y(0) = 1$

3. $y' - e^x y = \cos x$, $y(0) = 1$

4. $y' - e^x y = e^x$, $y(0) = 1$

5. $y' + y \cos x = 0$, $y(\pi) = -1$

6. $y' + y \cos x = e^{-\sin x}/x$, $y(\pi) = 0$

7. $xy' = 4y$, $y(1) = 3$

8. $xy' = 4x$, $y(1) = 3$

9. $xy' = 4y - \cos x^3$, $y(1) = 3$

10. $dz/dx = Qz$, Q a constant

11. $dz/dx = Qx$, Q a constant

12. $y' + y \cos x = \sin x$

13. $y' + \pi y^2 = 0$

14. $y' + \pi y = e^{-\pi x} + \cos x$, $y(0) = 0$

15. $y' + xy = \cos x$

16. $y' + 2y = 4x^2 - 3e^{2x}$

17. $y' - 2y = 4x^2 - 3e^{2x}$, $y(0) = 4$

18. $y' + 5y = \cos 5x + e^{-5x}$

19. $P' - kP = -E \cos \omega t$, ω constant
    (A model for a population subject to a periodic variation between emigration and immigration. See section 2.2.)

20. $R \, dq/dt + q/C = 0$, R, C constant

21. $R \, dq/dt + q/C = -E \sin \omega t$, R, C, E, ω constant

22. $R \, dq/dt + q/C = -E$, R, C, E constant

    The preceding three problems are models for the charge q on a capacitor in an RC circuit. See exercises 3–4, page 77, of the chapter exercises for chapter 2.

23. $dy/dt + ky = Q$, k, Q constant
    (A model for decay of a radioactive isotope being bombarded by a neutron beam. See exercise 2, page 77, of the chapter exercises for chapter 2.)

24. $P' = aP - sP^2$, a, s constants

25. $P' = aP$, a constant

26. You are given two functions $H(t)$ and $P(t)$ with the following properties:
    (i) $H(t)$ is a solution of $y' + a(t)y = 0$.
    (ii) $H(0) = 3$.
    (iii) $P(t)$ is a solution of $y' + a(t)y = f(t)$.
    (iv) $P(0) = -1$.
    Express each of the following in terms of $H(t)$ and $P(t)$. If you lack sufficient information, explain what is missing.
    (a) A general solution of $y' + a(t)y = f(t)$.
    (b) The solution of $y' + a(t)y = f(t)$, $y(0) = 5$.

27. You are given the following information:
    (i) A solution of $y' + f(x)y = 0$ is $S(x)$.
    (ii) A solution of $y' + f(x)y = g(x)$ is $R(x)$.
    (iii) $S(0) = 2$, $R(0) = -3$.

(a) Find a general solution of $y' + f(x)y = g(x)$.

(b) Solve the initial-value problem $y' + f(x)y = g(x)$, $y(0) = 5$.

Express your answers in terms of the functions R and S.

28. Suppose that f is a function that satisfies the differential equation $f' + xf = x^4$ and that $f(1) = 6$. The function $z(x)$ satisfies the equation $z' + xz = 0$. Write a solution of the initial-value problem

$$y' + xy = x^4, \quad y(1) = 0,$$

in terms of the functions f and z. Explain your reasoning.

29. (a) Solve the heat-loss model initial-value problem

$$\frac{dT}{dt} = -\frac{Ak}{cm}(T - T_{out}), \quad T(0) = T_i,$$

when T_{out} is constant.

(b) Show that your solution exhibits the following behavior: If $T_i > T_{out}$, then the solution decays to a steady state T_{out}, and if $T_i < T_{out}$, then the solution grows to the same steady state. Is this behavior physically reasonable?

30. (a) Solve the emigration model initial-value problem

$$P' = kP - E, \quad P(0) = P_i,$$

when E is constant.

(b) Show that your solution exhibits the following behavior: If $P_i < E/k$, then the solution decays, and if $P_i > E/k$, then the solution grows. What happens if $P_i = E/k$? Is this behavior physically reasonable?

31. Several of the exercises in chapter 2 offer you a choice of four initial-value problems as possible models of situations described there. Using uniform notation, the four equations are

(a) $c'(t) + kc(t) = d$

(b) $c'(t) + kc(t) = -d$

(c) $c'(t) - kc(t) = d$

(d) $c'(t) - kc(t) = -d$

Assume the constants k, d are positive. The situations being modeled include the following:

Radioactive decay with bombardment. A radioactive isotope is decaying at a rate proportional to its mass $c(t)$; the proportionality constant is k. A neutron beam bombarding the sample creates a new isotope at a rate of d g/s. (See exercise 2, page 77.)

Chemical reaction. A particular chemical reaction produces a chemical at a rate proportional to the concentration $c(t)$ of that chemical; the proportionality constant is k. Diluting the mixture reduces the concentration of the chemical at a rate of d g/cm^3-sec. (See exercise 8, page 78.)

Money in the bank. An amount of money $c(t)$ is earning interest, continuously compounded, at a daily rate of k (or $100k$ percent). Regular deposits are made at a rate of d dollars per day. (See exercise 11, page 54.)

Find a general solution of each of these equations. From the behavior of the general solution, identify which equation might be a reasonable model for each situation listed. Use your intuitive ideas about these situations to guide your selection.

32. A population of yeast spores increases due to reproduction by a fixed proportion R each day. The spores are harvested at a constant rate of H spores per day. Exercise 6 of section 2.2 asks you to model this situation. One model you might obtain is

$$y' = Ry - H, \quad y(0) = y_i,$$

where y is the population of spores. Find a solution of this initial-value problem and use it to predict the long-term fate of this population. Do the relative values of the parameters R, H, y_i matter?

33. Use the model for the population of Ireland during the potato famine, $P' = 0.015P - 0.209$, $P(1847) = 8$, to predict the date when the island would have become deserted. (Recall that t is measured in years, P in millions of people.)

34. Use separation of variables to derive a general solution of the linear, constant-coefficient equation $b_1 y' + b_0 y = 0$, $b_1 \neq 0$. Show that this solution is the same as that given by the characteristic equation method.

35. Exercise 2, page 77, develops a model for the mass y of a radioactive isotope being bombarded by a neutron beam,

$$\frac{dy}{dt} + ky = Q, \quad y(0) = y_i,$$

where k is the decay constant of the isotope and Q is the (constant) rate of production of a new isotope due to the neutron beam.

(a) Find a particular solution of this equation.

(b) Find a general solution of this equation.

(c) Solve the governing initial-value problem.

(d) What becomes of the mass of the isotope after a long time has passed?

36. Exercise 4, page 77, develops a model for the charge q on the capacitor in an RC circuit subject to an imposed voltage drop E,

$$R\frac{dq}{dt} + \frac{1}{C}q = -E, \quad q(0) = q_i.$$

The resistance R and capacitance C in the circuit are constant. Assume E is constant as well.

(a) Find a particular solution of this equation.

(b) Find a general solution of this equation.

(c) Solve the governing initial-value problem.

(d) What becomes of the charge on the capacitor after a long time has passed?

37. Repeat the analysis of the RC circuit of exercise 36 when the imposed voltage drop varies sinusoidally, $E = A \sin \omega t$, where the amplitude A and frequency ω are constant. (Do not confuse the amplitude A with an undetermined coefficient!)

38. Repeat the analysis of the RC circuit of exercise 36 when the imposed voltage drop decays exponentially, $E = Ae^{-\alpha t}$, where A and $\alpha > 0$ are constant.

39. Solve the heat-loss model

$$\frac{dT}{dt} = -\frac{Ak}{cm}(T - T_{out}(t)), \quad T(0) = T_i$$

when the outside temperature T_{out} varies periodically between a high L and a low ℓ,

$$T_{out}(t) = \ell + L\frac{1 + \cos(\pi t/12)}{2}.$$

Either use DELAB or confirm your results with it. In addition, verify that this expression for $T_{out}(t)$ varies as advertised. If t is measured in hours, what is the period of T_{out}?

40. Verify that

$$y = \frac{e^{4x}}{4x} + \frac{C}{x}$$

is a general solution of

$$xy' + y = e^{4x}.$$

For which values of x is this solution valid? Does DELAB provide the same solution? Does it give the domain upon which the solution is defined?

41. Argue that the solutions of $y' = -ky$ decay to zero when k is a positive constant. (This is the isotope decay equation derived in exercise 1 of the chapter 2 exercises.) Use the uniqueness theorem 3, page 185, to argue that this decay is *asymptotic*; $y(t) \neq 0$ for finite t unless $y(t) \equiv 0$ for *all* t. Is this result consistent with the analytic solution of this equation?

4.8 ■ CHAPTER PROJECTS

1. **Integrating factor.** An *integrating factor* can find a general solution of a linear, nonhomogeneous equation in one operation. Here is the basic idea. The nonhomogeneous equation $xy' + y = e^{4x}$ is easy to solve once we notice that its left side is the derivative of a product: $xy' + y = (xy)'$. Since the differential equation can be written $(xy)' = e^{4x}$, an antiderivative and simplification yield

$$\int (xy)' \, dx = \int e^{4x} \, dx,$$

$$xy = \frac{e^{4x}}{4} + C,$$

$$y(x) = \frac{e^{4x}}{4x} + \frac{C}{x},$$

a *general solution* of $xy' + y = e^{4x}$.

But not every equation can be written as a derivative of a product; e.g.,

$$y' + xy = -3x. \tag{4.22}$$

To obtain such a derivative, multiply the entire equation by an **integrating factor** $\mu(x)$,

$$\mu(x)y' + x\mu(x)y = -3x\mu(x),$$

and choose μ to make the *left side equal to the derivative* $(\mu(x)y)'$. Since $(\mu(x)y)' = \mu(x)y' + \mu'(x)y$, this condition requires

$$\mu(x)y' + x\mu(x)y = \mu(x)y' + \mu'(x)y.$$

The coefficients of y' in this equation are equal. The coefficients of y are equal if

$$\mu' = x\mu.$$

Now solve this simple first-order equation to find μ. Then use the integrating factor μ to solve (4.22) by converting it to the form $(\mu y)' = -3x\mu(x)$ and integrating.

(a) Carry out these details to create an example of the use of an integrating factor.

(b) Choose several linear, nonhomogeneous, first-order equations from the text or invent several (e.g., $y' + y\cos x = \pi \cos x$). Find a general solution using an integrating factor.

(c) Find an integrating factor for the heat-loss equation $T' = -(Ak/cm)(T - T_{out}(t))$. Does the form of the integrating factor seem to depend on the form of the linear differential operator ($\mathcal{L}T = T' + (Ak/cm)T$ in this case) or on the form of the forcing term $((Ak/cm)T_{out}(t))$?

(d) Solve the general linear, first-order equation $y' + g(x)y = f(x)$ using an integrating factor. Show that the resulting general solution is identical with that obtained by variation of parameters (4.11).

2. **Integrating factors and homogeneous solutions.** General solutions obtained by integrating factors (see the preceding project) always seem to contain a term such as C/μ that appears to be the homogeneous solution part of the general solution $y_p + Cy_h$. For example, an integrating factor for the heat-loss equation $T' = -(Ak/cm)(T - T_{out}(t))$ is $\mu = e^{Akt/cm}$. The general solution of the homogeneous version of this equation, $T' + AkT/cm = 0$, is $Ce^{-Akt/cm} = C/\mu$.

Use one or both of the following approaches to show that this observation is no accident: If μ is an integrating factor for a linear equation, then $1/\mu$ is a solution of the corresponding homogeneous equation.

(a) Find the integrating factor $\mu(x)$ for the general linear equation $y' + g(x)y = f(x)$. (Try some examples first.) Differentiate and use the fundamental theorem of calculus to show that $1/\mu(x)$ is a solution of $y' + g(x)y = 0$. Explicitly state any conditions that must be imposed on f or g.

(b) The integrating factor for $y' + g(x)y = f(x)$ is found by requiring that $\mu(y' + g(x)y) = (\mu y)'$. This equality holds if $\mu' = g(x)\mu$. Use $\mu' = g(x)\mu$ to show that $y_h = 1/\mu(x)$ satisfies $y_h' + g(x)y_h = 0$ and, hence, that $1/\mu$ is a homogeneous solution.

Note that this approach never needs an explicit expression for μ. It only needs to know the differential equation of which μ is a solution. But how do you show $\mu \neq 0$?

3. **Heat loss and emigration with variable forcing.** Consider the heat-loss model when the outside temperature T_{out} varies with time,

$$T'(t) = -\frac{Ak}{cm}(T(t) - T_{out}(t)), \quad T(0) = T_i.$$

Find an expression for the solution of this initial-value problem. Use that expression to determine a relation between T_i and $T_{out}(t)$ and other conditions, if necessary, that guarantee that the solution decreases for all time. Are these conditions physically reasonable?

Repeat this process for the emigration model when the emigration rate $E(t)$ varies with time,

$$P' = kP - E(t), \quad P(t_0) = P_i,$$

but now seek conditions that make the solution *increasing* for all time.

4. **Numerical stability and difference equations.** As described in section 3.1, a numerical method is stable if the values $y_0, y_1, \ldots$ that it generates when applied to $y' = ky$, $k < 0$, satisfy $y_n \to 0$ as n grows.

When a numerical method is applied to the linear equation $y' = ky$, the result is a *difference equation* for the successive approximate values. For example, applying Euler to $y' = ky$ yields

$$y_{n+1} = y_n + ky_n\Delta t. \tag{4.23}$$

One motivation for the name *difference equation* is that this equation was derived by approximating a derivative by a difference quotient.

Such constant-coefficient linear difference equations can be solved by methods analogous to those used for constant-coefficient linear differential equations. The solutions obtained by such an approach can then be analyzed to determine the range of values of Δt for which the method is numerically stable.

(a) Show that (4.23) can be written in the form

$$b_1 y_{n+1} + b_0 y_n = 0.$$

State a reasonable definition for *constant-coefficient, linear, first-order difference equation*. State a definition for *homogeneous difference equation*. State a definition for *general solution* of a constant-coefficient, homogeneous, linear, first-order difference equation. What sort of difference equation is (4.23)?

(b) Show that difference equations of the form $b_1 y_{n+1} + b_0 y_n = 0$ can be solved by an approach analogous to characteristic equations: assume a solution of the form $y_n = r^n$, r an unknown constant, substitute, and solve for r.

(c) Find a general solution of (4.23). Use it to determine the range of values of Δt for which the Euler method is numerically stable. (*Hint*: What relation must r satisfy to guarantee that $y_n \to 0$?) That is, argue that if Δt is chosen appropriately, then *every* solution of (4.23)—every Euler approximation to a solution of $y' = ky$, $k < 0$—will decay as n grows.

(d) Repeat this analysis for the *implicit Euler method*

$$y_{n+1} = y_n + f(t_{n+1}, y_{n+1})\Delta t.$$

(e) Repeat this analysis for the Heun method.

(f) Repeat this analysis for the general second-order method (3.12).

(g) Why is the concept of a *general solution* of a difference equation important to the analysis of numerical stability?

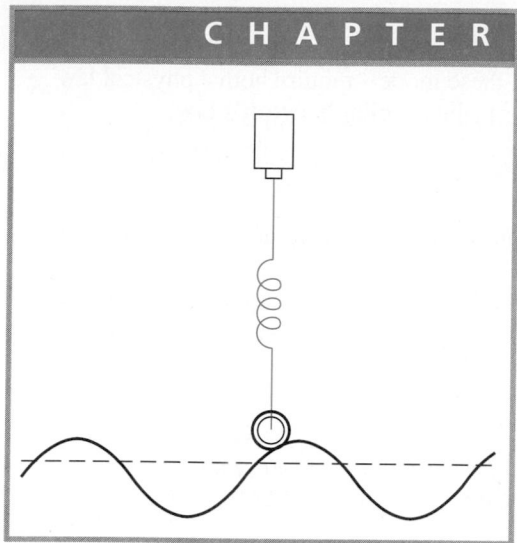

5

Two-dimensional Models: Oscillating Systems

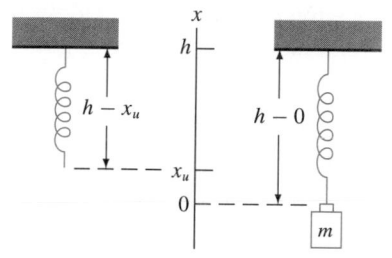

FIGURE 5.1   A vertical spring-mass system showing the location of the free end of the spring at $x = x_u$ before the mass is added and after, when the mass hangs in equilibrium at $x = 0$.

Oscillations are ubiquitous. They are seen in the natural repetitions of life cycles, in the jiggle of an automobile suspension, and in the rhythmic slamming of an old screen door.

We will not be so ambitious as to study biological clocks and their cycles, but we will model an intellectual precursor to an automobile suspension, a spring-mass system, as well as an electric circuit and a pendulum. In some cases, the behavior and structure of these models will be very similar to that of a population of rabbits and foxes but the "species" will be position and velocity of a mass on a spring, for example.

Sometimes the most useful mathematical form of these models will be a system of two first-order equations, sometimes it will be a single second-order equation, depending on the perspective and the mathematical tools to be used.

5.1 ■ SPRINGS AND MASSES

Figure 5.1 shows a typical **spring-mass system**. In the absence of an external disturbance, the mass should hang quietly from the spring, the downward pull of its weight just balanced by the upward pull of the slightly stretched spring. If the mass is pulled farther downward, for instance, and released, we expect it to oscillate up and down with slowly decreasing amplitude until it has returned to its equilibrium position.

Which factors determine the frequency of those oscillations? Their amplitude? The speed of the decay back to the rest position? We will derive several models that begin to answer these questions.

5.1.1 An Undamped System

Consistent with previous experience, these models require both a physical law and an experimental fact. The former is the familiar **Newton's law**,

PHYSICAL LAW

$$F_t = ma,$$

where F_t is the total force acting on the mass m in figure 5.1 and a is its acceleration.

Introducing a signed quantity such as acceleration requires a coordinate axis. The axis shown in figure 5.1 takes upward as positive, lets x denote the position of the mass, and chooses its *origin as the rest position* of the mass.

> The choice of an origin for a coordinate system is arbitrary. Two different points on the coordinate axes will still be the same absolute distance apart regardless of the location of the origin. However, some choices of origin are more convenient than others. Hindsight and experience are the best guides to good choices.

EXPERIMENTAL FACT

The experimental fact relates force in the spring to extension.

Hooke's law   The force exerted by a spring is proportional to the extension from its unweighted length, and it opposes the extension.

The constant of proportionality k is called the **spring constant**. The extension of the spring is just its current, weighted length minus its original, unweighted length.

> The net extension of the spring is current length minus original length so that *positive* extension corresponds to a *longer* spring.

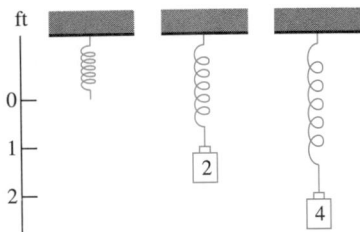

FIGURE 5.2   A spring that offers a resisting force of 2 lb when it is extended 1 ft.

Figure 5.2 illustrates Hooke's law. A 2-lb weight extends the spring 1 ft from its unstretched position, while a 4-lb weight stretches it 2 ft, twice as far. From the conceptual equation,

$$\text{force exerted by spring} = k \times \text{extension of spring},$$

we compute the spring constant

$$k = \frac{2 \text{ lb}}{1 \text{ ft}} = 2 \text{ lb/ft.}$$

We could equally well compute $k = (4 \text{ lb})/(2 \text{ ft}) = 2$ lb/ft.

Stop and Think

5.1 How far would this spring extend if a weight of 9 lb were added to it? A weight of 2,000 lb?

5.2 The spring is extended 3 in. What is the weight of the mass suspended from it?

5.3 What is the connection between Hooke's law and the following observation? On a fisherman's spring scale, increments of equal weight are equally spaced.

We require a mathematical formulation of Hooke's law. If z denotes the net extension of the spring from its *unweighted* length, then downward displacement of the end of the spring from its unweighted position corresponds to positive z. (Negative z would correspond to shortening the spring by pushing the lower end up.) Hence,

$$F_s = kz,$$

where F_s is the force exerted by the spring on the mass and z is the net extension of the spring from its unweighted length.

This equation certainly restates the experimental fact: The force of the spring is proportional to its extension, and it acts in the direction opposite to the direction of the extension. For example, the spring force acts upward (positive) if we pull the spring downward from its unweighted position.

Stop and Think

5.4 Hooke's law says that the graph of force vs. extension is a straight line, regardless of the force applied. How is the slope of that line related to the spring constant k? How does the slope of the force-extension graph for a stronger spring—say, from an automobile suspension—compare with a weak spring like that on a screen porch door?

5.5 Of course, real springs can't support an unlimited force. How should the graph of force vs. extension be modified to reflect that fact? How should the mathematical statement of Hooke's law be modified?

Using $F_s = kz$ requires an expression for the extension z. When the mass is at an arbitrary coordinate $x(t)$, the lower end of the spring has moved from its unweighted position x_u to x. Hence,

$$z = (h - x(t)) - (h - x_u) = x_u - x(t),$$

where $h - x(t)$ is the current (at time t) length and $h - x_u$ is the original, unweighted length; see figures 5.1 and 5.3.

To find x_u, calculate the extension z caused by placing the mass at the origin,

$$z = (h - 0) - (h - x_u) = x_u.$$

The length of the spring after adding the weight is $h - 0$; the length before is $h - x_u$. (See figure 5.1.)

When the mass is at rest, its weight and the force of the spring are in equilibrium. Hence, the spring must be exerting a force mg exactly balancing the force of gravity $-mg$ acting on the mass. Since $F_s = kz = kx_u = mg$, we have

$$x_u = \frac{mg}{k}.$$

As expected, x_u is positive; the end of the spring was above the origin before the weight was added.

Therefore, the force exerted by the spring on the mass when it is at x is

HOOKE'S LAW

$$F_s = k(x_u - x),$$

where $x_u = mg/k$. This is the form of Hooke's law we shall use; it relates the force exerted by the spring to the position of the mass.

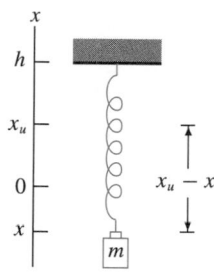

FIGURE 5.3 The extension of the vertical spring is $z = x_u - x$.

Extension = current length
        − original length

Now use these laws of Newton and Hooke to model the situation shown in figure 5.1. Neglect internal friction in the spring, air resistance, etc. and consider only the force of the spring and the force of gravity acting on the mass.

Since acceleration is the first derivative of velocity, Newton's law is just

NEWTON'S LAW

$$F_t = mv'(t),$$

where $v(t) = x'(t)$ is the velocity of the mass at time t and $x(t)$ is its position. The forces acting on the mass are those of the spring and of gravity: $F_t = F_s + F_g$. From $F_s = k(x_u - x)$ and $F_g = -mg$, we obtain

Total force on mass = spring force + gravity

$$mv'(t) = k(x_u - x(t)) - mg.$$

Substituting $x_u = mg/k$ into this last equation gives the differential equation we sought,

Law + fact ⇒ equation

$$mv'(t) + kx(t) = 0.$$

The relation $mv' + kx = 0$ is but *one* equation involving *two* unknowns. One resolution is to add a second equation that defines v in terms of x, leading to the *system of first-order differential equations*

$$x' = v, \tag{5.1}$$
$$mv' = -kx. \tag{5.2}$$

Another resolution is substituting $v = x'$ in order to eliminate v, leading to a *single second-order differential equation*

$$mx'' + kx = 0. \tag{5.3}$$

Both the system (5.1–5.2) and the single equation (5.3) are equivalent statements of the combination of Newton's law and Hooke's law.

Stop and Think

5.6 Verify this claim of equivalence. That is, show that substituting (5.1) into (5.2) yields the second-order equation (5.3) and that (5.3) coupled with the definition $v \equiv x'$ yields the system (5.1–5.2).

5.7 If the position *function* $x(t)$ is known, then the velocity function can be found by differentiation, $v(t) = dx(t)/dt$. If a single *value* of the position x is known, can the corresponding value of velocity be found by differentiation? Does each value of position x correspond to a single value of velocity v?

One initial condition was needed to eliminate the single arbitrary constant in the general solution of a single first-order equation. By analogy, *two* first-order equations like (5.1–5.2) ought to require *two* initial conditions to specify the motion of the mass uniquely, as should a single *second*-order equation like (5.3).

What might those two initial conditions be? Physical intuition suggests that they are the initial position and the initial velocity of the mass. For example, we expect the motion to be much different if it is begun by lifting the mass up slightly and releasing it from rest than if it is begun by giving the mass a strong upward push from the equilibrium position.

Stop and Think **5.8** What values of $x(0)$ and $v(0)$ might correspond to "lifting the mass up slightly and releasing it from rest"? To "giving the mass a strong upward push from the equilibrium position"?

If x_i denotes initial position and v_i denotes initial velocity, then such initial conditions would take the form

Initial conditions

$$x(0) = x_i, \quad v(0) = x'(0) = v_i.$$

Although based only on physical intuition, this argument for two initial conditions is correct. An initial-value problem that models the situation shown in figure 5.1 can be written as a system of two first-order equations,

UNDAMPED SPRING-MASS MODEL: SYSTEM

$$\begin{aligned} x' &= v, \\ mv' &= -kx, \end{aligned} \tag{5.4}$$

$$x(0) = x_i, \quad v(0) = v_i, \tag{5.5}$$

or as a single second-order equation,

UNDAMPED SPRING-MASS MODEL: SINGLE EQUATION

$$mx''(t) + kx(t) = 0, \tag{5.6}$$

$$x(0) = x_i, \quad x'(0) = v_i. \tag{5.7}$$

These models are called **undamped** because, as we shall see, they predict that the mass will oscillate forever once started.

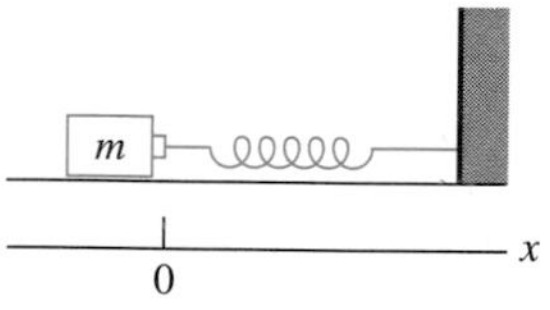

5.1.2 A Damped System

Now turn the system of figure 5.1 on its side, as shown in figure 5.4, to consider a different problem. The coordinate axis is drawn so that $x = 0$ corresponds to the location of the mass when the spring is unstretched. The new effect to consider here is friction between the mass and the surface. Friction is an example of *damping*, an action that tends to exhaust the energy of the system and bring it eventually to rest.

FIGURE 5.4 A horizontal spring-mass system; the origin is located at the rest point of the free end of the unstretched spring.

Stop and Think **5.9** The model describing the vertical spring-mass system of figure 5.1 is called *undamped*, suggesting that it includes no effects that eventually would bring the oscillating mass to a halt. In fact, such systems will not oscillate forever. What physical effects have been omitted from the models (5.4–5.5) and (5.6–5.7) to justify calling them *undamped*?

The simplest experimental observation is the following:

Experimental fact. The force of friction is proportional to velocity and acts to oppose the motion.

If F_f is the force of friction exerted by the surface on the mass in figure 5.4 and $v(t) = x'(t)$ is the velocity of the mass, then a mathematical formulation of the experimental fact is

Experimental fact

$$F_f = -pv(t).$$

The constant of proportionality $p > 0$ is called the **friction**, or **damping coefficient**. The negative sign ensures that the force of friction opposes the motion; when $v > 0$, then $F_f < 0$ and vice versa.

> That friction should oppose the motion is obvious; that its magnitude should be proportional to velocity is perhaps less clear.
>
> Experience does suggest that the force of friction increases with velocity. (It also depends on the finish of the surfaces in contact, on the force pressing them together, and on the area in contact.) Otherwise, sliding a heavy box across the floor would not become more difficult as we tried to push it faster. (Getting it started is another matter because the force of static friction exceeds that of dynamic friction.)
>
> Choosing the force of friction to be proportional to velocity leads ultimately to a linear differential equation. Although a more complex relation between force and velocity might better fit experimental observations, it would also make the resulting differential equation nonlinear and more difficult to analyze. The relation $F_f = -pv$, like many others, is a compromise between an accurate fit to data and analytic tractability.
>
> The constant of proportionality p has units of force/velocity; e.g.,
>
> $$\text{lb/(ft/s)} = \text{lb·s/ft} \quad \text{or} \quad \text{N/(m/s)} = \text{N·s/m} = \text{kg/s}$$
>
> (because $1\,\text{N} = 1\,\text{kg·m/s}^2$).

Since $x = 0$ is the position of the mass when the spring is unstretched, the extension of the spring in this coordinate system is the *negative* of the position x of the mass. Hence, the spring force is

Hooke's law in a horizontal coordinate system

$$F_s = -kx.$$

The rest of the analysis parallels the derivation of the undamped model (5.4–5.5) or (5.6–5.7). Ignoring all horizontal forces but that of the spring and of friction means that $F_t = F_f + F_s$. (Gravity does not appear because it acts vertically. This derivation accounts for horizontal forces.) Using

- Newton's law $F_t = ma = mv'$,
- the friction force relation $F_f = -pv$, and
- the spring force relation $F_s = -kx$,

we find

Total force on mass = friction force + spring force

$$mv'(t) = -pv(t) - kx(t).$$

Once again, the presence of two unknowns x and v can be accommodated either by adding a second equation $x' = v$ or by substituting $v = x'$ to eliminate the unknown v. Of course, two initial conditions are appropriate in either case.

The result is either a system of two first-order equations,

DAMPED SPRING-MASS MODEL: SYSTEM

$$x' = v,$$
$$mv' = -kx - pv,$$
$$x(0) = x_i, \quad v(0) = v_i,$$

(5.8)

or a single second-order equation,

DAMPED SPRING-MASS MODEL: SINGLE EQUATION

$$mx''(t) + px'(t) + kx(t) = 0,$$

(5.9)

$$x(0) = x_i, \quad x'(0) = v_i.$$

(5.10)

These models are called **damped** because friction can be shown to slow the motion.

5.1.3 A Forced System

Suppose the vertical spring-mass system of figure 5.1 is hung from a ceiling. An energetic grandmother begins rocking vigorously in her chair on the floor directly above. Her motion will certainly cause the mass to move. How is its motion related to the frequency and the strength (amplitude) of the disturbances shaking the ceiling? The model derived here explores the effect of such external forcing functions.

Consider the situation shown in figure 5.5, where the motion of grandmother's rocking chair imparts a displacement $h(t)$ to the top of the spring at time t. We proceed as in the first derivation, neglecting air resistance and considering only the force of gravity and the force of the spring.

When the ceiling—the upper end of the spring—was stationary, the force of the spring on the mass was computed by finding the extension of the spring. Once again, we compare the current length of the spring with its original unstretched length to find its net extension and, hence, the force it exerts on the mass.

The position of the top of the spring at $t = 0$ is just $h(0)$, and the coordinate of the lower end of the unstretched spring is $x_u - mg/k$, as before. Hence, the unstretched length of the spring is

$$h(0) - x_u = h(0) - \frac{mg}{k}.$$

At any instant t, the top of the spring will be at position $h(t)$ while the mass will be at $x(t)$. Thus, the length of the spring at time t is $h(t) - x(t)$, and the net extension of the spring is

$$z = (h(t) - x(t)) - \left(h(0) - \frac{mg}{k} \right).$$

From the Hooke's law relation $F_s = kz$, the net force exerted by the spring on the mass is thus

$$F_s = k \left(h(t) - h(0) - x(t) + \frac{mg}{k} \right).$$

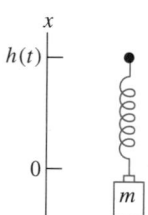

FIGURE 5.5 The position of the upper end of the vertical spring-mass system varies with time; the coordinate of the upper end of the spring at time t is $h(t)$.

Extension = current length − original length.

Hooke's law

The rest of the derivation is straightforward. Since we are considering only the forces of the spring and of gravity, the total force on the mass is

Total force = gravity + spring

$$F_t = F_g + F_s.$$

Using $F_t = mv'(t)$ yields

$$mv'(t) = -mg + k\left(h(t) - h(0) - x(t) + \frac{mg}{k}\right)$$

$$= k(h(t) - h(0)) - kx(t)$$

or

$$mv'(t) + kx(t) = k(h(t) - h(0)).$$

The model for the spring-mass system of figure 5.5 is completed by adding the same set of initial conditions as before grandmother's rocker appeared. The model can be written as a system that displays both position and velocity,

FORCED SPRING-MASS
MODEL: SYSTEM

$$x' = v,$$
$$mv' = -kx + k(h(t) - h(0)), \tag{5.11}$$
$$x(0) = x_i, \quad v(0) = v_i, \tag{5.12}$$

or, as a single equation in position alone,

FORCED SPRING-MASS
MODEL: SINGLE EQUATION

$$mx''(t) + kx(t) = k(h(t) - h(0)), \tag{5.13}$$
$$x(0) = x_i, \quad x'(0) = v_i. \tag{5.14}$$

Either of these models is called **forced** because the movement of the rocker can force the system into motion through the displacement of the upper end of the spring, even if the mass itself is initially in its equilibrium position.

5.1.4 Examples

Suppose we are given the spring shown in figure 5.2. We observe from that figure that adding a 2-lb weight stretches the spring one foot; its spring constant is $k = (2 \text{ lb})/(1 \text{ ft}) = 2 \text{ lb/ft}$.

■ **EXAMPLE 1**   *A 6-lb weight is suspended from the spring of figure* 5.2. *Write an initial-value problem that governs the motion of the mass if it is started from its equilibrium position with a downward velocity of* 6 *in./s. Neglect air resistance.*

The undamped model (5.6–5.7),

$$x'' + kx = 0, \quad x(0) = x_i, \quad x'(0) = v_i,$$

of section 5.1.1 applies; the coordinate system is shown in figure 5.1. We have seen that the spring constant is $k = 2$ lb/ft. The mass is $m = 6/32 =$

3/16 slug. The initial position is $x(0) = 0$. The initial velocity is $v(0) = -6$ in./s $= -0.5$ ft/s.

The governing initial-value problem is

$$\frac{3}{16}x''(t) + 2x(t) = 0, \quad x(0) = 0, \quad x'(0) = -0.5. \ \blacksquare$$

The ugly name for the unit of mass, the "slug," may be the best argument for the metric system! Loosely, one slug is the mass of an object that weighs 32 lb on the surface of the earth: $F_g = mg = 1$ slug $\cdot$ 32 ft/s $= 32$ lb.

Stop and Think **5.10** Write the model derived in the preceding example using a system of two first-order equations as in (5.4).

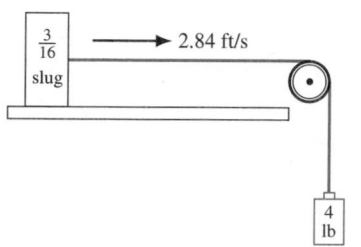

FIGURE 5.6 The falling 4-lb weight pulls the mass across the table at a constant speed of 2.84 ft/s.

■ **EXAMPLE 2** *The mass of the preceding example is placed on a table as in figure 5.6. After some experimentation, we discover that the falling weight of 4 lb shown there pulls the mass of 3/16 slug at a steady speed of 2.84 ft/s. The spring of the preceding example is then connected horizontally between this mass and a fixed anchor as in figure 5.4. The spring is stretched 9 in., and the mass is released. Write an initial-value problem that governs the resulting motion. Include the effects of friction.*

The damped model (5.9–5.10),

$$x'' + px' + kx = 0, \quad x(0) = x_i, \quad x'(0) = v_i,$$

applies. Use the coordinate system of figure 5.4. We know that $m = 3/16$ slug and $k = 2$ lb/ft. The data of the experiment shown in figure 5.6 reveal that the friction coefficient is

$$p = (4 \text{ lb})/(2.84 \text{ ft/s}) = 1.41 \text{ lb-s/ft}.$$

The initial position corresponds to stretching the spring 9 in.; i.e., $x(0) = 0.75$ ft. If the mass is simply released, its initial velocity is zero: $x'(0) = 0$.

The governing initial-value problem is

$$\frac{3}{16}x''(t) + 1.41x'(t) + 2x(t) = 0, \quad x(0) = 0.75, \quad x'(0) = 0. \ \blacksquare$$

Stop and Think **5.11** Write the model derived in the preceding example using a system of two first-order equations as in (5.8).

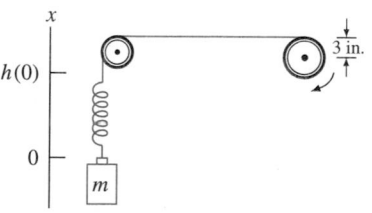

FIGURE 5.7 The turntable placed on its side oscillates the upper end of the vertical spring. It is shown here at $t = 0$.

■ **EXAMPLE 3** *The spring of the first example and a 4-lb weight are suspended as shown in figure 5.7. The end of the cable supporting the spring is connected 3 in. from the center of an old 45-rpm phonograph turntable placed vertically. Figure 5.7 shows the position of the assembly at $t = 0$; the mass is at rest. Neglect air resistance and write an initial-value problem that governs the motion of the mass.*

The forced model (5.13–5.14),

$$mx''(t) + kx(t) = k(h(t) - h(0)), \quad x(0) = x_i, \quad x'(0) = v_i,$$

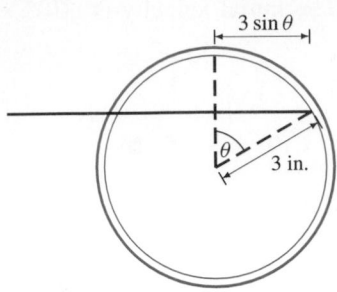

FIGURE 5.8 After a brief interval of time, the turntable of the previous figure has rotated through an angle θ to the position shown here. The end of the cable connected to the turntable has moved 3 sin θ in. to the right.

of section 5.1.3 obviously applies. The mass is $m = 1/8$ slug, and the spring constant is $k = 2$ lb/ft. The challenge here is finding the position $h(t)$ of the top of the spring at time t.

Figure 5.8 shows the turntable after it has turned through an angle θ. If we neglect the vertical displacement of the cable, then the end attached to the turntable has moved to the right 3 sin θ in. or $\frac{1}{4}$ sin θ ft; i.e., the upper end of the spring has been raised so that its position is $h(t) = h(0) + \frac{1}{4}$ sin θ. In general, $h(t) = h(0) + \frac{1}{4}$ sin $\theta(t)$, where $\theta(t)$ is the angular rotation of the turntable at time t.

The turntable is spinning at 45 rpm (revolutions per minute) or 90π rad/min $= 3\pi/2$ rad/s, since one revolution covers 2π rad. If t is measured in seconds, then $\theta(t) = 3\pi t/2$, and $h(t) = h(0) + \frac{1}{4} \sin(3\pi t/2)$.

We are told that figure 5.7 shows the position of the mass at $t = 0$, when the mass is at rest. Both initial position and initial velocity must be zero.

Because $k = 2$, the governing initial-value problem is

$$\frac{1}{8}x''(t) + 2x(t) = \frac{1}{2} \sin \frac{3\pi t}{2}, \quad x(0) = 0, \quad x'(0) = 0.$$

Stop and Think   **5.12** Write the model derived in the preceding example using a system of two first-order equations as in (5.11).

> We were careful to convert units of distance to feet and units of time to seconds in this model because we used $g = 32$ ft/s^2 to convert the given weight of 4 lb to a mass of 1/8 slug. The units in the parameter g dictate the choice of units for length and time.

Note that the initial location $h(0)$ of the upper end of the spring does not appear in this model. The important quantity is the net displacement of the end of the spring. ■

5.1.5 Preliminary Analysis

Analytic, numerical, and graphical tools suggest some of the sorts of behavior predicted by these three varieties of spring-mass models, undamped, damped, and forced. Subsequent chapters will develop in more detail the ideas introduced here.

Undamped

To begin an analysis of the undamped model, note that the differential equation (5.6) provides a relation between acceleration and position,

$$x''(t) = -\frac{k}{m}x(t).$$

When the mass is above the equilibrium position—i.e., when $x(t) > 0$—then acceleration is negative and vice versa. The magnitude of acceleration is symmetric across the point $x = 0$. If the mass is not somehow prevented from successively passing back and forth across the equilibrium position, its motion could be symmetric and repetitious.

The relation $x'' = -(k/m)x$ also tells us that, give or take a constant, x must be a function whose second derivative is its own negative, such as a sine or cosine. Inspired by the characteristic equation method, we might try guessing that $x(t) = \sin rt$, where r is an unknown constant.

Substitution shows that $r = \sqrt{k/m}$. A parallel attack beginning with $x(t) = \cos rt$ yields the same result, leading us to conclude that $\sin \sqrt{k/m}\, t$ and $\cos \sqrt{k/m}\, t$ are solutions of $mx'' + kx = 0$. The spring-mass equation has periodic solutions that oscillate with frequency $\sqrt{k/m}$.

Stop and Think

5.13 Carry out the details of substituting $x(t) = \sin rt$ into $mx'' + kx = 0$ to verify that $r = \sqrt{k/m}$. Repeat with $x(t) = \cos rt$.

5.14 Use this same process to show that the system form of the model (5.4) has the same solutions, $x(t) = \sin \sqrt{k/m}\, t$ and $x(t) = \cos \sqrt{k/m}\, t$. What is the velocity expression that corresponds to each of these position functions?

5.15 Neither *linear* nor *homogeneous* nor *general solution* has been defined for the scalar second-order equation (5.6). Nonetheless, you can make educated guesses. Do you think (5.6) is linear or nonlinear? Homogeneous or nonhomogeneous? What might be the form of its general solution?

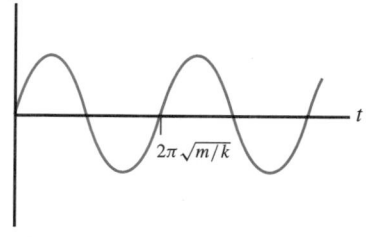

FIGURE 5.9   A graph of $\sin \sqrt{k/m}\, t$, one solution of $mx'' + kx = 0$.

Figure 5.9 shows a graph of $\sin \sqrt{k/m}\, t$. Its shape is consistent with the concavity required by $x'' = -(k/m)x$. When $x(t) > 0$, then $x''(t) < 0$; i.e., the graph is concave down. The graph is concave up when $x(t) < 0$. When the graph crosses the horizontal axis, there is an inflection point.

Stop and Think

5.16 What is the *period* (time between peaks) of the oscillations shown in figure 5.9? Does increasing the mass lead to faster or slower oscillations? Does making the spring stiffer (less deflection from the same force) lead to faster or slower oscillations?

5.17 What is the period of the spring-mass system described in example 1? What is its frequency of oscillation; that is, how many complete cycles does it execute per unit time?

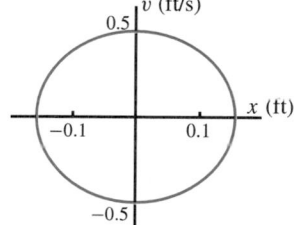

FIGURE 5.10   Phase plane diagram of the solution of the undamped system $x' = v$, $3v'/16 = -2x$, of example 1.

MATLAB

DELAB can find analytic solutions of second-order equations like $mx'' + kx = 0$. For guidance, select **Help, Textbook**, then go to chapter 5, analytic solution.

Does the solution of $mx'' + kx = 0$ provided by DELAB contain any new terms that were not analyzed in the previous discussion?

The phase plane diagram of figure 5.10 shows another representation of the periodic solution of the undamped spring-mass system $x' = v$, $3v'/16 = -2x$ of example 1. Its plot of velocity v vs. position x is a closed curve because the motion is periodic.

Stop and Think

5.18 Locate the point on the phase plane in figure 5.10 that corresponds to the initial conditions of example 1, $x(0) = 0$, $v(0) = -0.5$.

5.19 How would this phase plane diagram change if the initial values were $x(0) = 0$, $v(0) = +0.5$? If they were $x(0) = 0$, $v(0) = -1.5$? $x(0) = 0.1$, $v(0) = 0$?

5.20 In which direction is the closed curve of figure 5.10 traversed as time increases, clockwise or counterclockwise? Use physical intuition or the governing system of equations, $x' = v$, $3v'/16 = -2x$, to decide.

5.21 Real spring-mass systems will eventually come to rest as internal friction, air resistance, and other effects dissipate their initial energy. How would such a gradual slowing down be represented in a phase plane diagram?

MATLAB

The graphical tools in DELAB can draw phase plane diagrams. For guidance, select **Help, Textbook**, then go to chapter 5, figure 5.10.

Damped

Substitution shows that neither $\sin \sqrt{k/m}\, t$ nor $\cos \sqrt{k/m}\, t$ is a solution of the second-order damped spring-mass equation (5.9). How do solutions of this equation behave?

Stop and Think   **5.22** Verify by direct substitution that neither $\sin \sqrt{k/m}\, t$ nor $\cos \sqrt{k/m}\, t$ is a solution of $mx'' + px' + kx = 0$.

5.23 Try guessing a solution of $mx'' + px' + kx = 0$ that has the form $x(t) = \sin rt$, r a constant. What goes wrong? Does that same problem occur with $x(t) = \cos rt$?

With no ready analytic solution at hand, we could use a numerical method to find the behavior of the solution of a specific model, say that of the damped assembly considered in example 2. Rewritten as a system of two first-order equations, it is

$$x' = v',$$

$$\tfrac{3}{16} v' = -2x - 1.41v,$$

$$x(0) = 0.75, \quad v(0) = 0.$$

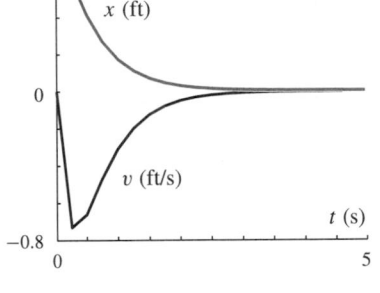

FIGURE 5.11 Numerical approximation to the solution of $x' = v$, $3v'/16 = -2x - 1.41v$, $x(0) = 0.75$, $v(0) = 0$, the damped model of example 2.

The characteristic equation method can also find an analytic solution of a constant coefficient equation like that of example 2. See chapter 6.

Figure 5.11 shows a graph of the numerical approximation to the solution of this initial-value problem. (It was obtained using fourth-order Runge-Kutta with $\Delta t = 0.25$.) Surprisingly, it shows no oscillations whatsoever!

Stop and Think   **5.24** Why aren't the curves in figure 5.11 smoother? They seem to be made up of short line segments.

MATLAB

Use MATLAB's DELAB to produce your own version of figure 5.11. For guidance, select **Help, Textbook**, then go to chapter 5, figure 5.11.

MATLAB (CONTINUED)

Figure 5.11 used $\Delta t = 0.25$ with the fourth-order Runge-Kutta method. Choosing a larger value of Δt, say $\Delta t = 0.5$, seems to produce a much different picture. (Try it!) Why can you be reasonably confident that the approximate solution obtained with the smaller value of Δt is the more accurate one? What additional numerical experiments could you conduct to build confidence in the approximate solution shown in figure 5.11?

Stop and Think **5.25** The system in example 1 oscillates. The system in example 2 does not. The governing equations are nearly the same: $3x''/16 + 2x = 0$ in example 1, $3x''/16 + 1.41x' + 2x = 0$ in example 2. What *physical* effect accounts for the *mathematical* difference between the two governing equations? Could such a physical effect prevent oscillation?

Forced

Numerical tools provide a picture of the response of the forced spring-mass system modeled in example 3. Figure 5.12 was obtained using fourth-order Runge-Kutta with $\Delta t = 0.2$.

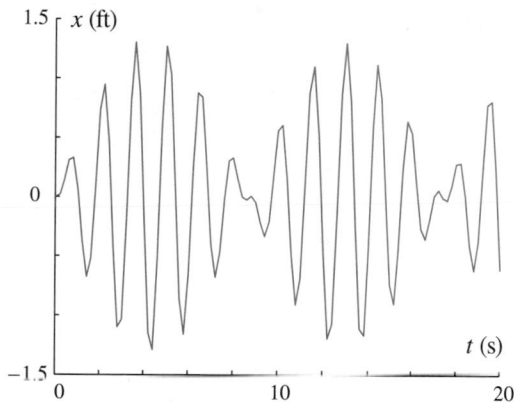

FIGURE 5.12 Numerical approximation to the solution of $x' = v$, $v'/8 = -2x + (\sin 3\pi t/2)/2$, $x(0) = 0$, $v(0) = 0$, the forced model of example 3.

Figure 5.12 seems full of surprises. Two different oscillations appear, one with a relatively short period that seems roughly consistent with the forcing frequency, the other with a longer period that has no obvious explanation. The amplitude of the motion of the mass is about 1.5 *feet* even though the top of the spring is only moving about 3 *inches* in each direction. The analytic tools of the next chapter will help to explain some of these features.

Stop and Think **5.26** Estimate the period of the fast and the slow oscillations appearing in figure 5.12. Is the slow oscillation consistent with the period of the forcing function (wiggling the top of the spring) or with the natural vibrations of the spring-mass system without forcing (when the top of the spring is held stationary)? Could the slow oscillation be due to beating, the source of the wavering sound you hear when two slightly out-of-tune musical instruments play the same note simultaneously?

MATLAB

Use the numerical tools in DELAB to draw your own version of figure 5.12. Convince yourself that the choice of $\Delta t = 0.2$ with fourth-order Runge-Kutta is producing a reliable approximate solution.

5.1.6 Exercises

EXERCISE GUIDE	
To gain experience . . .	**Try exercises**
Deriving spring-mass models	7–9, 10(b–c), 11, 12(c), 13
Writing and using spring-mass models	1, 6, 10(d)
With force relations (Hooke, damping, etc.)	2, 10(a), 12(a–b)
With solutions of spring-mass models	3–4, 12(e), 14, 17
Analyzing and interpreting solution behavior	5, 12(d), 15–16, 18–19

1. When a mass of 3 kg is suspended from a spring, the end of the spring moves 0.24 m. A force of 0.36 newton is required to drag a mass of 7 kg across a tabletop at a speed of 0.4 m/s.

   (a) Suppose a 2-kg mass is suspended vertically from the spring (as in figure 5.1) and that the mass is set in motion by pulling it down 4 cm and releasing it from rest. Write an initial-value problem whose solution will describe the position of the mass at any time t.

   (b) Suppose the spring is connected to the 7-kg mass on the tabletop as shown in figure 5.4 and the mass is set in motion by giving it a speed of 0.2 m/s to the right. Write an initial-value problem whose solution will describe the position of the mass at any time t.

   (c) The upper end of the suspended system of part (a) is now connected to the turntable of figure 5.7. Write an initial-value problem whose solution will describe the position of the mass at any time t if the mass is set in motion as in part (a).

2. Does a rubber band obey Hooke's law? Conduct an experiment using coins or similar objects as weights.

3. The undamped spring-mass equation can be written

$$x'' = -\frac{k}{m}x.$$

Since x has to be the negative of its second derivative, the text concludes that x could be a sine or a cosine function. Determine the precise form of those functions by assuming that $x = \sin rt$ and substituting into the differential equation. Obtain a (characteristic) equation for r and solve it to determine a functional form for x. Repeat for the cosine function.

4. The text argues that the undamped spring-mass equation $mx'' + kx = 0$ has the solutions $\sin \sqrt{k/m}\, t$ and $\cos \sqrt{k/m}\, t$.

   (a) Verify this claim by substituting these proposed solutions into the differential equation.

   (b) Show that $C_1 \sin \sqrt{k/m}\, t + C_2 \cos \sqrt{k/m}\, t$ is also a solution for any arbitrary constants C_1, C_2. (You are verifying the *principle of superposition* for this second-order equation.)

   (c) Solve the undamped spring-mass model

$$mx'' + kx = 0, \quad x(0) = x_i, \quad x'(0) = x_i,$$

by choosing C_1, C_2 in the solution of part (b) to satisfy the given initial conditions.

5. The text claims that the undamped spring-mass equation $mx'' + kx = 0$ has $\sin \sqrt{k/m}\, t$ as a solution.

(a) Sketch a graph of this function for $0 \leq t \leq 2\pi/\sqrt{k/m}$.

(b) How many cycles of the sine function does your graph contain?

(c) What is the period of this function? What does this model predict to be the period of oscillation of the mass?

(d) Does making the spring stiffer cause the mass to oscillate faster or slower? What happens if the mass is changed? If your car bounces too rapidly on a rough road, should you replace its springs with stiffer ones or weaker ones? If you can not change the springs, should you ask your passengers to get out and walk or should you fill the trunk with rocks?

6. The forced spring-mass equation (5.13) has the form

$$mx'' + \cdots = k(h - \cdots .$$

(a) What are the units of kh?

(b) Someone hands you the equation $6x'' + 2x = A \sin \omega t$. What might it model? Describe the physical situation as specifically as you can. What does A represent?

7. The text derived the model

$$mx'' + kx = k(h(t) - h(0)), \quad x(0) = x_i, \quad x'(0) = v_i,$$

for the forced, undamped vertical spring-mass system. It also considered a spring-mass system without forcing that was damped by friction. It obtained the model

$$mx'' + px' + kx = 0, \quad x(0) = x_i, \quad x'(0) = v_i.$$

In fact, even a freely suspended mass is subject to damping due to air resistance and internal resistance in the spring. These damping forces can be modeled as being proportional to velocity and acting in the opposite direction, just like the force of friction acting on the mass sliding horizontally.

(a) Compare the undamped, forced model with the damped model without forcing. Predict what a model of a forced, damped system would look like.

(b) Verify your prediction by deriving such a model. Specifically, consider the system shown in figure 5.5 and suppose that the mass is subject to air resistance which is proportional to its velocity and acts in the opposite direction.

8. Figure 5.1, which shows the vertical spring-mass system, places the mass below the fixed end of the spring. Place the fixed end of the spring below the mass, like the spring in a stationary pogo stick, and derive a model for this situation.

9. Figure 5.4, which shows the horizontal spring-mass system, has the fixed end of the spring to the right of the mass. Place the fixed end of the spring to the left of the mass and derive a model for this situation.

10. A shock absorber in an automobile can be thought of as a damped version of the vertical spring-mass system of figure 5.1. The mass of the car is m, and it is attached to a piston moving in a chamber filled with hydraulic fluid. See figures 5.13, 5.14. The hydraulic fluid exerts a force on the mass that is proportional to its velocity and *opposes* its motion.

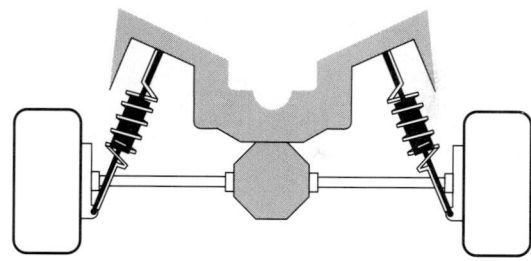

FIGURE 5.13 A view of a sport car's rear suspension, showing the damping shock absorbers inside coil springs.

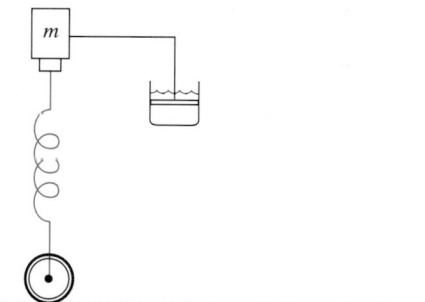

FIGURE 5.14 A schematic model of an automobile shock absorber system. The lower end of the spring is attached to a wheel of the car. The piston moves in oil to damp the motion of the mass m, which represents the automobile.

(a) Write a mathematical expression for the force the hydraulic fluid exerts on the moving mass.

(b) Use the expression in part (a) to derive a model of the automobile shock absorber system shown in figure 5.14. Be sure to define a coordinate system.

(c) Derive a model for this same system if the fixed (lower) end of the spring in figure 5.14 is moved to the position $h(t)$ at time t by driving the car over a rough road.

(d) What other model(s) considered in this section have the same mathematical structure as the one you just derived? Would a model of this form be suitable if we wanted to include the effects of air resistance or internal friction in the spring? Would you expect any differences in the relative values of the damping coefficients when you compare a shock absorber model with one involving air resistance?

11. Suppose the anchor for the horizontal spring-mass system of figure 5.4 is replaced by the turntable of figure 5.7, as shown in figure 5.15. Derive a model for this situation. (Neglect the vertical displacement of the end of the spring, as in the last example in this section.)

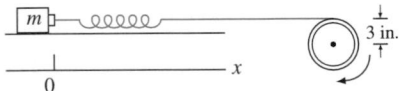

FIGURE 5.15   The vertical turntable of figure 5.7 is now oscillating the end of the horizontal spring-mass system that was formerly fixed. The spring is attached to the turntable 3 in. from its center.

12. A 12-in. hacksaw blade is clamped in a horizontal position. The height of its free end above the worktable is measured as quarters are placed on that end. The situation is shown in figure 5.16, and the corresponding data appear in table 5.1.

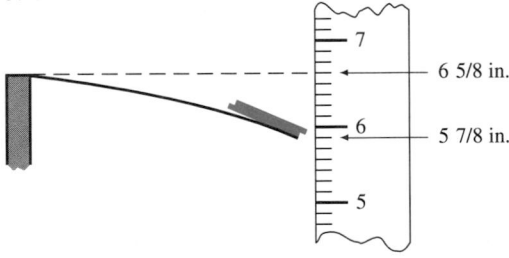

FIGURE 5.16   A hacksaw blade deflects from its horizontal position (shown dashed) when two quarters are laid on its free end.

(a) Is it reasonable to suppose that the displacement of the end of the hacksaw blade is proportional to the force applied? Is such a relationship exact here?

(b) If the displacement is proportional to the force applied, find the proportionality constant.

TABLE 5.1

Number of Quarters	Height above Table
0	6 5/8 in.
1	6 1/4 in.
2	5 7/8 in.
3	5 1/2 in.
4	5 3/16 in.

(c) Derive a model for the position of the end of the hacksaw blade when three quarters are taped to it and it is released from rest from a position 6 in. above the worktable. Be sure to define a coordinate system.

(d) Use the model equation to determine the period of the oscillations of the quarters taped to the end of the hacksaw blade.

(e) Use a numerical method of your choice from DELAB to approximate the solution of the initial-value problem that models the motion of the tip of the hacksaw blade. Is the period the same as part (d)?

13. The left end of the hacksaw blade of figure 5.16 is now being oscillated gently about its old fixed position of 6 5/8 in. above the worktable. Specifically, its height as a function of time is

$$h(t) = 6.625 + 0.5 \sin \omega t,$$

where the frequency ω is constant.

(a) How far above its old fixed position does the left end of the hacksaw blade move?

(b) Two quarters are taped to the right end of the blade. Derive a model for the position of this end of the blade if the quarters are at rest before the left end of the blade begins to move. (See exercise 12.)

14. Examples 1 and 2 derive spring-mass models of the form $3x''/16 + px' + 2x = 0$. The equation of example 1 has $p = 0$ and its solutions oscillate, that of example 2 has $p = 1.41$ and its solutions appear not to oscillate. Write this equation as a first-order system. Then use a numerical method of your choice from DELAB to answer the following:

(a) Do solutions of $x' = v$, $3v'/16 + px' + 2x = 0$ exhibit oscillations for small positive values of p?

(b) If the answer to the previous question is *yes*, estimate the value of p at which oscillations in the solutions of $x' = v$, $3v'/16 + px' + 2x = 0$ disappear and pure decay sets in.

15. The following argument shows that *undamped spring-mass systems conserve energy*. It uses the governing equations $x' = v$, $mv' = -kx$.

(a) Show that multiplying $mv' = -kx$ by v leads to $mvv' + kxx' = 0$.

(b) Argue that $\int mvv' dt = mv^2/2 + C$, $\int kxx' dt = kx^2/2 + C$, and, hence, that

$$\frac{mv^2}{2} + \frac{kx^2}{2} = C$$

for some constant C. Recall that the first expression is the kinetic energy of the mass, the second the potential energy stored in the spring.

(c) Evaluate C in terms of the initial values $x(0)$, $v(0)$. Explain why conservation of energy has been proven.

16. Generalize the argument of the previous exercise to show that *damped spring-mass systems do not conserve energy*. Begin with the governing system $x' = v$, $v' = -kx - pv$ and derive

$$\frac{mv^2}{2} + \frac{kx^2}{2} = C - p \int_0^t v^2 \, dt.$$

Evaluate the constant C and explain why dissipation of energy has been proven.

17. Figure 5.11, page 214, shows an approximate solution of the damped spring-mass initial-value problem of example 2. The figure suggests that solutions of the governing equation might be decaying exponentials. The equation itself, $3x''/16 + 1.41x' + 2x = 0$, has constant coefficients, and we have been able to find exponential solutions of *first*-order differential equations by the method of characteristic equations.

Try the same approach here, even though this equation is of second order. Assume a solution of the form $x = e^{rt}$, r a constant, substitute, and find values of r. Have you *proven* that *every* solution of $3x''/16 + 1.41x' + 2x = 0$ involves decaying exponentials?

18. If the top of the spring in example 3 were held stationary, then the forced system modeled there would reduce to an undamped system like that of example 1. The governing equation would be $x''/8 + 2x = 0$. What is the period of the oscillations predicted by this equation? Do oscillations of this period appear in figure 5.12, page 215, the plot of the approximate solution of the forced system?

19. Use a numerical method of your choice from DELAB to explore the behavior of a forced system like that of example 3, page 211, as the forcing frequency varies. Specifically, suppose the speed of the turntable in that example can be varied. Write the governing equation as

$$\frac{1}{8}x'' + 2x = \frac{1}{2}\sin \omega t$$

and study the variation in the amplitude of the motion with forcing frequency ω. (Note that ω has units of rad/s; $\omega/2\pi$ has the familiar engineering units of hertz or cycles per second.) What value of ω seems to produce the most violent response? How is that value related to the frequency of the natural oscillations of the system without forcing, which is modeled by $x''/8 + 2x = 0$?

5.2 ■ THE PENDULUM

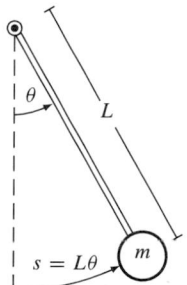

FIGURE 5.17 A diagram of a pendulum.

The pendulum in a grandfather clock swings to and fro with comforting regularity, releasing the escapement mechanism at precisely set intervals so the clock can keep time accurately. What determines the period (the elapsed time) of an oscillation? Is it the size of the weight? The length of the pendulum? How hard the weight is pushed when the clock is started? A model of the motion of a pendulum will answer some of those questions, at least to the extent that the mathematical model accurately represents reality.

Figure 5.17 shows a diagram of a pendulum. A mass m is suspended from an arm of length L, whose mass we will neglect. The arm is attached to a pivot which we will assume is frictionless. The angle of the arm from the vertical is denoted θ, and it is measured counterclockwise in *radians*. (A positive angle is motion to the right; negative, to the left.)

As figure 5.18 suggests, the instantaneous motion of the mass is tangent to its circular path. The displacement of the mass is the distance $s = L\theta$ it has traveled along that circular path.

If θ is measured in radians, then the distance described by an angle θ along the circumference of a circle of radius L is just $L\theta$; e.g., if $\theta = 2\pi$, then $s = 2\pi L$, the circumference of the circle.

Since L is constant, the velocity of the mass is $v = Ld\theta/dt$ and its acceleration is $a = Ld^2\theta/dt^2 = L\theta''$. We emphasize that velocity and acceleration are *tangent* to the path of the mass.

To derive a model, use Newton's law,

THE LAW

$$F_t = ma,$$

and a familiar experimental fact:

THE FACT

Experimental fact.   The force of gravity is proportional to mass.

But since all motion is tangent to the path of the mass, we must consider the components of acceleration and gravity force in the tangential direction.

In the counterclockwise-is-positive orientation of figure 5.18, the force of gravity acting *tangent* to the path of the mass is $F_g = -mg \sin\theta$. We have just argued that the acceleration of the mass in the tangential direction is $L\theta''$. Equating total force in Newton's law to the force of gravity, the only force acting on the mass, yields

$$mL\theta'' = -mg \sin\theta$$

or

$$L\theta'' + g \sin\theta = 0. \tag{5.15}$$

We can impose two initial conditions, the angle θ_i and the angular velocity ω_i of the pendulum at the instant of release.

A common notation for angular velocity is ω, similar to v for linear velocity. Angular velocity is usually measured in radians per unit time; an angular velocity of 2π rad/min is 1 rev/min (rpm).

Using that notation, Newton's law can be written $F_t = mL\omega'$ and the pendulum equation (5.15) can be written instead as the system

$$\theta' = \omega,$$
$$L\omega' = -g \sin\theta.$$

The complete model for the pendulum is

$$L\theta'' + g \sin\theta = 0, \quad \theta(0) = \theta_i, \quad \theta'(0) = \omega_i.$$

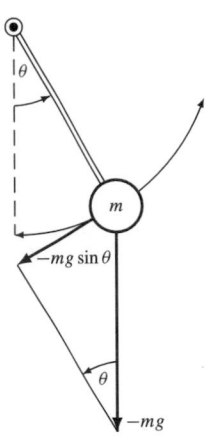

FIGURE 5.18   The gravity force $-mg \sin\theta$ is tangent to the path of the mass when the pendulum is at angle θ.

PENDULUM MODEL

Stop and Think

5.27   Write the pendulum model as a system of first-order differential equations.

5.28   Compare the pendulum system you just wrote with the spring-mass model (5.1–5.2). Identify similarities and differences in the mathematical forms.

Since the mass m of the pendulum bob appears nowhere in the model, we conclude that the frequency of the ticks in a grandfather clock are independent of the mass of the bob. But the length of the arm and the acceleration of gravity do affect the motion.

If you study a clock's pendulum, you will often find an adjustment screw below the mass to permit small changes in the length of the pendulum arm.

Since the acceleration of gravity g also affects the motion of a pendulum, grandfather clocks would not be good time keepers on the space shuttle, even if size were not an issue.

Although we have not formally defined *second-order linear equation*, your experience with first-order equations should persuade you that the pendulum equation $L\theta'' + g \sin\theta = 0$ is probably not linear. But we can linearize this equation using the approximation $\sin\theta \approx \theta$; e.g.,

$$\sin(0.1 \text{ rad}) = 0.0998334\ldots.$$

Then we obtain the approximate model

APPROXIMATE PENDULUM MODEL

$$L\theta'' + g\theta = 0, \quad \theta(0) = \theta_i, \quad \theta'(0) = \omega_i.$$

The basis for the approximation $\sin\theta \approx \theta$ is the same as that of linear stability analysis: We hope we can neglect powers of small terms. In this case, we are neglecting the terms $\theta^3, \theta^5, \ldots$ in the Taylor polynomial expansion of $\sin\theta$,

$$\sin\theta = \theta - \frac{1}{3!}\theta^3 + \frac{1}{5!}\theta^5 + \cdots.$$

This approximate pendulum model has exactly the same mathematical form as the undamped spring-mass model (5.6–5.7),

$$mx'' + kx = 0, \quad x(0) = x_i, \quad x'(0) = v_i.$$

The exercises ask you to make precise the analogy between the two.

5.2.1 Preliminary Analysis

Like the undamped spring-mass equation, the approximate pendulum equation can be written to suggest what its solutions might be:

$$\theta'' = -\frac{g}{L}\theta.$$

Give or take a constant, θ must be a function whose second derivative is its own negative, such as a sine or cosine. A little more guessing leads to the conclusion that $\sin\sqrt{g/L}\,t$ and $\cos\sqrt{g/L}\,t$ are solutions of $L\theta'' + g\theta = 0$. So the approximate pendulum equation has periodic solutions that oscillate with frequency $\sqrt{g/L}$.

Ignoring friction in the pivot and simplifying the original model equation has permitted determining (at least approximately) the frequency of oscillation of a simple pendulum.

MATLAB

Ask DELAB to find an analytic solution of the approximate pendulum equation $L\theta'' + g\theta = 0$. Does it provide a solution that contains oscillating terms consistent with the previous discussion?

Can DELAB find a solution of the pendulum equation $L\theta'' + g\sin\theta = 0$? What does that result suggest to you about the importance of graphical and numerical tools?

Of course, frictionless pivots exist only as an ideal. Exercise 8 asks you to show that incorporating the damping effects of friction lead to the *damped pendulum model*

DAMPED PENDULUM MODEL
$$L\theta'' + p\theta' + g\sin\theta = 0, \quad \theta(0) = \theta_i, \quad \theta'(0) = \omega_i,$$

and its linear approximation

APPROXIMATE DAMPED
PENDULUM MODEL
$$L\theta'' + p\theta' + g\theta = 0, \quad \theta(0) = \theta_i, \quad \theta'(0) = \omega_i.$$

Stop and Think   **5.29** Write each of these models as first-order systems.

5.2.2 Exercises

EXERCISE GUIDE	
To gain experience . . .	**Try exercises**
With mechanical analogies	1
Deriving pendulum models	7–9
With solutions of pendulum models	2–4, 10
With the accuracy of the approximate model	5(a), 6
Analyzing and interpreting solution behavior	5(b)

1. The text claims that the approximate pendulum model

$$L\theta'' + g\theta = 0, \quad \theta(0) = \theta_i, \quad \theta'(0) = \omega_i,$$

and the undamped spring-mass model (5.6–5.7),

$$mx'' + kx = 0, \quad x(0) = x_i, \quad x'(0) = v_i,$$

are analogous. Verify this claim by finding the analog of $L, g, \theta, \theta', \theta'', \theta_i$, and ω_i. Explain the physical significance of each analogy. Summarize your results in a table.

2. The approximate pendulum equation can be written

$$\theta'' = -\frac{g}{L}\theta.$$

Since θ has to be the negative of its second derivative, the text concludes that θ could be a sine or a cosine function. Determine the precise form of the sine function by assuming that $\theta = \sin rt$ and substituting into the differential equation. Obtain a (characteristic) equation for r and solve it to determine a functional form for θ. Repeat for the cosine function.

3. The text argues that the approximate pendulum equation $L\theta'' + g\theta = 0$ has the solutions $\sin\sqrt{g/L}\,t$ and $\cos\sqrt{g/L}\,t$.

(a) Verify this claim by substituting these proposed solutions into the differential equation.

(b) Show that $C_1\sin\sqrt{g/L}\,t + C_2\cos\sqrt{g/L}\,t$ is also a solution for any arbitrary constants C_1 and C_2. (You are verifying the *principle of superposition* for this second-order equation.)

(c) Solve the approximate pendulum model

$$L\theta'' + g\theta = 0, \quad \theta(0) = \theta_i, \quad \theta'(0) = \omega_i,$$

by choosing C_1 and C_2 in the solution of part (b) to satisfy the given initial conditions.

(d) Do the initial values θ_i and ω_i affect the frequency of the pendulum's oscillations? Is the way we start a clock's pendulum important?

4. The text claims that the approximate pendulum equation $L\theta'' + g\theta = 0$ has $\sin\sqrt{g/L}\,t$ as a solution.

   (a) Sketch a graph of this function for $0 \le t \le 2\pi/\sqrt{g/L}$.

   (b) How many cycles of the sine function does your graph contain?

   (c) What is the period of this function? What does this model predict to be the period of oscillation of the pendulum?

   (d) Does lengthening its pendulum cause a grandfather clock to run faster or slower? If the clock were moved to the moon where the gravity constant g is smaller, would the clock run faster or slower?

5. A typical grandfather clock might have a pendulum about 30 in. long. There is usually room for it to swing about 6 in. on either side of vertical.

   (a) How accurate is the approximation $\sin\theta \approx \theta$ for such a pendulum? (A Taylor polynomial can provide a general expression for the error.)

   (b) Use the results of the previous problem to determine the period of the oscillations of such a pendulum. Is this period consistent with your observations?

6. For what range of values of θ is the approximation $\sin\theta \approx \theta$ accurate to within 1%? 5%? What are these limiting angles in degrees?

7. The text derived the pendulum model using radians to measure the angular deviation of the pendulum from vertical.

   (a) Carefully derive the same model with angular deviation measured in *degrees*. (In the interest of good notation, you should give the angle measured in degrees a name other than θ, such as ϕ.)

   (b) Show that your model and the text's are equivalent by changing the dependent variable in your model to an angle measured in radians; e.g., substitute $\phi = 180\,\theta/\pi$.

8. The text neglected friction in the pivot bearing when it derived the pendulum model. Assume that the force of friction in the pivot is proportional to angular velocity θ' and acts to oppose the motion. Derive the **damped pendulum model**

   $$L\theta'' + p\theta' + g\sin\theta = 0, \quad \theta(0) = \theta_i, \quad \theta'(0) = \omega_i.$$

   What precisely does the term $p\theta'$ represent?

9. The previous problem asks you to derive a model for a pendulum that includes friction in the pivot. Find the corresponding approximate model.

10. Does the exact pendulum equation $L\theta'' + g\sin\theta$ support oscillations? Give an opinion supported by numerical evidence from DELAB using a method (as well as values of L and g) of your choice. Compare the period of any oscillations you observe with those predicted by the corresponding approximate pendulum equation $L\theta'' + g\theta = 0$.

5.3 ■ *RLC* CIRCUITS

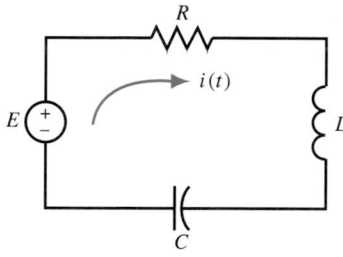

FIGURE 5.19 A series *RLC* circuit with an independent voltage source E.

The mechanical spring-mass systems of section 5.1 have a precise electrical analog, an *RLC* circuit, a circuit containing a resistance R, an inductance L, and a capacitance C. Figure 5.19 illustrates a series *RLC* circuit, which also includes an *imposed emf* (electromotive force) or an *independent voltage source* E. The charge on the capacitor in this circuit is the analog of the displacement of a mass attached to a spring.

We begin with a brief summary of the important physical ideas.

Current is the rate of flow of charged particles, and current flows between two points in a circuit because of a potential difference between those two points. The unit of charge is the *coulomb* (C). The unit of current is the *ampere* (A), 1 C/s.

Electric potential is the ability to move charged particles, and potential differences are measured in *volts* (V). For example, the imposed voltage E in figure 5.19 could be a battery. It lifts or pumps positively charged particles from its negative (lower potential) end to its positive (higher potential) end. The charged particles at its positive end are prepared to flow "downhill" toward its negatively charged end.

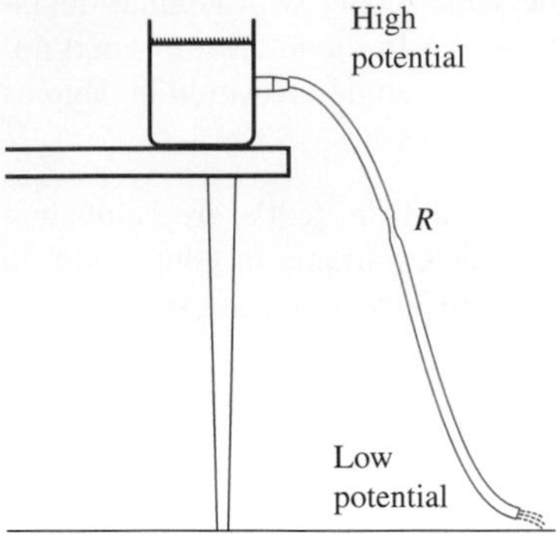

High
potential

R

Low
potential

FIGURE 5.20   The bucket of water on
the table is like a charged battery.

The situation is like placing a bucket of water on top of a table, for
electrical potential is analogous to mechanical potential energy. The water in
the bucket in figure 5.20 flows downhill toward the lower end of the hose,
the lower potential part of this fluid circuit. Lifting the bucket onto the table,
raising its potential, is like charging a battery.

The narrow constriction in the hose is a resistance. It slows the flow of
water and dissipates its potential energy just as the resistor *R* in the circuit
of figure 5.19 slows the flow of charged particles, causing a potential drop.
The faster the rate of flow, the greater the potential drop. A precise statement
follows.

Experimental fact 1.   The potential drop across a resistor is proportional to the rate of flow.

The proportionality constant is called the **resistance** *R*. If *q* denotes
charge, then dq/dt is the rate of flow of charged particles, or **current** *i*. The
potential change or voltage drop v_R across a resistor can be written as

$$v_R = R\frac{dq}{dt} = iR.$$

The potential that resistors dissipate appears as heat and light energy in toasters,
lightbulbs, etc.

The capacitor *C* in figure 5.19 is also a cause of a potential drop. It
consists of two conducting plates separated by an insulator, or *dielectric*. The
dielectric is a kind of dam for the flow of charged particles. When too much
charge builds up on one side of the dam, charged particles back up through the
circuit, trying to reach the other side.

Experimental fact 2.   The potential drop across a capacitor is proportional to the charge that has accumulated on the capacitor.

The reciprocal of the constant of proportionality is the **capacitance** *C*.
The potential change or voltage drop v_C across a capacitor can be written as

$$v_C = \frac{1}{C}q.$$

Like a dam holding water in a river, capacitors can store charge. When Ben Franklin
flew his kite in a lightning storm, he stored the charge that traveled down the
string of his kite in a Leyden jar, a capacitor.

Inductors are coils of wire. They resist the flow of charged particles
through a kind of feedback mechanism. Changing currents in the coil induce
magnetic fields that in turn induce an opposing current in the inductor. It is a
bit like turning some of the flow in a hose back on itself, but it is the *rate of
change* of the flow that provokes the resistance.

Experimental fact 3.   The potential drop across an inductor is proportional to the rate of change of current in the inductor.

The constant of proportionality is the **inductance** L. The potential change or voltage drop v_L across an inductor can be written as

$$v_L = L\frac{di}{dt}.$$

Since changing electric currents produce magnetic fields, magnetic compasses are useless near electric motors. Indeed, it is precisely those magnetic fields that turn the flow of charged particles into mechanical energy. Conversely, the charged particles that flow in a video display tube sometimes can be affected by the magnetic field from a nearby, powerful loudspeaker.

These three experimental facts are summarized in table 5.2.

TABLE 5.2 **Voltage drops across a resistor, a capacitor, and an inductor, as well as the units used to measure the size of each.**

Circuit Element	Units	Voltage Drop
Resistance R	Ohm (Ω)	$v_R = R\,dq/dt$
Capacitance C	Farad (F)	$v_C = q/C$
Inductance L	Henry (H)	$v_L = L\,di/dt$

The basic physical law, the analog of Newton's law, is actually a statement of conservation of energy:

Kirchhoff's voltage law. The sum of the voltage changes around a closed loop is zero.

To use Kirchhoff's voltage law, we simply traverse the loop once in a chosen direction, accounting for each of the voltage changes we encounter.

Traveling the circuit from the negative side of the battery to its positive side, the voltage *drop* is negative $(-E)$ because the battery can pump positively charged particles from its negative side to its positive side. Recalling the water analogy of figure 5.20, we can think of the battery as picking up charged particles from the floor (negative potential) and placing them on the table (positive potential) so that they can flow through the hose. The three experimental facts in table 5.2 tell us how to account for the voltage drops at the resistor, inductor, and capacitor.

Traveling the circuit of figure 5.19 in a clockwise direction and summing voltage drops, Kirchhoff's voltage law yields

$$-E + v_R + v_L + v_C = 0.$$

Substituting from the experimental facts, we have

$$-E + Ri + L\frac{di}{dt} + \frac{1}{C}q = 0.$$

Since current is the rate of change of charge $(i = dq/dt)$, we can write a system of first-order differential equations,

$$\frac{dq}{dt} = i,$$

$$L\frac{di}{dt} = E - Ri - \frac{1}{C}q.$$

Stop and Think   **5.30** Compare this *RLC* system with the spring-mass model (5.1–5.2). Identify similarities and differences in the mathematical forms.

Alternatively, we can substitute $i = q'$ and replace the first-order system with a single second-order equation

$$Lq'' + Rq' + \frac{1}{C}q = E.$$

The mechanical analogy is now evident. Comparing this *RLC* circuit equation with the equation for a damped, forced spring-mass system,

$$mx'' + px' + kx = k(h(t) - h(0)),$$

we see that resistance R is the analog of the friction coefficient p, charge q is the analog of displacement x, and so on. The exercises ask you to pursue this analogy in detail.

The bookkeeping for Kirchhoff's voltage law is like that for Newton's law. When we modeled spring-mass systems in section 5.1, we accumulated all the forces on the mass and set them equal to mass times acceleration. Experimental facts told us how forces due to friction and the spring depended on position and velocity. Here Kirchhoff's voltage law told us how to accumulate the various voltage drops around the circuit (they sum to zero), and the experimental facts told us how to express those voltage drops in terms of charge and current.

Adding a specification of initial charge q_i on the capacitor and initial current i_i completes the model:

SERIES *RLC* CIRCUIT MODEL
(CHARGE)
$$Lq'' + Rq' + \frac{1}{C}q = E, \quad q(0) = q_i, \quad q'(0) = i_i.$$

Stop and Think   **5.31** Since the charge on the capacitor is proportional to the voltage drop across it, an alternate form of the initial condition on charge is $q(0) = Cv_c(0)$.

Since $q' = i$, a derivative of the charge equation yields an equation for the current in this loop,

$$Li'' + Ri' + \frac{1}{C}i = E'.$$

Reasonable initial conditions for this equation are the current at $t = 0$, $i(0) = i_i$, and its rate of change at $t = 0$. From the third experimental fact, we can express $i'(0)$ in terms of the initial voltage drop across the inductor,

$$Li'(0) = v_L(0).$$

SERIES *RLC* CIRCUIT MODEL (CURRENT)

Hence, a model for the current in the series *RLC* circuit is

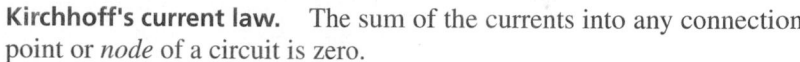

$$Li'' + Ri' + \frac{1}{C}i = E', \quad i(0) = i_i, \quad i'(0) = v_L(0)/L.$$

The two models—charge and current—have the same mathematical form.

To gain a feel for the behavior of these two similar equations, eliminate sources and resistance by setting $E = 0$ and $R = 0$. Then the current equation, for example, can be written

$$i'' + \frac{1}{LC}i = 0.$$

Either by analogy with the solutions $\sin\sqrt{k/m}\,t$ and $\cos\sqrt{k/m}\,t$ of the spring-mass equation $mx'' + kx = 0$ or by directly substituting $\sin rt$ or $\cos rt$ into $i'' + i/LC = 0$, we find that $r = 1/\sqrt{LC}$. Hence, the *RLC* circuit equation with $R = 0 = E$ has periodic solutions that oscillate with frequency $1/\sqrt{LC}$.

Figure 5.21 illustrates a different configuration of these three components, a source-free, parallel *RLC* circuit, the dual of the series *RLC* circuit with $E = 0$. The exercises ask you to derive a governing equation for this circuit using the following statement of conservation of charge.

Kirchhoff's current law. The sum of the currents into any connection point or *node* of a circuit is zero.

The equation you obtain there is

$$Cv'' + \frac{1}{R}v' + \frac{1}{L}v = 0.$$

It is obviously analogous to the series current and charge equations derived before.

> The circuit in figure 5.19 is a *series* circuit because current flows serially from one component to the next; the current flow in each component is the same. The circuit in figure 5.21 is a *parallel* circuit because the current flows in parallel, or simultaneously, through all the components; the voltage drop across all components is the same.

> The voltage source in figure 5.19 is called *independent* because the voltage drop across its terminals is independent of the current flowing through it.

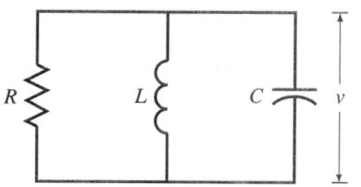

FIGURE 5.21 A source-free, parallel *RLC* circuit.

5.3.1 Exercises

EXERCISE GUIDE	
To gain experience . . .	Try exercises
With electrical and mechanical analogies	1
Deriving circuit models	5–6, 8
With solutions of circuit models	2–4
Analyzing and interpreting solution behavior	7–8

1. The text claims there is an obvious analogy between *RLC* circuits modeled by the equation

$$Lq'' + Rq' + \frac{1}{C}q = E$$

and forced, damped spring-mass systems modeled by

$$mx'' + px' + kx = k(h(t) - h(0)).$$

Verify this claim by finding the analogs of L, R, C, E, q, q'. Explain the physical significance of each analogy. Summarize your results in a table.

2. With $R = 0$, the series *LC* current equation can be written

$$i'' = -\frac{1}{LC}i.$$

Since i has to be the negative of its second derivative, the text concludes that i could be a sine or a cosine function. Determine the precise form of those functions by assuming that $i = \sin rt$ and substituting into the differential equation. Obtain a (characteristic) equation for r and solve it to determine a functional form for i. Repeat for the cosine function.

3. The text argues that the equation $Li'' + i/C = 0$ governing the current in a series *LC* circuit without forcing has the solutions $\sin(t/\sqrt{LC})$ and $\cos(t/\sqrt{LC})$.

(a) Verify this claim by substituting these proposed solutions into the differential equation.

(b) Show that $C_1 \sin(t/\sqrt{LC}) + C_2 \cos(t/\sqrt{LC})$ is also a solution for any arbitrary constants C_1, C_2. (You are verifying the *principle of superposition* for this second-order equation.)

(c) Solve the model for an *LC* circuit without forcing,

$$Li'' + \frac{1}{C}i = 0, \quad i(0) = i_i, \quad i'(0) = v_L(0)/L,$$

by choosing C_1, C_2 in the solution of part (b) to satisfy the given initial conditions.

4. The text claims that the equation $Li'' + i/C = 0$ for an *LC* circuit without forcing has $\sin(t/\sqrt{LC})$ as a solution.

(a) Sketch a graph of this function for $0 \le t \le 2\pi\sqrt{LC}$.

(b) How many cycles of the sine function does your graph contain?

(c) What is the period of this function? What does this model predict to be the period of oscillation of the charge on the capacitor in this circuit?

(d) Would increasing the size of the capacitor cause a graph of the oscillations in current versus time to be pushed tighter together or to be more spread out? What if the inductor is enlarged, increasing L by adding more turns to the coil?

(e) Suppose the inductor is actually the magnet coil in a speaker. How would the changes described in part (d) affect the sound you hear?

(The following three problems first appeared in the chapter exercises for chapter 2.)

5. Use the experimental facts in the text and Kirchhoff's voltage law to derive a model for the charge q on the capacitor in the *RC* circuit shown in figure 5.22. These steps will lead you to the desired model:

(a) Write a mathematical statement of each experimental fact.

(b) Write Kirchhoff's voltage law in terms of v_R, v_C.

(c) Combine the results of the preceding two steps to derive the governing differential equation

$$R\frac{dq}{dt} + \frac{1}{C}q = 0.$$

Complete the model for a series *RC* circuit by imposing an appropriate initial condition.

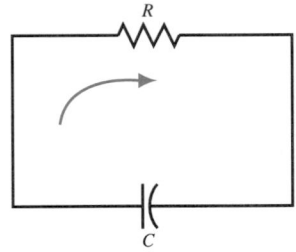

FIGURE 5.22   A source-free, series *RC* circuit.

6. Suppose the *RC* circuit of the preceding problem now includes a voltage source (imposed emf) E. Derive the following model for this circuit:

$$R\frac{dq}{dt} + \frac{1}{C}q = E, \quad q(0) = q_i.$$

See figure 5.23.

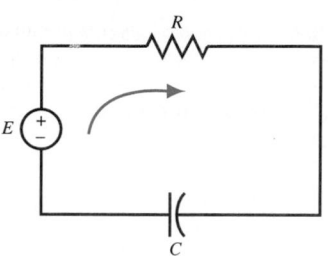

FIGURE 5.23 A series RC circuit with a voltage source E.

7. Analyze the models of the RC circuits in the two preceding problems by answering the following questions for each:

(a) By studying the sign of dq/dt in the differential equation, determine whether a given initial charge q_i on the capacitor will grow or decay. (Both R and C are positive constants.)

(b) By studying the solution of each initial-value problem, determine whether a given initial charge q_i on the capacitor will grow or decay.

(c) If charge stops flowing in either circuit, what is the steady-state charge on the capacitor?

8. The preceding problems deal with RC circuits. Figure 5.24 shows an unusual RC circuit, Ben Franklin's kite flying in

a lightning storm. The lightning arises from a voltage difference between the atmosphere and the kite. Current can flow down the wet kite string to the Leyden jar, a kind of capacitor, on the ground. In principle, you could calculate the capacitance of the Leyden jar and the resistance of the kite string. Derive a model that could help determine the potential difference that caused the lightning bolt. Describe how you would use the model to estimate this potential difference.

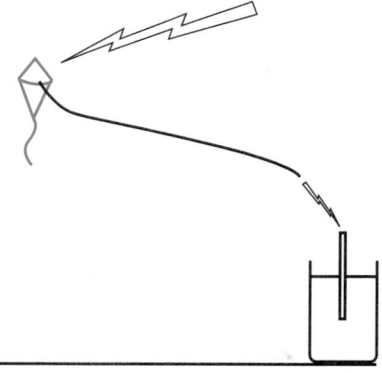

FIGURE 5.24 A schematic diagram of Ben Franklin's kite flying in a lightning storm; charge accumulates in the Leyden jar at the lower right.

5.4 ■ CHAPTER EXERCISES

EXERCISE GUIDE	
To gain experience . . .	**Try exercises**
Deriving and analyzing mechanical models	1–2, 6–7
Deriving and analyzing electrical models	3
Analogies among models	4–5

1. The model (5.6–5.7),

$$x'' + kx = 0, \quad x(0) = x_i, \quad x'(0) = v_i,$$

for the vertical spring-mass system was derived using a coordinate system whose origin is at the rest position of the mass; see figure 5.1, page 203.

(a) Introduce an alternate coordinate system y whose origin is at the lower end of the spring *before* the weight

is suspended from it. Derive the model

$$my''(t) + ky(t) = -mg, \quad y(0) = y_i, \quad y'(0) = v_i,$$

where $y(t)$ is the position of the mass in the new coordinate system at time t, y_i is its initial position, and v_i is its initial velocity.

(b) Since the model in part (a) and the model (5.6–5.7) describe the same motion, you should be able to convert one initial-value problem into the other. Show that you can by finding a relation between x, the coordinate of figure 5.1, and y, the coordinate used in part (a). (*Hint:* The origin of the y-axis is located at x_u in the x coordinate system.)

Substitute the expression for y into the model of part (a) and show that it reduces to (5.6–5.7). Reverse the process and obtain the model of part (a) by substituting into (5.6–5.7).

2. Figure 5.25 illustrates a conceptual model of an automobile suspension. The mass is the car, and the "fixed" end of the spring is connected to a wheel traversing a sinusoidal road. Derive a model for the vertical motion of the car if it is traveling this road at speed S.

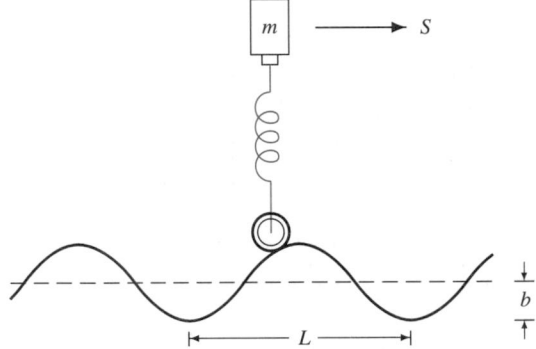

FIGURE 5.25   This spring-mass system is a conceptual model of an automobile suspension. The mass m represents the automobile. The bumps in the sinusoidal road are L units apart, and each is b units above their average level. The car is moving with speed S.

3. As directed in the following, use

Kirchhoff's current law. The sum of the currents into any connection point, or *node*, of a circuit is zero.

and the experimental facts given in the text to derive the equation governing the parallel RLC circuit of figure 5.21, page 227.

(a) Use Kirchhoff's current law to sum the currents i_R, i_C, i_L through the resistor, capacitor, and inductor where they enter the upper node in figure 5.21, page 227.

(b) Show that the experimental facts governing resistors, capacitors, and inductors can also be written

$$i_R = \frac{1}{R}v, \quad i_C = C\frac{dv}{dt},$$

$$i_L = \frac{1}{L}\left(\int_0^t v(\tau)\,d\tau - v(0)\right),$$

where v is the voltage drop across each of the three components.

(c) Combine the law with the experimental facts to obtain the governing differential equation

$$Cv'' + \frac{1}{R}v' + \frac{1}{L}v = 0.$$

4. Someone hands you the equation $L\theta'' + g\theta = 0$ and says, "This really could be a model of a spring-mass system. The mass is L, the spring constant is g, and θ is the distance the mass has moved from its equilibrium, not an angle." Is such an assertion reasonable? If so, some term in the equation must represent the force of the spring on the mass. Which term is it? Does that term seem to provide a physically reasonable representation of the force exerted by a spring stretched to various lengths?

5. Repeat the previous exercise using the exact pendulum equation $L\theta'' + g\sin\theta = 0$.

6. Figure 5.26 shows a schematic diagram of the passenger-seat suspension system in an automobile; m is the mass of the passenger, k is the spring constant of the spring in the seat, M is the mass of the car, and K is the spring constant of the car's suspension. Derive a model for the positions of the passenger and the automobile.

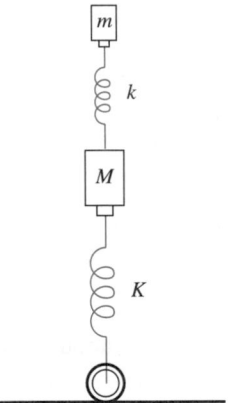

FIGURE 5.26   A schematic diagram of the passenger-seat suspension system in an automobile.

7. The lower end of the spring in figure 5.26 represents the point where the car's suspension contacts the road. Make the model of the preceding problem more realistic by allowing the coordinate of that point to vary with time.

5.5 ■ CHAPTER PROJECTS

1. **Nonlinear spring.** The idealization of Hooke's law that force is proportional to extension does not remain valid for large displacements. For sufficiently large extensions, the resisting force increases more rapidly than linearly. For example, as the coils of a spring are stretched nearly straight, the spring will resist even the smallest additional displacement with a substantially increased force. The last entry in table 5.1 of exercise 12, page 218, is an example of this phenomena. (What would that entry be if force remained proportional to displacement?) A common modification to accommodate this increase in force is the nonlinear force-extension law

$$F_s = kz + az^3,$$

where a is typically much smaller than k.

(a) What sign should a have? Sketch the force-extension relationship for both positive and negative values of z and argue that spring force is approximately proportional to extension for small values of z.

(b) Derive a model similar to (5.6–5.7), page 207, for the vertical, undamped spring-mass system using this force-extension relationship.

(c) Although *linear* is not yet defined for second-order differential equations, guess an appropriate definition by analogy with the first-order case. Does the differential equation obtained in part (b) appear to be linear or nonlinear?

(d) Using your choice of a numerical method from DELAB and parameter values m, k, and a, explore whether nonlinear spring-mass systems can support oscillations. Does the period of any oscillations you observe appear to depend upon amplitude? Is the period of oscillations you observe close to the period predicted by the corresponding linear equation, $mx'' + kx = 0$?

2. **Bridge cables.** Common wisdom holds that the Tacoma Narrows bridge collapsed because of linear resonant vibrations induced by high winds, much as in the classic advertisement, "Is it real or is it Memorex?" in which a glass shatters as it resonates to a singer's voice. (See section 6.7 for more about resonance.) Recent work by Lazer and McKenna ([16]) suggests instead that bridges can collapse because of the nonlinear response of the cables that suspend and supposedly stabilize the roadbed.

The response of the springs we have considered is linear; $F_s = kz$ means that spring force is proportional to extension. Compression and extension are resisted equally, with only the sign of the spring force changing. A cable, however, is a *nonlinear spring*. It resists only extension. That is, its spring force law is

$$F_s = kz^+ = \begin{cases} kz, & z \geq 0 \\ 0, & z < 0. \end{cases}$$

(The superscript $+$ is defined by $z^+ = z$ if $z \geq 0$, $z^+ = 0$ if $z < 0$.)

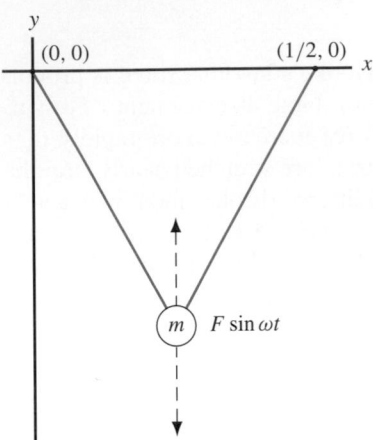

FIGURE 5.27   A mass m suspended from cables of equal unstretched length ℓ_0.

Figure 5.27 shows a mass m suspended from two cables, a simple prototype of a bridge suspension system. When the cables are modeled so that they resist only extension, the governing equations are

$$m\frac{d^2x}{dt^2} = -k\left(\sqrt{x^2 - y^2} - \ell_0\right)^+ \frac{x}{\sqrt{x^2 + y^2}}$$

$$+ k\left(\sqrt{(1/2 - x)^2 + y^2} - \ell_0\right)^+ \frac{1/2 - x}{\sqrt{(1/2 - x)^2 + y^2}}$$

$$- p\frac{dx}{dt}, \tag{5.16}$$

$$m\frac{d^2y}{dt^2} = -k\left(\sqrt{x^2 - y^2} - \ell_0\right)^+ \frac{y}{\sqrt{x^2 + y^2}}$$

$$- k\left(\sqrt{(1/2 - x)^2 + y^2} - \ell_0\right)^+ \frac{y}{\sqrt{(1/2 - x)^2 + y^2}}$$

$$- p\frac{dy}{dt} + F \sin \omega t - mg. \tag{5.17}$$

Here k is the coefficient in the spring force law of the cable, p is a damping coefficient, ℓ_0 is the unstretched length of the cables, and F and ω are the magnitude and frequency, respectively, of a vertical force applied directly to the mass.

This system exhibits an extraordinarily rich range of responses. In anticipation of further studies of its behavior in a later project, justify each of the terms in the governing equations (5.16–5.17). In addition, supply appropriate initial conditions.

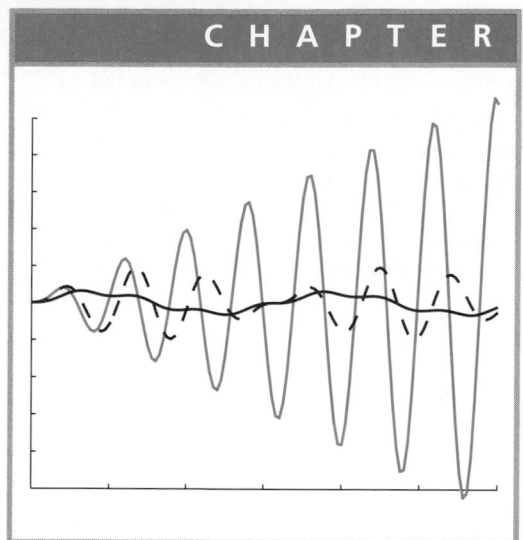

6

Analytic Tools for Two Dimensions

Analyzing the models of oscillatory phenomena developed in the previous chapter demands the mathematical concepts and solution procedures described here. Much of the work is a natural generalization of first-order ideas. This chapter concentrates on analytic tools for a single second-order equation; chapter 8 provides an analogous development for systems of first-order equations.

6.1 ▪ BASIC DEFINITIONS

6.1.1 Linear, Homogeneous, Etc.

The series of definitions that follow apply to second-order equations. Many of these ideas are straightforward generalizations of their first-order analogs.

Definition 1. *The function u is a **solution** on the interval $x_0 \leq x \leq x_1$ of the second-order differential equation $d^2y/dx^2 = f(x, y, dy/dx)$ if u is twice continuously differentiable and if*

$$\frac{d^2u}{dx^2} = f\left(x, u, \frac{du}{dx}\right) \quad \text{for } x_0 \leq x \leq x_1.$$

*Further, u is a **solution of the initial-value problem***

$$\frac{d^2y}{dx^2} = f\left(x, y, \frac{dy}{dx}\right), \quad y(x_0) = y_i, \quad y'(x_0) = z_i,$$

on the interval $x_0 \leq x \leq x_1$ if u is a solution of the differential equation and if

$$u(x_0) = y_i, \quad \left.\frac{du}{dx}\right|_{x=x_0} = z_i.$$

A second-order equation in the form $u'' = f(x, u, u')$ can be written easily as a first-order system. Define a second variable $v = u'$. Then use $u'' = v'$ to write the equivalent system

$$u' = v$$
$$v' = f(x, u, v).$$

Stop and Think   **6.1** If u were the position of a mass suspended from a spring, what would v represent? Compare the two forms of the spring-mass model, (5.1–5.2) and (5.3), page 206.

6.2 Write the damped pendulum equation $L\theta'' + p\theta' + g\sin\theta = 0$ as a first-order system.

Definition 2. *A **linear, second-order equation** is a second-order differential equation that can be written in the form*

LINEAR EQUATION

$$a_2(x)y'' + a_1(x)y' + a_0(x)y = r(x). \tag{6.1}$$

Of the unknown and its derivatives up to order two, exactly one can appear in any term of the equation, and that unknown must appear raised to the first power only, not as the argument of a nonlinear function. The expression $r(x)$ is sometimes called the *forcing function*. A second-order equation that is not linear is **nonlinear**.

Note the parallel with the general form of the linear first-order equation,

$$a_1(x)y' + a_0(x)y = r(x).$$

Exercise 20 asks you to develop the concept of a linear second-order differential operator.

Stop and Think   **6.3** Give an example of a first-order equation that is linear; nonlinear.

Recall some definitions from section 4.1: A differential equation is **homogeneous** if the identically zero function is a solution. Otherwise, the equation is **nonhomogeneous**. For the linear equation $a_2(x)y'' + a_1(x)y' + a_0(x)y = r(x)$, the *corresponding homogeneous equation* is $a_2(x)y'' + a_1(x)y' + a_0(x)y = 0$.

Stop and Think   **6.4** Give an example of a first-order equation that is homogeneous; nonhomogeneous.

6.5 Verify that the "corresponding homogeneous equation" $a_2(x)y'' + a_1(x)y' + a_0(x)y = 0$ is indeed homogeneous.

A second-order linear equation has **constant coefficients** if the coefficients a_2, a_1, and a_0 in the general form (6.1) are all constants. Otherwise, it has **variable coefficients**.

Stop and Think   **6.6** Give an example of a first-order equation that has constant coefficients; variable coefficients.

■ **EXAMPLE 1**   *Classify the following as linear or nonlinear, homogeneous or nonhomogeneous, and constant or variable coefficient.*

- The undamped, unforced spring-mass equation of section 5.1.1,

Linear, homogeneous, constant coefficient

$$mx''(t) + kx(t) = 0,$$

is linear; choose $a_2 = m$, $a_1 = 0$, $a_0 = k$, $r = 0$, and identify the unknown x in the preceding equation with the dependent variable y in the general form (6.1) and the independent variable t with x in (6.1). This equation is homogeneous because $x(t) = 0$ is a solution, and it has constant coefficients $a_2 = m$, $a_1 = 0$, $a_0 = k$.

- The damped, unforced spring-mass equation of section 5.1.2,

Linear, homogeneous, constant coefficient

$$mx''(t) + px'(t) + kx(t) = 0,$$

is linear, homogeneous, and has constant coefficients $a_2 = m$, $a_1 = p$, $a_0 = k$.

- The damped, forced spring-mass equation of section 5.1.3,

Linear, nonhomogeneous, constant coefficient

$$mx''(t) + px'(t) + kx(t) = k(h(t) - h(0)),$$

is linear (choose $a_2 = m$, $a_1 = p$, $a_0 = k$, $r = k(h(t) - h(0))$), and it has constant coefficients $a_2 = m$, $a_1 = p$, $a_0 = k$. However, it is nonhomogeneous because substituting the trial solution $x(t) = 0$ in this equation yields

$$0 = k(h(t) - h(0)),$$

an equality that is clearly not valid unless $k = 0$ (no spring) or $h(t) \equiv h(0)$ (no movement of the upper end of the spring).

- The more general spring-mass equation derived in exercise 15 of section 5.1.6,

Nonlinear, homogeneous

$$mx'' + kx + ax^3 = 0,$$

is certainly homogeneous, for the trial solution $x(t) = 0$ reduces this equation to an equality.

It is nonlinear because of the ax^3 term. Comparing the given equation with the general linear equation (6.1) suggests that choosing $a_2(t) = m$, $a_1(t) = 0$, $a_0(t) = ax^2$. However, the function a_0 in (6.1) can depend *only* on the independent variable, which is t in this equation. The choice $a_0 = ax^2$ causes a_0 to depend upon the dependent variable x. Similarly, we can not choose $r(t) = ax^3$. The term ax^3 can not be associated with any other term in (6.1) because the unknown x is raised to a power other than the first.

Since *constant coefficient* is defined only for linear equations, it has no meaning here.

• The pendulum equation

$$L\theta'' + g \sin\theta = 0,$$

derived in section 5.2, is homogeneous because substituting $\theta = 0$ reduces it to an identity. It is nonlinear because the general linear form (6.1) certainly cannot accommodate the sine of the dependent variable. ■

6.1.2 The Form of General Solutions

We might expect that second-order equations have two arbitrary constants in their solutions. Roughly speaking, one appears for each antiderivative needed to solve the equation. A **second-order initial-value problem** couples a second-order equation with *two* pieces of initial data. One is the value of the solution, and the other is the value of its derivative at a specified time. (Other types of initial data can occur, but we will seldom encounter such situations.)

For example, the forced, damped spring-mass system model of section 5.1,

$$mx''(t) + px'(t) + kx(t) = k(h(t) - h(0)),$$
$$x(0) = x_i, \quad x'(0) = v_i,$$

is a linear second-order initial-value problem. We will see that the general solution (which we have yet to define) of this differential equation involves two arbitrary constants that are determined by the two initial conditions.

Recall that a general solution of a linear *first*-order homogeneous equation is just a constant multiple of any nontrivial solution of the homogeneous equation. The arbitrary constant is determined by an initial condition, if one is given.

Stop and Think   **6.7** Is the population equation $P' - kP = 0$ linear? If so, write its general solution.

For example, $y = e^{-x}$ is a nontrivial solution of the first-order equation

$$y'(x) + y(x) = 0.$$

A general solution of this homogeneous equation is

$$y_g(x) = Ce^{-x},$$

where C is an arbitrary constant.

The situation is more complicated for second-order equations. The linear homogeneous second-order equation

$$y'' + y' - 12y = 0$$

has two nontrivial solutions

$$y_1(x) = e^{-4x}, \quad y_2(x) = e^{3x}.$$

(They can be found by the characteristic equation method.) It is tempting to say simply, "A general solution of the *second*-order homogeneous equation $y'' + y' - 12y = 0$ is

$$y_h(x) = C_1 e^{-4x} + C_2 e^{3x},$$

a sum of constant multiples of *two* nontrivial solutions of the given equation."

But another pair of nontrivial solutions of this differential equation is e^{-4x} and $8e^{-4x}$, and an equally good candidate for a general solution might be

$$C_1 e^{-4x} + C_2 8 e^{-4x}.$$

You might object that the nontrivial solutions used here are not "different" in the same way that e^{-4x} and e^{3x} are different. But then you need to be precise about the meaning of *different*.

One sign that the two functions in this last solution are not sufficiently different is the observation that the sum $C_1 e^{-4x} + C_2 8 e^{-4x}$ can be written

$$(C_1 + 8C_2)e^{-4x},$$

an expression that really contains only a single arbitrary constant, $C_1 + 8C_2$. The two functions e^{-4x} and $8e^{-4x}$ are not different. They are constant multiples of one another, and any sum containing them will reduce to one involving only a single arbitrary constant. But with only one arbitrary constant we can, in general, satisfy only one initial condition.

Stop and Think   **6.8** Are e^{-4x} and e^{2-4x} solutions of $y'' + y' - 12y = 0$? Are they different in the sense that $C_1 e^{-4x} + C_2 e^{2-4x}$ can not be collapsed to a form that involves only a single arbitrary constant?

Constructing a general solution of a linear homogeneous second-order equation requires two nontrivial solutions that are *different in the sense that the two arbitrary constants will not collapse into one*. This kind of "different" is called *linearly independent*.

Definition 3. *Two functions $y_1(x)$ and $y_2(x)$ are **linearly independent** on the interval $a \le x \le b$ if*

$$C_1 y_1(x) + C_2 y_2(x) = 0, \quad a \le x \le b,$$

*implies $C_1 = C_2 = 0$. Otherwise, they are **linearly dependent** on $a \le x \le b$.*

The expression $C_1 y_1 + C_2 y_2$ is called a **linear combination** of y_1 and y_2. If the interval is obvious from the context, then explicit reference to it can be omitted.

The condition
$$C_1 y_1 + C_2 y_2 = 0 \implies C_1 = C_2 = 0$$
is just a statement that the two functions can not be multiples of one another. (Read $\implies$ as *implies*.) Unfortunately, the idea of avoiding functions that are constant multiples of one another does not generalize easily to more than two functions, but this definition of linear independence does.

■ **EXAMPLE 2**   *Show that the functions $y_1(x) = e^{-4x}$, $y_2(x) = e^{3x}$ are linearly independent on $-\infty < x < \infty$.*

To establish this claim, assume the contrary: assume that the two functions are linearly *dependent*. Then there are constants C_1 and C_2, not both zero, satisfying

$$C_1 e^{-4x} + C_2 e^{3x} = 0, \quad -\infty < x < \infty.$$

Suppose $C_2 \neq 0$. Then simple division immediately yields

$$\frac{C_1}{C_2} = -e^{7x}.$$

Although the quotient on the left is constant, the exponential on the right is certainly not constant as x ranges over all real numbers. A similar contradiction occurs if we assume $C_1 \neq 0$. Hence, e^{-4x} and e^{3x} are linearly independent over the entire real line. ■

■ **EXAMPLE 3**   *Show that the functions e^{-4x} and $8e^{-4x}$ are linearly dependent on $-\infty < x < \infty$.*

We must exhibit nonzero constants C_1 and C_2 satisfying

$$C_1 e^{-4x} + C_2 8 e^{-4x} = 0$$

for $-\infty < x < \infty$. Dividing this last expression by e^{-4x} yields

$$C_1 = -8C_2.$$

Choosing $C_1 = -8$, $C_2 = 1$, for example, certainly satisfies $C_1 e^{-4x} + C_2 8 e^{-4x} = 0$ for all real x. Since these values of C_1 and C_2 are nonzero, the functions e^{-4x} and $8e^{-4x}$ are linearly dependent on $-\infty < x < \infty$. ■

Stop and Think

6.9 Does the same argument establish that e^{-4x} and e^{2-4x} are linearly dependent on $-\infty < x < \infty$?

6.10 Make up a similar example of linearly dependent solutions of $y'' + y' - 12y = 0$ using e^{3x}.

A pair of linearly independent solutions of a homogeneous linear equation are the building blocks of a *general solution*, a flexible, all-purpose solution that includes *every* possible solution of the differential equation (providing $a_2 \neq 0$).

The homogeneous solution is a linear combination of independent solutions of the homogeneous equation.

HOMOGENEOUS SOLUTION

Definition 4. *If y_1 and y_2 are linearly independent solutions of a homogeneous linear second-order equation on a given interval, then the **general solution of the homogeneous equation** on that interval, denoted y_h, is an arbitrary linear combination of those two solutions:*

$$y_h(x) = C_1 y_1(x) + C_2 y_2(x).$$

Rather than the cumbersome phrase *general solution of the homogeneous equation*, y_h is often (but inaccurately) called the **homogeneous general solution** or just the **homogeneous solution**.

Stop and Think **6.11** The combination $C_1y_1 + C_2y_2$ is called a general *solution* of the linear homogeneous differential equation, but is it obvious that this combination is actually a *solution*? Does calling it a solution make it a solution?

Since y_1 and y_2 are linearly independent, neither of them can be the trivial solution. If one were identically zero, then the constant that multiplies it in the linear combination $C_1y_1 + C_2y_2 = 0$ could be any nonzero quantity, thereby rendering the pair linearly dependent.

■ **EXAMPLE 4** *Construct a general solution of the homogeneous equation $y'' + y' - 12y = 0$ that is valid on the interval $-\infty < x < \infty$. Verify that the general solution actually solves $y'' + y' - 12y = 0$.*

Find the values of the arbitrary constants in the general solutions that recover the solutions $y \equiv 0$, $y = 12e^{-4x}$, and $y = -e^{3x}$. Find the values of the arbitrary constants that give the solution satisfying the initial conditions $y(0) = -1$, $y'(0) = -10$.

Example 2 established that e^{-4x} and e^{3x} are linearly independent on $-\infty < x < \infty$. Substitution shows that each function is a solution of $y'' + y' - 12y = 0$. Hence, a general solution of this homogeneous equation is $y_h = C_1e^{-4x} + C_2e^{3x}$.

Substitute to verify that $y_h = C_1e^{-4x} + C_2e^{3x}$ solves $y'' + y' - 12y = 0$:

$$y_h'' + y_h' - 12y_h$$
$$= (C_1e^{-4x} + C_2e^{3x})'' + (C_1e^{-4x} + C_2e^{3x})' - 12(C_1e^{-4x} + C_2e^{3x})$$
$$= C_1(16e^{-4x} - 4e^{-4x} - 12e^{-4x}) + C_2(9e^{3x} + 3e^{3x} - 12e^{3x}) = 0.$$

To recover the trivial solution $y \equiv 0$ from y_h, choose $C_1 = C_2 = 0$. To recover $y = 12e^{-4x}$, choose $C_1 = 12$, $C_2 = 0$. To recover $y = -e^{3x}$, choose $C_1 = 0$, $C_2 = -1$.

To satisfy the initial conditions $y(0) = -1$, $y'(0) = -10$, require that y_h satisfy

$$y_h(0) = C_1 + C_2 = -1,$$
$$y_h'(0) = -4C_1 + 3C_2 = -10.$$

Solving the pair of equations $C_1 + C_2 = -1$, $-4C_1 + 3C_2 = -10$ yields $C_1 = 1$, $C_2 = -2$. The solution of the initial-value problem $y'' + y' - 12y = 0$ is $y(x) = e^{-4x} - 2e^{3x}$. ■

Stop and Think **6.12** Is $C_1e^{-4x} + C_28e^{-4x}$ a general solution of $y'' + y' - 12y = 0$?

To extend the notion of the all-encompassing general solution to nonhomogeneous linear equations, add one more term.

General solution of *nonhomogeneous equation = particular solution + homogeneous solution*

Definition 5. *Let y_p be any solution of a nonhomogeneous linear second-order equation. If y_h is a general solution of the corresponding homogeneous equation, then*

GENERAL SOLUTION

$$y_g(x) = y_p(x) + y_h(x)$$

*is a **general solution of the nonhomogeneous equation**. The interval of validity of this general solution is the smaller of the intervals upon which y_p is a particular solution of the nonhomogeneous equation and y_h is a general solution of the homogeneous equation.*

The solution y_p of the nonhomogeneous equation is called a **particular solution**, as in "any particular solution."

■ **EXAMPLE 5**   *Construct a general solution of the nonhomogeneous equation*

$$\frac{1}{8}x''(t) + 2x(t) = \frac{1}{2}\sin\frac{3\pi t}{2}$$

governing the forced spring-mass system of example 3 of section 5.1, and construct a general solution of the corresponding homogeneous equation,

$$\frac{1}{8}x''(t) + 2x(t) = 0.$$

We will steal a few first-order ideas to solve this second-order equation.

Begin with the homogeneous equation $x''/8 + 2x = 0$. Either by using the characteristic equation method to be developed in section 6.3 or by guessing solutions of the form $x = \sin rt$ and $x = \cos rt$ for some constant r, we find two solutions,

$$x_1(t) = \sin 4t, \quad x_2(t) = \cos 4t.$$

Stop and Think   **6.13** Substitute x_1, x_2 into the equation and verify that they are indeed solutions.

To construct a general solution of the homogeneous equation, first show that $\sin 4t$ and $\cos 4t$ are linearly independent on the interval of interest for these initial-value problems, $t \geq 0$. Assume there are constants C_1 and C_2 such that

$$C_1 \sin 4t + C_2 \cos 4t = 0$$

for $t \geq 0$. Since the left-hand side of this last expression is identically zero, its derivative is identically zero as well,

$$4C_1 \cos 4t - 4C_2 \sin 4t = 0.$$

These two equations for C_1 and C_2,

$$C_1 \sin 4t + C_2 \cos 4t = 0,$$
$$4C_1 \cos 4t - 4C_2 \sin 4t = 0,$$

certainly have the solution $C_1 = C_2 = 0$, as you can verify by direct substitution. The theory of linear equations guarantees that *there is no solution other than this trivial solution if the determinant of coefficients*

$$W = \begin{vmatrix} \sin 4t & \cos 4t \\ 4\cos 4t & -4\sin 4t \end{vmatrix}$$

is not zero. But this determinant is

Value of 2×2 determinant = product of main diagonal − product of off-diagonal entries

$$W = (\sin 4t)(-4 \sin 4t) - (4 \cos 4t)(\cos 4t)$$
$$= -4 \sin^2 4t - 4 \cos^2 4t = -4 \neq 0, \quad t \geq 0.$$

Consequently, the only constants C_1, C_2 satisfying $C_1 \sin 4t + C_2 \cos 4t = 0$ for all $t \geq 0$ are $C_1 = C_2 = 0$. Hence, $\sin 4t$, $\cos 4t$ are linearly independent on $t \geq 0$.

Stop and Think

6.14 To see that the determinant W does not arise mysteriously, solve the system

$$C_1 \sin 4t + C_2 \cos 4t = 0,$$
$$4C_1 \cos 4t - 4C_2 \sin 4t = 0$$

for C_1 by multiplying the first equation by $-4 \sin 4t$ (the coefficient of C_2 in the second equation) and the second equation by $\cos 4t$ (the coefficient of C_2 in the first equation). Subtract the second equation from the first to obtain $W C_1 = 0$. Hence, $C_1 = 0$ if $W \neq 0$.

6.15 Repeat a variant of these steps to show that $C_2 = 0/W$.

Since these two functions are linearly independent, a general solution of the homogeneous equation $x''/8 + 2x = 0$ is

$$x_h(t) = C_1 \sin 4t + C_2 \cos 4t$$

for $t \geq 0$.

To derive a particular solution, adapt undetermined coefficients to find that

Guess $x_p(t) = A \sin(3\pi t/2)$, substitute, and solve for A.

$$x_p(t) = \frac{16}{64 - 9\pi^2} \sin \frac{3\pi t}{2} \approx -0.644 \sin \frac{3\pi t}{2}$$

solves the nonhomogeneous equation $x''/8 + 2x(t) = (1/2) \sin(3\pi t/2)$. Hence, for $t \geq 0$, a general solution of this nonhomogeneous equation is

$$x_g(t) = x_p(t) + x_h(t)$$

$$\approx -0.644 \sin \frac{3\pi t}{2} + C_1 \sin 4t + C_2 \cos 4t. \quad ■$$

MATLAB

Use MATLAB's DELAB to find an analytic solution of one of the linear equations appearing in example 1. Does it return a general solution?

6.1.3 Superposition

Is a general solution really a solution? Why is the homogeneous part of a general solution multiplied by an arbitrary constant but not the particular solution? Answers lie in an important property of linear equations that was introduced in section 4.1, the principle of superposition.

Superposition: a sum of solutions solves the linear equation that has the sum of the forcing terms.

Theorem 1 (Principle of superposition). *If $y(x)$ is a solution of the linear equation*

$$a_2(x)y'' + a_1(x)y' + a_0(x)y = r(x)$$

with forcing term $r(x)$ and if $z(x)$ is a solution of the same differential expression with forcing term $s(x)$,

$$a_2(x)z'' + a_1(x)z' + a_0(x)z = s(x),$$

then the linear combination $u(x) = C_1 y(x) + C_2 z(x)$ is a solution of the differential equation with forcing term $C_1 r(x) + C_2 s(x)$,

$$a_2(x)u'' + a_1(x)u' + a_0(x)u = C_1 r(x) + C_2 s(x).$$

Proof. Substitute and evaluate $a_2(x)u'' + a_1(x)u' + a_0(x)u$. ■

Stop and Think   **6.16** Carry out the details of the proof of the principle of superposition.

When we constructed a general solution of the homogeneous equation $x''/8 + 2x = 0$, we formed a linear combination of the two functions $\sin 4t$ and $\cos 4t$, each a solution of the equation with a forcing term of zero. Consequently, the principle of superposition asserts that the general solution

$$u = C_1 \sin 4t + C_2 \cos 4t$$

satisfies the equation

$$\frac{1}{8}u'' + 2u = C_1 \cdot 0 + C_2 \cdot 0 = 0;$$

i.e., the general solution $C_1 \sin 4t + C_2 \cos 4t$ satisfies the homogeneous equation. *The principle of superposition shows that a general solution is indeed a solution!*

Constructing a general solution of the nonhomogeneous equation

$$\frac{1}{8}x'' + 2x = \frac{1}{2}\sin\frac{3\pi t}{2}$$

follows a similar pattern. The general solution $x_h(t)$ of the homogeneous equation satisfies an equation with zero forcing term, and the particular solution $x_p(t)$ satisfies an equation with forcing term $(1/2)\sin(3\pi t/2)$. The principle of superposition guarantees that the linear combination

$$u = x_p(t) + x_h(t),$$

which is the general solution of the nonhomogeneous equation, satisfies the nonhomogeneous equation

$$\frac{1}{8}u''(t) + 2u(t) = \frac{1}{2}\sin\frac{3\pi t}{2} + 0 = \frac{1}{2}\sin\frac{3\pi t}{2};$$

i.e., the general solution $x_p + x_h$ of the nonhomogeneous equation actually satisfies the nonhomogeneous equation.

■ **EXAMPLE 6** *The preceding discussion used the principle of superposition to show that the general solution $x_h(t) = C_1 \sin 4t + C_2 \cos 4t$ of the unforced spring-mass equation*

$$\frac{1}{8}x''(t) + 2x(t) = 0$$

is actually a solution of this equation. Verify by direct substitution that the principle of superposition is correct.

Substituting the general solution $x_h(t) = C_1 \sin 4t + C_2 \cos 4t$ into the differential equation yields

$$\frac{1}{8}x_h''(t) + 2x_h(t)$$

$$= C_1\left(-\frac{1}{8}16\sin 4t + 2\sin 4t\right) + C_2\left(-\frac{1}{8}16\cos 4t + 2\cos 4t\right)$$

$$= C_1 \cdot 0 + C_2 \cdot 0 = 0.$$

The principle of superposition is correct. *Adding two solutions of a homogeneous equation produces a solution of the homogeneous equation.* ■

■ **EXAMPLE 7** *Again verify by direct substitution that the principle of superposition is correct, this time using the general solution*

$$x_g(t) = \frac{16}{64 - 9\pi^2}\sin\frac{3\pi t}{2} + x_h(t)$$

of the forced spring-mass equation

$$\frac{1}{8}x''(t) + 2x(t) = \frac{1}{2}\sin\frac{3\pi t}{2},$$

where $x_h(t) = C_1 \sin 4t + C_2 \cos 4t$ is the general solution of the homogeneous equation $x''/8 + 2x = 0$. Recall that $16/(64 - 9\pi^2) \approx -0.644$.

Substituting the general solution x_g into the left side of the differential equation yields

$$\frac{1}{8}x_g''(t) + 2x_g(t)$$

$$= \frac{0.644 \cdot 9\pi^2}{8 \cdot 4}\sin\frac{3\pi t}{2} - 2 \cdot 0.644\sin\frac{3\pi t}{2} + \frac{1}{8}x_h''(t) + 2x_h(t)$$

$$= \frac{1}{2}\sin\frac{3\pi t}{2} + 0 = \frac{1}{2}\sin\frac{3\pi t}{2}.$$

The principle of superposition is correct again: Adding a solution of a *non*-homogeneous equation to the solution of the corresponding homogeneous equation produces a solution of the *non*homogeneous equation. ■

Just as with first-order equations, a general solution of a linear differential equation is the starting point for finding a solution of an initial-value problem.

■ **EXAMPLE 8** *Solve the initial-value problem modeling the forced spring-mass system in example 3 of section 5.1,*

$$\frac{1}{8}x''(t) + 2x(t) = \frac{1}{2}\sin\frac{3\pi t}{2}, \quad x(0) = 0, \quad x'(0) = 0.$$

We already know that a general solution of the differential equation is

$$x_g(t) = \frac{16}{64 - 9\pi^2}\sin\frac{3\pi t}{2} + C_1\sin 4t + C_2\cos 4t$$

$$\approx -0.644\sin\frac{3\pi t}{2} + C_1\sin 4t + C_2\cos 4t.$$

The first initial condition requires

$$x_g(0) = C_2 = 0.$$

To apply the second initial condition, compute $x_g'(t)$ and use $C_2 = 0$:

$$x_g'(t) = \frac{16}{64 - 9\pi^2}\frac{3\pi}{2}\cos\frac{3\pi t}{2} + 4C_1\cos 4t.$$

We must choose C_1 to satisfy

$$x_g'(0) = \frac{16}{64 - 9\pi^2}\frac{3\pi}{2} + 4C_1 = 0,$$

or $C_1 = -6\pi/(64 - 9\pi^2) \approx 0.759$.

Hence, the solution of the given initial-value problem is

$$x(t) = \frac{16}{64 - 9\pi^2}\sin\frac{3\pi t}{2} - \frac{6\pi}{64 - 9\pi^2}\sin 4t$$

$$\approx -0.644\sin\frac{3\pi t}{2} + 0.759\sin 4t. \quad ■$$

The principle of superposition shows why a general solution of a linear equation contains so many different solutions of that equation, one for each possible set of initial conditions. The particular solution component ensures equality with the forcing function in the differential equation, and the homogeneous solution terms supply the arbitrary constants that can be chosen to satisfy the initial conditions.

Does a general solution contain *every* solution of a linear initial-value problem? The answer is *yes* (if $a_2 \neq 0$), but proof requires a uniqueness theorem for linear second-order equations analogous to theorem 3, page 185, for first-order equations, an issue not taken up here. A uniqueness theorem would guarantee that the one solution constructed by a proper choice of constants in a general solution is the *only* solution of the given initial-value problem.

6.1.4   Steady States and Transients

As with first-order equations, a **steady state** or **equilibrium** of a differential equation is any constant solution. An equilibrium y_{ss} is *asymptotically stable* (or frequently just *stable*) if small perturbations of it die out, that is, if solutions whose initial values are near y_{ss} approach y_{ss} in the limit of large time:

Definition 6. *The equilibrium y_{ss} is **asymptotically stable** if there are some $\delta_1, \delta_2 > 0$ such that every solution having $y_{ss} - \delta_1 \leq y(0) \leq y_{ss} + \delta_1$, $-\delta_2 \leq y'(0) \leq \delta_2$ satisfies $\lim_{t \to \infty} y(t) = y_{ss}$.*

In the context of a first-order system which is equivalent to a given second-order equation, an equilibrium is sometimes called a *singular point*.

Stop and Think

6.17 Is a steady-state solution of a linear equation a particular solution of that equation?

6.18 Suppose a second-order differential equation has a steady state $y_{ss} = k$ for some constant k. What is a solution of that differential equation subject to the initial conditions $y(0) = k$, $y'(0) = 0$?

6.19 Loosely, a steady state is stable if solutions that start near it eventually approach it. The definition of stability includes two conditions, $y_{ss} - \delta_1 \leq y(0) \leq y_{ss} + \delta_1$ and $-\delta_2 \leq y'(0) \leq \delta_2$. Explain why each makes precise the notion of "starting near" the steady-state solution of a second-order equation.

The **transient** component of a solution is that part of the solution (if any) whose limit is zero for large values of the independent variable; that is, if y_{ss} is a stable steady state approached by the solution $y(t)$, then $y(t) - y_{ss}$ is the transient part of $y(t)$.

An equation can also have an oscillating "steady state" or periodic solution. A periodic solution is a solution that is a periodic function of the independent variable. Periodic solutions, like steady states, can be *stable* or *unstable*, depending upon whether small perturbations of the periodic solution decay or grow with time. A transient can decay to a periodic solution.

■ **EXAMPLE 9**   *The solution of the initial-value problem*

$$y''(t) + 2y'(t) + 5y(t) = -9.8,$$
$$y(0) = 0, \quad y'(0) = 0.04,$$

is

$$y(t) = -1.96 + e^{-t}(1.96 \cos 2t + \sin 2t).$$

Use the solution to find a steady state of the differential equation. Does this solution approach that steady state? If so, find the transient part of this solution.

If this equation is to have a *constant* steady state y_{ss}, then y_{ss} must satisfy

$$y_{ss}'' + 2y_{ss}' + 5y_{ss} = 5y_{ss} = -9.8$$

because the derivative of a constant is zero. Hence, $y_{ss} = -9.8/5 = -1.96$. Using the solution $y(t)$, observe that

$$y(t) - y_{ss} = e^{-t}(1.96\cos 2t + \sin 2t) \to 0.$$

Hence, this solution approaches $y_{ss} = -1.96$, and the transient is $y_{tran} = y(t) - y_{ss} = e^{-t}(1.96\cos 2t + \sin 2t)$. ■

Is the steady state $y_{ss} = -1.96$ of the linear equation $y'' + 2y' + 5y = -9.8$ stable? If its general solution can be written in the form $y_g = -1.96 + y_h$ and $y_h \to 0$, then the answer is *yes*.

There is enough information in the previous example to construct a general solution and answer the stability question. Since $y_{ss} = -1.96$ is a particular solution of this differential equation, the remaining terms $e^{-t}(1.96\cos 2t + \sin 2t)$ in the solution of the initial-value problem should be a solution of the corresponding homogeneous equation. Perhaps the coefficient 1.96 of $\cos 2t$ and the coefficient 1 of $\sin 2t$ began life as arbitrary constants, before they were assigned these values to satisfy the initial conditions. If so, a general solution of the equation could be $y_g = -1.96 + e^{-t}(C_1\cos 2t + C_2\sin 2t)$.

Stop and Think   **6.20** Verify that $y_g = -1.96 + e^{-t}(C_1\cos 2t + C_2\sin 2t)$ is indeed a solution of $y''(t) + 2y'(t) + 5y(t) = -9.8$. What else must you test to verify that it is a *general* solution?

The constants C_1, C_2 are determined by the initial conditions, but no matter what values they assume, $\lim_{t\to\infty} y_g(t) = -1.96$.

Stop and Think   **6.21** Why does $\lim_{t\to\infty} y_g(t) = -1.96$?

The behavior of a general solution determines the behavior of every solution.

The equilibrium is stable to perturbations of any size; solutions need not start near $y_{ss} = -1.96$ to approach it. Analyzing the behavior of the all-purpose general solution has determined the behavior of every solution of this linear differential equation, regardless of the initial conditions that solution satisfies.

■ **EXAMPLE 10**   *The nonlinear pendulum equation $L\theta'' + g\sin\theta = 0$ has a steady state $\theta_{ss} = -\pi$. Is it stable?*

Stop and Think   **6.22** What is the position of the pendulum when $\theta = -\pi$? Do you expect this steady state to be stable or unstable?

Since this equation is nonlinear, a general solution approach is *not* appropriate. Instead, employ a linear stability analysis, as in section 3.3, page 118.

Consider the perturbed solution

solution = steady state
+ perturbation

$$\theta(t) = \theta_{ss} + q(t)$$

where $q(0)$ is small (and so is $q'(0)$). The goal is to decide if the perturbation q grows or decays with time.

The behavior of q is determined by the requirement that $\theta(t)$ solve the differential equation. To invoke that condition, substitute $\theta(t) = \theta_{ss} + q(t)$ into $L\theta'' + g \sin \theta = 0$. Since θ_{ss} is a constant, we have

$$\begin{aligned} \theta''(t) &= \theta_{ss}'' + q''(t) \\ &= (-\pi)'' + q''(t) \\ &= q''(t). \end{aligned} \tag{6.2}$$

For the second term in the differential equation, a Taylor expansion of $\sin x$ about x_0 (see appendix theorem 3) sets the stage for linearization of the equation for $q(t)$. A general Taylor expansion of the sine function is

$$\sin x = \sin x|_{x=x_0} + \frac{d \sin x}{dx}\bigg|_{x=x_0} (x - x_0) + \frac{d^2 \sin x}{dx^2}\bigg|_{x=x_0} \frac{(x - x_0)^2}{2} + \cdots .$$

In this case, $x = -\pi + q$ and $x_0 = -\pi$. Hence,

$$\begin{aligned} \sin(\theta_{ss} + q(t)) &= \sin(-\pi) + \cos(-\pi)q(t) - \sin(-\pi)(q(t))^2 + \cdots \\ &= -q(t) + \cdots . \end{aligned} \tag{6.3}$$

Stop and Think **6.23** What is the lowest power of q appearing in the terms $+ \cdots$ in the last equation? Under what conditions on q are those powers of q smaller than q itself?

Combining equations (6.2) and (6.3), the result of the substitution is

$$\begin{aligned} L\theta'' + g \sin \theta &= 0, \\ L(\theta_{ss} + q)'' + g \sin(\theta_{ss} + q) &= 0, \\ Lq'' - gq + \cdots &= 0, \end{aligned}$$

and the approximate, linearized equation for the perturbation q is

Linearized stability equation

$$Lq'' - gq = 0.$$

Stop and Think **6.24** Why is $Lq'' - gq = 0$ an approximate equation? What was omitted? Why was it omitted?

The (linearized) answer to the stability question now comes down to the behavior of the general solution of this linear, constant coefficient, homogeneous differential equation for q. If the solutions of $Lq'' - gq = 0$ grow, then the steady state $\theta_{ss} = -\pi$ is unstable. Subsequent sections in this chapter will develop tools for answering that stability question. ■

Numerical solution of $Lq'' - gq = 0$ could establish that the steady state $\theta_{ss} = -\pi$ is unstable to certain perturbations, those that produce approximate numerical solutions that seem to grow. An *analytic* study of the general solution is required to settle the stability question for *all* solutions of $Lq'' - gq = 0$.

The analytic tool to be developed is characteristic equations. That approach shows that a general solution of $Lq'' - gq = 0$ is $q_g = C_1 e^{\sqrt{g/L}\,t} + C_2 e^{-\sqrt{g/L}\,t}$. Hence, q grows without bound and θ_{ss} is unstable unless $C_1 = 0$, a condition that is satisfied only by initial conditions of the form $q(0) = q_i$, $q'(0) = -q_i\sqrt{g/L}$. Can you explain how such a special perturbation might actually bring the pendulum back to rest in the vertical position?

MATLAB

Use the analytic solution tool in MATLAB's DELAB to find a general solution of $Lq'' - gq = 0$.

6.1.5 Exercises

<table>
<tr><th colspan="2">EXERCISE GUIDE</th></tr>
<tr><th>To gain experience ...</th><th>Try exercises</th></tr>
<tr><td>With linear versus nonlinear</td><td>1–11, 19–20</td></tr>
<tr><td>With homogeneous versus nonhomogeneous</td><td>1–11, 15(a–b), 18(a)</td></tr>
<tr><td>With constant versus variable coefficient</td><td>1–11</td></tr>
<tr><td>Checking solutions of equations</td><td>13</td></tr>
<tr><td>With particular solutions</td><td>12</td></tr>
<tr><td>With homogeneous solutions</td><td>15(b), 18(a, d)</td></tr>
<tr><td>Using general solutions</td><td>15(c), 18(b)</td></tr>
<tr><td>Solving initial-value problems</td><td>15(c), 18(b)</td></tr>
<tr><td>Using the principle of superposition</td><td>16, 19</td></tr>
<tr><td>With steady-state and transient solutions</td><td>15(a–b, d), 18(a)</td></tr>
<tr><td>With periodic solutions</td><td>18(c)</td></tr>
<tr><td>With equations as models</td><td>14, 17</td></tr>
</table>

Classify the following ten second-order differential equations as linear or nonlinear and as homogeneous or nonhomogeneous. Identify each linear equation as having constant or variable coefficients. Justify each classification.

1. $3x''(t)/16 + 2x(t) = 0$ (See example 1 of section 5.1.)

2. $3x''(t)/16 + 1.41x'(t) + 2x(t) = 0$ (See example 2 of section 5.1.)

3. $my''(t) + ky(t) = -mg$, m, k, g constants (See exercise 6 of the chapter 5 exercises.)

4. $my''(t) + py'(t) = -mg$, m, p, g constants

5. $my''(t) + py'(t) = -ky(t)$, m, p, k constants

6. $x''(t)/8 + 2x(t) = (1/2)x(t)\sin(3\pi t/2)$

7. $x''(t)/8 + 2x(t) = (\pi t/2)\sin x(t)$

8. $t^3 x''(t) + 2x(t) = (1/2)\sin(3\pi t/2)$

9. $x''(t) + x'(t)\cos 4t = 2(x(t) - 1)$

10. $x''(t) + x'(t)\cos 4t = 2(x(t) - 1)^2$

11. Verify that the two spring-mass model equations considered in example 5,

$$\frac{1}{8}x''(t) + 2x(t) = \frac{1}{2}\sin\frac{3\pi t}{2},$$

$$\frac{1}{8}x''(t) + 2x(t) = 0,$$

are indeed linear. Verify that the first equation is nonhomogeneous and that the second is homogeneous. Do these equations have constant or variable coefficients?

12. Verify that

$$x_p(t) = \frac{16}{64 - 9\pi^2}\sin\frac{3\pi t}{2} \approx -0.644\sin\frac{3\pi t}{2}$$

is indeed a particular solution of the forced spring-mass equation

$$\frac{1}{8}x''(t) + 2x(t) = \frac{1}{2}\sin\frac{3\pi t}{2}.$$

13. Verify by direct substitution that

$$x(t) = -0.644\sin\frac{3\pi t}{2} + 0.759\sin 4t$$

is a solution of the initial-value problem

$$\frac{1}{8}x''(t) + 2x(t) = \frac{1}{2}\sin\frac{3\pi t}{2}, \quad x(0) = 0, \quad x'(0) = 0.$$

14. Identify a particular physical situation of which the initial-value problem

$$y''(t) + 2y'(t) + 5y(t) = -9.8,$$
$$y(0) = 0, \quad y'(0) = 0.04,$$

of example 8 might be a model. Be sure to specify units.

15. Example 9 considered the initial-value problem

$$y''(t) + 2y'(t) + 5y(t) = -9.8,$$
$$y(0) = 0, \quad y'(0) = 0.04,$$

whose solution is

$$y(t) = -1.96 + e^{-t}(1.96\cos 2t + \sin 2t).$$

That example found that the transient component of its solution is

$$y_{\text{tran}}(t) = e^{-t}(1.96\cos 2t + \sin 2t)$$

while the equation has the steady state $y_{\text{ss}} = -1.96$.

(a) Does the steady state satisfy the homogeneous or the nonhomogeneous differential equation? Could the steady-state solution serve as a particular solution of the nonhomogeneous equation? Give a set of initial values that will yield precisely this steady-state solution.

(b) Does the transient component satisfy the homogeneous or the nonhomogeneous differential equation?

(c) Show that a general solution of this initial-value problem is

$$y(t) = -1.96 + e^{-t}(C_1\cos 2t + C_2\sin 2t).$$

Derive the solution of this initial-value problem from this general solution.

(d) Use the general solution of part (c) to verify that the steady state $y_{\text{ss}} = -1.96$ is stable.

16. The **forced response** y_{for} of a second-order linear initial-value problem

$$a_2(x)y'' + a_1(x)y' + a_0(x)y = r(x),$$
$$y(0) = y_i, \quad y'(0) = z_i, \tag{6.4}$$

is defined to be the solution of

$$a_2(x)y''_{\text{for}} + a_1(x)y'_{\text{for}} + a_0(x)y_{\text{for}} = r(x),$$
$$y_{\text{for}}(0) = 0, \quad y'_{\text{for}}(0) = 0,$$

and the free response y_{free} is the solution of

$$a_2(x)y''_{\text{free}} + a_1(x)y'_{\text{free}} + a_0(x)y_{\text{free}} = 0,$$
$$y_{\text{free}}(0) = y_i, \quad y'_{\text{free}}(0) = z_i.$$

Use the principle of superposition to argue that the solution of the initial-value problem (6.4) can be written as the sum of its forced and free components,

$$y(x) = y_{\text{for}}(x) + y_{\text{free}}(x).$$

17. Identify a particular physical situation of which the initial-value problem

$$y''(t) + 2y'(t) + 5y(t) = 185\sin 4t,$$
$$y(0) - 9, \quad y'(0) - -47,$$

might be a model. Be sure to specify units.

18. Consider the initial-value problem

$$y''(t) + 2y'(t) + 5y(t) = 185\sin 4t,$$
$$y(0) = 9, \quad y'(0) = -47,$$

whose solution is

$$y(t) = 8\cos 4t - 11\sin 4t + e^{-t}(\cos 2t - \sin 2t).$$

The transient component of this solution is

$$y_{\text{tran}}(t) = e^{-t}(\cos 2t - \sin 2t).$$

This equation also has the periodic solution

$$y_{\text{ps}}(t) = 8\cos 4t - 11\sin 4t.$$

(a) Does the transient component satisfy the homogeneous or the nonhomogeneous differential equation?

(b) Show that a general solution of this initial-value problem is

$$y(t) = 8\cos 4t - 11\sin 4t + e^{-t}(C_1\cos 2t - C_2\sin 2t).$$

Derive the solution of this initial-value problem from this general solution.

(c) Does the periodic solution satisfy the homogeneous or the nonhomogeneous differential equation? Could it serve as a particular solution of the nonhomogeneous equation? Give a set of initial values that will yield precisely this periodic solution.

19. Recall the often-repeated calculus slogan, "The derivative of a sum is the sum of the derivatives." Does this saying apply to second derivatives? Explain why the principle of superposition is a consequence of this statement. (The principle of superposition follows from the linearity of the derivative operation.)

20. In parallel with the ideas of section 4.1.2, introduce the notion of a second-order differential operator and develop an alternate definition of *second-order linear differential equation* analogous to definition 4, page 139, for first-order equations. Show that a second-order equation that is linear in the sense of definition 2, page 234, defines a linear second-order differential operator; i.e., show that $\mathcal{L}y \equiv a_2(x)y'' + a_1(x)y' + a_0(x)y$ defines a linear second-order differential operator.

6.2 ■ TESTING LINEAR INDEPENDENCE

This section develops a simple linear independence test for solutions of linear equations. To begin, recall the process used in example 5, page 240, of the previous section to demonstrate that two functions are linearly independent.

To show that $\sin 4t$, $\cos 4t$ are linearly independent for $t \geq 0$, example 5 finds two constants C_1, C_2 satisfying

$$C_1\sin 4t + C_2\cos 4t = 0, \quad t \geq 0.$$

If the only choice is $C_1 = C_2 = 0$, then $\sin 4t$, $\cos 4t$ are linearly independent on $t \geq 0$.

Since C_1, C_2 are chosen so that $C_1\sin 4t + C_2\cos 4t$ is the identically zero function, the derivative of this expression is also identically zero,

$$4C_1\cos 4t - 4C_2\sin 4t = 0, \quad t \geq 0.$$

The pair of equations formed from the linear combination and its derivative,

$$C_1\sin 4t + C_2\cos 4t = 0, \tag{6.5}$$

$$4C_1\cos 4t - 4C_2\sin 4t = 0, \tag{6.6}$$

certainly has the solution $C_1 = C_2 = 0$. If there are no other values of C_1 and C_2 satisfying this pair of equations, then $\sin 4t$, $\cos 4t$ are linearly independent.

The theory of linear equations asserts that (6.5–6.6) has *exactly one* solution if the determinant of coefficients

$$W = \begin{vmatrix} \sin 4t & \cos 4t \\ 4\cos 4t & -4\sin 4t \end{vmatrix}$$

is not zero. To evaluate this determinant, subtract the product of the off-diagonal entries from the product of those on the diagonal:

$$W = (\sin 4t)(-4\sin 4t) - (4\cos 4t)(\cos 4t)$$

$$= -4\sin^2 4t - 4\cos^2 4t = -4 \neq 0, \quad t \geq 0.$$

Hence, the obvious solution $C_1 = C_2 = 0$ of (6.5–6.6) is the only solution of this pair of equations, and we conclude that $\sin 4t$, $\cos 4t$ are linearly independent on $t \geq 0$.

You may be more familiar with these determinant ideas in the context of Cramer's rule for solving systems of linear equations. This result states that if $W \neq 0$, then the *unique* solution of the pair of equations (6.5–6.6) is

$$C_1 = \frac{R_1}{W}, \quad C_2 = \frac{R_2}{W}.$$

Here R_1 and R_2 are the determinants obtained by replacing the first and second columns of W, respectively, by the zero terms appearing on the right side of (6.5–6.6),

$$R_1 = \begin{vmatrix} 0 & \cos 4t \\ 0 & -4\sin 4t \end{vmatrix}, \quad R_2 = \begin{vmatrix} \sin 4t & 0 \\ 4\cos 4t & 0 \end{vmatrix}.$$

But the columns of zeros in these determinants force $R_1 = R_2 = 0$, and the only solution of (6.5–6.6) is $C_1 = C_2 = 0$. See appendix, section 6.

This approach extends quite nicely to pairs of solutions of second-order equations. We describe that extension here.

Let $y_1(x)$, $y_2(x)$ be two functions that are continuously differentiable on the interval $a \leq x \leq b$. Define the **Wronskian** $W(y_1, y_2)$ of y_1, y_2 by

DEFINITION OF THE WRONSKIAN

$$W(y_1, y_2) = \begin{vmatrix} y_1(x) & y_2(x) \\ y_1'(x) & y_2'(x) \end{vmatrix}$$

(after the Polish mathematician H. Wronski). We may expand the determinant to write

EXPANDED WRONSKIAN

$$W(y_1, y_2) = y_1(x)y_2'(x) - y_2(x)y_1'(x).$$

The Wronskian of two functions is evidently a function of x.

To establish that the two functions y_1, y_2 are linearly independent on the interval $a \leq x \leq b$, we must show that the *only* constants C_1, C_2 satisfying

$$C_1 y_1(x) + C_2 y_2(x) = 0, \quad a \leq x \leq b,$$

are $C_1 = C_2 = 0$.

Proceed exactly as before, with y_1, y_2 replacing $\sin 4t$, $\cos 4t$. Since the linear combination $C_1 y_1 + C_2 y_2$ is identically zero on the interval $[a, b]$, its derivative must be zero there as well,

$$\frac{d}{dx}(C_1 y_1(x) + C_2 y_2(x)) = C_1 y_1'(x) + C_2 y_2'(x) = 0, \quad a \leq x \leq b.$$

We now have a pair of simultaneous linear equations for the unknown constants C_1, C_2,

$$C_1 y_1(x) + C_2 y_2(x) = 0,$$
$$C_1 y_1'(x) + C_2 y_2'(x) = 0,$$

the analog of (6.5–6.6). The first equation is the original linear combination, and the second is its derivative. This pair of equations has no solution other than $C_1 = C_2 = 0$ if its determinant of coefficients is not zero on $[a, b]$; that is, $C_1 = C_2 = 0$ if

$$W(y_1, y_2) \neq 0, \quad a \leq x \leq b,$$

because the Wronskian is exactly the determinant of coefficients of this pair of linear equations. This argument proves the first property of the Wronskian.

$W \neq 0$ guarantees linear independence.

Theorem 2. *The pair of continuously differentiable functions y_1, y_2 is linearly independent on the interval $a \leq x \leq b$ if $W(y_1, y_2) \neq 0$ there.*

■ **EXAMPLE 11**   *Show that the pair of functions $\sin 4t$, $\cos 4t$ is linearly independent on $t \geq 0$.*

Since these two functions are certainly continuously differentiable, we can immediately compute the Wronskian,

$$W(\sin 4t, \cos 4t) = \sin 4t (\cos 4t)' - \cos 4t (\sin 4t)'$$

$$= -4 \sin^2 4t - 4 \cos^2 4t = -4 \neq 0, \quad t \geq 0.$$

Hence, theorem 2 guarantees that $\sin 4t$, $\cos 4t$ are linearly independent on $t \geq 0$. ■

Note that this theorem provides no information if the Wronskian happens to be zero at one or more points scattered through the interval. The following two results together rectify this deficiency when the functions in question are solutions of the same second-order homogeneous linear differential equation. Further, they show that we need to check the Wronskian only at a single point of the interval.

The Wronskian of two solutions is either always zero or never zero.

Theorem 3. *Let y_1, y_2 be solutions of the second-order homogeneous linear differential equation*

$$y''(x) + P(x)y' + Q(x)y = 0, \quad a \leq x \leq b.$$

Assume that y_1, y_2 are defined and twice continuously differentiable on the interval $[a, b]$ and that the coefficient $P(x)$ is continuous on $[a, b]$. Then either $W(y_1, y_2) \neq 0$ for every x in $[a, b]$ or $W(y_1, y_2) = 0$ for every x in $[a, b]$.

Technically, the requirement that y_1, y_2 be twice continuously differentiable is redundant; see the definition of a solution of a second-order equation, definition 1, page 233.

The essence of the proof of this theorem is the observation that W (as a function of x) is a solution of the first-order linear equation

$$\frac{dW}{dx} + P(x)W = 0.$$

Hence, W may be written

$$W = C \exp\left(\int_a^x P(s)\, ds \right),$$

where C is an arbitrary constant. Since the exponential is never zero, either $W \neq 0$ on $[a, b]$ because $C \neq 0$ or $W \equiv 0$ on $[a, b]$ because $C = 0$. The exercises lead you through this proof step by step.

Theorem 3 asserts that *if the Wronskian of two solutions of a linear equation is zero at one point of the interval of interest, then it is zero throughout that interval.* However, we have yet to show that a zero Wronskian implies linear dependence. The next theorem disposes of that case.

If the Wronskian of two solutions is zero at one point, then the solutions are linearly dependent.

Theorem 4. *Let y_1, y_2 be twice continuously differentiable solutions of the linear equation*

$$y''(x) + P(x)y' + Q(x)y = 0, \quad a \leq x \leq b.$$

Let the coefficient P be continuous on $[a, b]$. If $W(y_1, y_2) = 0$ at any point of $[a, b]$, then y_1, y_2 are linearly dependent on $[a, b]$.

The proof of this theorem is outlined in the exercises as well.

Together, these theorems provide the following linear independence test. Recall that the functions in question must be twice continuously differentiable solutions of $y''(x) + P(x)y' + Q(x)y = 0$ and that the coefficient P must be continuous on the interval of interest.

Linear independence test. Two solutions of the linear equation $y''(x) + P(x)y' + Q(x)y = 0$ are linearly *independent* on their domain of definition if their Wronskian *is not zero* for at least one point of the interval. The two solutions are linearly *dependent* if their Wronskian *is zero* for at least one point of the interval.

■ **EXAMPLE 12** *Use the linear independence test to show that e^{-4x}, e^{3x} are linearly independent solutions of*

$$y'' + y' - 12y = 0, \quad -\infty < x < \infty.$$

Substitution immediately verifies that these two functions are solutions of this equation on the given interval.

This differential equation is certainly of the form required by the linear independence test; take $P(x) = -1$, $Q(x) = -12$.

The Wronskian of these two solutions is

$$W(e^{-4x}, e^{3x}) = e^{-4x}(e^{3x})' - e^{3x}(e^{-4x})' = 7e^{-x}.$$

Since $e^{-x} \neq 0$ for $-\infty < x < \infty$, the solutions e^{-4x}, e^{3x} are linearly independent there. ■

■ **EXAMPLE 13**   *Use the linear independence test to show that* e^{-4x}, $8e^{-4x}$ *are linearly dependent solutions of*

$$y'' + y' - 12y = 0, \quad -\infty < x < \infty.$$

Obviously, both functions are solutions of the given equation. Their Wronskian is

$$W(e^{-4x}, 8e^{-4x}) = e^{-4x}(8e^{-4x})' - 8e^{-4x}(e^{-4x})' = 0.$$

Hence, these two solutions are linearly dependent because $W = 0$ for all x. ■

■ **EXAMPLE 14**   *Two solutions of*

$$x^2 y'' + xy' - 4y = 0$$

are x^2 and x^{-2}. Are they linearly independent on any nontrivial intervals?

To apply the linear independence test, the given equation must be put into the form $y''(x) + P(x)y' + Q(x)y = 0$. To perform the necessary division by x^2, we can consider only $x \neq 0$. We obtain

$$y'' + x^{-1}y' - 4x^{-2}y = 0.$$

We now have the form $y''(x) + P(x)y' + Q(x)y = 0$ with $P(x) = x^{-1}$ and $Q(x) = -4x^{-2}$. This form of P is certainly continuous for $x \neq 0$.

Evaluating the Wronskian leads to

$$\begin{aligned} W(x^2, x^{-2}) &= x^2(x^{-2})' - x^{-2}(x^2)' \\ &= -4x^{-1} \neq 0, \quad x \neq 0. \end{aligned}$$

Hence, x^2 and x^{-2} are linearly independent solutions of $x^2 y'' + xy' - 4y = 0$ on any interval that excludes $x = 0$. ■

6.2.1 Exercises

EXERCISE GUIDE	
To gain experience . . .	**Try exercises**
Verifying linear independence	1, 2–3, 12–13
Constructing general solutions	2–3
Verifying linearity of equations	4
Constructing general solutions	5
With proofs of Wronskian relations and linear independence	6–11

1. Determine whether each of the given pairs of functions is linearly independent for all x. When the two functions are not linearly independent for all x, give an interval where they are linearly independent, if possible. If they are not linearly independent, explain why.

   (a) $\cos \omega x$, $\sin \omega x$, ω a positive constant

   (b) $\cos 2x$, $\cos x$

   (c) $\cos 2x$, $\sin 2(x - \pi/4)$

   (d) e^{rx}, e^{sx}, r, s constants, $r \neq s$

   (e) e^{rx}, xe^{rx}, r constant

   (f) e^{rx}, $e^{r(x-1)}$, r constant

In exercises 2–3,

   (i) Determine whether the two functions are linearly independent or linearly dependent solutions of the given differential equation on the interval shown.

   (ii) Explain the independence/dependence behavior you observe.

   (iii) Write a general solution of the differential equation each time you find a linearly independent pair of solutions.

2. (a) $y'' + y = 0$; $\sin x$, $\cos x$ on $-\infty < x < \infty$

   (b) $y'' + y = 0$; $\sin x$, $\cos(x - \pi/2)$ on $-\infty < x < \infty$

   (c) $y'' + y = 0$; $\sin x$, $\cos(x - \pi/4)$ on $-\infty < x < \infty$

3. (a) $y'' - y = 0$; e^x, e^{-x} on $-\infty < x < \infty$

   (b) $y'' - y = 0$; e^x, e^{1-x} on $-\infty < x < \infty$

   (c) $y'' - y = 0$; e^{-x}, e^{1-x} on $-\infty < x < \infty$

4. Verify that the differential equation

$$y''(x) + P(x)y' + Q(x)y = 0$$

appearing in theorem 3 is indeed linear. Can the general linear second-order differential equation

$$a_2(x)y'' + a_1(x)y' + a_0(x)y = r(x)$$

always be written in this form?

5. Example 14 considers the solutions x^2 and x^{-2} of

$$x^2 y'' + xy' - 4y = 0.$$

Write a general solution of this equation for $x \neq 0$. Can you guess a general solution of the nonhomogeneous equation

$$x^2 y'' + xy' - 4y = -4?$$

(*Hint*: Does this equation have a steady state?)

6. To contribute to the proof of theorem 3, show by completing the following steps that the Wronskian $W(y_1, y_2)$ satisfies the differential equation

$$\frac{dW}{dx} + P(x)W = 0$$

as a function of x. Recall that y_1, y_2 are twice continuously differentiable solutions of

$$y''(x) + P(x)y' + Q(x)y = 0.$$

   (a) Show directly from the definition of W that $dW/dx = y_1 y_2'' - y_2 y_1''$.

   (b) Each of the functions y_1 and y_2 satisfies $y'' + Py' + Qy = 0$. Write out the mathematical statements of these facts, multiply the equation for y_1 by y_2'' and that for y_2 by y_1''. Subtract the two expressions and show that the sum reduces to $dW/dx + PW = 0$.

   (c) The preceding steps derived the differential equation for W from the definition of W. Describe a series of steps like these that would allow you to work backwards, establishing that W is a solution of $dW/dx + PW = 0$ by directly substituting in this equation.

7. The proof of theorem 3 depends upon the claim that a solution of

$$\frac{dW}{dx} + P(x)W = 0$$

is

$$W = C \exp\left(\int_a^x P(s)\, ds \right).$$

   (a) Verify by direct substitution that the expression $W = C \exp\left(\int_a^x P(s)\, ds \right)$ is indeed a solution of $dW/dx + PW = 0$.

   (b) Use a solution method you learned earlier to derive the solution $W = C \exp\left(\int_a^x P(s)\, ds \right)$ from $dW/dx + PW = 0$.

8. The hypotheses of theorem 3 require that the two solutions y_1 and y_2 be twice continuously differentiable and that the coefficient $P(x)$ be continuous. Explain why each of these conditions is imposed.

9. Combine the ideas of the preceding three exercises to write out a complete proof of theorem 3.

10. This exercise outlines a proof of theorem 4. Suppose that y_1, y_2 are two solutions of

$$y''(x) + P(x)y' + Q(x)y = 0$$

and that their Wronskian $W(y_1, y_2)$ is zero throughout the interval $[a, b]$ upon which the solutions y_1, y_2 are defined.

   (a) Argue that if both y_1 and y_2 were identically zero, then they would be linearly dependent.

(b) Now consider the case when at least one of the two is not identically zero; suppose it is y_1. Argue that

$$\frac{W}{y_1^2} = \frac{d}{dx}\left(\frac{y_2}{y_1}\right) = 0$$

and, hence, that y_2/y_1 is constant on any subinterval of $[a, b]$ where $y_1 \neq 0$. Does this complete the proof that y_1 and y_2 are linearly dependent? If not, add the necessary details.

11. State precisely how the linear independence test follows from the three theorems stated in this section.

12. Suppose y_1, y_2 are twice continuously differentiable solutions of

$$y''(x) + P(x)y' + Q(x)y = 0$$

on some interval $0 \leq x \leq x_1$, that P is continuous on that interval, and that $y_1(0) = 1$, $y_1'(0) = 0$, $y_2(0) = 0$, $y_2'(0) = 1$. Show that y_1, y_2 are linearly independent on $0 \leq x \leq x_1$.

13. Suppose y_1, y_2 are twice continuously differentiable solutions of

$$y''(x) + P(x)y' + Q(x)y = 0$$

on some interval $0 \leq x \leq x_1$, that P is continuous on that interval, and that $y_1(0) = a$, $y_1'(0) = b$, $y_2(0) = c$, $y_2'(0) = d$. Give a condition involving the four constants a, b, c, d which guarantees that y_1, y_2 are linearly independent on $0 \leq x \leq x_1$.

6.3 ■ CHARACTERISTIC EQUATIONS: REAL ROOTS

6.3.1   The Possibilities

A constant-coefficient, homogeneous, first-order linear equation such as the population equation

$$P' - kP = 0$$

can be solved by finding its *characteristic equation*: Assume a solution of the form $P(t) = e^{rt}$, r a constant, substitute into the differential equation, and divide by $e^{rt} \neq 0$ to obtain the characteristic equation

$$r - k = 0.$$

Then use $r = k$ to write a nontrivial solution $P(t) = e^{kt}$ of $P' = kP$. A general solution is $P_g(t) = Ce^{kt}$, where C is an arbitrary constant.

Stop and Think   **6.25** Why must r be a constant in the assumed solution form e^{rt}?

The same procedure applies to constant-coefficient, homogeneous, second-order, linear equations, but the second derivative leads to a quadratic characteristic equation. Quadratic equations with real coefficients have roots of three types, real and distinct, real and repeated, and complex. Each type of characteristic root leads to a distinct type of general solution of the original differential equation.

The characteristic roots tell all there is to know about the behavior of the homogeneous solution of the differential equation. Real characteristic roots produce exponential solutions, growing or decaying according to whether the root is positive or negative. Complex roots produce oscillating solutions, products of sines and cosines with exponentials. Those oscillations grow or decay according to whether the exponential growth rate is positive or negative.

The general constant-coefficient, homogeneous, second-order, linear differential equation has the form

CONSTANT-COEFFICIENT
DIFFERENTIAL EQUATION

$$b_2 y'' + b_1 y' + b_0 y = 0, \tag{6.7}$$

where the coefficients b_2, b_1, b_0 all are constants. Assuming a solution of the form $y = e^{rx}$ and substituting into this equation yields

Substitute $y = e^{rx}$.

$$b_2 r^2 e^{rx} + b_1 r e^{rx} + b_0 e^{rx} = 0.$$

The exponential e^{rx} is never zero; divide it out to obtain the **characteristic equation**

CHARACTERISTIC EQUATION

$$b_2 r^2 + b_1 r + b_0 = 0.$$

The roots of the characteristic equation are called **characteristic roots** of the differential equation.

■ **EXAMPLE 15**   *Find the characteristic equation for the constant-coefficient, homogeneous, second-order linear differential equation*

$$y'' + y' - 12y = 0.$$

Use that characteristic equation to find two solutions of this differential equation.

A careful reader could argue that this request is carelessly worded. The example should ask for two *nontrivial* solutions.

The equation $y'' + y' - 12y = 0$ fits the general form (6.7) with $b_2 = 1$, $b_1 = 1$, and $b_0 = -12$. Assuming $y = e^{rx}$, substituting in the differential equation, and dividing out the nonzero factor e^{rx} leads to

Substitute $y = e^{rx}$.
Divide to find the characteristic equation.

$$r^2 e^{rx} + r e^{rx} - 12 e^{rx} = 0,$$
$$r^2 + r - 12 = 0.$$

Solve the characteristic equation $r^2 + r - 12 = 0$ by factoring,

Find characteristic roots.

$$r^2 + r - 12 = (r + 4)(r - 3) = 0.$$

Its roots are $r = -4, 3$, and two solutions of the differential equation $y'' + y' - 12y = 0$ are

Write homogeneous solutions.

$$y_1 = e^{-4x}, \quad y_2 = e^{3x}. \ ■$$

Stop and Think   **6.26** Are these solutions linearly independent? If so, use them to write a general solution of $y'' + y' - 12y = 0$. Does DELAB obtain the same general solution?

The characteristic equation $b_2 r^2 + b_1 r + b_0 = 0$ of the second-order, constant-coefficient differential equation (6.7) is a quadratic equation with real coefficients. Consequently, its roots fall into one of three categories:

Three categories of characteristic roots

1. Two *distinct real roots*,

2. A *single repeated real root*, or

3. A pair of *complex conjugate roots*.

Recall that a *complex number* is one of the form $\alpha + i\beta$, where α, β are real numbers and $i^2 = -1$; e.g., $3 - 2i$, $-1 + i$, and πi are all complex numbers. The *real part* of the complex number $z = \alpha + i\beta$ is α; we write $\operatorname{Re} z = \alpha$. The *imaginary part* of z is β; we write $\operatorname{Im} z = \beta$. If $z = 3 - 2i$, then $\operatorname{Re} z = 3$ and $\operatorname{Im} z = -2$.

The *complex conjugate* of $\alpha + i\beta$ is $\alpha - i\beta$. The complex conjugate of $3 - 2i$ is $3 + 2i$, and the complex conjugate of πi is $-\pi i$.

Examples of these three cases follow:

1. $r^2 + r - 12 = (r + 4)(r - 3) = 0$ has the *distinct real roots* $r = -4$, 3 (as seen above in example 15).
2. $r^2 + 8r + 16 = (r + 4)(r + 4) = 0$ has the single *repeated real root* $r = -4$ (as seen below in example 16).
3. $r^2 + 8r + 20 = 0$ has the pair of *complex conjugate roots* $r = -4 + 2i$, $-4 - 2i$ (as seen below in example 17).

In the last case, solve $r^2 + 8r + 20 = 0$ using the quadratic formula:

$$r = \frac{-8 \pm \sqrt{8^2 - 4 \cdot 20}}{2} = -4 \pm 2i.$$

Note that the negative *discriminant* (the term inside the square root),

$$8^2 - 4 \cdot 20 = -16,$$

guarantees complex roots of this equation because $\sqrt{-16} = 4i$.

In case 1, when the characteristic equation $b_2 r^2 + b_1 r + b_0 = 0$ has two *distinct real roots* $r_1, r_2, r_1 \neq r_2$, then two solutions of $b_2 y'' + b_1 y' + b_0 y = 0$ are

$$e^{r_1 x}, \quad e^{r_2 x}.$$

These two solutions will be seen to be linearly independent; they are all we need to construct a general solution of the differential equation (6.7).

In case 2, when there is only *one real root*, we can not hope to construct a general solution until we have somehow found a second, linearly independent solution. In case 3, when there is a pair of *complex conjugate roots*, we again have two solutions that can be shown to be linearly independent, but we have the problem of interpreting an exponential with a complex exponent.

Resolving these two issues will occupy most of this section and the next. Notice that these problems never arose with first-order equations because a linear characteristic equation (e.g., $r - k = 0$ from the simple population equation) can have only one real solution. Of course, only one nontrivial solution of a homogeneous first-order differential equation such as $P' - kP = 0$ is needed to construct a general solution.

■ **EXAMPLE 16**   *Find the characteristic equation and one solution of*

$$y'' + 8y' + 16y = 0.$$

Assume $y = e^{rx}$ and substitute in $y'' + 8y' + 16y = 0$ to obtain

Substitute $y = e^{rx}$.

$$r^2 e^{rx} + 8r e^{rx} + 16 e^{rx} = 0.$$

Dividing out the exponential factor yields the characteristic equation,

Characteristic equation

$$r^2 + 8r + 16 = (r + 4)^2 = 0.$$

As illustrated in case 2, this characteristic equation has the single real root $r = -4$. Hence, one solution of $y'' + 8y' + 16y = 0$ is

Write one homogeneous solution.

$$y_1 = e^{-4x}.$$

We need other techniques to obtain a second solution. ■

■ **EXAMPLE 17**   *Find the characteristic equation of*

$$y'' + 8y' + 20y = 0.$$

Assume $y = e^{rx}$ and substitute in $y'' + 8y' + 20y = 0$ to obtain

Substitute $y = e^{rx}$.

$$r^2 e^{rx} + 8re^{rx} + 20e^{rx} = 0.$$

Dividing out the exponential factor yields the characteristic equation,

Characteristic equation

$$r^2 + 8r + 20 = 0.$$

As illustrated in case 3, this characteristic equation has the pair of complex conjugate roots $r = -4 \pm 2i$. The corresponding homogeneous solutions are $e^{(-4\pm 2i)x}$. The next section will develop an interpretation for these complex exponential functions. ■

6.3.2 Distinct Real Roots

When the characteristic equation $b_2 r^2 + b_1 r + b_0 = 0$ has distinct real roots $r_1, r_2, r_1 \neq r_2$, then the two solutions

Distinct, real characteristic roots produce two linearly independent solutions.

$$y_1 = e^{r_1 x}, \quad y_2 = e^{r_2 x}$$

are linearly independent. The general solution of the constant-coefficient, homogeneous equation $b_2 y'' + b_1 y' + b_0 y = 0$ is

General solution

$$y_g(x) = C_1 e^{r_1 x} + C_2 e^{r_2 x}.$$

The exercises ask you to verify the claim of linear independence.
    To illustrate, the two solutions

$$y_1 = e^{-4x}, \quad y_2 = e^{3x}$$

of

$$y'' + y' - 12y = 0$$

that were exhibited in example 15, page 257, follow immediately from the distinct real roots $r = -4, 3$ of its characteristic equation

$$r^2 + r - 12 = (r + 4)(r - 3) = 0.$$

Example 12, page 253, verified that these two solutions are indeed linearly independent. Hence, a general solution of $y'' + y' - 12y = 0$ is

$$y_g(x) = C_1 e^{-4x} + C_2 e^{3x}.$$

■ **EXAMPLE 18**   *Find a general solution of*

$$6y'' - 12y' - 90y = 0$$

that is valid for all x.

Following the standard pattern, assume $y = e^{rx}$, substitute, and divide out the exponential to obtain the characteristic equation

Characteristic equation and roots

$$6r^2 - 12r - 90 = (3r + 9)(2r - 10) = 0.$$

Its roots are $r = -3, 5$. Hence, two solutions of $6y'' - 12y' - 90y = 0$ are

Homogeneous solutions

$$y_1 = e^{-3x}, \quad y_2 = e^{5x}.$$

To test for linear independence, evaluate the Wronskian

Test linear independence.

$$W(e^{-3x}, e^{5x}) = e^{-3x}(e^{5x})' - e^{5x}(e^{-3x})'$$
$$= 5e^{2x} + 3e^{2x} = 8e^{2x}.$$

Since $8e^{2x} \neq 0$ for all x, the solutions e^{-3x}, e^{5x} are linearly independent for all x. A general solution of $6y'' - 12y' - 90y = 0$ valid for all x is

General solution

$$y_g(x) = C_1 e^{-3x} + C_2 e^{5x}. ■$$

To summarize:

Distinct real roots. If the characteristic equation for a constant-coefficient, homogeneous, second-order linear equation has the *distinct real roots* r_1, r_2, then a pair of linearly independent solutions is

$$e^{r_1 x}, \quad e^{r_2 x}.$$

A general solution is $C_1 e^{r_1 x} + C_2 e^{r_2 x}$.

The exercises ask you to verify the claim that $e^{r_1 x}$ and $e^{r_2 x}$ are linearly independent.

6.3.3 Repeated Real Roots

Example 16 showed that the characteristic equation

$$r^2 + 8r + 16 = (r+4)^2 = 0$$

of the homogeneous differential equation

$$y'' + 8y' + 16y = 0$$

has the single repeated real root $r = -4$. Consequently, this differential equation has one solution,

$$y_1 = e^{-4x}.$$

How can we find a second linearly independent solution from which to construct a general solution?

To understand why the characteristic equation has provided only one solution, factor the derivative operators appearing in the differential equation. Let D denote the first derivative operator d/dx and recall that D^2, two applications of the first derivative operator, is just the second derivative operator d^2/dx^2.

$$y'' + 8y' + 16y = D^2 y + 8Dy + 16y$$
$$= (D+4)(D+4)y = 0.$$

Stop and Think **6.27** Carefully work through the meaning of this notation:

$$(D+4)(D+4)y = (D+4)(y'+4y) = D(y'+4y) + 4(y+4) = \cdots.$$

What happens when we substitute the solution $y_1 = e^{-4x}$ into this expression? The two factors $(D+4)$ appearing in the equation play different roles. The second one is boxed to emphasize the difference:

$$(D+4)\boxed{(D+4)}e^{-4x} = (D+4)\boxed{(D+4)e^{-4x}}$$
$$= (D+4)\boxed{(-4e^{-4x} + 4e^{-4x})}$$
$$= (D+4)\boxed{0} = 0.$$

The zero appears in the preceding box because e^{-4x} satisfies

$$(D+4)e^{-4x} = 0.$$

That is, e^{-4x} is a solution of the original second-order differential equation because it satisfies the *first*-order differential equation $(D+4)e^{-4x} = 0$. The left-hand factor $(D+4)$ in the differential equation never had any work to do because the right-hand one reduced e^{-4x} to zero entirely on its own.

This observation suggests that we could obtain a second solution y_2 if we could construct y_2 so that $\boxed{(D+4)y_2}$ is a nontrivial function (rather than the zero function obtained previously) which the left-hand factor $(D+4)$ could reduce to zero. That is, if we let

$$s(x) = \boxed{(D+4)y_2(x)}, \tag{6.8}$$

then we would like to choose y_2 so that

$$y_2'' + 8y_2' + 16y_2 = (D+4)\boxed{(D+4)y_2(x)}$$
$$= (D+4)s(x) = 0.$$

The last line tells us that $s(x)$ should be a solution of $Ds + 4s = s' + 4s = 0$; one choice is $s(x) = e^{-4x}$.

Stop and Think   **6.28**   Verify that one solution of $s' + 4s = 0$ is e^{-4x}. List all other possible solutions of $s' + 4s = 0$. Is *every one* of these solutions an acceptable choice for $s(x)$?

Using $s = e^{-4x}$ in (6.8) provides a first-order differential equation for the second solution y_2,

$$e^{-4x} = \boxed{(D+4)y_2(x)},$$

or

$$y_2' + 4y_2 = e^{-4x}.$$

This first-order, nonhomogeneous, linear, constant-coefficient equation can be solved by the method of undetermined coefficients. The solution y_2 has the form

$$y_2 = Axe^{-4x}.$$

> The first guess for the particular solution is $y_2 = Ae^{-4x}$, but it satisfies the homogeneous equation. Hence, we multiply it by x.

Since the actual value of A is unimportant here, we can simply take the second solution to be

$$y_2 = xe^{-4x}.$$

The exercises ask you to verify by direct substitution that y_2 is indeed a solution of $y'' + 8y' + 16y = 0$. They also ask you to show that xe^{-4x} is linearly independent of e^{-4x} for all x.

Stop and Think   **6.29**   Justify the claim that "the actual value of A is unimportant here." (*Hint*: The goal of this analysis is a general solution of a linear, homogeneous differential equation.)

The pattern suggested by this analysis is one of multiplying the first solution by the independent variable to obtain the second. The exercises ask you to verify in general that:

Repeated real roots.   If the characteristic equation for a constant-coefficient, homogeneous, second-order linear equation has the single *repeated real root* r, then a pair of linearly independent solutions is

$$e^{rx}, \quad xe^{rx}.$$

A general solution is $C_1 e^{rx} + C_2 x e^{rx}$.

■ **EXAMPLE 19**   *Find a general solution of*

$$4y'' - 12y' + 9y = 0.$$

To find the characteristic equation, substitute the assumed solution $y = e^{rx}$ into the differential equation. We obtain

Characteristic equation

$$4r^2 - 12r + 9 = 4\left(r - \tfrac{3}{2}\right)^2 = 0.$$

The characteristic equation evidently has the single repeated real root $r = 3/2$. The preceding discussion indicates that two solutions of $4y'' - 12y' + 9y = 0$ are

Two homogeneous solutions

$$y_1 = e^{3x/2}, \quad y_2 = x e^{3x/2}.$$

To assess their linear independence, evaluate the Wronskian:

Test linear independence.

$$W(e^{3x/2}, x e^{3x/2}) = e^{3x/2}(x e^{3x/2})' - x e^{3x/2}(e^{3x/2})' = e^{3x} \neq 0$$

for all x. Hence, these two solutions are linearly independent for all x. A general solution of $4y'' - 12y' + 9y = 0$ is

General solution

$$y_g = C_1 e^{3x/2} + C_2 x e^{3x/2}. ■$$

■ **EXAMPLE 20**   *Solve the initial-value problem*

$$4y'' - 12y' + 9y = 0, \quad y(0) = 0, \quad y'(0) = -3.$$

From the previous example, we have the general solution

Find general solution.

$$y_g = C_1 e^{3x/2} + C_2 x e^{3x/2}.$$

Use initial conditions
to evaluate C_1, C_2.

The initial condition $y(0) = 0$ forces $C_1 = 0$. Then calculate

$$y_g' = C_2 \left(\frac{3x e^{3x/2}}{2} + e^{3x/2} \right).$$

The second initial condition $y'(0) = -3$ forces $C_2 = -3$. The solution of the initial-value problem is

Solution of initial-value problem

$$y(x) = -3x e^{3x/2}. ■$$

6.3.4 Exercises

EXERCISE GUIDE	
To gain experience ...	Try exercises
Using the characteristic equation method	1–17, 23
Finding equations, given solutions	11–17
Verifying linear independence of solutions	1–10, 18–19, 20–21(b)
Verifying given functions are solutions	20–22(a)
Finding general solutions	1–10, 18–19, 20–21(c), 23
Solving initial-value problems	1–10
With the foundations of the characteristic equation method	11–17, 22(b), 23–24
Understanding repeated roots	22, 24–26

In exercises 1–10,

(i) Find the characteristic equation of the given differential equation.

(ii) Find two linearly independent solutions of the differential equation.

(iii) Verify that these solutions are linearly independent for all values of the independent variable.

(iv) Write a general solution of the differential equation.

(v) Solve the differential equation subject to the given initial conditions.

(vi) Confirm the general solution you obtain using the analytic solution tool in DELAB.

1. $2y'' - 2y' - 4y = 0$, $y(0) = 1$, $y'(0) = -4$

2. $y'' = y$, $y(0) = 0$, $y'(0) = 2$

3. $y'' = -2y' + y$, $y(0) = 1$, $y'(0) = -1$

4. $x''(t) + x'(t) - 6x(t) = 0$, $x(1) = 6$, $x'(1) = 0$

5. $2w''(x) - 3w'(x) + w(x) = 0$, $w(0) = 2$, $w'(0) = 1$

6. $8z(t) - 10z'(t) - 3z''(t) = 0$, $z(0) = 2$, $z'(0) = 1$

7. $4y'' - 8y' + 4y = 0$, $y(0) = -3$, $y'(0) = 4$

8. $x''/3 - 2x' - 9x = 0$, $x(0) = 0$, $x'(0) = 1$

9. $12u'' + 5u' - 2u = 0$, $u(1) = 1$, $u'(1) = -1$

10. $y'' + 2ay' + a^2y = 0$, $y(0) = m$, $y'(0) = n$, a, m, n constants

In exercises 11–17, write a constant-coefficient, second-order differential equation that has the given pair of functions as solutions. Write the most general equation you can.

11. e^x, e^{-x}

12. e^{2x}, e^{-3x}

13. e^x, xe^x

14. $-e^{-x}$, xe^{-x}

15. e^{x+1}, e^{2x-1}

16. e^x, π

17. e^x

18. Verify by direct substitution that

$$y_1 = e^{-4x}, \quad y_2 = e^{3x}$$

are solutions of the differential equation

$$y'' + y' - 12y = 0$$

of example 15. Show that y_1, y_2 are linearly independent. Construct a general solution of this differential equation.

19. If the characteristic equation

$$b_2r^2 + b_1r + b_0 = 0$$

of the general constant-coefficient, homogeneous equation

$$b_2y'' + b_1y' + b_0y = 0$$

has the distinct real roots $r_1, r_2, r_1 \neq r_2$, then the differential equation has two solutions,

$$y_1 = e^{r_1x}, \quad y_2 = e^{r_2x}.$$

Use the linear independence test to verify that these two solutions are linearly independent and, hence, that a general solution of this differential equation is

$$y_g(x) = C_1 e^{r_1 x} + C_2 e^{r_2 x}.$$

Is the condition $r_1 \neq r_2$ essential for linear independence? Is the condition that the roots be real essential? (*Hint:* Example 18 demonstrates linear independence of solutions when the roots of the characteristic equation are $r = -3, 5$. Can you use it as a guide to the general situation considered here?)

20. The discussion at the end of this section considers the constant-coefficient, homogeneous equation

$$y'' + 8y' + 16y = 0,$$

whose characteristic equation has the single repeated real root -4. Hence, one solution of this equation is $y_1 = e^{-4x}$. The text claims that a second linearly independent solution is $y_2 = xe^{-4x}$.

(a) Verify by substitution in $y'' + 8y' + 16y = 0$ that $y_2 = xe^{-4x}$ is indeed a solution.

(b) Verify that these two solutions are linearly independent for all x.

(c) Write a general solution of $y'' + 8y' + 16y = 0$.

21. The discussion at the end of this section claims

If the characteristic equation for a constant-coefficient, homogeneous, second-order linear equation has the single repeated real root r, then a pair of linearly independent solutions is e^{rx}, xe^{rx}. A general solution is $C_1 e^{rx} + C_2 xe^{rx}$.

Verify this claim by completing the following steps for the general equation

$$b_2 y'' + b_1 y' + b_0 y = 0.$$

You may wish to generalize your solution of the preceding exercise, which asks for much the same analysis in the context of a specific equation.

(a) Verify by direct substitution that y_2 is a solution of $b_2 y'' + b_1 y' + b_0 y = 0$. (*Hint:* Use the quadratic equation to find r in terms of the coefficients b_i. Remember that r is a repeated root. What does that fact tell you about the discriminant $b_1^2 - 4b_2 b_0$?)

(b) Verify that y_1 and y_2 are linearly independent for all x.

(c) Write a general solution of $b_2 y'' + b_1 y' + b_0 y = 0$.

22. (a) Verify by direct substitution that

$$y_1 = e^{3x/2}, \quad y_2 = xe^{3x/2}$$

are solutions of

$$4y'' - 12y' + 9y = 0,$$

the differential equation considered in example 19.

(b) Using $D \equiv d/dx$, write the differential equation $4y'' - 12y' + 9y = 0$ in the factored form used in the text,

$$4 \left(D - \tfrac{3}{2} \right) \boxed{\left(D - \tfrac{3}{2} \right) y} = 0.$$

Mimic the analysis on pages 261–262, preceding example 19, to answer the following questions:

(i) When $y = e^{3x/2}$, what expression does the factor $\left(D - \tfrac{3}{2} \right)$ outside the box operate upon? Is that expression a solution of $Dy - 3y/2 = dy/dx - 3y/2 = 0$?

(ii) When $y = xe^{3x/2}$, what expression does the factor $\left(D - \tfrac{3}{2} \right)$ outside the box operate upon? Is that expression a solution of $Dy - 3y/2 = dy/dx - 3y/2 = 0$?

23. By studying the discriminant $b_1^2 - 4b_2 b_0$ of the characteristic equation $b_2 r^2 + b_1 r + b_0 = 0$, give conditions on the coefficients b_2, b_1, b_0 that guarantee that the constant-coefficient differential equation

$$b_2 y'' + b_1 y' + b_0 = 0$$

has two linearly independent solutions of the form

(a) $e^{r_1 x}, e^{r_2 x}$,

(b) e^{rx}, xe^{rx}.

For each case, write out the general solution in terms of the coefficients b_2, b_1, b_0.

24. (a) Find the undetermined coefficient A in the assumed form

$$y_2 = Axe^{-4x}.$$

of a particular solution of

$$y_2' + 4y_2 = e^{-4x},$$

the first-order equation we encountered when we were trying to find a second solution of the equation

$$y'' + 8y' + 16y = 0,$$

whose characteristic equation has a single repeated real root.

(b) Verify that y_2 is a solution of $y'' + 8y' + 16y = 0$ for *any* choice of the constant A, including the value you find. This last argument justifies taking $A = 1$ in the discussion in the text.

25. The differential equation

$$y'' + 8y' + 16y = 0$$

has the single repeated characteristic root $r = -4$, and one solution is obviously $y_1 = e^{-4x}$. The last part of this section argues that a second solution is xe^{-4x}. This exercise outlines an alternate approach, known as *reduction of order*, to deriving that second solution.

(a) Assume that a second solution has the form $y_2 = u(x)e^{-4x}$, where $u(x)$ is an unknown function. Sub-

stitute y_2 into the differential equation and show that it reduces to $u'' = 0$.

(b) Argue that one solution of $u'' = 0$ is $u(x) = x$ and, hence, that a second solution of the differential equation is $y_2(x) = xe^{-4x}$.

26. Repeat the preceding exercise for the general equation $b_2 y'' + b_1 y' + b_0 y = 0$ with a single repeated characteristic root. (*Hint:* What is the value of the discriminant $b_1^2 - 4b_2 b_0$ when the characteristic equation has a single repeated root?)

6.4 ■ CHARACTERISTIC EQUATIONS: COMPLEX ROOTS

Constant-coefficient, linear, homogeneous differential equations have exponential solutions if their characteristic roots are real and distinct or real and repeated. This section determines the form of the solution when the characteristic roots are complex conjugates.

6.4.1 Complex Conjugate Roots

The characteristic equation

$$r^2 + 8r + 20 = 0$$

for the differential equation

$$y'' + 8y' + 20y = 0$$

has the pair of complex conjugate roots $r = -4 + 2i, -4 - 2i$. The corresponding solutions of $y'' + 8y' + 20y = 0$ are $e^{(-4+2i)x}$, $e^{(-4-2i)x}$. How can these complex exponentials be expressed in terms of real-valued functions? The principle of superposition provides the answer.

To convert complex exponential solutions to real-value solutions, recall the Taylor series (see appendix, section 3) about the origin for the exponential, the cosine, and the sine

$$e^x = 1 + x + \frac{x^2}{2} + \frac{x^3}{6} + \frac{x^4}{24} + \frac{x^5}{120} + \cdots,$$

$$\cos x = 1 - \frac{x^2}{2} + \frac{x^4}{24} + \cdots,$$

$$\sin x = x - \frac{x^3}{6} + \frac{x^5}{120} + \cdots.$$

Now let β be a real number, and write the series for the exponential with $x = i\beta$,

$$e^{i\beta} = 1 + i\beta + \frac{(i\beta)^2}{2} + \frac{(i\beta)^3}{6} + \frac{(i\beta)^4}{24} + \frac{(i\beta)^5}{120} + \cdots$$

$$= 1 - \frac{\beta^2}{2} + \frac{\beta^4}{24} + \cdots + i\left(\beta - \frac{\beta^3}{6} + \frac{\beta^5}{120} + \cdots\right)$$

$$= \cos\beta + i\sin\beta.$$

The last equality follows from comparing the preceding series for $e^{i\beta}$ with those for $\cos x$ and $\sin x$ using $x = \beta$.

This relationship is called **Euler's formula**,

EULER'S FORMULA

$$e^{i\beta} = \cos\beta + i\sin\beta.$$

■ **EXAMPLE 21** *Use Euler's formula to write two linearly independent, real-valued solutions of*

$$y'' + 8y' + 20y = 0.$$

Earlier work found that the characteristic equation $r^2 + 8r + 20 = 0$ has the complex conjugate roots $-4 \pm 2i$. Hence, the differential equation has the complex exponential solutions

Complex exponential solutions

$$z_1 = e^{(-4+2i)x}, \quad z_2 = e^{(-4-2i)x}.$$

Use Euler's formula with $\beta = 2x$ to rewrite z_1 and z_2:

Apply Euler's formula.

$$z_1 = e^{(-4+2i)x} = e^{-4x}e^{i(2x)} = e^{-4x}(\cos 2x + i\sin 2x),$$

$$z_2 = e^{(-4-2i)x} = e^{-4x}e^{i(-2x)} = e^{-4x}(\cos 2x - i\sin 2x).$$

Since both z_1 and z_2 are solutions of the homogeneous linear differential equation $y'' + 8y' + 20y = 0$, the principle of superposition guarantees that *any linear combination $C_1 z_1 + C_2 z_2$ is again a solution.* Is there a linear combination that would eliminate the imaginary terms from z_1 and z_2? Indeed, the sum of these two expressions is real-valued:

Sum of homogeneous solutions

$$z_1 + z_2 = 2e^{-4x}\cos 2x.$$

The function $2e^{-4x}\cos 2x$ is still a solution of $y'' + 8y' + 20y = 0$ because it is a *linear combination of solutions of this homogeneous linear equation.*

Similarly, any constant multiple of a solution of a homogeneous linear equation is again a solution of that equation. Multiply this new solution by $1/2$ to obtain a real-valued solution of $y'' + 8y' + 20y = 0$:

Constant multiple of a homogeneous solution

$$y_1 = e^{-4x}\cos 2x.$$

Another linear combination gives a purely imaginary solution,

Sum of homogeneous solutions

$$z_1 - z_2 = 2ie^{-4x}\sin 2x.$$

Using the same reasoning as in the preceding paragraph, multiply this homogeneous solution by $1/2i$ to obtain a second real-valued solution of $y'' + 8y' + 20y = 0$,

Constant multiple of a homogeneous solution

$$y_2 = e^{-4x} \sin 2x.$$

The exercises ask you to verify by direct substitution that y_1 and y_2 are indeed solutions of the differential equation.

To show that y_1, y_2 are linearly independent solutions, evaluate their Wronskian,

$$W(e^{-4x} \cos 2x,\ e^{-4x} \sin 2x)$$

Test linear independence.

$$= e^{-4x} \cos 2x\,(e^{-4x} \sin 2x)' - e^{-4x} \sin 2x\,(e^{-4x} \cos 2x)'$$

$$= 2e^{-8x}(\cos^2 2x + \sin^2 2x) = 2e^{-8x} \neq 0$$

for all x. Hence, for all x, a real-valued general solution of $y'' + 8y' + 20y = 0$ is

Real-valued general solution

$$y_g = C_1 e^{-4x} \cos 2x + C_2 e^{-4x} \sin 2x.$$

MATLAB

Does DELAB obtain a real-valued or a complex-valued general solution of $y'' + 8y' + 20y = 0$?

Obviously (meaning the details are left to the exercises), this analysis can be carried over without change to the general case:

Complex conjugate roots. A constant-coefficient, homogeneous, linear differential equation whose characteristic equation has the complex conjugate roots $r = \alpha + i\beta,\ \alpha - i\beta$ has the general solution

$$y_g = C_1 e^{\alpha x} \cos \beta x + C_2 e^{\alpha x} \sin \beta x.$$

The *real part* of the characteristic root appears in the *exponential*, and the *complex part* appears as the angular frequency in the *cosine* and *sine*.

The constant β in both $\cos \beta x$ and $\sin \beta x$ is called the **angular frequency** or **circular frequency** because it measures the number of radians by which the angle βx in either the cosine or the sine increases when x increases by one unit.

The quotient $\beta/2\pi$ is the **cyclic frequency**, or simply **frequency**; it measures the number of complete cycles executed by the cosine or sine when x increases by one unit; one *hertz* (after the German physicist H. R. Hertz), abbreviated Hz, is one cycle per second.

The inverse $2\pi/\beta$ of cyclic frequency is the **period**, the time required to complete one full cycle. See section 6.5.1.

6.4.2 More Examples

Examples of equations with both real and complex characteristic roots follow.

■ **EXAMPLE 22** *Solve the initial-value problem*

$$\frac{3}{16}x''(t) + 1.41x'(t) + 2x(t) = 0, \quad x(0) = 0.75, \quad x'(0) = 0,$$

which models the damped spring-mass system of example 2, page 211.

The usual assumption $x(t) = e^{rt}$ leads to the characteristic equation

Characteristic equation

$$\frac{3}{16}r^2 + 1.41r + 2 = 0.$$

The quadratic formula reveals that its roots are approximately $r = -1.87$, -5.65. Hence, two linearly independent solutions of the differential equation are $e^{-1.87t}$, $e^{-5.65t}$, and a general solution is

General solution

$$x_g(t) = C_1 e^{-1.87t} + C_2 e^{-5.65t}.$$

To solve the initial-value problem, use the initial conditions to determine C_1, C_2. These conditions require

Apply initial conditions.

$$x_g(0) = C_1 + C_2 = 0.75,$$
$$x_g'(0) = -1.87C_1 - 5.65C_2 = 0.$$

Solving this pair of simultaneous equations yields

Solve for C_1, C_2.

$$C_1 = 1.12, \quad C_2 = -0.37.$$

The solution of the given initial-value problem is

Solution of initial-value problem

$$x(t) = 1.12e^{-1.87t} - 0.37e^{-5.65t}. \quad ■$$

It appears that this spring-mass system does not oscillate and that it approaches the steady state $x_{ss} = 0$. The solution itself is the transient component of the response.

Such a system, one that decays to rest without oscillating, is called **overdamped**. See figure 6.1. *Overdamped* systems are signaled by *negative, real* characteristic roots.

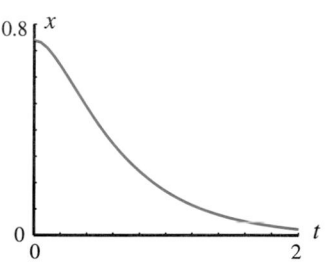

FIGURE 6.1 Position vs. time for the overdamped spring-mass system of example 22.

■ **EXAMPLE 23** *Consider the spring-mass system of the preceding example with a slightly stronger spring, one with a spring constant $k = 2.67$ lb/ft. The governing initial-value problem is*

$$\frac{3}{16}x''(t) + 1.41x'(t) + 2.67x(t) = 0, \quad x(0) = 0.75, \quad x'(0) = 0.$$

Find its solution.

Assume $x = e^{rt}$, substitute, and divide out e^{rt} to find the characteristic equation

Characteristic equation

$$\frac{3}{16}r^2 + 1.41r + 2.67 = 0.$$

Since its discriminant

$$(1.41)^2 - 4\frac{3}{16}2.67$$

is zero, this characteristic equation has the single repeated real root $r = -3.76$. Two linearly independent solutions of the differential equation are thus $e^{-3.76t}$, $te^{-3.76t}$. A general solution is

General solution

$$x_g(t) = C_1 e^{-3.76t} + C_2 te^{-3.76t}.$$

The initial conditions lead to two equations for the constants C_1, C_2,

Apply initial conditions.

$$x_g(0) = C_1 = 0.75,$$
$$x'_g(0) = -3.76C_1 + C_2 = 0.$$

Their solution is $C_1 = 0.75$, $C_2 = 2.82$.
The solution of the given initial-value problem is

Write solution of initial-value problem.

$$x(t) = 0.75e^{-3.76t} + 2.82te^{-3.76t}. \quad ■$$

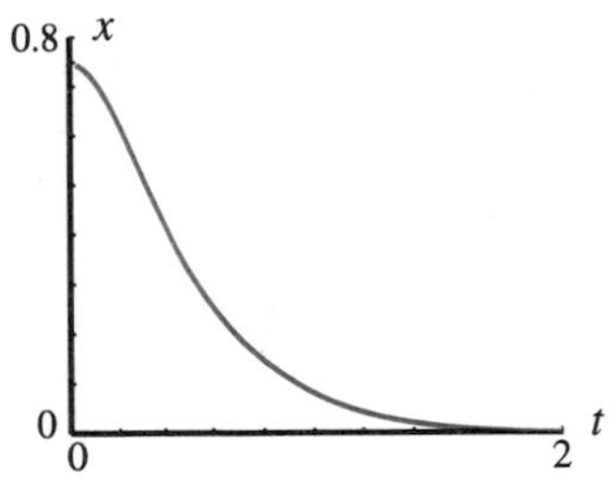

FIGURE 6.2 Position vs. time for the critically damped spring-mass system of example 23.

This solution also decays to the steady state $x_{ss} = 0$ without oscillating. See figure 6.2. A system such as this with a *single negative (real) characteristic root* is called **critically damped** because it separates overdamped systems such as that of example 22 from damped, oscillating systems like the one in the next example.

■ **EXAMPLE 24**   *Now consider the spring-mass system of the last example with a still stronger spring, one with spring constant $k = 4$ lb/ft. The governing initial-value problem is*

$$\frac{3}{16}x''(t) + 1.41x'(t) + 4x(t) = 0, \quad x(0) = 0.75, \quad x'(0) = 0.$$

Find its solution.

The usual argument yields the characteristic equation

Characteristic equation

$$\frac{3}{16}r^2 + 1.41r + 4 = 0.$$

Its roots are $r = -3.76 + 2.68i, -3.76 - 2.68i$.
Rather than write the complex exponential form of the solutions arising from these roots and form linear combinations to obtain real-valued solutions, proceed directly to the real-valued solutions. The real part -3.76 of the root

governs the exponential, and the imaginary part 2.68 is the frequency in the cosine and sine terms. The general solution is

General solution

$$x_g(t) = e^{-3.76t}(C_1 \cos 2.68t + C_2 \sin 2.68t).$$

The initial conditions require that C_1, C_2 be the solution of

Apply initial conditions.

$$x_g(0) = C_1 = 0.75,$$
$$x'_g(0) = -3.76C_1 + 2.68C_2 = 0.$$

The solution of this pair of equations is $C_1 = 0.75$, $C_2 = 1.05$. The solution of the given initial-value problem is

Write solution of initial-value problem.

$$x(t) = e^{-3.76t}(0.75 \cos 2.68t + 1.05 \sin 2.68t). \quad ■$$

The solution oscillates because of the cosine and sine terms, but the exponential factor still decays to the steady state $x_{ss} = 0$. See figure 6.3. Such a system is called **underdamped**. *Underdamped* systems are signaled by *complex conjugate characteristic roots with negative real part.*

Stop and Think

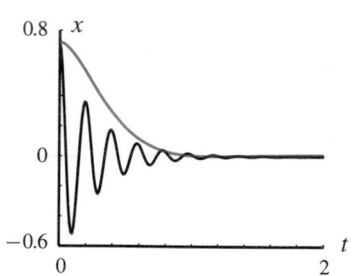

FIGURE 6.3 Position vs. time for the underdamped spring-mass system of example 24 (blue) and a similar system with a stronger spring (black).

6.30 Underdamped spring-mass systems are supposed to oscillate as they decay. Why does the plot of the solution of example 24 in figure 6.3 (blue curve) not show any oscillations? (*Hint:* The solution of $3x''/16 + 1.41x' + 200x = 0$, $x(0) = 0.75$, $x'(0) = 0$, is also plotted there (blue curve). That solution function is

$$x(t) \approx e^{-3.76t}(0.75 \cos 32.4t + 0.0869 \sin 32.4t).)$$

6.31 The text says that underdamped spring-mass systems—those with decaying oscillations—require "complex conjugate characteristic roots with negative real part." Which aspect of this behavior comes from the complex roots? Which aspect from the negative real part?

6.32 Spring-mass systems that are overdamped, critically damped, or underdamped all have characteristic roots with negative real part. What common behavior is guaranteed by that negative real part?

6.33 Spring-mass systems that are overdamped, critically damped, or underdamped all possess the steady state $x_{ss} = 0$. Is that steady state stable or unstable?

6.34 Describe a phase plane plot—a graph of $v = x'$ vs. x—for an underdamped spring-mass system.

■ **EXAMPLE 25** *Solve the initial-value problem*

$$\frac{3}{16}x''(t) + 2x(t) = 0, \quad x(0) = 0, \quad x'(0) = -0.5,$$

which models the undamped spring-mass system of example 1, section 5.1.

The characteristic equation is

Characteristic equation

$$\frac{3}{16}r^2 + 2 = 0.$$

It has the pure imaginary roots $r = \pm\sqrt{32/3}\,i \approx \pm 3.27i$. Since the real part of these complex roots is zero, the general solution contains no exponential term. However, the imaginary part 3.27 enters as the frequency in the cosine and sine terms. The general solution of this differential equation is

General solution

$$x_g(t) = C_1 \cos 3.27t + C_2 \sin 3.27t.$$

The initial conditions require

Apply initial conditions.

$$x_g(0) = C_1 = 0,$$
$$x'_g(0) = 3.27C_2 = -0.5.$$

The solution of this pair of equations is obviously $C_1 = 0$, $C_2 = -0.15$. The solution of the given initial-value problem is

Solution of initial-value problem

$$x(t) = -0.15 \sin 3.27t. \quad ■$$

This solution exhibits no decay whatsoever; this system is **undamped**. It has a periodic solution with period $2\pi/3.27 = 1.92$ s. See figure 6.4. *Undamped* systems are signaled by *pure imaginary characteristic roots*.

Equivalently, the circular frequency of this solution is 3.27 rad/s. Its cyclic frequency is $3.27/2\pi = 0.52$ Hz. In other words, it takes 1.92 s for the mass to complete a round-trip, bouncing down and then back up to its starting position, or it completes 0.52 round-trip every second.

Stop and Think

6.35 Deduce from each of figures 6.1–6.4 the initial values that govern the solution being plotted. Confirm your deduction by referring to the relevant example.

6.36 List the four types of spring-mass systems—overdamped, critically damped, underdamped, and undamped. Label each type with a description of the characteristic roots that identify it. Argue that each type of spring-mass system can be identified by the value of the discriminant of its characteristic equation.

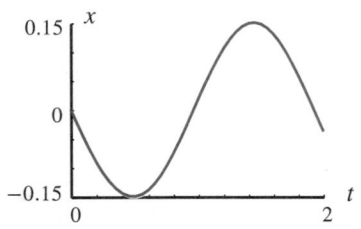

FIGURE 6.4   Position vs. time for the undamped spring-mass system of example 25.

MATLAB

Use the analytic solution tool in DELAB to verify the analytic solutions exhibited in each of the preceding examples. Use the graphical tool to produce your own versions of figures 6.1–6.4 and to verify your conjectures about the nature of phase plots of underdamped and undamped spring-mass systems.

6.4.3   Summary of Characteristic Equations

The constant-coefficient, homogeneous, second-order linear equation

$$b_2 y'' + b_1 y' + b_0 y = 0$$

can be solved by assuming $y = e^{rx}$, r a constant. Substituting this proposed solution in the differential equation and dividing by the exponential leaves the *characteristic equation*

CHARACTERISTIC EQUATION

$$b_2 r^2 + b_1 r + b_0 = 0.$$

Since the coefficients b_i are real, this quadratic equation may have

TYPES OF CHARACTERISTIC ROOTS

1. Distinct real roots r_1, r_2 if the discriminant $b_1^2 - 4b_2b_0 > 0$,
2. A single real root r if the discriminant $b_1^2 - 4b_2b_0 = 0$,
3. Complex conjugate roots $\alpha \pm i\beta$ if the discriminant $b_1^2 - 4b_2b_0 < 0$.

Stop and Think   **6.37** Explain why the discriminant of the characteristic equation determines the type of characteristic root.

LINEARLY INDEPENDENT HOMOGENEOUS SOLUTIONS

Real-valued, linearly independent solutions of $b_2 y'' + b_1 y' + b_0 y = 0$ **are**

Distinct real roots r_1, r_2: $y_1 = e^{r_1 x}$, $y_2 = e^{r_2 x}$.

Repeated real root r: $y_1 = e^{rx}$, $y_2 = xe^{rx}$.

Complex conjugate roots $\alpha \pm i\beta$: $y_1 = e^{\alpha t} \sin \beta t$, $y_2 = e^{\alpha t} \cos \beta t$.

6.4.4 Exercises

EXERCISE GUIDE	
To gain experience . . .	**Try exercises**
Using the characteristic equation method	1–16
Verifying linear independence of solutions	1–16, 28(d)
Verifying that given functions are solutions	26(b)
Deriving real-valued solutions from complex-valued solutions	26–29
Finding general solutions	1–23, 24–25(a), 27(a), 28(e)
Solving initial-value problems	17–23, 24(c), 25(d), 27(b)
Finding steady states	30–31
Analyzing solution behavior	24(b–d, f), 25(b, c), 33–34
With the foundations of the characteristic equation method	32–33

In exercises 1–16,

   (i) Find the characteristic equation of the given differential equation. If the concept of characteristic equation is not appropriate, explain why and go on to the next exercise.

   (ii) Find two linearly independent solutions of the differential equation.

   (iii) Verify that these solutions are linearly independent for all values of the independent variable.

   (iv) Write a general solution of the differential equation.

   (v) Confirm the general solution you obtain using the analytic solution tool in DELAB.

1. $y'' - 9y = 0$

2. $4x''(t) + 8x'(t) + 5x(t) = 0$

3. $w''(x) + 9w(x) = 0$

4. $2z''(t) - 7z'(t) + 3z(t) = 0$

5. $y'' - 4y' + 40y = 0$

6. $y'' - 4y' + 40y = \cos \pi x$

7. $x'' + x' - 6x = 0$

8. $9y'' + 36y' + 4y = 0$

9. $9y'' + 36y' + 4y^2 = 0$

10. $mx''(t) + kx(t) = 0$, m, k constants

11. $mx''(t) + kx(t) = k[h(t) - h(0)]$, m, k constants

12. $L\theta'' + g\theta = 0$, L, g constants

13. $L\theta'' + g \sin \theta = 0$, L, g constants

14. $mx'' + px' + kx = 0$, m, p, k constants, $p^2 - 4mk > 0$

15. $mx'' + px' + kx = 0$, m, p, k constants, $p^2 - 4mk = 0$

16. $mx'' + px' + kx = 0$, m, p, k constants, $p^2 - 4mk < 0$

In exercises 17–23,
    (i) Write a general solution of the differential equation.
    (ii) Solve the initial-value problem.

17. $u'' - 4u' + 40u = 0$, $u(\pi/6) = 1$, $u'(\pi/6) = -1$

18. $9x'' + 36x' + 4x = 0$, $x(0) = 2$, $x'(0) = 3$

19. $4y'' + 8y' + 5y = 0$, $y(\pi) = 0$, $y'(\pi) = -4$

20. $z'' + z' - 6z = 0$, $z(0) = 4$, $z'(0) = -6$

21. $x'' + 9x = 0$, $x(\pi) = 3$, $x'(\pi) = -9$

22. $2w'' - 7w' + 3w = 0$, $w(0) = 8$, $w'(0) = 12$

23. $x'' - 9x = 0$, $x(0) = 18$, $x'(0) = 18$

24. In the undamped spring-mass system model derived in section 5.1,

$$mx''(t) + kx(t) = 0, \quad x(0) = x_i, \quad x'(0) = v_i,$$

the parameters m, k in the differential equation are positive constants.

(a) Find a general solution of the differential equation.

(b) Argue that your solution is purely oscillatory, neither growing nor decaying with time.

(c) Find the angular frequency and period of these oscillations.

(d) Argue that the period of the oscillations in this spring-mass system depends only on the mass m and the spring constant k, not upon the initial conditions; i.e., the period of the oscillations is independent of how the motion is initiated.

(e) Find the solution of this initial-value problem.

(f) Find the maximum displacement of the mass from its equilibrium position. Does it reduce to the value you expect when the initial velocity v_i is zero?

25. In the damped spring-mass system model derived in section 5.1,

$$mx''(t) + px'(t) + kx(t) = 0,$$
$$x(0) = x_i, \quad x'(0) = v_i,$$

the parameters m, p, k in the differential equation are positive constants.

(a) Find a general solution of this differential equation. If you need to consider different ranges of the parameters m, p, k, state them clearly.

(b) Argue that your solution is decaying toward a steady state of zero, regardless of the values of m, p, k. Which parameter(s) in the equation determine(s) the rate of decay? Are they the one(s) you expect?

(c) For what ranges of the parameters does the solution of this equation exhibit oscillations? Are the underlying oscillations of the same frequency as those of the undamped spring-mass model? (See exercise 24.)

(d) Find the solution of the initial-value problem.

26. The first example in this section begins with two complex exponential solutions

$$z_1 = e^{(-4+2i)x}, \quad z_2 = e^{(-4-2i)x}$$

of the differential equation

$$y'' + 8y' + 20y = 0.$$

It then obtains two new solutions

$$y_1 = e^{-4x} \cos 2x, \quad y_2 = e^{-4x} \sin 2x$$

as linear combinations of z_1, z_2.

(a) Write y_1 and y_2 as linear combinations of z_1, z_2.

(b) Verify by substitution that y_1, y_2 are indeed solutions of $y'' + 8y' + 20y = 0$. Why is this direct verification actually unnecessary?

27. The last example in this section shows that the differential equation

$$\tfrac{3}{16}x''(t) + 2x(t) = 0,$$

from the undamped spring-mass model of example 1, section 5.1, has the characteristic equation

$$\tfrac{3}{16}r^2 + 2 = 0.$$

The roots of this equation are $r = \pm\sqrt{32/3}\,i \approx \pm 3.27i$. Consequently, $3x''/16 + 2x = 0$ has the linearly independent solutions

$$z_1 = e^{3.27it}, \quad z_2 = e^{-3.27it}.$$

(a) Write a general solution of $3x''/16 + 2x = 0$ in terms of the complex exponential solutions z_1, z_2.

(b) Choose the constants in your answer to part (a) so that your solution satisfies the initial conditions

$$x(0) = 0, \quad x'(0) = -0.5.$$

(Your constants will be complex numbers.)

(c) Use Euler's formula to reduce your (apparently) complex-valued solution of this initial-value problem to that obtained in example 25 in the text,

$$x(t) = -0.15 \sin 3.27t.$$

28. The differential equation

$$\tfrac{3}{16}x''(t) + 1.41x'(t) + 4x(t) = 0$$

considered in example 24 has the characteristic equation

$$\tfrac{3}{16}r^2 + 1.41r + 4 = 0.$$

Its roots are $r = -3.76 + 2.68i$, $-3.76 - 2.68i$. The example proceeded from these roots directly to a real-valued general solution of the differential equation. This exercise asks you to derive that real-valued solution in more detail.

(a) Write a general solution of $3x''/16 + 1.41x' + 4x = 0$ in terms of complex exponentials.

(b) Choose a set of values for the arbitrary constants in your general solution that reduces it, via Euler's formula, to the real-valued function $e^{-3.76t} \cos 2.68t$.

(c) Choose another set of values for the arbitrary constants in your general solution that reduces it, via Euler's formula, to the real-valued function $e^{-3.76t} \sin 2.68t$.

(d) Show that the solutions obtained in parts (b) and (c) are linearly independent.

(e) Write a general solution of $3x''/16 + 1.41x' + 4x = 0$ using these real-valued solutions.

29. Suppose that the characteristic equation for a homogeneous, constant-coefficient, second-order, linear differential equation has the complex conjugate roots $r = \alpha \pm i\beta$. Using the preceding exercise as a guide, argue that a general solution of such a differential equation is

$$e^{\alpha x}(C_1 \cos \beta x + C_2 \sin \beta x).$$

30. Find a steady state, if one exists, of each of the initial-value problems in exercises 17–23.

31. Suppose you are asked to find steady states, if they exist, for a series of constant-coefficient initial-value problems. Can you determine if a steady state exists simply by looking at the roots of the characteristic polynomial? What roots must the polynomial possess to give a (constant) steady-state solution? Could you predict the steady state of an initial-value problem without forcing from knowledge of the roots of the characteristic polynomial alone? Justify your reasoning. Try your ideas out on some of the initial-value problems of exercises 17–23.

32. The characteristic equation for the general homogeneous, constant-coefficient, linear, second-order equation

$$b_2 y'' + b_1 y' + b_0 y = 0$$

is

$$b_2 r^2 + b_1 r + b_0 = 0.$$

The discriminant of this quadratic equation is

$$d = b_1^2 - 4b_2 b_0.$$

To justify the statements in the summary at the end of this section, argue that the sign of d determines which of three cases (real distinct, single repeated, or complex conjugate roots of $b_2 r^2 + b_1 r + b_0 = 0$) occurs. Specifically, show each of the following.

(a) If $d > 0$, then the characteristic equation has real, distinct roots.

(b) If $d = 0$, then the characteristic equation has a single repeated real root.

(c) If $d < 0$, then the characteristic equation has a pair of complex conjugate roots.

33. Suppose the discriminant of a characteristic equation is negative. (See the preceding exercise.) What condition(s) on the coefficients of the homogeneous differential equation guarantee(s) the solution will decay to a steady state of zero?

34. Example 23 considers the initial-value problem

$$\tfrac{3}{16}x''(t) + 1.41x'(t) + 2.67x(t) = 0,$$
$$x(0) = 0.75, \quad x'(0) = 0,$$

and claims that it models the same spring-mass system as does the initial-value problem

$$\tfrac{3}{16}x''(t) + 1.41x'(t) + 2x(t) = 0,$$
$$x(0) = 0.75, \quad x'(0) = 0,$$

but for a slightly stiffer spring. Justify this claim. How is the term *stiffer spring* reflected in relative values of spring constants?

6.5 ■ ANALYZING MODELS WITHOUT FORCING

Much of the point of an analytic solution technique like characteristic equations is the insight it provides into solution behavior. This section explores some of those insights for several different systems that lack external forcing.

6.5.1   Spring-mass Systems

Recall initial-value problems such as

$$mx''(t) + px'(t) + kx(t) = 0, \quad x(0) = x_i, \quad x'(0) = v_i,$$

that were derived in section 5.1 as models of spring-mass systems without external forcing. Here m is the mass, x_i is the initial position in a coordinate system whose origin is at the rest position of the mass, and v_i is the initial velocity given to the mass. The spring constant k is the proportionality constant in Hooke's law, the experimental observation that force exerted by the spring is proportional to its extension. The damping coefficient p serves a parallel purpose as the proportionality constant relating friction or other damping forces to velocity.

The differential equation $mx''(t) + px'(t) + kx(t) = 0$ is simply a statement of Newton's law. The term mx'' is mass times acceleration. It is balanced by the sum of the force of friction and the force of the spring, $-px' - kx$.

Undamped Spring-mass Systems

In an undamped system, the damping coefficient p is zero. A typical model is then

UNDAMPED SPRING-MASS
MODEL

$$mx''(t) + kx(t) = 0, \quad x(0) = x_i, \quad x'(0) = v_i.$$

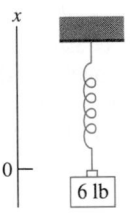

FIGURE 6.5   The vertical spring-mass system of example 26.

Model

■ **EXAMPLE 26**   *A spring suspended vertically extends* 1 ft *when a* 2-lb *weight is added to it. (Hence, its spring constant is* $k = 2$ lb/1 ft $= 2$ lb/ft.) *A* 6-lb *weight is suspended from this spring, and the mass is started in motion from its equilibrium position with a downward velocity of* 0.5 ft/s. *(See figure 6.5.) Will the mass continue oscillating or will the oscillations decay to zero? If the mass continues oscillating, determine the period of the oscillations of the mass.*

Example 1 of section 5.1 showed that the governing initial-value problem is

$$\frac{3}{16}x''(t) + 2x(t) = 0, \quad x(0) = 0, \quad x'(0) = -0.5.$$

Example 25, page 271, solved this initial-value problem. Briefly, the constant-coefficient differential equation $3x''/16 + 2x = 0$ has the characteristic equation

Characteristic equation

$$\frac{3}{16}r^2 + 2 = 0,$$

which has the pure imaginary roots $r = \pm\sqrt{32/3}\, i \approx \pm 3.27i$. The general solution of $3x''/16 + 2x = 0$ is

General solution

$$x_g(t) = C_1 \cos 3.27t + C_2 \sin 3.27t.$$

The initial conditions $x(0) = 0$, $x'(0) = -0.5$, require $C_1 = 0$, $C_2 = -0.15$. The solution of the initial-value problem is

Solution of initial-value problem

$$x(t) = -0.15 \sin 3.27t.$$

Because the roots of the characteristic equation are pure imaginary, the solution exhibits no exponential decay (or growth). It oscillates steadily with period $2\pi/3.27 = 1.92$ s, as illustrated in figure 6.6. ■

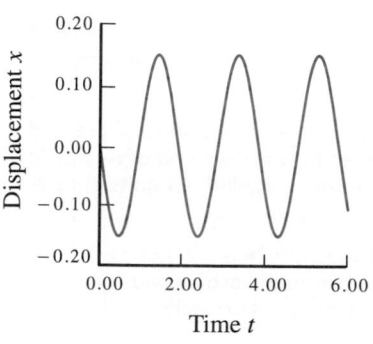

FIGURE 6.6   The steady oscillations of the solution of the undamped spring-mass model $3x''/16 + 2x = 0$, $x(0) = 0$, $x'(0) = -0.5$.

> **MATLAB**
>
> Does the analytic solution tool in ODETOOL correctly find a general solution of $3x''/16 + 2x(t) = 0$? Of the general equation $mx'' + kx = 0$? Is the solution of the latter wrong or simply not real-valued?

Do *all* undamped spring-mass systems oscillate periodically without decay? How do the initial conditions affect the response? Can they cause oscillations to grow or decay?

■ **EXAMPLE 27**   *Answer the preceding questions by showing that the general solution of the undamped spring-mass equation*

$$mx''(t) + kx(t) = 0$$

never contains exponential factors to cause growth or decay.

The characteristic equation of $mx''(t) + kx(t) = 0$ is

Characteristic equation

$$mr^2 + k = 0.$$

Since $k > 0$ and $m > 0$, its roots are pure imaginary,

Characteristic roots

$$r = \pm\sqrt{-\frac{k}{m}} = \pm i\frac{k}{m}.$$

The general solution of $mx''(t) + kx(t) = 0$ is

General solution

$$x_g(t) = C_1 \cos \omega_n t + C_2 \sin \omega_n t,$$

where $\omega_n = \sqrt{k/m}$.

Regardless of the values of the constants C_1 and C_2 (which are determined by the initial conditions), the solution will be a sum of the simple periodic functions $\cos \omega_n t$ and $\sin \omega_n t$. No exponential decay occurs because the roots of the characteristic equation are pure imaginary. The roots of $mr^2 + k = 0$ will always be pure imaginary as long as m and k are positive. ■

The frequency $\omega_n = \sqrt{k/m}$ appearing in the general solution $x_g(t) = C_1 \cos \omega_n t + C_2 \sin \omega_n t$ is the **natural (angular) frequency** of the undamped spring-mass system with mass m and spring constant k. The **natural (cyclic) frequency** is $\omega_n/2\pi$. The **natural period**, or the time for one complete cycle, is $2\pi/\omega_n$.

> Recall that the term *angular* (or *circular*) frequency is used because the angle $\omega_n t$, the argument of the cosine and sine terms in this solution, passes through an angle of ω_n radians in one unit of time. For example, the natural frequency of the system appearing in example 26 is $\sqrt{2/(3/16)} \approx 3.27$ rad/s, or a little more than π rad/s.
>
> Recall that the term *cyclic* frequency is used because $\omega_n/2\pi$ is the number of round-trips the mass makes in one unit of time. (Each round-trip, or complete cycle, corresponds to an angle of 2π rad.)
>
> Angular frequency is measured in radians per unit of t—e.g., in radians per second if t is measured in seconds. Frequency is measured in cycles per unit of t—e.g., in cycles per second, or hertz, if t is measured in seconds. Period has units of time, such as seconds.
>
> The natural cyclic frequency of the system of example 26 is $3.27/2\pi \approx 0.52$ Hz (cycles/s); about one-half of a complete cycle is executed each second. Compare this cyclic frequency with the angular frequency of approximately π rad/s. How many radians correspond to a complete cycle?
>
> When the natural frequency $\omega_n/2\pi$ has units of cycles per second, or hertz, the period $2\pi/\omega_n$ has units of seconds (per cycle). We saw in example 26 that the natural period of that system is $2\pi/3.27 = 1.92$ s, or about 2 s. If the mass is executing about 1/2 cycle each second, it should require about 2 s for one complete cycle.
>
> Common notation for angular frequency is ω, as in $\cos \omega t$. Common notation for cyclic frequency is f, as in $\cos 2\pi f t$.

■ **EXAMPLE 28**   *Show that an undamped spring-mass system always oscillates at its natural frequency, regardless of how the motion is started. Show that the graph of those oscillations has the shape of a simple sinusoid.*

The motion of a mass m suspended from a spring with spring constant k in the absence of damping is governed by the initial-value problem

$$mx''(t) + kx(t) = 0, \quad x(0) = x_i, \quad x'(0) = v_i.$$

Example 27 showed that the general solution of $mx''(t) + kx(t) = 0$ is

$$x_g(t) = C_1 \cos \omega_n t + C_2 \sin \omega_n t,$$

where $\omega_n = \sqrt{k/m}$ is the natural frequency.

The initial conditions $x(0) = x_i$, $x'(0) = v_i$ require that $C_1 = x_i$, $C_2 = v_i/\omega_n$. Hence, the solution of this initial-value problem is

$$x(t) = x_i \cos \omega_n t + \frac{v_i}{\omega_n} \sin \omega_n t. \tag{6.9}$$

The undamped spring-mass system always oscillates at its natural frequency ω_n, but it is not yet clear that the graph of the sum of the sine and cosine terms is simply a sinusoid.

Stop and Think **6.38** Check that the solution (6.9) actually satisfies the initial conditions.

To argue that (6.9) is, in fact, a sinusoid, we must combine the two terms in (6.9) into a single coefficient multiplying a sine function. If we could find an angle ϕ and a positive constant A such that

$$x_i = A \sin \phi, \quad \frac{v_i}{\omega_n} = A \cos \phi, \tag{6.10}$$

then a simple trigonometric identity would suffice, for (6.9) would become

$$x(t) = A \sin \phi \cos \omega_n t + A \cos \phi \sin \omega_n t$$
$$= A \sin(\omega_n t + \phi).$$

This form of the solution of the initial-value problem is certainly a simple sinusoid of frequency ω_n. It has simply been shifted by the **phase angle** ϕ. The factor A is evidently the maximum displacement or **amplitude** of the oscillations. See figure 6.7.

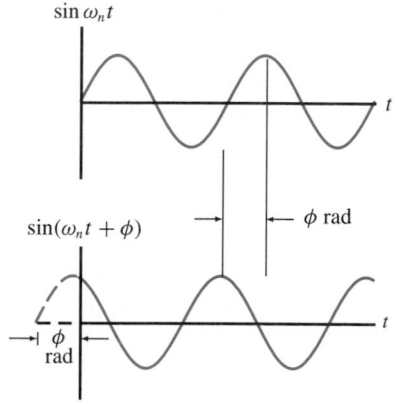

FIGURE 6.7 The upper graph is $\sin \omega t$. The lower graph is $\sin(\omega t + \phi)$—that is, $\sin \omega t$ shifted left by ϕ radians. The corresponding time shift left is ϕ/ω.

The graph of $\sin(\omega t + \phi)$ is just the graph of $\sin \omega t$ shifted to the *left* by ϕ radians. When $t = 0$, $\sin \omega t = 0$, but $\sin(\omega t + \phi) = \sin \phi$. The phase shift ϕ has given the graph of $\sin(\omega t + \phi)$ a "head start" of ϕ radians.

However, we must still show that we can actually find a phase angle ϕ and a positive amplitude A satisfying $x_i = A \sin \phi$, $v_i/\omega_n = A \cos \phi$. To find A, square and sum these two expressions:

$$x_i^2 + \left(\frac{v_i}{\omega_n} \right)^2 = A^2 \cos^2 \phi + A^2 \sin^2 \phi = A^2.$$

Hence,

AMPLITUDE

$$A - \sqrt{x_i^2 + \left(\frac{v_i}{\omega_n} \right)^2}. \tag{6.11}$$

Since the motion was begun with nontrivial initial conditions, either $x_i \neq 0$ or $v_i \neq 0$. Assuming that the latter holds, divide the first expression in (6.10) by the second to obtain

PHASE ANGLE

$$\tan \phi = \frac{\omega_n x_i}{v_i}. \tag{6.12}$$

Since the inverse tangent relation is not single valued, we have two choices for ϕ in the interval $0 \leq \phi < 2\pi$, each of which satisfies this last relation. We choose the one consistent with the two equations (6.10) and with $A > 0$. (The next example illustrates this process.)

Regardless of the (nontrivial) values of the initial displacement and velocity, the motion described by $A \sin(\omega_n t + \phi)$ has a simple sinusoidal graph with frequency ω. ■

Example 28 illustrates an *alternate form of the general solution* of the undamped spring-mass equation $mx'' + kx = 0$. That alternate general solution,

PHASE ANGLE-AMPLITUDE FORM OF GENERAL SOLUTION

$$x_g(t) = A \sin(\omega_n t + \phi),$$

is called the **phase angle-amplitude** form of the general solution. The amplitude A is determined by $A = \sqrt{x_i^2 + (v_i/\omega_n)^2}$. The phase angle ϕ is chosen to satisfy simultaneously $\tan \phi = \omega_n x_i / v_i$, equations (6.10), and the condition $A > 0$.

The expression $A \sin(\omega_n t + \phi)$ has arbitrary constants A, ϕ in place of the usual C_1, C_2. While expressing A, ϕ in terms of the initial conditions is harder than finding C_1, C_2, the phase angle-amplitude form of the general solution immediately provides the amplitude of the motion and its phase relative to a sine function.

■ **EXAMPLE 29**   *Find the solution of the initial-value problem*

$$\frac{3}{16} x''(t) + 2x(t) = 0, \quad x(0) = 0, \quad x'(0) = -0.5,$$

in phase angle-amplitude form.

The work of example 26 shows that $\omega_n = \sqrt{2/\frac{3}{16}} \approx 3.27$. We need only find values A and ϕ that make the alternate general solution

Phase angle-amplitude form of general solution

$$x_g(t) = A \sin(3.27t + \phi)$$

of this differential equation satisfy the initial conditions $x(0) = 0$, $x'(0) = -0.5$. These require

Apply initial conditions.

$$A \sin \phi = 0, \quad 3.27A \cos \phi = -0.5.$$

These equations are the analog of (6.10).

Solving each of the preceding equations for the trig function, squaring, and adding gives

Solve for A.

$$A^2 = \left(\frac{-0.5}{3.27} \right)^2$$

Hence, $A = 0.5/3.27 \approx 0.15$. (Note that A is chosen to be positive.) To find ϕ, divide $A \sin \phi = 0$ by $3.27A \cos \phi = -0.5$ to obtain

Solve for ϕ.

$$\tan \phi = 0.$$

Either $\phi = 0$ or $\phi = \pi$. Both choices satisfy $A \sin \phi = 0$, but only $\phi = \pi$ satisfies $3.27A \cos \phi = -0.5$.

Stop and Think

6.39 Show that $\phi = 0$ does *not* satisfy both $A \sin \phi = 0$ and $3.27A \cos \phi = -0.5$.

Hence, $\phi = \pi$, and the solution of the initial-value problem is

Write solution of initial-value problem.

$$x(t) = 0.15 \sin(3.27t + \pi).$$

This solution is the same as that obtained in example 26, for

$$\sin(3.27t + \pi) = -\sin 3.27t. \quad \blacksquare$$

> MATLAB
>
> Use the graphical tools in DELAB to plot the solution of this initial-value problem. Identify the phase angle $\phi = \pi$ on that plot.

Damped Spring-mass Systems

To incorporate damping and to generalize the findings of examples 22–25 at the end of the last section, consider the damped spring-mass model

Damped spring-mass model

$$mx''(t) + px'(t) + kx(t) = 0, \quad x(0) = x_i, \quad x'(0) = v_i.$$

The positive constant p is the damping coefficient.

The behavior of any nontrivial solution of $mx''(t) + px'(t) + kx(t) = 0$ is determined by its characteristic equation,

Characteristic equation

$$mr^2 + pr + k = 0.$$

Its roots are

Characteristic roots

$$r_1, r_2 = \frac{-p \pm \sqrt{p^2 - 4mk}}{2m}.$$

Evidently, three cases arise, depending upon the sign of the **discriminant** $p^2 - 4mk$. If p is small enough, then the discriminant is negative. The characteristic roots have an imaginary part, and the solution of $mx''(t) + px'(t) + kx(t) = 0$ contains oscillatory terms. That is, the mass oscillates if the damping is not too large. The spring-mass system is *underdamped*.

If the damping coefficient is large enough to make $p^2 - 4mk$ nonnegative, then the characteristic roots are real and the solution of the differential equation contains no oscillatory terms. The mass can not oscillate if the damping is sufficiently large. Such a system is *overdamped*.

More precisely, the equation $mx'' + px' + kx = 0$ is

underdamped if $p^2 - 4mk < 0$,

critically damped if $p^2 - 4mk = 0$, or

overdamped if $p^2 - 4mk > 0$.

Stop and Think

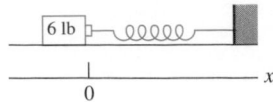

FIGURE 6.8   The horizontal spring-mass system of example 30.

Characteristic equation

The discriminant determines the type of characteristic roots.

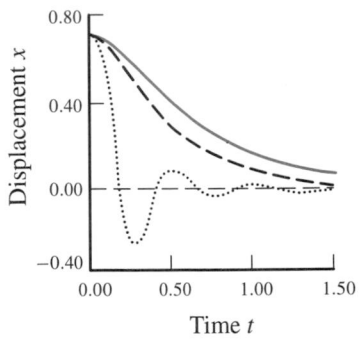

FIGURE 6.9   Solutions of the damped spring-mass model
$3x''/16 + 1.41x' + kx = 0$,
$x(0) = 0.75$, $x'(0) = 0$ for $k = 2$
(overdamped: solid blue curve),
$k = 8/3$ (critically damped: dashed curve), and $k = 200$ (underdamped: dotted curve).

Governing differential equation

6.40 Describe the nature of the characteristic roots associated with a spring-mass equation in each of the cases of underdamping, overdamping, and critical damping. Describe the behavior of the solutions of the equation in each case.

■ **EXAMPLE 30**   *Example 2 of section 5.1 showed that the horizontal spring-mass system illustrated in figure 6.8 is modeled by the initial-value problem*

$$\frac{3}{16}x''(t) + 1.41x'(t) + 2x(t) = 0, \quad x(0) = 0.75, \quad x'(0) = 0.$$

(The weight of 6 lb has mass of 3/16 slug. The spring constant is 2 lb/ft, and the friction coefficient is 1.41 lb·s/ft.) Is the force of friction between the mass and the table sufficient to overdamp this system?

This initial-value problem was solved in example 22 of the preceding section, but the complete solution is not needed. To answer the damping question, simply examine the sign of the discriminant of the characteristic equation of $3x''/16 + 1.41x' + 2x = 0$,

$$\frac{3}{16}r^2 + 1.41r + 2 = 0.$$

The discriminant is

$$(1.41)^2 - 4\left(\frac{3}{16}\right)2 = \frac{1}{2} > 0.$$

Hence, the characteristic roots are real and distinct. The system is overdamped. The initial conditions have nothing whatsoever to do with whether or not the system is damped.

Indeed, in the course of solving this initial-value problem in example 22, we found that the general solution of the differential equation is

$$x_g(t) = C_1 e^{-1.87t} + C_2 e^{-5.65t}.$$

The absence of oscillatory terms is simply a reflection of the real characteristic roots dictated by the positive discriminant.

The graph of the solution of this initial-value problem is the solid blue curve in figure 6.9. It decays to zero without oscillating, although with a negative initial velocity of sufficient magnitude, the solution would cross the x-axis once before it decayed to zero.   ■

■ **EXAMPLE 31**   *Can the spring in the system of the preceding example be replaced by one that will allow the system to oscillate? If so, specify a range of acceptable spring constant values.*

If the mass and friction coefficient are fixed and the spring has constant k, then the system is governed by the differential equation

$$\frac{3}{16}x'' + 1.41x' + kx = 0.$$

The system will oscillate if this differential equation has a characteristic equation with complex roots.

The discriminant of the characteristic equation is

The discriminant determines the type of characteristic roots.

$$(1.41)^2 - 4 \left(\frac{3}{16} \right) k = 2 - \frac{3k}{4}.$$

A negative discriminant leads to complex roots. Hence, the system will oscillate if $2 - 3k/4 < 0$, or if

The discriminant is negative if . . .

$$k > 8/3.$$

The dotted curve in figure 6.9 illustrates the decaying oscillations typical of an underdamped system ($k = 200$ lb/ft).

The system is critically damped if $k = 8/3$. Displacement decays to zero without oscillating, as the dashed curve of figure 6.9 shows. Compare these conclusions with the solutions of the initial-value problems in examples 22 and 24 of the last section, in which this system is considered with springs having $k = 8/3 \approx 2.67$ and $k = 4$ lb/ft. ■

■ **EXAMPLE 32** *Show that every damped spring-mass system achieves a steady state of zero as $t \to \infty$. In other words, show that the rest state is a stable equilibrium of a damped system.*

The characteristic roots of the damped spring-mass equation $mx'' + px' + kx = 0$ are

$$r_1, r_2 = \frac{-p \pm \sqrt{p^2 - 4mk}}{2m}.$$

Consider the three possible cases: underdamped, overdamped, and critically damped.

If the system is underdamped, then these roots are a complex conjugate pair. They may be written

Underdamped characteristic roots

$$r_1, r_2 = \alpha \pm i\beta,$$

where

$$\alpha = -\frac{p}{2m}, \quad \beta = \frac{\sqrt{4mk - p^2}}{2m}.$$

The corresponding general solution of $mx'' + px' + kx = 0$ is

$$x_g(t) = e^{\alpha t} (C_1 \cos \beta t + C_2 \sin \beta t).$$

Since $\alpha < 0$, $\lim_{t \to \infty} x_g(t) = 0$ for any choice of C_1, C_2; i.e., for any choice of initial values x_i, v_i, the solution of the initial-value problem modeling the underdamped spring-mass system tends toward a steady state of zero.

If the system is overdamped, then the differential equation has the general solution

$$x_g(t) = C_1 e^{r_1 t} + C_2 e^{r_2 t},$$

where r_1, r_2 are the distinct real roots

Overdamped characteristic roots

$$r_1, r_2 = \frac{-p \pm \sqrt{p^2 - 4mk}}{2m}.$$

Evidently, both roots are negative, and for any set of initial values, the solution of the differential equation must again decay to a steady state of zero, regardless of the initial conditions.

Stop and Think   **6.41** Obviously, the root $-p - \sqrt{\cdots}$ is negative. Show that the root with the positive sign is also negative; argue that $\sqrt{p^2 - 4mk} < p$ as long as $p^2 - 4mk > 0$ and $m, k > 0$.

If the system is critically damped, then the single repeated characteristic root is $r = -p/2m$, and the general solution of the differential equation is

$$x_g(t) = C_1 e^{rt} + C_2 t e^{rt}.$$

Since $r < 0$, $\lim_{t \to \infty} x_g(t) = 0$.

Stop and Think   **6.42** Use l'Hôpital's rule to show that $\lim_{t \to \infty} t e^{rt} = 0$ when $r < 0$.

Hence, damped spring-mass systems decay to a zero steady state.   ■

Stop and Think   **6.43** The analysis of the last example reveals that the general solution of a damped spring-mass equation decays to zero. It also shows that *the general solution is never identically zero no matter how large t grows* (unless $C_1 = C_2 = 0$, the case of no motion whatsoever). Do real spring-mass systems actually oscillate forever like solutions of these differential equations? Do the oscillations eventually become so small that they are invisible to an observer? Or do real systems actually stop moving? Explain this apparent paradox.

The analysis of the last example also shows that the transient response of a model of a damped spring-mass system without forcing is simply the solution itself.

6.5.2   The Pendulum

The nonlinear pendulum equation

$$L\theta'' + g \sin \theta = 0$$

derived in section 5.2 *can not* be solved by the characteristic equation method, but its linear approximation

$$L\theta'' + g\theta = 0 \tag{6.13}$$

can. Since (6.13) is identical mathematically with the undamped spring-mass equation, no new analysis is required for the linear pendulum equation.

Stop and Think   **6.44** Why can't the characteristic equation method find solutions of the nonlinear pendulum equation?

6.45 Make a table to convert pendulum notation to spring-mass notation.

For example, the characteristic equation of $L\theta'' + g\theta = 0$ is

$$Lr^2 + g = 0 \tag{6.14}$$

and its general solution is

$$\theta_g = C_1 \sin \omega_n t + C_2 \cos \omega_n t, \tag{6.15}$$

where the natural frequency is $\omega_n = \sqrt{g/L}$. Like an undamped spring-mass system, an undamped pendulum will oscillate forever.

Stop and Think
6.46 Find the roots of the characteristic equation (6.13) and verify the form of the general solution (6.15).

By analogy, the notions of over- and underdamping can be extended to the damped linear pendulum equation, $L\theta'' + p\theta' + g\theta = 0$. That work is left to the reader.

Stop and Think
6.47 Using physical intuition, describe the motion of an underdamped pendulum. Of an overdamped pendulum.

6.48 Give conditions on the pendulum parameters L, p, and g that determine whether a pendulum is overdamped, critically damped, or underdamped.

6.5.3 *RLC* Circuits

Section 5.3 showed that the series *RLC* circuit illustrated in figure 6.10 is modeled by the initial-value problem

$$Lq'' + Rq' + \frac{1}{C}q = E, \quad q(0) = q_i, \quad q'(0) = i_i.$$

Here q is the charge on the capacitor C, and E is a voltage source that can vary with time. The initial charge q_i on the capacitor can also be expressed in terms of the initial voltage drop $v_C(0)$ across the capacitor: $q_i = Cv_C(0)$. The initial current in the circuit i_i is dq/dt evaluated at $t = 0$.

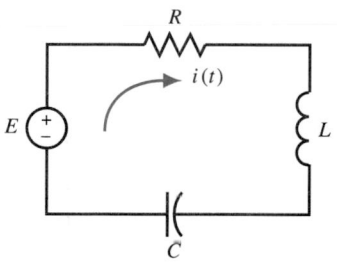

FIGURE 6.10 A series *RLC* circuit with an independent voltage source E.

This model has precisely the same mathematical form as the spring-mass models analyzed in the first part of this section,

$$mx'' + px' + kx = f(t), \quad x(0) = x_i, \quad x'(0) = v_i.$$

From the mathematical analogy between spring-mass systems and series *RLC* circuits, we can derive physical analogies, and we can adapt the analysis of one physical problem to the other. This transfer of understanding between two apparently different physical situations illustrates the power of mathematical abstraction.

The mathematical analogy follows from a comparison of the governing differential equations. For example, mass m corresponds to inductance L because these parameters are the coefficients of the second derivative terms in their respective equations. These analogies are summarized in table 6.1.

TABLE 6.1   **Mathematical analogies between the spring-mass model** $mx'' + px' + kx = f(t), x(0) = x_i, x'(0) = x_i,$ **and the series** *RLC* **circuit model** $Lq'' + Rq' + q/C = E, q(0) = q_i, q'(0) = i_i$

Spring-mass System	Series *RLC* Circuit
Mass m	Inductance L
Friction coefficient p	Resistance R
Spring constant k	Reciprocal capacitance $1/C$
Force $f(t)$ applied to spring	Voltage source E
Displacement x of mass	Charge q on capacitor
Velocity x'	Current $q' = i$
Acceleration x''	Rate of change of current di/dt
Initial position x_i of mass	Initial charge q_i on capacitor
Initial velocity v_i	Initial current i_i

Likewise, we can construct physical analogies from the mathematical parallels. For example, the inertia term mx'' in the spring-mass equation, the "mass times acceleration" expression from Newton's law, is analogous to the inductive potential term Lq'' in the series *RLC* circuit equation. Additional physical analogies are listed in table 6.2.

TABLE 6.2   **Physical analogies between the spring-mass model** $mx'' + px' + kx = f(t), x(0) = x_i, x'(0) = x_i,$ **and the series** *RLC* **circuit model** $Lq'' + Rq' + q/C = E, q(0) = q_i, q'(0) = i_i$

Spring-mass System	Series *RLC* Circuit
Inertia mx''	Inductive potential $Lq'' = Li'$
Damping force px'	Voltage drop $Rq' = Ri$ across resistor
Restoring force kx	Capacitive potential q/C

The ideas developed at the beginning of this section—underdamped, overdamped, critically damped, etc.—apply equally well to *RLC* circuits. As with the brief discussion of the undamped pendulum in section 6.5.2, the analogies guide the development of these concepts, but most of the details are left to the exercises.

■ **EXAMPLE 33**   *The series RLC circuit of figure 6.10 has its inductance and capacitance fixed. The fixed resistance R is replaced by a potentiometer, a variable resistor. For what range of values of R is the response of the circuit underdamped?*

From the work at the start of this section, a spring-mass system is underdamped when $p^2 - 4mk < 0$. Using the analogies of table 6.1, we immediately conclude that the circuit is underdamped if $R^2 - 4L(1/C) < 0$ or if

$$R < 2\sqrt{\frac{L}{C}}. \tag{6.16}$$

Alternatively, argue from the definition: a spring-mass equation is underdamped if its characteristic roots are complex, for then it exhibits decaying oscillations. Apply the same criterion to the characteristic roots of $Lq'' + Rq' + q/C = 0$, the equation governing the series RLC circuit. The discriminant of the characteristic equation $Lr^2 + Rr + 1/C = 0$ is $R^2 - 4L/C$. Complex characteristic roots occur when $R^2 - 4L/C < 0$. Hence, the circuit is underdamped if (6.16) holds, the condition derived from the mathematical analogies. ■

■ **EXAMPLE 34**   *What is the natural frequency of an undamped series RLC circuit?*

Since *natural frequency* is a concept associated with an undamped system, we should first decide what makes a series RLC circuit undamped. Consulting the tables reveals that resistance R is analogous to the friction or damping coefficient p; each is the coefficient of the first derivative term in its respective differential equation. Hence, a series RLC circuit is undamped when $R = 0$.

The natural frequency of the undamped spring-mass equation $mx'' + kx = 0$ is the real factor in the imaginary characteristic root: $\omega_n = \sqrt{k/m}$. Since the spring constant k is analogous to the reciprocal capacitance $1/C$ and m is analogous to L, the natural (circular) frequency of the undamped series RLC circuit is $\omega_n = 1/\sqrt{LC}$.

Again, a direct attack is equally effective. The characteristic roots of $Lq'' + q/C = 0$ are $\pm 1/\sqrt{-LC}$. Hence, the natural frequency, the frequency of the sine and cosine terms in the general solution of this equation, is $1/\sqrt{LC}$. ■

Oscillations in electrical circuits often occur on a much faster time scale, commonly microseconds, than mechanical oscillations, although the 60-hertz oscillations of the power lines in homes, offices, and schools are a ubiquitous exception.

The inductors and capacitors that appear in household electronic devices such as stereos and televisions have values that include the prefix *milli-* (10^{-3}, abbreviated m), *micro-* (10^{-6}, abbreviated μ), or *pico-* (10^{-12}, abbreviated p); e.g., one might find an inductor of 1 mH = 10^{-3} H and a capacitor of 1 μF = 10^{-6} F.

■ **EXAMPLE 35**   *Suppose the elements in the circuit of figure 6.10 have the values $R = 200\,\Omega$, $L = 100\,\mu$H, $C = 100\,$pF, $E \equiv 0$. (The circuit has no voltage source.) What is the natural frequency of the corresponding undamped circuit? What is the frequency of the oscillations in the damped circuit? How much time is required for the amplitude of the oscillations in the charge on the capacitor to die to one-half its initial value?*

The natural frequency of the undamped ($R = 0$) circuit is

$$\omega_n = \frac{1}{\sqrt{LC}} = \frac{1}{\sqrt{100 \times 10^{-6} \cdot 100 \times 10^{-12}}}$$

$$= 1 \times 10^7 \text{ rad/s} = 10 \text{ rad/}\mu\text{s}.$$

The corresponding cyclic frequency is 1.5915 cycles/μs.

The frequency of the oscillations in the damped circuit is the imaginary part of the complex characteristic root of $Lr^2 + Rr + 1/C = 0$. Using the given circuit parameters, those roots are

$$r = -\frac{R}{2L} \pm i\sqrt{\frac{1}{LC} - \left(\frac{R}{2L}\right)^2} = -2 \times 10^6 \pm i9.798 \times 10^6.$$

Hence, the relatively small damping of 200 Ω has reduced the frequency only slightly, to 9.798×10^6 rad/s, or 9.798 rad/μs. The corresponding cyclic frequency is 1.5594 cycles/μs.

The real part of this characteristic root, $-R/2L = -2 \times 10^6$, controls the exponential decay of these oscillations. The oscillations will have one-half their initial amplitude when

$$e^{-2 \times 10^6 t_{1/2}} = \frac{1}{2}.$$

Using logarithms, we solve for $t_{1/2}$, the time for the amplitude to decay by one-half:

$$-2 \times 10^6 t_{1/2} = -\ln 2,$$

$$t_{1/2} = (2 \times 10^{-6}) \ln 2 \approx 1.3863 \times 10^{-6} \text{ s} = 1.3863 \ \mu\text{s}.$$

This decay is illustrated in figure 6.11. Is that graph consistent with these calculations? ■

Compare this calculation of decay time with that of the half-life of a decaying population or a radioactive compound, as in the exercises for section 2.1.

The vertical axis in figure 6.11 plots the voltage drop across the capacitor rather than charge on the capacitor because voltage drop is easily measured by attaching oscilloscope probes to either side of the capacitor. From experimental fact 2 of section 5.3, the relation between capacitor voltage and charge is $v_C = q/C$.

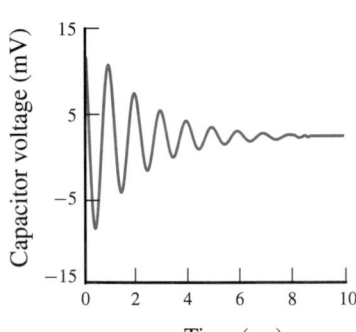

FIGURE 6.11 Capacitor voltage versus time for the series *RLC* circuit of figure 6.10 without forcing; $R = 200 \,\Omega$, $L = 100 \,\mu$H, $C = 100$ pF. Initial conditions are $v_C(0) = 10$ mV (corresponding to $q(0) = 10^{-12}$ C) and $i(0) = q'(0) = -1 \,\mu$A.

6.5.4 Exercises

EXERCISE GUIDE	
To gain experience ...	**Try exercises**
With phase angle–amplitude solutions	1, 13–14, 16, 19–22
With relations among amplitude, phase, period, initial conditions, etc.	2–4, 15, 16–18, 23, 25, 31

EXERCISE GUIDE (Continued)	
To gain experience ...	**Try exercises**
Analyzing and interpreting damped response	4, 5(a), 6–11, 24, 25, 31, 32
Analyzing and interpreting periodic response	2–3, 5(b–c), 8, 12, 15–16, 18, 23, 30, 32(a)
With analogies among various physical systems	22, 26–29

1. Derive the phase angle-amplitude form of the solution of the initial-value problem modeling each of the undamped systems described here. Assume a general solution of the form $x_g(t) = A \sin(\omega t + \phi)$, and determine the constants A and ϕ directly from the initial conditions.

   (a) A mass of 600 g is suspended from a spring that extended 4 cm when a mass of 100 g was hung from it. The 600-g mass is set in motion by releasing it from rest 6 cm above its equilibrium position.

   (b) The spring of part (a) is now set horizontally, attached to a fixed anchor as in figure 6.8. A 2-kg air puck is attached to the free end of the spring. (An air puck rides on a cushion of air so that it can slide with essentially no friction.) The puck is pushed 7 cm toward the fixed anchor and released.

2. Consider the system described in part (a) of exercise 1.

   (a) Suppose the mass is released from rest from an arbitrary position x_i. How does the amplitude of the resulting motion depend upon x_i? How does the phase angle depend upon x_i? Do these relationships seem physically reasonable?

   (b) Suppose the mass is started in motion from its equilibrium position with a velocity v_i. How does the amplitude of the resulting motion depend upon v_i? How does the phase angle depend upon v_i? Do these relationships seem physically reasonable?

3. The air-puck system described in part (b) of exercise 1 is set in motion from its equilibrium position with an unknown initial velocity.

   (a) The period of its oscillations is observed to be 6.39 s. Can you determine the initial velocity of the puck?

   (b) The maximum displacement of the puck is observed to be 14.1 cm. Can you determine its initial velocity?

4. The system described in part (b) of exercise 1 is not truly undamped, although that exercise suggested that you could model it that way. In fact, during a period of 4 min, the maximum displacement of the oscillations of the puck was observed to decrease from 10 cm to 5.6 cm. What is the damping coefficient?

5. Example 27 shows that the general solution of the undamped model equation

$$mx''(t) + kx(t) = 0$$

contains no exponential factors to cause either growth or decay.

   (a) Explain why no solution of the initial-value problem

$$mx''(t) + kx(t) = 0,$$
$$x(0) = x_i, \quad x'(0) = v_i,$$

   will exhibit growth or decay, regardless of the choice of initial values x_i, v_i.

   (b) Show that all solutions of this initial-value problem oscillate with period $2\pi \sqrt{m/k}$.

   (c) What happens when $x_i = v_i = 0$?

6. A 12-kg mass is suspended from a spring. Air resistance and internal friction in the spring resist the motion of the mass with a force whose magnitude is 0.02 times the velocity of the mass. Find the range of spring-constant values for which the mass will *not* oscillate.

7. A mass of 4 kg is suspended from a spring in a liquid that offers a resistance force whose magnitude is eight times the velocity of the mass.

   (a) Does decreasing the magnitude of the spring constant cause the mass to oscillate slower or faster?

   (b) What range of values of the spring constant will prevent the mass from oscillating?

8. You are given the characteristic roots r_1, r_2 of a model of a spring-mass system. Give a step-by-step recipe (an algorithm) for determining from r_1, r_2 whether the system is damped or undamped. If it is damped, determine whether it is underdamped, critically damped, or overdamped.

9. In example 2, page 211, of section 5.1 we derived a model for a horizontal spring-mass system without forcing,

$$\frac{3}{16}x''(t) + 1.41x'(t) + 2x(t) = 0,$$

$$x(0) = 0.75, \quad x'(0) = 0.$$

In example 22, page 269, of section 6.4 we found that the solution of this initial-value problem is

$$x(t) = 1.12e^{-1.90t} - 0.37e^{-5.62t}.$$

(a) In what way(s) does the form of this solution tell you that this system is overdamped? Verify that it is indeed overdamped.

(b) Does the position $x(t)$ ever become negative? (*Hint:* Examine the local extrema of the function x.)

(c) Review the physical situation that this initial-value problem models. Might you be able to start the mass moving in such a way that its position could be negative at some time? Describe such a set of initial conditions qualitatively.

(d) Provide a mathematical statement for the set of initial conditions you described in part (c). Solve the given differential equation subject to those initial conditions and show that you do indeed obtain negative values of $x(t)$. (If you don't obtain the desired results, revise your guess about an appropriate set of initial conditions.)

(e) For a given initial position x_i, give the range of initial velocities that will always yield negative values of $x(t)$ at some time.

10. Example 32 argues that any solution of the overdamped spring-mass system model

$$mx''(t) + px'(t) + kx(t) = 0,$$

$$x(0) = x_i, \quad x'(0) = v_i,$$

must decay to zero because the real, distinct characteristic roots of $mx'' + px' + kx = 0$ are both negative. Justify the claim that those roots, which are defined by

$$r_1, r_2 = \frac{-p \pm \sqrt{p^2 - 4mk}}{2m},$$

are both negative.

11. Example 32 claims that the critically damped spring-mass system model

$$mx''(t) + px'(t) + kx(t) = 0,$$

$$x(0) = x_i, \quad x'(0) = v_i,$$

possesses a stable steady state of zero because its general solution

$$x_g(t) = C_1 e^{rt} + C_2 t e^{rt}$$

decays to zero. (The repeated characteristic root r is $-p/2m$.) Verify that $\lim_{t \to \infty} x_g(t) = 0$.

12. Show that the general solution

$$x_g(t) = C_1 \cos \omega_n t + C_2 \sin \omega_n t$$

of

$$mx''(t) + kx(t) = 0$$

does indeed have period $2\pi/\omega_n = 2\pi/\sqrt{k/m}$; i.e., show that $x_g(t + 2\pi/\omega_n) = x_g(t)$ for all t.

13. To make more plausible the claim of example 28 that the general solution

$$x_g(t) = C_1 \cos \omega_n t + C_2 \sin \omega_n t$$

can be replaced by the phase angle-amplitude form

$$x(t) = A \sin(\omega_n t + \phi),$$

expand $A \sin(\omega_n t + \phi)$ using the sum-of-angles identity for the sine function. Find C_1, C_2 in terms of A, ϕ by equating the two forms of the general solution.

Express C_1, C_2 in terms of A in the four special cases $\phi = 0, \pi/2, \pi, 3\pi/2$. Can the phase angle-amplitude form duplicate behavior that involves just a sine or just a cosine in the familiar general solution form $x_g = C_1 \cos \omega_n t + C_2 \sin \omega_n t$?

The next exercise asks you to go the other way, finding A, ϕ in terms of C_1, C_2.

14. Express the arbitrary constants A, ϕ in the phase angle amplitude form of the general solution

$$x_g(t) = A \sin(\omega t + \phi)$$

of the undamped spring-mass equation $mx'' + kx = 0$ in terms of the arbitrary constants C_1, C_2 that appear in the usual form of the general solution

$$x_g(t) = C_1 \cos \omega t + C_2 \sin \omega t.$$

(The previous exercise asks you to go the other way, finding C_1, C_2 in terms of A, ϕ.)

15. Show that the maximum value of the expression

$$C_1 \cos \omega t + C_2 \sin \omega t$$

is $\sqrt{C_1^2 + C_2^2}$. That is, find the amplitude of this typical solution expression in terms of the constants C_1, C_2.

16. Example 28 solves the initial-value problem

$$mx''(t) + kx(t) = 0, \quad x(0) = x_i, \quad x'(0) = v_i,$$

by finding the constants A, ϕ in the phase angle-amplitude form of the general solution

$$x(t) = A\sin(\omega_n t + \phi).$$

Those constants are determined by

$$A = \sqrt{x_i^2 + \left(\frac{v_i}{\omega_n}\right)^2}, \quad \tan\phi = \frac{\omega_n x_i}{v_i},$$

where $\phi, 0 \le \phi < 2\pi$, is chosen to satisfy $\tan\phi = \omega_n/v_i$, equation (6.10), and the condition $A > 0$.

(a) Does the formula $A = \sqrt{x_i^2 + (v_i/\omega_n)^2}$ give a reasonable value for amplitude when $v_i = 0$? (Recall that $\sqrt{x^2} = |x|$.)

(b) What are possible values of ϕ when $v_i = 0$, $x_i \ne 0$? Does the sign of x_i determine ϕ uniquely? Is the resulting solution physically reasonable?

(c) What are possible values of ϕ when $x_i = 0$, $v_i \ne 0$? Does the sign of v_i determine ϕ uniquely? Is the resulting solution physically reasonable?

(*Hint*: See example 29.)

17. Figure 6.7 shows that the graph of $\sin(\omega t + \phi)$ is just the graph of $\sin\omega t$ shifted to the left by ϕ radians; i.e., the graph of $\sin(\omega t + \phi)$ starts ϕ radians before the graph of $\sin\omega t$. But the horizontal axis in figure 6.7 is measured in units of time. By how many *time* units is the graph of $\sin(\omega t + \phi)$ ahead of the graph of $\sin\omega t$? That is, find a formula that translates a phase shift of ϕ radians into units of time. (*Hint*: A phase shift of π radians corresponds to a phase shift of one-half period. Can you establish a similar proportionality relationship for arbitrary phase angles?)

18. An undamped, vertical spring-mass system is set in motion by a variety of combinations of initial displacements and velocities. Each choice is required to produce a maximum displacement of M units from the equilibrium position of the mass.

(a) Given the initial position, prescribe an initial velocity that will produce the desired maximum displacement. Is the choice of initial velocity unique? Given M, is there a maximum initial displacement?

(b) As the initial displacement is increased to larger and larger positive values and the initial velocity is given the corresponding positive value you found in part (a), will the mass reach its peak displacement sooner or later? (*Hint*: Examine the phase angle-amplitude form.) Is your analytic conclusion consistent with your physical intuition?

19. Show that the underdamped problem considered in example 24 of section 6.4,

$$\frac{3}{16}x''(t) + 1.41x'(t) + 4x(t) = 0,$$

$$x(0) = 0.75, \quad x'(0) = 0,$$

can be solved by beginning with a general solution in phase angle-amplitude form

$$x_g(t) = Ae^{-3.76t}\sin(2.68t + \phi)$$

and using the initial conditions $x(0) = 0.75$, $x'(0) = 0$, to determine the arbitrary constants A and ϕ. Explain the presence of the numbers 3.76 and 2.68 in this general solution. Show that your phase angle-amplitude solution is equivalent to that obtained from the familiar form of the general solution,

$$x(t) = e^{-3.76t}(0.75\cos 2.68t + 1.05\sin 2.68t).$$

20. Show that the general underdamped initial-value problem

$$mx''(t) + px'(t) + kx(t) = 0, \quad x(0) = x_i, \quad x'(0) = 0,$$

can be solved by beginning with a general solution in the phase angle-amplitude form

$$x_g(t) = Ae^{\alpha t}\sin(\beta t + \phi),$$

where A, ϕ are arbitrary constants to be determined by the initial conditions and α, β are determined by the characteristic equation.

21. The preceding exercise describes a phase angle-amplitude form for a general solution of an underdamped initial-value problem. Find the values of the arbitrary constants A, ϕ in that solution in terms of the arbitrary constants C_1, C_2 that appear in the usual general solution form

$$x_g(t) = e^{\alpha t}(C_1\cos\beta t + C_2\sin\beta t).$$

22. Derive the phase angle-amplitude form of the general solution of the linear undamped pendulum equation $L\theta'' + g\theta = 0$ by mimicking this section's analysis of the undamped spring-mass equation $mx'' + kx = 0$. Check your results by substituting the analogous pendulum variables into the spring-mass solution expressions (6.11) and (6.12), page 279.

23. Suppose the pendulum of a grandfather clock is required to execute a complete swing every 2 s.

(a) What should be the length of the pendulum arm?

(b) If you based your length calculation on the linear pendulum equation, check the accuracy of your result by solving the nonlinear pendulum equation using a numerical method of your choice from DELAB. (First write the single equation as a first-order system.) How accurate is the linear prediction of the period? Does the period observed in the nonlinear pendulum appear to depend on the amplitude of the oscillations?

24. Because of friction in the pivot, the pendulum in a grandfather clock would come to a halt were it not for the energy imparted through the weight and escapement mechanism. Without the weight and escapement, the simplest linear model for these decaying oscillations would be $L\theta'' + p\theta' + g\theta = 0$, the damped pendulum equation. Suppose the oscillations of the pendulum in a well-made clock diminished to half of the initial amplitude in about 10 min. Estimate the damping coefficient p. (*Hint*: See example 35. Be sure to use consistent units of time.)

25. Find conditions on the parameters R, L, C that guarantee that a series *RLC* circuit such as that of figure 6.10 is

    (a) underdamped,

    (b) critically damped,

    (c) overdamped.

   The voltage drop v across a *parallel RLC* circuit such as that shown in figure 6.12 is governed by

$$Cv'' + \frac{1}{R}v' + \frac{1}{L}v = 0.$$

Exercises 26–32 refer to such circuits.

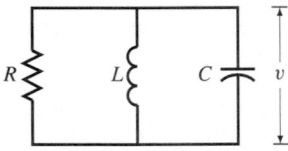

FIGURE 6.12   A parallel *RLC* circuit.

26. Construct a table similar to table 6.1 illustrating the mathematical analogies between a parallel *RLC* circuit and a spring-mass system.

27. Construct a table similar to table 6.2 illustrating the physical analogies between a parallel *RLC* circuit and a spring-mass system.

28. Construct a table similar to table 6.1 illustrating the mathematical analogies between a parallel *RLC* circuit and a *series RLC* circuit.

29. Construct a table similar to table 6.2 illustrating the physical analogies between a parallel *RLC* circuit and a *series RLC* circuit.

30. Find the natural frequency of an undamped parallel *RLC* circuit.

31. Find conditions on the parameters R, L, C that make a parallel *RLC* circuit

    (a) underdamped,

    (b) critically damped,

    (c) overdamped.

32. The oscillations of the charge on the capacitor in a series *RLC* circuit without a voltage source are governed by $Lq'' + Rq' + q/C = 0$.

    (a) Write the frequency of the oscillations in this circuit in terms of the natural frequency of the undamped circuit. Which of the three circuit parameters R, L, C cause the two frequencies to differ?

    (b) Find an expression for the time required for the amplitude of the oscillations in this circuit to decrease to one-half their initial value. Which of the three circuit parameters R, L, C control this characteristic decay time?

6.6 ■ UNDETERMINED COEFFICIENTS

6.6.1 Overview

Section 4.4 showed that the method of undetermined coefficients can find particular solutions of *constant-coefficient, linear*, first-order equations whose forcing terms are polynomials, sines, cosines, or exponentials. (See table 4.1, page 168, for a quick review.) Those ideas extend immediately to constant-coefficient, second-order, linear equations

$$b_2 y'' + b_1 y' + b_0 y = f(x)$$

with forcing terms $f(x)$ from the same family.

The essential idea of undetermined coefficients is to assume a trial particular solution that involves the same sorts of functions as appear in the forcing term. For example, if the forcing term is

$$\sin 2\pi t,$$

then the trial particular solution is

$$A \cos 2\pi t + B \sin 2\pi t.$$

The coefficients A and B are determined by substituting the trial solution into the differential equation.

If part of a trial particular solution is a solution of the *homogeneous* equation, then the coefficient of that part will disappear when it is substituted into the differential equation. Hence, its coefficient will remain undetermined. For first-order equations, this difficulty can be circumvented by multiplying the first trial solution by the independent variable. See the examples of section 4.4 for illustrations of this process.

In this situation lies the only real difference between undetermined coefficients for first-order equations and second-order equations. Multiplying a trial solution of a second-order equation by the independent variable may yet leave a solution of the homogeneous equation in the trial solution. In that case, we simply multiply by the independent variable one more time.

The rules for efficiently guessing a form of the particular solution (or, to be more elegant, constructing a trial solution) are summarized in table 6.3. It includes several natural extensions of the first-order guidelines of, e.g., forcing terms which are the product of a polynomial and an exponential.

TABLE 6.3 **Particular solutions via undetermined coefficients for the constant-coefficient, second-order linear equation** $b_2 y'' + b_1 y' + b_0 y = f(x)$

$b_2 y'' + b_1 y' + b_0 y = f(x)$	
Forcing Term $f(x)$	**Trial Particular Solution**
$a_n x^n + \cdots + a_1 x + a_0$	$A_n x^n + \cdots + A_1 x + A_0$
$(a_n x^n + \cdots + a_1 x + a_0)e^{qx}$	$(A_n x^n + \cdots + A_1 x + A_0)e^{qx}$
$(a_n x^n + \cdots + a_1 x + a_0) \cos px$	$(A_n x^n + \cdots + A_1 x + A_0) \cos px$
$\quad + (b_n x^n + \cdots + b_1 x + b_0) \sin px$	$\quad + (B_n x^n + \cdots + B_1 x + B_0) \sin px$
$ae^{qx} \cos px + be^{qx} \sin px$	$Ae^{qx} \cos px + Be^{qx} \sin px$

If the assumed form of the particular solution solves the corresponding homogeneous equation, multiply the assumed form by x. Repeat if necessary.

Lowercase letters a, a_i, b, b_i are constants given in the forcing function. The coefficients to be determined are denoted by uppercase letters A, A_i, B, B_i.

In simplest terms, the important points for efficiently forming trial particular solutions are as follows:

- Guess a trial particular solution of the same form as the forcing function.

- If the forcing term involves a sine *or* a cosine, the trial solution should contain *both* a sine *and* a cosine.
- If any part of the trial solution is a solution of the homogeneous differential equation, multiply the trial solution by the independent variable. Repeat if necessary.

The idea of undetermined coefficients is often useful because it suggests which terms will appear in a particular solution of a given model equation. The drudgery of finding the actual value of the undetermined coefficients is often better left to software like MATLAB.

6.6.2 Examples

■ **EXAMPLE 36**   *Find a particular solution of*

$$x''(t) = -g,$$

the equation governing the position of a rock moving vertically; g is the constant of gravitation. (See exercise 5 of the chapter 1 exercises.)

This equation is certainly a constant-coefficient, linear, second-order equation. Since its forcing term is a constant (i.e., a polynomial of degree $n = 0$), table 6.3 directs us to use the trial solution

The first trial particular solution . . .

$$x_p(t) = A_0,$$

where A_0 is a coefficient to be determined. But substituting x_p into $x'' = -g$ immediately yields

. . . fails.

$$x_p'' = (A_0)'' = 0.$$

The trial solution is a solution of the homogeneous equation; A_0 can not be determined.

Abandoning table 6.3 for the moment, solve $x'' = -g$ directly by two successive antiderivatives,

$$x'(t) = -\int g\, dt = -gt + C_1,$$

$$x(t) = \int (-gt + C_1)\, dt = -gt^2/2 + C_1 t + C_2,$$

where C_1, C_2 are arbitrary constants. Substituting this expression for $x(t)$ into the left side of $x'' = -g$ reveals

$$x''(t) = \left(-gt^2/2\right)'' + (C_1 t)'' + (C_2)'' = -g + 0 + 0.$$

That is, a particular solution of $x'' = -g$ is

$$x_p(t) = -gt^2/2.$$

The other terms in the solution, $C_1t + C_2$, satisfy the homogeneous equation $x''(t) = 0$.

This same particular solution arises from the guidelines of table 6.3. Begin by finding the solutions of the homogeneous version of this equation,

$$x''(t) = 0.$$

Since its characteristic equation $r^2 = 0$ has the repeated real root $r = 0$, a pair of linearly independent solutions of $x''(t) = 0$ is

Linearly independent homogeneous solutions

$$x_1 = e^{0t} = 1, \quad x_2 = te^{0t} = t.$$

We see immediately that the first trial solution $x_p(t) = A_0$ is just a multiple of the homogeneous solution $x_1 = 1$. Following table 6.3, multiply the trial solution by the independent variable to obtain a new trial solution,

Second trial particular solution

$$x_p(t) = A_0t.$$

But this trial solution is just a multiple of the second homogeneous solution $x_2 = t$. Hence, multiply again by the independent variable to obtain yet another trial solution,

Third trial solution

$$x_p(t) = A_0t^2,$$

that, at last, does not contain a solution of the homogeneous differential equation.

Substituting this trial solution into the original differential equation $x'' = -g$ yields

Substitute the trial solution.

$$x_p''(t) - (A_0t^2)'' - 2A_0 - -g$$

or

Find the undetermined coefficient.

$$A_0 = -g/2.$$

Hence, the method of undetermined coefficients yields a particular solution of $x'' = -g$,

Particular solution

$$x_p(t) = -gt^2/2.$$

With no additional effort, we can couple this particular solution with the homogeneous solutions found above to obtain the general solution

General solution

$$x_g(t) = -gt^2/2 + C_1t + C_2,$$

the solution obtained by antidifferentiation. ■

This example motivates the strategy of multiplying a trial particular solution that satisfies the homogeneous equation by the independent variable. Do you see any similarities between this example and example 28, page 165, of section 4.4, which motivated the same strategy for first-order equations?

The next example treats a similar situation, but for a differential equation that can not be solved by antidifferentiation.

■ **EXAMPLE 37**   *Find a particular solution of*

$$y'' + 4y' + 4y = 8e^{-2x}.$$

To find linearly independent solutions of the corresponding homogeneous equation,

$$y'' + 4y' + 4y = 0,$$

find the roots of its characteristic equation,

$$r^2 + 4r + 4 = (r + 2)^2 = 0.$$

Since the characteristic equation has the repeated real root $r = -2$, linearly independent solutions of the homogeneous equation are

Linearly independent homogeneous solutions

$$y_1 = e^{-2x}, \quad y_2 = xe^{-2x}.$$

Because the forcing term in $y'' + 4y' + 4y = 8e^{-2x}$ is an exponential, table 6.3 suggests a trial particular solution of the same form,

First trial particular solution

$$y_p(x) = A_0 e^{-2x}.$$

But this trial solution satisfies the homogeneous equation $y'' + 4y' + 4y = 0$ because it is a constant multiple of the homogeneous solution $y_1 = e^{-2x}$. Multiply by the independent variable to obtain a new trial solution,

Second trial solution

$$y_p(x) = A_0 x e^{-2x}.$$

Now we have a multiple of the homogeneous solution $y_2 = xe^{-2x}$. One more multiplication of the trial solution by x yields

Third trial solution

$$y_p(x) = A_0 x^2 e^{-2x}.$$

We accept this expression as our final trial solution because it contains no solutions of the homogeneous differential equation.

Substituting the trial particular solution $y_p(x) = A_0 x^2 e^{-2x}$ into the non-homogeneous equation $y'' + 4y' + 4y = 8e^{-2x}$ yields

Substitute the trial solution.

$$y_p'' + 4y_p' + 4y_p$$
$$= A_0 e^{-2x}(2 - 8x + 4x^2) + 4A_0 e^{-2x}(2x - 2x^2) + 4A_0 x^2 e^{-2x}$$
$$= 2A_0 e^{-2x} = 8e^{-2x}.$$

Find the undetermined
coefficient.

The last equality yields $2A_0 = 8$, or $A_0 = 4$. A particular solution of

$$y'' + 4y' + 4y = 8e^{-2x}$$

is

Particular solution

$$y_p(x) = 4x^2 e^{-2x}.$$

Using the homogeneous solutions y_1, y_2, we can also write a general solution of $y'' + 4y' + 4y = 8e^{-2x}$,

General solution

$$y_g(x) = 4x^2 e^{-2x} + C_1 e^{-2x} + C_2 x e^{-2x}. \quad \blacksquare$$

■ **EXAMPLE 38**   *Find a general solution of*

$$y'' + 4y' + 4y = 6xe^{-2x}.$$

The homogeneous version of this equation is $y'' + 4y' + 4y = 0$. We have already found linearly independent solutions of that homogeneous equation,

Linearly independent
homogeneous solutions

$$y_1 = e^{-2x}, \quad y_2 = xe^{-2x}.$$

Since the forcing term in $y'' + 4y' + 4y = 6xe^{-2x}$ is a polynomial of first-order multiplying an exponential, table 6.3 directs us to consider a trial particular solution that is also a product of a first-order polynomial with the same exponential,

First trial particular solution

$$y_p(x) = (A_1 x + A_0)e^{-2x}.$$

But both the terms in this trial particular solution are solutions of the homogeneous equation $y'' + 4y' + 4y = 0$:

$$A_0 e^{-2x} = A_0 y_1 \quad \text{and} \quad A_1 x e^{-2x} = A_1 y_2.$$

To obtain a new trial solution, multiply the first guess by x:

Second trial solution

$$y_p(x) = x(A_1 x + A_0)e^{-2x} = (A_1 x^2 + A_0 x)e^{-2x}.$$

The last term in this trial solution, $A_0 x e^{-2x}$, is still a solution of the homogeneous differential equation, so we must multiply again by x to produce yet another trial solution,

Third trial solution

$$y_p(x) = x(A_1 x^2 + A_0 x)e^{-2x} = (A_1 x^3 + A_0 x^2)e^{-2x}.$$

This trial solution contains no solutions of the homogeneous equation $y'' + 4y' + 4y = 0$.

Substituting the trial particular solution

$$y_p(x) = (A_1 x^3 + A_0 x^2)e^{-2x}$$

into the differential equation $y'' + 4y' + 4y = 6xe^{-2x}$ and simplifying yields

Substitute the trial solution.

$$6A_1xe^{-2x} + 2A_0e^{-2x} = 6xe^{-2x}.$$

Equating coefficients of xe^{-2x} and e^{-2x} on both sides of this last expression gives two equations for the two undetermined coefficients A_0, A_1,

Equate coefficients.

$$\begin{aligned} e^{-2x} &: \quad 2A_0 = 0, \\ xe^{-2x} &: \quad 6A_1 = 6. \end{aligned}$$

The solution of this pair of simultaneous equations is obviously $A_0 = 0$, $A_1 = 1$. A particular solution of $y'' + 4y' + 4y = 6xe^{-2x}$ is thus

Particular solution

$$y_p(x) = x^3e^{-2x}.$$

A general solution of $y'' + 4y' + 4y = 6xe^{-2x}$ is

General solution

$$y_g(x) = x^3e^{-2x} + C_1e^{-2x} + C_2xe^{-2x}. \quad ■$$

■ **EXAMPLE 39**   *Find a particular solution of*

$$y'' + 4y' + 4y = 18xe^{-2x} - \sin \pi x - 2e^{-2x}.$$

The homogeneous version of this equation is $y'' + 4y' + 4y = 0$. Linearly independent solutions of this homogeneous equation are

Linearly independent homogeneous solutions

$$e^{-2x}, \quad xe^{-2x}.$$

The forcing term $18xe^{-2x} - \sin \pi x - 2e^{-2x}$ can be separated into two groups, those involving the exponential e^{-2x} and those involving the $\sin \pi x$ term. To simplify the algebra, consider two separate problems, one for each group of forcing terms:

Separate into two problems.

$$\begin{aligned} \text{Problem 1:} &\quad y'' + 4y' + 4y = 18xe^{-2x} - 2e^{-2x}, \\ \text{Problem 2:} &\quad y'' + 4y' + 4y = -\sin \pi x. \end{aligned}$$

The principle of superposition guarantees that a sum of particular solutions of problems 1 and 2 will be a solution of the original equation, $y'' + 4y' + 4y = 18xe^{-2x} - \sin \pi x - 2e^{-2x}$.

The forcing term in problem 1, $(18x - 2)e^{-2x}$, is a polynomial of first-order multiplying the exponential e^{-2x}. Since this is precisely the same form as the forcing term in the previous example, we can mimic that analysis exactly. In particular, we find that our initial trial solution

First trial particular solution for problem 1

$$y_p(x) = (A_1x + A_0)e^{-2x}$$

must be multiplied twice by the independent variable before we obtain the final trial solution, which is free of homogeneous solutions,

Final trial solution for problem 1

$$y_p(x) = (A_1 x^3 + A_0 x^2)e^{-2x}.$$

Substituting $y_p(x) = (A_1 x^3 + A_0 x^2)e^{-2x}$ into problem 1 yields a slight variant of an expression in the previous example,

Substitute into problem 1.

$$6A_1 x e^{-2x} + 2A_0 e^{-2x} = 18x e^{-2x} - 2e^{-2x}.$$

Equating coefficients of xe^{-2x} and e^{-2x} yields two simultaneous equations for the two undetermined coefficients A_0, A_1,

Equate coefficients.

$$e^{-2x}: \quad 2A_0 = -2,$$
$$xe^{-2x}: \quad 6A_1 = 18.$$

The solution of this pair of equations is $A_0 = -1$, $A_1 = 3$. A particular solution of problem 1 is

Particular solution of problem 1

$$y_{p1}(x) = (3x^3 - x^2)e^{-2x}.$$

To find a particular solution of problem 2, table 6.3 suggests the trial solution

Trial particular solution for problem 2

$$y_p(x) = A \cos \pi x + B \sin \pi x.$$

Since none of the terms in this trial solution is a solution of the homogeneous equation $y'' + 4y' + 4y = 0$, we can substitute the trial solution directly into problem 2. A little algebra yields

Substitute into problem 2.

$$[(4 - \pi^2)A + 4\pi B] \cos \pi x + [-4\pi A + (4 - \pi^2)B] \sin \pi x = -\sin \pi x.$$

Equating coefficients of $\cos \pi x$, $\sin \pi x$ gives two equations for the two undetermined coefficients A, B,

Equate coefficients.

$$\cos \pi x: \quad (4 - \pi^2)A + 4\pi B = 0,$$
$$\sin \pi x: \quad -4\pi A + (4 - \pi^2)B = -1.$$

The solution of this system is

$$A = \frac{4\pi}{(4 + \pi^2)^2}, \quad B = -\frac{4 - \pi^2}{(4 + \pi^2)^2}.$$

A particular solution of problem 2 is thus

Particular solution of problem 2

$$y_{p2}(x) = \frac{4\pi}{(4 + \pi^2)^2} \cos \pi x - \frac{4 - \pi^2}{(4 + \pi^2)^2} \sin \pi x.$$

The principle of superposition allows us to combine the particular solutions y_{p1} and y_{p2} of problems 1 and 2 to obtain a particular solution of the original equation, $y'' + 4y' + 4y = 18xe^{-2x} - \sin \pi x - 2e^{-2x}$,

The particular solution of the original problem is the sum of particular solutions 1 and 2.

$$y_p(x) = y_{p1}(x) + y_{p2}(x)$$

$$= (3x^3 - x^2)e^{-2x} + \frac{4\pi}{(4 + \pi^2)^2} \cos \pi x - \frac{4 - \pi^2}{(4 + \pi^2)^2} \sin \pi x. \quad ■$$

■ **EXAMPLE 40**   *Solve the initial-value problem*

$$\frac{1}{8}x''(t) + 2x(t) = \frac{1}{2} \sin \frac{3\pi t}{2}, \quad x(0) = 0, \quad x'(0) = 0,$$

that governs the forced spring-mass system of example 3, page 211, of section 5.1.

The characteristic equation of this differential equation is

$$\frac{1}{8}r^2 + 2 = 0.$$

Its roots are $r = \pm\sqrt{-16} = \pm 4i$. Linearly independent solutions of the homogeneous equation $x''/8 + 2x = 0$ are

Linearly independent homogeneous solutions

$$\cos 4t, \quad \sin 4t.$$

Table 6.3 suggests the trial particular solution

Trial particular solution

$$x_p(t) = A \cos \frac{3\pi t}{2} + B \sin \frac{3\pi t}{2}.$$

Since these particular cosine and sine functions are not solutions of the homogeneous equation $x''/8 + 2x = 0$, we can substitute this trial solution directly into the given nonhomogeneous equation:

$$\left(-\left(\frac{3\pi}{2}\right)^2 \frac{1}{8} + 2\right) A \cos \frac{3\pi t}{2} + \left(-\left(\frac{3\pi}{2}\right)^2 \frac{1}{8} + 2\right) B \sin \frac{3\pi t}{2}$$

Substitute trial solution.

$$= \frac{1}{2} \sin \frac{3\pi t}{2}.$$

Equating coefficients of $\cos(3\pi t/2)$ and $\sin(3\pi t/2)$ yields

Equate coefficients.

$$A = 0, \quad B = \frac{16}{64 - 9\pi^2}.$$

A particular solution of the given nonhomogeneous equation is

Particular solution

$$x_p(t) = \frac{16}{64 - 9\pi^2} \sin \frac{3\pi t}{2}.$$

Combining this particular solution with the homogeneous solutions found before allows us to construct the general solution

General solution

$$x_g(t) = \frac{16}{64 - 9\pi^2} \sin \frac{3\pi t}{2} + C_1 \cos 4t + C_2 \sin 4t.$$

The first initial condition requires

Find C_1.

$$x(0) = C_1 = 0.$$

Using $C_1 = 0$ and imposing the second initial condition lead to

Find C_2.

$$x'(0) = \frac{24\pi}{64 - 9\pi^2} + 4C_2 = 0 \implies C_2 = \frac{-6\pi}{64 - 9\pi^2}.$$

The solution of the given initial-value problem is

$$x(t) = \frac{16}{64 - 9\pi^2} \sin \frac{3\pi t}{2} - \frac{6\pi}{64 - 9\pi^2} \sin 4t$$

Solution of initial-value problem

$$\approx -0.64 \sin \frac{3\pi t}{2} + 0.76 \sin 4t. \ \blacksquare$$

Evidently, the motion of this undamped system neither grows nor decays with time.

Stop and Think   **6.49** Explain why characteristic equations and undetermined coefficients combine to provide an easy answer to the question, "Which frequencies will appear in solutions of

$$\frac{1}{8} x''(t) + 2x(t) = \frac{1}{2} \sin \frac{3\pi t}{2} ?"$$

■ **EXAMPLE 41**   *Suppose the angular frequency of the forcing function in the differential equation of the preceding example is changed from $3\pi/2$ to 4. Find the position of the mass as a function of time if it is started with the initial conditions $x(0) = x'(0) = 0$.*

We are asked to solve the initial-value problem

$$\frac{1}{8} x''(t) + 2x(t) = \frac{1}{2} \sin 4t, \quad x(0) = 0, \quad x'(0) = 0.$$

Since only the forcing term has changed from the last example, we know that linearly independent solutions of the homogeneous equation $x''/8 + 2x = 0$ are

Linearly independent homogeneous solutions

$$\cos 4t, \quad \sin 4t.$$

Begin finding a particular solution with the usual trial solution for a sine or cosine forcing term,

First trial particular solution

$$x_p(t) = A \cos 4t + B \sin 4t.$$

But for the change in angular frequency, this trial solution is the same as that of the last example.

Obviously, both terms in this trial solution are multiples of the homogeneous solutions $\cos 4t$, $\sin 4t$. We can create a new trial solution by multiplying by the independent variable,

Second trial solution

$$x_p(t) = At \cos 4t + Bt \sin 4t.$$

Since none of the terms in this trial solution is a solution of $x''/8 + 2x = 0$, we can substitute it into the nonhomogeneous differential equation $x''/8 + 2x = (\sin 4t)/2$:

Substitute trial solution.

$$\frac{1}{8}x_p''(t) + 2x_p(t) = -A \sin 4t + B \cos 4t = \frac{1}{2}\sin 4t. \qquad (6.17)$$

Equating coefficients of the $\sin 4t$ and $\cos 4t$ terms yields equations for A and B,

Equate coefficients.

$$-A = \tfrac{1}{2},$$
$$B = 0.$$

Obviously, $A = -1/2$, $B = 0$, and a particular solution of $x''/8 + 2x = (\sin 4t)/2$ is

Particular solution

$$x_p(t) = -\frac{1}{2}t \sin 4t.$$

A general solution of $x''/8 + 2x = (\sin 4t)/2$ is

General solution

$$x_g(t) = -\frac{1}{2}t \sin 4t + C_1 \cos 4t + C_2 \sin 4t.$$

The initial conditions $x(0) = x'(0) = 0$ require

Find C_1, C_2.

$$x_g(0) = C_1 = 0,$$
$$x_g'(0) = 4C_2 = 0.$$

Hence, the solution of the given initial-value problem is

Solution of initial-value problem

$$x(t) = -\frac{1}{2}t \sin 4t. \quad ■$$

The mass in this system *never* oscillates periodically. Indeed, the factor t in this solution causes the displacement to swing to ever greater extremes until finally the spring breaks from overextension. This system is in *resonance*, a phenomenon studied in the next section.

■ **EXAMPLE 42**   *Determine an appropriate trial particular solution for*

$$y'' - y' - 6y = 3x \sin 2x - 5 \cos 3x + x^3(e^{-2x} + e^{2x}) + \pi - x^4.$$

As in example 39, decompose the given problem into subproblems, each of which can be solved individually. The principle of superposition guarantees that the sum of particular solutions of the subproblems is a particular solution of the original equation.

Form each subproblem by collecting in each all forcing terms of similar type—i.e., all exponentials with the same exponent, all sines and cosines with the same frequency, all polynomials, and so forth. One such set of subproblems is as follows:

Separate into subproblems.

- All sine, cosine terms with angular frequency 2:

$$\text{Problem 1:} \quad y'' - y' - 6y = 3x \sin 2x.$$

- All sine, cosine terms with angular frequency 3:

$$\text{Problem 2:} \quad y'' - y' - 6y = -5 \cos 3x.$$

- All exponentials with exponent $-2x$:

$$\text{Problem 3:} \quad y'' - y' - 6y = x^3 e^{-2x}.$$

- All exponentials with exponent $2x$:

$$\text{Problem 4:} \quad y'' - y' - 6y = x^3 e^{2x}.$$

- All polynomials:

$$\text{Problem 5:} \quad y'' - y' - 6y = \pi - x^4.$$

To construct trial solutions for each of these subproblems, we need to know the solutions of the homogeneous equation

$$y'' - y' - 6y = 0.$$

The characteristic equation $r^2 - r + 6 = (r - 3)(r + 2) = 0$ reveals that homogeneous solutions are

Linearly independent homogeneous solutions

$$e^{3x}, \quad e^{-2x}.$$

For problem 1, a trial solution is

Trial solution for problem 1

$$y_{p1}(x) = (A_1 x + A_0) \cos 2x + (B_1 x + B_0) \sin 2x.$$

None of the terms in this expression is a solution of the homogeneous equation $y'' - y' - 6y = 0$. Both the cosine and the sine in the trial solution are multiplied by a first-order polynomial because the first-order polynomial $3x$ multiplies the sine term in problem 1.

For problem 2, a trial solution is

Trial solution for problem 2

$$y_{p2}(x) = C \cos 3x + D \sin 3x.$$

31. Verify that the solution

$$y_g(x) = x^3 e^{-2x} + C_1 e^{-2x} + C_2 x e^{-2x}$$

of

$$y'' + 4y' + 4y = 6xe^{-2x}$$

obtained in example 38 is indeed a *general* solution. If you completed the previous exercise, can you use some of its analysis here? Explain.

32. Consider the undamped, forced spring-mass system model

$$mx''(t) + kx(t) = A \sin \omega t, \quad x(0) = x_i, \quad x'(0) = v_i,$$

where A, ω are fixed parameters. (Recall that m is mass, k is the spring constant, and x_i, v_i are the initial position and velocity.)

(a) Assume $\omega \neq \sqrt{k/m}$ and solve the initial-value problem.

(b) Assume $\omega = \sqrt{k/m}$ and solve the initial-value problem.

(c) How does the behavior of the solutions obtained in parts (a) and (b) differ? How does the value of ω affect the solution procedure you use?

33. (a) Solve the damped, forced spring-mass system model

$$mx''(t) + px'(t) + kx(t) = A \sin \omega t,$$
$$x(0) = x_i, \quad x'(0) = v_i,$$

where A, ω are fixed parameters. (Recall that m is mass, p is the damping or friction coefficient, k is the spring constant, and x_i, v_i are the initial position and velocity.)

(b) The previous exercise considered the undamped ($p = 0$) version of this model, and two cases arose, $\omega = \sqrt{k/m}$ and $\omega \neq \sqrt{k/m}$. Do such considerations arise here? Does the solution of this problem remain bounded for all time?

34. Example 39 obtained the particular solution

$$y_p(x) = (3x^3 - x^2)e^{-2x} + \frac{4\pi}{(4 + \pi^2)^2} \cos \pi x$$
$$- \frac{4 - \pi^2}{(4 + \pi^2)^2} \sin \pi x$$

of

$$y'' + 4y' + 4y = 18xe^{-2x} - \sin \pi x - 2e^{-2x}$$

by decomposing the original problem into two subproblems:

Problem 1: $y'' + 4y' + 4y = 18xe^{-2x} - 2e^{-2x}$,
Problem 2: $y'' + 4y' + 4y = -\sin \pi x$.

Verify by direct substitution that y_p, which is a sum of a particular solution of problem 1 and a particular solution of problem 2, is indeed a solution of the full problem. Which part of y_p is a solution of problem 1? Of problem 2? Do these parts each yield the expected forcing terms when they are substituted into the left side of the equation? Show how your calculations support your answer to this last question.

35. Example 42 broke the problem of finding the form of a particular solution of

$$y'' - y' - 6y = 3x \sin 2x - 5 \cos 3x + x^3 (e^{-2x} + e^{2x}) + \pi - x^4$$

into a series of subproblems:

Problem 1: $y'' - y' - 6y = 3x \sin 2x$
Problem 2: $y'' - y' - 6y = -5 \cos 3x$
Problem 3: $y'' - y' - 6y = x^3 e^{-2x}$
Problem 4: $y'' - y' - 6y = x^3 e^{2x}$
Problem 5: $y'' - y' - 6y = \pi - x^4$

Find a particular solution of the original differential equation by finding a particular solution of each of these subproblems and summing. You may use as much of the information in example 42 as you find useful.

36. In its search for a particular solution of

$$x''(t) = -g,$$

example 36 proposed three forms for a trial solution of $x'' = -g$. They are

$$x_{p1}(t) = A_0,$$
$$x_{p2}(t) = A_0 t,$$
$$x_{p3}(t) = A_0 t^2.$$

Confirm the wisdom of finally using x_{p3} by direct substitution of the first two trial solutions in $x''(t) = -g$. Explain why these two trial solutions are unsatisfactory. Show that the third trial solution is not a solution of the homogeneous equation $x'' = 0$.

37. In its search for a particular solution of

$$y'' + 4y' + 4y = 8e^{-2x},$$

example 37 proposed three forms for a trial solution of $y'' + 4y' + 4y = 8e^{-2x}$:

$$y_{p1}(x) = A_0 e^{-2x},$$
$$y_{p2}(x) = A_0 x e^{-2x},$$
$$y_{p3}(x) = A_0 x^2 e^{-2x}.$$

Confirm the wisdom of finally using y_{p3} by direct substitution of y_{p1} and y_{p2} in $y'' + 4y' + 4y = 8e^{-2x}$. Explain why these two trial solutions are unsatisfactory. Show that x_{p3} is not a solution of the homogeneous equation $y'' + 4y' + 4y = 0$.

38. In its search for a particular solution of

$$y'' + 4y' + 4y = 6xe^{-2x},$$

example 38 proposed three forms for a trial solution of $y'' + 4y' + 4y = 6xe^{-2x}$:

$$y_{p1}(x) = (A_1x + A_0)e^{-2x},$$

$$y_{p2}(x) = (A_1x^2 + A_0x)e^{-2x},$$

$$y_{p3}(x) = (A_1x^3 + A_0x^2)e^{-2x}.$$

Confirm the wisdom of finally using y_{p3} by direct substitution of the first two trial solutions in $y'' + 4y' + 4y = 6xe^{-2x}$. Explain why these two trial solutions are unsatisfactory. Show that the third trial solution is not a solution of the homogeneous equation $y'' + 4y' + 4y = 0$.

39. Example 39 found that a particular solution of

$$y'' + 4y' + 4y = 18xe^{-2x} - \sin \pi x - 2e^{-2x}$$

is

$$y_p(x) = (3x^3 - 2x^2)e^{-2x} + \frac{4\pi}{(4 + \pi^2)^2} \cos \pi x$$

$$- \frac{4 - \pi^2}{(4 + \pi^2)^2} \sin \pi x.$$

Find a general solution of this differential equation.

40. Example 40 found that a solution of the initial-value problem

$$\frac{1}{8}x''(t) + 2x(t) = \frac{1}{2} \sin \frac{3\pi t}{2}, \quad x(0) = 0, \ x'(0) = 0,$$

is

$$x(t) = \frac{16}{64 - 9\pi^2} \sin \frac{3\pi t}{2} - \frac{6\pi}{64 - 9\pi^2} \sin 4t$$

$$\approx -0.64 \sin \frac{3\pi t}{2} + 0.76 \sin 4t.$$

Use this solution as a starting point for the construction of a general solution of the differential equation. How does the general solution you obtain compare with that obtained in example 40? Explain the differences and similarities. Is one general solution better than the other in any sense?

41. Table 6.3, which summarizes the form of the trial particular solution for various forcing functions, does not include the hyperbolic sine or cosine among its forcing terms. Correct this omission by adding entries for forcing terms of the form

(a) $a \cosh px + b \sinh px$

(b) $(a_nx^n + \cdots + a_1x + a_0) \cosh px$
$+ (b_nx^n + \cdots + b_1x + b_0) \sinh px$

(c) $ae^{qx} \cosh px + be^{qx} \sinh px$

(Recall that $\cosh px = (e^{px} + e^{px})/2$, $\sinh px = (e^{px} - e^{px})/2$.) Do these entries add any really new information?

6.7 ■ ANALYZING MODELS WITH FORCING

In systems subject to external forcing, an important question is the relation between the frequency of the external forcing and the amplitude and frequency of the response. For example, a high-amplitude response is desirable in a radio tuner circuit, where the forcing is the broadcast signal, but undesirable in an automobile suspension, where the forcing is a bump in the road. Undetermined coefficients is one of the tools that supports such analyses.

6.7.1 Spring-mass Systems

Example 3, page 211, of section 5.1 derived the initial-value problem

Forced spring-mass model

$$\frac{1}{8}x''(t) + 2x(t) = \frac{1}{2} \sin \frac{3\pi t}{2}, \quad x(0) = 0, \quad x'(0) = 0, \tag{6.18}$$

as a model of a vertical spring-mass system whose upper end is being oscillated with angular frequency $3\pi/2$ rad/s. (See figure 6.13.) Example 40 of

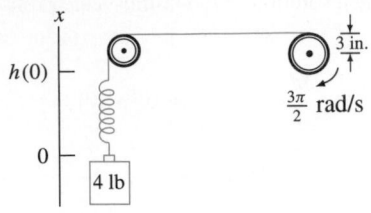

FIGURE 6.13 This spring-mass system is forced by oscillating the upper end of the spring at a frequency of $3\pi/2$ rad/s.

section 6.6 found that the solution of this initial-value problem is

$$x(t) = \frac{16}{64 - 9\pi^2} \sin \frac{3\pi t}{2} - \frac{6\pi}{64 - 9\pi^2} \sin 4t$$

$$\approx -0.64 \sin \frac{3\pi t}{2} + 0.76 \sin 4t. \qquad (6.19)$$

The term involving $\sin(3\pi t/2)$ evidently arises from the external forcing. To explain the other term, which has angular frequency 4 rad/s, note that the natural frequency of this system is

$$\omega_n = \sqrt{\frac{2}{1/8}} = 4 \text{ rad/s.}$$

Although the $\sin 4t$ term may have been stimulated by the external forcing, it represents the natural oscillations of the system.

Graphs of this solution and of the forcing term are shown in figure 6.14. Note that the forcing frequency $3\pi/2 \approx 4.7$ rad/s and the natural frequency 4 rad/s are not too different. (The corresponding periods are $2\pi/(3\pi/2) = 4/3 = 1.33$ s for the forcing and $2\pi/4 = 1.57$ s for the natural vibrations.) Rapid oscillations at about those frequencies do seem to appear in figure 6.14, but there is a much slower oscillation as well, the long-period pulsations in the amplitude of the motion.

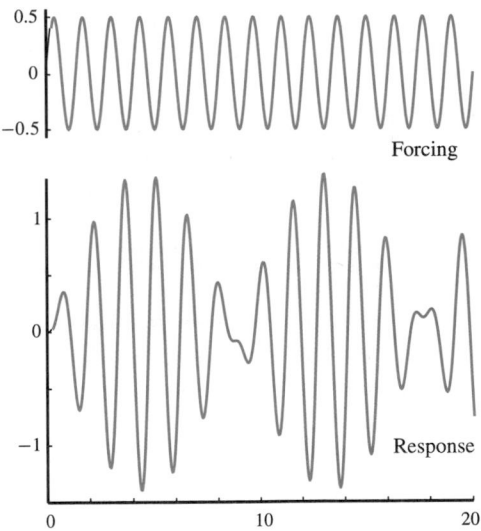

FIGURE 6.14 A plot of the forcing term (upper curve) and the displacement (lower curve) of the forced spring-mass system shown in figure 6.13.

How does the forcing frequency affect the amplitude of the response? What determines the balance between natural frequency and forced frequency in the system's response? What is the source of the low frequency pulsation?

To begin answering these questions, consider this same system with the forcing frequency changed from $3\pi/2$ rad/s to 4 rad/s, the natural frequency of this system. Example 41 of section 6.6 used undetermined coefficients to solve the governing initial-value problem,

$$\frac{1}{8}x''(t) + 2x(t) = \frac{1}{2}\sin 4t, \quad x(0) = 0, \quad x'(0) = 0. \tag{6.20}$$

Recall that the first guess for a particular solution was the linear combination

Trial particular solution

$$x_p(t) = A\cos 4t + B\sin 4t.$$

Since both of the terms in this proposed particular solution solved the homogeneous equation $x''/8 + 2x = 0$, we multiplied by t. We finally obtained the particular solution

Final particular solution

$$x_p(t) = -\frac{1}{16}t\sin 4t.$$

The factor t in this solution appears because the forcing frequency of 4 rad/s is identical to this undamped system's natural frequency. Because of this added factor, the magnitude of the solution grows without bound, in contrast to the solution (6.19) of the system (6.18), whose forcing frequency is different from its natural frequency.

The undamped system (6.20) is said to be in **resonance** because it is being forced at its natural frequency. The natural frequency of an *undamped* system is also called its **resonant frequency**. Evidently, the *response of an undamped system in resonance grows without bound.*

RESONANT FREQUENCY (UNDAMPED)

> Although the solution of the mathematical model grows without bound for all time, displacement is limited in reality by two factors. At very large displacements, the simple linear relationship of Hooke's law fails, and the governing differential equation must be replaced by one containing a nonlinear spring force-displacement relationship; see project 1 of the chapter 5 projects. Ultimately, the spring breaks because it is stretched too far.

The previous analysis shows that the response of a system can change dramatically as the forcing frequency varies. The next example explores this phenomenon in more detail.

■ **EXAMPLE 43**   *Examine the behavior of the response of the undamped, forced system*

$$\frac{1}{8}x''(t) + 2x(t) = F\sin\omega t, \quad x(0) = 0, \quad x'(0) = 0,$$

as the forcing frequency ω varies. Do not allow resonance to occur.

Forbidding resonance is equivalent to requiring $\omega \neq \omega_n$, where

$$\omega_n = \sqrt{\frac{k}{m}} = \sqrt{\frac{2}{1/8}} = 4 \text{ rad/s.}$$

To find a solution of this initial-value problem, mimic the analysis of example 40 of the previous section.

The method of undetermined coefficients suggests a particular solution of the form

Trial particular solution

$$x_p(t) = A \cos \omega t + B \sin \omega t.$$

Since $\omega \neq \omega_n$, none of the terms in this trial particular solution are solutions of the homogeneous equation $x''/8 + 2x = 0$; no added factor of t is required. Substituting into $x''/8 + 2x = \sin \omega t$ leads to

Find undetermined coefficients.

$$A = 0, \quad B = \frac{8F}{16 - \omega^2}.$$

Hence, a general solution of $x''/8 + 2x = F \sin \omega t$ is

Write general solution.

$$x_g(t) = \frac{8F}{16 - \omega^2} \sin \omega t + C_1 \cos 4t + C_2 \sin 4t.$$

Stop and Think   **6.50** Confirm that x_g is indeed a general solution.

Find C_1, C_2.

The initial conditions $x(0) = x'(0) = 0$ require $C_1 = 0$, $C_2 = -2\omega F/(16 - \omega^2)$. The solution of the initial-value problem is

$$x(t) = \frac{8F}{16 - \omega^2} \sin \omega t - \frac{2\omega F}{16 - \omega^2} \sin 4t$$

Write solution of initial-value problem.

$$= \frac{F}{16 - \omega^2} (8 \sin \omega t - 2\omega \sin 4t). \tag{6.21}$$

Stop and Think   **6.51** Confirm that $x(t)$ is a solution of the given initial-value problem.

The magnitude of the extreme displacement—the amplitude—is controlled by the leading factor

Amplitude factor

$$\frac{F}{16 - \omega^2}.$$

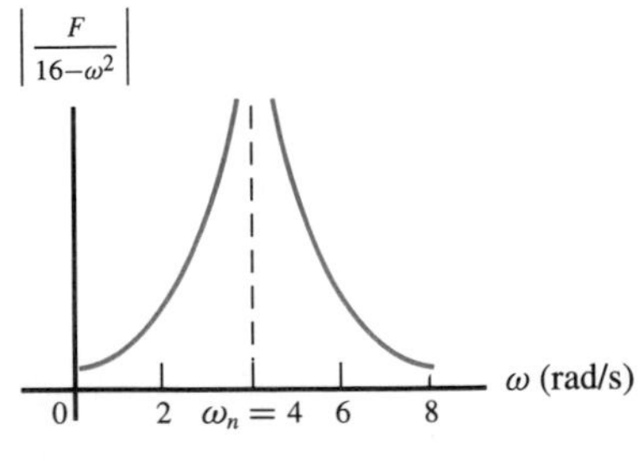

It is no surprise that the amplitude of the response of this linear system is proportional to the amplitude F of the forcing function.

But the forcing frequency also has a profound effect: as ω approaches the resonant frequency of 4 rad/s, the amplitude of the response grows without bound. At resonance ($\omega = 4$) the solution (6.21) is no longer valid, and we must turn to the solution

$$x(t) = -\frac{1}{16} t \sin 4t$$

FIGURE 6.15   A plot of the amplitude of the displacement of the undamped system $x''/8 + 2x = F \sin \omega t$ considered in example 43 as the forcing frequency ω approaches 4 rad/s, the natural frequency of this system.

discussed earlier, which grows with time because of the factor t multiplying the sine function. Figure 6.15 shows the qualitative behavior of the maximum displacement as ω approaches the resonant frequency $\omega_n = 4$ with F fixed. Figure 6.16 shows plots of the displacement for three different forcing frequencies, $\omega = 1, 3, 3.9$ rad/s, all with forcing amplitude $F = 1$. Note how

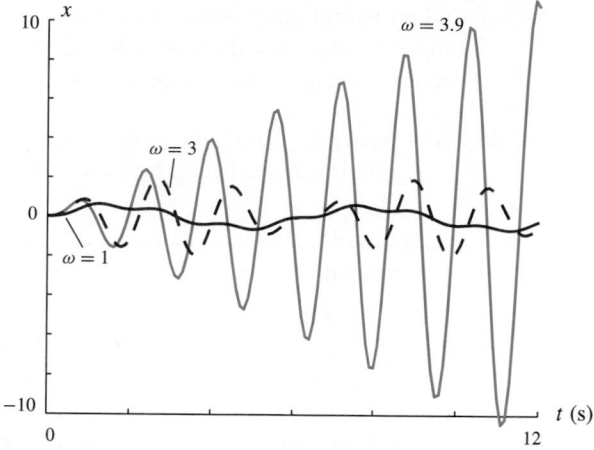

FIGURE 6.16 The response of $x''/8 + 2x = \sin \omega t$, $x(0) = x'(0) = 0$, for $\omega = 1$ rad/s (solid black curve), $\omega = 3$ rad/s (dashed black curve), and $\omega = 3.9$ rad/s (blue curve). Note that the amplitude of the response increases as ω approaches the resonant frequency $\omega_n = 4$ rad/s, although the forcing amplitude remains constant at 1.

the amplitude of the response grows as the forcing frequency approaches the resonant frequency, 4 rad/s.

The solution (6.21) also explains the low-frequency pulsations seen in figure 6.14. Analytically, the source of those pulsations is seen most easily by manipulating trig identities in an expression for the velocity,

$$v(t) = x'(t)$$

$$= \frac{8F\omega}{16 - \omega^2}(\cos \omega t - \cos 4t)$$

$$= \frac{-16F\omega}{16 - \omega^2} \sin\left(\frac{\omega - 4}{2}t\right) \sin\left(\frac{\omega + 4}{2}t\right).$$

The identity is

$$\cos a - \cos b = \cos\left(\frac{a+b}{2} + \frac{a-b}{2}\right) - \cos\left(\frac{a+b}{2} - \frac{a-b}{2}\right)$$

$$= -2 \sin \frac{a-b}{2} \sin \frac{a+b}{2}.$$

Velocity variations exhibit two characteristic frequencies. One is a fast frequency $(\omega + 4)/2$ that differs very little from either the forcing frequency or the natural frequency when the two are nearly equal. The other frequency $(\omega - 4)/2$ is slower, approaching zero as the forcing frequency ω approaches the natural frequency of 4 rad/s.

It is this slow variation, known as *beating*, that accounts for the pulsations in amplitude seen in figures 6.14 and 6.17. (You can hear these pulsations when two slightly out-of-tune musical instruments play the same note.) In those figures, $\omega = 3\pi/2$ and the difference in frequencies is $(\omega - 4)/2 \approx 0.36$ rad/s, corresponding to a period of about 17.6 s.

Stop and Think   **6.52** Verify the calculation of the period of the beating oscillations. Is that period consistent with figures 6.14 and 6.17? ■

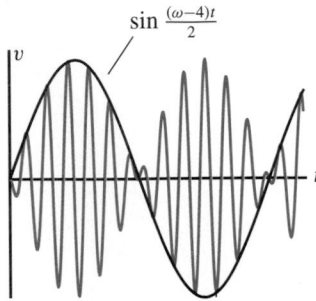

$\sin \dfrac{(\omega-4)t}{2}$

FIGURE 6.17 Beating at a frequency of $(\omega - 4)t/2$ in the velocity of the forced spring-mass system of example 43. The blue curve is velocity. The black curve is the pulsating amplitude envelope due to beating. It is proportional to $\sin(\omega - 4)t/2$.

MATLAB

Use the numerical tools in DELAB to find an approximate solution of $x''/8 + 2x = \sin \omega t$, $x(0) = x'(0) = 0$, for values of ω near the resonant frequency of 4 rad/s. (First write the single equation as a first-order system.) Do these solutions exhibit unbounded growth? If not, does the amplitude appear to increase as $\omega \to 4$, consistent with figure 6.15? If the solutions appear to grow without bound, it that growth due to numerical error or is it consistent with behavior predicted by the differential equation?

■ **EXAMPLE 44**   *Can the solution of the forced, underdamped system*

$$\frac{3}{16}x''(t) + 1.41x'(t) + 8x(t) = F \sin \omega t, \quad x(0) = 0, \quad x'(0) = 0,$$

grow without bound? What value of the forcing frequency ω will produce the largest displacements over time?

The condition $k > 8/3$ derived in example 31, page 282, can be used to verify that this system is indeed underdamped. Alternatively, the presence of complex characteristic roots confirms that it is underdamped.

The characteristic roots of $3x''/16 + 1.41x' + 8x = 0$ are $-3.76 \pm 5.34i$. Hence, the homogeneous part of the general solution of this differential equation is

General solution of homogeneous equation

$$e^{-3.76t}(C_1 \cos 5.34t + C_2 \sin 5.34t).$$

Obviously, this part of the solution decays with time for any values of C_1, C_2. It does not produce unbounded growth, regardless of the forcing frequency (or initial conditions), and its contributions to the maximum displacement are negligible after a sufficient period of time has passed. (For example, at $t = 10$ s, $e^{-3.76t} \approx 4.38 \times 10^{-14}$.)

To examine the contributions of the particular solution to the displacement, use undetermined coefficients and assume a particular solution of the form

Trial particular solution

$$x_p(t) = B \cos \omega t + C \sin \omega t.$$

After the usual substitution in the differential equation, we find that

Find undetermined coefficients.

$$B = -\frac{1.41\omega^2 F}{2\omega^2 + D^2}, \quad C = \frac{DF}{2\omega^2 + D^2},$$

where $D = k - 3\omega^2/16$.

Our real interest here is in the amplitude, or maximum displacement, of the particular solution x_p. Just as in example 28, page 278, of section 6.5, we could represent this particular solution in the phase angle-amplitude form

Phase angle-amplitude form of particular solution

$$x_p(t) = A\sin(\omega t + \phi).$$

The same calculations that derived the amplitude expression (6.11) from the solution (6.9) considered in example 28 lead as well to the amplitude A of this particular solution,

Amplitude of particular solution

$$A = \sqrt{B^2 + C^2} = \frac{F}{\sqrt{2\omega^2 + D^2}}.$$

Stop and Think **6.53** Derive the amplitude expression A.

We can now answer the question, "What value of forcing frequency ω makes the amplitude A of the response a maximum?" by seeking the maximum of A as a function of ω. Using the chain rule, calculate

$$\frac{dA}{d\omega} = -\frac{F}{2(2\omega^2 + D^2)^{3/2}}\left(4\omega + 2D\frac{dD}{d\omega}\right).$$

Since only the term in the large parentheses can make $dA/d\omega$ zero, evaluate it to find

$$4\omega + 2D\frac{dD}{d\omega} = \omega\left(\frac{9\omega^2}{64} - 2\right).$$

Hence, $dA/d\omega = 0$ when $\omega = 0$, a value we discard (why?), and when $9\omega^2/64 - 2 = 0$, or

Forcing frequency for maximum amplitude

$$\omega = \sqrt{\frac{128}{9}} = \frac{8\sqrt{2}}{3} \approx 3.77 \text{ rad/s.} \ \blacksquare$$

The particular solution whose amplitude we have just examined is, in fact, the *periodic solution* of this underdamped system, and the homogeneous solution is its transient. We have shown that this limiting periodic state has maximum amplitude at a forcing frequency of 3.77 rad/s. The forcing frequency at which a periodic solution of an *under*damped system has its maximum is called the **resonant frequency** of that system.

RESONANT FREQUENCY (UNDERDAMPED)

Stop and Think **6.54** The resonant frequency of an underdamped system is the forcing frequency that provides the periodic solution of largest amplitude. Is the definition of *resonant frequency* consistent with that for undamped systems?

6.55 Does it make sense to speak of the resonant frequency of an *over*damped system?

Note that the resonant frequency of this damped system is *not* the natural frequency of the corresponding undamped system. That natural frequency is

$$\omega_n = \sqrt{\frac{8}{3/16}} \approx 6.53 \text{ rad/s}$$

while we have seen that the resonant frequency is 3.77 rad/s. Without damping, the resonant frequency would be ω_n.

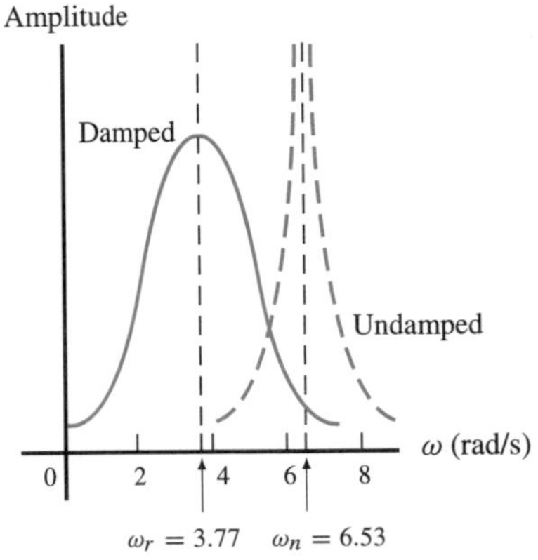

Figure 6.18 shows the variation of the amplitude of the forced response with frequency for the damped system of example 44 and for its undamped analog. The resonant frequency in the damped case is $\omega_r = 3.77$ rad/s, as calculated earlier, while the resonant frequency without damping is the natural frequency $\omega_n = 6.53$ rad/s. With damping, the amplitude of the response reaches a finite maximum when the system is forced at its resonant frequency. Without damping, the amplitude of the response grows without bound as the forcing frequency approaches the resonant (natural) frequency.

The exercises ask you to show that the resonant frequency of the underdamped spring-mass system

$$mx'' + px' + kx = F \sin \omega t$$

is

FIGURE 6.18   Amplitude of response versus forcing frequency for the system of example 44 (solid curve on the left) and its undamped analog (dashed curve on the right).

$$\omega_r = \sqrt{\frac{k}{m} - \frac{p^2}{2m^2}}.$$

That is, the amplitude of the periodic solution of this equation is a maximum when $\omega = \omega_r$.

Adjusting the forcing frequency for maximum amplitude is the complement of tuning a radio, in which the natural frequency of the tuner circuit is adjusted for maximum amplitude at the fixed forcing frequency of the desired station. Resonance in the moderately damped tuner circuit provides the strongest signal.

Resonance was long believed to account for the collapse of the Tacoma Narrows Bridge in 1940. The bridge began twisting and undulating in moderate winds that might have excited certain resonant modes. Ultimately, the amplitude of the response grew so large that the bridge collapsed. However, more recent research [16] suggests that nonlinearity rather than resonance may have been the culprit; see project 5 at the end of this chapter.

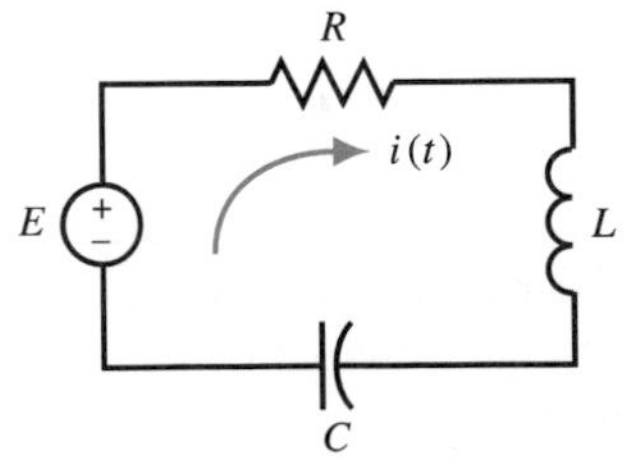

6.7.2   *RLC* Circuits

A varying voltage source in a series *RLC* circuit (see figure 6.19) can also produce resonance effects analogous to those seen in spring-mass systems.

Example 35, page 287, of section 6.5 explored the damped response of the circuit of figure 6.19 without forcing. The next example compares the response when a voltage source imposes a forcing frequency.

FIGURE 6.19   A series *RLC* circuit with an independent voltage source *E*.

■ **EXAMPLE 45**   *Suppose the RLC circuit of example 35 of section 6.5 ($R = 200\,\Omega$, $L = 100\,\mu H$, $C = 100\,pF$) contains an independent voltage source $E = V \sin \omega t$ with $V = 10\,mV$, $\omega = 2\,rad/\mu s$. What is the amplitude of the periodic response to the forcing that appears after the transient dies away? About how much time must pass before that periodic solution establishes itself?*

To answer the second question first, recall from example 35 that the transient decays to half its amplitude in about $1.4\,\mu s$. So 2 or 3 μs should be enough time for the transient to be effectively dissipated.

Stop and Think   **6.56** Where do you look to find the number that controls the decay of the transient?

To determine the amplitude of the periodic solution, first find a particular solution of

$$Lq'' + Rq' + q/C = V \sin \omega t$$

that does not involve any homogeneous solution terms. That particular solution will be the desired periodic solution.

Stop and Think   **6.57** Why are the solutions of the homogeneous equation $Lq'' + Rq' + q/C = 0$ not periodic functions?

Since amplitude of response is the issue, seek a particular solution in the phase angle-amplitude form

Phase angle-amplitude form of particular solution

$$q_p = A \sin(\omega t + \phi),$$

where A, ϕ are undetermined coefficients. (The exercises use the more conventional form $q_p = B_1 \cos \omega t + B_2 \sin \omega t$.)

Substituting in the differential equation, expanding $\sin(\omega t + \phi)$ using the sum-of-angles identity, and equating coefficients of $\sin \omega t$ and $\cos \omega t$ lead, respectively, to

Substitute to find equations defining A, ϕ.

$$A\left[\left(\frac{1}{C} - L\omega^2\right)\cos\phi - R\omega\sin\phi\right] = V,$$

$$\left(\frac{1}{C} - L\omega^2\right)\sin\phi + R\omega\cos\phi = 0. \tag{6.22}$$

From the second equation, we have

$$\tan\phi = -\frac{R\omega}{1/C - L\omega^2}.$$

We conclude that

Determining ϕ ...

$$\cos\phi = \pm\frac{1/C - L\omega^2}{D}, \qquad \sin\phi = \mp\frac{R\omega}{D}, \tag{6.23}$$

where

$$D = \sqrt{\left(\frac{1}{C} - L\omega^2\right)^2 + R^2\omega^2}.$$

Stop and Think **6.58** Carry out the details of these manipulations.

Using these expressions for $\cos\phi$ and $\sin\phi$ in (6.22) yields

Amplitude value
$$A = \pm\frac{V}{D}.$$

To ensure $A > 0$, choose the positive sign above (and hence the positive sign with the cosine and the negative sign with the sine function in (6.23)) if $V > 0$. All the signs are reversed if $V < 0$.

For the given circuit parameters $R = 200\,\Omega$, $L = 100\,\mu\text{H}$, $C = 100\,\text{pF}$, and $V = 10\,\text{mV}$, we calculate

$$A = 1.0408 \times 10^{-12}\,\text{C}.$$

This charge on a 100-pF capacitor corresponds to a voltage drop across the capacitor of 10.408 mV, because $v_C = q/C$.

The graph on the right in figure 6.20 (which was computed using a fourth-order Runge-Kutta method with $\Delta t = 0.001\,\mu\text{s}$) illustrates this periodic response with its amplitude of approximately 10 mV. That same graph shows that the transient has decayed to insignificance in about 3 μs.

Compare the decay of the transient shown in the graph on the left in figure 6.20 with the emergence of the periodic solution in the graph on the right. Note the appearance of the higher frequency unforced oscillations in the early part of the forced response, when the forced oscillations and the natural oscillations are struggling for dominance. ■

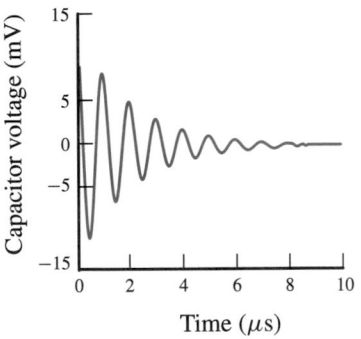 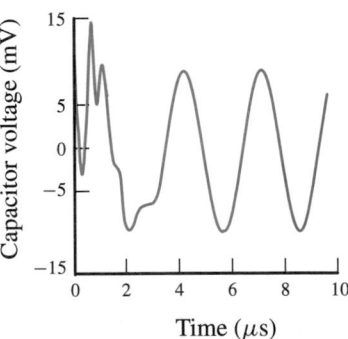

FIGURE 6.20 Capacitor voltage versus time for the series *RLC* circuit of figure 6.19 without forcing (left graph) and with forcing (right graph); $R = 200\,\Omega$, $L = 100\,\mu\text{H}$, $C = 100\,\text{pF}$. The forcing function for the graph on the right is $E = V\sin\omega t$ with $V = 10\,\text{mV}$, $\omega = 2\,\text{rad}/\mu\text{s}$. Initial conditions are $v_C(0) = 10\,\text{mV}$ (corresponding to $q(0) = 10^{-12}\,\text{C}$) and $i(0) = q'(0) = -1\,\mu\text{A}$.

■ **EXAMPLE 46** *Increase the resistance in the circuit of the previous example to R = 4,500 Ω. What is the phase relationship between the forcing function and the periodic response?*

Because the forcing amplitude $V = 10\,\text{mV}$ is positive, the analysis of the preceding example and equation (6.23) together force the phase angle ϕ to satisfy

Determining ϕ...

$$\cos\phi = \frac{1/C - L\omega^2}{D}, \quad \sin\phi = -\frac{R\omega}{D},$$

where D is as before. For the given circuit parameters, we find

$$\cos\phi = 0.7295, \quad \sin\phi = -0.6839.$$

Since ϕ must lie in the fourth quadrant, we have $\phi = 5.5300$ rad, or about 317°.

When $t = 0$, the argument of the periodic response $A\sin(\omega t + \phi)$ is about 317° ahead of the argument of the forcing function $V\sin\omega t$. At an angle of 317°, the sine function in the periodic solution will have already passed through its first maximum at 90° ($\pi/2$ rad), its second zero at 180° (π rad), and its first minimum at 270° ($3\pi/2$ rad). The periodic solution will be climbing back toward zero on its way to another maximum when the forcing function is only beginning its climb from zero toward its first maximum.

Figure 6.21 compares the periodic response (right graph) with the forcing function (left graph) to illustrate this phase difference. ■

Stop and Think **6.59** The 317° phase difference shown in figure 6.21 is nearly 90% of a full period of 360°. Do the graphs show a time difference between peaks that is about 90% of the period of the forcing function?

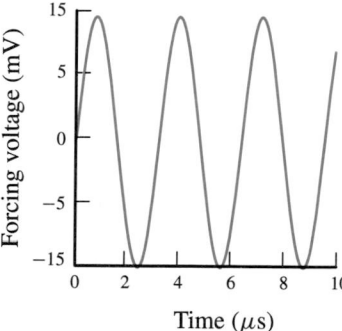 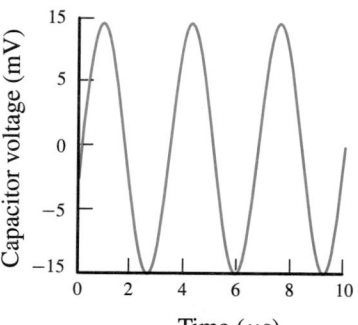

FIGURE 6.21 Voltage source (left graph) and periodic response (capacitor voltage—right graph) versus time for the *RLC* circuit of figure 6.19 with $R = 4,500\,\Omega$, $L = 100\,\mu\text{H}$, $C = 100\,\text{pF}$. The forcing function is $E = V\sin\omega t$ with $V = 10\,\text{mV}$, $\omega = 2\,\text{rad}/\mu\text{s}$.

6.7.3 Exercises

EXERCISE GUIDE	
To gain experience . . .	**Try exercises**
Relating physical systems to equations	1
Finding resonant frequency	2–4, 13(c), 14, 18
With relations among system parameters, resonance, and amplitude	3–5, 14, 17, 22–23, 27
Analyzing solution behavior as forcing frequency varies	7, 13(c), 15–16, 18
Solving forced equations	8–9, 11, 18, 22–23
Finding maxima of solutions	10, 13(b–c), 18
With phase angle-amplitude solutions	11–12, 19, 24–28

1. *Completely* describe a physical situation of which each of the following initial-value problems could be a model.

   (a) $2x'' + 3x' + x = 0$, $x(0) = 1$, $x'(0) = -2$

   (b) $2x'' + 3x' + x = 5 \sin \omega t$, $x(0) = 1$, $x'(0) = -2$

2. Find the resonant frequency of each of the following systems.

   (a) A mass of 600 g is suspended from a spring that extended 4 cm when a mass of 100 g was hung from it. The 600-g mass is set in motion by releasing it from rest 6 cm above its equilibrium position.

   (b) The spring of part (a) is now set horizontally, attached to a fixed anchor as in figure 6.8, page 282. A 2-kg air puck is attached to the free end of the spring. (An air puck rides on a cushion of air so that it can slide with essentially no friction.) The puck is pushed 7 cm toward the fixed anchor and released.

   Do you have enough information in each case to determine the resonant frequency? Are you given any extra information? List both missing and extraneous information, if any.

3. Consider the system described in part (a) of exercise 2.

   (a) Suppose the mass is released from rest from an arbitrary position x_i. How does the resonant frequency of the system depend upon x_i?

   (b) Suppose the mass is started in motion from its equilibrium position with a velocity v_i. How does the resonant frequency of the system depend upon v_i?

4. An air-puck system similar to the one described in part (b) of exercise 2 is set in motion from its equilibrium position with an unknown initial velocity.

   (a) The maximum displacement of the puck is observed to be 14.1 cm. Can you determine the resonant frequency of the system?

   (b) The period of its oscillations is observed to be 6.39 s. Can you determine the resonant frequency of the system?

5. The system described in part (b) of exercise 2 is not truly undamped, although that exercise suggests that you could model it that way. In fact, during a period of 4 min, the maximum displacement of the oscillations of the puck was observed to decrease from 10 cm to 5.6 cm. What is the system's resonant frequency?

6. Following the style of example 43, examine the behavior of the response of the general undamped, forced system

   $$mx''(t) + kx(t) = F \sin \omega t, \quad x(0) = 0, \quad x'(0) = 0,$$

   as the forcing frequency ω varies. Do not allow resonance to occur.

7. Following the style of example 43, examine the behavior of the response of the general damped, forced system

   $$mx''(t) + px' + kx(t) = F \sin \omega t,$$
   $$x(0) = x_i, \quad x'(0) = v_i,$$

   as the forcing frequency ω varies. Does the form of the solution change at resonance?

8. Example 43 solves the initial-value problem

$$\frac{1}{8}x''(t) + 2x(t) = F \sin \omega t,$$

$$x(0) = 0, \quad x'(0) = 0,$$

of which the problem solved in example 40 of section 6.6,

$$\frac{1}{8}x''(t) + 2x(t) = \frac{1}{2}\sin\frac{3\pi t}{2},$$

$$x(0) = 0, \quad x'(0) = 0,$$

is a special case. Show that the solution obtained in example 43 reduces to the solution found in example 40, section 6.6,

$$x(t) = \frac{16}{64 - 9\pi^2}\sin\frac{3\pi t}{2} - \frac{6\pi}{64 - 9\pi^2}\sin 4t,$$

when the parameters F, ω appearing in the first equation are given the appropriate values. Similarly, show that the undetermined coefficients used to find the particular solution and the arbitrary constants in the general solution reduce to the appropriate values.

9. Supply the details of the derivation of the solution

$$x(t) = \frac{F}{16 - \omega^2}(8 \sin \omega t - 2\omega \sin 4t)$$

of the initial-value problem

$$\frac{1}{8}x''(t) + 2x(t) = F \sin \omega t, \quad x(0) = 0, \quad x'(0) = 0,$$

considered in example 43.

10. Example 43 suggests that extreme displacements of the solution of the forced initial-value problem

$$\frac{1}{8}x''(t) + 2x(t) = F \sin \omega t,$$

$$x(0) = 0, \quad x'(0) = 0,$$

occur at those times t for which $x'(t) = 0$, where

$$x(t) = \frac{F}{16 - \omega^2}(8 \sin \omega t - 2\omega \sin 4t)$$

is the solution of this initial-value problem. Find an equation whose solutions are these values of t. In the case $F = 1/2$, $\omega = 3\pi/2$, find the first two such values of t graphically, numerically, or by any other means available from MATLAB. What are the corresponding extreme values of displacement?

11. Find the solution of the system of example 43,

$$\frac{1}{8}x''(t) + 2x(t) = F \sin \omega t,$$

$$x(0) = 0, \quad x'(0) = 0,$$

with both the particular and homogeneous parts of the solution expressed in phase angle-amplitude form. That is, express the homogeneous part in a form such as

$$A \sin(\omega_n t + \phi)$$

and the particular part in a form such as

$$B \sin(\omega t + \theta),$$

where B, θ are the amplitude and phase angle, respectively, of the particular solution.

12. Example 44 claims that a particular solution

$$x_p(t) = B \cos \omega t + C \sin \omega t$$

of the underdamped differential equation

$$\frac{3}{16}x''(t) + 1.41x'(t) + 4x(t) = F \sin \omega t$$

can be represented in the phase angle-amplitude form

$$x_p(t) = A \sin(\omega t + \phi)$$

with amplitude

$$A = \frac{F}{\sqrt{2\omega^2 + D^2}},$$

where $D = k - 3\omega^2/16$. It suggests that the calculations are identical with those of example 28.

(a) Verify this claim by deriving the given expression for A.

(b) Obtain an expression for the phase angle ϕ. Does the response represented by the particular solution x_p given above lead (reach a peak before) or lag (reach a peak after) the forcing function? Does it seem physically reasonable that the parameters you find appearing in ϕ should control the phase relationship between the forcing and the response?

13. Consider a forced, damped system described by the differential equation

$$\frac{3}{16}x''(t) + 1.41x'(t) + kx(t) = F \sin \omega t.$$

(This equation with $k = 8$ was considered in example 44.)

(a) For what range of values of the spring constant k is this system *underdamped*? Assume in part (b) that k is restricted to that range of values.

(b) Find the amplitude of the particular solution of this system.

(c) Show that for $k > 8/3$ this system can exhibit resonance; i.e., show that there is a value of the forcing frequency ω for which the amplitude of the periodic solution of this equation exhibits a local maximum. Find the corresponding resonant frequency. Show that for $k \leq 8/3$, the amplitude does not exhibit a local maximum. Do the values of k for which resonance does not occur correspond to the system being nearly overdamped or very far from being overdamped? Is this relation physically reasonable?

14. The resonant frequency of a damped, forced spring-mass system is the forcing frequency which gives the periodic solution of the equation its largest amplitude. Is the resonant frequency of a damped, forced system the same as the frequency of the decaying oscillations in the unforced system? Answer using information obtained in example 44.

15. Using a numerical method of your choice from DELAB, conduct some numerical experiments to estimate the amplitude of the periodic solution of

$$\frac{1}{8}x'' + 2x = \sin \omega t$$

for ω equal to 0.1, 0.5, 0.9, 1.1, 2, and 4 times the resonant frequency of this system. (First write the single equation as a first-order system.) Plot the amplitude values versus frequency and comment on the connection with the graph of figure 6.15. How could you obtain similar results using the analytic solution of this problem? What might happen were you to force this system at precisely its resonant frequency?

16. Repeat exercise 15 for a system with the same mass and spring constant but with a damping coefficient one-quarter that required to critically damp the system. Can you force this system exactly at its resonant frequency for an extended period of time?

17. Why do we not speak of the resonant frequency of an *over*damped system?

18. Find the value of forcing frequency ω that maximizes the amplitude of the periodic response of the forced parallel *RLC* circuit modeled by

$$Cv'' + \frac{1}{R}v' + \frac{1}{L}v = F \sin \omega t.$$

19. Find the phase angle between the forcing function and the periodic response of the parallel *RLC* circuit of exercise 18.

20. Can the amplitude of the response of a forced series *RLC* circuit be made arbitrarily large by reducing the resistance in the circuit while holding all other parameters constant? What if both the resistance and the forcing frequency can be varied?

21. Referring to the series *RLC* circuit of figure 6.19 with $R = 200 \, \Omega$, $L = 100 \, \mu\text{H}$, $C = 100 \, \text{pF}$, the text in example 45 says glibly, "So 2 or 3 μs should be enough time for the transient to be effectively dissipated." But "effectively" is not very precise.

(a) How much time must elapse before the amplitude of the transient has decayed to 1% of its original size?

(b) How much time must elapse before the amplitude of the transient has decayed to δ times its original size, where $0 < \delta < 1$?

(c) Answer the preceding question for the general *RLC* circuit governed by $Lq'' + Rq' + q/C = 0$.

22. The text finds the amplitude of the periodic response of the series *RLC* circuit equation

$$Lq'' + Rq' + q/C = V \sin \omega t$$

by seeking a particular solution in the phase angle-amplitude form $q_p = A \sin(\omega t + \phi)$. Instead, derive the amplitude as follows.

(a) Assume an undetermined coefficients solution of the form $q_p = B_1 \cos \omega t + B_2 \sin \omega t$. Under what conditions on R, L, and/or C can you be sure that this proposed particular solution contains no solutions of the corresponding homogeneous equation?

(b) Find B_1, B_2.

(c) Argue that the amplitude of this solution, its maximum value, is $\sqrt{B_1^2 + B_2^2}$, thereby verifying the amplitude expression obtained in the text.

23. Example 45 derives the amplitude of the periodic response of

$$Lq'' + Rq' + q/C = V \sin \omega t$$

using the phase angle-amplitude form $A \sin(\omega t + \phi)$ of the particular solution. Supply the details of that derivation.

24. Example 46 determines that the periodic response of a certain series *RLC* circuit leads the forcing function by 317°. That circuit has $R = 4,500 \, \Omega$, $L = 100 \, \mu\text{H}$, and $C = 100 \, \text{pF}$, and it is subject to a voltage source $E = V \sin \omega t$ with $V = 10 \, \text{mV}$ and $\omega = 2 \, \text{rad}/\mu\text{s}$. The response and the forcing are shown in figure 6.21. Answer the question the text asks:

The 317° phase difference is about 90% of a full period of 360°. Do the graphs show a time difference between peaks that is about 90% of the period of the forcing function?

Compute the period of the forcing function and compare it with the time differences between peaks you read from the graphs of figure 6.21.

25. Could example 45 have concluded that the phase angle of the periodic response is $\phi = -43°$ rather than $\phi = 317°$? Verify that the equations defining ϕ are satisfied by this choice. Would one then say that the periodic response *lags* the forcing by 43°? Sketch a version of figure 6.21 and show this phase difference.

26. Example 45 concludes that the periodic response of the circuit considered there leads the forcing by 317°. Express this lead in seconds. Is the time lead you obtain consistent with that shown in figure 6.21?

27. Find the amplitude of the periodic response of the series *RLC* circuit considered in example 46. That circuit has

$R = 4,500\,\Omega$, $L = 100\,\mu H$, and $C = 100\,pF$, and it is subject to a voltage source $E = V \sin \omega t$ with $V = 10\,mV$ and $\omega = 2\,rad/\mu s$. Compare that amplitude with the one shown in the appropriate graph of figure 6.21. Which of the parameters R, L, C, ω, V affect the amplitude? Does doubling the strength V of the forcing voltage change the amplitude? How about doubling the frequency of the voltage source?

28. Find the phase angle between the periodic response and the forcing function in the series *RLC* circuit of example 45 with $R = 200\,\Omega$, $L = 100\,\mu H$, and $C = 100\,pF$. The voltage source is $E = V \sin \omega t$ with $V = 10\,mV$ and $\omega = 2\,rad/\mu s$. Which of the parameters R, L, C, ω, V affect the phase angle? Does doubling the strength V of the voltage source change the phase angle? Does doubling the frequency of the voltage source change the phase angle?

> Audio purists worry when phase varies with frequency because sound quality suffers. The apparent location of an instrument could change as the notes it plays rise or fall in frequency.

6.8 ■ LINEAR VERSUS NONLINEAR

In what ways are nonlinear models of oscillatory phenomena different from linear models? How does nonlinearity change the relations among frequency, amplitude, initial conditions, and the other parameters of the problem? What relationships do we find between a nonlinear equation and the linear approximations that arise in the study of the stability of equilibrium points?

A natural vehicle for such comparisons is the nonlinear pendulum model of section 5.2,

NONLINEAR PENDULUM
MODEL

$$L\theta'' + p\theta' + g \sin \theta = 0, \quad \theta(0) = \theta_i, \quad \theta'(0) = \omega_i, \tag{6.24}$$

and its linear approximation,

LINEAR PENDULUM MODEL

$$L\theta'' + p\theta' + g\theta = 0, \quad \theta(0) = \theta_i, \quad \theta'(0) = \omega_i. \tag{6.25}$$

Recall that L is the length of the pendulum arm, p is a damping coefficient (see exercise 8, page 223, of section 5.2.), g is the acceleration of gravity, θ is the angle in radians of the pendulum from the vertical, θ_i is the initial angular displacement of the pendulum, and ω_i is its initial angular velocity.

Stop and Think
6.60 In order to explicitly display the angular velocity $\omega = \theta'$, write each of these pendulum models as first-order systems.

6.8.1 Amplitude and Period

What is the relation between the amplitude and the period of the oscillations of an undamped ($p = 0$) pendulum? For simplicity, always start the pendulum by displacing it to one side and releasing it from rest: $\omega_i = 0$.

In this case, the solution of the linear initial-value problem (6.25) is

$$\theta(t) = \theta_i \cos \sqrt{\frac{g}{L}} t.$$

We make two observations:

1. The natural frequency $\omega_n = \sqrt{g/L}$ is independent of initial position θ_i.

2. The amplitude θ_i of the response is independent of length L and the acceleration of gravity g.

The upper graph of figure 6.22 illustrates the first observation for a pendulum of length $L = 0.9930$ m. The solid black curve is the graph of angular displacement for $\theta_i = 0.15$ rad, and the blue dashed curve is that for $\theta_i = 1$ rad. The periods, and hence the frequencies, of these two curves are identical.

What is the relation between period and amplitude for the nonlinear equation $L\theta'' + g \sin \theta = 0$? Unfortunately, analytic solution tools fail, but fourth-order Runge-Kutta, for example, enables us to draw the graphs shown in the lower half of figure 6.22.

The solid black curves in figure 6.22, those corresponding to the smaller amplitude, seem to have the same frequency, but the blue dashed curve in the lower graph clearly has a longer period. With each swing, it passes through the vertical or $\theta = 0$ position slightly later than the motion described by the solid black curve. *The period of the nonlinear pendulum has increased with amplitude, while that of the linear pendulum remained fixed.*

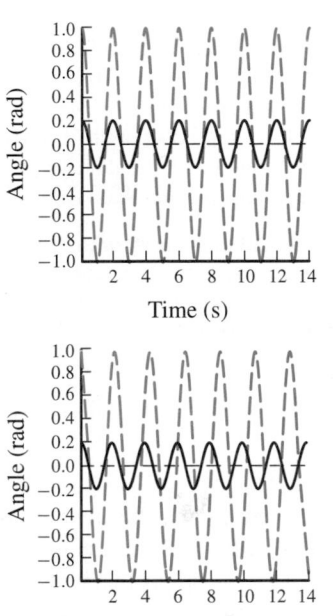

FIGURE 6.22   Graphs of solutions of the linear (equation (6.25)—upper plot) and the nonlinear (equation (6.24)—lower plot) pendulum equations for $\theta_i = 0.15$ rad (solid black curve) and $\theta_i = 1$ rad (dashed blue curve). In both cases, $L = 0.9930$ m, $\omega_i = 0$.

MATLAB

Write the single equation as a first-order system, then use DELAB's numerical solution tools to generate your own version of figure 6.22.

Why is the difference in predicted period between the two equations pronounced at the larger amplitude but not at the smaller amplitude? At the smaller amplitude of 0.15 rad ($\approx 8.6°$), the approximation $\sin \theta \approx \theta$ that connects the nonlinear pendulum equation

$$L\theta'' + g \sin \theta = 0$$

and the linear pendulum equation

$$L\theta'' + g\theta = 0$$

is more nearly valid than it is at 1 rad ($\approx 57°$); $\sin 0.15 = 0.14944$, while $\sin 1 = 0.84147$. As the nonlinearity asserts itself, predictions based on the solution of a linear approximation lose validity.

Stop and Think **6.61** What is the percent error in the approximation $\sin\theta \approx \theta$ at the two initial values considered here, $\theta_i = 0.15$ rad and $\theta_i = 1.0$ rad?

We leave to exercise 8 the task of conducting a parallel investigation of the relation between amplitude and the parameters L and g.

6.8.2 Steady States and Stability

What are the steady states of the nonlinear pendulum equation? They are the constant solutions θ_{ss} of

$$L\theta_{ss}'' + p\theta_{ss}' + g\sin\theta_{ss} = g\sin\theta_{ss} = 0.$$

From $\sin\theta_{ss} = 0$, we obtain

$$\theta_{ss} = 0, \pm\pi, \pm 2\pi, \ldots .$$

Stop and Think **6.62** Show that the *undamped* pendulum equation $L\theta'' + g\sin\theta = 0$ exhibits the same steady states.

This endless list of possible steady states is at first surprising, for the only apparent stationary position of the pendulum is hanging straight down. But figure 6.23 explains the list.

Steady-state values such as $\theta_{ss} = 0, \pm 2\pi, \ldots$, which involve zero or *even* multiples of π, correspond to the familiar straight-down position, perhaps after the pendulum has passed through one or more complete revolutions around its pivot. (A full revolution increments θ by 2π. The increment is positive if the revolution is counterclockwise, and negative if the rotation is clockwise.)

The equilibria at *odd* multiples of π occur when the pendulum is positioned straight *up*. Knowing that this inverted position is difficult to attain in practice suggests that these steady states are unstable, although their straight-down counterparts ought to be stable. The linear stability analysis that follows confirms our intuition in several cases. (Compare these examples with the analysis of the stability of the steady states of the logistic equation $P' = aP - sP^2$ in section 3.3.)

Begin by analyzing the stability of a steady state that ought to be unstable. For simplicity, ignore damping.

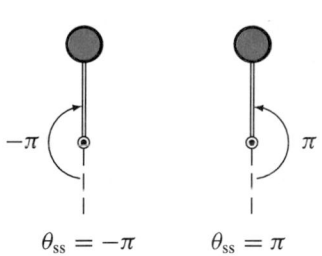

FIGURE 6.23 Some possible steady states of a pendulum.

Identify the steady state.

Perturb the steady state.

■ **EXAMPLE 47**   *Determine the stability of the equilibrium $\theta_{ss} = \pi$ of the undamped ($p = 0$) nonlinear pendulum equation in (6.24).*

What becomes of a small perturbation of the steady state $\theta_{ss} = \pi$? Intuition suggests that a perturbation of the straight-up state should grow. Confirming that intuition mathematically will establish that this steady state is unstable.

Call this small perturbation $q(t)$; it is defined by

$$\theta(t) = \pi + q(t),$$

where $\theta(t)$ is a solution of the nonlinear pendulum equation $L\theta'' + g\sin\theta = 0$.

The steady state plus its perturbation must satisfy the governing equation because that sum describes the motion of the pendulum. That is, every possible motion of the pendulum is a solution of the governing equation.

Substituting into the differential equation leads to

Substitute into differential equation.

$$L(\pi + q(t))'' + g \sin(\pi + q(t)) = 0,$$
$$Lq''(t) + g \sin(\pi + q(t)) = 0.$$

To find a *linear* (but approximate) equation for q, expand $\sin x$ in a Taylor polynomial about $x = \pi$:

$$\sin(\pi + q) = \sin x|_{x=\pi} + \left(\frac{d}{dx} \sin x \bigg|_{x=\pi} \right) q + \cdots$$

Expand nonlinearity.

$$= \sin \pi + (\cos \pi)q + \cdots$$

$$= -q + \cdots .$$

The linearizing approximation is

Linearize.

$$\sin(\pi + q) \approx -q.$$

The resulting *linear* equation for the perturbation q is

Linear equation for perturbation

$$Lq'' - gq = 0,$$

only slightly different from the undamped approximate pendulum equation. But that negative sign before the second term has profound consequences!

The characteristic roots of this equation are $\pm\sqrt{g/L}$, and the corresponding general solution is

$$q = C_1 e^{\sqrt{g/L}\,t} + C_2 e^{-\sqrt{g/L}\,t}.$$

Except in the very special circumstances that render $C_1 = 0$, the perturbation q grows exponentially.

Here is an example of the power of a general solution. It represents *every* possible form the perturbation q could take.

Stop and Think   **6.63** Verify that the characteristic roots of $Lq'' - gq = 0$ are $\pm\sqrt{g/L}$. Verify the general solution and the claims about its exponential growth.

6.64 What special initial conditions are required to make $C_1 = 0$? Is a random disturbance of an upward-pointing pendulum likely to produce these initial conditions?

Since all but a very special form of perturbation of the steady state $\theta_{ss} = \pi$ will grow, we conclude that it is unstable. ■

Now consider a straight-down steady state, one our intuition suggests should be stable.

■ **EXAMPLE 48**   *Determine the stability of the equilibrium* $\theta_{ss} = 2\pi$ *of the damped* $(p > 0)$ *pendulum.*

Proceeding as in the last example, we wish to determine the fate of the small perturbation q of the steady state $\theta_{ss} = 2\pi$. Hence, we define q by

Perturb the steady state.

$$\theta(t) = 2\pi + q(t),$$

where $\theta(t)$ is a solution of the nonlinear pendulum equation $L\theta'' + p\theta' + g \sin \theta = 0$.

Substituting θ into the differential equation leads to an equation for the unknown perturbation q,

Substitute into differential equation.

$$L(2\pi + q(t))'' + p(2\pi + q(t))' + g \sin(2\pi + q(t)) = 0,$$
$$Lq''(t) + pq'(t) + g \sin(2\pi + q(t)) = 0.$$

Now expand $\sin x$ in a Taylor polynomial about $x = 2\pi$:

$$\sin(2\pi + q) = \sin x|_{x=2\pi} + \left(\frac{d}{dx} \sin x \bigg|_{x=2\pi} \right) q + \cdots$$

Expand nonlinearity.

$$= \sin 2\pi + (\cos 2\pi)q + \cdots$$
$$= q + \cdots .$$

The approximation

Linearize.

$$\sin(2\pi + q) \approx q$$

gives a linear equation for q. This approximation fails if the perturbation q becomes large, but by then stability is already out of the question.

The resulting *linear* equation for the perturbation q is

Linear equation for perturbation

$$Lq'' + pq' + gq = 0, \tag{6.26}$$

just the familiar (damped) approximate pendulum equation. From the characteristic equation $Lr^2 + pr + g = 0$, the characteristic roots of this differential equation are

$$r = \frac{-p \pm \sqrt{p^2 - 4Lg}}{2L}.$$

These roots determine the behavior of the perturbation q and, hence, the stability of the steady state.

If $p^2 - 4Lg \geq 0$ (overdamped or critically damped pendulum), then both roots are real and negative.

Stop and Think   **6.65** Verify this assertion. In particular, why is $-p + \sqrt{p^2 - 4Lg} < 0$ when $p > 0$?

Hence, if $p^2 - 4Lg \geq 0$, every solution of the linearized perturbation equation (6.26) is a decaying exponential, regardless of initial conditions.

Stop and Think   **6.66** Use the general solution of (6.26) to verify this assertion.

If $p^2 - 4Lg < 0$ (underdamped pendulum), then the characteristic roots are complex conjugates with negative real part $-p/2L$.

Stop and Think   **6.67** Verify this assertion, particularly the claim that the real part of the characteristic roots is negative.

Hence, if $p^2 - 4Lg < 0$, every solution of the linearized perturbation equation (6.26) is a periodic function multiplied by a decaying exponential, regardless of initial conditions.

Stop and Think   **6.68** Use the general solution of (6.26) to verify this assertion.

Therefore, all nontrivial solutions of the linearized perturbation equation (6.26) decay to zero, regardless of the (positive) values of L, p, and g. The steady state θ_{ss} is *stable*. ■

Note that both of these examples follow the same pattern as the first-order linear stability analysis of section 3.3:

Linear stability analysis

1. Define the perturbed solution $\theta(t) = \theta_{ss} + q(t)$.

2. Substitute θ and simplify.

3. If necessary, linearize by *assuming* that the perturbation q remains small for all time and by eliminating terms involving powers of q greater than one.

4. Examine the resulting (possibly approximate) linear differential equation for $q(t)$ to determine if it grows (θ_{ss} is unstable) or decays (θ_{ss} is stable).

Stop and Think   **6.69** The approximate, *linear* pendulum equation seems to govern the perturbations of the *stable* steady states of the original nonlinear pendulum equation. Which linear equation governs the stability of unstable steady states?

The linearizations
$$\sin(2\pi + p) \approx p, \quad \sin(\pi + p) \approx -p$$
used in these two examples are also consequences of familiar trig identities and an old friend, $\sin p \approx p$. Taylor polynomial expansions were used in these examples because they can be applied to a wide variety of nonlinearities. Of course, $\sin p \approx p$ was first derived using a Taylor polynomial.

6.8.3 Exercises

EXERCISE GUIDE	
To gain experience . . .	**Try exercises**
Identifying linear and nonlinear equations	1, 5, 6–7(c–d), 11(b)

<table>
<tr><td colspan="2">**EXERCISE GUIDE (Continued)**</td></tr>
<tr><td>**To gain experience ...**</td><td>**Try exercises**</td></tr>
<tr><td>Solving linear equations</td><td>2–3, 5</td></tr>
<tr><td>With validity of linear models</td><td>6(b), 11(b), 12–13</td></tr>
<tr><td>Relating period, amplitude, etc. to system parameters</td><td>3–4, 6–10, 11(a), 12–13(a–b)</td></tr>
<tr><td>With damped pendulum model</td><td>17–19</td></tr>
<tr><td>Determining stability of equilibria</td><td>14–18, 20–24</td></tr>
<tr><td>Analyzing and interpreting solution behavior</td><td>6–9, 14–15, 18, 21</td></tr>
</table>

1. Verify that the pendulum equation

$$L\theta'' + g\sin\theta = 0$$

is nonlinear and that its approximate counterpart

$$L\theta'' + g\theta = 0$$

is linear. Are these equations homogeneous or nonhomogeneous?

2. Derive the solution

$$\theta(t) = \theta_i \cos\sqrt{\frac{g}{L}}t$$

of the linear initial-value problem $L\theta'' + g\theta = 0$, $\theta(0) = \theta_i$, $\theta'(0) = 0$, thereby verifying the solution formula used in the text.

3. Solve the general initial-value problem of the approximate pendulum model,

$$L\theta'' + g\theta = 0, \quad \theta(0) = \theta_i, \quad \theta'(0) = \omega_i.$$

Argue that the frequency of this solution is *always* $\sqrt{g/L}$, regardless of the initial values θ_i, ω_i.

4. The text makes the following observations about the solution

$$\theta(t) = \theta_i \cos\sqrt{\frac{g}{L}}t$$

of the initial-value problem $L\theta'' + g\theta = 0$, $\theta(0) = \theta_i$, $\theta'(0) = 0$:

(i) The natural frequency $\omega_n = \sqrt{g/L}$ is independent of initial position θ_i.

(ii) The amplitude θ_i of the response is independent of length L and the acceleration of gravity g.

Verify each claim.

5. The text claims, "Unfortunately, analytic solution tools fail" when applied to the nonlinear equation $L\theta'' + g\sin\theta = 0$. Demonstrate this assertion by attempting to solve this equation by the characteristic equation method. To what types of equations is this method applicable? What fails here?

6. The text makes two observations about the behavior of the solution of the linear pendulum problem $L\theta'' + g\theta = 0$, $\theta(0) = \theta_i$, $\theta'(0) = 0$:

(i) The natural frequency $\omega_n = \sqrt{g/L}$ is independent of initial position θ_i.

(ii) The amplitude of the response is independent of length L and the acceleration of gravity g.

(a) Construct a parallel set of observations for the initial-value problem $L\theta'' + g\theta = 0$, $\theta(0) = 0$, $\theta'(0) = \omega_i$.

(b) Describe the physical difference between this problem and the one considered in the text.

(c) Is your first observation valid for the corresponding nonlinear problem? Write the equation as a first-order system, then use DELAB to investigate numerically.

(d) Is your second observation valid for the corresponding nonlinear problem? Write the equation as a first-order system, then use DELAB to investigate numerically.

7. Repeat parts (a) and (b) of exercise 6 for the initial-value problem $L\theta'' + g\theta = 0$, $\theta(0) = \theta_i$, $\theta'(0) = \omega_i$.

8. The text observes that the frequency of the response of the linear pendulum equation $L\theta'' + g\theta = 0$ is independent of the initial position θ_i. It then goes on to show that the frequency of the response of the nonlinear pendulum equation $L\theta'' + g\sin\theta = 0$ is dependent upon θ_i.

Using DELAB, conduct a parallel numerical investigation to determine if the nonlinear pendulum equation obeys the second property the text observed for the linear equation:

The amplitude θ_i of the response is independent of length L and the acceleration of gravity g.

9. A pendulum executing small oscillations is transported to the moon, where the force of gravity is less than that on the earth. What becomes of the period of the pendulum's oscillations? What becomes of the amplitude of those oscillations? What can you say if the pendulum is executing large oscillations?

10. The numerical examples graphed in figure 6.22 were computed for a pendulum with length 0.9930 m. Why was that unusual length chosen?

11. The upper graph in figure 6.22 shows solutions of

$$0.9930\theta'' + 9.8\theta = 0, \quad \theta(0) = \theta_i, \quad \theta'(0) = 0,$$

for $\theta_i = 0.15$ and $\theta = 1$.

(a) Read the period of the solutions from the graph. Are the values you read consistent with the solution of this initial-value problem?

(b) Describe the pendulum modeled by this initial-value problem. Does the choice of the value of θ_i affect the validity of this model?

12. The *dashed blue* curves in figure 6.22 show the response of the approximate pendulum equation $L\theta'' + g\theta = 0$ (upper graph) and the nonlinear pendulum equation $L\theta'' + g\sin\theta = 0$ (lower graph) when the mass is released from rest from an angle of 1 rad. (Recall $L = 0.9930$ m.)

(a) For the motion shown, how good or bad is the approximation $\sin\theta \approx \theta$ that relates these two equations?

(b) What difference in period, if any, can you detect in the graph between the responses of the two equations?

(c) Are your answers to the preceding two questions consistent?

13. The *solid black* curves in figure 6.22 show the response of the approximate pendulum equation $L\theta'' + g\theta = 0$ (upper graph) and the nonlinear pendulum equation $L\theta'' + g\sin\theta = 0$ (lower graph) when the mass is released from rest from an angle of 0.15 rad. (Recall $L = 0.9930$ m.)

(a) For the motion shown, how good or bad is the approximation $\sin\theta \approx \theta$ that relates these two equations?

(b) What difference in period, if any, can you detect in the graph between the responses of the two equations?

(c) Are your answers to the preceding two questions consistent?

14. If perturbations of a steady state can be kept arbitrarily close to it by making the initial perturbation sufficiently small, then the equilibrium is called **neutrally stable**.

(There is no requirement that the perturbation return to the steady state.) Neutral stability is a mathematical ideal, and a linearized prediction of neutral stability may or may not be an accurate characterization of the behavior of the full nonlinear equation.

Show that the steady state $\theta_{ss} = 0$ of the undamped nonlinear pendulum equation $L\theta'' + g\sin\theta = 0$ is neutrally stable. Is this result consistent with your intuition?

15. Determine the stability of the equilibrium $\theta_{ss} = -\pi$ of the pendulum equation $L\theta'' + g\sin\theta = 0$. Are your results consistent with your intuition?

16. Find the equilibrium points of the approximate pendulum equation $L\theta'' + g\theta = 0$. Determine their stability. (*Hint:* See the description of *neutral stability* in exercise 14.)

17. Find the equilibrium points of the *damped* approximate pendulum equation $L\theta'' + p\theta' + g\theta = 0$. Determine their stability.

18. The text shows that the equilibrium points of the damped pendulum equation $L\theta'' + p\theta' + g\sin\theta = 0$ are $\theta_{ss} = 0, \pm\pi, \pm2\pi, \dots$.

(a) Which of these correspond to a straight-up position of the pendulum? To a straight-down position? Are there any other equilibrium positions?

(b) Determine the stability of *all* of the straight-up equilibria. Are your results consistent with your intuition?

(c) Determine the stability of *all* of the straight-down equilibria. Are your results consistent with your intuition?

19. Examine the derivation of the undamped pendulum equation in section 5.2. What physical effects might the damping term $p\theta'$ represent in the damped pendulum equation $L\theta'' + p\theta' + g\sin\theta = 0$? When the damped equation is written in this form, which of the parameters L, g or the mass m of the pendulum bob might the coefficient p involve?

20. In conducting its linear stability analysis of the equilibrium point $\theta_{ss} = 2\pi$ of the nonlinear pendulum equation, the text expands $\sin x$ in a Taylor polynomial about $x = 2\pi$:

$$\sin(2\pi + q) = \sin x\big|_{x=2\pi} + \left(\frac{d}{dx}\sin x\bigg|_{x=2\pi}\right)q + \cdots$$
$$= \sin 2\pi + (\cos 2\pi)q + \cdots.$$

(a) What is the lowest power of q that appears in the terms represented by the ellipsis $\cdots$?

(b) Are these terms really smaller than the term involving q? Specifically compare the term involving q with the first term represented by the ellipsis $\cdots$ when $q = 0.1$.

(c) If you kept any of the terms represented by the ellipsis $\cdots$, would the resulting equation in q be linear? Explain.

21. Linearized stability analyses of the undamped pendulum equation $L\theta'' + g\sin\theta$ lead to one or the other of the equations

$$Lq'' + gq = 0, \quad Lq'' - gq = 0$$

for the perturbation q of the steady state in question.

(a) Find the general solution of each of these equations.

(b) Argue that the solutions of the first are always

bounded.

(c) Argue that the solutions of the second grow exponentially unless q and q' obey a special initial condition. State that condition.

In exercises 22–24, find the equilibrium points of the given equation and determine their stability.

22. $x'' + x^2 - 1 = 0$

23. $x'' + x^2 - x = 0$

24. $x'' + x^3 - x = 0$

6.9 ■ CHAPTER EXERCISES

EXERCISE GUIDE	
To gain experience . . .	**Try exercises**
With steady-state and periodic solutions	1–2
Determining linear independence	3–5
Constructing general solutions	4–14
Solving initial-value problems	10–12
Analyzing and interpreting solution behavior	15–18
With extensions of characteristic equations and undetermined coefficients to higher orders	19–24
Analyzing damped systems	25–28, 30(b), 32–33, 41
With relations among system parameters and amplitude, frequency, etc.	29–31, 35–38
With forcing and resonance	33–38, 41 42
Determining stability of equilibria	39–40

1. Argue that a steady-state solution of a differential equation can *always* be used as a particular solution.

2. Argue that a periodic solution of a differential equation can *always* be used as a particular solution.

3. Determine whether each of the given pairs of functions is linearly independent for all x. When the two functions are not linearly independent for all x, give an interval where they are linearly independent, if possible. If they are not linearly independent, explain why.

(a) x, x^2

(b) $1 + x, 1 - x$

(c) $\ln|x|, \ln x^2$

In exercises 4 and 5,

(i) Determine whether the two functions are linearly independent or linearly dependent solutions of the given differential equation on the interval shown.

(ii) Explain the independence/dependence behavior you observe.

(iii) Write a general solution of the differential equation each time you find a linearly independent pair of solutions.

4. (a) $y'' + xy' = 0$; $\int_1^x e^{-s^2/2}\,ds$, 1 on $1 \le x$

(b) $y'' + xy' = 0$; $\int_1^x e^{(1-s^2)/2}\,ds$, π on $1 \leq x$

(c) $y'' + xy' = 0$; $\int_1^x e^{-s^2/2}\,ds$, $\int_1^x e^{(1-s^2)/2}\,ds$ on $1 \leq x$

5. (a) $x^2 y'' - 3xy' + 3y = 0$; x, x^3 on $x \leq -2$

   (b) $x^2 y'' - 3xy' + 3y = 0$; $x(1-x)(1+x)$, x^3 on $x \leq -2$

   (c) $x^2 y'' - 3xy' + 3y = 0$; x, $x(4+x^2)$ on $x \leq -2$

In exercises 6–14, find a general solution of the given equation or solve the given initial-value problem.

6. $4x''(t) + 8x'(t) + 5x(t) = t(t-2)$

7. $y'' - 9y = e^{-3x}$

8. $y'' + 4y' + 40y = -2x^2 + e^{-2x}$

9. $y'' - 4y' + 40y = e^{2x} - e^{-2x}$

10. $z'' + 9z = 2\cos 4t$, $z(\pi) = 3$, $z'(\pi) = -9$

11. $z'' + 9z = t\sin 3t$, $z(\pi) = 3$, $z'(\pi) = -9$

12. $2w'' - 7w' + 3w = 6e^{3t}$, $w(0) = 8$, $w'(0) = -12$

13. $2w'' - 7w' + 3w = -1$

14. $y'' + 4y' + 4y = 16e^{2x} + 12\sin 2x$

15. Find conditions on the parameters α, β that guarantee that all solutions of
$$x'' + \alpha x' + \beta x = 0$$
are

   (a) Purely oscillatory,

   (b) Decaying to zero for large t,

   (c) Not oscillating,

   (d) Growing in amplitude for all t.

What might α, β represent if this differential equation arose in a spring-mass model? Are all of the values you found physically reasonable in such a setting?

16. Find initial values x_i, v_i that guarantee the solution of
$$x'' - 12x = 0, \quad x(0) = x_i, \quad x'(0) = v_i,$$
is

   (a) Decaying to zero,

   (b) Growing without bound,

   (c) Constant for all t.

Could this initial-value problem be a model of a spring-mass system?

17. Find conditions on the parameters a, b that guarantee
$$x'' + ax' + bx = 4\sin 2t$$
has solutions that

   (a) Have constant amplitude for all time,

   (b) Have period π in the limit as $t \to \infty$,

   (c) Are oscillatory and include a component with period 4,

   (d) Have amplitude growing linearly with time.

What might a, b represent if this differential equation arose in a spring-mass model? Are all of the values you found physically reasonable in such a setting?

18. Find a linear, constant-coefficient, homogeneous differential equation having the given pair of functions as solutions. For each function, find an initial-value problem of which it is a solution.

   (a) $\cos x$, $2\sin x$

   (b) $3x$, $2e^{-x}$

Use characteristic equations and undetermined coefficients to find general solutions of the third- and fourth-order equations in exercises 19–24.

19. $y''' - 4y' = 0$

20. $y''' + y'' - y - 1 = 0$

21. $y''' - 3y'' + 4y' - 12y = 0$

22. $y^{iv} - 16y = 0$

23. $y''' - 4y' = 48e^{4t}$

24. $y^{iv} - 16y = -30\cos t$

25. A weight of 32 lb is suspended from a spring so that it hangs in a container of light oil. The oil resists the motion of the weight with a force equal in magnitude to three times the velocity of the weight. The spring exerts a resisting force of 4 lb for every foot it is extended. The weight is raised 1 ft and released. How much time is required for the amplitude of the motion to decrease by 50%?

26. Suppose the weight in the system described in exercise 25 can be varied arbitrarily without affecting the damping force exerted by the oil. (For example, a hollow container could be filled with varying amounts of lead shot.) For what range of *weights* will this system be underdamped? Overdamped? Critically damped? Is this variation with weight physically reasonable?

27. A 2-kg mass is suspended from a spring that extended 7 cm when the mass was hung from it. That mass is suspended in a variety of different fluids, ranging from air to heavy motor oil. Argue that after a disturbance of its position, the mass will always tend to return to its original equilibrium position, *no matter what* the numerical value of the damping constant for the various fluids.

28. The differential equation

$$2x'' - x' + 4x = 0$$

is proposed as a model of a damped spring-mass system. What is wrong with it? What is the effect of the proposed damping term in this equation?

29. A certain spring extends 4 cm when a mass of 100 g is suspended from it. A 600-g mass is hanging at rest from this spring when it is suddenly struck from below. The mass moves upward 13 cm. What initial velocity was given to it by this blow? Would the mass have moved just as far if the blow had been directed downward?

30. A horizontal spring-mass system such as that of figure 5.4, page 207, can be taken as a simplified model of the energy-absorbing barriers that are now placed around bridge abutments on some interstate highways. The fixed end of the spring is attached to the bridge abutment on the right, and the mass is that of the car and the restraint system combined.

   (a) If the speed limit were raised from 55 mi/h to 65 mi/h, how would the spring have to be changed to limit maximum displacement from a collision at the higher speed to that at the lower speed? (Ignore damping.)

   (b) How would the presence of damping alter your conclusion in part (a)? Should such systems be underdamped or overdamped?

31. Under what initial conditions is the maximum displacement of an undamped spring-mass system independent of both the mass and the spring constant?

32. A screen door on a porch is often pulled shut by an attached spring. It is usually set into motion from its equilibrium (closed) position by pushing it open. Friction in the hinges and between the door and the door frame damps the motion. Should such a system be overdamped, underdamped, or critically damped? In theory, will an overdamped screen door ever shut tightly?

33. Describe key differences between the response of an undamped spring-mass system and a damped system when each is forced at its resonant frequency.

34. You are given the characteristic roots r_1, r_2 of a model of a spring-mass system. Give a step-by-step recipe (an algorithm) for determining from r_1, r_2 whether resonance is possible. If resonance can occur, determine the resonant frequency of the system.

35. The resonant frequency of an underdamped spring-mass system is the value of the forcing frequency ω that makes the amplitude of the periodic solution of

$$mx'' + px' + kx = F \sin \omega t$$

a maximum.

   (a) Find a particular solution (of frequency ω) of this equation and show that its amplitude A is F/D, where

$$D = \sqrt{(k - m\omega^2)^2 + p^2\omega^2}.$$

   (b) Argue that the resonant frequency must be the value of ω that makes D a *minimum*. Find that minimum and show that the resonant frequency is

$$\omega_r = \sqrt{\frac{k}{m} - \frac{p^2}{2m^2}}.$$

   (*Hint*: It is easier to minimize D^2 than D.)

   (c) How does this resonant frequency ω_r compare with the natural frequency of the *undamped* system as damping decreases?

   (d) Use the formula for ω_r to check the resonant frequency value found in example 44, page 312, of section 6.7.

36. Using the result of part (a) of the preceding exercise and the ideas outlined there, show that the forcing frequency that gives the *velocity* of the mass its maximum amplitude is the undamped natural frequency $\omega_n = \sqrt{k/m}$. How does this frequency compare with the resonant frequency— the forcing frequency that produces maximum amplitude of displacement—found in the preceding exercise when damping is small?

37. Using the amplitude expression derived in the text, find the forcing frequency ω that maximizes the amplitude of the periodic response (capacitor voltage or capacitor charge) of a series *RLC* circuit with voltage source $E = V \sin \omega t$.

38. Show that the periodic response of the *current* in a series *RLC* circuit has maximum amplitude when the voltage source $E = V \sin \omega t$ has frequency $\omega = 1/\sqrt{LC}$.

39. Generalize the text's analysis of the stability of the equilibria $\theta_{ss} = \pi, 2\pi$ of $L\theta'' + g \sin \theta = 0$ to show that

   (a) the equilibrium points $0, \pm 2\pi, \pm 4\pi, \ldots$ of $L\theta'' + g \sin \theta = 0$ are neutrally stable (see exercise 14, p. 328).

   (b) the equilibrium points $\pm \pi, \pm 3\pi, \ldots$ of $L\theta'' + g \sin \theta = 0$ are unstable.

40. Determine the stability of the nontrivial equilibrium solution of the predator-prey (fox-rabbit) equations

$$F' = -(d_F - \alpha R)F,$$
$$R' = (b_R - \beta F)R$$

that were derived in section 5.4. Can you recognize the linear equation that governs perturbations of this equilibrium? Have you seen a similar equation in other contexts? Can you suggest a linear population model that is valid near the equilibrium?

41. The quantity $2\sqrt{mk}$ is sometimes called the *critical damping* value for the system $mx'' + px' + kx = 0$, and the ratio $d = p/2\sqrt{mk}$ is called the *damping ratio*. Explain these names by showing that the system is underdamped, critically damped, or overdamped according to whether $d < 1$, $d = 1$, $d > 1$. Show that the resonant frequency of an underdamped system can be written $\omega_n\sqrt{1 - d^2}$, where $\omega_n = \sqrt{k/m}$ is the natural frequency.

42. Use DELAB to conduct a numerical study of the approach to resonance in an undamped spring-mass system. What is the effect of the pulsation due to beating as the forcing and natural frequencies become more nearly equal? What would you experience if you were riding on the mass? Would the experience be different if moderate damping were added?

6.10 ■ CHAPTER PROJECTS

1. **Basic definitions for equations of order n.** A linear differential equation of order n is one of the form

$$a_n(x)\frac{d^n y}{dx^n} + \cdots + a_1(x)\frac{dy}{dx} + a_0(x)y = r(x).$$

   (a) For such an equation, define *solution*, *homogeneous*, *nonhomogeneous*, *initial-value problem*, *solution of initial-value problem*.

   (b) Complete the definition: *The functions* $y_1, \ldots, y_n$ *are* linearly independent *on the interval* .... Define *general solution* of a linear, homogeneous equation of order n. Define *general solution* of a linear, nonhomogeneous equation.

   (c) Formulate an analog of the Wronskian condition for the linear independence of n solutions of a linear, homogeneous differential equation of order n. Conjecture what the analogs of theorems 2–4 in section 6.2 might be. (These relate linear independence and the Wronskian.)

   (d) State and prove a principle of superposition for linear equations of order n.

2. **Solution methods for constant-coefficient equations of order n.** Consider the constant-coefficient linear equation of order n,

$$b_n y^{(n)} + \cdots + b_1 y' + b_0 y = r(x),$$

   where $b_n, \ldots, b_0$ are (real) constants.

   (a) Develop the characteristic equation solution method for the homogeneous form of this equation. Summarize your method in a form similar to the second-order summary on page 273. What linear independence results are needed to put this method on a rigorous foundation?

   (b) Develop the method of undetermined coefficients for this equation when the forcing term $r(x)$ is a polynomial, a linear combination of sines and cosines, or an exponential. Provide a table similar to table 6.3 for these cases.

3. **The van der Pol equation.** The van der Pol equation

$$\frac{d^2 x}{dt^2} - \epsilon\left(1 - x^2\right)\frac{dx}{dt} + x = 0$$

is a classic source of nonlinear oscillatory phenomena.

(a) B. van der Pol first described and analyzed this equation in the mid-1920's in terms of "negative resistance" in vacuum tube circuits. (See [26], for example.) Explain this terminology in terms of a hypothetical mechanical system or electrical circuit. Specifically, show the variation in the damping force coefficient with amplitude.

(b) Write the van der Pol equation as a system, then use a numerical method of your choice from DELAB to show that this equation achieves the same periodic limiting state for a wide range of initial conditions. Consider $\epsilon = 0.1, 1.0, 10.0$.

(c) Compare the shape of the periodic oscillations for $\epsilon = 0.1$ with those for $\epsilon = 10$. Explain why the former are reasonably sinusoidal. The latter are called *relaxation oscillations*. Suggest why that name is appropriate.

4. **Period of a nonlinear spring.** Project 1 of the chapter 5 projects asks for a derivation of a model of a spring-mass system with the nonlinear force-extension relation

$$F_s = kz + az^3,$$

where $a > 0$ is much smaller than k. Using DELAB, determine the variation with a of the period of such a system (undamped). (Keep k, initial displacement, and mass fixed.) For fixed spring parameters, determine the variation of the period with amplitude. What are the corresponding relations for a linear spring-mass system?

5. **More bridge cables.** Project 2 of chapter 5 introduced in (5.16–5.17) a prototype model for the nonlinear response of the cables in a suspension bridge such as the Golden Gate Bridge. This project explores the great variety of periodic responses of this system, some of which may account for the collapse of the Tacoma Narrows Bridge. In the following, use $m = 1$, $g = 9.8$, $k = 10$, $p = 0.05$, $\omega = 1$, $F = 8.5$, and $\ell_0 = 0.3135$.

(a) Write the system (5.16–5.17) of two second-order equations as a system of four first-order equations.

(b) Using DELAB, show that the mass simply oscillates vertically if the initial conditions do not displace it to the left or right. Plot y versus x so that you can see the motion of the mass. (Try a second- or fourth-order Runge-Kutta or an adaptive method. Allow the system to run long enough for the transient to decay.)

(c) Now impose initial conditions that displace the mass to the side. Compare the long-time motion you find with that of others who complete this same project. Figure 6.24 shows an example of the motion that can result. Do you find what might be chaotic motion, that is, trajectories without any apparent regular pattern?

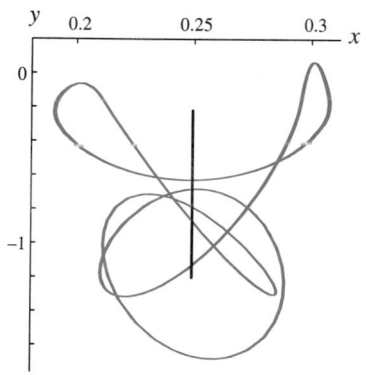

FIGURE 6.24 A trajectory from the prototype bridge suspension model (5.16–5.17), courtesy of P.J. McKenna.

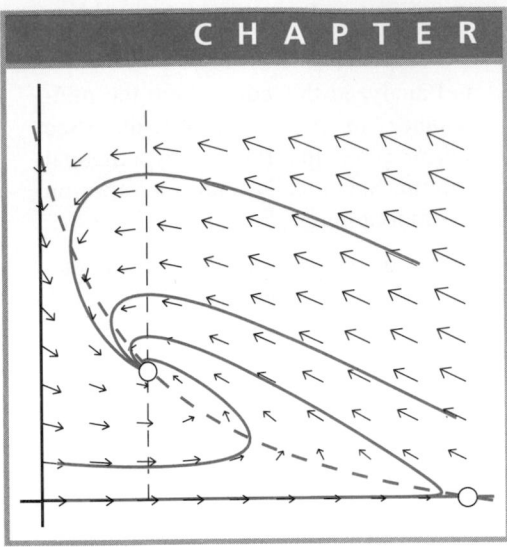

Graphical Tools for Two Dimensions

Direction fields and phase lines provide vivid illustrations of stability and related phenomena for first-order equations. In section 2.4, phase plane diagrams were briefly introduced to provide similar visualizations of behavior in such two-dimensional systems as models of epidemics and competition.

This chapter explores phase plane diagrams in more detail. The graphical tools it develops complement the numerical and analytic tools studied earlier, and, in turn, the study of the phase plane initiated here will be complemented by the analytic tools for first-order systems introduced in chapter 8.

7.1 ■ THE PHASE PLANE

Understanding models of the behavior of two populations, for example, obviously requires information about both species. Perhaps less obviously, but for the same reason, understanding the behavior of a device like a pendulum or a spring-mass system also requires information about two variables, position and velocity. A snapshot of a pendulum, for example, captures only its position. From that photo alone, an observer can not determine the direction of its motion.

Stop and Think   **7.1** Would it be equally difficult to determine the direction of motion of a spring-mass system from a snapshot?

Position and velocity information can be presented in two separate graphs, position versus time and velocity versus time, as in figure 7.1. But in the absence of forcing, it is really the relation between velocity and position that matters. For example, a pendulum with *positive* displacement and *positive* velocity will continue to swing even farther from its equilibrium, while one with *negative* position and *positive* velocity must be moving toward its equilibrium.

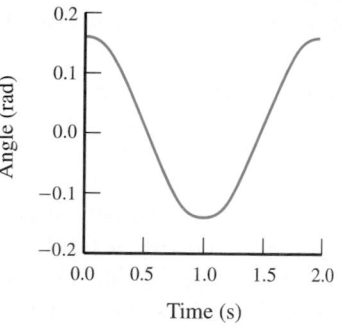
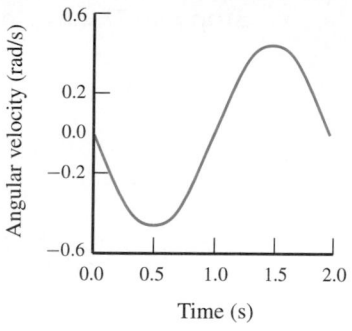

FIGURE 7.1 Graphs of angular position θ (left graph) and angular velocity $\omega = \theta'$ (right graph) versus time for the nonlinear pendulum model $L\theta'' + g \sin\theta = 0$, $\theta(0) = 0.15$, $\theta'(0) = 0$ with $L = 0.9930$ m.

For a second-order equation such as the pendulum equation, the natural way to capture the relation between velocity and position is in a graph of velocity versus position. In general, a plot of the derivative of the dependent variable (velocity) against the dependent variable (position) in a second-order equation is known as a **phase plane**. For a system of two first-order equations, the phase plane is a graph of one dependent variable against the other. A curve in the phase plane that represents the evolution of the two variables, say θ and ω, with time is called a **trajectory**.

For example, a phase plane plot of the pendulum equation

$$L\theta'' + g \sin\theta = 0$$

is a graph of $\omega = \theta'$ versus θ—that is, a graph of velocity versus position. Of course, angular velocity ω is more visible when this second-order equation is written as a first-order system,

$$\theta' = \omega,$$
$$L\omega' = -g \sin\theta.$$

The initial conditions $\theta(0) = 0.15$, $\theta'(0) = 0$ become $\theta(0) = 0.15$, $\omega(0) = 0$.

Recall that second-order equations can be converted to first-order systems by introducing a new dependent variable—ω in this case—to replace the first derivative of the unknown. The first equation in the system is just the definition of the new dependent variable; see section 6.1, p. 234.

Figure 7.2 shows the phase plane plot of ω versus θ that corresponds to the graphs of ω versus t and θ versus t shown in figure 7.1. Notice that *a simple closed curve in the phase plane corresponds to a periodic solution.* Each time the trajectory is traced, position and velocity return to the same values. Because time never appears explicitly in the pendulum equations $\theta' = \omega$, $L\omega' = -g \sin\theta$, the time elapsed in traversing a section of the trajectory depends only on θ and ω, not on any external influence that could change from cycle to cycle. We could choose any point on the trajectory to determine initial values of θ and $\theta' = \omega$ and always retrace this same trajectory.

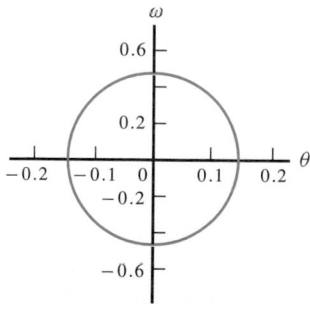

FIGURE 7.2 The phase plane plot of angular velocity ω versus angular position θ corresponding to the velocity- and position-versus-time graphs of figure 7.1.

Stop and **Think**

7.2 Identify the points in the phase plane plot of figure 7.2 that correspond to $t = 0$, $0.5, \ldots$.

7.3 Could a single trajectory in the phase plane correspond to more than one initial-value problem? (*Hint:* How many different sets of initial conditions could produce the trajectory shown in figure 7.2?)

Equations such as the pendulum equation, in which the only functions of time are the unknowns themselves, are called **autonomous**. The phase plane perspective is appropriate for autonomous systems because the response of the system depends upon its state—the current values of the dependent variables—and not upon time. The phase plane picture is invariant in time.

Another autonomous system is the predator-prey (fox-rabbit) model of section 2.4.1,

$$F' = -(d_F - \alpha R)F,$$
$$R' = (b_R - \beta F)R.$$

The *unforced* spring-mass equation is also autonomous.

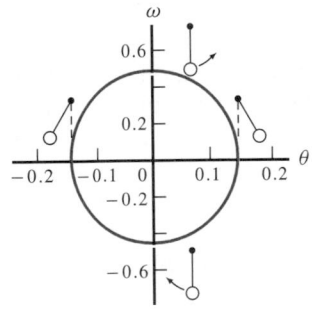

FIGURE 7.3 The phase plane of figure 7.2 marked with the position of the pendulum. The trajectory is traversed in the clockwise direction as time increases.

Which direction on the trajectory of figure 7.2 corresponds to increasing time, clockwise or counterclockwise? Starting at $\theta = 0.15$ rad, $\omega = 0$ rad/s corresponds to being on the positive θ-axis at the three o'clock position. Releasing the pendulum causes it to drop; θ will decrease. At the same time, its velocity will *decrease* in the sense of moving from zero toward negative values; ω is negative because θ is decreasing with time. The trajectory is traced in the clockwise direction.

Figure 7.3 redraws the trajectory with diagrams of the pendulum added to emphasize the connection between location in the phase plane and motion of the pendulum.

Naturally, these physical arguments are reflected in the governing differential equations,

$$\begin{aligned} \theta' &= \omega, \\ L\omega' &= -g \sin \theta. \end{aligned} \tag{7.1}$$

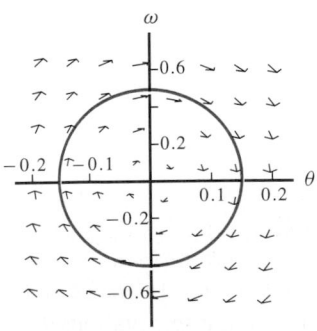

FIGURE 7.4 The phase diagram of the undamped pendulum with direction field arrows defined by the system (7.1).

If $\pi > \theta > 0$, then $L\omega' = -g \sin \theta$ forces $\omega' < 0$. If initially $\omega = 0$ and $\omega' < 0$ (because $\theta > 0$), then ω must become negative, forcing the trajectory from its starting point on the positive θ-axis into the fourth quadrant.

A direction field diagram provides yet another way to see the connection between the governing differential equations (7.1) and the trajectory shown in figures 7.2 and 7.3. A **direction field diagram** (or **vector field** or **flow field** diagram) for a first-order system like (7.1) is a plot of the direction defined by the pair of values (θ', ω') at each point of the $\omega\theta$-plane. It is the two-dimensional analog of a phase line diagram for a single first-order equation.

Think of following the point $(\theta(t), \omega(t))$ that is drawing the trajectory as it moves across the phase plane. It begins at (θ_i, ω_i). The velocity (vector) of the point is $(\theta'(t), \omega'(t))$, and it is that vector—the direction in which the point is about to move—that defines the arrows in figure 7.4. Of course, the velocity values $\theta'(t)$ and $\omega'(t)$ are given by the right side of the system (7.1).

These direction field diagrams are sometimes called *flow diagrams* because they depict the flow of the trajectories or they are called *vector fields* because they show the direction field vectors.

In many engineering disciplines, $(\theta(t), \omega(t))$ is called the *state vector* of the pendulum, *state* because it includes both position and velocity—all the information needed to describe the state or motion of the pendulum—and *vector* because it represents a vector from the origin of the phase plane to the current point on the trajectory.

> MATLAB
>
> Watch the motion of a trajectory as MATLAB draws it using the phase plane tool in the graphical toolbox in DELAB. Note how it follows the flow of the vector field. For guidance, select **Help, Textbook**, then go to chapter 7, figure 7.4.

Stop and Think   **7.4** Determine the signs of the derivatives in (7.1) and use that information to predict the direction of the arrows in the direction field of figure 7.4; e.g., argue that the arrows in the first quadrant must point down and to the right ("southeast") because $\theta, \omega > 0$ force $\theta' = \omega > 0$ ("east") and $\omega' = -(g/L)\sin\theta < 0$ ("south").

At a steady state, both θ and ω are constant. Hence, the trajectory of a steady state is simply a point in the phase plane. The equilibrium points of the pendulum equation must all lie on the horizontal axis in the phase plane, because the velocity ω is zero at such a point.

Stop and Think   **7.5** Speculate: If a steady state—stable or unstable—is just a single point in the phase plane, how should the behavior of trajectories near such a point distinguish stable from unstable steady states?

As a specific example, consider the damped pendulum equation

$$\theta' = \omega,$$
$$L\omega' = -g\sin\theta - p\omega. \tag{7.2}$$

Since steady states are constant solutions of the governing equations, they can be found by solving the differential equations with the derivatives set to zero:

$$\theta'_{ss} = 0 = \omega_{ss},$$
$$L\omega'_{ss} = 0 = -g\sin\theta_{ss} - p\omega_{ss}.$$

The first equation confirms the observation of the preceding paragraph: $\omega_{ss} = 0$. From the second equation, we find $\sin\theta_{ss} = 0$ and

$$\theta_{ss} = 0, \pm\pi, \pm 2\pi, \dots .$$

These equilibrium or **stationary** points are shown in figure 7.5.

What becomes of trajectories that start near these stationary points? The stability analysis of section 6.8, amplified by some of the exercises, provides

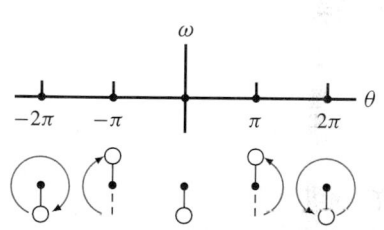

FIGURE 7.5   The equilibrium points of the nonlinear damped pendulum system (7.2) lie on the θ-axis in the phase plane.

the answer. The stationary points $\theta_{ss} = 0, \pm 2\pi, \ldots, \omega_{ss} = 0$, which correspond to the pendulum's usual straight-down rest position, are stable. Perturbations of these equilibria result in decaying oscillations around the equilibrium point. That is, a trajectory starting near one of the stationary points $\theta_{ss} = 0, \pm 2\pi, \ldots, \omega_{ss} = 0$, approaches it over time.

Put differently, any motion of the pendulum must come to rest as friction in the pivot or air resistance exhausts its energy. These stable equilibria are sometimes called **attracting points** or **attractors** because they draw to them trajectories that start near enough.

Stop and Think

7.6 Suppose the pendulum is underdamped. Sketch how the solution trajectory might approach a stable equilibrium point.

7.7 Suppose the pendulum is overdamped. Sketch how the solution trajectory might approach a stable equilibrium point.

Figure 7.6 shows the phase plane for the (under)damped pendulum system (7.2). Note the instability of the stationary points $\theta_{ss} = \pm\pi, \pm 3\pi, \ldots$, $\omega_{ss} = 0$, consistent with our physical intuition about the instability of the inverted position.

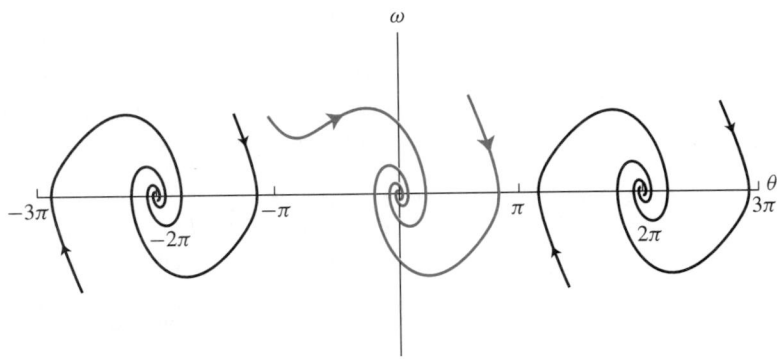

FIGURE 7.6   The phase plane for the damped pendulum system (7.2).

Stop and Think

7.8 How do you "note the instability of the stationary points $\theta_{ss} = \pm\pi, \pm 3\pi, \ldots$, $\omega_{ss} = 0$"? What is it in figure 7.6 that demonstrates the instability of these stationary points?

7.9 Mark figure 7.6 with diagrams of the pendulum's position as in figure 7.3.

7.10 Study the trajectory that begins with θ slightly larger than π and $\omega = 0$; it spirals toward the equilibrium point $(\theta_{ss}, \omega_{ss}) = (2\pi, 0)$. How many times does the pendulum loop around its pivot in the course of traversing this trajectory?

7.11 How would figure 7.6 change if the pendulum were *over*damped?

The behavior shown in figure 7.6 is consistent with the stability analysis of section 6.8 and with physical intuition. Such figures can be generated easily using the graphical tools in MATLAB's DELAB. But the primary point is that much of the solution behavior shown there can be derived directly from the governing differential equations (7.2). The exercises request such analyses.

If the pendulum is undamped, then the spirals of figure 7.6 are replaced by elliptically shaped curves surrounding $\theta = 0, \pm 2\pi, \ldots$ as in figure 7.7. These oscillations are called *neutrally stable* because a perturbation leads simply to another, nearby oscillation. The motion neither decays nor grows; see exercise 14, page 328 of section 6.8.

Indeed, these oscillations are neutrally stable even to the relatively large perturbations shown in this figure. Since any point on a trajectory could serve as an initial point, the distance between points on the trajectory and the equilibrium gives an idea of the size of the perturbation. For example, the elliptically shaped curve closest to the origin in figure 7.7 could have been started from the point on the positive θ-axis with $\theta = 1$, $\omega = 0$. Of course, the change from $\theta = 0$ to $\theta = 1$ is hardly a small perturbation.

If the closed curve of figure 7.2 were scaled to fit these axes, it would represent a small perturbation of the equilibrium at $\theta = 0$. That curve would be a tiny ellipse around the origin passing through the point $\theta = 0.15$, $\omega = 0$. (The curve in figure 7.2 appears more circular than the closed curves in figure 7.7 because of the relative scaling of the axes.)

The stationary points $\theta = \pm \pi, \ldots$ are unstable, with or without damping, because they correspond to attempting to stop the pendulum in an inverted position. Section 6.8 showed that small perturbations of these equilibria grow with time.

The trajectory labeled U (for *unstable*) in figure 7.7 begins close to $\theta = -\pi$, $\omega = 0$; it is a small perturbation of that unstable equilibrium point. Although it eventually returns to its starting point, it also cycles far from that equilibrium point, demonstrating the instability of the equilibrium $\theta = \pi$.

Physically, this motion corresponds to releasing the pendulum from rest from the position shown in figure 7.8, a negative angular displacement just short of vertical. The pendulum swings down and back up the other side, not quite reaching the vertical position $\theta = \pi$ before it returns to its starting point. The trajectory in figure 7.7 likewise falls just short of $\theta = \pi$.

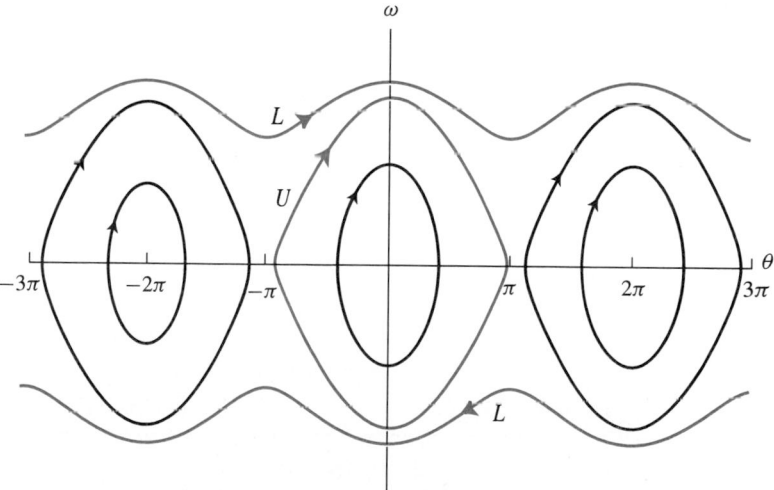

FIGURE 7.7 The phase plane for the undamped pendulum system (7.1).

FIGURE 7.8   The position of the pendulum to start the trajectory labeled U in figure 7.7. The pendulum is displaced just to the left of vertical, to a negative angle slightly smaller in magnitude than $-\pi$.

The curves labeled L (for *looping*) in figure 7.7 represent continual looping motion of the pendulum. The pendulum has been displaced and pushed so hard that it spins continuously around its pivot like a propeller. Exercise 18 asks you to decide which of the curves labeled L represents clockwise motion and which represent counterclockwise motion.

MATLAB

As you read this chapter, keep DELAB close at hand so that you can expand your own understanding of the graphical tools being used.

Any autonomous system such as the ones mentioned in this section can be written in the general form

$$y' = f(y, z),$$
$$z' = g(y, z),$$

where prime (′) denotes differentiation with respect to an independent variable such as t. For example, in the pendulum equation, $y = \theta$ and $z = \omega$.

The stationary points or equilibria of such systems are simply solutions (y_{ss}, z_{ss}) of

$$f(y_{ss}, z_{ss}) = 0,$$
$$g(y_{ss}, z_{ss}) = 0.$$

A phase plane trajectory can be thought of as a solution of the single first-order equation

$$\frac{dy}{dz} = \frac{f(y, z)}{g(y, z)},$$

provided $g \neq 0$.

7.1.1 Exercises

EXERCISE GUIDE	
To gain experience . . .	**Try exercises**
Relating second-order equations to autonomous systems	1–2, 10–12(a)
Finding stationary points	3
Drawing phase plane plots	4–7(a–c), 9, 10–12(b)
Analyzing phase plane plots	4–7(d–e), 8–9, 10–12(b), 13–22

1. Under what conditions is the damped pendulum system

$$\theta' = \omega,$$
$$L\omega' = -g\theta - p\omega$$

underdamped? Overdamped?

2. Identify the dependent variables y, z and the functions f, g so that each of the following can be written in the autonomous form

$$y' = f(y, z),$$
$$z' = g(y, z).$$

If necessary, convert second-order equations to first-order systems.

(a) Undamped pendulum system

$$\theta' = \omega,$$
$$L\omega' = -g \sin \theta.$$

(b) Damped pendulum equation $L\theta'' + p\theta' + g \sin \theta = 0$.

(c) Approximate pendulum equation $L\theta'' + g\theta = 0$.

(d) Unforced spring-mass equation $mx'' + px' + kx = 0$.

(e) Predator-prey system

$$F' = -(d_F - \alpha R)F,$$
$$R' = (b_R - \beta F)R.$$

(f) SIR system

$$S' = -bIS,$$
$$I' = bIS - rI.$$

(g) Competition equations

$$y' = y(1 - y - ez),$$
$$z' = rz(1 - z - fy).$$

3. Find the stationary points of each of the equations listed in the previous exercise.

4. Solve the approximate pendulum model

$$L\theta'' + g\theta = 0, \quad \theta(0) = \theta_i, \quad \theta'(0) = 0,$$

with $L = 0.9930$ m, $g = 9.8$ m/s^2, $\theta_i = 0.15$ rad.

(a) Write formulas for $\theta(t)$ and $\omega(t) = \theta'(t)$. Use these to construct a phase plane plot of ω versus θ. (You will be plotting a curve defined by parametric equations with t as the parameter.)

(b) Mark the times $t = 0.5, 1.0, 1.5, 2.0$ s on the trajectory. Which direction, clockwise or counterclockwise, corresponds to t increasing?

(c) Solve the θ and ω equations for the trigonometric function each contains. Square and add to show that ω and θ satisfy an equation whose graph is a simple closed curve. Identify that curve.

(d) Mark the amplitude or maximum value of $\theta(t)$ on your phase plane graph. How is the amplitude related to the initial conditions?

(e) Compare your results with figure 7.2, a trajectory of the solution of the nonlinear pendulum equation $L\theta'' + g \sin \theta = 0$ subject to the same initial conditions.

(f) Confirm your analysis with a phase plane diagram drawn by DELAB.

5. Repeat exercise 4 using arbitrary values of the parameters L, g, and θ_i.

6. Repeat exercise 4 using the initial conditions $\theta(0) = 0$, $\theta'(0) = \omega_i$ and arbitrary values of the parameters L, g, and ω_i. What value(s) must ω_i have if the motion is to have a specified amplitude?

7. Repeat exercise 4 using the general initial conditions $\theta(0) = \theta_i$, $\theta'(0) = \omega_i$ and arbitrary values of the parameters L, g, θ_i, and ω_i. What value(s) must θ_i and ω_i have if the motion is to have a specified amplitude?

8. Referring to the pendulum trajectory shown in figure 7.2, the text claims

> We could choose any point on the trajectory as initial values of θ and $\theta' = \omega$ and always retrace this trajectory.

Read from the graph the initial conditions appropriate for starting on the trajectory at

(a) The 3 o'clock position (positive θ-axis).

(b) The 6 o'clock position (negative ω-axis).

(c) The 9 o'clock position (negative θ-axis).

(d) The 12 o'clock position (positive ω-axis).

9. The text argues that the trajectory of figure 7.2 is traced in the clockwise direction starting from $\theta = 0.15$, $\omega = 0$, the 3 o'clock position in the phase plane.

(a) What happens when the pendulum is started with initial conditions that correspond to the 9 o'clock position? Argue both from your physical intuition about the behavior of the pendulum and from the governing equations,

$$\theta' = \omega,$$
$$L\omega' = -g \sin \theta.$$

(b) Starting on the phase plane trajectory of figure 7.2 at the 12 o'clock position corresponds to beginning the motion with the pendulum hanging straight down. What is the sign of the initial velocity of the pendulum? In which direction must the pendulum move as it begins its motion, to the right or to the left? In which direction is the trajectory traced, clockwise or counterclockwise? Argue both from your physical intuition about the behavior of the pendulum and from the governing equations.

(c) Answer the questions posed in part (b) when the starting point is the 6 o'clock position on the trajectory.

10. (a) Write a first-order system corresponding to the approximate pendulum equation $L\theta'' + g\theta = 0$.

(b) Draw a complete phase plane diagram for this system. Identify both stationary points and periodic solutions. Is it reasonable physically to consider large values of θ and ω?

11. (a) Letting $v = x'$, write a first-order system corresponding to the undamped spring-mass equation $mx'' + kx = 0$.

(b) Draw a complete phase plane diagram for this system. Identify both stationary points and periodic solutions. Are there physical limitations on the size of the values of x and v you can consider?

12. (a) Letting $v = x'$, write a first-order system corresponding to the damped spring-mass equation $mx'' + px' + kx = 0$.

(b) Draw a complete phase plane diagram for this system. Identify stationary points, attractors, and periodic solutions, if any. Are there physical limitations on the size of the values of x and v you can consider?

Exercises 13–18 ask about figure 7.7, the phase plane diagram of the undamped pendulum. In answering each question provide both a *physical* and a *mathematical* justification for your answer. Confirm your results using DELAB. Base your mathematical arguments on the governing equations,

$$\theta' = \omega,$$
$$L\omega' = -g \sin \theta.$$

13. Are the elliptically shaped curves that represent oscillations around the stable equilibrium points $\theta = 0, \pm 2\pi$ traversed in a clockwise or a counterclockwise direction?

14. The trajectory labeled U is a perturbation of the unstable equilibrium at $\theta = -\pi$. Is it traversed in a clockwise or a counterclockwise direction?

15. Suppose the unstable equilibrium at $\theta = -\pi$ were perturbed to a point slightly to the left of $-\pi$—i.e., $\theta(0) < -\pi$. Sketch the resulting trajectory and give the direction in which it is traversed.

16. Suppose the unstable equilibrium at $\theta = \pi$ were perturbed to a point slightly to the left of π—i.e., $\theta(0) < \pi$. How would the resulting trajectory compare with the curve labeled U?

17. Suppose the unstable equilibrium at $\theta = \pi$ were perturbed to a point slightly to the right of π—i.e., $\theta(0) > \pi$. Sketch the resulting trajectory and give the direction in which it is traversed.

18. The text claims that the trajectories labeled L represent looping motion of the pendulum; it is spinning continuously in one direction about its pivot.

(a) Justify this claim.

(b) In which direction is the pendulum spinning on the upper trajectory labeled L? On the lower trajectory?

(c) Might you find a trajectory like the upper one labeled L with valleys so deep they nearly touched the θ-axis at odd multiples of π? If so, how would the initial conditions for such a trajectory compare with those for the one shown in figure 7.7? Could such a trajectory actually cross the θ-axis?

Exercises 19–22 ask about figure 7.6, the phase plane diagram of the damped pendulum. In answering each question provide both a *physical* and a *mathematical* justification for your answer. Confirm your results using DELAB. Base your mathematical arguments on the governing equations,

$$\theta' = \omega,$$
$$L\omega' = -g \sin \theta - p\omega.$$

19. Are the spiral trajectories around the stable equilibrium points $\theta = 0, \pm 2\pi$ being traversed in a clockwise or a counterclockwise direction?

20. Sketch a trajectory that begins on the θ-axis slightly to the left of $\theta = \pi$. Such a trajectory represents a small perturbation of the unstable equilibrium at $\theta = \pi$.

21. Figure 7.7, the phase plane for the undamped pendulum, includes two trajectories marked L. Sketch their analogs for the damped pendulum.

22. Argue on physical grounds that no trajectory in the phase plane of figure 7.6 for the damped pendulum can be a closed curve unless it is a single point.

23. Use the undamped pendulum system (7.1) to draw direction field arrows on the phase plane diagram of figure 7.7 for the undamped pendulum. Confirm your results using DELAB.

24. Use the damped pendulum system (7.2) to draw direction field arrows on the phase plane diagram of figure 7.6 for the damped pendulum. Confirm your results using DELAB.

7.2 ■ NULLCLINES AND LOCAL LINEARIZATION

The last section illustrated the connection between the phase plane direction field and the behavior predicted by various models of a pendulum. Since tools like DELAB can easily draw the flow diagram for a given two-dimensional system, the important challenges are understanding and predicting the connections between the form of the differential equations and the kinds of behavior those diagrams depict. This section explores some of those connections with particular attention to the geometry of the trajectories and behavior near equilibrium points.

> These sorts of questions about the generic response of a system are often lumped into the category of *qualitative behavior*, as opposed to the quantitative or numerical behavior that is observed for specific sets of parameter values and initial conditions.

7.2.1 Nullclines

Important points on the phase plane are steady states, those points at which both components of the solution of the governing equations are constant. Equilibrium points are located at the intersection of **nullclines**, curves in the phase plane along which one or the other of the solution components is constant.

Stop and Think   **7.12** Why would steady states occur at points in the phase plane where nullclines intersect?

For example, consider the underdamped linear pendulum system

$$\theta' = \omega,$$
$$\omega' = -\theta - \tilde{p}\omega. \tag{7.3}$$

> In (7.3), $\tilde{p} = p/L$. For convenience, we have chosen L so that $g/L = 1\ \text{s}^{-2}$. In fact, by an appropriate choice of time units, one can *always* write the damped linear pendulum system in this form; see exercise **18**.

Stop and Think   **7.13** For what values of $\tilde{p}$ is the system (7.3) underdamped? Is (7.3) underdamped for relatively small or relatively large values of $\tilde{p}$?

The dependent variable θ is constant along the nullcline defined by $\theta' = 0$, that is, along the line

$\theta' = 0$ nullcline

$$\omega_{\theta'=0} = 0,$$

while ω is constant along the nullcline defined by $\omega' = 0$, that is, along the line $-\theta - \tilde{p}\omega = 0$ or

$\omega' = 0$ nullcline

$$\omega_{\omega'=0} = -(1/\tilde{p})\theta.$$

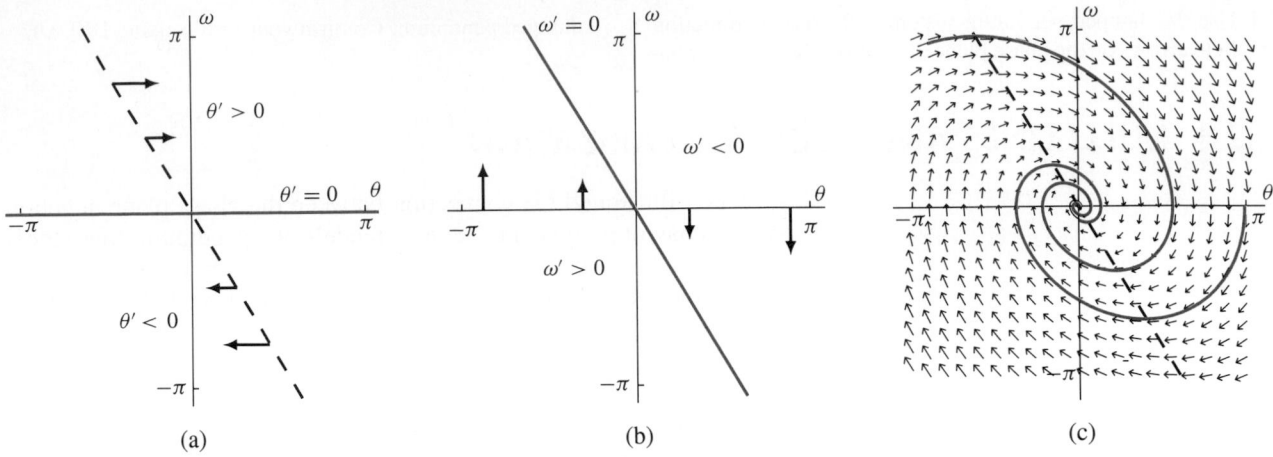

(a)   (b)   (c)

FIGURE 7.9   Nullclines and corresponding flow directions for the damped pendulum system (7.3).

These nullclines are shown in figure 7.9. Note that their intersection locates an equilibrium point, where both θ and ω are constant.

Figure 7.9(a) shows the nullcline $\theta' = 0$ along which θ is constant. Since θ can not change sign without crossing its nullcline, the sign of θ' can be determined by testing just a single point in each of the two halves of the phase plane defined by the nullcline. Obviously, $\theta' = \omega$ is positive above the nullcline and negative below it.

Figure 7.9(b) shows the nullcline $\omega' = 0$ along which ω is constant. Testing $\omega' = -\theta - \tilde{p}\omega$ for very large values of $|\omega|$, one positive and the other negative, verifies the sign behavior of ω' shown there.

Stop and Think   **7.14** Verify the sign behavior of ω' shown in figure 7.9(b) by using very large values of $|\theta|$, one positive and the other negative. (Recall that $\tilde{p} > 0$.)

7.15 The $\omega' = 0$ nullcline in figure 7.9 is drawn with a rather steep slope. Is that steepness consistent with the system (7.3) being *under*damped?

Figure 7.9(a) shows direction arrows attached to the $\omega' = 0$ nullcline. Those arrows are horizontal (ω is not changing along them) because at any point on that nullcline $\omega' = 0$. Those vectors increase in magnitude with distance from the θ axis—with increasing ω—because $\theta' = \omega$.

Stop and Think   **7.16** Provide parallel arguments for the direction and size of the vectors attached to the $\theta' = 0$ nullcline in figure 7.9(b). Why are they vertical? Why does their magnitude increase with distance from the ω axis—with increasing θ?

7.17 Show that the direction vectors in figure 7.9(c) are consistent with the nullcline information in parts (a) and (b).

7.18 Give a physical description of the motion of the pendulum described by each of the trajectories in figure 7.9(c).

> **MATLAB**
>
> Set $\tilde{p} = 0.6$ in (7.3) and use DELAB to draw your own version of figure 7.9(c), including nullclines. How does the value of $\tilde{p}$ affect the diagram? For guidance, select **Help, Textbook**, then go to chapter 7, figure 7.9.

Figure 7.9(c) is a composite of the information obtained from each of the nullclines in parts (a) and (b). The vector field shown there suggests that the flow is clockwise with a circular character. How does one explain the inward spiral shown there (and in figure 7.6, page 338)? Some possible arguments include the following:

- **Physical intuition** says the underdamped trajectory should spiral toward the rest point at the origin as the pendulum dissipates its energy with oscillations of decreasing amplitude.

- **MATLAB** phase plane diagrams of underdamped versions (e.g., $\tilde{p} = 0.1$) of (7.3) show spirals. (Use the graphical tools in DELAB.)

- **The general solution** of the second-order equation equivalent to (7.3) is driven to zero by decaying exponentials.

Stop and Think　**7.19** Derive the second-order equation $\theta'' + \tilde{p}\theta' + \theta = 0$. Use its characteristic equation to verify the claim about decaying exponentials in the general solution.

7.20 Is any kind of damping, over- or under-, sufficient to guarantee decaying exponentials in the general solution of $\theta'' + \tilde{p}\theta' + \theta = 0$?

- **The vector** from the origin of the phase plane to the solution point $(\theta(t), \omega(t))$, the so-called *state vector*, is decreasing in length as it rotates around the origin tracing out the trajectory.

That last geometrical argument requires justification. Figure 7.10 shows a hypothetical trajectory and the state vector $\mathbf{r}(t)$ from the origin to the point $(\theta(t), \omega(t))$ that is tracing out the trajectory. Is the length of that vector, $|\mathbf{r}(t)| = \sqrt{\theta^2(t) + \omega^2(t)}$, decreasing with time?

The behavior of the length of $\mathbf{r}(t)$ will be determined by the differential equations (7.3) that define its components $\theta(t)$ and $\omega(t)$. To avoid the bother of the square root, consider instead the *square* of the length,

$$s(t) = \theta^2(t) + \omega^2(t).$$

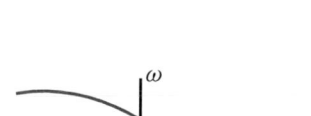
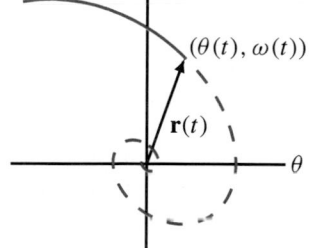

FIGURE 7.10　The state vector tracing out a possible phase plane trajectory of the underdamped pendulum system (7.3).

Showing that $s'(t) < 0$ will establish that $|\mathbf{r}(t)|$ is shrinking with time; the derivative itself is easy to compute,

$$s'(t) = 2\theta\theta'(t) + 2\omega\omega'(t).$$

Stop and Think　**7.21** Verify the expression for $s'(t)$.

Of course, the differential equations (7.3) that define the components of $\mathbf{r}(t)$ provide expressions for $\theta'(t)$ and $\omega'(t)$. Using (7.3), we find

$$s'(t) = 2\theta\omega + 2\omega(-\theta - \tilde{p}\omega) = -2\tilde{p}\omega^2 \le 0. \tag{7.4}$$

The (square of the) length of the state vector is decreasing with time (unless $\omega(t) \equiv 0$, in which case the pendulum is at rest). The trajectory in figure 7.10 is indeed drawing closer to the equilibrium point at the origin.

Stop and Think

7.22 What does this analysis predict about the trajectories of an undamped pendulum? Is that prediction physically reasonable?

Figure 7.11 illustrates the sorts of trajectory we expect when we combine the nullcline information of figure 7.9 with the analysis that showed trajectories coming closer to the origin: trajectories are spiraling clockwise in toward the steady state $\theta_{ss} = 0$, $\omega_{ss} = 0$.

> The analysis has *not* shown that the trajectories actually circle the origin; that part of the result is based on physical intuition and observation of solution behavior.
>
> Confirmation of trajectories' circling the origin is left to the exercises. One argument suggested in the exercises builds on the geometric perspective begun here, that of representing the trajectory in polar coordinates. The work shown here dealt only with the radial coordinate; demonstrating that the trajectory actually circles the origin requires showing that the angular coordinate increases without bound.

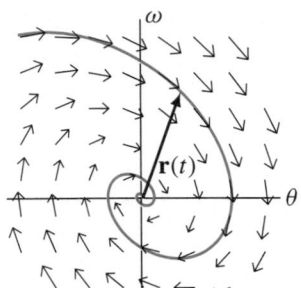

FIGURE 7.11 Trajectories of the underdamped pendulum system (7.3) spiral toward the origin because the length of the state vector decreases with time.

7.2.2 Local Linearization

Now we will conduct a similar nullcline analysis of the nonlinear pendulum system

$$\begin{aligned}\theta' &= \omega, \\ \omega' &= -\sin\theta - \tilde{p}\omega.\end{aligned} \tag{7.5}$$

After analyzing the nonlinear system, it will be apparent that the linear approximation has provided a local view—a kind of zoomed in, close-up picture—of one part of the nonlinear system's more complicated behavior. This simplified close-up is an example of *local linearization*, the ability of linear approximations to provide a description of the behavior of a nonlinear system in one locality or neighborhood.

> Local linearization is a fact of everyday life. For example, we use flat maps to find our way around a city, even though the surface of the earth is curved. Of course, the spherical earth is very well approximated locally by its tangent plane, the flat map.
>
> The approximation $\sin\theta \approx \theta$ is the linearizing approximation that replaces the nonlinear system (7.5) with the linear system (7.3) .

The $\theta' = 0$ nullcline of (7.5) is the same straight line as for the linear system (7.3),

$\theta' = 0$ nullcline

$$\omega_{\theta'=0} = 0,$$

but the $\omega' = 0$ nullcline is more complicated. It is the sine curve

$\omega' = 0$ nullcline

$$\omega_{\omega'=0} = -(1/\tilde{p})\sin\theta.$$

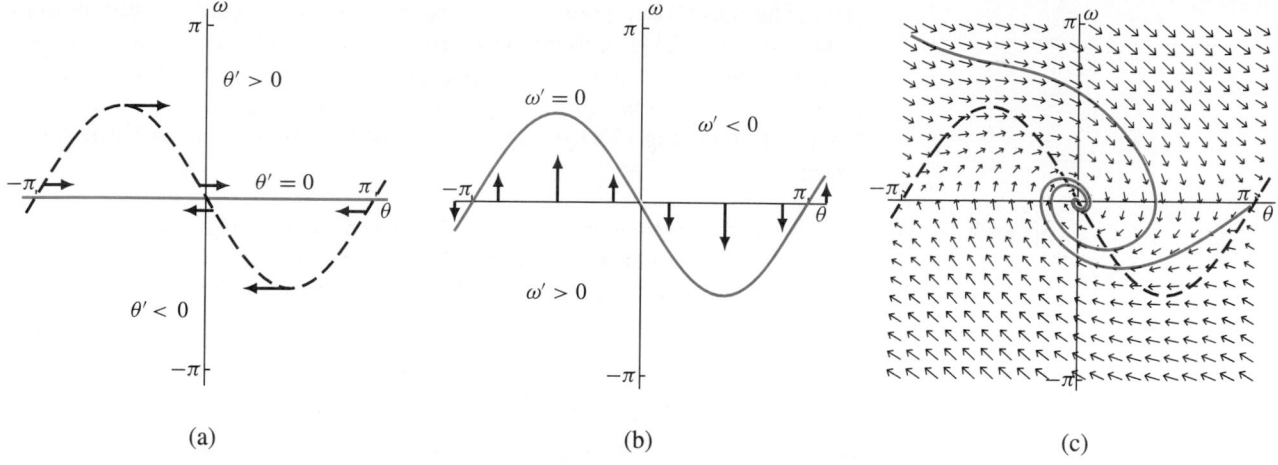

(a)   (b)   (c)

FIGURE 7.12 Nullclines and corresponding flow directions for the damped nonlinear pendulum system (7.5).

Figure 7.12 displays these two nullclines and the vector field associated with them. There is an infinity of equilibrium points at the intersections of the two nullclines,

$$\theta_{ss} = 0, \pm\pi, \pm2\pi, \ldots, \quad \omega_{ss} = 0.$$

Compare with the steady states listed in section 6.8.2, page 323.

Stop and Think

7.23 Describe the position of the pendulum for each of these steady states.

7.24 Verify the sign behavior of θ' shown in figure 7.12(a).

7.25 Verify the sign behavior of ω' shown in figure 7.12(b). (*Hint*: Test the sign of $\omega' = -\sin\theta - \tilde{p}\omega$ for very large values of $|\omega|$, one positive and the other negative.)

7.26 Explain why the direction field vectors emanating from the sine curve in figure 7.12(a) are horizontal. (*Hint*: The sine curve is the $\omega' = 0$ nullcline.) Explain their variation in magnitude with ω. (*Hint*: How should $|\theta'|$ change with ω?)

7.27 Explain why the direction field vectors emanating from the horizontal axis in figure 7.12(b) are vertical. (*Hint*: The horizontal axis is the $\theta' = 0$ nullcline.) Explain their variation in magnitude with θ. (*Hint*: How should $|\omega'|$ change with θ?)

Zoom in on the origin in figure 7.6 or 7.12(c). Get close enough and you will see 7.9(c), the flow field of the linearized version of the nonlinear system (7.3). All of the analysis of the preceding section applies, at least to the extent that the underlying linearizing approximation $\sin\theta \approx \theta$ is valid.

MATLAB

Set $\tilde{p} = 0.6$ in (7.5) and use DELAB to draw your own version of figure 7.12(c). Zoom in on that diagram until you see the linearized phase plane. For guidance, select **Help, Textbook**, then go to chapter 7, figure 7.12(c).

The equilibrium point $\theta_{ss} = 0$, $\omega_{ss} = 0$ is stable; nearby solutions spiral toward it in a clockwise direction, as in figure 7.11. The assertion of stability is a consequence of linear approximation; e.g., one could use linear stability analysis as in section 6.8 or one could study the length of the state vector, ultimately invoking (7.4) via the approximation $\sin\theta \approx \theta$, as requested in exercise 21.

Stop and Think   **7.28** Might the behavior at the other straight-down equilibrium points $\theta_{ss} = \pm 2\pi$, $\pm 4\pi$, ... be similar? Could figure 7.11 be a plausible close-up view of the phase diagram of the nonlinear system (7.5) near one of these equilibria?

What of the behavior near one of the straight-up equilibria, say $\theta_{ss} = \pi$, $\omega_{ss} = 0$? The phase diagram of figure 7.12(c) indicates that the trajectories of nearby solutions will move away. Physical intuition and the linear stability analysis of section 6.8 suggests as well that such motion should be away from this unstable equilibria. The details of the arguments are left to the exercises; the key is to define $\phi = \theta - \pi$ (the angular deviation from the straight-up position) in order to derive the local linear approximation

$$\begin{aligned} \phi' &= \omega, \\ \omega' &= \phi - \tilde{p}\omega. \end{aligned} \tag{7.6}$$

Figure 7.12 shows a complete phase plane diagram for the nonlinear system. The linear system phase plane of figure 7.11 is a close-up of the behavior near the origin. Exercise 8 asks you to find a linear approximation that gives the close-up near the unstable equilibrium $\theta_{ss} = \pi$, $\omega_{ss} = 0$.

Stop and Think   **7.29** Explain the physical behavior represented by trajectories that start near the unstable steady state $(\pi, 0)$. Where does the pendulum settle when this steady state is disturbed? Where do these nearby trajectories settle?

7.30 Should the local linear analysis that applies to $\theta_{ss} = \pi$, $\omega_{ss} = 0$ also apply to $\theta_{ss} = -\pi$, $\omega_{ss} = 0$?

The formalism of local linear approximation for a general autonomous system of two first-order equations

$$\begin{aligned} y' &= f(y, z), \\ z' &= g(y, z) \end{aligned}$$

is not hard to write down. Suppose there is an equilibrium at (y_{ss}, z_{ss}). To study perturbations of this equilibrium, define $u = y - y_{ss}$ and $v = z - z_{ss}$. Some simple manipulations involving first-order Taylor approximations (see exercise 22) lead to the linear system

$$\begin{aligned} u' &= f_y(y_{ss}, z_{ss})u + f_z(y_{ss}, z_{ss})v, \\ v' &= g_y(y_{ss}, z_{ss})u + g_z(y_{ss}, z_{ss})v. \end{aligned}$$

evaluated at (y_{ss}, z_{ss}). A more general study of such linear, constant-coefficient systems will occupy most of chapter 8, where we will see that the coefficients of the right side of this system, the **Jacobian matrix**

$$\begin{pmatrix} f_y(y_{ss}, z_{ss}) & f_z(y_{ss}, z_{ss}) \\ g_y(y_{ss}, z_{ss}) & g_z(y_{ss}, z_{ss}) \end{pmatrix},$$

usually determine completely the behavior of the system near the equilibrium point (y_{ss}, z_{ss}).

7.2.3 Epidemics

The SIRS system (2.31), page 72,

SIRS SYSTEM

$$\begin{aligned} S' &= -bIS + g(P - S - I), \\ I' &= bIS - rI, \end{aligned} \tag{7.7}$$

which was derived to model an epidemic in which some of those who recover from the disease again become susceptible, exhibits two steady states

$$S_{ss}^1 = \frac{r}{b}, \quad I_{ss}^1 = \frac{g(P - r/b)}{r + g} \quad \text{and} \quad S_{ss}^2 = P, \quad I_{ss}^2 = 0,$$

providing $r/b < P$. (See exercise 12, page 76.) The second steady state is more desirable than the first, for in it no one is infected.

Stop and Think   **7.31** What is the source of the requirement $r/b < P$?

7.32 Explain why "no one is infected" in the second steady state.

What does a nullcline analysis suggest about the stability of these steady states? That is, which of these steady states is likely to be observed and which is similar to an inverted pendulum, disappearing as soon as it is disturbed?

The $I' = 0$ nullclines are the axis $I \equiv 0$ and

$$S_{I'=0} = r/b.$$

The $S' = 0$ nullcline is the curve

$$I_{S'=0} = \frac{g(P - S)}{bS + g}. \tag{7.8}$$

To sketch the $S' = 0$ nullcline on the IS phase plane, note that the function $I_{S'=0}(S)$ defined by (7.8) satisfies $I_{S'=0}(0) = P$, $I_{S'=0}(P) = 0$, $dI_{S'=0}(S)/dS < 0$ (decreasing), and $d^2I_{S'=0}(S)/dS^2 > 0$ (concave up). (See exercise 9.)

Figure 7.13 illustrates these nullclines, the direction field vectors associated with each, and the corresponding phase plane diagram. The two steady states appear at the intersection of the nullclines.

Stop and Think   **7.33** Verify the sign of I' shown in figure 7.13(a).

7.34 Verify the sign of S' shown in figure 7.13(b).

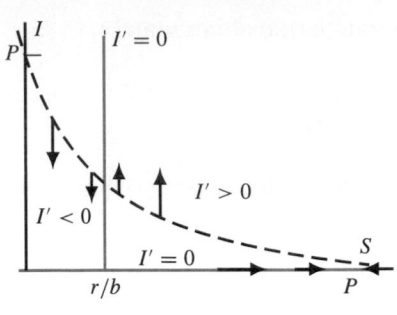
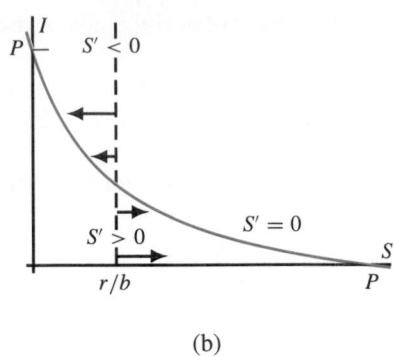
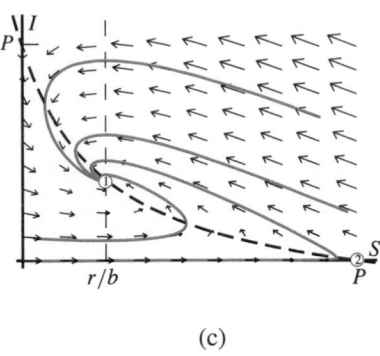

(a)   (b)   (c)

FIGURE 7.13 Nullclines ($I' = 0$ in (a), $S' = 0$ in (b)), corresponding flow directions, the two steady states (part (c)), and typical trajectories of the SIRS system (7.7) with $r/b < P$.

7.35 Justify the orientation and magnitude of the direction vectors shown in figure 7.13(a).

7.36 Justify the orientation and magnitude of the direction vectors shown in figure 7.13(b).

7.37 Show that the direction vectors in figure 7.13(c) are consistent with the nullcline information in parts (a) and (b).

7.38 Provide a physical description of the course of the epidemic described by each of the four trajectories in figure 7.13(c).

7.39 The trajectories in figure 7.13(c) appear to cross the $S' = 0$ nullcline with a vertical slope, the $I' = 0$ nullcline with a horizontal slope. Is that observation a coincidence? Can it be generalized?

The phase diagram of figure 7.13(c), which is based on the nullcline information shown in parts (a) and (b), suggests that the first steady state, in which some portion of the population remains infected, may be stable. Indeed, that steady state seems to be showing spiral behavior similar to the damped pendulum, figure 7.9. The analytic tools developed in the next chapter will make that analogy explicit.

MATLAB

To draw a phase plane similar to that of figure 7.13, apply DELAB's graphical tools to (7.7) with the parameter values $b = 2$, $r = 1$, $g = 1$, and $P = 2$.

The second steady state shown in 7.13(c), in which the population is free of infection, appears to be unstable except to perturbations of a special form, those that remain on the S-axis. Indeed, in the special case $I_i = 0$, $P > S_i > 0$, the system (7.7) is easy to solve. (See exercise 10.) One finds

$$I(t) \equiv 0, \quad S(t) = (S_i - P)e^{-gt} + P \to P.$$

However, the vector field of 7.13(c) shows that perturbations of this steady state with $I_i \neq 0$ will move away from it.

The phase plane behavior displayed by the SIRS system and by the pendulum system hardly exhaust the variety of behavior that two-dimensional systems can exhibit. The exercises explore both new and analogous phase plane geometries exhibited by predator-prey, competition, and other systems.

7.2.4 Exercises

1. Arguing from first principles as in the text, construct the analog of the nullcline phase plane diagrams in figure 7.9 for the *undamped* pendulum system

$$\theta' = \omega,$$
$$\omega' = -\theta.$$

Using DELAB, show that figure 7.9 looks more and more like your diagram for the undamped system as $\tilde{p} \to 0$.

2. Consider the undamped spring-mass system

$$x' = v,$$
$$v' = -x.$$

(Think of choosing $k/m = 1$ or see exercise 20.)

(a) Construct the analog of the nullcline diagrams in figure 7.9 for this system.

(b) Argue that $x^2 + v^2$ is constant on any trajectory. (*Hint*: $2xx' = d(x^2)/dt$.) Add several representative trajectories to your diagram, give the direction in which they are traversed, and explain their physical significance.

3. Consider the underdamped spring-mass system

$$x' = v,$$
$$v' = -x - \tilde{p}v.$$

(Think of choosing $k/m = 1$ or see exercise 20.)

(a) Construct the analog of the nullcline diagrams in figure 7.9 for this system.

(b) Argue that $x^2 + v^2$ is *decreasing* on any trajectory. (*Hint*: $2xx' = d(x^2)/dt$.) Explain the geometric significance in the phase plane of $x^2 + v^2$. Add several representative trajectories to your diagram, give the direction in which they are traversed, and explain their physical significance. Confirm with DELAB.

4. Describe the changes that occur in the nullcline diagrams of parts (a) and (b) of figure 7.9 as the damping coefficient $\tilde{p}$ increases. Confirm your results with DELAB.

5. Figure 7.9(c) illustrates the nullcline analysis of the damped linear pendulum system (7.3).

(a) In the first quadrant, the direction field vectors appear to point down and to the right. Analyze the differential equations directly and show that they predict exactly this behavior.

(b) Provide the analogous descriptions and arguments for the other three quadrants.

6. Consider the nonlinear pendulum system (7.5) near the equilibrium $\theta_{ss} = 2\pi$, $\omega_{ss} = 0$. Conduct a nullcline analysis and comment on the behavior of trajectories near $(2\pi, 0)$. Which *linear* phase plane diagram is a good approximation to the behavior you observe?

7. Consider the nonlinear pendulum system (7.5) near the equilibrium $\theta_{ss} = 2\pi$, $\omega_{ss} = 0$.

(a) Let $\phi = \theta - 2\pi$. (Since ϕ is the deviation from the straight-down position, it is analogous to θ, less the one complete revolution represented by $\theta = 2\pi$.) Substitute $\theta = \phi + 2\pi$ in (7.5), use an appropriate linearizing approximation, and derive a linear system just like (7.3).

(b) Use that linear system to conclude that this equilibrium is stable and that figure 7.11 is an accurate local representation of its phase diagram.

(c) Confirm this local analysis using a phase plane diagram of the full nonlinear system produced by DE-LAB.

8. The text claims that the system (7.6),

$$\phi' = \omega,$$
$$\omega' = \phi - \tilde{p}\omega,$$

is a linear approximation to the angular deviation ϕ of a damped pendulum from its straight-up equilibrium $\theta_{ss} = \pi$, $\omega_{ss} = 0$.

(a) Derive (7.6) from the nonlinear damped pendulum system (7.5) using the substitution $\theta = \pi + \phi$ and an appropriate linear approximation. How does the mathematical form of this system differ from that of the damped linear pendulum system (7.3)?

(b) Conduct a nullcline analysis similar to that which produced figure 7.9. Comment on the similarities and differences between your diagram and figure 7.9.

(c) Use arguments similar to those surrounding (7.4) to show that the length of the state vector of this system is *increasing* with time. What are the stability implications of that result?

(d) Use DELAB to confirm the results of your phase plane analysis of (7.6).

(e) Are the results of this analysis consistent with your physical intuition?

(f) Show that precisely the same analysis applies to the equilibrium $\theta_{ss} = -\pi$, $\omega_{ss} = 0$.

9. The text makes the following assertions about the $S' = 0$ nullcline (7.8),

$$I_{S'=0} = \frac{g(P - S)}{bS + g}$$

of the SIRS system (7.7):

(a) $I_{S'=0}(0) = P$, $I_{S'=0}(P) = 0$

(b) $dI_{S'=0}(S)/dS < 0$

(c) $d^2 I_{S'=0}(S)/dS^2 > 0$

Verify each assertion.

10. The text says that the SIRS system (7.7) is easy to solve in the special case $I_i = 0$, $P > S_i > 0$. It claims that

$$I(t) \equiv 0, \quad S(t) = (S_i - P)e^{-gt} + P \to P.$$

Verify this claim, first by showing $I(t) \equiv 0$, then by solving the remaining first-order equation for $S(t)$.

11. Use the phase plane tool in DELAB to investigate the behavior near the stable steady state of the SIRS system (7.7) (steady state 1 in figure 7.13) for large and small values of the reinfection parameter g. It what ways is the behavior near this steady state like that of a damped pendulum? In what ways is it different?

12. Conduct a nullcline analysis of the predator-prey system of example 14, page 68, of section 2.4,

$$F' = -(0.04 - 0.0011R)F,$$
$$R' = (0.06 - 0.0009F)R.$$

Graph the nullclines, determine the direction field along them, identify steady state(s), and speculate upon their stability. Use the phase plane tool in DELAB to confirm your analysis. It what ways is the behavior near the steady state(s) like that of a pendulum? In what ways is it different?

13. Repeat the previous exercise for the modified predator-prey system

$$F' = -(0.04 - 0.0011R)F,$$
$$R' = (0.06 - 0.0009F - 0.0009R)R.$$

The prey (rabbit) equation includes a logistic growth term; prey can not grow without bound, even in the absence of predators. How has the addition of logistic growth of the prey changed the phase plane behavior near the steady state?

14. Conduct a nullcline analysis of the competition system of example 15, page 73, of section 2.4,

$$y' = y(1 - y - 2z),$$
$$z' = z(1 - z - 2y).$$

Graph the nullclines, determine the direction field along them, identify steady state(s), and speculate upon their stability. Use the phase plane tool in DELAB to confirm your analysis. Compare the phase plane diagram with the time plots of figure 2.16, page 74. Confirm the phase plane diagram of figure 2.17, page 74. Describe the physical situation being depicted.

15. Conduct a nullcline analysis of a variant of the competition system of example 15, page 73, of section 2.4,

$$y' = y(1 - y - 2z),$$
$$z' = z(1 - z - 0.5y).$$

Graph the nullclines, determine the direction field along them, identify steady state(s), and speculate upon their stability. Use the phase plane tool in DELAB to confirm your analysis. Compare the phase plane diagram with the time plots of figure 2.16, page 74, and with the phase plane diagram of figure 2.17, page 74. Describe the physical situation being depicted and differences with example 15.

16. Conduct a nullcline analysis of a variant of the competition system of example 15, page 73, of section 2.4,

$$y' = y(1 - y - 0.5z),$$
$$z' = z(1 - z - 2y).$$

Graph the nullclines, determine the direction field along them, identify steady state(s), and speculate upon their stability. Use the phase plane tool in DELAB to confirm your analysis. Compare the phase plane diagram with the time plots of figure 2.16, page 74, and with the phase plane diagram of figure 2.17, page 74. Describe the physical situation being depicted.

17. Conduct a nullcline analysis of a variant of the competition system of example 15, page 73, of section 2.4,

$$y' = y(1 - y - 0.5z),$$
$$z' = z(1 - z - 0.5y).$$

Graph the nullclines, determine the direction field along them, identify steady state(s), and speculate upon their stability. Use the phase plane tool in DELAB to confirm your analysis. Compare the phase plane diagram with the time plots of figure 2.16, page 74, and with the phase plane diagram of figure 2.17, page 74. Describe the physical situation being depicted.

18. Argue as follows that the damped linear pendulum system

$$\theta' = \omega,$$
$$L\omega' = -g\theta - p\omega \qquad (7.9)$$

can always be written in a simpler form that amounts to choosing $g/L = 1$,

$$\theta' = \omega,$$
$$\omega' = -\theta - \tilde{p}\omega, \qquad (7.10)$$

by choosing to measure time in appropriate units.

(a) Define a new, dimensionless time unit by $\tau \equiv t\sqrt{g/L}$. For any function $f(t)$, define $F(\tau) = f(\tau/\sqrt{g/L}) = f(t)$. Use the chain rule to show that

$$\frac{df(t)}{dt} = \frac{dF(\tau)}{d\tau}\frac{d\tau}{dt} = \sqrt{g/L}\,\frac{df(\tau)}{d\tau}.$$

How is this new time scale related to the natural frequency of the undamped pendulum? Why is τ dimensionless?

(b) Introduce the notation $\dot{F}(\tau) \equiv dF/d\tau$; recall that $f'(t) \equiv df/dt$. Define $\Theta(\tau) \equiv \theta(\tau/\sqrt{g/L})$ and $\Omega(\tau) \equiv \omega(\tau/\sqrt{g/L})$. Use the derivative result from the first step to show that (7.9) can be written using the τ time units as

$$\sqrt{g/L}\,\dot{\Theta} = \Omega,$$
$$\sqrt{g/L}\,L\dot{\Omega} = -g\Theta - p\Omega.$$

(c) Show that this system is equivalent to the single second-order equation

$$\ddot{\Theta} + (P/L)\dot{\Theta} + \Theta = 0.$$

Find an expression for P.

(d) *Redefine* angular velocity in the τ time scale by $\Omega = d\Theta/d\tau$. Find the first-order system equivalent to $\ddot{\Theta} + (P/L)\dot{\Theta} + \Theta = 0$. Show that it has the "$g/L = 1$" form of (7.10) when θ, ω, and $\tilde{p}$ in (7.10) are appropriately defined. Give those definitions.

19. Repeat exercise 18 for the damped *non*linear pendulum equation; that is, show that a proper choice of time scale is equivalent to choosing $g/L = 1$. Compare the time scales used in the linear and the nonlinear problems.

20. Adapt the arguments of exercise 18 to show that the damped spring-mass system

$$x' = v,$$
$$mv' = -kx - pv$$

can always be written in the simpler form (that amounts to choosing $k/m = 1$)

$$x' = v,$$
$$v' = -x - \tilde{p}v$$

by introducing an appropriate time scale. Interpret that time scale physically. Define $\tilde{p}$.

21. Show that the (square of the) length of the state vector $\mathbf{r}(t)$ of the damped nonlinear pendulum system (7.5) is decreasing with time, at least so long as the linear approximation $\sin\theta \approx \theta$ is valid. (*Hint*: Compute $s'(t)$ from the differential equations, then approximate to obtain (7.4).)

22. Suppose that the general autonomous system of two first-order equations

$$y'(t) = f(y(t), z(t)),$$
$$z'(t) = g(y(t), z(t))$$

has an equilibrium at (y_{ss}, z_{ss}). After defining $u(t) = y(t) - y_{ss}$ and $v(t) = z(t) - z_{ss}$, the text claims that the linear system

$$u' = f_y(y_{ss}, z_{ss})u + f_z(y_{ss}, z_{ss})v,$$
$$v' = g_y(y_{ss}, z_{ss})u + g_z(y_{ss}, z_{ss})v$$

(approximately) governs perturbations of that equilibrium. Verify that claim as follows.

(a) Substitute $y = u + y_{ss}$, $z = v + z_{ss}$ into the original system to find the nonlinear system satisfied by u, v.

(b) Use first-order Taylor expansions (see section A.3, page 625) to argue that

$$f(u + y_{ss}, v + z_{ss}) \approx f_y(y_{ss}, z_{ss})u + f_z(y_{ss}, z_{ss})v.$$

Obtain a similar result for $g(u + y_{ss}, v + z_{ss})$. Write the resulting system of linear equations that govern $u(t)$ and $v(t)$. Why is it true that $f(y_{ss}, z_{ss}) = 0$ and $g(y_{ss}, z_{ss}) = 0$? Why is this linear system only an approximation?

7.3 ■ LIMIT CYCLES AND STABILITY

The local linear analysis of the last section demonstrated that trajectories that start near the pendulum's unstable straight-up steady state $\theta_{ss} = \pi$, $\omega_{ss} = 0$ leave it. The global picture showed that eventually such trajectories spiral in toward one of the stable, straight-down steady states like $\theta_{ss} = 0$, $\omega_{ss} = 0$; e.g., see figure 7.12, page 347.

Some trajectories that leave unstable states can also spiral toward a closed curve in the phase plane known as a **limit cycle**, one that is approached forward (or backward) in time by nearby trajectories. Such closed curves correspond to *periodic solutions* of the governing differential equations. Limit cycles are remarkable because they represent *self-sustained oscillations* in autonomous systems, systems that lack periodic forcing.

One source of such behavior is a series RLC circuit, as in figure 7.14, whose resistance R varies with current. The arguments of section 5.3 show that the equation governing current i is

$$Li'' + R(i)i' + \frac{1}{C}i = 0.$$

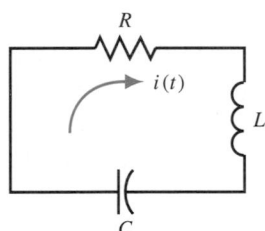

FIGURE 7.14   A series *RLC* circuit.

Some circuit devices actually produce negative resistance—they supply energy to the circuit—for small currents but become true resistors for larger currents. One simple expression that reflects such behavior is

$$R(i) = \epsilon(i^2 - i_0^2).$$

The threshold current for the transition from negative to positive resistance is i_0. Taking $i_0 = 1$ for simplicity, the resulting circuit equation is

VAN DER POL EQUATION

$$Li'' + \epsilon(i^2 - 1)i' + \frac{1}{C}i = 0, \tag{7.11}$$

the *van der Pol equation*.

The Dutch electrical engineer B. van der Pol introduced this differential equation during his studies in the 1920's of triode vacuum tube circuits. See [26].

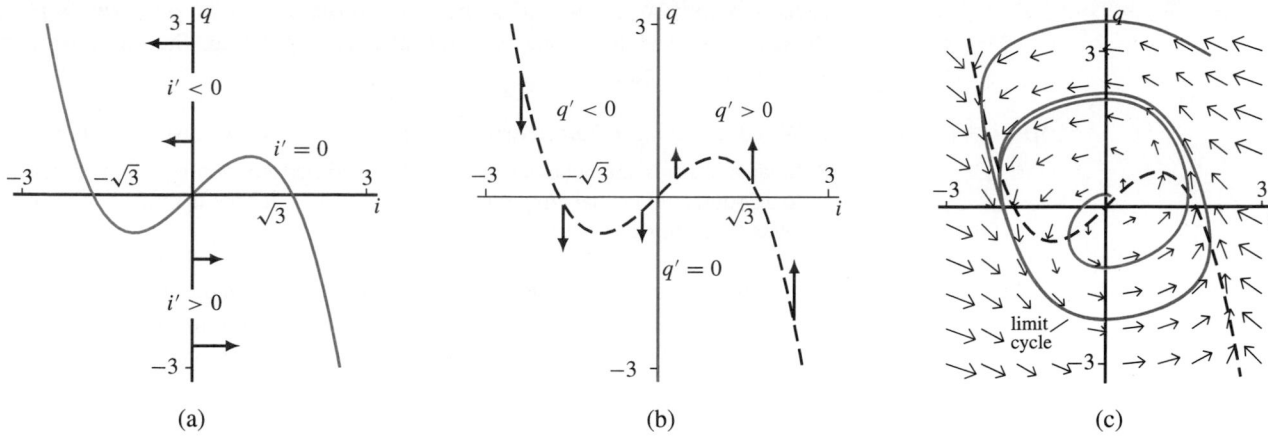

(a)　　　　　　　　(b)　　　　　　　　(c)

FIGURE 7.15　Nullclines and corresponding flow directions for the van der Pol system (7.12).

Although the second-order equation (7.11) could be written as a system by using the usual substitution ($i' = y$, $Ly' = \cdots$), it is more useful physically to let the second variable be q, the charge on the capacitor. Recall that

$$i \equiv \frac{dq}{dt}, \quad q = \int i \, dt.$$

To introduce the variable q, integrate (7.11):

$$\int \left(Li'' + \epsilon(i^2 - 1)i' + \frac{1}{C}i \right) dt = Li' + \epsilon \left(\frac{i^3}{3} - i \right) + \frac{1}{C}q = k,$$

for some arbitrary constant k.

Choose $k = 0$ and set $L = 1$, $C = 1$ for simplicity. Then a first-order system equivalent to (7.11) is

VAN DER POL SYSTEM

$$
\begin{aligned}
i' &= -\epsilon(i^2/3 - 1)i - q, \\
q' &= i.
\end{aligned}
\tag{7.12}
$$

Stop and Think　**7.40**　Compute i'' from the first equation of (7.12) and verify that (7.11) and (7.12) are equivalent.

7.41　Verify that $\int (Li'' + \epsilon(i^2 - 1)i' + i/C) \, dt = Li + \epsilon \left(i^3/3 - i \right) + q/C$.

The nullcline analysis summarized in figure 7.15 suggests a vaguely circular counterclockwise flow in the phase plane. The nullclines intersect in a single equilibrium point at the origin.

To test the stability of the equilibrium $i_{ss} = 0$, substitute the perturbed solution $i(t) = i_{ss} + p(t) = p(t)$ into the governing equation (7.11) and linearize by eliminating powers of the small perturbation p:

$$
\begin{aligned}
i'' + \epsilon(i^2 - 1)i' + i &= p'' + \epsilon(p^2 - 1)p' + p \\
&\approx p'' - \epsilon p' + p = 0.
\end{aligned}
\tag{7.13}
$$

Exercise 5 asks you to verify that the characteristic roots of $p'' - \epsilon p' + p = 0$ always include at least one whose real part is positive, thereby ensuring instability of the steady state $i_{ss} = 0$.

Stop and Think

7.42 Why does having a characteristic root with positive real part ensure instability?

7.43 In what ways is the behavior of this linearized perturbation equation similar to that obtained in analyzing the stability of $\theta_{ss} = \pi$ for the damped nonlinear pendulum? In what ways is it different?

The trajectory starting near the origin of the phase plane in figure 7.15(c) illustrates this instability. The surprise is its limiting behavior, its approach to the *limit cycle*, the closed curve highlighted there. A **limit cycle** is a trajectory that closes back on itself in finite time.

MATLAB

Choose $\epsilon = 1$ and use DELAB to draw your own version of figure 7.15(c). How does the shape of the limit cycle change with ϵ? For guidance, select **Help, Textbook**, then go to chapter 7; select figure 7.4 for a refresher on phase planes, figure 7.1 for a refresher on nullclines.

When this circuit is perturbed from its rest state, it quickly begins oscillating periodically. Furthermore, those periodic oscillations appear to be *stable*, for any trajectory that starts near the limit cycle returns to it as time passes.

A formal definition of **stable limit cycle** could be given in terms of a sufficiently small distance of the initial point from the limit cycle leading to the trajectory's eventual approach to within an arbitrary close distance of the limit cycle. However, the intuitive notion "start close to get arbitrarily close in large time" is sufficient at this stage.

What causes such behavior? Figure 7.15(c) suggests that there is some sort of balance between the repulsion of the unstable equilibrium at the origin and the generally inward flow of the direction field far from the unstable equilibrium. That intuitive idea can be made rigorous in the Poincarè-Bendixson theorem, which we state informally (and without proof) here.

EXISTENCE OF LIMIT CYCLES

Theorem 1 (Informal Poincarè-Bendixson theorem). *Suppose there is a bounded region $\mathcal{R}$ in the phase plane with two properties.*

1. *$\mathcal{R}$ contains a single unstable equilibrium point of a system of two differential equations.*

2. *No solution trajectory of this system leaves $\mathcal{R}$ once it has entered it.*

Then that system has a periodic solution lying entirely in $\mathcal{R}$.

That periodic solution is a limit cycle. The limit cycle may be approached as $t \to \infty$ (stable limit cycle) or as $t \to -\infty$ (unstable limit cycle).

The Poincarè-Bendixson theorem is among the first contributions to the flowering of the geometric theory of differential equations that began at the end of the nineteenth century. See [13] for a proof.

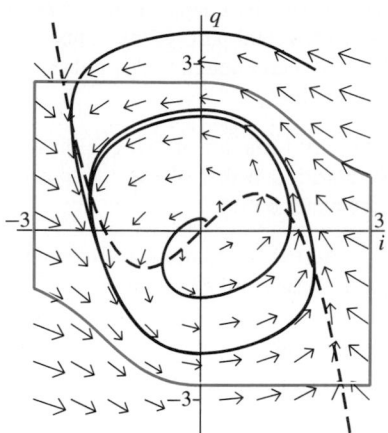

FIGURE 7.16 A region surrounding an unstable equilibrium point from which the flow of the van der Pol system (7.12) can not escape.

How might one apply the Poincarè-Bendixson theorem to the van der Pol system (7.12)? Figure 7.16 suggests one way the region $\mathcal{R}$ that encloses the equilibrium point at the origin might be constructed. The flow can never leave this region; it must trap any trajectories that enter.

The sides of this region are vertical lines that intersect the i-axis beyond $|i| = \sqrt{3}$ (at $|i| = 3$ in the figure). The nullcline analysis of figure 7.15 shows that the flow is always inward across these vertical lines that are on the appropriate side (above on the right, below on the left) of the cubic $i' = 0$ nullcline.

The top of the region consists of two pieces. One is a portion of a solution curve starting on the right side of the region (at $i = 3$, $q = 1$ in figure 7.16). Since no other trajectory can cross this one, nearby trajectories either remain above it or below it. This portion of the bounding region ends when it crosses the $q' = 0$ nullcline, the vertical axis.

The balance of the top of the bounding region is a horizontal line from the $i = 0$ end of the bounding trajectory to the left side of the bounding region ($i = -3$ in the figure). The nullcline analysis ensures that all flows are inward along this line as well.

Stop and Think　**7.44** Why can't the top of the region $\mathcal{R}$ in figure 7.16 be a simple horizontal line?

> This intuitive discussion assumes a great deal about solution behavior that is evident from figure 7.15 but requires careful proof. For example, how can we be sure that the solution trajectory that begins at $i = 3$, $q = 1$ actually reaches the vertical axis $i = 0$ in finite time? Such arguments are the hard part of applying the Poincarè-Bendixson theorem.

Stop and Think　**7.45** Argue from the nullcline analysis summarized in figure 7.15 that the flows along the straight sides of the bounding region $\mathcal{R}$ in figure 7.16 are indeed inward.

7.46 If the van der Pol system (7.12) obeys a uniqueness theorem analogous to theorem 5, page 188, which was proved for a single scalar equation, then there is never more than one solution of (7.12) satisfying a given set of initial conditions $i(0) = i_i$, $q(0) = q_i$. Why would such a uniqueness theorem guarantee that trajectories of (7.12) can not cross at an isolated point?

7.3.1 Exercises

EXERCISE GUIDE	
To gain experience . . .	Try exercises
With nullclines and flow fields	1, 6, 7(b, d), 8(b, d)
Examining stability of steady states	5, 7(c), 8(c)
Finding limit cycles	6, 7(e), 8(e)
With mathematical and physical analogies	3–4, 7(a), 8(a)

1. Verify each of the indicated features of figure 7.15, a null-cline and phase plane diagram for the van der Pol system (7.12),

$$i' = -\epsilon(i^2/3 - 1)i - q,$$
$$q' = i.$$

   (a) The shape of the $i' = 0$ nullcline and the flow field directions shown in figure 7.15(a).

   (b) The shape of the $q' = 0$ nullcline and the flow field directions shown in figure 7.15(b).

   (c) The orientation shown in figure 7.15(c) of the flow field direction arrows in the four regions defined by the intersecting $i' = 0$ and $q' = 0$ nullclines.

2. What should be the slope of a trajectory of the van der Pol system (7.12) as it crosses the $i' = 0$ nullcline? The $q' = 0$ nullcline? Verify that the trajectories in figure 7.15(c) cross the nullclines with the correct slope.

3. Suppose the van der Pol equation (7.11), $Li'' + \epsilon(i^2 - 1)i' + i/C = 0$, were a model of displacement in a spring-mass system rather than current in an *RLC* circuit.

   (a) Write (7.11) in the usual spring-mass notation and identify the ways, if any, in which the behavior of the inertia (mx''), damping (px'), and spring force (kx) terms in this equation differ from those in the standard spring-mass equation.

   (b) Use your understanding of this "van der Pol spring-mass system" to explain how self-sustained oscillations like those seen in the limit cycle of figure 7.15(c) might arise.

4. Suppose the van der Pol equation (7.11), $Li'' + \epsilon(i^2 - 1)i' + i/C = 0$, were a model of a pendulum rather than an *RLC* circuit.

   (a) Write (7.11) in the usual pendulum notation and identify the ways, if any, in which the behavior of the inertia $(L\theta'')$, damping $(p\theta')$, and gravity $(g\theta)$ terms in this equation differ from those in the standard linear pendulum equation.

   (b) Use your understanding of this "van der Pol pendulum system" to explain how self-sustained oscillations like those seen in the limit cycle of figure 7.15(c) might arise.

5. Working as follows, verify the details of the stability analysis of the steady state $i_{ss} = 0$ of the van der Pol equation (7.11), $Li'' + \epsilon(i^2 - 1)i' + i/C = 0$.

   (a) Substitute $i(t) = i_{ss} + p(t)$ to derive (7.13), $p'' - \epsilon p' + p = 0$, the approximate linear equation governing the perturbation $p(t)$.

   (b) Verify that at least one of the characteristic roots of (7.13) always has positive real part. (Why is just one root with a positive real part enough to claim instability of an equilibrium point?)

   (c) Does the linear stability behavior of this equilibrium point bear any similarities to that of the equilibrium point $\theta_{ss} = \pi$ of the pendulum?

6. Use the graphical tools in DELAB to draw a vector field diagram and enough trajectories of the system

$$y' = z + y(1 - y^2 - z^2),$$
$$z' = -y + z(1 - y^2 - z^2)$$

to exhibit a stable limit cycle. Describe the oscillations represented by the limit cycle. On a copy of the phase diagram, draw a region $\mathcal{R}$ that could be used to invoke the Poincarè-Bendixson theorem and prove that your observations are not a numerical artifact.

7. Suppose a funky pendulum has been constructed so that its damping coefficient depends on angle; it is modeled by $\theta'' + p(\theta)\theta' + \theta = 0$ with $p(\theta) = d(\theta^2 - \theta_0^2)$ for some fixed angle θ_0 and coefficient $d > 0$.

   (a) Describe the behavior of this damping term. Is it always damping—slowing—the motion?

   (b) Introduce angular velocity $\omega = \theta'$ and write the governing second-order equation as a pair of first-order equations. Conduct a nullcline analysis of this system similar to that in figure 7.15, parts (a) and (b). How does the value of θ_0 affect the nullcline diagrams?

   (c) Find the steady state(s) and determine their stability. How do the values of d and θ_0 affect this analysis?

   (d) Use the graphical tools in DELAB to draw a vector field diagram, say for $d = 1$, $\theta_0 = 0.5$. Confirm that it is consistent with your nullcline and linear stability analyses.

   (e) Let DELAB draw enough trajectories to convince you that this system has a (stable) limit cycle. On a copy of the phase diagram, speculate on the construction of a region $\mathcal{R}$ that could be used to invoke the Poincarè-Bendixson theorem.

8. This exercise continues the investigation of the funky pendulum begun in exercise 7.

   (a) Careful inspection of the funky pendulum's governing equation, $\theta'' + p(\theta)\theta' + \theta = 0$ with $p(\theta) = d(\theta^2 - \theta_0^2)$, reveals that it has the form of a van der Pol equation. Introduce the new variable $V = \int \theta \, dt$ (compare with *q* in the *RLC* circuit equation) and show that you obtain the system

$$\theta' = -d(\theta^2/3 - \theta_0^2)\theta - V,$$
$$V' = \theta, \tag{7.14}$$

which is analogous to the van der Pol system (7.12).

> Using $\sin\theta \approx \theta$, it can be shown that *V* is proportional to the work done against the force of gravity along the path of the pendulum; i.e., *V* approximates potential energy. The potential energy stored by displacing the pendulum can be released through velocity, just as charge stored in a capacitor can be released through current.

(b) Conduct a nullcline analysis of (7.14) analogous to that of figure 7.15, parts (a) and (b).

(c) Mimicking the text's analysis of the van der Pol system (7.12), find the steady state(s) of (7.14) and determine their stability. How do the values of *d* and θ_0 affect this analysis? Should the results of this analysis be any different from the analysis of the θ-ω system of exercise 7(b)?

(d) Use the graphical tools in DELAB to draw a vector field diagram, say for $d = 1$, $\theta_0 = 0.5$. Confirm that it is consistent with your nullcline and linear stability analyses. Why is this picture different from that drawn in exercise 7(d) for the θ-ω system derived there?

(e) Let DELAB draw enough trajectories to convince you that (7.14) has a (stable) limit cycle. On a copy of the phase diagram, speculate on the construction of a region $\mathcal{R}$ that could be used to invoke the Poincarè-Bendixson theorem. Compare with figure 7.16.

7.4 ■ CHAPTER EXERCISES

EXERCISE GUIDE	
To gain experience ...	**Try exercises**
With *autonomous*, *linear*, etc.	6(a), 7(b)
Finding equilibrium points	5(e–f), 6(b–c), 7(c)
Drawing and analyzing flow diagrams	1–2, 4, 7(g, j), 8–10, 11(f), 12
Analyzing stability of equilibria	3, 4(b), 7(d, f, j), 8–9, 11(c–d)
With nullcline analysis	3, 5–6, 7(e)
With limit cycles	6(d), 7(h–j), 11(c–f)
With phase planes in polar coordinates	8–11

1. Figure 7.12, page 347, the phase plane for the damped nonlinear pendulum system (7.5), provides little information about the behavior of trajectories near the unstable equilibrium point ($\theta_{ss} = \pi$, $\omega_{ss} = 0$).

(a) Use the graphical tools in DELAB to rectify this deficiency. What becomes of perturbations of this equilibrium? What is the physical interpretation of the phase plane behavior you observe?

(b) How does the phase portrait change as the damping parameter $\tilde{p}$ is varied through positive values? Could the difference between an underdamped and an overdamped system be detected from the phase portrait alone? How might you interpret *underdamped* and *overdamped* in this nonlinear setting? (Recall that these terms were defined for linear equations.)

2. Suppose the previous problem is posed for the equilibrium point ($\theta_{ss} = -\pi$, $\omega_{ss} = 0$). Give a physical argument that allows you to carry over the results for the point ($\theta_{ss} = \pi$, $\omega_{ss} = 0$) with only simple changes. Give the mathematical (and rigorous) version of that symmetry argument.

3. Introduce the charge $q = \int i\, dt$ and write an *i-q* system of two first-order equations that is equivalent to the *RLC* series equation $Li'' + Ri' + i/C = 0$, as in the derivation of (7.12) from (7.11). Conduct a nullcline analysis of this system and sketch a flow field for it. Find the steady state of this system and analyze its stability. To what other physical systems is this *RLC* circuit analogous? Do those systems exhibit the same behavior as you have identified here?

4. The damped nonlinear pendulum system (7.5) has an unstable steady state at the point $(\theta_{ss} = \pi, \omega_{ss} = 0)$ (figure 7.12(c), page 347). The SIRS system (7.7) has an unstable steady state at the point $(S_{ss} = P, I_{ss} = 0)$ (figure 7.13(c), page 350).

   (a) Demonstrate that both steady states share a common property: in spite of their instability, each can be approached by a trajectory that carefully follows a specific direction. Identify (a linear approximation to) that direction in each case.

   (b) Use the graphical tools in DELAB to sketch flow diagrams and sample trajectories for each system near the unstable equilibrium. Describe similarities and differences in the flow diagram seen in each case. (Zoom in to concentrate on local behavior.) If the two phase diagrams were printed on rubber paper, might you be able to deform one so that it looked like the other?

   > Such equilibria are called *saddle points* for reasons that should be clear from the geometry of the flow diagrams. They will studied in more detail in chapter 8.

5. A general autonomous system of two first-order equations can be written in the form

$$y' = f(y, z),$$
$$z' = g(y, z).$$

   (a) Write the equation that defines the $y' = 0$ nullcline(s).

   (b) What is the slope of a phase plane trajectory as it crosses a $y' = 0$ nullcline?

   (c) Write the equation that defines the $z' = 0$ nullcline(s).

   (d) What is the slope of a phase plane trajectory as it crosses a $z' = 0$ nullcline?

   (e) Write the equations that determine the equilibrium points.

   (f) Do equilibrium points lie on the intersection of $y' = 0$ and $z' = 0$ nullclines?

6. Chapter 8 defines a constant-coefficient, linear system of two first-order equations as one of the form

$$y' = ay + bz,$$
$$z' = cy + dz,$$

   where $a, \ldots, d$ are constants.

   (a) Is such a system autonomous?

   (b) Find the nullclines of this system. What type of curve are they?

(c) How many equilibrium points might such a system have?

(d) Suppose such a system has one unstable equilibrium point. Could the Poincarè-Bendixson theorem possibly guarantee the existence of a limit cycle?

7. Consider a general system of the form

$$u' = -g(u) - v,$$
$$v' = u$$

for some differentiable function g.

   (a) Show that with a proper choice of u, v, and g, the van der Pol system (7.12) is just a special case of this general equation.

   (b) Show that this system is equivalent to the second-order equation $u'' + g'(u)u' + u = 0$. Is this equation linear or nonlinear?

   (c) Suppose $g(0) = 0$. Show that this system has a steady state $u_{ss} = 0$, $v_{ss} = 0$.

   (d) Conduct a linear stability analysis of the steady state $u_{ss} = 0$ using the equivalent second-order equation $u'' + g'(u)u' + u = 0$. What additional information do you need about g in order to complete this analysis?

   (e) Suppose $g(u) = u(u - a)(b - u)$ for some constants $a < 0 < b$. Conduct a nullcline analysis analogous to that shown in figure 7.15, page 355, parts (a) and (b).

   (f) Use $g(u) = u(u - a)(b - u)$ to complete the analysis of the stability of $u_{ss} = 0$ begun in part (d).

   (g) Sketch a complete flow diagram analogous to that in figure 7.15(c).

   (h) Speculate upon the construction of a region $\mathcal{R}$ that could be used to invoke the Poincarè-Bendixson theorem; compare with figure 7.16, page 357. What do you conclude about the possibility of such a system supporting a limit cycle oscillation?

   (i) Use the graphical tools in DELAB to confirm your analysis for some specific values of a and b of your choice.

   (j) Consider a more general form of g. Require that $g(u) = 0$ for $u = a$, 0, b for some constants $a < 0 < b$. Impose as few additional requirements as necessary to ensure that the preceding analysis—the apparent existence of a limit cycle around the unstable equilibrium at the origin of the phase plane—holds for a general class of functions g. Describe the character of these requirements in a sketch. Express them analytically as well.

8. Combine the two equations in the undamped pendulum system

$$\theta' = \omega,$$
$$\omega' = -\theta$$

to show that $\theta^2 + \omega^2 = C$. (*Hint*: $2\theta\theta' = d(\theta^2)/dt$.) In the phase plane, what is the geometric significance of this formula—that is, what is the shape of the corresponding trajectory? How is C related to the initial values of θ and ω? Describe the physical behavior that corresponds to these trajectories.

> This analysis shows that the steady state ($\theta_{ss} = 0$, $\omega_{ss} = 0$) is *neutrally stable*; perturbations of this state are periodic solutions that orbit around it, neither returning too close to it nor departing too far from it.

9. Repeat exercise 8 for the predator-prey system

$$F' = -(d_F - \alpha R)F,$$
$$R' = (b_R - \beta F)R,$$

linearized about the steady state $F_{ss} = b_R/\beta$, $R_{ss} = d_F/\alpha$. (*Hint*: Combine the approximate equations to obtain an expression of the form $mFF' + nRR' = 0$ for appropriate constants m, n; the resulting trajectory is *not* a circle.)

10. Previous analyses of the underdamped linear pendulum have shown that it swings continuously with decreasing amplitude, behavior that should correspond to a trajectory spiraling inward around the steady state ($\theta_{ss} = 0$, $\omega_{ss} = 0$). One way to study that behavior in the phase plane is through the polar coordinate $(r(t), \phi(t))$ representation of that trajectory. The radial coordinate is defined by $r^2 = \theta^2 + \omega^2$, the angular coordinate by $\tan\phi = \omega/\theta$. (The polar coordinate angle is denoted ϕ rather than the usual θ to avoid confusion with the pendulum angle.)

Equation (7.4) shows that the trajectory of any damped linear pendulum system is indeed approaching the point $(0, 0)$ because its radial coordinate $r(t)$ (the length of the state vector $\mathbf{r}(t)$) is decreasing with time. The following argument shows that the trajectory is actually a spiral if the system is underdamped.

(a) Write the second-order equation equivalent to the first-order linear pendulum system $\theta' = \omega$, $\omega' = -\theta - \tilde{p}\omega$. Find the range of values of $\tilde{p}$ for which this system is underdamped.

(b) Using its definition and the governing differential equation, show that the angular coordinate ϕ satisfies

$$\frac{d\phi}{dt} = -1 - \tilde{p}\frac{\theta\omega}{r^2}.$$

(c) By finding its critical points, argue that the function $f(\theta, \omega) = \theta\omega/r^2 = \theta\omega/(\theta^2 + \omega^2)$ satisfies $|f| \le 1/2$ for all θ, ω. Conclude that $\phi' \le -1 + \tilde{p}/2$.

(d) Show that the angular coordinate ϕ decreases without bound if $\tilde{p}$ is in the range of values for which the system is underdamped. If ϕ is decreasing, in which direction must the trajectory be winding?

11. Exercise 6 of section 7.3 asked you to use the graphical tools in DELAB to exhibit a stable limit cycle of the system

$$y' = z + y(1 - y^2 - z^2),$$
$$z' = -y + z(1 - y^2 - z^2) \tag{7.15}$$

and to construct a plausible region $\mathcal{R}$ that could be used to invoke the Poincarè-Bendixson theorem. Converting the phase plane representation of this system to polar coordinates permits a rigorous description of that region.

(a) Define the polar coordinate (r, θ) by $r^2 = y^2 + z^2$, $\tan\theta = z/y$. Show that (7.15) is equivalent to the (uncoupled) system

$$r' = r(1 - r^2),$$
$$\theta' = -1. \tag{7.16}$$

(*Hint*: $rr' = yy' + zz'$. (Why?) Try multiplying and adding the equations in (7.15) to obtain an equation involving r'.)

(b) Show that a solution of (7.16) is

$$r = (1 + ke^{-2t})^{-1/2}, \qquad \theta = -t + \theta_i. \tag{7.17}$$

Extra: Does the uniqueness theorem 5, page 188, apply to each of the equations in (7.16)? If so, is this solution *general* in the sense that every solution of (7.16) can be written in the form (7.17) for an appropriate choice of the constants k and θ_i?

(c) Describe the behavior of the solution (7.17) as $t \to \infty$. Does the system (7.16) have a limit cycle? Is it stable? What is its period? Its amplitude? Convert your description to the original y, z variables.

(d) Analyze the stability of the steady state $r_{ss} = 0$ of the radial equation $r' = r(1 - r^2)$. What other steady states does this equation have? What is their stability and significance?

(e) Show that for any $\delta > 0$, the region

$$\mathcal{R} = \{r \mid |r| < 1 + \delta\}$$

satisfies the hypotheses of the Poincarè-Bendixson theorem, thereby providing a second rigorous verification of the presence of the limit cycle $r = 1$.

(f) Use the graphical tools in DELAB to draw a flow diagram and sample trajectories of (7.15). Are these consistent with your analysis? Mark typical (r, θ) coordinates on a copy of your diagram, particularly in the vicinity of the limit cycle. Would a flow diagram of the polar coordinate system (7.16) look the same or different?

12. Under suitable conditions, the uniqueness theorem for a single first-order equation prohibits a first-order initial-value problem from having more than one distinct solu-

tion. Similar theorems are available for second-order equations: Under suitable conditions, a second-order initial-value problem can have no more than one distinct solution. What does such a theorem say about the possibility of two distinct trajectories in the phase plane meeting in a common point?

7.5 ■ CHAPTER PROJECTS

1. **Oscillations in autocatalytic chemical reactions.** The simple chemical reaction scheme

$$
\begin{aligned}
2X + Y &\rightleftharpoons 3X, \\
A &\rightarrow Y, \\
X &\rightleftharpoons B
\end{aligned}
\tag{7.18}
$$

involving the four species X, Y, A, and B can support oscillations in the concentrations of X and Y. The first step is *autocatalytic* in X; i.e., X makes more of itself (while consuming Y as well). The third reaction disposes of X, effectively removing it from the reaction network, while the second reaction adds Y, a substance that promotes the formation of X. How might these competing influences—the feedback in X and the relative amounts of A and B—balance one another?

This reaction scheme can be modeled by the system [21]

$$
\begin{aligned}
x' &= x^2 y - x + b, \\
y' &= a - x^2 y.
\end{aligned}
\tag{7.19}
$$

Lowercase letters represent concentrations of the corresponding chemical species.

(a) Argue that the system (7.19) is a plausible model of the reaction scheme (7.18). One approach is to regard reaction notation such as $A \rightarrow Y$ as defining a birth-like phenomena with per capita birth rate of unity; $2X + Y$ requires an interaction between two X's and one Y. (Or consult a chemistry text.)

(b) Locate the steady state(s) of this system. Conduct a nullcline analysis and speculate upon possible behavior near the steady state.

(c) Consider $b = 0.1$. Using the graphical tools in DELAB, explore the phase plane behavior of (7.19) for several different values of a, say $a = 0.9$ and $a = 0.7$. What do you observe? How does the stability of the steady state appear to change? What global behavior emerges as a is decreased?

(d) Make a copy of a phase plane diagram for a version of (7.19) that exhibits a limit cycle. On it draw a possible bounding region $\mathcal{R}$ that could be used to invoke the Poincarè-Bendixson theorem. Argue from the differential equations (7.19) that the region you drew has the required properties. Does the stability of the enclosed equilibrium point appear to be of the type required by the Poincarè-Bendixson theorem? What can you conclude about the existence of a limit cycle?

(e) Provide a physical interpretation of your observations.

2. **Structural instability in predator-prey models.** A family of predator-prey models can be written in the form

$$F' = \phi(F, R)F,$$
$$R' = \rho(F, R)R. \tag{7.20}$$

Here F is the predator (fox) population, R the prey (rabbit) population, and ϕ and ρ are their respective per capita birth rates. In the familiar model (2.24–2.25) derived in section 2.4.1, we found

$$\phi(F, R) \equiv \phi_0(F, R) = -(d_F - \alpha R),$$
$$\rho(F, R) \equiv \rho_0(F, R) = (b_R - \beta F). \tag{7.21}$$

Exercise 9 of the chapter exercises shows that the nontrivial steady state of this form of the predator-prey system is neutrally stable; that is, perturbations of this state are periodic solutions that orbit around it, neither returning too close to it nor departing too far from it.

This project explores the uncertain nature of neutral stability. The family of predator-prey systems (7.20) is *structurally unstable* in the sense that small changes in the form of the model—small changes in the growth rate functions $\phi(F, R)$ and $\rho(F, R)$—lead to dramatic changes in the response of the system.

(a) Suppose that the fractions of the fox and rabbit populations that reproduce per unit time are given by $\phi(F, R)$ and $\rho(F, R)$, respectively. Derive the governing equations (7.20) from appropriate conservation laws.

(b) Consider the familiar predator-prey model, in which ϕ and ρ are defined by (7.21). Find the nontrivial steady state of this system, conduct a nullcline analysis, demonstrate the neutral stability of the nontrivial steady state as in exercise 9 of the chapter exercises, and provide appropriate phase plane diagrams. Give a physical interpretation of the behavior of the populations near the steady state.

(c) Suppose the birth rate terms ϕ_0, ρ_0 of the familiar predator-prey model are perturbed by the addition of one term,

$$\phi(F, R) = \phi_1(F, R) \equiv \phi_0(F, R),$$
$$\rho(F, R) = \rho_1(F, R) \equiv \rho_0(F, R) + cR.$$

Provide a physical interpretation of this change in the case $c < 0$.
To be concrete, consider the parameter values $d_F = 0.04$, $\alpha = 0.0011$, $b_R = 0.06$, $\beta = 0.009$, $c = -0.0001$. Find the nontrivial steady state

of this system, conduct a nullcline analysis, and explore the stability of the nontrivial steady state using the graphical tools in DELAB. Support your analysis with appropriate phase plane diagrams.

Give a physical interpretation of the behavior of the populations near the steady state. How has the character of the equilibrium changed from that of the familiar case (7.21)?

(d) Now perturb the familiar birth rate terms ϕ_0, ρ_0 in a slightly different way,

$$\phi(F, R) = \phi_2(F, R) \equiv \phi_0(F, R) - s_F F,$$
$$\rho(F, R) = \rho_2(F, R) \equiv \rho_0(F, R) - (aR - c)R.$$

Provide a physical interpretation of the added terms in the case s_F, a, $c > 0$.

Let d_F, α, b_R, and β be as before. Set $s_F = 0.0001$, $a = 0.00005$, $c = 0.005$. Find the nontrivial steady state of this system, conduct a nullcline analysis, and explore the stability of the nontrivial steady state using the graphical tools in DELAB. Support your analysis with appropriate phase plane diagrams.

Give a physical interpretation of the behavior of the populations near the steady state. How has the character of the equilibrium changed from that of the familiar case (7.21)? What new *global* behavior has emerged? What is its physical significance?

(e) Two apparently small modifications of the familiar predator-prey birth rate terms have dramatically changed the behavior of the system. Summarize the fate of the neutrally stable oscillations for each modification of the birth rate terms; that is, describe the structural instability of the predator-prey system defined by (7.21). Explain why this system would be an unsatisfactory model for a field biologist.

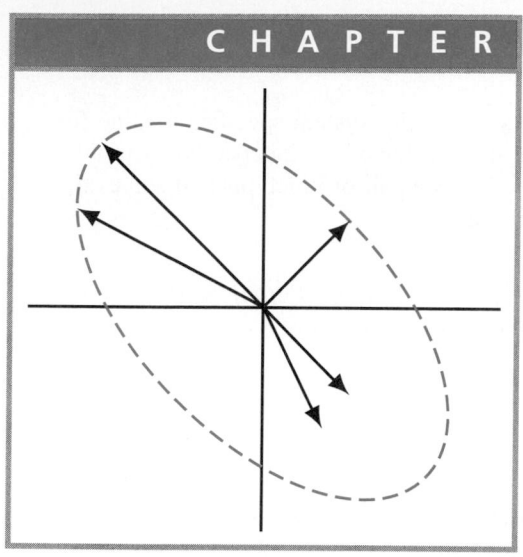

Analytic Tools for Higher Dimensions

Stability and local behavior of many sorts of models are determined by systems of linear differential equations with constant coefficients. This chapter studies such systems, particularly their application to understanding the local behavior of nonlinear systems.

8.1 ■ BASIC DEFINITIONS: SYSTEMS

Systems of first-order equations arise naturally in many models, the predator-prey system

$$F' = -(d_F - \alpha R)F,$$
$$R' = (b_R - \beta F)R$$

of section 2.4.1 being but one example. Of course, single equations of order 2 (or higher) can be converted to systems of first-order equations by introducing a new variable to account for the first derivative; e.g., introduce angular velocity $\omega = \theta'$ to convert the damped pendulum equation $L\theta'' + p\theta + g\sin\theta = 0$ to the first-order system

$$\theta' = \omega,$$
$$L\omega' = -g\sin\theta - p\omega. \tag{8.1}$$

Systems are the most general framework for the study of differential equations because they incorporate higher-order equations as well as systems themselves.

The generic first-order system of two differential equations has the form

$$y' = f(t, y, z), \tag{8.2}$$
$$z' = g(t, y, z), \tag{8.3}$$

where y and z are the **dependent variables** and t is the **independent variable**. A *solution* of a system is a pair of continuously differentiable functions

$(y(t), z(t))$ that reduce both equations to an identity when substituted for the corresponding dependent variables.

An **initial-value problem** for a first-order system specifies a value for each of the dependent variables at a fixed value of the independent variable, and a *solution of an initial-value problem* is a pair of functions that solves the system and satisfies the initial conditions.

More precisely:

Definition 1. *The pair of functions $(u(t), v(t))$ is a **solution** on the interval $t_0 \leq t \leq t_1$ of the first-order system (8.2–8.3) if u, v are continuously differentiable there and if*

$$u' = f(t, u, v),$$
$$v' = g(t, u, v), \tag{8.4}$$

*for $t_0 \leq t \leq t_1$. Further, (u, v) is a **solution of the initial-value problem***

$$y' = f(t, y, z), \quad y(t_0) = y_0,$$
$$z' = g(t, y, z), \quad z(t_0) = z_0, \tag{8.5}$$

if (u, v) is a solution of the system (8.2–8.3) and, in addition,

$$u(t_0) = y_0, \quad v(t_0) = z_0.$$

The equations $y(t_0) = y_0, z(t_0) = z_0$, are the **initial conditions**.

8.1.1   Vector Notation

Vector notation provides a compact way to write these systems and their solutions. Systems of two equations have solutions with two components, for example, $u(t)$ and $v(t)$ for the system (8.4) in the definition or $\theta(t)$ and $\omega(t)$—angular position and velocity—for the pendulum system (8.1). In two dimensions, vectors likewise have two components; one example is the state vector **r** introduced in section 7.2 during the discussion of the phase plane of the pendulum. The first or horizontal component is $\theta(t)$. The second or vertical component is $\omega(t)$. Both are illustrated in figure 8.1.

The solution of the pendulum system (8.1) can be written as $\mathbf{r}(t) = (\theta(t), \omega(t))$, equating the vector with the coordinates of its tip. The ordered-pair notation $(\cdot, \cdot)$ separates the first component from the second. The solution of (8.4) can be written $\mathbf{u}(t) = (u(t), v(t))$, the initial conditions as $\mathbf{u}(0) = (y_0, z_0)$.

Boldface notation distinguishes the vector **u** from the single function u. Both $\mathbf{r}(t)$ and $\mathbf{u}(t)$ are called **solution vectors**. They are vector-valued functions. The components of these vector functions are *scalar*-valued functions; for a given t, the output of a vector function is an ordered pair in two dimensions (or an ordered triple in three dimensions or an ordered n-tuple in n dimensions) of numbers or *scalars*.

A matrix is a rectangular array of numbers or scalar-valued functions. (See section A.5, page 629.) Matrices with two rows and one column, so-called 2×1 matrices or *column vectors*, provide alternative notation for a

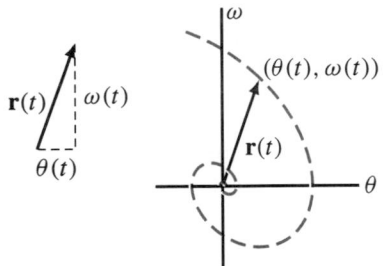

FIGURE 8.1 Two representations of solutions $\theta(t)$ and $\omega(t)$ of the damped pendulum system 8.1, one as defining the horizontal and vertical components of the state vector $\mathbf{r}(t)$ and the other as coordinates of the point $(\theta(t), \omega(t))$ on the solution trajectory at the tip of the state vector.

solution vector, as in

$$\mathbf{r}(t) = \begin{pmatrix} \theta(t) \\ \omega(t) \end{pmatrix}, \quad \mathbf{u}(t) = \begin{pmatrix} u(t) \\ v(t) \end{pmatrix}, \quad \text{or} \quad \mathbf{u}(0) = \begin{pmatrix} y_0 \\ z_0 \end{pmatrix}.$$

Defining the additional matrix (or vector) functions

$$\mathbf{P}(\mathbf{r}) = \begin{pmatrix} \omega(t) \\ -(g/L)\sin\theta(t) - (p/L)\omega(t) \end{pmatrix}, \quad \mathbf{f}(t, \mathbf{r}) = \begin{pmatrix} f(t, u, v) \\ g(t, u, v) \end{pmatrix}$$

permits writing the systems (8.1) and (8.4) in the very compact forms

$$\mathbf{r}'(t) = \mathbf{P}(\mathbf{r}), \quad \mathbf{u}'(t) = \mathbf{f}(t, \mathbf{r}).$$

In the same spirit, the initial-value problem (8.5) can be written $\mathbf{y}'(t) = \mathbf{f}(t, \mathbf{y})$, $\mathbf{y}(0) = \mathbf{y}_0$, where $\mathbf{y}_0 = (y_0, z_0)$.

The power of vector notation is its easy extension to systems of more than two equations. For example, definition 1 can easily be extended to define the solution of an initial-value problem $\mathbf{y}' = \mathbf{f}(t, \mathbf{y})$, $\mathbf{y}(0) = \mathbf{y}_0$, for a system of n equations by letting $\mathbf{y}$ and $\mathbf{f}$ be $n \times 1$ column vectors whose components are functions with the appropriate properties.

The *transpose* of a matrix, denoted ^{T}, is obtained by swapping its rows and columns; e.g., the transpose of a 1×2 *row vector* is a 2×1 column vector; e.g.,

$$\mathbf{r}(t) = \begin{pmatrix} \theta(t) \\ \omega(t) \end{pmatrix} = \begin{pmatrix} \theta(t) & \omega(t) \end{pmatrix}^T. \tag{8.6}$$

It takes less space on the page to write $\mathbf{r}(t) = \begin{pmatrix} \theta(t) & \omega(t) \end{pmatrix}^T$ than to write it as a column vector.

The various forms of notation are used interchangeably: an ordered pair like $(\theta(t), \omega(t))$ to emphasize a point moving on the $\omega\theta$ plane or a vector $\mathbf{r}(t)$, written either as $\mathbf{r}(t) = \begin{pmatrix} \theta(t) & \omega(t) \end{pmatrix}^T$ or as a column vector as in (8.6), to emphasize orientation or distance from the origin.

Stop and Think

8.1 Write the vector that points from the origin up 5 units and to the left 4 units in three different ways: as the coordinates of its tip, as a column vector (matrix), and as the transpose of a row vector (matrix). What is the length of this vector? How far is the point in question from the origin?

8.2 Write the vector that points from the origin to the point with x coordinate $\cos t$ and y coordinate $-\sin t$ in three different ways: as the coordinates of its tip, as a column vector (matrix), and as the transpose of a row vector (matrix). What path does this vector trace over time? (*Hint*: How far is the point in question from the origin?) Of what system of differential equations is this the solution vector? (*Hint*: $(\cos t)' = -\sin t$, $(-\sin t)' = -\cos t$.)

■ **EXAMPLE 1** *Write the spring-mass equation $mx'' + kx = 0$ as a first-order system. Use a solution of the original second-order equation to write a solution of the equivalent system. Write everything in matrix-vector notation.*

Introduce the additional dependent variable $v = x'$; v is the velocity of the mass. Then an equivalent system is

$$x' = v,$$
$$v' = -(k/m)x \qquad (8.7)$$

or $\mathbf{x}'(t) = \mathbf{f}(\mathbf{x})$, where

$$\mathbf{x}(t) = \begin{pmatrix} x(t) \\ v(t) \end{pmatrix} \quad \text{and} \quad \mathbf{f}(\mathbf{x}) = \begin{pmatrix} v(t) \\ -(k/m)x(t) \end{pmatrix}.$$

One solution of the single second-order scalar equation $mx'' + kx = 0$ is $x(t) = \sin\sqrt{k/m}\,t$. Using $v = x'$, a solution of the system is $x(t) = \sin\sqrt{k/m}\,t$ and

$$v(t) = \left(\sin\sqrt{\frac{k}{m}}\,t\right)' = \sqrt{\frac{k}{m}}\,\cos\sqrt{\frac{k}{m}}\,t$$

or

$$\mathbf{x}(t) = \begin{pmatrix} \sin\sqrt{k/m}\,t \\ \sqrt{k/m}\,\cos\sqrt{k/m}\,t \end{pmatrix}$$

or

$$\mathbf{x}(t) = \left(\sin\sqrt{k/m}\,t \quad \sqrt{k/m}\,\cos\sqrt{k/m}\,t \right)^T$$

or

$$\mathbf{x}(t) = \left(\sin\sqrt{k/m}\,t,\ \sqrt{k/m}\,\cos\sqrt{k/m}\,t \right). \quad ■$$

Stop and Think

8.3 The spring-mass system also has the trivial solution $x(t) = 0$, $v(t) = 0$. Write this solution in each of the four forms exhibited in example 1.

8.4 Explain the difference between the zero in the expression $x(t) = 0$ and the zero in $\mathbf{x}(t) = \mathbf{0}$.

■ **EXAMPLE 2**   *Section 2.4 derived the epidemic model*

$$S' = -bIS,$$
$$I' = bIS - rI,$$
$$R' = rI.$$

Write this system in vector notation.

Define the 3×1 column vectors

$$\mathbf{y}(t) = \begin{pmatrix} S(t) \\ I(t) \\ R(t) \end{pmatrix}, \quad \mathbf{f}(\mathbf{y}(t)) = \begin{pmatrix} -bI(t)S(t) \\ -bI(t)S(t) - rI(t) \\ rI(t) \end{pmatrix}.$$

Then these epidemic equations may be written $\mathbf{y}'(t) = \mathbf{f}(\mathbf{y}(t))$. ■

8.1.2 Linear, Homogeneous, Etc.

Informally, the system (8.2–8.3) of two first-order differential equations is *linear* if f and g are linear in the dependent variables; i.e., if f, g can be written in the form

$$f(t, y, z) = k_1(t)y + k_2(t)z + F(t),$$
$$g(t, y, z) = \ell_1(t)y + \ell_2(t)z + G(t).$$

The dependent variables in a linear system can appear raised only to the first power and multiplied only by a function of the independent variable.

The general linear system of two equations,

LINEAR SYSTEM, TWO EQUATIONS

$$\begin{aligned} y' &= k_1(t)y + k_2(t)z + F(t), \\ z' &= \ell_1(t)y + \ell_2(t)z + G(t), \end{aligned} \tag{8.8}$$

can be written as the matrix system

MATRIX FORM OF LINEAR SYSTEM

$$\mathbf{y}' = \mathbf{A}(t)\mathbf{y} + \mathbf{f}(t)$$

by introducing the vector functions

$$\mathbf{y}(t) = \begin{pmatrix} y(t) \\ z(t) \end{pmatrix}, \quad \mathbf{f}(t) = \begin{pmatrix} F(t) \\ G(t) \end{pmatrix},$$

and the matrix function

$$\mathbf{A}(t) = \begin{pmatrix} k_1(t) & k_2(t) \\ \ell_1(t) & \ell_2(t) \end{pmatrix}.$$

The notation $\mathbf{A}(t)\mathbf{y}(t)$ employs the "row times column" definition of matrix multiplication (see section A.5, page 630):

$$\mathbf{A}(t)\mathbf{y}(t) = \begin{pmatrix} k_1(t) & k_2(t) \\ \ell_1(t) & \ell_2(t) \end{pmatrix} \begin{pmatrix} y(t) \\ z(t) \end{pmatrix} = \begin{pmatrix} k_1(t)y(t) + k_2(t)z(t) \\ \ell_1(t)y(t) + \ell_2(t)z(t) \end{pmatrix}.$$

Stop and Think   8.5 Verify the following matrix multiplication:

$$\begin{pmatrix} 1 & 2 \\ 3 & 4 \end{pmatrix} \begin{pmatrix} y(t) \\ z(t) \end{pmatrix} = \begin{pmatrix} y(t) + 2z(t) \\ 3y(t) + 4z(t) \end{pmatrix}.$$

Matrix notation leads directly to the general definition of *linear* system.

Definition 2. *The system of n first-order equations $\mathbf{y}'(t) = \mathbf{F}(t, \mathbf{y}(t))$ is **linear** if and only if it can be written in the form*

LINEAR FIRST-ORDER SYSTEM

$$\mathbf{y}' = \mathbf{A}(t)\mathbf{y} + \mathbf{f}(t)$$

*for some $n \times n$ matrix $\mathbf{A}$ and some $n \times 1$ column vector $\mathbf{f}$, each of which depends only on the independent variable t. Otherwise, the system is **nonlinear**.*

A 2×2 system is *homogeneous* if it has the trivial solution $y \equiv 0, z \equiv 0$.

Trivial solution

Homogeneous system

In general, the **trivial solution** is the vector $\mathbf{y} = \mathbf{0} = \begin{pmatrix} 0 & \cdots & 0 \end{pmatrix}^T$, all of whose components are zero. The system $\mathbf{y}' = \mathbf{f}(t, \mathbf{y})$ is **homogeneous** if and only if it possesses the trivial solution. Otherwise, it is **nonhomogeneous**.

Stop and Think   **8.6** Justify or give a counterexample: the system (8.2–8.3) is homogeneous if and only if $f(t, 0, 0) = 0$ and $g(t, 0, 0) = 0$ for all t.

8.7 Is it correct to say that a system of two first-order equations is homogeneous if and only if $(0, 0)$ is a steady state?

8.8 Show that the linear system $\mathbf{y}'(t) = \mathbf{A}(t)\mathbf{y}(t)$ is homogeneous.

■ **EXAMPLE 3**   *Show that the predator-prey system*

$$V' = -(d_V - \alpha R)V,$$
$$R' = (b_R - \beta V)R$$

is nonlinear and homogeneous. (To avoid confusion with the forcing function $F(t)$ in (8.8) and elsewhere in this section, $V(t)$ denotes the fox (varmint!) population.)

Define

$$\mathbf{y}(t) = \begin{pmatrix} V(t) \\ R(t) \end{pmatrix}, \quad \mathbf{P}(\mathbf{y}) = \begin{pmatrix} -(d_V - \alpha R) & 0 \\ 0 & (b_R - \beta V) \end{pmatrix}.$$

Then this predator-prey system can be written in the form $\mathbf{y}' = \mathbf{P}(\mathbf{y})\mathbf{y}$. However, the coefficient matrix $\mathbf{P}(\mathbf{y})$ is a function of the unknown fox and rabbit populations V and R, violating the definition of linear system.

To test homogeneity, substitute the proposed solution $\mathbf{y} = \mathbf{0} = \begin{pmatrix} 0 & 0 \end{pmatrix}^T$ into $\mathbf{y}' = \mathbf{P}(\mathbf{y})\mathbf{y}$ and test the equality:

$$\mathbf{0}' \; ? =? \; \mathbf{P}(\mathbf{0})\mathbf{0}$$

$$\mathbf{0} \; = \; \mathbf{0} \; \checkmark$$

Hence, the predator-prey system is homogeneous. ■

Stop and Think   **8.9** Write out the details of the matrix calculation $\mathbf{0}' = \mathbf{P}(\mathbf{0})\mathbf{0} = \mathbf{0}$.

8.10 Define a matrix $\mathbf{P}$ so that the epidemic system of example 2 can be written in the form $\mathbf{y}' = \mathbf{P}\mathbf{y}$. Is this system linear or nonlinear? Homogeneous or nonhomogeneous?

Constant-coefficient system

A 2×2 linear system has **constant coefficients** if the coefficients $k_1(t)$, $k_2(t)$, $\ell_1(t)$, $\ell_2(t)$ in (8.8) are constant. Otherwise, it has **variable coefficients**. More generally, the linear system of n equations $\mathbf{y}' = \mathbf{A}\mathbf{y} + \mathbf{f}$ has constant coefficients if the coefficient matrix $\mathbf{A}$ is constant.

■ **EXAMPLE 4**   *Are the pendulum (8.1) and spring-mass (8.7) systems linear or nonlinear, homogeneous or nonhomogeneous, constant or variable coefficient?*

The pendulum system (8.1) is nonlinear because of the term $\sin\theta$. It is homogeneous because substituting the solution pair $(\theta, \omega) = (0, 0)$ reduces both equations to identities. Since *constant coefficient* is defined only for linear systems, that concept has no meaning for this system.

The spring-mass system (8.7) is linear. Either make the identifications $y = x$, $z = v$, $k_1 = 0$, $k_2 = 1$, $\ell_1 = -k/m$, $\ell_2 = 0$ or write it in the matrix-vector form $\mathbf{x}' = \mathbf{A}\mathbf{x}$ with

$$\mathbf{x}(t) = \begin{pmatrix} x(t) \\ v(t) \end{pmatrix}, \quad \mathbf{A} = \begin{pmatrix} 0 & 1 \\ -(k/m) & 0 \end{pmatrix}.$$

It is a constant-coefficient system because each of the coefficients k_1, k_2, ℓ_1, ℓ_2 is constant or because the coefficient matrix $\mathbf{A}$ is constant. It is homogeneous because it possesses the trivial solution $(x, v) = (0, 0)$ or $\mathbf{x} = \mathbf{0}$.

Note that the forced spring-mass system

$$x' = v,$$
$$v' = -(k/m)x - A \sin \pi t$$

is nonhomogeneous because substituting $(x, v) = (0, 0)$ fails to yield equality in *all* equations:

$$0 = 0,$$
$$0 \neq 0 - A \sin \pi t.$$

However, this forced system is linear and has constant coefficients, just like its homogeneous counterpart in the preceding paragraph. ■

Stop and Think   **8.11** Define $\mathbf{f}(t)$ so that this forced system can be written in the form $\mathbf{x}' = \mathbf{A}\mathbf{x} + \mathbf{f}(t)$. Does the definition of the coefficient matrix $\mathbf{A}$ change?

8.1.3   Linear Independence

Linearly independent functions   **Definition 3.** *Two vector functions*

$$\mathbf{y}_1 = \begin{pmatrix} y_1(t) & z_1(t) \end{pmatrix}^T \quad and \quad \mathbf{y}_2 = \begin{pmatrix} y_2(t) & z_2(t) \end{pmatrix}^T$$

*are **linearly independent** on the interval $t_0 \leq t \leq t_1$ if and only if the equality $C_1\mathbf{y}_1 + C_2\mathbf{y}_2 = \mathbf{0}$, $t_0 \leq t \leq t_1$, or component by component,*

$$C_1 y_1(t) + C_2 y_2(t) = 0,$$
$$C_1 z_1(t) + C_2 z_2(t) = 0, \quad t_0 \leq t \leq t_1,$$

*holds in the single case $C_1 = C_2 = 0$. Otherwise, $\mathbf{y}_1$ and $\mathbf{y}_2$ are **linearly dependent**.*

Stop and Think   **8.12** Is it correct to say that two vectors are linearly independent if they are not parallel?

A sum of multiples of two vectors

LINEAR COMBINATION

$$C_1 \mathbf{y}_1 + C_2 \mathbf{y}_2 = \begin{pmatrix} C_1 y_1 + C_2 y_2 \\ C_1 z_1 + C_2 z_2 \end{pmatrix}$$

is called a *linear combination* of $\mathbf{y}_1$ and $\mathbf{y}_2$. Two solutions of a linear system (homogeneous or nonhomogeneous) are *linearly independent* if the only linear combination that yields the trivial solution is the one with $C_1 = C_2 = 0$.

Paralleling the study in chapter 6 of the Wronskian for second-order equations, Cramer's rule reveals that the solutions $\mathbf{y}_1(t)$ and $\mathbf{y}_2(t)$ are linearly independent on $t_0 \le t \le t_1$ if the determinant

WRONSKIAN

$$W(t) = \begin{vmatrix} y_1 & y_2 \\ z_1 & z_2 \end{vmatrix} = \det\left(\mathbf{y}_1(t) \quad \mathbf{y}_2(t) \right)$$

is not zero on that interval. The determinant $W(t)$ is known as the **Wronskian**. It is the determinant of coefficients of the simultaneous equations $C_1\mathbf{y}_1 + C_2\mathbf{y}_2 = \mathbf{0}$ that must be solved for the unknowns C_1, C_2 in order to test linear independence.

As with the Wronskian of two solutions of a single second-order equation, under suitable conditions, either $W(t)$ is identically zero or it is never zero on the interval of definition of the solutions $\mathbf{y}_1$ and $\mathbf{y}_2$. Hence, we have:

Linear independence test.   The solutions $\mathbf{y}_1$ and $\mathbf{y}_2$ are *linearly independent* if their Wronskian $W(t)$ is not zero for at least one point in their interval of definition.

■ **EXAMPLE 5**   *Let*

$$\mathbf{B} = \begin{pmatrix} 0 & 1 \\ 16 & 0 \end{pmatrix}.$$

Solutions of the linear system $\mathbf{y}' = \mathbf{B}\mathbf{y}$ *are*

$$\mathbf{y}_1 = \begin{pmatrix} e^{4t} \\ 4e^{4t} \end{pmatrix}, \quad \mathbf{y}_2 = \begin{pmatrix} e^{-4t} \\ -4e^{-4t} \end{pmatrix}.$$

Show that they are linearly independent on $-\infty < t < \infty$.

To verify linear independence, evaluate the Wronskian

$$W = \det\left(\mathbf{y}_1(t) \quad \mathbf{y}_2(t) \right) = \begin{vmatrix} y_1 & y_2 \\ z_1 & z_2 \end{vmatrix} = \begin{vmatrix} e^{4t} & e^{-4t} \\ 4e^{4t} & -4e^{-4t} \end{vmatrix} = -8 \ne 0.$$

Since $W \ne 0$ for at least one point of $-\infty < t < \infty$ (in fact, for *every* point!), the solutions $\mathbf{y}_1$, $\mathbf{y}_2$ are linearly independent for all t. Verification that $\mathbf{y}_1$, $\mathbf{y}_2$ are indeed solutions of $\mathbf{y}' = \mathbf{B}\mathbf{y}$ is left to the exercises. ■

Any linear combination of two solutions of a linear, homogeneous system is again a solution of that system. To verify this claim, suppose $\mathbf{y}'_1 = \mathbf{A}\mathbf{y}_1$ and $\mathbf{y}'_2 = \mathbf{A}\mathbf{y}_2$. Then test that $C_1\mathbf{y}_1 + C_2\mathbf{y}_2$ solves $\mathbf{y}' = \mathbf{A}\mathbf{y}$:

$$\begin{aligned}
(C_1\mathbf{y}_1 + C_2\mathbf{y}_2)' \ ? &= ? \ \mathbf{A}(C_1\mathbf{y}_1 + C_2\mathbf{y}_2) \\
(C_1\mathbf{y}_1 + C_2\mathbf{y}_2)' &= C_1\mathbf{y}'_1 + C_2\mathbf{y}'_2 \\
&= C_1\mathbf{A}\mathbf{y}_1 + C_2\mathbf{A}\mathbf{y}_2 \\
&= \mathbf{A}(C_1\mathbf{y}_1 + C_2\mathbf{y}_2) \quad \checkmark
\end{aligned}$$

Suppose $\mathbf{y}_1(t) = \left(y_1(t) \quad z_1(t) \right)^T$ and $\mathbf{y}_2(t) = \left(y_2(t) \quad z_2(t) \right)^T$ are linearly independent solutions of the linear, homogeneous system

$$y' = k_1(t)y + k_2(t)z,$$
$$z' = \ell_1(t)y + \ell_2(t)z$$

and that C_1, C_2 are arbitrary constants. Then the **general solution of the homogeneous linear system** or, more briefly, the **homogeneous solution** is a solution $\mathbf{y}_h$ of the form

Homogeneous solution

$$\mathbf{y}_h = \begin{pmatrix} y_h \\ z_h \end{pmatrix} = C_1\mathbf{y}_1 + C_2\mathbf{y}_2 = \begin{pmatrix} C_1 y_1 + C_2 y_2 \\ C_1 z_1 + C_2 z_2 \end{pmatrix}.$$

If the two solutions $\mathbf{y}_1$, $\mathbf{y}_2$ are linearly independent, then the matrix formed from these two column vectors is called a **fundamental matrix** of that system. The corresponding solution vectors are said to form a set of **fundamental solutions** of the system.

Fundamental matrix

$$\mathbf{Y}(t) = \left(\mathbf{y}_1(t) \quad \mathbf{y}_2(t) \right) = \begin{pmatrix} y_1(t) & y_2(t) \\ z_1(t) & z_2(t) \end{pmatrix}.$$

Informally, a fundamental matrix is a general solution of $\mathbf{y}' = \mathbf{A}\mathbf{y}$ without the arbitrary constants; its columns list the *fundamental solutions*, the building blocks of a general solution of the homogeneous system.

The Wronskian is the determinant of the fundamental matrix: $W = \det \mathbf{Y}$. Since the solutions are linearly independent, $\det \mathbf{Y} \neq 0$, a condition that guarantees that the fundamental matrix $\mathbf{Y}$ is *invertible* or *nonsingular*; $\mathbf{Y}^{-1}$ exists. See section A.5, pages 632–634.

A general solution of a homogeneous system is a linear combination of fundamental solutions. Using the fundamental matrix $\mathbf{Y}(t) = \left(\mathbf{y}_1(t) \quad \mathbf{y}_2(t) \right)$, we can write in matrix-vector form a general solution of the homogeneous equation or, using the usual shorter phrase, the

---

Homogeneous solution

$$\mathbf{y}_h(t) = C_1\mathbf{y}_1(t) + C_2\mathbf{y}_2(t) = \mathbf{Y}(t)\mathbf{c},$$

---

where $\mathbf{c}$ is a vector of arbitrary constants,

$$\mathbf{c} = \begin{pmatrix} C_1 \\ C_2 \end{pmatrix}.$$

Stop and Think   **8.13** Write out the details of the matrix multiplication $\mathbf{y}(t)\mathbf{c}$ to exhibit the two components of the homogeneous solution and the two arbitrary constants.

A given solution $\mathbf{y}_p = \begin{pmatrix} y_p & z_p \end{pmatrix}^T$ of a nonhomogeneous linear system

$$y' = k_1(t)y + k_2(t)z + F(t),$$
$$z' = \ell_1(t)y + \ell_2(t)z + G(t)$$

is called a **particular solution**. A **general solution of a nonhomogeneous system** is a solution $\mathbf{y}_g$ of the form

General solution of nonhomogeneous system

$$\mathbf{y}_g = \begin{pmatrix} y_g \\ z_g \end{pmatrix} = \mathbf{y}_p + C_1\mathbf{y}_1 + C_2\mathbf{y}_2 = \begin{pmatrix} y_p + C_1y_1 + C_2y_2 \\ z_p + C_1z_1 + C_2z_2 \end{pmatrix}.$$

The general solution construction is applied in parallel to both components of the solution.

If $\mathbf{y}_p$ is a particular solution of the nonhomogeneous system $\mathbf{y}' = \mathbf{A}(t)\mathbf{y} + \mathbf{f}(t)$, then a general solution is

---

General solution

$$\mathbf{y}_g = \mathbf{y}_p + \mathbf{y}_h = \mathbf{y}_p + C_1\mathbf{y}_1 + C_2\mathbf{y}_2.$$

Stop and Think   **8.14** With $\mathbf{y}_h$, $\mathbf{y}_p$, and $\mathbf{y}_g$ defined as above, complete the indicated equality:

(a) $\mathbf{y}'_h - \mathbf{A}\mathbf{y}_h =$
(b) $\mathbf{y}'_p - \mathbf{A}\mathbf{y}_p =$
(c) $\mathbf{y}'_g - \mathbf{A}\mathbf{y}_g =$

■ **EXAMPLE 6**   *The linear system* $\mathbf{y}' = \mathbf{B}\mathbf{y} + \mathbf{f}(t)$, *where*

$$\mathbf{B} = \begin{pmatrix} 0 & 1 \\ 16 & 0 \end{pmatrix}, \quad \mathbf{f}(t) = \begin{pmatrix} 0 \\ 8t \end{pmatrix},$$

or

$$y' = z,$$
$$z' = 16y + 8t$$

is equivalent to the second-order equation $y'' - 16y = 8t$. *Use that equivalence to find a general solution of this equation.*

Stop and Think   **8.15** Verify that the two forms of this system are equivalent to one another and to $y'' - 16y = 8t$.

Linearly independent solutions of the homogeneous equation $y'' - 16y = 0$ are e^{4t} and e^{-4t}. Using $z = y'$, we can construct from them two pairs of solutions of the homogeneous system $\mathbf{y}' = \mathbf{B}\mathbf{y}$ or

$$y' = z,$$
$$z' = 16y.$$

The resulting linearly independent homogeneous solutions of the system are

$$\mathbf{y}_1 = \begin{pmatrix} e^{4t} \\ 4e^{4t} \end{pmatrix}, \quad \mathbf{y}_2 = \begin{pmatrix} e^{-4t} \\ -4e^{-4t} \end{pmatrix}.$$

The linear independence of these solutions was established in example 5. A general solution of the homogeneous system is

$$\mathbf{y}_h = C_1\mathbf{y}_1 + C_2\mathbf{y}_2.$$

A fundamental matrix for this system is

$$\mathbf{Y} = \begin{pmatrix} \mathbf{y}_1 & \mathbf{y}_2 \end{pmatrix} = \begin{pmatrix} y_1 & y_2 \\ z_1 & z_2 \end{pmatrix} = \begin{pmatrix} e^{4t} & e^{-4t} \\ 4e^{4t} & -4e^{-4t} \end{pmatrix}.$$

To obtain a particular solution of the nonhomogeneous system, note that a particular solution of the equivalent second-order equation $y'' - 16y = 8t$ is $-t/2$. Choosing $y_p = -t/2$ and again using $z = y'$ gives $z_p = -1/2$. Hence, we have the particular solution vector

$$\mathbf{y}_p = \begin{pmatrix} y_p \\ z_p \end{pmatrix} = \begin{pmatrix} -t/2 \\ -1/2 \end{pmatrix}.$$

Combining this solution pair with the homogeneous solution above yields a general solution

$$\mathbf{y}_g = \mathbf{y}_p + C_1\mathbf{y}_1 + C_2\mathbf{y}_2 = \mathbf{y}_p + \mathbf{Yc} = \begin{pmatrix} -t/2 + C_1e^{4t} + C_2e^{-4t} \\ -1/2 + 4C_1e^{4t} - 4C_2e^{-4t} \end{pmatrix}$$

of the original system $\mathbf{y}' = \mathbf{By} + \mathbf{f}(t)$. ■

MATLAB

DELAB can find solutions of first-order systems. For guidance, select **Help, Textbook**, then go to chapter 8, example 6.

Most of the previous discussion has been limited to systems of two equations. The extension to systems of n equations is straightforward. The key definition is linear independence of $n \times 1$ column vectors.

LINEARLY INDEPENDENT VECTORS

Definition 4. *The vector functions* $\mathbf{y}_1(t), \ldots, \mathbf{y}_j(t)$ *are* **linearly independent** *on the interval* $t_0 \le t \le t_1$ *if and only if the single linear combination that is zero throughout the interval,*

$$C_1\mathbf{y}_1(t) + \cdots + C_j\mathbf{y}_j(t) = \mathbf{0}, \quad t_0 \le t \le t_1,$$

is the one with $C_1 = \cdots = C_j = 0$. *Otherwise,* $\mathbf{y}_1, \ldots, \mathbf{y}_j$ *are* **linearly dependent**.

For a linear system of n equations, the Wronskian of a set of n solutions is

WRONSKIAN

$$W(t) = \det \begin{pmatrix} \mathbf{y}_1 & \cdots & \mathbf{y}_n \end{pmatrix}$$

Those solutions are linearly independent if $\det W(t) \neq 0$ for at least one t in the interval upon which they are defined.

If $\mathbf{y}_1, \ldots, \mathbf{y}_n$ are linearly independent solutions of $\mathbf{y}' = \mathbf{A}(t)\mathbf{y} + \mathbf{f}(t)$, then a fundamental matrix is

FUNDAMENTAL MATRIX

$$\mathbf{Y}(t) = \begin{pmatrix} \mathbf{y}_1 & \cdots & \mathbf{y}_n \end{pmatrix},$$

the matrix whose i-th column is $\mathbf{y}_i(t)$, $i = 1, \ldots, n$. A general solution of the homogeneous system $\mathbf{y}' = \mathbf{A}(t)\mathbf{y}$ is

HOMOGENEOUS SOLUTION

$$\mathbf{y}_h = \mathbf{Y}\mathbf{c},$$

where $\mathbf{c}$ is a vector of arbitrary constants $\begin{pmatrix} C_1 & \cdots & C_n \end{pmatrix}^T$. If $\mathbf{y}_p$ is a particular solution of the nonhomogeneous system $\mathbf{y}' = \mathbf{A}(t)\mathbf{y} + \mathbf{f}(t)$, then a *general solution* is

GENERAL SOLUTION

$$\mathbf{y}_g = \mathbf{y}_p + \mathbf{Y}\mathbf{c}.$$

A **principle of superposition** can also be stated for linear systems but that task is left to the exercises. Of course, superposition is the key to showing that a general solution is indeed a solution!

8.1.4 Exercises

EXERCISE GUIDE	
To gain experience . . .	**Try exercises**
Writing second-order equations as systems	1, 3–4, 16–18
With linear versus nonlinear systems	2, 5–6
With homogeneous versus nonhomogeneous systems	2, 5–6, 9–10
With constant- versus variable-coefficient systems	2, 5–6
Verifying solutions of systems	7, 8(a), 11–12, 16, 17–18(e, g)
With linear independence	11, 16(b), 19–21, 27
Using general solutions and fundamental solutions	8(b), 11, 17–18(e, h, i), 27
With steady states	10
Using and developing superposition	12–15, 22
With generalizations to higher order	24–26

1. Write each of the following second-order equations as an equivalent first-order system. If initial conditions are given, write an equivalent initial-value problem for the system. When appropriate, give a physical interpretation of the new variable you introduce. Classify each of the resulting systems as linear or nonlinear, homogeneous or nonhomogeneous, and variable or constant coefficient, if appropriate.

(a) $4y'' - 16y' + 2y = 0$

(b) $4y'' - 16y' + 2y = e^{2t}$, $y(0) = 1$, $y'(0) = -2$

(c) $4y'' + 16y'(1 - y^2) = y'$

(d) The damped spring-mass equation $mx'' + px' + kx = 0$, $x(0) = x_0$, $x'(0) = v_0$.

(e) The forced, undamped spring-mass equation $mx'' + kx = A \sin \omega t$.

(f) The series RLC equation $Lq'' + Rq' + q/C = E(t)$.

(g) The approximate pendulum equation $L\theta'' + g\theta = 0$, $\theta(0) = \theta_0$, $\theta'(0) = \omega_0$.

(h) The damped pendulum equation $L\theta'' + p\theta' + g \sin \theta = 0$.

(i) The generic linear second-order equation $a_2 y'' + a_1 y' + a_0 y = h(t)$.

2. Classify each of the following systems as linear or nonlinear, homogeneous or nonhomogeneous, and variable or constant coefficient (if appropriate).

(a) The unsimplified SIR system

$$S' = -bIS,$$
$$I' = bIS - rI,$$
$$R' = rI.$$

(b) The SIR system

$$S' = -bIS,$$
$$I' = bIS - rI.$$

(c) The unsimplified SIRS system

$$S' = -bIS + gR,$$
$$I' = bIS - rI,$$
$$R' = rI - gR.$$

(d) The SIRS system

$$S' = -bIS + g(P - S - I),$$
$$I' = bIS - rI.$$

(e) The competition system

$$y' = y(1 - y - ez),$$
$$z' = rz(1 - z - fy).$$

(f) The autocatalytic chemical reaction system

$$x' = x^2 y - x + b,$$
$$y' = a - x^2 y.$$

(g) The modified predator-prey model

$$F' = -(d_F - \alpha R)F - s_F F,$$
$$R' = (b_R - \beta F)R - (aR - c)R.$$

(h) The limit cycle system

$$y' = z + y(1 - y^2 - z^2),$$
$$z' = -y + z(1 - y^2 - z^2).$$

(i) The van der Pol system

$$i' = -\epsilon(i^2/3 - 1)i - q,$$
$$q' = i.$$

(j) The general system

$$u' = -g(u) - v,$$
$$v' = u,$$

with $g(0) = 0$ and with $g(0) \neq 0$.

3. Write a first-order system that is equivalent to the generic second-order equation $y'' = f(t, y, y')$.

4. Verify that the linear second-order equation

$$a_2(t)y'' + a_1(t)y' + a_0(t)y = h(t)$$

is equivalent to the first-order system

$$y' = z,$$
$$a_2(t)z' = -a_0(t)y - a_1(t)z + h(t)$$

as follows:

(a) Suppose $y(t)$ is a solution of the second-order equation. Show that $(y(t), y'(t))$ solves the first-order system.

(b) Suppose $(y(t), z(t))$ solves the first-order system. Show that $y(t)$ solves the second-order equation.

5. Classify each of the following equations as
   (i) Linear or nonlinear.
   (ii) Homogeneous or nonhomogeneous.
   (iii) Constant or variable coefficient.

If necessary, convert second-order equations to first-order systems.

(a) $4y'' - 16y' + 2y = 0$

(b) $4y'' - 16y' + 2y = e^{2t}$, $y(0) = 1$, $y'(0) = -2$

(c) $4y'' + 16y'(1 - y^2) = y'$

(d) The damped spring-mass equation $mx'' + px' + kx = 0$.

(e) The forced, damped spring-mass equation $mx'' + px' + kx = A \sin \omega t$.

(f) The forced, *undamped* spring-mass equation $mx'' + kx = A \sin \omega t$.

(g) The predator-prey system
$$F' = -(d_F - \alpha R)F,$$
$$R' = (b_R - \beta F)R.$$

(h) The damped pendulum system
$$\theta' = \omega,$$
$$L\omega' = -g \sin \theta - p\omega.$$

(i) The forced pendulum equation $L\theta'' + p\theta' + g \sin \theta = A \sin \pi t$.

(j) The series RLC equation $Lq'' + Rq' + q/C = E(t)$.

(k) The approximate pendulum equation $L\theta'' + g\theta = 0$.

6. Example 6 considers the systems
$$y' = z,$$
$$z' = 16y + 8t$$
and
$$y' = z,$$
$$z' = 16y.$$
Verify that both systems are linear, that the first is nonhomogeneous, that the second is homogeneous, and that both have constant coefficients.

7. Verify by substitution that
$$\mathbf{y}_1 = \begin{pmatrix} e^{4t} \\ 4e^{4t} \end{pmatrix}, \quad \mathbf{y}_2 = \begin{pmatrix} e^{-4t} \\ -4e^{-4t} \end{pmatrix}$$
are both solutions of the system
$$y' = z,$$
$$z' = 16y,$$
as claimed in example 6. Explain how you could find a solution of the equivalent second-order equation $y'' - 16y = 0$ from each of these solutions.

8. Example 6 obtains the general solution
$$\mathbf{y}_g = \begin{pmatrix} -t/2 + C_1 e^{4t} + C_2 e^{-4t} \\ -1/2 + 4C_1 e^{4t} - 4C_2 e^{-4t} \end{pmatrix}$$
of the linear system
$$y' = z,$$
$$z' = 16y + 8t.$$

(a) Verify by direct substitution that $\mathbf{y}_g$ is indeed a solution of this system.

(b) Use this general solution to solve this system subject to the initial conditions $\mathbf{y}(0) = \begin{pmatrix} 2 & 3/2 \end{pmatrix}^T$.

9. Show that the linear system
$$y' = k_1(t)y + k_2(t)z + F(t),$$
$$z' = \ell_1(t)y + \ell_2(t)z + G(t)$$
is homogeneous if and only if $F = G = 0$.

10. Show that a homogeneous system has the steady state $(0, 0)$.

11. For the system
$$y' = 2z,$$
$$z' = -2y,$$

(i) Verify that each of the following vector functions is a solution of the system.

(ii) Determine which pairs are linearly independent.

(iii) Use each linearly independent pair to write a general solution of this system.

(iv) Use each linearly independent pair to write a fundamental matrix of this system.

(a) $\mathbf{y}_1 = \begin{pmatrix} \sin 2t & \cos 2t \end{pmatrix}^T$

(b) $\mathbf{y}_2 = \begin{pmatrix} \cos 2(t - \pi/2) & \sin 2(t - \pi/2) \end{pmatrix}^T$

(c) $\mathbf{y}_3 = \begin{pmatrix} \sin 2(t - \pi/4) & \cos 2(t - \pi/4) \end{pmatrix}^T$

12. Suppose $\mathbf{y}_1$ and $\mathbf{y}_2$ are solutions of the homogeneous system
$$y' = k_1(t)y + k_2(t)z,$$
$$z' = \ell_1(t)y + \ell_2(t)z.$$
Verify by substitution that the combination $C_1\mathbf{y}_1 + C_2\mathbf{y}_2$ is also a solution of the homogeneous system. Does it matter whether $\mathbf{y}_1$ and $\mathbf{y}_2$ are linearly independent?

13. The text claims

Any linear combination of two solutions of a linear, homogeneous system is again a solution of that system.

(a) Prove that this claim is correct.

(b) Is the result of this exercise the same as the result of exercise 12? Explain.

14. (a) Show that a constant multiple of a solution of a linear homogeneous system is again a solution of that system.

(b) Is this result different from the result of the preceding two exercises or is it a special case? Explain.

15. Suppose y_h is a solution of the homogeneous system

$$y' = k_1(t)y + k_2(t)z,$$
$$z' = \ell_1(t)y + \ell_2(t)z,$$

and that y_p is a particular solution of the nonhomogeneous system

$$y' = k_1(t)y + k_2(t)z + F(t),$$
$$z' = \ell_1(t)y + \ell_2(t)z + G(t).$$

Verify by substitution that the combination $y_g = y_p + y_h$ is also a solution of the nonhomogeneous system. Does it matter whether or not y_h is nontrivial?

16. Example 1 uses the solution $x = \sin\sqrt{k/m}\,t$ of the spring-mass equation $mx'' + kx = 0$ to construct a solution of the equivalent system

$$x' = v,$$
$$v' = -(k/m)x.$$

(a) Find a solution of $mx'' + kx = 0$ that is linearly independent of $\sin\sqrt{k/m}\,t$. Use it to construct another solution of this system.

(b) Show that the solution you construct in part (a) and the solution

$$\left(\,x \quad v\,\right)^T = \left(\,\sin\sqrt{\frac{k}{m}}\,t \quad \sqrt{\frac{k}{m}}\cos\sqrt{\frac{k}{m}}\,t\,\right)^T$$

obtained in the text are linearly independent.

(c) Write a fundamental matrix for this system.

17. (a) Find a nontrivial solution of the second-order homogeneous equation $x'' + 4x = 0$.

(b) Write this second-order equation as a first-order system. Use the solution you just found to write a solution of this first-order system.

(c) Find a *general* solution of $x'' + 4x = 0$.

(d) Use the general solution you just found to write a general solution of the equivalent first-order system of part (b).

(e) Verify that you have a general solution of the system. Do you evaluate a determinant that looks like the Wronskian of two solutions of the second-order equation $x'' + 4x = 8t$?

(f) Find a particular solution of $x'' + 4x = 8t$.

(g) Write $x'' + 4x = 8t$ as a first-order system. Use the particular solution you just found to write a particular solution of this first-order system.

(h) Combine the analysis of the preceding steps to write a general solution of the nonhomogeneous system.

(i) Use the preceding analysis to write a fundamental matrix for this system. Write a general solution using the fundamental matrix.

18. Repeat exercise 17 for the homogeneous equation $x'' - 9x = 0$ and the nonhomogeneous equation $x'' - 9x = -10\sin t$.

19. The text claims, "Cramer's rule reveals that the solutions $y_1(t)$ and $y_2(t)$ are linearly independent on $t_0 \le t \le t_1$ if the determinant

$$\begin{vmatrix} y_1 & y_2 \\ z_1 & z_2 \end{vmatrix}$$

is not zero on that interval." Prove this assertion.

20. The text states that the linear independence of the solutions $y_1(t)$ and $y_2(t)$ of a linear system can be tested by examining the determinant

$$\begin{vmatrix} y_1 & y_2 \\ z_1 & z_2 \end{vmatrix}.$$

Show that if the linear system was obtained from a linear, second-order equation, than this determinant is just the Wronskian of two solutions of the second-order equation.

21. (a) One solution of a linear, homogeneous system satisfies the initial conditions $y_1(0) = \left(\,1 \quad 0\,\right)^T$. Another satisfies the initial conditions $y_2(0) = \left(\,0 \quad 1\,\right)^T$. Show that the two solutions must be linearly independent.

(b) Give a general relation between two sets of initial conditions that will guarantee that the corresponding solutions of a homogeneous system are linearly independent.

22. State a *principle of superposition* for linear systems. It might begin

Let $\mathbf{y}_j$ be solutions of the nonhomogeneous system

$$y' = k_1(t)y + k_2(t)z + F_j(t),$$
$$z' = \ell_1(t)y + \ell_2(t)z + G_j(t)$$

for $j = 1, 2$. Then for any constants C_1, C_2, the combination $\mathbf{y}_c = C_1\mathbf{y}_1 + C_2\mathbf{y}_2 \ldots$.

23. The system

$$y' = -y/2 + 3z/2,$$
$$z' = 3y/2 - z/2$$

has the solutions $\mathbf{y}_1 = \begin{pmatrix} e^t & e^t \end{pmatrix}^T$, $\mathbf{y}_2 = \begin{pmatrix} e^{-2t} & -e^{-2t} \end{pmatrix}^T$. Classify this system as linear or nonlinear, homogeneous or nonhomogeneous, constant or variable coefficient. Are these solutions linearly independent? If appropriate and possible, write both a general solution and a fundamental matrix for this system. Write this system and these solutions in vector notation as well.

24. The text suggests in passing that an equation of *any* order can be reduced to a first-order system.

    (a) By introducing the variables $y_1 = u$, $y_2 = u'$, and $y_3 = u''$, find a system of three first-order equations in y_1, y_2, y_3 that are equivalent to the linear third-order equation

    $$b_3 u^{(3)} + b_2 u'' + b_1 u' + b_0 u = h(t).$$

    (b) Repeat these same steps for the generic third-order equation

    $$u^{(3)} = f(t, u, u', u'').$$

    (c) Does one equation include the other as a special case?

25. Repeat exercise 24 for the fourth-order equations

    $$b_4 u^{(4)} + b_3 u^{(3)} + b_2 u'' + b_1 u' + b_0 u = h(t)$$

    and

    $$u^{(4)} = f(t, u, u', u'', u^{(3)}).$$

26. Repeat exercise 24 for the *n*-th order equations

    $$b_n u^{(n)} + \cdots + b_2 u'' + b_1 u' + b_0 u = h(t)$$

    and

    $$u^{(n)} = f(t, u, u', u'', \ldots, u^{(n-1)}).$$

8.2 ■ CONSTANT-COEFFICIENT, HOMOGENEOUS SYSTEMS

We seek a general solution of the generic constant-coefficient, homogeneous linear system $\mathbf{y}' = \mathbf{B}\mathbf{y}$, $\mathbf{B}$ a constant matrix. Written explicitly in the case of two first-order equations, the problem is

$$y' = ay + bz,$$
$$z' = cy + dz,$$

or

$$\mathbf{y}' = \mathbf{B}\mathbf{y}, \quad \text{with} \quad \mathbf{y} = \begin{pmatrix} y \\ z \end{pmatrix} \quad \text{and} \quad \mathbf{B} = \begin{pmatrix} a & b \\ c & d \end{pmatrix}.$$

These are the simplest nontrivial systems of first-order equations, and they arise naturally in the local analysis of the stability of equilibria of nonlinear systems.

Earlier experience with constant-coefficient equations immediately suggests guessing exponential solutions e^{rt} and substituting to find the parameter r. That approach works again, but the need to find a solution *vector*—a solution with two (or more) components—requires finding coefficients for the exponential as well. Those coefficients are determined by the properties of $\mathbf{B}$; they form a vector (one coefficient for each component of the solution) whose geometric characteristics determine the phase plane picture of this equation and, hence, the local behavior of any nonlinear system of which it is a valid linear approximation.

8.2.1 Motivation

Example 6, page 374, of section 8.1 constructed the linearly independent solutions

$$\mathbf{y}_1 = \begin{pmatrix} e^{4t} \\ 4e^{4t} \end{pmatrix}, \quad \mathbf{y}_2 = \begin{pmatrix} e^{-4t} \\ -4e^{-4t} \end{pmatrix}$$

of the homogeneous, constant-coefficient system

$$\mathbf{y}' = \mathbf{B}\mathbf{y} \quad \text{with} \quad \mathbf{B} = \begin{pmatrix} 0 & 1 \\ 16 & 0 \end{pmatrix}. \tag{8.9}$$

That example started from the equivalent second-order scalar equation $y'' - 16y = 0$. Can we develop an approach that does not rely on an equivalent second-order equation, a crutch that may not always be available? Specifically, can we extend the ideas of characteristic equations to first-order systems?

These solutions certainly involve exponentials of the form e^{rt}. For example, the solution $\mathbf{y}_1$ involves e^{4t}—obviously e^{rt} with $r = 4$—but the two components of this solution have different coefficients,

$$\mathbf{y}_1 = \begin{pmatrix} e^{4t} \\ 4e^{4t} \end{pmatrix} = \begin{pmatrix} 1 \\ 4 \end{pmatrix} e^{4t}.$$

Since the exponential is the same but the coefficients of the two components are different, a reasonable general form for the solution might be

$$\mathbf{y} = \begin{pmatrix} y \\ z \end{pmatrix} = \begin{pmatrix} pe^{rt} \\ qe^{rt} \end{pmatrix} = \begin{pmatrix} p \\ q \end{pmatrix} e^{rt} = \mathbf{p}e^{rt}, \tag{8.10}$$

where the scalar coefficients p, q (or the vector coefficient $\mathbf{p}$) and the growth rate r are constants to be determined by substituting into the system. The following example explores this process.

Stop and Think   **8.16** To test your understanding of the notation, write out the scalar equations for the components: $y = \cdots$, $z = \cdots$.

■ **EXAMPLE 7**   *Use the assumed solution form* $\mathbf{y} = \mathbf{p}e^{rt}$ *defined in (8.10) to find linearly independent solutions of the homogeneous, constant-coefficient system (8.9).*

Substituting the guess $\mathbf{y} = \mathbf{p}e^{rt}$ into the system yields the two scalar equations

Substitute $\mathbf{y} = \mathbf{p}e^{rt}$.

$$rpe^{rt} = qe^{rt},$$
$$rqe^{rt} = 16pe^{rt}.$$

Dividing by the exponential and collecting terms leaves

Equations for p, q in terms of r

$$-rp + q = 0, \tag{8.11}$$
$$16p - rq = 0. \tag{8.12}$$

For any value of r, this system always has the trivial solution $p = q = 0$. But choosing $p = q = 0$ gives only $\mathbf{y} = \mathbf{0}$, the trivial solution of the differential equation, hardly a building block for a general solution.

Note that the system of simultaneous equations (8.11–8.12) can be written compactly in matrix notation by using the 2×2 *identity matrix*

2 × 2 identity matrix

$$\mathbf{I} = \begin{pmatrix} 1 & 0 \\ 0 & 1 \end{pmatrix}.$$

Since

$$\mathbf{B} - r\mathbf{I} = \begin{pmatrix} 0 & 1 \\ 16 & 0 \end{pmatrix} - r \begin{pmatrix} 1 & 0 \\ 0 & 1 \end{pmatrix} = \begin{pmatrix} -r & 1 \\ 16 & -r \end{pmatrix},$$

$(\mathbf{B} - r\mathbf{I})\mathbf{p} = \mathbf{0}$ is equivalent to (8.11–8.12); $(\mathbf{B} - r\mathbf{I})$ is just the coefficient matrix of those equations.

The key question!

Here is the key question: Are there values of r for which the pair of equations (8.11–8.12) can be solved for *nontrivial p and q*? Trying to solve for p using Cramer's rule (see section A.6, page 634) leads to

$$p \det(\mathbf{B} - r\mathbf{I}) = \begin{vmatrix} 0 & 1 \\ 0 & -r \end{vmatrix} = 0,$$

where $\det(\mathbf{B} - r\mathbf{I})$ is the determinant of coefficients,

$$\det(\mathbf{B} - r\mathbf{I}) = \begin{vmatrix} -r & 1 \\ 16 & -r \end{vmatrix} = r^2 - 16.$$

To argue without appealing to Cramer, solve (8.11–8.12) for p by multiplying (8.11) by $-r$, (8.12) by -1, and adding. The result is $(r^2 - 16)p = \det(\mathbf{B} - r\mathbf{I})\,p = 0$.

If $\det(\mathbf{B} - r\mathbf{I}) \neq 0$, then $p = 0/\det(\mathbf{B} - r\mathbf{I}) = 0$, the unwanted trivial solution. If $\det(\mathbf{B} - r\mathbf{I}) = 0$, then $0 \cdot p = 0$, an indeterminate form that permits a nontrivial solution $p \neq 0$.

Requiring that the determinant of coefficients $\det(\mathbf{B} - r\mathbf{I})$ *be zero* gives an equation for r,

Characteristic equation by requiring nontrivial p, q

$$\det(\mathbf{B} - r\mathbf{I}) = r^2 - 16 = 0,$$

the *characteristic equation* of this system of differential equations. Of course, $r^2 - 16 = 0$ is the characteristic equation of the equivalent second-order equation $y'' - 16y = 0$ as well.

The solutions of the characteristic equation $r^2 - 16 = 0$ are

Characteristic values

$$r = \pm 4.$$

The corresponding values of p and q must satisfy (8.11–8.12) with either $r = 4$ or $r = -4$.

When $r = 4$, the pair of equations (8.11–8.12) becomes

Equations for p, q when $r = 4$

$$\begin{aligned} -4p + q &= 0, \\ 16p - 4q &= 0. \end{aligned}$$

But the second equation is redundant; it is the first multiplied by -4. Hence, we can determine p and q only to within a constant multiple:

q in terms of p when r = 4

$$q = 4p, \quad p \text{ arbitrary.}$$

Arbitrarily choosing $p = p_1 = 1$ (any nonzero value will do) yields $q_1 = 4$ or $\mathbf{p}_1 = \begin{pmatrix} 1 & 4 \end{pmatrix}^T$. The corresponding solution of the system of differential equations (8.9) is

Solution for r = 4

$$\mathbf{y}_1 = \mathbf{p}_1 e^{4t} = \begin{pmatrix} 1 \\ 4 \end{pmatrix} e^{4t} = \begin{pmatrix} e^{4t} \\ 4e^{4t} \end{pmatrix}.$$

Similarly, the values of p and q corresponding to the second characteristic value $r = -4$ must satisfy

Equations for p, q when r = −4

$$4p + \quad q = 0,$$
$$16p + 4q = 0.$$

The second equation is again a multiple of the first, but now

q in terms of p when r = −4

$$q = -4p, \quad p \text{ arbitrary.}$$

Choosing $p = p_2 = 1$ leads to $q_2 = -4$ or $\mathbf{p}_2 = \begin{pmatrix} 1 & -4 \end{pmatrix}^T$. The corresponding solution of the differential equations (8.9) is

Solution for r = −4

$$\mathbf{y}_2 = \mathbf{p}_2 e^{-4t} = \begin{pmatrix} 1 \\ -4 \end{pmatrix} e^{-4t} = \begin{pmatrix} e^{-4t} \\ -4e^{-4t} \end{pmatrix}.$$

Example 5 verified the linear independence of these solutions by showing that their Wronskian $W(t) = \det \begin{pmatrix} \mathbf{y}_1 & \mathbf{y}_2 \end{pmatrix}$ is nonzero.

A general solution of the homogeneous system (8.9) is

$$\mathbf{y}_g = C_1 \mathbf{y}_1 + C_2 \mathbf{y}_2 = C_1 \begin{pmatrix} 1 \\ 4 \end{pmatrix} e^{4t} + C_2 \begin{pmatrix} 1 \\ -4 \end{pmatrix} e^{-4t}. \quad ■$$

Based on this one example, we define the *characteristic equation* of the constant-coefficient system $\mathbf{y}' = \mathbf{B}\mathbf{y}$ to be

CHARACTERISTIC EQUATION
OF $\mathbf{y}' = \mathbf{B}\mathbf{y}$

$$\det(\mathbf{B} - r\mathbf{I}) = 0.$$

The validity of this definition will become clear.

8.2.2 Geometry: Eigenvalues and Eigenvectors

The general solution developed in the last example has the form

Any solution is a combination of two vectors.

$$\mathbf{y}_g = C_1\mathbf{p}_1 e^{4t} + C_2\mathbf{p}_2 e^{-4t}.$$

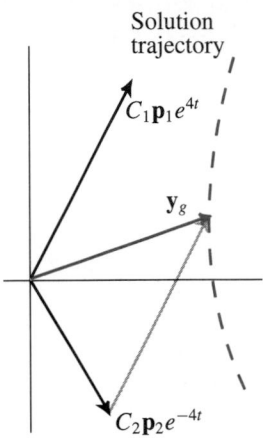

FIGURE 8.2 The general solution vector of (8.9) is the sum of two components whose directions are determined by $\mathbf{p}_1$ and $\mathbf{p}_2$.

Suppose C_1 and C_2 have been determined by initial conditions. Then at any instant of time, the location of the solution in the phase plane could be determined by combining two vectors. *Any solution is a combination of these two vectors!* Their directions are $\mathbf{p}_1$ and $\mathbf{p}_2$, and their lengths are $C_1 e^{4t}$ and $C_2 e^{-4t}$, respectively, as shown in figure 8.2.

The vectors $\mathbf{p}_1$ and $\mathbf{p}_2$ evidently play a key role in determining the behavior of the system. They also have a natural geometric interpretation in terms of the coefficient matrix of the system $\mathbf{y}' = \mathbf{B}\mathbf{y}$.

To see that connection, substitute one of these solution vectors, say, $\mathbf{y}_2 = \mathbf{p}_2 e^{-4t}$, into $\mathbf{y}' = \mathbf{B}\mathbf{y}$ and cancel the exponential factor:

$$\mathbf{y}'_2 = -4\mathbf{p}_2 e^{-4t} = \mathbf{B}\mathbf{p}_2 e^{-4t},$$

$$-4\mathbf{p}_2 = \mathbf{B}\mathbf{p}_2,$$

$$-4\begin{pmatrix} 1 \\ -4 \end{pmatrix} = \begin{pmatrix} 0 & 1 \\ 16 & 0 \end{pmatrix}\begin{pmatrix} 1 \\ -4 \end{pmatrix} = \begin{pmatrix} -4 \\ 16 \end{pmatrix}.$$

Evidently, $\mathbf{B}$ and $\mathbf{p}_2$ enjoy a special relationship: $\mathbf{B}\mathbf{p}_2$ is parallel to $\mathbf{p}_2$ (but four times longer and in the opposite direction, due to the factor -4).

Stop and Think **8.17** Show that $\mathbf{B}$ and $\mathbf{p}_1 = \begin{pmatrix} 1 & 4 \end{pmatrix}^T$ are related in a similar way: $\mathbf{B}\mathbf{p}_1$ is parallel to $\mathbf{p}_1$ but the change in length is 4.

This relationship arises in solving *any* constant-coefficient, homogeneous, linear system $\mathbf{y}' = \mathbf{B}\mathbf{y}$. Begin with the solution form

Form of characteristic equation solution

$$\mathbf{y} = \mathbf{p}e^{rt}.$$

Since $\mathbf{y}' = r\mathbf{p}e^{rt}$, substituting in $\mathbf{y}' = \mathbf{B}\mathbf{y}$ leads to

Substitute y.

$$r\mathbf{p}e^{rt} = \mathbf{B}\mathbf{p}e^{rt}.$$

Canceling the nonzero exponential leaves

Divide by e^{rt}.

$$\mathbf{B}\mathbf{p} = r\mathbf{p}.$$

Solving $\mathbf{B}\mathbf{p} = r\mathbf{p}$ is asking for a (nonzero) vector $\mathbf{p}$ whose *direction* is unaffected when it is multiplied by the matrix $\mathbf{B}$, while its *magnitude* is altered by the factor r. The nonzero vector $\mathbf{p}$ is called a **characteristic vector** or **eigenvector** of the matrix $\mathbf{B}$, and r is its corresponding **characteristic value** or **eigenvalue**.

Using the identity matrix $\mathbf{I}$, this important relation can be written

---

Eigenvalue-eigenvector relation

$$\mathbf{Bp} = r\mathbf{p} \quad \text{or} \quad (\mathbf{B} - r\mathbf{I})\mathbf{p} = \mathbf{0}.$$

Finding the eigenvalues of the matrix $\mathbf{B}$ requires solving the *homogeneous* system of linear equations $(\mathbf{B} - r\mathbf{I})\mathbf{p} = \mathbf{0}$. That system *always* has the solution $\mathbf{p} \equiv 0$. From Cramer's rule, it can have nontrivial solutions $\mathbf{p} \neq 0$ *only when the determinant of coefficients is zero*—i.e., only when

---

Characteristic equation

$$\det(\mathbf{B} - r\mathbf{I}) = 0.$$

The solutions of the **characteristic equation** of the matrix $\mathbf{B}$ are the eigenvalues of $\mathbf{B}$. This expression is also the *characteristic equation of the system of differential equations* $\mathbf{y}' = \mathbf{By}$ because $\mathbf{B} - r\mathbf{I}$ is just the determinant of coefficients of the system defining $\mathbf{p} = \begin{pmatrix} p & q \end{pmatrix}^T$.

> The condition that leads to the characteristic equation is the complement of that encountered in studying the Wronskian. Here we want nontrivial solutions; there having only the trivial solution ensures linear independence.
>
> The Wronskian is the determinant of the fundamental matrix $\mathbf{Y}$. The columns of $\mathbf{Y}(t)$ are linearly independent if and only if the system of equations $\mathbf{Y}(t)\mathbf{c} = \mathbf{0}$ has only the trivial solution $\mathbf{c} = \mathbf{0}$. That system has only the trivial solution if $W = \det \mathbf{Y} \neq 0$. In contrast, we require here that $\det(\mathbf{B} - r\mathbf{I}) = 0$ to obtain *non*trivial solutions in addition to the trivial solution.

■ **EXAMPLE 8** *Write the eigenvalue-eigenvector relation and the characteristic equation for the matrix*

$$\mathbf{B} = \begin{pmatrix} -1/2 & 3/2 \\ 3/2 & -1/2 \end{pmatrix}.$$

Let $\mathbf{p} = \begin{pmatrix} p & q \end{pmatrix}^T$. Then the relation $\mathbf{Bp} = r\mathbf{p}$ is

$$\begin{pmatrix} -1/2 & 3/2 \\ 3/2 & -1/2 \end{pmatrix} \begin{pmatrix} p \\ q \end{pmatrix} = r \begin{pmatrix} p \\ q \end{pmatrix}$$

$$\begin{pmatrix} (-1/2 - r)p & 3q/2 \\ 3p/2 & (-1/2 - r)q \end{pmatrix} = \begin{pmatrix} 0 \\ 0 \end{pmatrix}$$

$$(\mathbf{B} - r\mathbf{I})\mathbf{p} = \mathbf{0}.$$

The characteristic equation of $\mathbf{B}$ comes from the requirement that this homogeneous system have nonzero solutions. It is $\det(\mathbf{B} - r\mathbf{I}) = (-1/2 - r)^2 - 9/4 = 0$. ■

Stop and Think   **8.18** Is it correct to say that an eigenvector **p** of **B** is a vector with the property that **p** and **Bp** are linearly dependent?

■ **EXAMPLE 9**   *Write the eigenvalue-eigenvector relation and the characteristic equation for the general 2×2 constant-coefficient matrix*

$$\mathbf{B} = \begin{pmatrix} a & b \\ c & d \end{pmatrix}.$$

Write $\mathbf{p} = \begin{pmatrix} p & q \end{pmatrix}^T$. Then the relation $\mathbf{Bp} = r\mathbf{p}$ is

$$\begin{pmatrix} a & b \\ c & d \end{pmatrix} \begin{pmatrix} p \\ q \end{pmatrix} = r \begin{pmatrix} p \\ q \end{pmatrix}$$

$$\begin{pmatrix} (a-r)p & bq \\ cp & dq \end{pmatrix} = \begin{pmatrix} 0 \\ 0 \end{pmatrix}$$

$$(\mathbf{B} - r\mathbf{I})\mathbf{p} = \mathbf{0}.$$

The characteristic equation of **B**, which arises from the requirement that this homogeneous system have nontrivial solutions, is

$$\det(\mathbf{B} - r\mathbf{I}) = (a-r)(d-r) - bc = 0. \quad ■$$

The next two examples demonstrate the connection between eigenvalues and eigenvectors of **B** and the solutions of $\mathbf{y}' = \mathbf{By}$. The first finds the eigenvalues and eigenvectors of a matrix **B** from its characteristic equation. The second finds a general solution of $\mathbf{y}' = \mathbf{By}$ using that information.

■ **EXAMPLE 10**   *Find the eigenvalues and eigenvectors of the matrix*

$$\mathbf{B} = \begin{pmatrix} -1/2 & 3/2 \\ 3/2 & -1/2 \end{pmatrix}.$$

Example 8 found the characteristic equation or *characteristic polynomial* of the matrix **B**,

Characteristic equation of coefficient matrix

$$\det(\mathbf{B} - r\mathbf{I}) = (-1/2 - r)^2 - 9/4$$
$$= r^2 + r - 2 = (r+2)(r-1) = 0.$$

The roots of this polynomial are $r = -2, 1$; that is, the eigenvalues of the matrix **B** are $r = -2, 1$.

To find the eigenvector corresponding to the eigenvalue $r_1 = 1$, return to the defining relation $(\mathbf{B} - r\mathbf{I})\mathbf{p} = \mathbf{0}$, substitute $r_1 = 1$, and solve for the components p, q of the eigenvector $\mathbf{p}_1$:

Substitute first eigenvalue to find eigenvector.

$$(\mathbf{B} - \mathbf{I})\mathbf{p}_1 = \begin{pmatrix} -3/2 & 3/2 \\ 3/2 & -3/2 \end{pmatrix} \begin{pmatrix} p \\ q \end{pmatrix} = \mathbf{0}.$$

Multiplying by 2/3 to clear the fractions leaves two equations,

$$-p + q = 0,$$
$$p - q = 0,$$

that are not independent; one of the unknowns is free to vary.

From the first equation,

$$q = p.$$

Arbitrarily choosing $p = 1$ yields $q = 1$. (The choice of p is arbitrary because an eigenvector contributes direction rather than magnitude to the solution, as illustrated in figure 8.2.) An eigenvector of **B** associated with the eigenvalue $r_1 = 1$ is

First eigenvector

$$\mathbf{p}_1 = \begin{pmatrix} 1 \\ 1 \end{pmatrix}.$$

To find the eigenvector of **B** associated with the eigenvalue $r_2 = -2$, substitute $r_2 = -2$ into the defining relation $(\mathbf{B} - r_2\mathbf{I})\mathbf{p}_2 = \mathbf{0}$ and solve for the components of **p**:

Substitute second eigenvalue to find eigenvector.

$$(\mathbf{B} + 2\mathbf{I})\mathbf{p}_2 = \begin{pmatrix} 3/2 & 3/2 \\ 3/2 & 3/2 \end{pmatrix} \begin{pmatrix} p \\ q \end{pmatrix} = \mathbf{0}.$$

A nontrivial solution of the resulting equations,

$$p + q = 0,$$
$$p + q = 0,$$

requires $q = -p \neq 0$; choose $p = 1, q = -1$. An eigenvector of **B** associated with the eigenvalue $r_2 = -2$ is

Second eigenvector

$$\mathbf{p}_2 = \begin{pmatrix} 1 \\ -1 \end{pmatrix}. \blacksquare$$

In two dimensions, eigenvectors and eigenvalues could also be found by testing many different vectors **x** until one turned up with the property that **Bx** was parallel to **x**. Such a vector would be an eigenvector **p** and the corresponding eigenvalue r would be the ratio of the length of **Bp** to **p**. Figure 8.3 illustrates this idea for the matrix **B** of example 10.

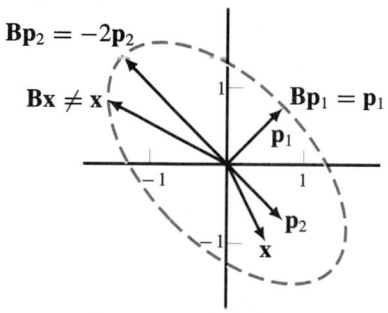

FIGURE 8.3 Because $\mathbf{p}_2$ is an eigenvector of **B** with eigenvalue $r_2 = -2$, the vector $\mathbf{Bp}_2$ is twice as long as $\mathbf{p}_2$ and points in the opposite direction. $\mathbf{Bp}_1$ is identical with $\mathbf{p}_1$ because $\mathbf{p}_1$ corresponds to the eigenvalue $r_1 = 1$. Since **x** is not an eigenvector, $\mathbf{Bx} \neq \mathbf{x}$.

MATLAB

MATLAB offers a vivid visual demonstration of this relationship. To see it for the matrix of the last example, type

```
eigshow( [ -1/2 3/2; 3/2 -1/2 ] )
```

at the prompt in the MATLAB command window. For more information, type **help eigshow**. For guidance in accessing this demonstration through DELAB, select **Help, Textbook**, then go to chapter 8, eigshow.

Summarizing the preceding example,

To find eigenvalues r_i and eigenvectors $\mathbf{p}_i$ of a matrix B:

1. Solve the characteristic polynomial

$$\det(\mathbf{B} - r\mathbf{I}) = 0$$

   for the eigenvalues r_i.

2. For each eigenvalue r_i, determine the components of the eigenvector $\mathbf{p}_i$ by finding a nontrivial solution of the homogeneous system $(\mathbf{B} - r_i\mathbf{I})\mathbf{p}_i = 0$. (One or more components of $\mathbf{p}_i$ will be arbitrary.)

MATLAB

Matrices are represented in MATLAB by enclosing their entries in square brackets [...]. Rows of a matrix are separated by semicolons; e.g., to enter the matrix of the last example, type

```
B = [-1/2 3/2; 3/2 -1/2]
```

The MATLAB command `eig` returns both the eigenvalues and eigenvectors of a matrix. At the prompt in the MATLAB command window, type

```
B = [-1/2 3/2; 3/2 -1/2]
[V,D] = eig(B)
```

The eigenvectors are the columns of V and the corresponding eigenvalues are on the diagonal of D. Type `help eig` for more information. For guidance in finding eigenvalues using DELAB, select `Help, Textbook`, then go to chapter 8, eigenvalues and eigenvectors.

For this example, verify that $\mathbf{B}\mathbf{p} = r\mathbf{p}$ for each eigenvalue r and its associated eigenvector $\mathbf{p}$. How do the eigenvectors given by MATLAB differ from the ones obtained in example 10?

What choice of p_i in example 10 would give the eigenvectors obtained by MATLAB?

The next example illustrates how the eigenvalues and eigenvectors of a coefficient matrix $\mathbf{B}$ are combined to construct a solution of the differential equation $\mathbf{y}' = \mathbf{B}\mathbf{y}$.

■ **EXAMPLE 11**   *Find a general solution of* $\mathbf{y}' = \mathbf{B}\mathbf{y}$ *where*

$$\mathbf{B} = \begin{pmatrix} -1/2 & 3/2 \\ 3/2 & -1/2 \end{pmatrix}.$$

Assume solutions of the form $\mathbf{y} = \mathbf{p}e^{rt}$. Substitute in the homogeneous differential equation and divide out e^{rt} to obtain the eigenvalue-eigenvector relation $r\mathbf{p} = \mathbf{B}\mathbf{p}$, or

$$(\mathbf{B} - r\mathbf{I})\mathbf{p} = \mathbf{0}.$$

Evidently, r must be a root of the characteristic polynomial

Characteristic polynomial

$$\det(\mathbf{B} - r\mathbf{I}) = 0;$$

i.e., r is an eigenvalue of $\mathbf{B}$ and $\mathbf{p}$ is a corresponding eigenvector.

Example 10 found that the eigenvalues and eigenvectors of the coefficient matrix $\mathbf{B}$ are

Eigenvalues and eigenvectors

$$r_1 = 1, \ \mathbf{p}_1 = \begin{pmatrix} 1 \\ 1 \end{pmatrix}, \quad \text{and} \quad r_2 = -2, \ \mathbf{p}_2 = \begin{pmatrix} 1 \\ -1 \end{pmatrix}.$$

Since each eigenvalue-eigenvector pair generates a solution vector $\mathbf{p}_i e^{r_i t}$, a candidate general solution is $C_1 \mathbf{p}_1 e^t + C_2 \mathbf{p}_2 e^{-2t}$. To convey the same information in different terms, a candidate fundamental matrix is

Fundamental matrix

$$\mathbf{Y}(t) = \begin{pmatrix} \mathbf{p}_1 e^t & \mathbf{p}_2 e^{-2t} \end{pmatrix}. \tag{8.13}$$

To test the linear independence of these solutions, evaluate the Wronskian. For convenience, test its value at $t = 0$:

$$W(\mathbf{p}_1 e^t, \mathbf{p}_2 e^{-2t})\big|_{t=0} = \det \mathbf{Y}(0) = \det \begin{pmatrix} \mathbf{p}_1 & \mathbf{p}_2 \end{pmatrix} = -2 \neq 0.$$

Hence, $\mathbf{Y}(t)$ is indeed a fundamental matrix. Furthermore, *the linear independence of the solutions is a direct consequence of the linear independence of the eigenvectors* of the coefficient matrix.

A general solution of this system is

General solution

$$\mathbf{y}_g(t) = \mathbf{Y}(t)\mathbf{c} = C_1 \begin{pmatrix} 1 \\ 1 \end{pmatrix} e^t + C_2 \begin{pmatrix} 1 \\ -1 \end{pmatrix} e^{-2t}. \ \blacksquare$$

Solving an initial-value problem demonstrates the importance of linearly independent eigenvectors.

■ **EXAMPLE 12** *Solve the system of example 11 subject to the initial condition*

$$\mathbf{y}(0) = \mathbf{y}_i = \begin{pmatrix} 1 \\ 5 \end{pmatrix}.$$

The previous example found the general solution $\mathbf{y}_g(t) = \mathbf{Y}(t)\mathbf{c}$, where $\mathbf{Y}(t)$ is the fundamental matrix (8.13). The initial condition demands that the vector $\mathbf{c}$ of arbitrary constants be chosen to satisfy

Impose initial conditions.

$$\mathbf{y}_g(0) = \mathbf{Y}(0)\mathbf{c} = \mathbf{y}_0.$$

Symbolically, $\mathbf{c} = \mathbf{Y}(0)^{-1}\mathbf{y}_0$; the inverse matrix $\mathbf{Y}(0)^{-1} = \begin{pmatrix} \mathbf{p}_1 & \mathbf{p}_2 \end{pmatrix}^{-1}$ exists precisely because the eigenvectors of the coefficient matrix are linearly independent, forcing $\det \mathbf{Y}(0) \neq 0$.

In practice, we solve the system of equations defining **c** by elimination rather than compute the inverse matrix and multiply. The system defining C_1, C_2 is

Solve for **c**.

$$\mathbf{Y}(0)\mathbf{c} = \begin{pmatrix} 1 & 1 \\ 1 & -1 \end{pmatrix} \mathbf{c} = \begin{pmatrix} C_1 & C_2 \\ C_1 & -C_2 \end{pmatrix} \mathbf{c} = \begin{pmatrix} 1 \\ 5 \end{pmatrix}.$$

Elimination yields $C_1 = 3$, $C_2 = -2$. The solution of the initial-value problem is

Use **c** to write the solution of the initial-value problem.

$$\mathbf{y}(t) = \mathbf{Y}(t) \begin{pmatrix} 3 \\ -2 \end{pmatrix} = 3 \begin{pmatrix} 1 \\ 1 \end{pmatrix} e^t + -2 \begin{pmatrix} 1 \\ -1 \end{pmatrix} e^{-2t}. \quad ■$$

MATLAB

The calculation $\mathbf{c} = \mathbf{Y}(0)^{-1}\mathbf{y}_0$ can be carried out in MATLAB using the **inv** command. In the MATLAB command window, type

```
Y0 = [1 1; 1 -1]
c = inv(Y0)*[1; 5]
```

Explicitly finding the inverse of $\mathbf{Y}(0)$ is numerically inefficient in spite of the conceptual simplicity of multiplying by the inverse matrix, but the extra work can be forgiven for tiny 2×2 matrices. (The same can be said of Cramer's rule.)

The structure of these solution formulas is enlightening. Compare the general solution with the solution of the initial-value problem:

$$\mathbf{y}_g(t) = C_1 \begin{pmatrix} 1 \\ 1 \end{pmatrix} e^t + C_2 \begin{pmatrix} 1 \\ -1 \end{pmatrix} e^{-2t}$$

$$\mathbf{y}(t) = \mathbf{Y}(t) \begin{pmatrix} 3 \\ -2 \end{pmatrix} = 3 \begin{pmatrix} 1 \\ 1 \end{pmatrix} e^t - 2 \begin{pmatrix} 1 \\ -1 \end{pmatrix} e^{-2t}.$$

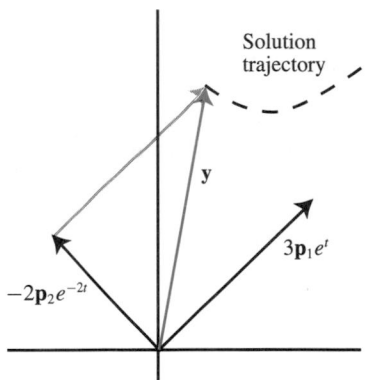

FIGURE 8.4 The solution trajectory is traced by a vector **y** that is the sum of two vectors whose lengths are changing exponentially, $3e^t$ and $-2e^{-2t}$.

Both solutions are linear combinations of two vectors, the eigenvectors $\mathbf{p}_1 = \begin{pmatrix} 1 & 1 \end{pmatrix}^T$ and $\mathbf{p}_2 = \begin{pmatrix} 1 & -1 \end{pmatrix}^T$. The relative size of each is determined initially by the constants C_1 and C_2; those values are $C_1 = 3$ and $C_2 = -2$ for the given initial conditions. As time passes, each of those vectors is scaled by an exponential factor whose growth or decay rate is determined by the corresponding eigenvalue ($r_1 = 1$ gives e^t, $r_2 = -2$ gives e^{-2t}). The eigenvectors determine direction; the eigenvalues determine behavior over time. Figure 8.4 illustrates these ideas; compare it with figure 8.2.

All the information that determines the behavior of a general solution of $\mathbf{y}' = \mathbf{B}\mathbf{y}$ is contained in the eigenvectors and eigenvalues of the coefficient matrix **B**. The eigenvectors **p** are just the special vectors whose direction remains unchanged when multiplied by **B**; that is, $\mathbf{B}\mathbf{p} = r\mathbf{p}$. The eigenvalue

r is the scaling, increasing or decreasing the length of **p** according to whether $|r| > 1$ or $|r| < 1$ and reversing its direction if $r < 0$.

Although linearly independent eigenvectors are the key to easy formation of a general solution, not every 2×2 matrix is blessed with two linearly independent eigenvectors; see exercise 30. However, *symmetric* matrices ($\mathbf{B} = \mathbf{B}^T$) and *Hermitian* matrices ($\mathbf{B} = \overline{\mathbf{B}}^T$) always have a complete set of linearly independent eigenvectors.

> Recall that an overbar denotes complex conjugate: $\overline{\alpha + i\beta} = \alpha - i\beta$. The complex conjugate of a matrix is the matrix of complex conjugates—i.e., if the entries of **B** are b_{ij}, then the entries of $\overline{\mathbf{B}}$ are $\overline{b}_{ij}$.

8.2.3 Complex Eigenvalues

The geometric interpretation of figure 8.3 fails when the eigenvalues of **B** are complex; that is, an eigenvector **p** is no longer defined by the property that **Bp** is parallel to **p** if the multiplier is a complex number. Nonetheless, the same pattern of computations can generate solutions of the differential equations if one step is added to obtain real-valued solutions.

■ **EXAMPLE 13**   *Find a real-valued fundamental matrix for the system* $\mathbf{y}' = \mathbf{B}\mathbf{y}$ *where*

$$\mathbf{B} = \begin{pmatrix} 0 & -1 \\ 16 & 0 \end{pmatrix}.$$

The characteristic polynomial of this matrix is

Characteristic polynomial

$$\det(\mathbf{B} - r\mathbf{I}) = \det \begin{pmatrix} -r & -1 \\ 16 & -r \end{pmatrix} = r^2 + 16 = 0.$$

Complex eigenvalues

Its roots, the eigenvalues of **B**, are evidently $r = \pm 4i$.

To find an eigenvector associated with $r_1 = 4i$, solve

Substitute first eigenvalue to find eigenvector.

$$(\mathbf{B} - 4i\mathbf{I})\mathbf{p} = \begin{pmatrix} -4i & -1 \\ 16 & -4i \end{pmatrix} \begin{pmatrix} p \\ q \end{pmatrix} = \mathbf{0}.$$

Both equations provide the same relation between p and q,

$$q = -4ip.$$

Choosing $p = 1$, an eigenvector associated with the eigenvalues $r_1 = 4i$ is

First eigenvector

$$\mathbf{p}_1 = \begin{pmatrix} 1 \\ -4i \end{pmatrix}.$$

Parallel calculations find the eigenvector associated with $r_2 = -4i$,

Use the second eigenvalue to find a second eigenvector.

$$\mathbf{p}_2 = \begin{pmatrix} 1 \\ 4i \end{pmatrix}.$$

After verifying the linear independence of the two eigenvectors $\mathbf{p}_1$ and $\mathbf{p}_2$, we can certainly form the general solution

Complex-valued general solution

$$C_1 \begin{pmatrix} 1 \\ -4i \end{pmatrix} e^{4it} + C_2 \begin{pmatrix} 1 \\ 4i \end{pmatrix} e^{-4it}. \tag{8.14}$$

How do we obtain a *real*-valued solution?

The answer hinges on these observations:

Complex conjugate pairs

1. Eigenvectors and eigenvalues of real matrices occur in complex conjugate pairs: If r, $\mathbf{p}$ is an eigenvalue-eigenvector pair for the real-valued matrix $\mathbf{B}$, then so is $\bar{r}$, $\bar{\mathbf{p}}$.

   The proof is simple. Suppose $\mathbf{Bp} = r\mathbf{p}$. Form the complex conjugate of both sides of the equation. Since $\mathbf{B}$ is real-valued, $\bar{\mathbf{B}} = \mathbf{B}$. Hence, $\mathbf{B}\bar{\mathbf{p}} = \bar{r}\bar{\mathbf{p}}$, and the conjugates form an eigenvalue-eigenvector pair as well.

   If $\mathbf{p}_1 e^{r_1 t}$ is one solution, then $\mathbf{p}_2 e^{r_2 t} = \overline{\mathbf{p}_1 e^{r_1 t}}$ is another.

Isolate real and imaginary parts.

2. If $z = \alpha + i\beta$ is a complex quantity, then linear combinations select real and imaginary parts,

$$\frac{z + \bar{z}}{2} = \alpha = \operatorname{Re} z, \quad \frac{z - \bar{z}}{2i} = \beta = \operatorname{Im} z.$$

Exercise 32 requests a verification.

Superposition of homogeneous solutions

3. A linear combination of homogeneous solutions is again a homogeneous solution.

   The first statement certainly holds in this case: $r_2 = -4i = \overline{4i} = \bar{r}_1$. And the eigenvectors are clearly complex conjugate pairs as well, $\mathbf{p}_2 = \bar{\mathbf{p}}_1$. Consequently, the choice of arbitrary constants $C_1 = C_2 = 1/2$ in the general solution (8.14) leads to the real-valued solution

$$\mathbf{y}_1 = \frac{1}{2} \left(\mathbf{p}_1 e^{4it} + \mathbf{p}_2 e^{-4it} \right) = \operatorname{Re} \left(\mathbf{p}_1 e^{4it} \right).$$

Using Euler's formula, $e^{4it} = \cos 4t + i \sin 4t$, we find

$$\mathbf{p}_1 e^{4it} = \begin{pmatrix} \cos 4t + i \sin 4t \\ 4 \sin 4t - 4i \cos 4t \end{pmatrix}.$$

Hence, the real part of this vector function is

First real fundamental solution

$$\mathbf{y}_1 = \operatorname{Re} \left(\mathbf{p}_1 e^{4it} \right) = \begin{pmatrix} \cos 4t \\ 4 \sin 4t \end{pmatrix}.$$

Likewise, the choice $C_1 = -C_2 = 1/2i$ in the general solution (8.14) leads to another real-valued solution,

Second real fundamental solution

$$\mathbf{y}_2 = \frac{1}{2i} \left(\mathbf{p}_1 e^{4it} - \mathbf{p}_2 e^{-4it} \right) = \operatorname{Im} \left(\mathbf{p}_1 e^{4it} \right) = \begin{pmatrix} \sin 4t \\ -4 \cos 4t \end{pmatrix}.$$

A real-valued general solution is

$$\mathbf{y}_g = C_1\mathbf{y}_1 + C_2\mathbf{y}_2 = C_1 \begin{pmatrix} \cos 4t \\ 4\sin 4t \end{pmatrix} + C_2 \begin{pmatrix} \sin 4t \\ -4\cos 4t \end{pmatrix}$$

and a real-valued fundamental matrix is

Real-valued fundamental matrix

$$\mathbf{Y}(t) = \begin{pmatrix} \cos 4t & \sin 4t \\ 4\sin 4t & -4\cos 4t \end{pmatrix}\mathbf{c}.$$

(Verification of linear independence is left to exercise 33.) ■

8.2.4 Summary

From the matrix perspective, the *behavior of solutions of* $\mathbf{y}' = \mathbf{By}$ *depends entirely on the eigenvalues and eigenvectors* of the constant coefficient matrix **B**.

Characteristic equation method for $\mathbf{y}' = \mathbf{By}$. If the $n \times n$ matrix **B** has n linearly independent eigenvectors $\mathbf{p}_i$ with associated eigenvalues r_i, $i = 1, \ldots, n$, then a *fundamental matrix* for

$$\mathbf{y}' = \mathbf{By}$$

is

$$\mathbf{Y}(t) = \begin{pmatrix} \mathbf{p}_1 e^{r_1 t} & \cdots & \mathbf{p}_n e^{r_n t} \end{pmatrix},$$

and a *general solution* of the homogeneous system is

$$\mathbf{y}_g(t) = \mathbf{Y}(t)\mathbf{c},$$

where **c** is the constant vector $\begin{pmatrix} C_1 & \cdots & C_n \end{pmatrix}^T$.

If **B** is a real matrix, then complex eigenvectors and eigenvalues occur in conjugate pairs $\mathbf{p}, \overline{\mathbf{p}}$ and $r, \overline{r}$. A pair of complex-valued columns

$$\mathbf{Y}(t) = \begin{pmatrix} \cdots & \mathbf{p}e^{rt} & \overline{\mathbf{p}}e^{\overline{r}t} & \cdots \end{pmatrix}$$

in the fundamental matrix can be replaced by the real columns

$$\mathbf{Y}(t) = \begin{pmatrix} \cdots & \text{Re}\,(\mathbf{p}e^{rt}) & \text{Im}\,(\mathbf{p}e^{rt}) & \cdots \end{pmatrix}.$$

If **B** does not have linearly independent eigenvectors, then these ideas will not produce a fundamental matrix. That possibility will not be explored in detail, but section 8.2.5 suggests how this problem is handled.

Stop and Think   **8.19** Rewrite the preceding summary for the specific case considered in example 11. (**B** has real eigenvalues.)

8.20 Rewrite the preceding summary for the specific case considered in example 13. (**B** has complex eigenvalues.)

■ **EXAMPLE 14**   *The temperature at both ends of an insulated rod of length 5 m is zero. Initially it is warmed throughout its length. Example 24 of section 9.6 shows that the temperature $U_i(t)$ as a function of time at the intermediate interior points $i = 1, \ldots, 4$ m from the left end are approximated by the system* $\mathbf{u}' = \mathbf{Au}$ *where*

$$\mathbf{u}(t) = \begin{pmatrix} U_1(t) \\ U_2(t) \\ U_3(t) \\ U_4(t) \end{pmatrix}, \quad \mathbf{A} = \begin{pmatrix} -2 & 1 & 0 & 0 \\ 1 & -2 & 1 & 0 \\ 0 & 1 & -2 & 1 \\ 0 & 0 & 1 & -2 \end{pmatrix}.$$

(*This system arises from approximating a partial differential equation for temperature in the rod as a function of time and position by a technique known as the method of lines.*)
    Will the rod cool as time passes? If so, how fast?

As we know, the characteristic equation method seeks solutions of $\mathbf{u}' = \mathbf{Au}$ of the form $\mathbf{u} = \mathbf{p}e^{rt}$; $\mathbf{p}$ is an eigenvector of $\mathbf{A}$ and r is the corresponding eigenvalue. Since every solution of this system is a linear combination of such solutions, the behavior over time is completely determined by the eigenvalues of $\mathbf{A}$.

The MATLAB command `eig` finds that the eigenvalues of $\mathbf{A}$ are $r_1 = -0.3820$, $r_2 = -1.3820$, $r_3 = -2.6180$, and $r_4 = -3.6180$. *Every* column of a fundamental matrix for this system is decaying. Hence, no matter what the initial conditions, this model predicts that the rod will cool to zero temperature.

The rate of decay is determined by these eigenvalues through the factors $e^{r_i t}$, $i = 1, \ldots, 4$. The slowest decaying term is $e^{r_1 t}$ since $r_1 = -0.3820$ is the least negative eigenvalue. ■

MATLAB

To obtain these eigenvalues, type

```
[V,D] = eig([-2 1 0 0; 1 -2 1 0; 0 1 -2 1; 0 0 1 -2])
```

at the prompt in the MATLAB command window. The eigenvalues appear on the diagonal of D. For more information, type `help eig`. To obtain these eigenvalues using DELAB, select the analytic or numerical eigenvalue tool.

■ **EXAMPLE 15**   *Find a general solution of* $\mathbf{y}' = \mathbf{By}$ *where*

$$\mathbf{B} = \begin{pmatrix} 2 & -1 \\ 1 & 2 \end{pmatrix}.$$

The characteristic equation of $\mathbf{B}$ is

$$\det(\mathbf{B} - r\mathbf{I}) = (2 - r)^2 + 1 = r^2 - 2r + 5 = 0.$$

Its roots, the eigenvalues of **B** are $r = 2 \pm i$.

To find the eigenvector corresponding to $r = 2 + i$, set $r = 2 + i$ in the eigenvalue relation $(\mathbf{B} - r\mathbf{I})\mathbf{p} = 0$. Since the two equations will be redundant, it suffices to solve just the first,

$$(2 - r)p - q = -ip + q = 0,$$

from which we obtain $q = ip$. Setting $p = 1$ yields the eigenvector $\mathbf{p} = \begin{pmatrix} 1 & -i \end{pmatrix}^T$. Real-valued linearly independent solutions of $\mathbf{y}' = \mathbf{B}\mathbf{y}$ are the real and imaginary parts of the complex-valued solution

$$\mathbf{p}e^{rt} = \begin{pmatrix} 1 \\ -i \end{pmatrix} e^{(2+i)t}$$

$$= \begin{pmatrix} 1 \\ -i \end{pmatrix} e^{2t}(\cos t + i \sin t)$$

$$= \begin{pmatrix} e^{2t} \cos t \\ e^{2t} \sin t \end{pmatrix} + i \begin{pmatrix} e^{2t} \sin t \\ -e^{2t} \cos t \end{pmatrix}.$$

From the real and imaginary parts, respectively, we obtain the solutions

$$\mathbf{y}_1 = \begin{pmatrix} e^{2t} \cos t \\ e^{2t} \sin t \end{pmatrix}, \quad \mathbf{y}_2 = \begin{pmatrix} e^{2t} \sin t \\ -e^{2t} \cos t \end{pmatrix}.$$

A candidate fundamental matrix is

$$\mathbf{Y}(t) = \begin{pmatrix} \mathbf{y}_1 & \mathbf{y}_2 \end{pmatrix}.$$

To test the linear independence of $\mathbf{y}_1$ and $\mathbf{y}_2$, evaluate the Wronskian $W(t) = \det \mathbf{Y}(t)$ at a convenient value of t, say $t = 0$:

$$W(0) = \det \mathbf{Y}(0) = \begin{vmatrix} 1 & 0 \\ 0 & -1 \end{vmatrix} = -1 \neq 0.$$

Hence, the solutions are independent, **Y** is indeed a fundamental matrix, and a general solution of this system is $\mathbf{y}_g = \mathbf{Y}(t)\mathbf{c}$, $\mathbf{c} = \begin{pmatrix} C_1 & C_2 \end{pmatrix}^T$. ■

Stop and Think   **8.21** Let $\mathbf{y}_g = \begin{pmatrix} y_g & z_g \end{pmatrix}^T$. Write out expressions for y_g and z_g.

8.2.5   Examples of Repeated Eigenvalues

If an eigenvalue is a repeated root of the characteristic equation, it may still be associated with two *linearly independent* eigenvectors. Then a general solution is easily constructed from those eigenvectors. The simplest 2×2 matrix with a repeated eigenvalue is the identity matrix, as the next example illustrates.

■ **EXAMPLE 16** *Find linearly independent solutions of* $\mathbf{y}' = \mathbf{I}\mathbf{y}$, *where* $\mathbf{I}$ *is the identity*

$$\mathbf{I} = \begin{pmatrix} 1 & 0 \\ 0 & 1 \end{pmatrix},$$

Substituting $\mathbf{y} = \mathbf{p}e^{rt}$ and dividing by the exponential leads to

System for $\mathbf{p} = \begin{pmatrix} p & q \end{pmatrix}^T$

$$(1-r)p = 0,$$
$$(1-r)q = 0.$$

Setting the determinant of coefficients to zero gives the characteristic equation of the system of differential equations or, equivalently, the characteristic polynomial of the coefficient matrix $\mathbf{I}$,

Characteristic equation

$$(r-1)^2 = 0.$$

It has the single repeated root $r = 1$.
To find p, q, set $r = 1$ in the algebraic system:

Find p, q for the repeated root $r = 1$.

$$0\,p = 0, \tag{8.15}$$
$$0\,q = 0. \tag{8.16}$$

Both p and q are completely arbitrary! One choice is $p_1 = 1$, $q_1 = 0$ and another is $p_2 = 0$, $q_2 = 1$, leading to two independent eigenvectors

$$\mathbf{p}_1 = \begin{pmatrix} 1 \\ 0 \end{pmatrix}, \quad \mathbf{p}_2 = \begin{pmatrix} 0 \\ 1 \end{pmatrix}$$

and the corresponding linearly independent solutions

$$\mathbf{y}_1 = \begin{pmatrix} 1 \\ 0 \end{pmatrix} e^t, \quad \mathbf{y}_2 = \begin{pmatrix} 0 \\ 1 \end{pmatrix} e^t. \quad \blacksquare$$

Of course, the system in this example consists of two *uncoupled* equations $y' = y$ and $z' = z$, for which we have just found nontrivial solutions. In spite of its simplicity, this example shows that the key to finding two independent solutions is the existence of two linearly independent eigenvectors corresponding to the repeated eigenvalue.

If there are not linearly independent eigenvectors associated with the repeated eigenvalue, the situation is more complicated. For a system of two equations with a single characteristic root and only one eigenvector the general remedy is

Repeated eigenvalue with one eigenvector

Given a repeated eigenvalue r with only the single eigenvector $\mathbf{p}$, one solution is of the form $\mathbf{p}e^{rt}$ and the second is of the form $(\mathbf{P}_1 + \mathbf{P}_2 t)e^{rt}$.

The next example illustrates this process.

■ **EXAMPLE 17** *Find linearly independent solutions of the system* $\mathbf{y}' = \mathbf{By}$ *with*

$$\mathbf{B} = \begin{pmatrix} 0 & 1 \\ -9 & 6 \end{pmatrix}.$$

Substituting $\mathbf{y} = \mathbf{p}e^{rt}$ and dividing by the exponential leads to the usual characteristic equation

Characteristic equation

$$\det(\mathbf{B} - r\mathbf{I}) = r^2 - 6r + 9 = (r-3)^2 = 0.$$

It has the single repeated root $r = 3$.
To find p, q, set $r = 3$ in the algebraic system

Find p, q for the repeated root $r = 3$.

$$-3p + q = 0, \tag{8.17}$$
$$-9p + 3q = 0 \tag{8.18}$$

and obtain $q = 3p$. Choosing $p = 1$ gives one solution of the system,

First solution

$$\mathbf{y}_1 = \mathbf{p}_1 e^{3t} = \begin{pmatrix} 1 \\ 3 \end{pmatrix} e^{3t}.$$

As directed above, seek a second solution in the form

Special form for second solution

$$\mathbf{y}_2 = \begin{pmatrix} P_1 + P_2 t \\ Q_1 + Q_2 t \end{pmatrix} e^{3t} = (\mathbf{P}_1 + \mathbf{P}_2 t)e^{3t}.$$

Substituting this assumed form into the system of differential equations yields

$$3P_1 e^{3t} + \boxed{t(P_2 e^{3t})'} + P_2 e^{3t} = Q_1 e^{3t} + \boxed{t(Q_2 e^{3t})} + 3Q_1 e^{3t} + \boxed{t(Q_2 e^{3t})'} + Q_2 e^{3t}$$

$$= -9P_1 e^{3t} + \boxed{t(-9P_2 e^{3t})} + 6Q_1 e^{3t} + \boxed{t(6Q_2 e^{3t})}.$$

Equate the coefficients of terms involving te^{3t} (boxed) and the coefficients of terms involving just e^{3t} to obtain two separate sets of equations. The equations coming from the boxed terms reduce to (8.17–8.18) with $p = P_2$ and $q = Q_2$; we conclude that $Q_2 = 3P_2$.
Using $P_2 = 1$ and $Q_2 = 3P_2 = 3$, the equations from the terms that are not boxed become

$$3P_1 - Q_1 = -1,$$
$$9P_1 - 3Q_1 = -3,$$

from which we conclude $Q_1 = 1 + 3P_1$. With $P_1 = 1$, a second solution of this system is

Second solution

$$\mathbf{y}_2 = \begin{pmatrix} P_1 + P_2 t \\ Q_1 + Q_2 t \end{pmatrix} e^{3t} = \begin{pmatrix} 1 + t \\ 4 + 3t \end{pmatrix} e^{3t}.$$

Verification of linear independence is left to the exercises. ■

The difficulty in example 17 is the presence of only one linearly independent eigenvector $\mathbf{p} = \begin{pmatrix} 1 & 3 \end{pmatrix}^T$.

MATLAB

Apply MATLAB's `eigshow` to the coefficient matrices of the last two examples. How does the difference in behavior show that one matrix has two linearly independent eigenvectors while the other has only a single eigenvector?

8.2.6 Exercises

EXERCISE GUIDE	
To gain experience . . .	**Try exercises**
Finding general solutions	1–18, 26–27
Solving initial-value problems	1–18
Verifying linear independence	13–15
Analyzing characteristic roots	16(b), 21, 38
Obtaining real-valued solutions from complex characteristic roots	1, 5–6, 9, 11, 17, 32–33
With the foundations of the characteristic equation method	16, 24
Finding eigenvalues and eigenvectors	13–18, 21, 29–31, 34
Finding fundamental matrices	13–18, 33
Finding general solutions	13–18, 21, 36
Analyzing solution behavior	21, 36
With eigenvalue ideas	21, 22, 34, 36–37

In exercises 1–12,

(i) Find a general solution of the given system.

(ii) Find a solution of the system that satisfies the given initial conditions.

(iii) As appropriate, check your analysis using MATLAB.

1. $y' = 2y + 4z, \; y(0) = 2$
$z' = -4y + 2z, \; z(0) = -2$

2. $y' = -y + 3z, \; y(0) = 0$
$z' = 3y - z, \; z(0) = 4$

3. $y' = 4y + 4z, \; y(0) = 4$
$z' = 3y - 4z, \; z(0) = 1$

4. $y' = 5y/2 + 3\sqrt{3}\,z/2, \; y(0) = 6$
$z' = 3\sqrt{3}\,y/2 - z/2, \; z(0) = 0$

5. $y' = -4y - 5z/2, \; y(0) = 0$
$z' = y - 2z, \; z(0) = 0$

6. $y' = -4y - 5z/2, \; y(0) = 1$
$z' = y - 2z, \; z(0) = -6$

7. $y' = y/2 + \sqrt{3}\,z/2, \; y(0) = 6$
$z' = \sqrt{3}\,y/2 - z/2, \; z(0) = 4$

8. $y' = z, \; y(0) = 1$
$z' = y, \; z(0) = 0$

9. $x' = v, \; x(0) = x_0$
$mv' = -kx, \; v(0) = 0$

10. $x' = v, \; x(0) = 0$
$mv' = -kx, \; v(0) = v_0$

11. $x' = v, \; x(0) = x_0$
$mv' = -kx - pv, \; v(0) = 0$

12. $\theta' = \omega, \; \theta(0) = \theta_0$
$L\omega' = -g\theta, \; \omega(0) = 0$

In exercises 13–18, for the given matrix **B**,

(i) Find the eigenvalues and associated eigenvectors of **B**.

(ii) Find linearly independent solutions of $\mathbf{y}' = \mathbf{By}$.

(iii) Find a fundamental matrix of $\mathbf{y}' = \mathbf{By}$.

(iv) Write a general solution of $\mathbf{y}' = \mathbf{By}$.

(v) Find a solution of $\mathbf{y}' = \mathbf{By}$, $\mathbf{y}(0) = \begin{pmatrix} 4 & -4 \end{pmatrix}^T$.

(vi) Confirm the eigenvalues and eigenvectors using MAT-LAB.

13. B $= \begin{pmatrix} 1 & 0 \\ 4 & 3 \end{pmatrix}$

14. B $= \begin{pmatrix} -2 & 0 \\ 2 & -1 \end{pmatrix}$

15. B $= \begin{pmatrix} 11 & 2 \\ 1 & 10 \end{pmatrix}$

16. B $= \begin{pmatrix} 1 & 2 \\ -1 & 1 \end{pmatrix}$

17. B $= \begin{pmatrix} 4 & 3 \\ -3 & 4 \end{pmatrix}$

18. B $= \begin{pmatrix} 3/2 & \sqrt{3}/2 \\ -\sqrt{3}/2 & 3/2 \end{pmatrix}$

19. Eigenvalues of the matrix

$$\mathbf{B} = \begin{pmatrix} 0 & 1 & 0 \\ 0 & 0 & 1 \\ 1 & 0 & 0 \end{pmatrix}$$

are $r = 1$, $(-1 \pm i\sqrt{3})/2$. Find the corresponding eigenvectors and write a real-valued general solution of $\mathbf{y}' = \mathbf{By}$. Check your work with MATLAB.

20. Consider the generic constant-coefficient system

$$y' = ay + bz,$$
$$z' = cy + dz.$$

(a) Show that its characteristic equation is

$$r^2 - (a + d)r + ad - bc = 0.$$

(b) Find conditions on the coefficients a, b, c, d that guarantee that the characteristic roots are

(i) Real and distinct.

(ii) Real and equal.

(iii) Complex conjugates.

21. In a setting similar to that described in example 14, the system $\mathbf{u}' = \mathbf{Au} + \mathbf{f}$,

$$\mathbf{u}(t) = \begin{pmatrix} U_1(t) \\ U_2(t) \\ U_3(t) \end{pmatrix}, \quad \mathbf{A} = 4 \begin{pmatrix} -2 & 1 & 0 \\ 1 & -2 & 1 \\ 0 & 1 & -2 \end{pmatrix},$$

$$\mathbf{f} = \begin{pmatrix} 4 \\ 0 \\ 12 \end{pmatrix},$$

models the temperature $U_i(t)$ as a function of time at three points interior to a thin, insulated rod whose ends are held at constant temperature. (See example 25 of section 9.6.) Using MATLAB as appropriate,

(a) Find the eigenvalues and corresponding eigenvectors of **B**.

(b) Write a general solution of the homogeneous system $\mathbf{u}' = \mathbf{Au}$.

(c) Find a steady-state solution $\mathbf{u}_{ss}$ of the nonhomogeneous system $\mathbf{u}' = \mathbf{Au} + \mathbf{f}$. Is $\mathbf{u}_{ss}$ a particular solution?

(d) Write a general solution of the nonhomogeneous system.

(e) Comment on the stability of the steady state $\mathbf{u}_{ss}$. To what temperature profile is the rod tending? (It is 2 m long. Its left and right ends are at $1°$ and $3°$, respectively; U_i, $i = 1, \dots, 3$ approximates the temperature at locations 0.5, 1.0, and 1.5 m from the left end.)

22. Show that the parameters p, q, r in the proposed solution

$$\mathbf{y} = \begin{pmatrix} p \\ q \end{pmatrix} e^{rt}$$

of the system of differential equations

$$y' = ay + bz,$$
$$z' = cy + dz$$

must satisfy

$$(a - r)p + bq = 0,$$
$$cp + (d - r)q = 0.$$

How is this expression related to the characteristic polynomial of the coefficient matrix of the system of differential equations?

23. The characteristic equation method for a 2×2 system is based on the following claim:

The system of linear algebraic equations

$$(a - r)p + bq = 0,$$
$$cp + (d - r)q = 0$$

has a nontrivial solution for p and q only if the determinant of coefficients is zero:

$$\begin{vmatrix} a - r & b \\ c & d - r \end{vmatrix} = (a - r)(d - r) - bc = 0.$$

(a) Use Cramer's rule to justify this statement.

(b) Justify the statement directly as follows. Eliminate q from the system of equations by multiplying the first equation by $d - r$, the second by b and subtracting, thereby obtaining $[(a - r)(d - r) - bc]p = 0$. Why must $(a - r)(d - r) - bc = 0$ if p is not to be zero? Carry out a parallel calculation for q. (*Hint:* See example 7, page 381.)

24. Prove that the real part and the imaginary part of a complex-valued solution of a linear homogeneous system with real coefficients are each solutions of that same system. Proceed as follows:

Suppose

$$\mathbf{y}_c = \begin{pmatrix} y_1 \\ z_2 \end{pmatrix} + i \begin{pmatrix} y_2 \\ z_2 \end{pmatrix}$$

is a complex-valued solution of the linear homogeneous system

$$y' = k_1(t)y + k_2(t)z,$$
$$z' = \ell_1(t)y + \ell_2(t)z,$$

where y_i, z_i, k_i, ℓ_i are all real-valued functions. Then $\mathbf{y}_1 = \begin{pmatrix} y_1 & z_1 \end{pmatrix}^T$ and $\mathbf{y}_2 = \begin{pmatrix} y_2 & z_2 \end{pmatrix}^T$ are real-valued solutions of the same system.

Are the two real-valued solutions necessarily linearly independent?

25. Prove that the given solutions of

$$y' = ay + bz,$$
$$z' = cy + dz$$

are linearly independent when the characteristic roots are as indicated.

(i) $\mathbf{p}_1 e^{r_1 t}$, $\mathbf{p}_2 e^{r_2 t}$ if the characteristic roots r_1, r_2 are *real and distinct*.

(ii) The *real part* and the *imaginary part* of the complex-valued solution $\mathbf{p}_1 e^{r_1 t}$ if r_1, r_2 are a *complex conjugate pair*.

Proceed as follows.

(a) Evaluate the Wronskian $W(t)$ of the solutions $\mathbf{p}_1 e^{r_1 t}$, $\mathbf{p}_2 e^{r_2 t}$. What conditions must you impose on $\mathbf{p}_1, \mathbf{p}_2$? How can you be sure these conditions are satisfied?

(b) Write $r_1, r_2 = \alpha \pm i\beta$, $p_1 = P_1 + iP_2$, $q_1 = Q_1 + iQ_2$, where all parameters are real. Use this notation to write the real part and the imaginary part of the solution $\mathbf{y} = \mathbf{p}_1 e^{r_1 t}$. Evaluate the Wronskian and proceed as in part (a).

26. Fill in the details of the computation of the second solution in example 17. Start with the paragraph that begins, "As directed above ..."

27. Find a general solution of

$$y' = y + 3z,$$
$$z' = -2z.$$

Then show that this system can be uncoupled by first solving the second equation for $z(t)$ and then using that solution as a forcing term in the first equation to find $y(t)$. Show that you obtain the same general solution.

28. Consider the second-order scalar equation $b_2 y'' + b_1 y' + b_0 y = 0$, b_i constant.

(a) Write the characteristic equation of this scalar equation. Write the first-order system equivalent to this equation. Show that their characteristic equations are identical.

(b) Show that the characteristic equation of the coefficient matrix of the first-order system is also identical to that of the scalar differential equation.

29. Verify the details omitted from example 13 by showing that $\mathbf{p}_2 = \begin{pmatrix} 1 & 4i \end{pmatrix}^T$ is indeed an eigenvector of the matrix

$$\mathbf{B} = \begin{pmatrix} 0 & -1 \\ 16 & 0 \end{pmatrix}$$

associated with the eigenvalue $r_2 = -4i$; that is, verify that $\mathbf{B}\mathbf{p}_2 = -4i\mathbf{p}_2$.

30. Show that the matrix

$$\mathbf{B} = \begin{pmatrix} 1 & 1 \\ 0 & 1 \end{pmatrix}$$

does not have two linearly independent eigenvectors. Find the (single) eigenvalue of $\mathbf{B}$ and show that any eigenvector $\mathbf{p} = \begin{pmatrix} p & q \end{pmatrix}^T$ associated with it must have $q = 0$. Hence, every eigenvector of this matrix is a scalar multiple of the vector $\begin{pmatrix} 1 & 0 \end{pmatrix}^T$.

31. Show that the diagonal matrix

$$\mathbf{B} = \begin{pmatrix} a & 0 \\ 0 & a \end{pmatrix},$$

$a \neq 0$, has two linearly independent eigenvectors even though it has the single repeated eigenvalue $r = a$. Show that the two independent eigenvectors can be chosen to be multiples of $\begin{pmatrix} 1 & 0 \end{pmatrix}^T$ and $\begin{pmatrix} 0 & 1 \end{pmatrix}^T$.

32. Using $\bar{z} = \alpha - i\beta$, verify the claim made in the text:

If $z = \alpha + i\beta$ is a complex quantity, then

$$\frac{z + \bar{z}}{2} = \alpha = \operatorname{Re} z, \quad \frac{z - \bar{z}}{2i} = \beta = \operatorname{Im} z.$$

33. Verify the result of example 13 that

$$\mathbf{Y}(t) = \begin{pmatrix} \cos 4t & \sin 4t \\ 4\sin 4t & -4\cos 4t \end{pmatrix} \mathbf{c}$$

is a fundamental matrix for the system $\mathbf{y}' = \mathbf{B}\mathbf{y}$ where

$$\mathbf{B} = \begin{pmatrix} 0 & -1 \\ 16 & 0 \end{pmatrix}.$$

Also show that $\mathbf{Y}' = \mathbf{B}\mathbf{Y}$.

34. MATLAB's **eigshow** demonstration is a visual tool for finding eigenvalues and eigenvectors of some 2×2 matrices. Choose from examples in the text matrices having distinct (nonzero) real eigenvalues, one repeated eigenvalue, and complex conjugate eigenvalues, respectively. (*Hint*: Try examples 8, 13, 17, and the identity matrix **I**.) Experiment, then suggest to what extent these three classes of matrices can be distinguished from one another by the behavior they exhibit in **eigshow**. (For help with **eigshow**, see the remarks following example 10.)

35. Construct a fundamental matrix for the system

$$\mathbf{y}'(t) = \begin{pmatrix} -4 & 2 \\ -3 & 3 \end{pmatrix} \mathbf{y}(t).$$

by finding the eigenvalues and eigenvectors of the coefficient matrix. Check your results using MATLAB.

Use this fundamental matrix to find the solution of this system subject to the initial condition $\mathbf{y}(0) = \begin{pmatrix} 1 & -2 \end{pmatrix}^T$.

36. Suppose you have conducted a linear stability analysis of an equilibrium point of a system of two nonlinear differential equations. You have found that the perturbations $\mathbf{y}(t)$ of that equilibrium satisfy the constant-coefficient system $\mathbf{y}' = \mathbf{B}\mathbf{y}$, where the real matrix **B** has two linearly independent eigenvectors. By examining the form of the general solution of this system, justify the following statements:

(a) The equilibrium is stable if the real part of *every* eigenvalue of **B** is negative.

(b) The equilibrium is unstable if the real part of *any* eigenvalue of **B** is positive.

(c) The equilibrium is neutrally stable if the real part of every complex eigenvalue of **B** is zero.

37. Given an eigenvalue r of the matrix **B**, why can Cramer's rule not be used to solve $(\mathbf{B} - r\mathbf{I})\mathbf{p} = 0$ for the components of the eigenvector **p**?

38. Verify that $\mathbf{Y}(t)\mathbf{c}$ is, in fact, a linear combination of the columns of $\mathbf{Y}(t)$ in the 2×2 case $\mathbf{Y} = \begin{pmatrix} \mathbf{y}_1 & \mathbf{y}_2 \end{pmatrix}$, where

$$\mathbf{y}_i = \begin{pmatrix} y_i \\ z_i \end{pmatrix}, \quad \mathbf{c} = \begin{pmatrix} C_1 \\ C_2 \end{pmatrix}.$$

Carry out the matrix multiplication and show that $\mathbf{Y}(t)\mathbf{c} = C_1\mathbf{y}_1 + C_2\mathbf{y}_2$.

8.3 ■ CONNECTIONS WITH THE PHASE PLANE

The explorations in chapter 7 of the phase plane of two-dimensional autonomous systems revealed a variety of possible behaviors in the vicinity of equilibrium points. Much of this local behavior can be fully analyzed using the tools of the previous section, the characteristic equation approach to representing solutions of constant-coefficient homogeneous linear systems $\mathbf{y}' = \mathbf{B}\mathbf{y}$ in terms of the eigenvalues and eigenvectors of the coefficient matrix **B**. A linear, homogeneous, constant-coefficient problem of the form $\mathbf{y}' = \mathbf{B}\mathbf{y}$ *always* arises as a linear approximation of an autonomous system in the neighborhood of an equilibrium point. In most cases, it is a good approximation too.

8.3.1   Autonomous Systems and the Jacobian

Recall that autonomous systems are those that do not depend explicitly upon the independent variable, usually t. Examples include the equations that model a spring-mass system without forcing, the pendulum, predator-prey interactions, epidemics, and many other situations.

Stop and Think   8.22   Write an example of an autonomous system of differential equations from each model mentioned.

Any autonomous system can be written in the general form

AUTONOMOUS SYSTEM

$$y' = f(y, z),$$
$$z' = g(y, z),$$

where prime ($'$) denotes differentiation with respect to an independent variable such as t. Equivalent vector notation is

$$\mathbf{y}' = \mathbf{f}(\mathbf{y}), \quad \text{where} \quad \mathbf{y} = \begin{pmatrix} y \\ z \end{pmatrix}, \quad \mathbf{f} = \begin{pmatrix} f(y, z) \\ g(y, z) \end{pmatrix}.$$

The *stationary points* or steady states or equilibria of such systems are simply solutions (y_{ss}, z_{ss}) of

STEADY STATE

$$\begin{aligned} f(y_{ss}, z_{ss}) &= 0, \\ g(y_{ss}, z_{ss}) &= 0 \end{aligned} \quad \text{or} \quad \mathbf{f}(\mathbf{y}_{ss}) = \mathbf{0}.$$

An equilibrium point is also called a *singular point* of $\mathbf{y}' = \mathbf{f}(\mathbf{y})$. Such a point is singular because no solution trajectory can pass through it without violating uniqueness.

■ **EXAMPLE 18**   *Argue that the pendulum system*

$$\begin{aligned} \theta' &= \omega, \\ L\omega' &= -g \sin \theta - p\omega \end{aligned} \tag{8.19}$$

is autonomous. Define the expressions required to write it in the vector form $\mathbf{y}' = \mathbf{f}(\mathbf{y})$. *Find its steady states in this notation.*

This system is autonomous because t appears explicitly nowhere in it. By defining

$$\mathbf{y}(t) = \begin{pmatrix} \theta(t) \\ \omega(t) \end{pmatrix}, \quad \mathbf{f}(\mathbf{y}) = \begin{pmatrix} \omega \\ -(g/L) \sin \theta - (p/L)\omega \end{pmatrix},$$

we can write it in the compact form $\mathbf{y}' = \mathbf{f}(\mathbf{y})$.

The steady states of this system are defined by

$$\mathbf{f}(\mathbf{y}_{ss}) = \begin{pmatrix} \omega_{ss} \\ -(g/L) \sin \theta_{ss} - (p/L)\omega_{ss} \end{pmatrix} = \mathbf{0}.$$

This vector equality is equivalent to two scalar equations,

$$\omega_{ss} = 0,$$
$$-(g/L)\sin\theta_{ss} - (p/L)\omega_{ss} = 0,$$

whose solutions are familiar: $\omega_{ss} = 0$, $\theta_{ss} = 0, \pm\pi, \pm2\pi, \ldots$ or

$$\mathbf{y}_{ss} = \begin{pmatrix} \theta_{ss} \\ \omega_{ss} \end{pmatrix} = \begin{pmatrix} 0 \\ \pm\pi \end{pmatrix}, \quad \begin{pmatrix} 0 \\ \pm2\pi \end{pmatrix}, \ldots . \quad ■$$

To study the behavior of solutions near such an equilibrium $\mathbf{y}_{ss}$, the standard technique of *linear approximation* is to introduce a perturbation $\mathbf{u}(t)$ defined by $\mathbf{y}(t) = \mathbf{y}_{ss} + \mathbf{u}(t)$, where $\mathbf{y}$ is a solution of the nonlinear system $\mathbf{y}' = \mathbf{f}(\mathbf{y})$; that is, the perturbation is defined by

$$\mathbf{y}(t) = \begin{pmatrix} y_{ss} + u(t) \\ z_{ss} + v(t) \end{pmatrix}$$

and

$$(y_{ss} + u(t))' = f(y_{ss} + u(t), z_{ss} + v(t)), \tag{8.20}$$
$$(z_{ss} + v(t))' = g(y_{ss} + u(t), z_{ss} + v(t)). \tag{8.21}$$

Alternatively, the perturbation $\mathbf{u}(t) = \mathbf{y}(t) - \mathbf{y}_{ss}$ is the difference between a solution $\mathbf{y}(t)$ that starts near the steady state and the steady state $\mathbf{y}_{ss}$ itself. "Starts near the steady state" means $\mathbf{u}(0)$ is a vector of small magnitude or $\mathbf{y}(0) \approx \mathbf{y}_{ss}$.

Linear approximation consists of expanding f and g on the right side of these differential equations in Taylor series (see section A.3, page 627) about the steady-state solution (y_{ss}, z_{ss}) and discarding terms involving powers greater than one of the (small) perturbations u and v.

For example, the Taylor expansion for f is

$$f(y_{ss} + u, z_{ss} + v) = f(y_{ss}, z_{ss}) + \frac{\partial f(y_{ss}, z_{ss})}{\partial y}u + \frac{\partial f(y_{ss}, z_{ss})}{\partial z}v + \cdots$$

This expansion can be simplified in two steps, one exact, the other approximate.

The exact simplification is $f(y_{ss}, z_{ss}) = 0$ from the definition of steady state.

The approximate simplification is ignoring whatever appears in $+\cdots$ in the Taylor expansion, the terms after the first partial derivatives that involve u^2, uv, v^2 and higher powers of these small terms. The justification for this simplification is that powers of small quantities (less than one in magnitude) are even smaller. The resulting *linear approximation* is

Linear approximation of f.

$$f(y_{ss} + u, z_{ss} + v) \approx \frac{\partial f(y_{ss}, z_{ss})}{\partial y}u + \frac{\partial f(y_{ss}, z_{ss})}{\partial z}v.$$

Stop and Think **8.23** Is it accurate to read this approximation as "rate of change of f with respect to y times change in y plus a similar expression in z"? Is u the change in y, v the change in z?

8.24 A classic example of linear approximation is "distance equals rate times time" when driving on a superhighway: even though distance covered is a relatively complex function (an integral over time) of the variable highway speed, that linear expression is a good approximation of the distance covered.

What are the analogs of distance, rate, and time in this linear approximation? Do there seem to be two rates and two times? Explain.

8.25 Use a Taylor expansion in one variable to show that $\sin\theta \approx \theta$ is a linear approximation valid for small θ.

Applying similar arguments to $g(y_{ss} + u(t), z_{ss} + v(t))$ leads finally to an approximate *constant-coefficient homogeneous linear system* that governs the perturbations $\mathbf{u} = \begin{pmatrix} u & v \end{pmatrix}^T$ in the vicinity of the steady state (y_{ss}, z_{ss}),

APPROXIMATE LINEAR SYSTEM

$$\begin{aligned} u' &= f_y(y_{ss}, z_{ss})u + f_z(y_{ss}, z_{ss})v \\ v' &= g_y(y_{ss}, z_{ss})u + g_z(y_{ss}, z_{ss})v. \end{aligned} \tag{8.22}$$

Stop and Think **8.26** Derive the second equation in this linear system, $v' = g_y u + g_z v$.

The (constant) coefficient matrix of the system (8.22),

JACOBIAN MATRIX

$$\mathbf{J}(y_{ss}, z_{ss}) = \begin{pmatrix} f_y(y_{ss}, z_{ss}) & f_z(y_{ss}, z_{ss}) \\ g_y(y_{ss}, z_{ss}) & g_z(y_{ss}, z_{ss}) \end{pmatrix},$$

is called the **Jacobian matrix** of the nonlinear autonomous system $\mathbf{y}' = \mathbf{f}(\mathbf{y})$ evaluated at the equilibrium $\mathbf{y}_{ss}$. Almost always, *the behavior of the linear system $\mathbf{u}' = \mathbf{Ju}$ near its equilibrium at the origin—behavior determined by the eigenvalues and eigenvectors of the Jacobian matrix $\mathbf{J}$—provides an accurate picture of the behavior of the full nonlinear system near its equilibrium point* (y_{ss}, z_{ss}).

Roughly, if the equilibrium of the linear system is not neutrally stable, then the linear approximation is a reliable guide to the local behavior of the nonlinear system.

Stop and Think **8.27** Why is the Jacobian a constant matrix?

A detailed picture of the behavior of a two-dimensional autonomous system near an equilibrium point $\mathbf{y}_{ss}$ depends upon an analysis of the linearized system $\mathbf{u}' = \mathbf{Ju}$ with the Jacobian matrix $\mathbf{J}$ evaluated at the equilibrium. In most cases, the behavior of the linear system $\mathbf{u}' = \mathbf{Ju}$ near the origin provides a good close-up view of the behavior of the nonlinear system $\mathbf{y}' = \mathbf{f}(\mathbf{y})$ near $\mathbf{y}_{ss}$.

■ **EXAMPLE 19**   *Find the Jacobian matrix $\mathbf{J}$ of the pendulum system (8.19) at the steady state $\mathbf{y}_{ss} = \begin{pmatrix} \pi & 0 \end{pmatrix}^T$. Find the eigenvalues of $\mathbf{J}$ and comment on their significance for the stability of this steady state.*

Stop and Think **8.28** What position of the pendulum corresponds to $\mathbf{y}_{ss} = \begin{pmatrix} \pi & 0 \end{pmatrix}^T$?

With $f(\theta, \omega) = \omega$, and $g(\theta, \omega) = -(g/L)\sin\theta - (p/L)\omega$, the Jacobian matrix of (8.19) is

Jacobian matrix

$$\mathbf{J} = \begin{pmatrix} f_\theta(\theta_{ss}, \omega_{ss}) & f_\omega(\theta_{ss}, \omega_{ss}) \\ g_\theta(\theta_{ss}, \omega_{ss}) & g_\omega(\theta_{ss}, \omega_{ss}) \end{pmatrix} = \begin{pmatrix} 0 & 1 \\ -(g/L)\cos\theta_{ss} & -p/L \end{pmatrix}. \quad (8.23)$$

Stop and Think **8.29** Verify the partial derivatives appearing in the Jacobian.

At the steady state $\theta_{ss} = \pi$, $\omega_{ss} = 0$, the Jacobian is

Jacobian evaluated at $\theta_{ss} = \pi$, $\omega_{ss} = 0$

$$\mathbf{J}(\pi, 0) = \begin{pmatrix} 0 & 1 \\ g/L & -p/L \end{pmatrix}.$$

The characteristic equation of $\mathbf{J}$ is

Characteristic equation of $\mathbf{J}$

$$\det(\mathbf{J} - r\mathbf{I}) = \det\begin{pmatrix} -r & 1 \\ g/L & -p/L - r \end{pmatrix} = r^2 + \frac{p}{L} - \frac{g}{L} = 0.$$

Its roots, the eigenvalues of $\mathbf{J}$, are

Eigenvalues of $\mathbf{J}$

$$r_\pm = \frac{-p \pm \sqrt{p^2 + 4gL}}{2L}.$$

They are real and of opposite sign: $r_- < 0 < r_+$.

Stop and Think **8.30** Verify the form of the characteristic equation of $\mathbf{J}$, its eigenvalues, and the assertion about the signs of the eigenvalues.

Perturbations $\mathbf{u} = \begin{pmatrix} u & v \end{pmatrix}^T$ of the steady state $\mathbf{y}_{ss} = \begin{pmatrix} \pi & 0 \end{pmatrix}^T$ are governed by the constant-coefficient homogeneous linear system $\mathbf{u}' = \mathbf{J}\mathbf{u}$. A general solution of this system has the form

General solution of $\mathbf{u}' = \mathbf{J}\mathbf{u}$

$$\mathbf{u} = C_1\mathbf{p}_+ e^{r_+ t} + C_2\mathbf{p}_- e^{r_- t}$$

where $\mathbf{p}_\pm$ are eigenvectors of $\mathbf{J}$ corresponding to the eigenvalues $r_\pm$. Since $r_+ > 0$, the magnitude of the perturbation $\mathbf{u}$ grows with time except for those special initial conditions that force $C_1 = 0$. Hence, the equilibrium $\mathbf{y}_{ss} = \begin{pmatrix} \pi & 0 \end{pmatrix}^T$ is unstable. ■

Stop and Think **8.31** If $\mathbf{p}_+ = \mathbf{0}$, then $\mathbf{u}$ would not grow with time regardless of the value of C_1. Why is $\mathbf{p}_+ \neq \mathbf{0}$?

This example illustrates one outcome of a stability analysis, real eigenvalues of the Jacobian. Just one positive eigenvalue guarantees instability, as in this case; both real eigenvalues must be negative to ensure stability.

Another possible outcome is a Jacobian whose eigenvalues are complex. In that case, the local solutions oscillate (spirals in the phase plane) because of the imaginary part of the eigenvalue. The direction of the spiral (in for stability, out for instability) is determined by the sign of the real part of the eigenvalue. If the real part is zero, then the spirals degenerate to closed curves (ellipses).

Stop and Think   **8.32** Which sign—positive or negative—of the real part of a complex eigenvalue corresponds to stability?

The balance of this section explores these two primary alternatives, beginning with real eigenvalues. In each case, *the form of the general solution of* $\mathbf{y}' = \mathbf{J}\mathbf{y}$ *determines the phase plane behavior.*

8.3.2   Real Eigenvalues

Begin with an autonomous system $\mathbf{y}' = \mathbf{f}(\mathbf{y})$. Find a steady state $\mathbf{y}_{ss}$. Use the Jacobian matrix $\mathbf{J}(\mathbf{y}_{ss})$ to construct a linear approximation of $\mathbf{y}' = \mathbf{f}(\mathbf{y})$ in the neighborhood of $\mathbf{y}_{ss}$.

Suppose the Jacobian matrix has real eigenvalues r_1, r_2 and corresponding linearly independent (nonparallel) eigenvectors $\mathbf{p}_1, \mathbf{p}_2$. Then the form of the general solution of the linear system $\mathbf{u}' = \mathbf{J}\mathbf{u}$ shows

GENERAL SOLUTION,
REAL EIGENVALUES

> *Any* **solution is a combination of two vectors,**
>
> $$\mathbf{u}_g = C_1\mathbf{p}_1 e^{r_1 t} + C_2\mathbf{p}_2 e^{r_2 t}. \qquad (8.24)$$

The eigenvectors $\mathbf{p}_1$, $\mathbf{p}_2$ of $\mathbf{J}$ define the directions. The contribution of the eigenvector $\mathbf{p}_i$ to the full solution $\mathbf{u}$ is controlled by the constant C_i, which is determined by initial conditions, and by the exponential factor $e^{r_i t}$, which may be growing or decaying, depending on the sign of r_i.

To emphasize the role of the eigenvalues and eigenvectors in determining the behavior of linear, homogeneous, constant-coefficient systems, recall a system whose general solution was constructed earlier.

■ **EXAMPLE 20**   *Analyze the behavior of the solutions of* $\mathbf{y}' = \mathbf{B}\mathbf{y}$ *with*

$$\mathbf{B} = \begin{pmatrix} -1/2 & 3/2 \\ 3/2 & -1/2 \end{pmatrix}.$$

The eigenvalues and eigenvectors of this coefficient matrix were found in example 10, page 386. Solutions of the system were constructed in example 11, page 388. The eigenvalues of $\mathbf{B}$ are $r_- = -2$, $r_+ = 1$. Corresponding eigenvectors are

$$\mathbf{p}_- = \begin{pmatrix} 1 \\ -1 \end{pmatrix}, \quad \mathbf{p}_+ = \begin{pmatrix} 1 \\ 1 \end{pmatrix}.$$

A general solution of $\mathbf{y}' = \mathbf{B}\mathbf{y}$ is (compare with (8.24))

$$\mathbf{y}_g = C_-\mathbf{p}_-e^{r_-t} + C_+\mathbf{p}_+e^{r_+t} = C_-\begin{pmatrix} 1 \\ -1 \end{pmatrix}e^{-2t} + C_+\begin{pmatrix} 1 \\ 1 \end{pmatrix}e^t. \quad (8.25)$$

The constants C_-, C_+ are determined by the given initial conditions. For example, the solution of $\mathbf{y}' = \mathbf{B}\mathbf{y}$ that satisfies the initial condition

$$\mathbf{y}(0) = \begin{pmatrix} 4 \\ -3 \end{pmatrix} \quad (8.26)$$

has $C_- = 3.5, C_+ = 0.5$.

Stop and Think **8.33** Verify $C_\pm$.

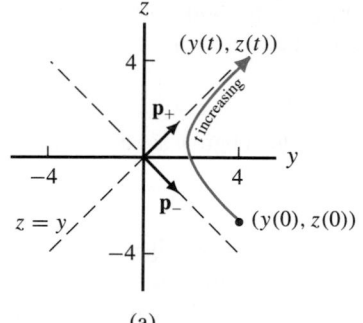

(a)

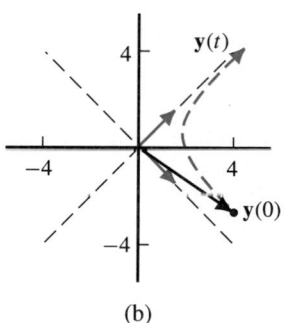

(b)

FIGURE 8.5 The solution trajectory of the system of example 20 that starts at $(4, -3)$ is asymptotic to the line $z = -y$, the direction of $\mathbf{p}_+$, the eigenvector corresponding to the positive eigenvalue $r_+ = 1$.

That is, the solution that starts at $(4, -3)$ is

$$\mathbf{y}(t) = 3.5\begin{pmatrix} 1 \\ -1 \end{pmatrix}e^{-2t} + 0.5\begin{pmatrix} 1 \\ 1 \end{pmatrix}e^t. \quad (8.27)$$

A phase plane diagram of this solution trajectory can be visualized in two ways:

- As a plot of the *parametric curve* $(y(t), z(t))$, where the functions y and z are defined by (8.27); i.e., by

$$y(t) = 3.5e^{-2t} + 0.5e^t,$$
$$z(t) = -3.5e^{-2t} + 0.5e^t.$$

- As the path in the plane traced by the (tip of the) *solution vector* or state vector $\mathbf{y}(t)$ defined by (8.27), $\mathbf{y}(t) = 3.5\mathbf{p}_-e^{-2t} + 0.5\mathbf{p}_+e^t$.

Of course, both approaches produce the same trajectory.

Parametric equation approach: The phase plane plot of the solution (8.27) begins in the fourth quadrant at the point $(y(0), z(0)) = (4, -3)$ because $y(0) = 3.5 + 0.5 = 4$, $z(0) = -3.5 + 0.5 = -3$.

As t increases, $e^{-2t} \to 0$ and $e^t \to \infty$. Hence, as t increases, the phase plane graph of the solution enters the first quadrant because $y \to \infty$ and $z \to -\infty$. Furthermore, for large t,

$$\frac{z(t)}{y(t)} = \frac{-3.5e^{-2t} + 0.5e^t}{3.5e^{-2t} + 0.5e^t} \approx \frac{0.5e^t}{0.5e^t} = 1.$$

That is, for large t, $z \approx y$: the phase plane plot of the solution is approaching the line $z = y$. This information is illustrated in figure 8.5(a).

MATLAB

Use the phase plane tool in DELAB to reproduce these trajectories and to experiment with different parameter values.

Vector approach: At $t = 0$, the tip of the solution vector $\mathbf{y}(t)$ points at

$$\mathbf{y}(0) = 3.5\mathbf{p}_- + 0.5\mathbf{p}_+ = 3.5 \begin{pmatrix} 1 \\ -1 \end{pmatrix} + 0.5 \begin{pmatrix} 1 \\ 1 \end{pmatrix} = \begin{pmatrix} 4 \\ -3 \end{pmatrix}.$$

As t increases, $\mathbf{y}(t) = 3.5\mathbf{p}_- e^{-2t} + 0.5\mathbf{p}_+ e^t$ grows along the direction of $\mathbf{p}_+$ because e^t is increasing. Simultaneously, it shrinks along the direction of $\mathbf{p}_-$ because e^{-2t} is decaying. The solution *grows* in the direction of the eigenvector associated with the *positive* eigenvalue, and it *shrinks* in the direction of the eigenvector associated with the *negative* eigenvalue. This behavior is shown in figure 8.5(b).

From either point of view, the equilibrium point at $(0,0)$ is *unstable* because most perturbations of this state grow. Indeed, perturbations grow in the direction of $\mathbf{p}_+$.

The only perturbations that do not grow are those whose initial conditions require $C_+ = 0$ in the general solution (8.25). Solutions that contain no component of the growing eigenvalue-eigenvector pair $r_+ = 1$, $\mathbf{p}_+ = \begin{pmatrix} 1 & -1 \end{pmatrix}^T$ can never grow.

Stop and Think   **8.34** Is it correct to say that the only perturbations which do not grow are those that lie along the line of the vector $\mathbf{p}_-$? ■

Saddle point

Equilibrium points at which the system has one positive eigenvalue and one negative eigenvalue are called **saddle points**. The reason for this suggestive name is clear from the flow field diagram of figure 8.6, which is drawn for the system of example 20. Saddle points are always unstable.

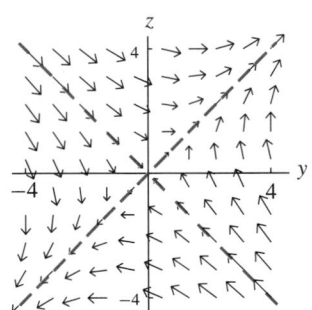

FIGURE 8.6 A typical saddle point instability arising from real eigenvalues of opposite sign, drawn here for the system of example 20.

■ **EXAMPLE 21**   *Example 19 found the Jacobian $\mathbf{J}$ of the damped pendulum system (8.19) at the equilibrium $\theta_{ss} = \pi$, $\omega_{ss} = 0$. Build on that work to analyze the behavior of the approximating system $\mathbf{u}' = \mathbf{J}\mathbf{u}$. Does the linear system accurately describe the behavior of the nonlinear system in the neighborhood of that equilibrium? Take $g = L = 1$ and $p = 2$.*

With $g = L = 1$ and $p = 2$, the work of example 19 yields the Jacobian of the nonlinear pendulum system (8.19) evaluated at $\theta_{ss} = \pi$, $\omega_{ss} = 0$,

$$\mathbf{J} = \begin{pmatrix} 0 & 1 \\ 1 & -2 \end{pmatrix}$$

and its eigenvalues

$$r_\pm = -1 \pm \sqrt{2}.$$

The usual analysis produces the eigenvectors

$$\mathbf{p}_\pm = \begin{pmatrix} 1 \\ -1 \pm \sqrt{2} \end{pmatrix}.$$

Stop and Think   **8.35** Verify that $\mathbf{p}_\pm$ are eigenvectors by showing that $\mathbf{J}\mathbf{p}_\pm = r_\pm\mathbf{p}_\pm$.

A general solution of the approximating linear system is (compare with (8.24))

$$\mathbf{u}_g = C_-\mathbf{p}_-e^{r_-t} + C_+\mathbf{p}_+e^{r_+t}.$$

Evidently, one component is decreasing with time and the other is increasing, the telltale sign of a saddle point instability.

The instability of the equilibrium $(\pi, 0)$ of the nonlinear pendulum is old news, but the linear analysis now permits a more detailed view of the nature of that instability. The eigenvector $\mathbf{p}_+$ associated with the positive eigenvalue is the dominant direction pulling trajectories that start near the equilibrium away from it. Only trajectories that start precisely on a line through the eigenvector $\mathbf{p}_-$ corresponding to the negative eigenvalue can return to the equilibrium point.

Figure 8.7 shows this behavior for the *linear* system $\mathbf{u}' = \mathbf{Ju}$. That figure certainly seems to be a plausible close-up view of the behavior near $(\pi, 0)$ shown in figure 8.8, a phase plane diagram of the *nonlinear* system (8.19). The linear analysis has identified the details of the behavior near the equilibrium. ■

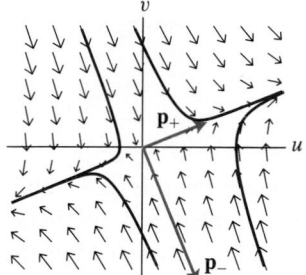

FIGURE 8.7 A phase diagram of the saddle point in a linear approximation to the critically damped pendulum system (8.19) near the equilibrium $\theta_{ss} = \pi$, $\omega_{ss} = 0$.

Stop and Think

8.36 The linear phase diagram of figure 8.7 is centered at the origin. But the equilibrium point of the nonlinear system is $(\pi, 0)$. Why are the two different?

8.37 Suppose figure 8.7, the approximate linear phase diagram, were to be pasted onto a blowup of figure 8.8, a phase diagram of the nonlinear pendulum system. Where would the origin of figure 8.7 be located on the blowup of figure 8.8? Which trajectory in figure 8.7 would match which trajectory in figure 8.8? Identify trajectories that appear in one figure but not in the other.

Different behavior occurs if both eigenvalues have the same sign; the equilibrium at the origin is then called a *node*. We will see that one component of the general solution (8.24) decays or grows faster than the other if the eigenvalues are distinct. Most trajectories enter or leave the origin tangent to the direction of the eigenvector having the eigenvalue of smaller magnitude.

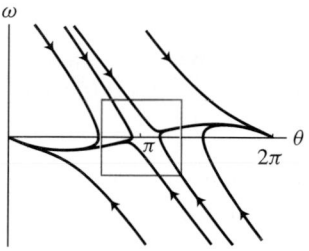

FIGURE 8.8 A phase diagram of the critically damped nonlinear pendulum system (8.19) with $g = L = 1$, $p = 2$; the blue box marks the closeup view shown in figure 8.7, the phase plane of the linear approximation $\mathbf{u}' = \mathbf{Ju}$.

■ **EXAMPLE 22** *Analyze the behavior at the origin of the system* $\mathbf{y}' = \mathbf{By}$ with

$$\mathbf{B} = \begin{pmatrix} -5 & -3 \\ -3 & -5 \end{pmatrix}.$$

The roots of the characteristic polynomial $\det(\mathbf{B} - r\mathbf{I}) = (-5 - r)^2 - 9 = 0$ are $r = -2, -8$. Corresponding eigenvectors are

Eigenvectors of $\mathbf{B}$

$$\mathbf{p}_{-2} = \begin{pmatrix} -1 \\ 1 \end{pmatrix}, \quad \mathbf{p}_{-8} = \begin{pmatrix} 1 \\ 1 \end{pmatrix}$$

and a general solution of $\mathbf{y}' = \mathbf{By}$ is (compare with (8.24))

General solution of $\mathbf{y}' = \mathbf{By}$

$$\mathbf{y}_g = C_1\mathbf{p}_{-2}e^{-2t} + C_2\mathbf{p}_{-8}e^{-8t} = C_1\begin{pmatrix} -1 \\ 1 \end{pmatrix}e^{-2t} + C_2\begin{pmatrix} 1 \\ 1 \end{pmatrix}e^{-8t}.$$

Stop and Think   **8.38** Verify the eigenvectors by computing **Bp**.

Evidently, all solutions decay to zero; the origin is a *stable node*. Calculating the ratio of the coefficients of the two eigenvectors (assuming $C_1 \neq 0$),

$$\frac{C_2 e^{-8t}}{C_1 e^{-2t}} = \frac{C_2}{C_1} e^{-6t} \to 0, \tag{8.28}$$

shows that the $\mathbf{p}_{-8}$ component is decaying much faster than its $\mathbf{p}_{-2}$ counterpart. After enough time has gone by, the $\mathbf{p}_{-2}$ component will dominate, regardless of the values of $C_1 \neq 0$ and C_2. These trajectories will enter the origin tangent to the $\mathbf{p}_{-2}$ direction, as figure 8.9 illustrates.

Stop and Think   **8.39** There are some trajectories that do not enter the origin tangent to the $\mathbf{p}_{-2}$ direction. Describe them. Where would they appear in figure 8.9? ■

Node

The situation analyzed in this example holds in general: the origin is a (proper) **node** of $\mathbf{y}' = \mathbf{B}\mathbf{y}$ if **B** has distinct real eigenvalues of the same sign. It is stable if the eigenvalues are negative, the case just considered, and unstable if the eigenvalues are positive. Most trajectories enter or leave the origin tangent to the direction of the eigenvector associated with the eigenvalue of smaller magnitude. See exercises 27 and 28.

Stop and Think   **8.40** "Most trajectories enter or leave the origin tangent to . . . ." Which trajectories do *not* enter or leave tangent to the direction of the eigenvector associated with the eigenvalue of smaller magnitude?

If the real eigenvalues are equal in magnitude but there are two linearly independent (nonparallel) eigenvectors, then the node is star-like; any direction of approach is possible. See exercise 29.

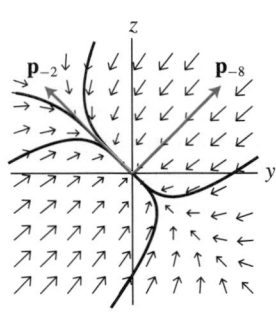

FIGURE 8.9   A stable node produced by the system of example 22.

■ **EXAMPLE 23**   *Analyze the stability of the steady state $\theta_{ss} = 0$, $\omega_{ss} = 0$ of the overdamped pendulum system (8.19). To be concrete, take $g = L = 1$ and $p = 4$.*

The Jacobian of the pendulum system is given in (8.23). Using $\theta_{ss} = 0$, $\omega_{ss} = 0$, $g = L = 1$ and $p = 4$, we obtain

$$\mathbf{J} = \begin{pmatrix} 0 & 1 \\ -1 & -4 \end{pmatrix}.$$

The eigenvalues of **J** are $r_{\pm} = -2 \pm \sqrt{3} < 0$, consistent with overdamping.

Stop and Think   **8.41** What property of these eigenvalues confirms that this pendulum is overdamped?

Corresponding eigenvectors are

$$\mathbf{p}_{\pm} = \begin{pmatrix} 1 \\ -2 \pm \sqrt{3} \end{pmatrix}.$$

Stop and Think **8.42** Write the characteristic polynomial of **J**. Find the eigenvalues and verify the expression for the eigenvectors.

Both eigenvalues are negative but r_- has larger magnitude: $r_- < r_+ < 0$. To emphasize that the behavior of solutions of the linearized system $\mathbf{u}' = \mathbf{J}\mathbf{u}$ depends on the eigenvalues but not on the length of the eigenvectors, we can replace $\mathbf{p}_\pm$ with parallel vectors of length unity. Dividing each of $\mathbf{p}_\pm$ by its length yields

$$\mathbf{p}_{\text{large}} = \frac{1}{2\sqrt{2+\sqrt{3}}}\mathbf{p}_-, \quad \mathbf{p}_{\text{small}} = \frac{1}{2\sqrt{2-\sqrt{3}}}\mathbf{p}_+.$$

The eigenvectors of unit length have been labeled *large* and *small* according to the relative magnitude of their corresponding eigenvalue.

Stop and Think **8.43** Verify the length calculations. Verify that $\mathbf{p}_{\text{large}}$, $\mathbf{p}_{\text{small}}$ are eigenvectors of **J** and that they have unit length.

Following (8.24), a general solution of the approximating linear system is

$$\mathbf{u}_g = C_1\mathbf{p}_{\text{large}}e^{(-2-\sqrt{3})t} + C_2\mathbf{p}_{\text{small}}e^{(-2+\sqrt{3})t}.$$

Stability is clear from the decaying exponentials. Paralleling example 22, all trajectories except those that start on the line through $\mathbf{p}_{\text{large}}$ enter the origin tangent to the direction of $\mathbf{p}_{\text{small}}$, as shown in figure 8.10. ■

Stop and Think **8.44** Verify the claim about the direction of approach to the origin in figure 8.9 using the appropriate analog of equation (8.28) from example 22.

8.45 Relabel the eigenvector directions in figure 8.10 as *fast* and *slow* according to their relative rate of decay. Which direction goes with which eigenvalue? Suppose the fast decay is very fast, almost instantaneous, and the slow decay is very slow, like ice melting. Use the difference in rates of decay to give an intuitive explanation for the way the trajectories in figure 8.10 enter the origin.

8.46 Compare the behavior shown in figure 8.10 with a close-up view of the origin of the pendulum phase plane in figure 7.9, page 344. Is the pendulum of figure 7.9 overdamped?

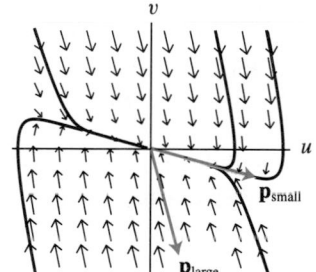

FIGURE 8.10 The steady state $\theta_{ss} = 0$, $\omega_{ss} = 0$ is a stable node for the overdamped pendulum system.

8.3.3 Complex Eigenvalues

Now suppose that the Jacobian matrix **J** of the approximating linear system $\mathbf{u}' = \mathbf{J}\mathbf{u}$ has complex conjugate eigenvalues

Complex conjugate eigenvalues

$$r_\pm = \alpha \pm i\beta$$

and associated complex conjugate eigenvectors $\mathbf{p}_\pm$. A real-valued general solution is constructed from the real part and the imaginary part of the complex-valued solutions $\mathbf{p}_\pm \, e^{(\alpha\pm i\beta)t}$. (See example 13, page 391, or example 15, page 394.)

A particular view of the structure of that general solution is useful here. Euler's formula introduces an exponential from the real part of the eigenvalue and bounded, periodic functions from the imaginary part,

$$e^{(\alpha \pm i\beta)t} = e^{\alpha t}(\cos t\beta t \pm \sin \beta t).$$

The complex-valued solutions $\mathbf{p}_\pm e^{(\alpha \pm i\beta)t}$ *always* contain the exponential factor $e^{\alpha t}$ multiplying either $\sin \beta t$ or $\cos \beta t$. Hence, if $\mathbf{J}$ has complex conjugate eigenvalues, the general solution of the linear system $\mathbf{u}' = \mathbf{J}\mathbf{u}$ shows that

GENERAL SOLUTION,
COMPLEX CONJUGATE
EIGENVALUES

> *Any* **solution is a combination of two vectors,**
>
> $$\mathbf{u}_g = e^{\alpha t}(C_1\mathbf{S}_1(\beta t) + C_2\mathbf{S}_2(\beta t)). \qquad (8.29)$$

The components of the vector functions $\mathbf{S}_1(\beta t)$ and $\mathbf{S}_2(\beta t)$ are linear combinations of $\sin \beta t$ and $\cos \beta t$; their length is bounded.

Stop and Think   **8.47** Compare the form of the general solution (8.29) in the case of complex conjugate eigenvalues with that for real eigenvalues, equation (8.24). Are both formed from two basic vectors? How do those vectors differ? Do both general solutions contain exponential factors?

How can such a general solution behave? Because their components are combinations of sines and cosines, the orientation of the vectors $\mathbf{S}_1(\beta t)$ and $\mathbf{S}_2(\beta t)$ will change but their lengths will remain bounded. (One example of these vectors is drawn in figure 8.11.) That is, *the general solution vector is the product of an exponential scaling factor $e^{\alpha t}$ and periodically oscillating vectors $\mathbf{S}_i(\beta t)$ of bounded length.*

If $\alpha \neq 0$, equilibria with complex conjugate eigenvalues are called **spiral points** because trajectories cycle around the origin as the $\mathbf{S}_i(\beta t)$ vectors vary, expanding or contracting with the exponential factor $e^{\alpha t}$. Clearly, a spiral point is **stable** if the real part α of the eigenvalues is positive, **unstable** if it is negative. A spiral point is sometimes called a *focus*.

If $\alpha = 0$, then the equilibrium is *neutrally stable* and the trajectories are ellipses around the origin—degenerate spirals that neither expand nor contract. Such an equilibrium is called a *center*.

Spiral point

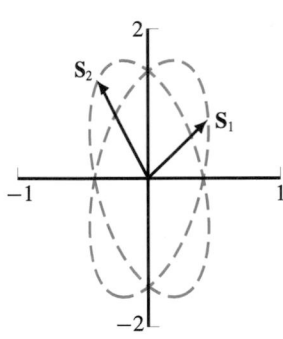

FIGURE 8.11 One cycle of the oscillating vectors $\mathbf{S}_i(\beta t)$ that appear in the general solution of $\mathbf{y}' = \mathbf{J}\mathbf{y}$ when $\mathbf{J}$ has complex conjugate eigenvalues; in example 24, $\beta = \sqrt{3}/2$.

■ **EXAMPLE 24**   *Analyze the stability of the steady state $\theta_{ss} = 0$, $\omega_{ss} = 0$ of the underdamped pendulum system (8.19). To be concrete, take $g = L = 1$ and $p = 1$.*

The Jacobian of the pendulum system is given in (8.23). Using $\theta_{ss} = 0$, $\omega_{ss} = 0$, $g = L = 1$ and $p = 1$, we obtain

$$\mathbf{J} = \begin{pmatrix} 0 & 1 \\ -1 & -1 \end{pmatrix}.$$

The eigenvalues of $\mathbf{J}$ are the complex conjugate pair $r_\pm = (-1 \pm i\sqrt{3})/2$. Corresponding (complex) eigenvectors are

$$\mathbf{p}_\pm = \begin{pmatrix} 1 \\ (-1 \pm i\sqrt{3})/2 \end{pmatrix}.$$

Stop and Think **8.48** Write the characteristic polynomial of $\mathbf{J}$. Find the eigenvalues and verify the expression for the eigenvectors. Is the pendulum indeed underdamped?

To form a real-valued general solution, identify the real and imaginary parts of $e^{r_+ t}\mathbf{p}_+$:

$$e^{r_+ t}\mathbf{p}_+ = e^{-t/2}\left(\cos(\sqrt{3}\,t/2) + i\sin(\sqrt{3}\,t/2)\right)\begin{pmatrix} 1 \\ (-1 + i\sqrt{3})/2 \end{pmatrix}$$

$$= e^{-t/2}\begin{pmatrix} \cos(\sqrt{3}\,t/2) \\ -\left(\cos(\sqrt{3}\,t/2) + \sqrt{3}\sin(\sqrt{3}\,t/2)\right)/2 \end{pmatrix}$$

$$+ i e^{-t/2}\begin{pmatrix} \sin(\sqrt{3}\,t/2) \\ \left(\sqrt{3}\cos(\sqrt{3}\,t/2) - \sin(\sqrt{3}\,t/2)\right)/2 \end{pmatrix}$$

$$= e^{-t/2}\left(\mathbf{S}_1(\sqrt{3}\,t/2) + i\mathbf{S}_2(\sqrt{3}\,t/2)\right).$$

The real part is $e^{-t/2}\mathbf{S}_1(\sqrt{3}\,t/2)$, the imaginary part is $e^{-t/2}\mathbf{S}_2(\sqrt{3}\,t/2)$. With these expressions for $\mathbf{S}_i(\sqrt{3}\,t/2)$, a real-valued general solution is given by (8.29) with $\alpha = -1/2$:

$$\mathbf{u}_g = e^{-t/2}\left(C_1\mathbf{S}_1(\sqrt{3}\,t/2) + C_2\mathbf{S}_2(\sqrt{3}\,t/2)\right).$$

One cycle of each of the oscillating vectors $\mathbf{S}_i(\sqrt{3}\,t/2)$ is plotted in figure 8.11. (They trace tilted ellipses.) Since those vectors are bounded, the exponential factor $e^{-t/2}$ forces $\mathbf{u}_g$ to zero for any values of the arbitrary constants C_i; the origin is stable.

The direction of rotation about the origin can be deduced from a study of the oscillating vectors $\mathbf{S}_i(\sqrt{3}\,t/2)$, but that information is also apparent from a nullcline analysis of the linearized system

$$u' = v,$$
$$v' = -u - v,$$

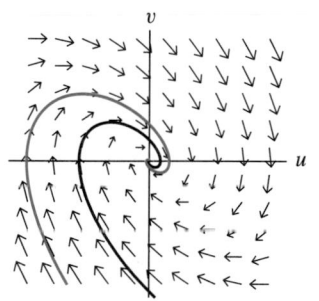

FIGURE 8.12 A stable spiral point resulting from a linear analysis of the equilibrium at the origin of an underdamped pendulum system.

or from the individual equations themselves; e.g., $u' = v$ forces flow field arrows to point toward the right in the upper half plane, and $v' = -u - v$ forces the flow field to point downward in the first quadrant. ■

Figure 8.12 exhibits the stable spiral point of the linear approximation.

Stop and Think **8.49** Why are no eigenvectors pictured in figure 8.12?

8.50 Is figure 8.12 a close-up picture near the origin of the nonlinear pendulum phase portrait of figure 7.9, page 344? Is the pendulum of figure 7.9 underdamped or overdamped?

8.51 The examples in this section show that the pendulum can exhibit three types of equilibria: a saddle point, a node, and a spiral point. Describe the circumstances under which each can appear and its physical significance.

8.3.4 Summary

> **Linear stability analysis of $\mathbf{y}' = \mathbf{f}(\mathbf{y})$.** Solve $\mathbf{f}(\mathbf{y}_{ss}) = \mathbf{0}$ to find a steady state $\mathbf{y}_{ss}$. Use the Jacobian matrix $\mathbf{J}(\mathbf{y}_{ss})$ to construct a linear approximation of $\mathbf{y}' = \mathbf{f}(\mathbf{y})$ in the neighborhood of $\mathbf{y}_{ss}$.

In most cases, the behavior of the linear system $\mathbf{u}' = \mathbf{J}\mathbf{u}$ near its equilibrium at the origin accurately depicts the behavior of the nonlinear system $\mathbf{y}' = \mathbf{f}(\mathbf{y})$ near its equilibrium $\mathbf{y}_{ss}$.

> **Stability for the linear system $\mathbf{u}' = \mathbf{J}\mathbf{u}$.** If the eigenvalues of $\mathbf{J}$ are
>
> - real and of opposite sign, the origin is a *saddle point*.
> - real, distinct, and of the same sign, the origin is a (proper) *node*.
> - complex conjugate with nonzero real part, the origin is a *spiral point*.
>
> A saddle point is unstable. A node is asymptotically stable if the eigenvalues are negative and unstable if they are positive. A spiral point is asymptotically stable if the real part of the eigenvalues is negative and unstable if it is positive.

These three types of behavior are illustrated in figure 8.13.

Under reasonable conditions (e.g., [20, p. 86]), a nonlinear system exhibits a saddle, a node, or a spiral in the neighborhood of an equilibrium point if the linear approximation around that point predicts such behavior. Saddle points are shown in figures 8.6 and 8.7, stable nodes in figures 8.9 and 8.10, and a stable spiral in figure 8.12.

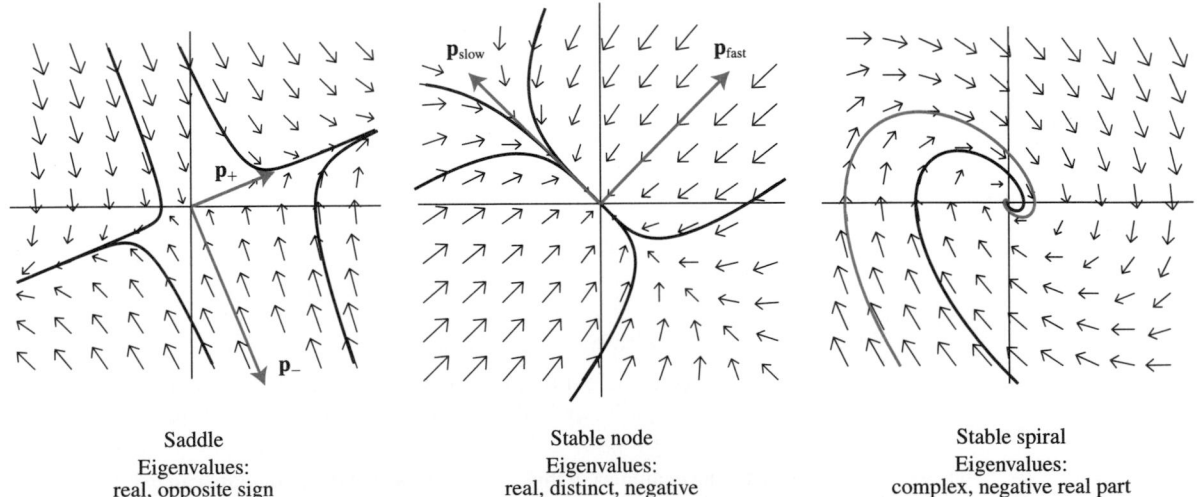

Saddle	Stable node	Stable spiral
Eigenvalues:	Eigenvalues:	Eigenvalues:
real, opposite sign	real, distinct, negative	complex, negative real part

FIGURE 8.13   A saddle point (always unstable), a stable node (unstable if the eigenvalues are positive), and a stable spiral point (unstable if the real part of the eigenvalues is positive).

MATLAB

Menu bar options in DELAB calculate the Jacobian of a nonlinear system at a selected equilibrium, determine its eigenvalues and eigenvectors, and display the phase plane of the linearized system. For guidance, select **Help, Textbook**, then go to chapter 8, linearization. Phase plane diagrams of linear or nonlinear systems can also be drawn directly; select **Graphical tools** in the equation window.

Two other situations arise. If the eigenvalues of **J** are real but identical, the equilibrium is an *improper node*. A single negative eigenvalue guarantees stability; a positive eigenvalue is unstable. However, the trajectories may approach or leave the origin at any angle, not just along the preferred directions of a conventional node, depending on the nature of the eigenvectors. See exercises 29 and 30 for examples. Similar behavior may or may not be present in a nonlinear system whose Jacobian has repeated real eigenvalues.

The other situation is that of pure imaginary eigenvalues. Eigenvalues with zero real part represent a transition between those with negative real part (stable) and those with positive real part (unstable). If the eigenvalues of the linear system are pure imaginary, then the origin is a neutrally stable center, but additional information is required to make a statement about the stability of the equilibrium of a nonlinear system whose Jacobian has pure imaginary eigenvalues. See exercises 31–34 at the end of this section.

The characteristic polynomial, and hence the eigenvalues, of the matrix

$$\mathbf{B} = \begin{pmatrix} a & b \\ c & d \end{pmatrix}$$

can be written in terms of two quantities

$$T = a + d \quad \text{and} \quad D = ad - bc.$$

(T is called the *trace* of **B**. Of course, D is its determinant.) Conditions can then be given in terms T and D, rather than the four components of **B**, that determine the nature of the origin as an equilibrium point of $\mathbf{y}' = \mathbf{B}\mathbf{y}$. See exercise 35.

8.3.5 Exercises

EXERCISE GUIDE	
To gain experience . . .	**Try exercises**
Classifying equilibria of linear systems	1–6, 11–18, 23–24, 26–30, 35
With *homogeneous*, *general solution*, etc.	7–8
With phase portraits and behavior near equilibria	10–18, 20–22, 25, 27–30
Calculating Jacobians and conducting linearized analysis	9, 19, 26, 31–34, 36
With eigenvectors and eigenvalues	1–6, 9, 14–18, 23–24, 27–30

In exercises 1–6, determine the nature of the equilibrium point at the origin for the system $y' = By$. As appropriate, use MATLAB or the results of the exercises from section 8.2 in which the eigenvalues and eigenvectors of these matrices were determined.

1. $B = \begin{pmatrix} 1 & 0 \\ 4 & 3 \end{pmatrix}$

2. $B = \begin{pmatrix} -2 & 0 \\ 2 & -1 \end{pmatrix}$

3. $B = \begin{pmatrix} 11 & 2 \\ 1 & 10 \end{pmatrix}$

4. $B = \begin{pmatrix} 1 & 2 \\ -1 & 1 \end{pmatrix}$

5. $B = \begin{pmatrix} 4 & 3 \\ -3 & 4 \end{pmatrix}$

6. $B = \begin{pmatrix} 3/2 & \sqrt{3}/2 \\ -\sqrt{3}/2 & 3/2 \end{pmatrix}$

7. Justify the statement, "By definition, a homogeneous system has the steady-state solution $y_{ss} \equiv 0$." Must a homogeneous system be linear to have such an equilibrium solution?

8. Verify that (8.25) is indeed a general solution of the system considered in example 20.

9. Examples 18 and 19 write the pendulum system in the general notation $y' = f(y)$, find its steady states, evaluate its Jacobian J at one of those steady states, determine the eigenvalues of J there, and comment on the stability predicted by the linearization. Carry out the same steps for each of the following systems. As appropriate, use DELAB to support or assist your analysis.

(a) The undamped spring-mass system $x' = v$, $v' = -(k/m)x$.

(b) The damped linear pendulum system $\theta' = \omega$, $\omega' = -(g/L)\theta - (p/L)\omega$.

(c) The specific predator-prey model $F' = -(d_F - \alpha R)F$, $R' = (b_R - \beta F)R$ with $d_F = 0.04$, $\alpha = 0.0011$, $b_R = 0.06$, $\beta = 0.009$.

(d) The general predator-prey model $F' = -(d_F - \alpha R)F$, $R' = (b_R - \beta F)R$.

(e) Relaxation oscillator (7.15) $y' = z + y(1 - y^2 - z^2)$, $z' = -y + z(1 - y^3 - z^3)$.

(f) The SIRS model $S' = -bIS + g(P - S - I)$, $I' = bIS - rI$ with $b = 2$, $g = 1$, $r = 1$, $P = 2$.

(g) The competition model $y' = y(1 - y - 2z)$, $z' = z(1 - z - 2y)$.

(h) The competition model $y' = y(1 - y - 2z)$, $z' = z(1 - z - 0.5y)$.

(i) The competition model $y' = y(1 - y - 0.5z)$, $z' = z(1 - z - 2y)$.

(j) The competition model $y' = y(1 - y - 0.5z)$, $z' = z(1 - z - 0.5y)$.

(k) The van der Pol system $i' = -\epsilon(i^2/3 - 1)i - q$, $q' = i$.

10. Sketch trajectories on the phase plane of figure 8.6 corresponding to the following initial conditions.

(a) $y(0) = 2$, $z(0) = -3$

(b) $y(0) = -4$, $z(0) = 3$

(c) $y(0) = 3$, $z(0) = 4$

(d) $y(0) = 1$, $z(0) = 1$

(e) $y(0) = -1$, $z(0) = 1$

(f) $y(0) = -1$, $z(0) = -1$

(g) $y(0) = 1$, $z(0) = -1$

11. Sketch a phase plane diagram for the system

$$y' = 4z,$$
$$z' = 4y.$$

Confirm your results with DELAB.

12. Sketch a phase plane diagram for the system

$$y' = -4y + 2z,$$
$$z' = -3y + 3z.$$

Confirm your results with DELAB.

13. Sketch a phase plane diagram for the system

$$y' = -y/2 + 3z/2,$$
$$z' = 3y/2 - z/2.$$

Confirm your results with DELAB.

14. Characterize the behavior near the origin and draw a phase plane diagram for the system $y' = By$ where B has the eigenvalues and eigenvectors

$$r_1 = -1, \quad p_1 = \begin{pmatrix} 2 \\ -1 \end{pmatrix}, \quad r_2 = 4, \quad p_2 = \begin{pmatrix} -1 \\ 2 \end{pmatrix}.$$

Is the equilibrium $(0, 0)$ stable or unstable?

15. (a) Characterize the behavior near the origin and draw a phase plane diagram for the system $\mathbf{y}' = \mathbf{By}$ where $\mathbf{B}$ has the eigenvalues and eigenvectors

$$r_1 = 1, \quad \mathbf{p}_1 = \begin{pmatrix} 2 \\ -1 \end{pmatrix}, \quad r_2 = -4, \quad \mathbf{p}_2 = \begin{pmatrix} -1 \\ 2 \end{pmatrix}.$$

Is the equilibrium $(0, 0)$ stable or unstable?

(b) Compare this phase plane diagram with that of the preceding exercise. How do the eigenvalues and eigenvectors differ? How are those differences reflected in the phase plane?

16. Characterize the behavior near the origin and draw a phase plane diagram for the system $\mathbf{y}' = \mathbf{By}$ where $\mathbf{B}$ has the eigenvalues and eigenvectors

$$r_1 = -5, \quad \mathbf{p}_1 = \begin{pmatrix} 2 \\ -1 \end{pmatrix}, \quad r_2 = -1, \quad \mathbf{p}_2 = \begin{pmatrix} -1 \\ 2 \end{pmatrix}.$$

Is the equilibrium $(0, 0)$ stable or unstable?

17. Characterize the behavior near the origin and draw a phase plane diagram for the system $\mathbf{y}' = \mathbf{By}$ where $\mathbf{B}$ has the eigenvalues and eigenvectors

$$r_1 = 1, \quad \mathbf{p}_1 = \begin{pmatrix} 2 \\ -1 \end{pmatrix}, \quad r_2 = 4, \quad \mathbf{p}_2 = \begin{pmatrix} -1 \\ 2 \end{pmatrix}.$$

Is the equilibrium $(0, 0)$ stable or unstable?

18. Characterize the behavior near the origin and draw a phase plane diagram for the system $\mathbf{y}' = \mathbf{By}$ where the real 2×2 matrix $\mathbf{B}$ has the following eigenvalues.

(a) $r_\pm = 2 \pm 2i$

(b) $r_\pm = -2 \pm 2i$

(c) $r_\pm = \pm 2i$

Is knowledge of eigenvectors required? Are there ambiguities in the phase diagram that can not be resolved without additional information?

19. Derive the linearized equation (8.22), $v' = g_y u + g_z v$, one of those that govern perturbations of a steady state of an autonomous system. That is, derive the second row of the Jacobian matrix of the system $\mathbf{y}' = \mathbf{f}(\mathbf{y})$.

20. Find parametric equations $(\theta(t), \omega(t))$ for the solution trajectories in the phase plane of the undamped pendulum system

$$\theta' = \omega,$$
$$\omega' = -(g/L)\theta.$$

Show that these trajectories are either ellipses or a single point. Find the time required for one circuit of such an ellipse. Does that time depend upon the size of the ellipse? What is the physical interpretation of these results?

21. Find parametric equations $(x(t), v(t))$ for the solution trajectories in the phase plane of the undamped spring-mass system

$$x' = v,$$
$$v' = -(k/m)x.$$

Show that these trajectories are either ellipses or a single point. Find the time required for one circuit of such an ellipse. Does that time depend upon the size of the ellipse? What is the physical interpretation of these results?

22. Figure 8.6 is the phase diagram of the saddle point for the system of example 20. Which line in that diagram might reasonably be labeled the stable axis? The unstable axis? How is the orientation of those axes related to the eigenvectors of the coefficient matrix of the differential equations?

23. Find the eigenvalues and eigenvectors of the matrix

$$\mathbf{B}_1 = \begin{pmatrix} -1/2 & -3/2 \\ -3/2 & -1/2 \end{pmatrix}.$$

Use that information to describe how solutions of the system $\mathbf{y}' = \mathbf{B}_1\mathbf{y}$ compare with solutions of the system studied in example 20. How would a phase diagram for this system differ from that of figure 8.6?

24. Verify that

$$\mathbf{p}_\pm = \begin{pmatrix} 1 \\ -1 \pm \sqrt{2} \end{pmatrix}$$

are eigenvectors of the Jacobian matrix

$$\mathbf{J} = \begin{pmatrix} 0 & 1 \\ 1 & -2 \end{pmatrix}$$

associated with the eigenvalues $r_\pm = -1 \pm \sqrt{2}$, as claimed in example 21. What information do these eigenvectors bring to the analysis of the phase behavior of $\mathbf{u}' = \mathbf{Ju}$?

25. Example 20 analyzed the saddle point of the system $\mathbf{y}' = \mathbf{By}$ with

$$\mathbf{B} = \begin{pmatrix} -1/2 & 3/2 \\ 3/2 & -1/2 \end{pmatrix}.$$

Using the general solution developed in that example, find a family of initial conditions that produce trajectories that head toward the origin. Mark the location of these special initial points on a copy of figure 8.6, a phase diagram of this system. Use that analysis to define the set of perturbations of the origin that will *not* return to it. Which seems more likely, a perturbation that returns to the origin or one that leaves it?

26. Consider the SIRS system

$$S' = -bIS + g(P - S - I),$$
$$I' = bIS - rI,$$

with $b = 2$, $g = 1$, $r = 1$, $P = 2$. Two equilibria of this system are displayed in figure 7.13(c).

(a) Conduct a linear analysis of the steady state marked 1 in figure 7.13(c). What sort of equilibrium point is it? What is the physical significance of this result? Is the linear analysis consistent with the nonlinear phase diagram?

(b) Conduct a linear analysis of the steady state marked 2 in figure 7.13(c). What sort of equilibrium point is it? What is the physical significance of this result? Is the linear analysis consistent with the nonlinear phase diagram?

27. Consider a system similar to that of example 22, page 409, $\mathbf{y}' = \mathbf{B}_1\mathbf{y}$ with

$$\mathbf{B}_1 = \begin{pmatrix} 5 & 3 \\ 3 & 5 \end{pmatrix}.$$

(a) Verify that $\mathbf{B}_1$ has eigenvalues $r = 2$, 8 and that associated eigenvectors are $\mathbf{p}_2 = \begin{pmatrix} -1 & 1 \end{pmatrix}^T$ and $\mathbf{p}_8 = \begin{pmatrix} 1 & 1 \end{pmatrix}^T$.

(b) Write a general solution of this system.

(c) Argue that the phase diagram of this system is exactly that of figure 8.9 with the flow field arrows reversed; that is, this system has an *unstable node* at the origin.

(d) Argue that all trajectories will emerge from the origin tangent to the $\mathbf{p}_2$ direction except those that start on the line through the origin defined by $\mathbf{p}_8$.

(e) What is the relation of this system to that of example 22, page 409, if time is reversed? That is, show that the change of independent variable $-t \leftarrow t$ converts one system into the other.

28. Suppose a 2×2 matrix $\mathbf{B}$ has two real eigenvalues $r_1 < r_2 < 0$. (Then it also has corresponding linearly independent eigenvectors.) Show that the system $\mathbf{y}' = \mathbf{B}\mathbf{y}$ has a stable node at the origin and determine the direction(s) in which trajectories approach the origin.

29. Suppose a 2×2 matrix $\mathbf{B}$ has a repeated real eigenvalue $r < 0$ with two corresponding linearly independent eigenvectors $\mathbf{p}_1$, $\mathbf{p}_2$. This exercise shows that the system $\mathbf{y}' = \mathbf{B}\mathbf{y}$ has a stable node at the origin. However, the configuration of that node is star-like—trajectories can approach the origin through any angle—rather than like that of figure 8.9, page 410, in which there are only a few directions of approach.

(a) Argue that all solutions of $\mathbf{y}' = \mathbf{B}\mathbf{y}$ approach the origin; i.e., the origin is a stable equilibrium point.

(b) Consider the (uncoupled) system $y' = -y$, $z' = -z$. Show that it has a stable, star-like node at the origin. Sketch a flow diagram.

(c) Consider the general situation: $\mathbf{B}$ has a repeated real eigenvalue $r < 0$ with two corresponding linearly independent eigenvectors $\mathbf{p}_1$, $\mathbf{p}_2$. Show that the flow diagram for $\mathbf{y}' = \mathbf{B}\mathbf{y}$ is essentially the same as for the uncoupled system of the previous step.

30. Suppose the 2×2 matrix $\mathbf{B}$ has a single eigenvalue $r < 0$. This exercise argues that the equilibrium of $\mathbf{y}' = \mathbf{B}\mathbf{y}$ at the origin is always stable.

(a) Suppose $\mathbf{B}$ has two linearly independent eigenvectors $\mathbf{p}_1$, $\mathbf{p}_2$ associated with the eigenvalue r. Write the form of the general solution and argue that it always decays to zero.

(b) Suppose $\mathbf{B}$ has only one independent eigenvector associated with r. Then the terms in the general solution contain factors of the form $(P_1 + P_2 t)e^{rt}$; e.g., see example 17, page 397. Argue that these terms decay to zero as well if $r < 0$, thereby completing the proof of stability.

31. Show that the origin is a center for the approximate undamped pendulum system $\theta' = \omega$, $\omega' = -(g/L)\theta$. Sketch the flow field and a few representative trajectories. Mark the direction of increasing time.

32. Show that the origin is a center for the approximate undamped pendulum system $\theta' = \omega$, $\omega' = -(g/L)\sin\theta$. Sketch the flow field and a few representative trajectories. Mark the direction of increasing time. What is the physical interpretation of this behavior?

33. Find the steady state in the first quadrant of the predator-prey system

$$F' = -(d_F - \alpha R)F,$$
$$R' = (b_R - \beta F)R.$$

Show that it is neutrally stable. Sketch the flow field and a few representative trajectories. Mark the direction of increasing time. What is the physical interpretation of this behavior?

34. Consider the system

$$y' = z,$$
$$z' = 2y - y^2 + az(y - 2)$$

in three cases, $a = 0, \pm 1$. Show that this system has an equilibrium at $(2, 0)$ in every case and that the corresponding linear system always exhibits a center (pure imaginary eigenvalues there). Use DELAB to explore the differences in the phase plane behavior of the *nonlinear* system near $(2, 0)$ in each case. Is the prediction by the linear approximation of a center at $(2, 0)$ a reliable indicator of the behavior of the nonlinear system?

Is there a similar difficulty at the equilibrium $(0, 0)$?

35. The text claims that the characteristic polynomial, and hence the eigenvalues, of the matrix

$$\mathbf{B} = \begin{pmatrix} a & b \\ c & d \end{pmatrix}$$

can be written in terms of two quantities

$$T = a + d \quad \text{and} \quad D = ad - bc,$$

the trace and the determinant of $\mathbf{B}$.

(a) Confirm this assertion by finding an expression for the eigenvalues of $\mathbf{B}$ in terms of T and D.

(b) Find conditions on T and D that guarantee that the origin is a saddle point of $\mathbf{y}' = \mathbf{B}\mathbf{y}$.

(c) Find conditions on T and D that guarantee that the origin is a stable node or an unstable node of $\mathbf{y}' = \mathbf{B}\mathbf{y}$.

(d) Find conditions on T and D that guarantee that the origin is a stable spiral or an unstable spiral of $\mathbf{y}' = \mathbf{B}\mathbf{y}$.

(e) Mark these regions on a plot of T vs. D.

36. Project 1 of chapter 7 considers the chemical reaction model

$$x' = x^2 y - x + b,$$
$$y' = a - x^2 y.$$

Fix $b = 0.1$. Find the equilibrium of this system and conduct a linear stability analysis of it. Describe the changes in the linear stability behavior with a. What behavior in the reaction would you observe as you varied the concentration a?

8.4 ■ NONHOMOGENEOUS SYSTEMS: VARIATION OF PARAMETERS

For a single first-order equation, the method of variation of parameters finds particular solutions of nonhomogeneous equations in the form $u(t)y_h(t)$, where y_h is a nontrivial solution of the associated homogeneous equation; see section 4.5. Here those ideas are extended to systems.

To find a particular solution of the nonhomogeneous equation

$$\mathbf{y}' = \mathbf{A}(t)\mathbf{y} + \mathbf{f}(t), \tag{8.30}$$

suppose two linearly independent solutions $\mathbf{y}_1, \mathbf{y}_2$ of the homogeneous system

$$\mathbf{y}' - \mathbf{A}(t)\mathbf{y}$$

are available. Form the fundamental matrix $\mathbf{Y}(t) = \begin{pmatrix} \mathbf{y}_1 & \mathbf{y}_2 \end{pmatrix}$ for that homogeneous system, and seek a particular solution in the form

Variation of parameters form of particular solution

$$\mathbf{y}_p(t) = \mathbf{Y}(t)\mathbf{u}(t),$$

where $\mathbf{u}(t)$ is to be determined by substituting into the nonhomogeneous system (8.30).

The name *variation of parameters* arises because the constant vector $\mathbf{c}$ in the general solution $\mathbf{Y}(t)\mathbf{c}$ of the homogeneous system has been replaced with a variable vector $\mathbf{u}(t)$.

Substituting this form of $\mathbf{y}_p(t)$ yields

$$\mathbf{y}_p' = \mathbf{Y}'(t)\mathbf{u}(t) + \mathbf{Y}(t)\mathbf{u}'(t) = \mathbf{A}(t)\mathbf{Y}(t)\mathbf{u}(t) + \mathbf{f}(t).$$

Now each of the columns of the fundamental matrix $\mathbf{Y}(t)$ is a solution of $\mathbf{y}' = \mathbf{A}\mathbf{y}$; that is, $\mathbf{Y}' = \mathbf{A}\mathbf{Y}$. So the previous expression reduces to the *basic equation of variation of parameters*, the equation defining the derivative of the unknown vector function $\mathbf{u}(t)$.

Basic equations of variation of parameters

$$\mathbf{Y}(t)\mathbf{u}'(t) = \mathbf{f}(t). \qquad (8.31)$$

In principle, it suffices to solve this system of equations for the components of $\mathbf{u}'$, integrate to obtain $\mathbf{u}$ (often a difficult chore), and write $\mathbf{y}_p(t) = \mathbf{Y}(t)\mathbf{u}(t)$ for the particular solution.

■ **EXAMPLE 25**   *Find a particular solution of* $\mathbf{y}'(t) = \mathbf{B}\mathbf{y}(t) + \mathbf{f}(t)$*, where*

$$\mathbf{B} = \begin{pmatrix} -4 & 2 \\ -3 & 3 \end{pmatrix}, \quad \mathbf{f}(t) = \begin{pmatrix} 10 \\ 5t \end{pmatrix}. \qquad (8.32)$$

First find a fundamental matrix of the homogeneous system $\mathbf{y}'(t) = \mathbf{B}\mathbf{y}(t)$. The eigenvalues of $\mathbf{B}$ are the roots of

Characteristic equation

$$\det(\mathbf{B} - r\mathbf{I}) = (-4 - r)(3 - r) + 6 = r^2 + r - 6 = (r + 3)(r - 2) = 0.$$

They are $r_1 = -3$, $r_2 = 2$. The usual manipulations find the corresponding eigenvectors: $\mathbf{p}_1 = \begin{pmatrix} 2 & 1 \end{pmatrix}^T$ and $\mathbf{p}_2 = \begin{pmatrix} 1 & 3 \end{pmatrix}^T$. Hence, a fundamental matrix is

Fundamental matrix

$$\mathbf{Y}(t) = \begin{pmatrix} 2e^{-3t} & e^{2t} \\ e^{-3t} & 3e^{2t} \end{pmatrix}.$$

Stop and Think   **8.52** Verify that $\mathbf{Y}$ is indeed a fundamental matrix of this system; i.e., verify that $\mathbf{Y}' = \mathbf{B}\mathbf{Y}$ and that the columns of $\mathbf{Y}$ are linearly independent.

Then the basic equation of variation of parameters, $\mathbf{Y}(t)\mathbf{u}'(t) = \mathbf{f}(t)$, leads to

Equations for $\mathbf{u}'$

$$\begin{pmatrix} 2e^{-3t} & e^{2t} \\ e^{-3t} & 3e^{2t} \end{pmatrix} \begin{pmatrix} u_1'(t) \\ u_2'(t) \end{pmatrix} = \mathbf{f}(t) = \begin{pmatrix} 10 \\ 5t \end{pmatrix}.$$

Solving this system by Cramer's rule requires calculation of the determinant of coefficients, which is just the Wronskian

$$W = \det \begin{pmatrix} 2e^{-3t} & e^{2t} \\ e^{-3t} & 3e^{2t} \end{pmatrix} = 5e^{-t}.$$

Then

Solve for $\mathbf{u}'$.

$$u_1' = \frac{1}{W} \det \begin{pmatrix} 10 & e^{2t} \\ 5t & 3e^{2t} \end{pmatrix} = 6e^{3t} - te^{3t},$$

$$u_2' = \frac{1}{W} \det \begin{pmatrix} 2e^{-3t} & 10 \\ e^{-3t} & 5t \end{pmatrix} = 2te^{-2t} - 2e^{-2t}.$$

Stop and Think   **8.53** Verify that these expressions are solutions of $\mathbf{Y}(t)\mathbf{u}'(t) = \mathbf{f}(t)$.

Integrating yields

$$u_1 = \left(\frac{19}{9} - \frac{t}{3}\right)e^{3t},$$

Integrate.

$$u_2 = \left(\frac{1}{2} - t\right)e^{-2t}.$$

A particular solution is

Write $\mathbf{y}_p$.

$$\mathbf{y}_p = \mathbf{Y}(t)\mathbf{u} = \begin{pmatrix} 2e^{-3t} & e^{2t} \\ e^{-3t} & 3e^{2t} \end{pmatrix}\begin{pmatrix} u_1(t) \\ u_2(t) \end{pmatrix} = \begin{pmatrix} 85/18 - 5t/3 \\ 65/18 - 10t/3 \end{pmatrix}.$$

A general solution is $\mathbf{y}_g = \mathbf{y}_p + \mathbf{Y}\mathbf{c}$. ■

Clearly, the integrations required by variation of parameters can become formidable. Much of the value of this method lies in the representation formula it provides for a particular solution.

To obtain such a representation, apply the fundamental theorem of calculus to the basic equation $\mathbf{u}'(t) = \mathbf{Y}(t)^{-1}\mathbf{f}(t)$ to write

Expression for $\mathbf{u}$

$$\mathbf{u}(t) = \int_a^t \mathbf{Y}(s)^{-1}\mathbf{f}(s)\, ds,$$

where a can be chosen arbitrarily. Hence, given a fundamental matrix $\mathbf{Y}(t)$, a general solution of $\mathbf{y}' = \mathbf{A}(t)\mathbf{y} + \mathbf{f}(t)$ is

General solution of
$\mathbf{y}' = \mathbf{A}(t)\mathbf{y} + \mathbf{f}(t)$

$$\mathbf{y}_g(t) = \mathbf{Y}(t)\int_a^t \mathbf{Y}(s)^{-1}\mathbf{f}(s)\, ds + \mathbf{Y}(t)\mathbf{c}.$$

The first term is a particular solution of the nonhomogeneous equation, and the second term is a general solution of the corresponding homogeneous system $\mathbf{y}' = \mathbf{A}(t)\mathbf{y}$.

A less general but less labor-intensive method for finding particular solutions, the method of undetermined coefficients, extends to constant-coefficient matrix systems. For example, if the forcing term includes e^t, then a particular solution would have the form $\mathbf{p}e^t$ for some undetermined coefficient vector $\mathbf{p}$. If $\mathbf{p}e^t$ could solve the homogeneous system, then an appropriate form for a particular solution would be $\mathbf{p}_1 te^t + \mathbf{p}_2$, where the vectors $\mathbf{p}_i$ are again determined by substituting in the nonhomogeneous system. See exercise 24 in this section as well as exercise 27 of the chapter exercises for chapter 8.

The proverbial alert reader will have noticed that variation of parameters can be applied to any linear system, constant- or variable-coefficient. But homogeneous solutions are the building blocks of variation of parameters, and we know how to find them only for constant-coefficient systems. Consequently, this discussion has been limited to constant-coefficient systems.

8.4.1 Exercises

EXERCISE GUIDE	
To gain experience …	**Try exercises**
Finding homogeneous solutions	1–11, 19–22
With variation of parameters and particular solutions	1–11, 12, 17–22
Finding and using general solutions	1–11, 12, 13(a), 14, 19–23
Solving initial-value problems	1–11, 13(b), 24
With the foundations of variation of parameters	15–16, 23
Extending undetermined coefficients to systems	24

In exercises 1–11,

(i) Find two linearly independent solutions $\mathbf{y}_1$, $\mathbf{y}_2$ of the homogeneous system.

(ii) Begin a variation of parameters solution of the nonhomogeneous system by finding u_1, u_2 in the assumed form $\mathbf{y}_p = \mathbf{Y}\mathbf{u}$. If necessary, express antiderivatives as definite integrals using the fundamental theorem of calculus.

(iii) Complete the variation of parameters solution by writing $\mathbf{y}_p$.

(iv) Write a general solution of the nonhomogeneous system.

(v) Find the solution of the nonhomogeneous system that satisfies the given initial conditions.

(vi) Confirm your results using MATLAB.

If a homogeneous solution is available from some other source (e.g., a previous exercise), use it rather than rederiving a solution.

1. $y' = -4y + 2z + 15t$, $y(0) = 3$
$z' = -3y + 3z - 20$, $z(0) = 4$

2. $y' = -y + 3z - 8e^{-2t}$, $y(0) = 0$
$z' = 3y - z + 12$, $z(0) = 4$

3. $y' = 2y + 4z - 2$, $y(0) = 2$
$z' = -4y + 2z + 4$, $z(0) = -2$

4. $y' = 4y + 4z - 24t$, $y(0) = 4$
$z' = 3y - 4z + 8\cos 2t$, $z(0) = 1$

5. $y' = 5y/2 + 3\sqrt{3}\,z/2 + 4$, $y(0) = 6$
$z' = 3\sqrt{3}\,y/2 - z/2 - 6t$, $z(0) = 0$

6. $y' = -4y - 5z/2 + 1$, $y(0) = 1$
$z' = y - 2z - 1$, $z(0) = -6$

7. $y' = -4y - 5z/2 + 1$, $y(0) = 0$
$z' = y - 2z - 1$, $z(0) = 0$

8. $x' = v$, $x(0) = x_0$
$mv' = -kx + F\sin\omega t$, $v(0) = 0$

9. $x' = v$, $x(0) = 0$
$mv' = -kx + F\sin\omega t$, $v(0) = v_0$

10. $x' = v$, $x(0) = x_0$
$mv' = -kx - pv + Fe^{-t}$, $v(0) = 0$

11. $\theta' = \omega$, $\theta(0) = \theta_0$
$L\omega' = -g\theta + 2\cos\pi t$, $\omega(0) = 0$

12. Example 25 obtains the particular solution

$$\mathbf{y}_p = \begin{pmatrix} 85/18 - 5t/3 \\ 65/18 - 10t/3 \end{pmatrix}$$

of the nonhomogeneous system

$$\mathbf{y}' = \begin{pmatrix} -4 & 2 \\ -3 & 3 \end{pmatrix}\mathbf{y} + \begin{pmatrix} 10 \\ 5t \end{pmatrix}.$$

Verify by substitution that $\mathbf{y}_p$ is indeed a particular solution of this system. Explain why it is *not* a general solution.

13. Example 25 claims that $\mathbf{y}_g = \mathbf{y}_p + \mathbf{Y}\mathbf{c}$ solves $\mathbf{y}' = \mathbf{B}\mathbf{y} + \mathbf{f}$, where $\mathbf{y}_p$ and $\mathbf{Y}$ are defined there, $\mathbf{B}$ and $\mathbf{f}$ by (8.32).

(a) Verify that $\mathbf{y}_g$ is indeed a general solution.

(b) Find the solution of this system that satisfies the initial conditions $\mathbf{y}(0) = \begin{pmatrix} 139/18 & 137/18 \end{pmatrix}^T$.

14. Example 25 obtains a particular solution

$$y_p = 85/18 - 5t/3,$$
$$z_p = 65/18 - 10t/3$$

of the nonhomogeneous system defined by (8.32) by setting to zero the constants of integration it finds in computing u_1, u_2 from

$$u_1' = 6e^{3t} - te^{3t},$$
$$u_2' = 2te^{-2t} - 2e^{-2t}.$$

Keep those constants (you might call them C_1, C_2) and show that you obtain a *general solution* of the nonhomogeneous system.

15. Why would variation of parameters fail if the homogeneous solutions were *not* linearly independent?

16. The text shows in general that substituting the particular solution form $\mathbf{y}_p(t) = \mathbf{Y}(t)\mathbf{u}(t)$ into $\mathbf{y}' = \mathbf{B}\mathbf{y} + \mathbf{f}$ leads to the system of equations

$$\mathbf{Y}(t)\mathbf{u}'(t) = \mathbf{f}(t)$$

for the derivative of the unknown vector function $\mathbf{u}(t)$. Verify this claim in the following special case by substituting directly into the system $\mathbf{y}' = \mathbf{B}\mathbf{y} + \mathbf{f}$ considered in example 25, where

$$\mathbf{B} = \begin{pmatrix} -4 & 2 \\ -3 & 3 \end{pmatrix}, \quad \mathbf{f}(t) = \begin{pmatrix} 10 \\ 5t \end{pmatrix}.$$

A fundamental matrix for this system is

$$\mathbf{Y}(t) = \begin{pmatrix} 2e^{-3t} & e^{2t} \\ e^{-3t} & 3e^{2t} \end{pmatrix}.$$

In exercises 17–18, use the given solutions $\mathbf{y}_1$, $\mathbf{y}_2$ of the homogeneous equation to construct a particular solution of the nonhomogeneous equation. Confirm your solution using DE-LAB.

17. $\mathbf{y}_1 = \begin{pmatrix} -e^{4t} \\ e^{4t} \end{pmatrix}$, $\quad \mathbf{y}_2 = \begin{pmatrix} e^{-4t} \\ e^{-4t} \end{pmatrix}$,

$$\mathbf{y}' = \begin{pmatrix} 0 & -4 \\ -4 & 0 \end{pmatrix}\mathbf{y} + \begin{pmatrix} 4t \\ -2e^{2t} \end{pmatrix}$$

18. $\mathbf{y}_1 = \begin{pmatrix} e^t \cos t \\ -e^t \sin t \end{pmatrix}$, $\quad \mathbf{y}_2 = \begin{pmatrix} e^t \sin t \\ e^t \cos t \end{pmatrix}$,

$$\mathbf{y}' = \begin{pmatrix} 1 & 1 \\ -1 & 1 \end{pmatrix}\mathbf{y} + \begin{pmatrix} 2e^t \\ -e^t \end{pmatrix}$$

In exercises 19–22, find a general solution of the given system. If you can use work from earlier exercises, do so. Confirm your results using DELAB.

19. $\mathbf{y}' = \begin{pmatrix} 1 & 0 \\ 4 & 3 \end{pmatrix}\mathbf{y} + \begin{pmatrix} 4e^{-2t} \\ 6e^{2t} \end{pmatrix}$

20. $\mathbf{y}' = \begin{pmatrix} -2 & 0 \\ 2 & -1 \end{pmatrix}\mathbf{y} + \begin{pmatrix} -6t \\ 2e^{3t} \end{pmatrix}$

21. $\mathbf{y}' = \begin{pmatrix} 11 & 2 \\ 1 & 10 \end{pmatrix}\mathbf{y} + \begin{pmatrix} 162 \\ -108t \end{pmatrix}$

22. $\mathbf{y}' = \begin{pmatrix} 1 & 2 \\ -1 & 1 \end{pmatrix}\mathbf{y} + \begin{pmatrix} 4\sqrt{2}e^{2t} \\ -4\sqrt{2}e^{-2t} \end{pmatrix}$

23. Use the general solution expression

$$\mathbf{y}_g(t) = \mathbf{Y}(t)\int_a^t \mathbf{Y}(s)^{-1}\mathbf{f}(s)\, ds + \mathbf{Y}(t)\mathbf{c}$$

for the system $\mathbf{y}' = \mathbf{A}(t)\mathbf{y} + \mathbf{f}(t)$ to write a solution of this equation subject to the initial condition $\mathbf{y}(0) = \mathbf{y}_0$. (Choose a to simplify the integral at $t = 0$.) How would the expression for $\mathbf{c}$ simplify if $\mathbf{Y}(0) = \mathbf{I}$?

24. Formulate an extension of the method of undetermined coefficients to systems and apply it to exercises 1–11 and 19–22, as assigned.

8.5 ■ CHAPTER EXERCISES

EXERCISE GUIDE	
To gain experience ...	**Try exercises**
Generalizing and extending concepts and methods	9, 19–25
Solving homogeneous systems	1–4, 15–20, 22–25
Finding particular solutions by variation of parameters	5–8, 10–14, 21
Finding general solutions	1–4, 5–8

EXERCISE GUIDE (Continued)	
To gain experience . . .	**Try exercises**
Solving initial-value problems	1–4, 5–8
Comparing analytic solutions with phase portraits	26
Extending undetermined coefficients to systems	27
With nullcline and phase plane analysis	28–32
Classifying equilibria of linear systems	33

In exercises 1–4,

(i) Find a general solution of the given system.

(ii) Find a solution of the system that satisfies the given initial conditions.

(iii) Confirm your results with DELAB.

1. $y' = z, \quad y(0) = 0$
$z' = y, \quad z(0) = 1$

2. $y' = -4y + 2z, \quad y(0) = 3$
$z' = -3y + 3z, \quad z(0) = 4$

3. $y' = -4z, \quad y(0) = 0$
$z' = -4y, \quad z(0) = 2$

4. $y' = y + z, \quad y(0) = 1$
$z' = -y + z, \quad z(0) = 2$

In exercises 5–8,

(i) Find two linearly independent solutions $\mathbf{y}_1$, $\mathbf{y}_2$ of the homogeneous system.

(ii) Begin a variation of parameters solution of the nonhomogeneous system by finding u_1, u_2 in the assumed form $\mathbf{y}_p = u_1(t)\mathbf{y}_1 + u_2(t)\mathbf{y}_2$. If necessary, express antiderivatives as definite integrals using the fundamental theorem of calculus.

(iii) Complete the variation of parameters solution by writing $\mathbf{y}_p$.

(iv) Write a general solution of the nonhomogeneous system.

(v) Find the solution of the nonhomogeneous system that satisfies the given initial conditions.

(vi) Confirm your results with DELAB.

If a homogeneous solution is available from some other source (e.g., a previous exercise), use it rather than rederive it.

5. $y' = y/2 + \sqrt{3}\,z/2 + 2t, \quad y(0) = 6$
$z' = \sqrt{3}\,y/2 - z/2 - 4t^2, \quad z(0) = 4$

6. $y' = z + 4\cos t, \quad y(0) = 1$
$z' = y, \quad z(0) = 0$

7. $y' = y + 3z + e^{2t}, \quad y(0) = 1$
$z' = 3y + z - 1, \quad z(0) = 0$

8. $y' = y - z - 2, \quad y(0) = 1$
$z' = 2y + 4z - 2e^{-2t}, \quad z(0) = 0$

9. Suppose you know two solutions $y_1(t)$, $y_2(t)$ of the homogeneous version of the second-order equation $y'' + P(t)y' + Q(t)y = f(t)$. Write this equation as a first-order system and develop a variation of parameters formula for a particular solution in terms of y_1, y_2.

Find a general solution of the systems given in exercises 10–25.

10. $\mathbf{y}' = \begin{pmatrix} 5 & -1 \\ -1 & 5 \end{pmatrix} \mathbf{y} + \begin{pmatrix} -12 \\ 2e-6t \end{pmatrix}$

11. $\mathbf{y}' = \begin{pmatrix} -7 & -3 \\ 6 & 2 \end{pmatrix} \mathbf{y} + \begin{pmatrix} 6t \\ -9 \end{pmatrix}$

12. $\mathbf{y}' = \begin{pmatrix} 11 & 6 \\ -30 & -16 \end{pmatrix} \mathbf{y} + \begin{pmatrix} 2e^t \\ -2 \end{pmatrix}$

13. $\mathbf{y}' = \begin{pmatrix} -9 & -4 \\ 20 & 9 \end{pmatrix} \mathbf{y} + \begin{pmatrix} e^{-t} \\ -e^t \end{pmatrix}$

14. $\mathbf{y}' = \begin{pmatrix} 0 & -4 \\ 4 & 0 \end{pmatrix} \mathbf{y} + \begin{pmatrix} \cos 4t \\ -\sin 4t \end{pmatrix}$

15. $\mathbf{y}' = \begin{pmatrix} -4 & -4 \\ 8 & 4 \end{pmatrix} \mathbf{y}$

16. $\mathbf{y}' = \begin{pmatrix} 1 & 1 \\ -2 & -1 \end{pmatrix} \mathbf{y}$

17. $\mathbf{y}' = \begin{pmatrix} -17 & -13 \\ 18 & 13 \end{pmatrix} \mathbf{y}$

18. $\mathbf{y}' = \begin{pmatrix} -8 & -5 \\ 9 & 4 \end{pmatrix} \mathbf{y}$

19. $\mathbf{y}' = \begin{pmatrix} -3 & -2 \\ 4 & 1 \end{pmatrix} \mathbf{y}$

20. $\mathbf{y}' = \begin{pmatrix} 2 & 2 & 0 \\ 2 & 1 & 1 \\ -7 & 2 & 3 \end{pmatrix} \mathbf{y}$

21. $\mathbf{y}' = \begin{pmatrix} 2 & 2 & 0 \\ 2 & 1 & 1 \\ -7 & 2 & 3 \end{pmatrix} \mathbf{y} + \begin{pmatrix} 0 \\ 0 \\ -1 \end{pmatrix}$

22. $\mathbf{y}' = \begin{pmatrix} 1 & 0 & -2 \\ 2 & 2 & 4 \\ 0 & 0 & 2 \end{pmatrix} \mathbf{y}$

23. $\mathbf{y}' = \begin{pmatrix} 0 & 1 & 0 \\ 0 & 0 & 1 \\ -12 & 13 & 0 \end{pmatrix} \mathbf{y}$

24. $\mathbf{y}' = \begin{pmatrix} 0 & 1 & 0 \\ 0 & 0 & 1 \\ 12 & 16 & 3 \end{pmatrix} \mathbf{y}$

25. $\mathbf{y}' = \begin{pmatrix} 8 & 2 & 12 \\ 5 & 2 & 9 \\ -5 & -1 & -8 \end{pmatrix} \mathbf{y}$

26. Choose a homogeneous system of two equations from the exercises of the previous sections. Solve the system, and confirm the correctness of the analytic solution by comparing it with a phase portrait generated using DELAB.

27. Formulate an extension of the method of undetermined coefficients to systems and apply it to exercises 10–14 and 21, as assigned.

28. Conduct a nullcline analysis of the linear system $\mathbf{y}' = \mathbf{By}$, where $\mathbf{B}$ is defined in example 20, page 406. Show that

your results are consistent with the phase portrait of this system shown in figure 8.6, page 408.

29. Conduct a nullcline analysis of the linear system $\mathbf{u}' = \mathbf{Ju}$ with

$$\mathbf{J} = \begin{pmatrix} 0 & 1 \\ 1 & -2 \end{pmatrix}$$

that was used in example 21, page 408, to study the stability of the equilibrium $\theta_{ss} = \pi$, $\omega_{ss} = 0$ of a damped pendulum system. Confirm the flow field shown in figure 8.7, page 409.

30. Conduct a nullcline analysis of the linear system $\mathbf{y}' = \mathbf{By}$, where $\mathbf{B}$ is defined in example 22, page 409. Show that your results are consistent with the phase portrait of this system shown in figure 8.9, page 410.

31. Conduct a nullcline analysis of the linear system $\mathbf{u}' = \mathbf{Ju}$ with

$$\mathbf{J} = \begin{pmatrix} 0 & 1 \\ -1 & -4 \end{pmatrix}$$

that was used in example 23, page 410, to study the stability of the equilibrium at the origin of an overdamped pendulum system. Confirm the flow field shown in figure 8.10, page 411.

32. Conduct a nullcline analysis of the linear system

$$u' = v,$$
$$v' = -u - v$$

that was used in example 24, page 412, to study the stability of the equilibrium at the origin of an underdamped pendulum system. Confirm the flow field shown in figure 8.12, page 413.

33. Use the eigenvalues of the coefficient matrix to classify the equilibrium at the origin for each of the systems in exercises 15–19, as assigned. Show that this classification is consistent with the behavior of the general solutions found in those exercises.

8.6 ■ CHAPTER PROJECTS

1. **Existence theorem for systems.** For a system of two equations, state and prove an existence theorem that is an extension of theorem 6, the existence theorem for a single equation. Show carefully how the bounding arguments in the proof of the theorem for a single equation must be modified to account for two equations.

2. **More autocatalytic reactions.** Return to project 1 of chapter 7 and augment its analysis as follows.

   (a) Find an expression for the equilibrium point in terms of the concentrations a, b. For convenience, fix $b = 0.1$. Linearize the system about this point and show that the eigenvalues of the Jacobian cross the imaginary axis as a is varied; that is, show that the eigenvalues of the Jacobian are complex conjugates and that their real part changes sign at a value of a known as a *bifurcation point*. (The ideas of exercise 35, page 419, might prove useful.)

(b) How does the nature of the equilibrium change as a varies through the bifurcation point? How does the phase portrait of the full nonlinear system change? Do you observe any dependence on the magnitude of the difference between a and the bifurcation point? What is the physical significance of that behavior?

This situation is an example of a *Hopf bifurcation*, the emergence of a closed orbit around an equilibrium point as the stability of the equilibrium reverses at a critical value of the bifurcation parameter a.

3. **More predator-prey.** Project 2 of chapter 7 considers three forms of the predator-prey model

$$F' = \phi_i(F, R)F,$$
$$R' = \rho_i(F, R)R,$$

$i = 0, 1, 2$, where the functions ϕ_i, ρ_i are defined there. Using the parameter values given in that project, find the steady states of each of these forms of the predator-prey model and conduct a linear analysis of their stability. How have the (relatively small) terms added to ϕ_i, ρ_i, $i = 1, 2$, changed the stability characteristics and type of equilibrium of the standard model ($i = 0$)?

4. **Maintaining a supply of bacteria.** A *chemostat* is a vessel containing a nutrient fluid in which a steady supply of bacteria or other organisms is maintained for laboratory studies. Nutrient enters the vessel at a steady rate, and effluent is removed at the same rate to preserve a constant volume of fluid. The effluent carries in it some of the bacteria and some of the nutrient. Bacteria growth increases with concentration of nutrient. Of course, the bacteria deplete the nutrient.

The goal is to balance these competing effects of increase and decrease so that the chemostat contains a constant concentration of bacteria.

Edelstein-Keshet [8] shows that a dimensionless model of the chemostat is

$$b' = a_1 \frac{n}{1 + n} b - b, \tag{8.33}$$

$$n' = -\frac{n}{1 + n} b - n + a_2. \tag{8.34}$$

Here $b(t)$, $n(t)$ are the concentrations of the bacteria and nutrient, respectively, $a_1 > 0$ is a dimensionless bacteria growth rate parameter, and $a_2 > 0$ is a dimensionless nutrient input flow rate parameter.

(a) Explain the physical origin of each of the terms appearing in (8.33–8.34).

(b) Find the steady states of (8.33–8.34). Which steady state corresponds to the purpose of the chemostat, maintaining a steady supply of bacteria? Which steady state is undesirable? How must the parameters a_1, a_2 be restricted to ensure that the steady states are biologically meaningful?

(c) Linearize the system about each of the steady states and determine the nature of those equilibrium points for the biologically meaningful values of a_1, a_2. Exhibit the phase plane behavior and discuss its physical significance. Does your analysis suggest that a chemostat would be a practical piece of laboratory apparatus?

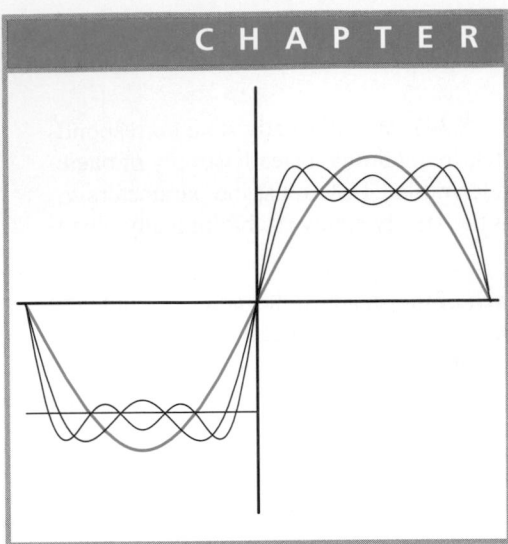

9

Diffusion Models and Boundary-Value Problems

Boundary-value problems use data about the solution at two different points, not just at a single initial point. Such problems arise naturally in models of diffusion, among many other important applications.

9.1 ■ DIFFUSION MODELS

Diffusion is the process that causes a dark drop of ink to spread from one end of a narrow pan of water to the other until all of the water is uniformly tinted by the diluted droplet. Likewise, diffusion is the process that carries heat through the walls of a house and pollution from one side of a quiet lake to another. It is central to many equilibrium phenomena in the natural world.

The rate at which a blob of ink spreads through water is determined by change in the concentration of ink between adjacent points in the fluid. When the ink is uniformly spread through the water, then there is no change in concentration of ink from one point to another. Diffusion is effectively halted, as shown in the right side of figure 9.1. When the droplet of ink first enters the water, there is a sharp concentration change at the boundary between the ink and the water. The ink diffuses rapidly toward the clear water.

FIGURE 9.1 When a drop of ink first enters the water, it diffuses rapidly at the boundary of the droplet, where the concentration gradient is greatest. Later, as shown in the diagram on the far right, diffusion is effectively halted because the concentration is uniform and the concentration gradient is zero.

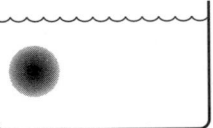

Rapid diffusion

No effective diffusion

The natural measure of the change in concentration from point to point is the derivative of concentration with respect to position, the **concentration gradient**. If $c(x)$ is the concentration at point x, then *the concentration gradient at x is $c'(x)$*. Since diffusion carries material from high concentration toward low concentration, the diffusive flow will be in the *positive x* direction if c' is *negative*, and vice versa.

One way to characterize diffusion is to find the rate of flow of ink (or salt or any other dissolved substance) through an imaginary unit surface placed in the fluid. The experimental fact that governs diffusion uses that idea.

Experimental fact. The rate of diffusion of a substance dissolved in a liquid through a unit area is proportional to the negative of its *concentration gradient*.

The constant of proportionality is the **diffusion coefficient**. If $c(x)$ denotes concentration, then this experimental fact can be written:

Flow rate due to diffusion through unit area at x is $-Dc'(x)$.

This idea is illustrated in figure 9.2.

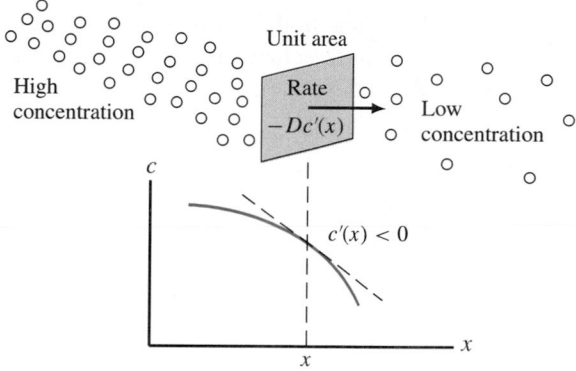

FIGURE 9.2 An illustration of the experimental fact governing diffusion. Ink (or any other dissolved substance) will diffuse from left to right because the concentration is negative at the point shown; concentration is higher on the left than on the right.

This experimental fact is sometimes known as **Fick's law**, but it is a "law" in the same sense as Hooke's law, which relates force in a spring to its extension. If the quantity that is diffusing is heat energy, then it is called **Fourier's law**.

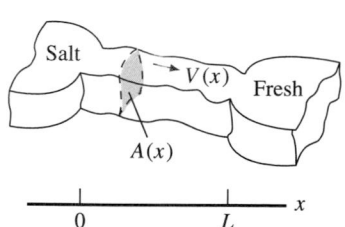

FIGURE 9.3 A schematic diagram of a channel connecting two bodies of water, salt on the left and fresh on the right. The length of the connecting channel is L. The area of the cross section at coordinate x is $A(x)$.

Now use these ideas to model the transfer of salt from the ocean to an estuary through a tidal inlet, as illustrated in figure 9.3. The ocean on the left contains salt in a concentration of c_0 mass per unit volume (say, g/m^3). Tidal flow can carry the salty water through the connecting channel at a velocity of $V(x)$ m/s into the fresher water on the right, where the salt concentration is c_L. Even without the flow of saltwater, diffusion would carry salt from the ocean through the channel toward the lower concentration in the estuary, just as the ink droplet in water spreads, seeking lower concentrations of itself.

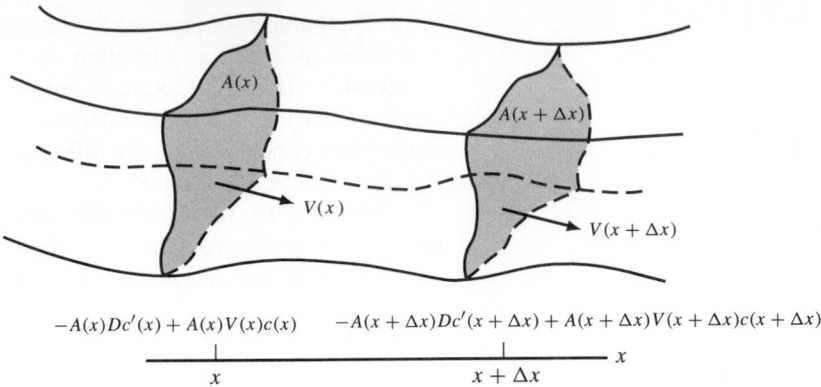

$$-A(x)Dc'(x) + A(x)V(x)c(x) \qquad -A(x+\Delta x)Dc'(x+\Delta x) + A(x+\Delta x)V(x+\Delta x)c(x+\Delta x)$$

FIGURE 9.4 At equilibrium, the flow of salt into this imaginary box in the channel equals the net flow out. This picture is drawn as if salt flows in through the left wall and out through the right wall.

We want to formulate a model that will describe the equilibrium distribution of salt in the connecting channel. The model should incorporate transport of salt due both to flow of water in the channel and to diffusion through the water in the channel. The governing law is that of conservation of mass (of salt, in this case).

CONSERVATION OF MASS

Law of conservation of mass. The net change in the mass of a fixed volume over a given period of time is the mass added less the mass removed.

Focus attention on a section of the channel between the coordinates x and $x + \Delta x$, as shown in figure 9.4. This imaginary box is known as a **control volume**. At equilibrium, the net change in the mass of salt in this box over time is zero. Consequently, the conservation law says

At equilibrium, the rate of flow *into* the control volume equals the rate of flow *out of* the control volume.

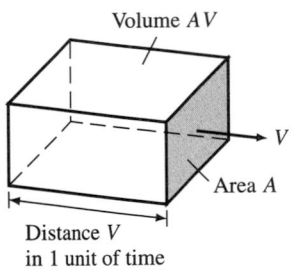

FIGURE 9.5 If water with velocity V flows through an area A, then in 1 unit of time a volume AV of water will pass through it.

The net volume of water per unit time that flows through either wall is the product of the area of the wall and the flow rate of the water in the channel, AV; see figure 9.5. The total rate at which *salt* enters due to this flow is just the product of the volume flow rate of water with the concentration of salt in the water, AVc. Hence, due to the flow in the channel, salt *enters* from the left at the rate

$$A(x)V(x)c(x),$$

and it *leaves* from the right at the rate

$$A(x + \Delta x)V(x + \Delta x)c(x + \Delta x).$$

(Our *enter* and *leave* labels suppose the water is flowing from left to right, as shown in figure 9.4.)

From the experimental fact, the rate of transport of salt *through a unit area* due to diffusion is $-Dc'$. Hence, the net rate of diffusion *in* through the left wall of the control volume is

$$-A(x)Dc'(x),$$

and the net rate *out* through the right wall is

$$-A(x + \Delta x)Dc'(x + \Delta x).$$

As figure 9.4 illustrates, the *total* flow rate *in* through the left is

$$-A(x)Dc'(x) + A(x)V(x)c(x),$$

and the *total out* through the right is

$$-A(x + \Delta x)Dc'(x + \Delta x) + A(x + \Delta x)V(x + \Delta x)c(x + \Delta x).$$

At equilibrium, the mass of salt in the control volume can not change over time. Hence, conservation of mass demands that the rate of salt flow in equal the rate of salt flow out:

Rate in = rate out

$$-A(x)Dc'(x) + A(x)V(x)c(x) = -A(x + \Delta x)Dc'(x + \Delta x)$$
$$+ A(x + \Delta x)V(x + \Delta x)c(x + \Delta x).$$

Collecting terms and dividing by Δx leads to

$$-D\frac{A(x + \Delta x)c'(x + \Delta x) - A(x)c'(x)}{\Delta x}$$
$$+ \frac{A(x + \Delta x)V(x + \Delta x)c(x + \Delta x) - A(x)V(x)c(x)}{\Delta x} = 0.$$

The limit as Δx goes to zero reveals a second-order differential equation

$$-D(A(x)c'(x))' + (A(x)V(x)c(x))' = 0.$$

A solution of this second-order equation should contain two arbitrary constants. Two conditions are needed to determine those two constants, and the natural ones to use are the concentrations c_0, c_L at the inlet and the outlet of the channel. Hence, the complete *diffusion model* is

DIFFUSION MODEL

$$-D(A(x)c'(x))' + (A(x)V(x)c(x))' = 0, \quad c(0) = c_0, \quad c(L) = c_L.$$

The diffusion model is an example of a *boundary-value problem*, and the auxiliary conditions

$$c(0) = c_0, \quad c(L) = c_L$$

are called *boundary conditions*. We solve boundary-value problems as you might expect: Find a general solution of the differential equation and use the

Evidently $c(0) = c_0$ because water with that concentration of pollutant is at the left end of the channel.

At the right end of the channel, the boundary condition must reflect the fact that all flow of the pollutant is prohibited. Setting $V = 0$ has already eliminated flow due to the movement of the water. Since the rate of flow of salt due to diffusion is given by $-Dc'$, there will be no diffusion if $c' = 0$. Hence, the boundary condition at $x = L$ is

$$c'(L) = 0,$$

and the complete model is the boundary-value problem

$$-D(A(x)c'(x))' = 0, \quad c(0) = c_0, \quad c'(L) = 0. \quad ■$$

A no-flow boundary condition like $c'(L) = 0$ is known as a *Neumann* boundary condition or a boundary condition of the *second kind*. A boundary condition such as $c(0) = c_0$ that specifies the value itself is a *Dirichlet* boundary condition or a boundary condition of the *first kind*.

■ **EXAMPLE 4**   *Suppose the bodies of water at either end of the channel in figure 9.3 have been cleared of pollutant. However, a microbe is digesting an otherwise harmless substance distributed throughout the water in the channel itself. The product of that digestion is a toxin that is being released at a concentration rate of $P(t)$ g/m^3·s (mass per unit volume per unit time) into the channel. Derive a model for the concentration of this toxic pollutant in the channel.*

Return to the control volume of figure 9.4. In this example, there are *three* ways to add pollutant to that imaginary box:

- Transport of pollutant through the wall of the box by the flowing water.
- Diffusion through the wall due to a nonzero concentration gradient.
- The added effect of release into the volume of the box at the concentration rate $P(t)$.

Since the control volume is at equilibrium, conservation of mass still demands that the rate at which pollutant is added to the control volume equals the rate at which it is lost.

Pollutant is released at a concentration rate of $P(t)$. Hence, the total mass of pollutant released into the control volume per unit time is the product of the volume of the box with the concentration rate P. Since the area of a side wall of the control volume is $A(x)$ and its width is Δx, its volume is about $A(x)\Delta x$. Hence, a mass of $A(x)P(x)\Delta x$ of salt is added to the control volume per unit time by the polluted bed and walls of the channel.

Combining this source with the flow and diffusion terms shown in figure 9.4, the "rate in equals rate out" conservation equation is

$$- A(x)Dc'(x) + A(x)V(x)c(x) + A(x)P(x)\Delta x$$
$$= -A(x + \Delta x)Dc'(x + \Delta x) + A(x + \Delta x)V(x + \Delta x)c(x + \Delta x).$$

Collecting terms and dividing by Δx leads to

$$-D\frac{A(x+\Delta x)c'(x+\Delta x)-A(x)c'(x)}{\Delta x}$$

$$+\frac{A(x+\Delta x)V(x+\Delta x)c(x+\Delta x)-A(x)V(x)c(x)}{\Delta x}=A(x)P(x).$$

The limit as Δx approaches zero yields the *diffusion equation with source term*

$$-D(A(x)c'(x))'+(A(x)V(x)c(x))'=A(x)P(x).$$

The large bodies of water at either end of the channel are free of pollutant. We can suppose that any pollutant which enters them from the channel is dispersed completely so that the ends of the channel are effectively pollution-free. Hence, reasonable boundary conditions are $c(0)=c(L)=0$. The complete model is

$$-D(A(x)c'(x))'+(A(x)V(x)c(x))'=A(x)P(x),\quad c(0)=c(L)=0.\ ■$$

9.1.1 Exercises

EXERCISE GUIDE	
To gain experience . . .	**Try exercises**
With concepts underlying the derivation of diffusion models	1–2
Classifying diffusion equations	4–5
Deriving models of mass diffusion	3, 6–7
Deriving models of thermal diffusion	8–14

1. The text derives the model for the distribution of salt in the channel as if the higher concentration of salt were on the left. Reverse the place of the ocean and the estuary and show that the model remains unchanged. The boundary values of c_0, c_L do not affect the derivation of the governing differential equation.

2. The text derives the model for the distribution of salt in the channel as if the higher concentration of salt were on the left; that is, as if diffusion were carrying the salt from left to right. It also supposes that the flow of water is in the same direction. Argue that neither the direction of diffusion nor the direction of water flow matters. The basic conservation balance,

$$-A(x)Dc'(x)+A(x)V(x)c(x)$$
$$=-A(x+\Delta x)Dc'(x+\Delta x)$$
$$+A(x+\Delta x)V(x+\Delta x)c(x+\Delta x),$$

still holds. If either the concentration gradient c' or the velocity V is such that one of the "in" terms on the left is actually removing salt from the control volume, then the sign of that term will be negative. Likewise, "out" terms on the right will change sign if they are actually contributing to the salt in the box rather than removing it.

3. The text derives the diffusion model

$$-D(A(x)c'(x))'+(A(x)V(x)c(x))'=0$$

under the assumption that the diffusion coefficient D is constant. Show that if the diffusion coefficient varies with position, then the governing equation is

$$-(D(x)A(x)c'(x))'+(A(x)V(x)c(x))'=0.$$

4. Is the diffusion equation

$$-D(A(x)c'(x))'+(A(x)V(x)c(x))'=0$$

linear or nonlinear? Homogeneous or nonhomogeneous? Under what conditions would it have constant coefficients?

5. Is the diffusion equation with source term

$$-D(A(x)c'(x))' + (A(x)V(x)c(x))' = A(x)P(x)$$

linear or nonlinear? Homogeneous or nonhomogeneous? Under what conditions would it have constant coefficients?

6. Suppose the cross section of the channel in example 1 is approximately semicircular throughout its length but that it is 90 m in diameter at the ocean and tapers roughly linearly to a diameter of 10 m at the lake. Furthermore, suppose water is flowing *from* the lake *to* the ocean at $62.5L^2/(45L - 40x)^2$ km/h, where x is distance from the end of the channel connected to the ocean. Write a model for the concentration of salt in the channel.

7. In Example 2 we took some shortcuts to derive a model for the situation shown in figure 9.6. Proceed from first principles, using the conservation law as in the text and the control volume shown in figure 9.7, to derive the model

$$xc''(x) + c'(x) = 0, \quad c(0) = c_0, \quad c(L) = c_L.$$

Exercises 8–14 ask you to derive models for the *equilibrium* temperature distribution in various bodies. The central experimental fact is as follows:

Experimental fact. The rate of diffusion of heat energy through a unit area is proportional to the negative of the *temperature gradient* at that point.

The constant of proportionality is the *thermal conductivity k*. If $T(x)$ denotes temperature, then

heat-flow rate through a unit area at x is $-kT'(x)$.

The basic conservation law is:

Law of conservation of heat energy. The net change in the heat energy of a body over a given period of time is the amount of heat energy added less the amount of heat energy removed.

At equilibrium, this conservation law demands that the rate at which heat is added to a control volume equal the rate at which heat is lost.

8. The ends of a thin insulated rod of length L are maintained at temperatures T_0, T_L. Suppose the rod has constant cross-sectional area A. Since the rod is thin and its sides are insulated, you can assume that heat flows only along its length. Derive a model for the temperature distribution in this rod as follows.

   (a) Argue that the rate at which heat energy flows through the left wall of the control volume shown in figure 9.8 is $-kAT'(x)$. Similarly, argue that heat-flow rate through the wall on the right is $-kAT'(x + \Delta x)$.

   (b) Why does the conservation law let you conclude that at equilibrium $-kAT'(x + \Delta x) + kAT'(x) = 0$?

   (c) Use a limiting argument to derive the differential equation $-kT''(x) = 0$.

   (d) Argue that the complete *temperature model* is

   $$-T''(x) = 0, \quad T(0) = T_0, \quad T(L) = T_L.$$

   (e) Can you find the temperature distribution in the rod? Does it seem reasonable?

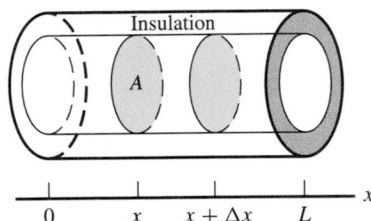

FIGURE 9.8 An insulated rod with cross-sectional area A. The control volume has its left end at x and its right end at $x + \Delta x$.

9. Repeat exercise 8 when the right end of the rod is perfectly insulated so that no heat energy can diffuse through it. Argue that the model is almost the same as before; only the boundary condition at $x = L$ is changed to

$$T'(L) = 0.$$

10. Repeat exercise 9 under each of the following conditions:

   (a) Cross-sectional area A varies with position. Obtain the model

   $$-(A(x)T'(x))' = 0, \quad T(0) = T_0, \quad T(L) = T_L.$$

   (b) Both cross-sectional area A and thermal conductivity k vary with position. Obtain the model

   $$-(k(x)A(x)T'(x))' = 0,$$
   $$T(0) = T_0, \quad T(L) = T_L.$$

   (c) Use the results of part (a) to write a model of the temperature distribution in a rod whose area varies uniformly from left to right so that the area at $x = L$ is twice that at $x = 0$.

11. The top and the bottom of a thin circular plate of thickness d and radius L are insulated; see figure 9.9. Its boundary is at temperature T_L. Because of symmetry and insulation, heat can flow only in the radial direction. Using a control volume, such as that in figure 9.7 derive the model

$$-(xT'(x))' = 0, \quad T'(0) = 0, \quad T(L) = T_L.$$

(*Hint*: The area of the surface of the control volume at radius x is $A(x) = 2\pi x d$.)

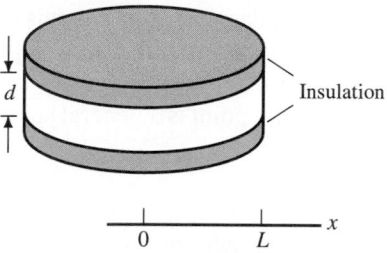

FIGURE 9.9   The top and bottom of a thin circular plate of thickness d and radius L are insulated.

12. Repeat exercise 11 when the outer edge of the disk is perfectly insulated so that no heat energy can diffuse through it. Argue that you obtain almost the same model as before; only the boundary condition at $x = L$ is changed to

$$T'(L) = 0.$$

Try to guess a solution of the boundary-value problem in this case. Does your guess seem physically reasonable? Do you seem to need more information?

13. By passing a current through the thin rod shown in figure 9.8, it is possible to heat it at a given rate. Suppose that heat energy per unit volume is being added at the rate $H(x)$. Derive the model

$$-kAT''(x) = H(x), \quad T(0) = T_0, \quad T(L) = T_L.$$

14. How is the preceding model changed if the left end of the bar is perfectly insulated so that no heat can diffuse through it?

9.2 ■ BOUNDARY-VALUE PROBLEMS: ANALYTIC TOOLS

The diffusion model

$$-D(A(x)c'(x))' + (A(x)c(x))' = 0, \quad c(0) = c_0, \quad c(L) = c_L,$$

derived in section 9.1 is an example of a **boundary-value problem**, a differential equation coupled with specifications of the solution at two different values of the independent variable.

Auxiliary conditions such as

$$c(0) = c_0, \quad c(L) = c_L,$$

or

$$c(0) = c_0, \quad c'(L) = 0,$$

are called **boundary conditions**. In contrast to initial conditions, which are all imposed at a single point, boundary conditions are imposed at two distinct points. Those points define the interval upon which the boundary-value problem must be solved. In the diffusion models of section 9.1, that interval is always $0 \le x \le L$.

Some typical boundary-value problems for the generic second-order differential equation $y'' = f(x, y, y')$ on the interval $a \le x \le b$ include

Typical boundary-value problems

$$y'' = f(x, y, y'), \quad y(a) = A, \quad y(b) = B,$$
$$y'' = f(x, y, y'), \quad y(a) = A, \quad y'(b) = B,$$
$$y'' = f(x, y, y'), \quad y(a) = A, \quad y(b) + dy'(b) = B,$$

$$\vdots$$

A **solution of a boundary-value problem** is a solution of the differential equation that satisfies the boundary conditions and that is defined throughout the interval at whose ends the boundary conditions are set. A formal definition is:

**SOLUTION OF
BOUNDARY-VALUE PROBLEM**

Definition 1. *The function $u(x)$ is a solution of the boundary-value problem $y'' = f(x, y, y')$, $y(a) = A$, $y'(b) = B$, if u is a solution of $y'' = f(x, y, y')$ on $a \leq x \leq b$, if $u(a) = A$, and if $u'(b) = B$.*

As such a definition suggests, solving a boundary-value problem is much like solving an initial-value problem: Find a solution formula containing two arbitrary constants, then choose the constants to satisfy the boundary conditions. For a *linear* differential equation, the starting point is a general solution, just as for an initial-value problem:

**Steps to solve a linear
boundary-value problem:**

1. Find a general solution of the differential equation.
2. Choose the arbitrary constants in the general solution to satisfy the boundary conditions.

The question of uniqueness, the number of solutions a boundary-value problem can possess, is addressed in project 2 of this chapter's projects.

■ **EXAMPLE 5**   *Solve the diffusion model*

$$-Dc'' + Vc' = 0, \quad c(0) = 28, \quad c(L) = 0,$$

derived in example 1. Here $D = 1.09 \times 10^{-9}$ m²/s, $V = 0.5556$ m/s, $L = 4{,}000$ m, and concentrations are measured in g/m³.

Find a general solution.

Use the characteristic equation method to solve the constant-coefficient, homogeneous differential equation

$$-Dc'' + Vc' = 0.$$

Its characteristic equation $-Dr^2 + Vr = 0$ has roots

$$r = 0, \quad r = V/D,$$

and the general solution of the differential equation is

$$c_g = C_1 + C_2 e^{Vx/D}.$$

**Determine C_1, C_2 from the
boundary conditions.**

At $x = 0$, the boundary condition requires

$$c_g(0) = C_1 + C_2 = 28.$$

The boundary condition at $x = L$ requires that

$$c_g(L) = C_1 + C_2 e^{VL/D} = 0.$$

To find C_1, C_2, solve these two simultaneous equations,

$$C_1 + C_2 = 28,$$
$$C_1 + C_2 e^{VL/D} = 0,$$

to find

$$C_1 = \frac{28 e^{VL/D}}{e^{VL/D} - 1}, \quad C_2 = \frac{-28}{e^{VL/D} - 1}.$$

The solution of the boundary-value problem is the general solution with C_1, C_2 replaced by these values:

Solution of boundary-value problem

$$c(x) = \frac{28 e^{VL/D}}{e^{VL/D} - 1} - \frac{28 e^{Vx/D}}{e^{VL/D} - 1}. \quad ■$$

MATLAB

The analytic tools in DELAB can find analytic solutions of many second-order boundary-value problems. For guidance, select `Help`, `Textbook`, then go to chapter 9, example 5.

■ **EXAMPLE 6**   *Describe the concentration profile given by the solution obtained in example 5.*

The given values of V, L, and D yield $VL/D \approx 2.04 \times 10^{12}$. Hence,

$$e^{VL/D} - 1 \approx e^{VL/D},$$

and the solution found in example 5 can be written as

$$c(x) = \frac{28 e^{VL/D}}{e^{VL/D} - 1} - \frac{28 e^{Vx/D}}{e^{VL/D} - 1} \approx 28 - 28 e^{-V(L-x)/D}.$$

Stop and Think

9.1 If $VL/D \approx 1.02 \times 10^{12}$, what is the error in the approximation $e^{VL/D} - 1 \approx e^{VL/D}$?

What can we learn from this approximate expression? When x is near zero, the second term is negligible; $e^{-VL/D} \approx 10^{-12}$. So $c \approx 28$ unless x is close enough to L to make $e^{-V(L-x)/D}$ significant, say, 0.1. If $e^{-V(L-x)/D} = 0.1$, then we find

$$x = L - \frac{D \ln 10}{V} \approx L - 4.52 \times 10^{-9}.$$

In other words, the concentration remains at its inlet level of 28 g/m³ throughout the length of the channel, dropping to its ending value of zero only in the last fraction of a millimeter. This precipitous drop is sketched in figure 9.10. ■

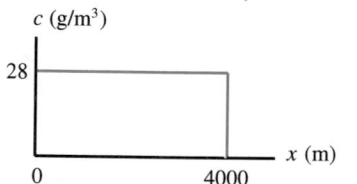

c (g/m³)

28

0   4000 x (m)

FIGURE 9.10 A graph of the solution of the boundary-value problem $-Dc'' + Vc' = 0$, $c(0) = 28$, $c(L) = 0$, analyzed in example 6. The rapid decline from $c = 28$ to $c = 0$ near $x = 4{,}000$ occurs much more precipitously than can be drawn here.

The sharp change in the solution at $x = 4{,}000$ that is displayed in figure 9.10 is an example of a *boundary layer*, a solution phenomenon we can not study here. Boundary layers arise in many settings in which there is a rapid transition between two regimes, such as that from the relatively high concentration in the channel to zero at its outlet.

■ **EXAMPLE 7**   *Example 3 proposes the model*

$$-D(A(x)c'(x))' = 0, \quad c(0) = c_0, \quad c'(L) = 0,$$

for diffusion in a channel of length L when the end at $x = L$ is dammed, thereby stopping diffusion. Solve this boundary-value problem when D and A are constant.

When D, A are constant, the differential equation simplifies to the constant-coefficient equation

$$c''(x) = 0.$$

Either from the characteristic equation $r^2 = 0$ with repeated root $r = 0$ or from direct integration, we find the general solution

Find a general solution.

$$c_g = C_1 + C_2 x.$$

Stop and Think   **9.2** What sort of curve is described by the equation $c_g = C_1 + C_2 x$? Is that shape consistent with $c'' = 0$?

The boundary conditions require

Determine C_1, C_2 from the boundary conditions.

$$c_g(0) = C_1 = c_0,$$
$$c_g'(L) = C_2 = 0.$$

Hence, the solution of the boundary-value problem is

$$c(x) = c_0.$$

The concentration is constant throughout the length of the channel.

Stop and Think   **9.3** Describe the geometry of the solution curve. Which component of the boundary-value problem $c'' = 0$, $c(0) = c_0$, $c'(L) = 0$, tells you that the solution is a straight line? That it is a horizontal line? That it is a horizontal line through c_0? ■

■ **EXAMPLE 8**   *Example 2 proposes the model*

$$xc''(x) + c'(x) = 0, \quad c'(0) = 0, \quad c(L) = c_L,$$

for the diffusion of salt in a circular pond of radius L. Solve this boundary-value problem.

Unfortunately, this second-order equation has variable coefficients, and we are equipped to solve only constant-coefficient, second-order equations.

Were we to skip ahead to section 11.2, however, we would discover that a general solution of this equation is

Find a general solution.

$$c_g = C_1 + C_2 \ln x.$$

MATLAB

Does DELAB find the same general solution?

Determine C_1, C_2 from the boundary conditions.

Accepting this general solution as given, find the constants C_1 and C_2. To evaluate the boundary condition at $x = 0$, compute

$$c_g'(x) = -\frac{C_2}{x},$$

a function that is not even defined at $x = 0$ *unless* $C_2 = 0$. If $C_2 = 0$, then

$$c_g'(0) = 0,$$

and the boundary condition is satisfied.

> We choose the constant C_2 not to satisfy the boundary condition explicitly but simply to permit the solution to be defined throughout the interval $0 \leq x \leq L$. What remains of the general solution does in fact satisfy the symmetry boundary condition $c'(0) = 0$.

The second boundary condition requires

$$c_g(L) = C_1 = c_L.$$

Hence, the solution of the boundary-value problem is

$$c(x) = c_L.$$

The salt concentration is constant throughout the pond. ■

■ **EXAMPLE 9**   *Example 4 proposes a model for the level of a toxin being released by microbes distributed throughout a channel; the channel connects two clean bodies of water. If the release rate of the toxin is a constant P, if the cross-sectional area of the channel is a constant, and if there is no flow in the channel, then the model reduces to*

$$-Dc''(x) = P, \quad c(0) = c(L) = 0.$$

Solve this boundary-value problem.

A general solution of this differential equation is readily obtained by direct integration but, for practice, exercise the constant-coefficient machinery of chapter 6.

Find a general solution.   As in example 7, the characteristic equation $r^2 = 0$ of the homogeneous differential equation $c''(x) = 0$ yields the general solution

$$c_h = C_1 + C_2 x.$$

To solve the forced equation $c''(x) = -P/D$ with P and D constant, use undetermined coefficients with the initial guess

$$c_p = B,$$

where B is the undetermined coefficient. But this guess satisfies the homogeneous equation. Multiplying by the independent variable leads to

$$c_p = Bx,$$

which again solves the homogeneous equation. Another multiplication leads to the final form of the particular solution,

$$c_p = Bx^2.$$

Substituting in $c''(x) = -P/D$ gives

$$2B = -\frac{P}{D},$$

and

$$c_p = -\frac{Px^2}{2D}.$$

The general solution of the differential equation is

$$c_g = C_1 + C_2 x - \frac{Px^2}{2D}.$$

The boundary conditions require

$$c_g(0) = C_1 = 0,$$

Determine C_1, C_2 from the boundary conditions

$$c_g(L) = C_2 L - \frac{PL^2}{2D} = 0.$$

We conclude that $C_1 = 0$, $C_2 = PL/2D$. The solution of the boundary-value problem is

$$c(x) = \frac{Px}{2D}(L - x).$$

The concentration of pollutant reaches a peak value of $PL^2/8D$ at $x = L/2$, the midpoint of the channel. See figure 9.11. ■

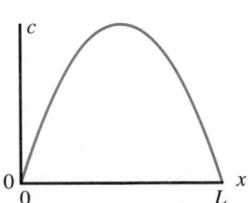

FIGURE 9.11   A graph of the solution of the boundary-value problem $-Dc'' = P$, $c(0) = c(L) = 0$, showing the concentration peak at the center of the channel.

MATLAB

Use DELAB to check the analysis in this example.

Stop and Think   **9.4** The differential equation considered in example 9 has the form $c'' = -P/D$, a constant. Which curves have constant second derivative? Is the solution $c(x) = Px(L-x)/2D$ one of these curves? Does $c'' = -P/D$ force the solution to be concave up or concave down? Does the solution function exhibit that concavity? Sketch the ways in which different boundary conditions (e.g., $c(0) = 0, c'(L) = 0$) select different pieces of this solution curve as the solution of the boundary-value problem on the interval $0 \le x \le L$.

9.2.1 Exercises

EXERCISE GUIDE	
To gain experience . . .	**Try exercises**
Finding general solutions	1–7, 18
Solving boundary-value problems	1–15
Analyzing and interpreting solution behavior	8–17

In exercises 1–7,

    (i) Find a general solution of the differential equation.

    (ii) Find a solution of the given boundary-value problem.

    (iii) Verify your analysis using DELAB.

1. $y'' - y = 0$, $y(0) = 1$, $y(2) = 0$

2. $y'' - y = 0$, $y(0) = 1$, $y'(2) = 4$

3. $y'' - y = \sin \pi x$, $y(0) = y(2) = 0$

4. $y'' - y = 0$, $y(0) = y(2) = 0$

5. $y'' - y = \sin \pi x$, $y(0) = 1$, $y'(2) = 4$

6. $y'' - 2y' + y = t^2$, $y(0) = -2$, $y(1) = 1$

7. $y'' - 2y' + y = e^t$, $y(0) = 0$, $y(1) = 1$

Each of exercises 8–14 displays a boundary-value problem based on a model for equilibrium temperature derived in the indicated exercise.

    (i) Solve the boundary-value problem and sketch a graph of the solution.

    (ii) Confirm the analytic solution using DELAB.

    (iii) Comment on the physical significance of the solution in light of the situation being modeled.

8. From section 9.1.1, exercise 8,

$$-T''(x) = 0, \quad T(0) = T_0, \quad T(L) = T_L.$$

9. From section 9.1.1, exercise 9,

$$-T''(x) = 0, \quad T(0) = T_0, \quad T'(L) = 0.$$

10. From section 9.1.1, exercise 11,

$$-(xT'(x))' = 0, \quad T'(0) = 0, \quad T(L) = T_L.$$

(*Hint*: See example 8 for a general solution.)

11. From section 9.1.1, exercise 12,

$$-(xT'(x))' = 0, \quad T'(0) = 0, \quad T'(L) = 0.$$

(*Hint*: See example 8 for a general solution.) Have you found one or many solutions of this boundary-value problem? Do you seem to need more information?)

12. From section 9.1.1, exercise 13 with $S(x) = x(L-x)$, k, A constant,

$$-kAT''(x) = x(L-x), \quad T(0) = T_0, \quad T(L) = T_L.$$

13. From section 9.1.1, exercise 14 with $S(x) = x(L-x)$, k, A constant,

$$-kAT''(x) = x(L-x), \quad T'(0) = 0, \quad T(L) = T_L.$$

14. From section 9.7, part (a) of chapter exercise 1,

$$T''(x) = 0, \quad T(0) = T_{in}, \quad T(L) = T_{out}.$$

15. Example 6 claims

$$e^{VL/D} - 1 \approx e^{VL/D}$$

when VL/D is large. How large must this quantity be to keep the error in this approximation under 1%? Do the values used in the text, $D = 1.09 \times 10^{-9}$ m^2/s, $V = 0.2778$ m/s, and $L = 4{,}000$ m, meet this criterion?

16. Verify the claim made in example 6, while analyzing the approximate solution $c \approx 28 - 28e^{-V(L-x)/D}$: If $e^{-V(L-x)/D} = 0.1$, then we find

$$x = L - \frac{D \ln 10}{V}.$$

17. Example 8 solved the boundary-value problem

$$xc''(x) + c'(x) = 0, \quad c'(0) = 0, \quad c(L) = c_L,$$

modeling the concentration of salt in the circular pond of figure 9.6. That figure also shows a hypothetical graph of the concentration profile in the pond. Redraw figure 9.6 with the correct concentration profile, as found in example 8.

18. Find a general solution of the differential equation considered in example 9,

$$-Dc''(x) = P,$$

by integration. Verify that you obtain the same general solution as in the example. (Recall that D, P are constants.)

19. Solve these versions of the toxin source model derived in example 4. Compare your solutions with those obtained by DELAB. Describe the situation each model represents.

(a) $-c''(x) = Px(L - x), c(0) = c(L) = 0$

(b) $-c''(x) + 4c'(x) = 4P, c(0) = c(L) = 0$

(c) $-c''(x) = Px(L - x), c(0) = c'(L) = 0$

9.3 ■ BOUNDARY-VALUE PROBLEMS: NUMERICAL METHODS

Numerical methods for approximating solutions of boundary-value problems can be developed using the same idea as for initial-value problems: approximate derivatives by difference quotients. Since boundary-value problems are constrained at two points, these approximation methods lead to systems of simultaneous equations rather than iterative schemes like the Euler method for initial-value problems.

> Because the underlying approximation is a difference quotient—$dy/dx \approx \Delta y/\Delta x$—the methods to be developed in this section are sometimes called *finite-difference methods*.

9.3.1 Approximating Derivatives

The numerical methods we consider are developed by approximating derivatives. The basic approximation is stopping short of the limit in the definition of derivative:

Forward-difference approximation
$$u'(x) = \lim_{\Delta x \to 0} \frac{u(x + \Delta x) - u(x)}{\Delta x} \quad \Rightarrow \quad u'(x) \approx \frac{u(x + \Delta x) - u(x)}{\Delta x}.$$

This approximation is illustrated on the left in figure 9.12; it is called a **forward-difference approximation** because the difference $u(x + \Delta x) - u(x)$ samples u *ahead* of the point x at which the approximation of u' is desired.

Another natural approximation is a **backward difference**,

Backward-difference approximation
$$u'(x) \approx \frac{u(x) - u(x - \Delta x)}{\Delta x}. \tag{9.1}$$

Stop and Think   **9.5** Explain the name *backward difference*.

That approximation is illustrated on the right in figure 9.12. A visual inspection of the forward and backward differences shown there suggests that neither

FIGURE 9.12 Forward-, centered-, and backward-difference approximations to the derivative $u'(x)$

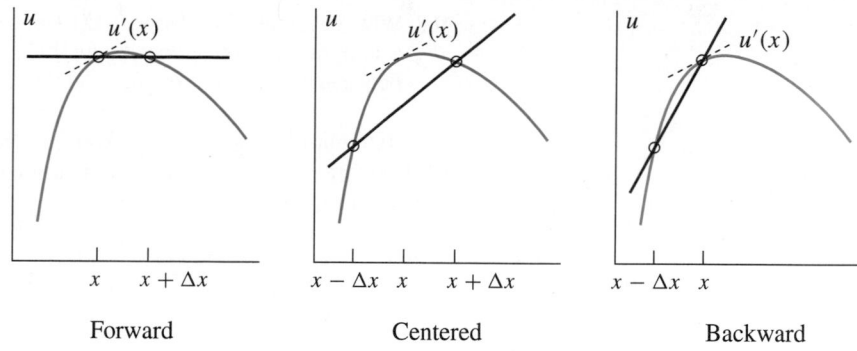

Forward   Centered   Backward

is a particularly good approximation to $u'(x)$, while the **centered difference**

CENTERED-DIFFERENCE
APPROXIMATION TO $u'(x)$

$$u'(x) \approx \frac{u(x + \Delta x) - u(x - \Delta x)}{2\Delta x} \qquad (9.2)$$

shown in the center of figure 9.12 is much better. Is that observation true in general?

Stop and Think   **9.6** Why does $2\Delta x$—not Δx—appear in the denominator of (9.2)?

An answer—a formal argument that centered differences are more accurate than forward or backward differences—comes from Taylor's theorem (see appendix section A.3):

$$u(x + \Delta x) = u(x) + u'(x)\Delta x + u''(x)\frac{\Delta x^2}{2!} + u'''(x)\frac{\Delta x^3}{3!} + \cdots . \qquad (9.3)$$

(Of course, $\cdots$ denotes terms involving even higher powers of Δx: Δx^4, etc.) Using this expansion in the forward-difference approximation yields

$$\frac{u(x + \Delta x) - u(x)}{\Delta x} = \frac{1}{\Delta x}\left(u'(x)\Delta x + u''(x)\frac{\Delta x^2}{2!} + u'''(x)\frac{\Delta x^3}{3!} + \ldots\right)$$

$$= u'(x) + u''(x)\frac{\Delta x}{2} + \cdots .$$

The error in the forward-difference approximation $u'(x) \approx (u(x + \Delta x) - u(x))/\Delta x$ is

Error in forward-difference
approximation

$$E_{\text{forward}} = \frac{u(x + \Delta x) - u(x)}{\Delta x} - u'(x) = u''(x)\frac{\Delta x}{2} + \cdots .$$

Since this error is dominated by a term involving Δx to the first power, the forward-difference approximation is said to be a *first-order* or *order one* approximation. Exercise 3 shows the backward difference is first-order, too.

Stop and Think   **9.7** According to this order terminology, a *second-order* approximation would be dominated by Δx^2. Why would such an approximation be better than one of first order?

9.8 What becomes of the forward-difference error term $-u''(x)\Delta x/2 + \cdots$ if u is a straight line? What does figure 9.12 show in this case? What is the relation between a forward difference and the slope of a line?

Now apply that same analysis to the centered-difference approximation (9.2). Using both the Taylor expansion (9.3) for $u(x + \Delta x)$ and its analog for $u(x - \Delta x)$ leads to

$$\frac{u(x + \Delta x) - u(x - \Delta x)}{2\Delta x}$$

$$= \frac{1}{2\Delta x}\left(u(x) + u'(x)\Delta x + u''(x)\frac{\Delta x^2}{2!} + u'''(x)\frac{\Delta x^3}{3!} + \cdots\right)$$

$$- \left(u(x) - u'(x)\Delta x + u''(x)\frac{\Delta x^2}{2!} + u'''(x)\frac{\Delta x^3}{3!} - \cdots\right)$$

$$= u'(x) + u'''(x)\frac{\Delta x^2}{3} + \cdots.$$

The error in the centered-difference approximation (9.2) is

Error in centered-difference approximation

$$E_{\text{centered}} = u'''(x)\frac{\Delta x^2}{3} + \cdots,$$

a *second-order* term. For small Δx, the second-order centered-difference approximation is indeed more accurate than the first-order forward-difference approximation.

Stop and Think

9.9 How small should Δx be to make the dominant term in the centered-difference error smaller than the dominant term in the forward-difference error?

9.10 Discuss: The conclusion that a centered difference is a more accurate approximation of $u'(x)$ than a forward difference needs an important qualifier: *if* the derivatives u'' and u''' are defined and smooth near x. Otherwise, an expression like that for E_{centered} could be meaningless.

To develop an approximation for the second derivative—the *derivative of the first derivative*—begin with a centered-difference approximation (9.2) of du'/dx on an interval whose *total length* is Δx,

$$u''(x) = \frac{du'(x)}{dx} \approx \frac{u'(x + \Delta x/2) - u'(x - \Delta x/2)}{\Delta x}. \tag{9.4}$$

The choice of the shorter interval depends on hindsight and the desire to ultimately involve only $u(x)$ and $u(x \pm \Delta x)$ in the approximation of $u''(x)$.

Now approximate the derivatives $u'(x \pm \Delta x/2)$ with centered differences, again on an interval whose length is Δx:

$$u'(x + \Delta x/2) \approx \frac{u(x + \Delta x) - u(x)}{\Delta x},$$

$$u'(x - \Delta x/2) \approx \frac{u(x) - u(x - \Delta x)}{\Delta x}. \tag{9.5}$$

Stop and Think **9.11** Verify both of these centered-difference approximation formulas.

Combining the approximations (9.4) and (9.5) gives a *centered-difference approximation to $u''(x)$*,

CENTERED-DIFFERENCE APPROXIMATION TO $u''(x)$

$$u''(x) \approx \frac{u(x - \Delta x) - 2u(x) + u(x + \Delta x)}{\Delta x^2}. \tag{9.6}$$

Exercise 5 asks you to show that this approximation is of second order.

9.3.2 Approximating Differential Equations

To use these derivative approximations to approximate a differential equation, consider

$$-c''(x) = \sin \pi x / L$$

on the interval $0 \leq x \leq L$. This equation might arise in a diffusion model with a source term, as in example 4.

Choose an integer n and let $\Delta x = L/n$. Define a uniform **mesh** or **grid**,

$$x_0 = 0, \, x_1 = \Delta x, \, x_2 = 2\Delta x, \ldots, x_{n-1} = (n-1)\Delta x, \, x_n = n\Delta x = L,$$

on the interval $0 \leq x \leq L$. That is, let $x_i = i \Delta x = iL/n, \, i = 0, \ldots, n$.

Stop and Think **9.12** How many points in total are in the grid? How many are on the *interior* of the interval? Which values of i correspond to points interior to the interval? Which values of i correspond to the endpoints of the interval?

9.13 Suppose the interval $1 \leq x \leq L$ is divided into n subintervals of equal length. Define $x_0, \ldots, x_n$.

We can approximate $u'(x_i)$ or $u''(x_i)$ at any of the interior grid points using the centered-difference formulas (9.2) and (9.6), respectively.

Since the differential equation $-c''(x) = \sin \pi x / L$ contains only the second derivative, we need only the approximation

$$c''(x_i) \approx \frac{c(x_i - \Delta x) - 2c(x_i) + c(x_i + \Delta x)}{\Delta x^2}$$

Centered-difference approximation to $c''(x_i)$

$$= \frac{c(x_{i-1}) - 2c(x_i) + c(x_{i+1})}{\Delta x^2}, \quad i = 1, \ldots, n - 1.$$

Stop and Think **9.14** How does the second line follow from the first?

9.15 Why is the index i limited to the range $i = 1, \ldots, n - 1$? What problem arises using this centered-difference approximation at $x_0 = 0$ or $x_n = L$?

If c_i is an approximation to $c(x_i)$, then the second-derivative approximation becomes

---

Centered-difference approximation to c_i''

$$c_i'' \approx \frac{c_{i-1} - 2c_i + c_{i+1}}{\Delta x^2}, \quad i = 1, \ldots, n - 1. \tag{9.7}$$

Approximating c'' at x_i requires three points, x_i and its right- and left-hand neighbors $x_{i\pm 1}$. Think of this *three-point stencil* sliding back and forth across the grid to approximate c'' at each of the grid points x_i interior to the interval $0 \le x \le L$, as shown in figure 9.13.

Now use the approximation (9.7) to c'' in the differential equation $-c''(x) = \sin \pi x / L$ at each of the *interior* grid points $x_1 = \Delta x$, $x_2 = 2\Delta x$, $\ldots, x_{n-1} = (n-1)\Delta x$. The differential equation is approximated by a *system of $n - 1$ simultaneous equations*,

Finite-difference approximation

$$-\frac{c_{i-1} - 2c_i + c_{i+1}}{\Delta x^2} = \sin \pi x_i / L, \quad i = 1, \ldots, n - 1.$$

With the fractions cleared, those $n - 1$ equations can be written

FINITE-DIFFERENCE APPROXIMATION

$$
\begin{aligned}
-c_0 + 2c_1 - c_2 &= \Delta x^2 \sin \pi x_1 / L \\
-c_1 + 2c_2 - c_3 &= \Delta x^2 \sin \pi x_2 / L \\
&\vdots \\
-c_{n-2} + 2c_{n-1} - c_n &= \Delta x^2 \sin \pi x_{n-1} / L.
\end{aligned}
\tag{9.8}
$$

Stop and Think

9.16 Show that each of these $n - 1$ equations corresponds to approximating $c''(x)$ at one of the interior grid points $x_1, \ldots, x_{n-1}$ interior to the interval $0 \le x \le L$. Why is there no equation corresponding to either of the endpoints, $x = 0$ ($i = 0$) or $x = L$ ($i = n$)?

9.17 Write out the analog of (9.8) for the more general differential equation $-c''(x) = f(x)$.

The system of simultaneous equations (9.8) constitutes a **finite-difference approximation** to $-c'' = \sin \pi x / L$ on $0 \le x \le L$. Note that there are $n + 1$ unknowns, the concentration values $c_0, c_1, \ldots, c_n$, but only $n - 1$ equations; there are two more unknowns than equations. Without two pieces of additional information—without boundary conditions—there is no hope of finding a unique solution for the c_i.

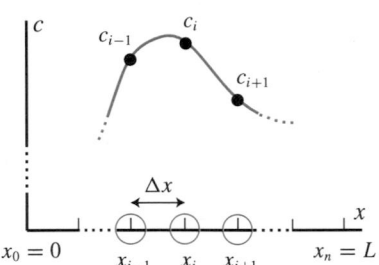

FIGURE 9.13 The three-point centered-difference stencil samples $c(x)$ at x_{i-1}, x_i, and x_{i+1} to approximate $c''(x_i)$.

Of course, an equal number of unknowns and equations does not guarantee the existence of a unique solution. For example, two equations in two unknowns could represent two parallel lines. Such a system has either no solution if the two lines are not collinear or an infinite number if they are.

9.3.3 Incorporating Boundary Conditions

Adding boundary conditions can provide the additional data needed to guarantee a unique solution of the system of simultaneous equations (9.8), a finite-difference approximation to $-c'' = \sin \pi x / L$ on $0 \le x \le L$.

■ EXAMPLE 10   *Write a finite-difference approximation to the solution of the boundary-value problem*

$$-c'' = \sin \pi x / L, \quad c(0) = c(L) = 0. \tag{9.9}$$

Stop and Think **9.18** Describe a physical situation for which this boundary-value problem might be a model.

The boundary conditions explicitly define two of the unknowns in the difference approximation:

$$c(0) = 0 \quad \Rightarrow \quad c_0 = 0,$$
$$c(L) = 0 \quad \Rightarrow \quad c_n = 0.$$

Use these relations to remove two of the unknowns in the difference equations (9.8); set $c_0 = 0$ in the first equation and $c_n = 0$ in the last.

Stop and Think **9.19** Explain why c_0 appears only in the first equation and c_n only in the last.

Now there are $n - 1$ equations in $n - 1$ unknowns $c_1, c_2, \ldots, c_{n-1}$:

$$2c_1 - c_2 \qquad\qquad\qquad = \Delta x^2 \sin \pi x_1 / L$$
$$-c_1 + 2c_2 - c_3 \qquad\qquad = \Delta x^2 \sin \pi x_2 / L$$
$$\vdots$$
$$-c_{n-2} + 2c_{n-1} = \Delta x^2 \sin \pi x_{n-1} / L.$$

Stop and Think **9.20** How would these equations change if the boundary conditions were $c(0) = 1$, $c(L) = 2$? (See exercise 6.)

These simultaneous equations—a finite-difference approximation to (9.9)—can be written in the compact matrix form

Finite-difference approximation to $-c'' = \sin \pi x / L$, $c(0) = c(L) = 0$

$$\mathbf{Ac = b}$$

by defining the square coefficient matrix

$$\mathbf{A} = \begin{pmatrix} 2 & -1 & 0 & \cdots & 0 & 0 & 0 \\ -1 & 2 & -1 & \cdots & 0 & 0 & 0 \\ & & & \ddots & & & \\ 0 & 0 & 0 & \cdots & 0 & -1 & 2 \end{pmatrix} \qquad (9.10)$$

and a vector of unknowns and a vector of source terms,

$$\mathbf{c} = \begin{pmatrix} c_1 \\ c_2 \\ \vdots \\ c_{n-1} \end{pmatrix}, \quad \mathbf{b} = \begin{pmatrix} \Delta x^2 \sin \pi x_1 / L \\ \Delta x^2 \sin \pi x_2 / L \\ \vdots \\ \Delta x^2 \sin \pi x_{n-1} / L \end{pmatrix}.$$

(See appendix A.5 to review matrix ideas.) The finite-difference approximation $\mathbf{Ac = b}$ was derived by incorporating the boundary conditions into the finite-difference approximation (9.8) to the differential equation $-c'' = \sin \pi x / L$. ■

Stop and Think **9.21** Write out the third row of **A**. The next-to-last row of **A**.

9.22 Write out the first three equations defined by $\mathbf{Ac} = \mathbf{b}$. The last two equations.

9.23 Is it correct to say that the unknown vector contains the approximate concentration values at the points of the grid *interior* to the interval $0 \leq x \leq L$?

■ **EXAMPLE 11**   *Use finite differences to find an approximate solution of the boundary-value problem*

$$-c'' = \sin \pi x/4, \quad c(0) = c(4) = 0, \tag{9.11}$$

using $\Delta x = 1$. Compare that approximation with the exact solution.

Stop and Think **9.24** What is the value of n corresponding to $\Delta x = 1$?

The boundary-value problem (9.11) is simply (9.9) with $L = 4$. From example 10, the finite-difference approximation to (9.11) using $\Delta x = 1$ is the system of three simultaneous equations defined by the matrix equation

$$\begin{pmatrix} 2 & 1 & 0 \\ 1 & 2 & 1 \\ 0 & 1 & 2 \end{pmatrix} \begin{pmatrix} c_1 \\ c_2 \\ c_3 \end{pmatrix} = \begin{pmatrix} \sin \pi/4 \\ \sin \pi/2 \\ \sin 3\pi/4 \end{pmatrix} = \begin{pmatrix} \sqrt{2}/2 \\ 1 \\ \sqrt{2}/2 \end{pmatrix}.$$

Stop and Think **9.25** Identify the matrix **A** and the vectors **c** and **b** defined by writing this matrix equation as $\mathbf{Ac} = \mathbf{b}$.

9.26 Explain the arguments $\pi/4$, $\pi/2$, and $3\pi/4$ appearing in the sine functions in **b**. Confirm that $\mathbf{b} = \begin{pmatrix} \sqrt{2}/2 & 1 & \sqrt{2}/2 \end{pmatrix}^T$.

9.27 Why are there three equations? To what points in the interval $0 \leq x \leq 4$ do they correspond?

9.28 Write out explicitly the three equations defined by $\mathbf{Ac} = \mathbf{b}$. Confirm that they are a correct finite-difference approximation to (9.11) when $\Delta x = 1$.

These three equations can be solved easily by *row elimination*; that is, by using the diagonal entry in each row to eliminate all occurrences of that unknown in the equation below it. The starting point is

Original system

$$\begin{pmatrix} 2 & -1 & 0 \\ -1 & 2 & -1 \\ 0 & -1 & 2 \end{pmatrix} \begin{pmatrix} c_1 \\ c_2 \\ c_3 \end{pmatrix} = \begin{pmatrix} \sqrt{2}/2 \\ 1 \\ \sqrt{2}/2 \end{pmatrix}.$$

Eliminating all occurrences of c_1 below the first equation requires multiplying the first row by $-(-1/2) = 1/2$ and adding it to the second. The result is

Eliminate c_1 below the diagonal.

$$\begin{pmatrix} 2 & -1 & 0 \\ 0 & 3/2 & -1 \\ 0 & -1 & 2 \end{pmatrix} \begin{pmatrix} c_1 \\ c_2 \\ c_3 \end{pmatrix} = \begin{pmatrix} \sqrt{2}/2 \\ 1 + \sqrt{2}/4 \\ \sqrt{2}/2 \end{pmatrix}.$$

Eliminating all occurrences of c_2 below the first equation requires multiplying the second row by $-(-1/(3/2)) = 2/3$ and adding it to the third. The result is

Eliminate c_2 below the diagonal.

$$\begin{pmatrix} 2 & -1 & 0 \\ 0 & 3/2 & -1 \\ 0 & 0 & 4/3 \end{pmatrix} \begin{pmatrix} c_1 \\ c_2 \\ c_3 \end{pmatrix} = \begin{pmatrix} \sqrt{2}/2 \\ 1 + \sqrt{2}/4 \\ 2(1+\sqrt{2})/3 \end{pmatrix}.$$

Stop and Think   **9.29** The system is now in *upper triangular form.* Explain that term.

9.30 State a general rule for computing the *multiplier*, the factor by which the diagonal row is multiplied before adding it to the row from which the diagonal unknown is to be eliminated; the multipliers are $-(-1/2)$ for the first row and $-(-1/(3/2))$ for the second. (Incidentally, the coefficient of the diagonal element is known as the *pivot*; the pivots here are 2 and $3/2$.)

Now solve for c_3, c_2, c_1 in that order, beginning with the last equation, a process known as *back substitution,*

$$4c_3/3 = 2(1+\sqrt{2})/3 \quad \Rightarrow \quad c_3 = (1+\sqrt{2})/2 \approx 1.2071,$$
$$3c_2/2 - c_3 = 1 + \sqrt{2}/4 \quad \Rightarrow \quad c_2 = 1 + \sqrt{2}/2 \approx 1.7071,$$
$$2c_1 - c_2 = \sqrt{2}/2 \quad \Rightarrow \quad c_1 = (1+\sqrt{2})/2 \approx 1.2071.$$

Stop and Think   **9.31** Carry out the details of the back substitution calculation of c_3, c_2, and c_1.

Hence, the finite-difference approximation using $\Delta x = 1$ to the solution of the boundary-value problem (9.11) is

$$c_1 \approx 1.2071, \quad c_2 \approx 1.7071, \quad c_3 \approx 1.2071,$$

or, in vector notation, $\mathbf{c} = \begin{pmatrix} 1.2071 & 1.7071 & 1.2071 \end{pmatrix}^T$.
From exercise 7, the exact solution of (9.11) is

$$c(x) = (4/\pi)^2 \sin \pi x/4. \tag{9.12}$$

The (absolute) error at each of the interior grid points is

$$e_1 = |c(x_1) - c_1| \approx 0.0608,$$
$$e_2 = |c(x_2) - c_2| \approx 0.0860,$$
$$e_3 = |c(x_3) - c_3| \approx 0.0608.$$

Of course, the error is zero at the boundary points. The exact and approximate solutions are shown in figure 9.14. ■

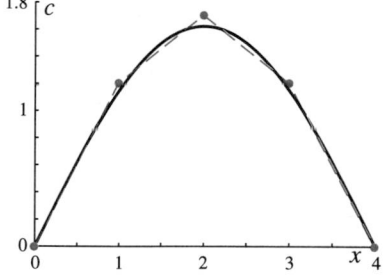

FIGURE 9.14 Finite-difference approximation using $\Delta x = 1$ (blue dots) and exact solution (black curve) of the solution of the boundary-value problem (9.11).

■ **EXAMPLE 12**   *Use finite differences to find an approximate solution of the boundary-value problem (9.11) of example 11, using $\Delta x = 0.5$. Compare that approximation with the exact solution and with the $\Delta x = 1$ approximation. Speculate on the order of the finite-difference approximation to the solution of (9.11).*

Stop and Think   **9.32** What value of n corresponds to $\Delta x = 0.5$? The value of $c(x)$ at the left end of the interval is approximated by c_i with $i = ?$ At the right end, $i = ?$

The mechanics are the same as for example 11: use the boundary conditions $c(0) = c(L) = 0$ to find $c_0 = c_8 = 0$. The finite-difference system (9.8) becomes seven equations in the seven unknowns $c_1, \ldots, c_7$.

Stop and Think   **9.33** The seven unknowns $c_1, \ldots, c_7$ approximate $c(x)$ at which values of x?

The matrix form of the finite-difference system is

$$
\begin{pmatrix}
2 & -1 & 0 & 0 & 0 & 0 & 0 \\
-1 & 2 & -1 & 0 & 0 & 0 & 0 \\
 & & \ddots & & & & \\
0 & 0 & 0 & 0 & -1 & 2 & -1 \\
0 & 0 & 0 & 0 & 0 & -1 & 2
\end{pmatrix}
\begin{pmatrix}
c_1 \\
c_2 \\
\vdots \\
c_6 \\
c_7
\end{pmatrix}
= \left(\frac{1}{2}\right)^2
\begin{pmatrix}
\sin \pi/8 \\
\sin \pi/4 \\
\vdots \\
\sin 3\pi/4 \\
\sin 7\pi/8
\end{pmatrix}.
$$

Stop and Think   **9.34** Write out in full the entries in the matrix form $\mathbf{Ac} = \mathbf{b}$ of this system. Show that the matrix $\mathbf{A}$ is *tridiagonal*; that is, show that $\mathbf{A}$ is zero everywhere except along its main diagonal (upper left to lower right) and along the diagonals just above and below the main diagonal.

Certainly, this system could be solved by row elimination, just as in example 11. However, MATLAB is quicker and more reliable. It finds the approximate values plotted in figure 9.15,

$$
\mathbf{c} = \begin{pmatrix} 0.6284 & 1.1612 & 1.5171 & 1.6421 & 1.5171 & 1.1612 & 0.6284 \end{pmatrix}^T.
$$

Stop and Think   **9.35** Locate the values $c_1, \ldots, c_7$ in figure 9.15.

FIGURE 9.15 Finite-difference approximation using $\Delta x = 0.5$ (blue dots) and exact solution (black curve) of the solution of the boundary-value problem (9.11).

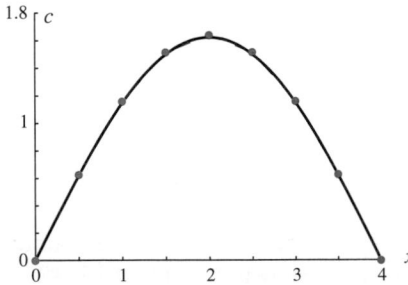

Comparison with the exact solution (9.12) shows that the maximum error occurs at $x = 2$,

$$
e_4 = |c(2) - c_4| \approx 0.0210.
$$

Stop and Think   **9.36** Verify that $i = 4$ is the index value corresponding to $x = 2$.

Compare this value with the maximum error $e_2 = 0.0860$ from example 11, in which $\Delta x = 1$:

$$\frac{\text{max error with } \Delta x = 1}{\text{max error with } \Delta x = 0.5} = \frac{0.0860}{0.0210} \approx 4.0952 \approx \left(\frac{1}{0.5}\right)^2.$$

The maximum error appears to vary with the square of Δx, suggesting that the finite-difference approximation to the solution of the boundary-value problem (9.11) might be of order two. ■

MATLAB

MATLAB was designed to solve linear systems like the finite-difference approximation $\mathbf{Ac} = \mathbf{b}$. For guidance in its application to finite-difference approximations, start DELAB, select `Help, Textbook`, then go to chapter 9, example 12.

Tridiagonal matrices like the coefficient matrix $\mathbf{A}$ in example 12 are examples of *sparse* matrices, matrices whose entries are predominantly zero. In computers, such matrices can be stored economically by keeping only the nonzero entries; e.g., $\mathbf{A}$ might be stored as three column vectors, one for the main diagonal and two for each of the secondary diagonals. Since $\mathbf{A}$ is *symmetric* across the main diagonal (that is, $\mathbf{A}^T = \mathbf{A}$), only one of those secondary diagonals need be stored. Besides the economical storage, computations are faster because there is no arithmetic involving values known to be zero.

Developing efficient storage schemes and fast, accurate numerical methods for solving finite-difference equations like those seen here, as well as the more complex ones that arise in more sophisticated models, are central concerns of the discipline of scientific computing. The results of this kind of work appear in everyday settings, including predicted storm tracks on TV weather reports and digital special effects such as the surface of the ocean in many scenes of the movie *Titanic*.

9.3.4 Derivative Boundary Conditions

If one of the boundary conditions involves the derivative of the unknown, then the two extra unknowns in the finite-difference approximation (9.8) are not automatically eliminated. The following example illustrates how a centered-difference approximation to a derivative in a boundary condition adds one more unknown but two more equations, thereby providing a unique solution of the difference approximation to the boundary-value problem.

■ **EXAMPLE 13** *Write a finite-difference approximation to the boundary-value problem*

$$-c''(x) = x(2L - x) + 2, \quad c(0) = 2, \quad c'(L) = 0. \qquad (9.13)$$

Stop and Think **9.37** Describe a physical situation for which (9.13) might be a model. In particular, describe the physical significance of the boundary condition $c'(L) = 0$.

To write the analog of (9.8) for $-c''(x) = x(2L - x) + 2$, apply the centered-difference approximation (9.6) to $c''(x)$ at $x = x_1, \ldots, x_{n-1}$. The $n - 1$ equations that result are

$$-c_0 + 2c_1 - c_2 \qquad\qquad\qquad = \Delta x^2(x_1(2L - x_1) + 2)$$

$$\vdots \qquad\qquad\qquad\qquad\qquad (9.14)$$

$$-c_{n-2} + 2c_{n-1} - c_n = \Delta x^2(x_{n-1}(2L - x_{n-1}) + 2).$$

Stop and Think   **9.38** Show that the first equation corresponds to approximating $-c''(x) = x(2L - x) + 2$ at $x = x_1$, the last to approximating it at $x = x_{n-1}$. Explain why each equation involves exactly three unknowns.

The unknown c_0 in the first equation in (9.14) can be evaluated immediately using the boundary condition $c(0) = 2$: $c_0 = 2$. Then that first equation becomes

$$2c_1 - c_2 = \Delta x^2(x_1(2L - x_1) + 2) + 2.$$

The "extra" unknown in the last equation is c_n, but it is not specified directly by the other boundary condition, $c'(L) = 0$. One possibility is to approximate the derivative in this boundary condition using a backward-difference approximation (9.1) to $c'(x_n)$,

Backward-difference approximation to $c'(L) = 0$

$$c'_n \approx \frac{c_n - c_{n-1}}{\Delta x},$$

as shown in figure 9.16. Then the boundary condition $c'(L) = 0$ becomes $c'_n \approx (c_n - c_{n-1})/\Delta x = 0$ or $c_n = c_{n-1}$; c_n is easily eliminated from the last of the equations (9.14).

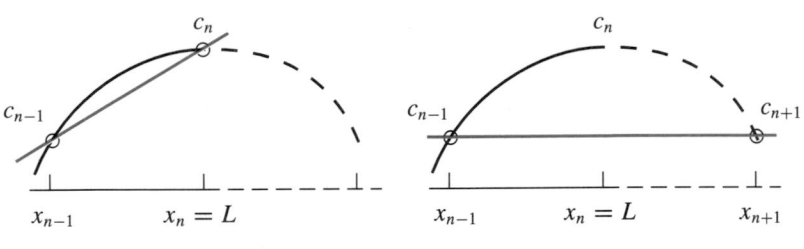

Backward difference   Centered difference

FIGURE 9.16   The boundary condition $c'(L) = 0$ can be approximated by a backward difference from x_n (left) or by a centered difference about x_n (right). The latter involves the fictitious point x_{n+1} and the fictitious unknown c_{n+1}.

Stop and Think   **9.39** Write the last equation in (9.14) after c_n is eliminated by using $c_n = c_{n-1}$, the backward-difference approximation to the boundary condition $c'(L) = 0$.

The drawback is that the backward-difference approximation to $c'(L)$ is only of first order while $c''(x)$ in the differential equation was approximated to second order. As figure 9.16 illustrates, using a more accurate centered-difference approximation to $c'(L)$ would require adding a fictitious grid point x_{n+1} outside the interval and yet another unknown, c_{n+1}. However, the situation is more promising than it first appears.

Using the centered-difference formula (9.2), the boundary condition $c'(L) = 0$ is approximated by

Centered-difference approximation to $c'(L) = 0$

$$c'_n \approx \frac{c_{n+1} - c_{n-1}}{2\Delta x} = 0.$$

The approximate boundary condition expresses the new unknown c_{n+1} in terms of one of the interior values,

$$c_{n-1} = c_{n+1}.$$

The extra unknown is eliminated, but there are still only $n - 1$ equations for the n unknowns $c_1, \ldots, c_n$.

Fortunately, one more equation can be added to the system (9.14), the equation that corresponds to approximating the differential equation $-c''(x) = x(2L - x) + 2$ at $x = x_n = L$, *the right endpoint of the interval*:

$$-c_{n-1} + 2c_n - c_{n+1} = \Delta x^2 (x_n(2L - x_n) + 2). \tag{9.15}$$

Stop and Think **9.40** Derive this equation by applying the centered-difference approximation (9.2) to $-c''(x_n)$.

The approximate boundary condition $c_{n-1} = c_{n+1}$ eliminates c_{n+1} from (9.15). Now there are n equations (corresponding to approximating $-c''(x) = x(2L - x) + 2$ at $x = x_1, \ldots, x_n$) in the n unknowns $c_1, \ldots, c_n$. We have the desired approximation:

Finite-difference approximation to $-c''(x) = x(2L - x) + 2$, $c(0) = 2$, $c'(L) = 0$

$$
\begin{aligned}
2c_1 - c_2 &= \Delta x^2 (x_1(2L - x_1) + 2) + 2 \\
&\ \ \vdots \\
-c_{n-2} + 2c_{n-1} - c_n &= \Delta x^2 (x_{n-1}(2L - x_{n-1}) + 2) \\
-2c_{n-1} + 2c_n &= \Delta x^2 (x_n(2L - x_n) + 2).
\end{aligned}
\tag{9.16}
$$

Stop and Think **9.41** Carefully derive the last equation in (9.16).

9.42 Explain why the first equation in (9.16) is different from the first equation in (9.14). What became of c_0? Why is there an extra term $+2$ on the right in this equation alone? ■

Approximating a derivative boundary condition

To summarize: A centered-difference approximation to a derivative boundary condition adds one unknown, the concentration at a fictitious grid point outside the interval (x_{n+1} in figure 9.16), but approximating the differential equation at the endpoint adds another equation. The derivative boundary condition eliminates the fictitious unknown c_{n+1}, thereby balancing the number of equations and unknowns.

■ **EXAMPLE 14**   *Find the approximate solution of (9.13) in the case L = 4, using the finite-difference approximation of example 13 with* $\Delta x = 1$.

In this case, the finite-difference approximation (9.16) reduces to four equations in four unknowns, $c_1, \ldots, c_4$, corresponding to approximating the differential equation at the four grid points $x_i = i\Delta x = i$, $i = 1, \ldots, 4$. To write (9.16) in the matrix form $\mathbf{Ac} = \mathbf{b}$, define the coefficient matrix

$$\mathbf{A} = \begin{pmatrix} 2 & -1 & 0 & 0 \\ -1 & 2 & -1 & 0 \\ 0 & -1 & 2 & -1 \\ 0 & 0 & -2 & 2 \end{pmatrix}$$

and the vectors

Ac = b approximates
$-c''(x) = x(8 - x) + 2$, $c(0) = 2$,
$c'(4) = 0$ with $\Delta x = 1$.

$$\mathbf{c} = \begin{pmatrix} c_1 \\ c_2 \\ c_3 \\ c_4 \end{pmatrix}, \quad \mathbf{b} = \begin{pmatrix} \Delta x^2(x_1(8 - x_1) + 2) + 2 \\ \Delta x^2(x_2(8 - x_2) + 2) \\ \Delta x^2(x_3(8 - x_3) + 2) \\ \Delta x^2(x_4(8 - x_4) + 2) \end{pmatrix} = \begin{pmatrix} 11 \\ 14 \\ 17 \\ 18 \end{pmatrix}.$$

Stop and Think   **9.43** Verify each of the entries in **A**, **c**, and **b**.

Using MATLAB to solve $\mathbf{Ac} = \mathbf{b}$ yields

$$\mathbf{c} = \begin{pmatrix} 51 & 91 & 117 & 126 \end{pmatrix}^T.$$

Both this approximate solution and the exact solution $c(x) = x^4/12 - 4x^3/3 - x^2 + 152x/3 + 2$ are plotted in figure 9.17. ■

Stop and Think   **9.44** The vector of absolute errors in the approximation **c** just found is $\mathbf{e} = \begin{pmatrix} 0.5833 & 1.0000 & 1.2500 & 1.3333 \end{pmatrix}^T$, as you can easily verify. Where does the maximum error occur? What might be the cause? How does the distribution of errors compare with that of example 11? Comment on the differences.

MATLAB

For guidance using MATLAB to solve $\mathbf{Ac} = \mathbf{b}$, select **Help, Textbook**, then go to chapter 9, example 14.

9.3.5 Summary

Key approximations for deriving finite-difference approximations to boundary-value problems are the centered-difference approximations

Centered-difference
approximation to $u'(x)$

$$u'(x) \approx \frac{u(x + \Delta x) - u(x - \Delta x)}{2\Delta x}$$

and

Centered-difference
approximation to $u''(x)$

$$u''(x) \approx \frac{u(x - \Delta x) - 2u(x) + u(x + \Delta x)}{\Delta x^2}.$$

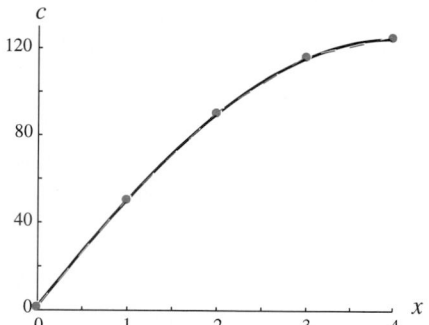

FIGURE 9.17 Finite-difference approximation using $\Delta x = 1$ (blue dots) and exact solution (black curve) of the solution of the boundary-value problem (9.13) with $L = 4$.

Divide the given interval $0 \le x \le L$ into n subintervals of length $\Delta x = L/n$. Define $x_i = i\,\Delta x$, $k = 0, \ldots, L$. Use the centered-difference formulas to approximate the differential equation $-c''(x) = \cdots$ at the interior points $x_1, \ldots, x_{n-1}$, thereby providing $n-1$ equations in $n+1$ unknowns $c_0, \ldots, c_n$, where c_i is the finite-difference approximation to $c(x_i)$.

The boundary conditions provide the extra information needed to determine the c_i, either directly through a condition like $c(0) = 0$ or indirectly through $c'(L) = 0$, as illustrated in example 13.

9.3.6 Exercises

EXERCISE GUIDE	
To gain experience . . .	**Try exercises**
Using and analyzing difference approximations	1–5
Studying errors in finite-difference approximations	3, 4(b), 8(d), 9–10, 12(iv), 13(b), 15
Finding exact solutions of boundary-value problems	6(c), 7, 12(iv)
Finding finite-difference approximations to boundary-value problems and their solutions	6(a, b, d), 11–14, 16
Analyzing models using difference approximations	8

1. For each of the following functions,

   (i) Estimate the derivative of the function at $x = 1$ using a forward difference, a backward difference, and a centered difference, each with a value of Δx of your choice.

   (ii) Compare your estimates with the exact value and comment on the accuracy of the approximation.

   (iii) Sketch a graph of the function and show the finite-difference approximations on it, as in figure 9.12.

   (iv) Use a Taylor expansion or a similar argument to explain the results you obtain.

   (a) $u(x) = x$

   (b) $u(x) = 4x - 3$

   (c) $u(x) = x^2$

   (d) $u(x) = x^2 + 2x - 3$

2. For each of the following functions,
   (i) Estimate the derivative of the function at $x = 1$ using a forward difference and a centered difference, first with a value of Δx of your choice, then with one-tenth that value.
   (ii) Compare the two sets of estimates with the exact value. By what factor has the error changed?
   (iii) Use a Taylor expansion to explain the relation between the change in the observed error and the change in Δx.

   (a) $u(x) = e^x$

   (b) $u(x) = \ln x$

   (c) $u(x) = \cos x$

3. Use a Taylor expansion to show that the backward-difference approximation (9.1) is of order one, as is its forward-difference relative. For what types of functions is this difference approximation exact? Explain your answer analytically and geometrically.

4. Derive and analyze an alternative approximation to $u''(x)$ as follows.

   (a) Apply the centered-difference approximation (9.2) twice, first to approximate $u''(x) = du'(x)/dx$, then to approximate the terms $u'(x \pm \Delta x)$ that appear in the approximation to du'/dx.

   (b) Use a Taylor expansion to determine the order of this approximation.

   (c) What difficulties might you experience were you to use this formula to approximate a differential equation?

5. Use a Taylor expansion to verify that the centered-difference approximation to $u''(x)$ given by (9.6) is indeed of second order. Explicitly compute the leading term in the error expression. For what types of functions is this approximation exact?

6. Consider the boundary-value problem

   $$-c'' = \sin \pi x/L, \quad c(0) = 1, \quad c(L) = 2,$$

   identical to that considered in example 10 except for the change in the boundary conditions.

   (a) Show that a finite-difference approximation can be written in the form $\mathbf{Ac} = \tilde{\mathbf{b}}$, where $\mathbf{A}$ and $\mathbf{c}$ are defined in example 10 and $\tilde{\mathbf{b}}$ is closely related to the vector $\mathbf{b}$ defined there. Explicitly exhibit $\tilde{\mathbf{b}}$.

   (b) Find a finite-difference approximation to the solution using $\Delta x = 1$. That is, solve $\mathbf{Ac} = \tilde{\mathbf{b}}$ in the case $\Delta x = 1$.

   (c) Compare the difference approximation with the exact solution.

   (d) Repeat part (a) for the general boundary conditions $c(0) = \alpha, c(L) = \beta$.

7. Several examples in the text compared exact solutions of boundary-value problems with finite-difference approximations. Find the exact solution of the indicated boundary-value problem. Compare your analytic solution with that given by DELAB.

   (a) Find the exact solution of (9.11), page 450, which was considered in example 11.

   (b) Find the exact solution of (9.13), page 453, with $L = 4$, which was considered in example 14.

8. Suppose the boundary-value problem

   $$-c'' = \sin \pi x/4, \quad c(0) = c(4) = 0,$$

   models diffusion of a pollutant in a channel connecting two bodies of water, as in figure 9.3, page 429. ($L = 4$ is the length of the channel in km, say.)

   (a) Explain why the rate of diffusion of pollutant out of the channel at its left and right ends is proportional to $c'(0)$ and $c'(L)$, respectively. Why would this value be important to you if you were monitoring the quality of the water in the large bays at either end of the channel?

   (b) The solution of this boundary-value problem was approximated in example 11 using $\Delta x = 1$. Use that approximate solution to estimate $c'(0)$ and $c'(L)$. Which difference formulas can you use? What is the order of the approximation to c'? Compare the approximate values with the exact values. Does the approximation overestimate or underestimate? Illustrate the approximations you use on a sketch like figure 9.14.

   (c) Repeat part (b) using the $\Delta x = 0.5$ finite-difference approximation of example 12.

   (d) Compare the errors in the $\Delta x = 1$ and $\Delta x = 0.5$ estimates of $c'(0), c'(L)$. Is the ratio of the errors consistent with the order of the difference formula you used to estimate these derivatives?

9. Examples 11 and 12 found finite-difference approximations to the solution of

   $$-c'' = \sin \pi x/4, \quad c(0) = c(4) = 0,$$

   using $\Delta x = 1$ and $\Delta x = 0.5$, respectively. Find the ratio of the errors in the $\Delta x = 0.5$ approximation to those of the $\Delta x = 1$ approximation at the grid points common to both. How do those ratios compare with the ratio of the Δx values? Speculate about the order of the finite-difference approximation developed in example 10.

10. Example 14 found a finite-difference approximation with $\Delta x = 1$ to the solution of (9.13) in the case $L = 4$. Use DELAB to find an approximation using $\Delta x = 0.5$. Find the ratio of the errors in the $\Delta x = 0.5$ approximation to those of the $\Delta x = 1$ approximation at the grid points common to both. How do those ratios compare with the ratio of the Δx values? Speculate about the order of the finite-difference approximation involved.

11. For each of the following differential equations on the interval $0 < x < L$, write the analog of the finite-difference approximation (9.8), page 448.

   (a) $-c''(x) = 2x$

   (b) $-c''(x) = f(x)$

   (c) $-c''(x) + 3c'(x) = f(x)$

   (d) $-c''(x) + 3c'(x) - 2c(x) = f(x)$

12. For each of the following boundary-value problems:

   (i) Write a finite-difference approximation.

   (ii) Set $\Delta x = 0.5$ and write out *all* of the equations in the approximation.

   (iii) Use DELAB to find an approximate solution for $\Delta x = 0.5$.

   (iv) Compare the approximate solution with the exact solution.

   (v) Repeat the previous two steps for $\Delta x = 0.25$. Is the reduction in error you observe consistent with a second-order approximation to the boundary-value problem?

   (a) $-c''(x) = x(2 - x)$, $c(0) = c(2) = 0$

   (b) $-c''(x) = x(4 - x) + 2$, $c(0) = 2$, $c'(2) = 0$

   (c) $-c''(x) + 3c'(x) = \sin \pi x/2$, $c(0) - c(2) = 0$

   (d) $-c''(x) + 3c'(x) - 2c(x) = \sin \pi x/4$, $c(0) = 0$, $c'(2) = 0$

13. Example 13 developed a finite-difference approximation to the boundary-value problem (9.13) using a centered-difference approximation for the boundary condition $c'(L) = 0$.

   (a) Develop the finite-difference approximation that results from using a (less accurate) backward-difference approximation to $c'(L) = 0$, as illustrated in figure 9.16.

   (b) Use DELAB to find this approximation for $L = 4$ and $\Delta x = 1$. Compare the accuracy of this approximation with the centered-difference result given in example 14. Comment on the differences.

14. Construct a finite-difference approximation to each of the following boundary-value problems. Write the approximation in the form $\mathbf{Ac} = \mathbf{b}$ and define $\mathbf{A}$, $\mathbf{c}$, and $\mathbf{b}$.

   (a) $-y''(x) + Q(x)y(x) = f(x)$, $y(a) = \alpha$, $y(b) = \beta$

   (b) $-y''(x) + Q(x)y(x) = f(x)$, $y'(a) = \alpha$, $y(b) = \beta$

   (c) $-y''(x) + Q(x)y(x) = f(x)$, $y(a) = \alpha$, $y'(b) = \beta$

   (d) $-y''(x) + Q(x)y(x) = f(x)$, $y(a) + ky'(a) = 0$, $y(b) = \beta$

   (e) $-y''(x) + Q(x)y(x) = f(x)$, $y(a) = \alpha$, $y(b) + ky'(b) = 0$

15. Use theorem 3, page 625, Taylor's theorem with remainder, to make more careful statements about the error in the forward- and centered-difference approximations to $u'(x)$.

   (a) For the forward-difference approximation $u'(x) \approx (u(x) - u(x - \Delta x))/\Delta x$, express the absolute error

   $$e(x) = \left| u'(x) - \frac{u(x) - u(x - \Delta x)}{\Delta x} \right|$$

   in the approximation at x in terms of the remainder R_2 in the Taylor expansion. Show that subject to appropriate hypotheses on u and its derivatives, the error $e(x)$ is bounded by a term proportional to the maximum of $|u''|$ on the interval whose endpoints are x and $x + \Delta x$.

   (b) For what class of functions is the forward-difference approximation exact? Justify your conclusion analytically and geometrically.

   (c) Repeat part (a) for the centered-difference approximation (9.2). Which derivative appears in the error bound?

   (d) Repeat part (b) for the centered-difference approximation.

16. Example 14 used DELAB to solve a finite-difference system $\mathbf{Ac} = \mathbf{b}$ of four simultaneous equations.

   (a) Use row elimination to solve that system. Confirm the solution obtained by DELAB.

   (b) How many row elimination steps would be required using the diagonal entry in row i if $\mathbf{A}$ were an arbitrary matrix? How many row elimination steps in total for an $n \times n$ matrix? How many steps per row are required when $\mathbf{A}$ is a tridiagonal finite-difference approximation? How many steps in total for an $n \times n$ tridiagonal matrix? What feature of the process of approximating the boundary-value problem ensures that $\mathbf{A}$ is tridiagonal?

9.4 ■ TIME-DEPENDENT DIFFUSION

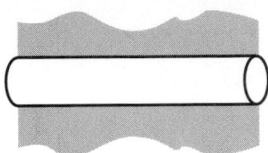

FIGURE 9.18   A thin rod whose sides are insulated.

The diffusion models considered so far in this chapter assume that the distribution of salt or pollutant or heat energy is in *equilibrium*; that is, the value of concentration of pollutant at a given point is constant in time. This section derives models that relax that assumption, permitting a study of the approach to equilibrium as well as the equilibrium itself.

The models derived here use several independent variables, those for spatial position as well as for time. Consequently, partial derivatives appear in the resulting models, which involve *partial differential equations*.

Section 9.1 studied the diffusion of salt and pollutants in water. The models developed here consider an analogous phenomenon, the diffusion of heat energy in a solid such as the thin, insulated rod shown in figure 9.18. The qualifiers *thin* and *insulated* mean that heat energy flow is effectively limited to the axial direction, the left-right direction in the figure.

The governing experimental fact is gradient-dependent diffusion, just as with substances dissolved in water:

EXPERIMENTAL FACT

Experimental fact.   The rate of diffusion of heat energy through a unit area is proportional to the negative of the *temperature gradient*.

The constant of proportionality is the *thermal conductivity k*. If $T(x)$ denotes temperature, then

$$\text{heat-flow rate through a unit area at } x = -k\frac{\partial T(x,t)}{\partial x}.$$

Figure 9.19 illustrates this idea.

The two-variable notation $T(x,t)$ indicates that temperature depends both on position x along the rod's axis and on time t. Consequently, the temperature gradient is denoted by the partial derivative $\partial T/\partial x$.

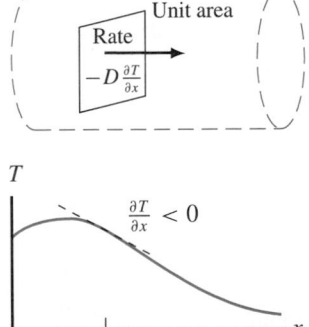

FIGURE 9.19   Heat is flowing through the unit area at the rate $-k\partial T(x,t)/\partial x$. The flow is from left to right because the temperature gradient is negative at that point.

The basic conservation law is as follows.

CONSERVATION LAW

Law of conservation of heat energy.   The net change in the heat energy of a body over a given period of time is the amount of heat energy added less the amount of heat energy removed.

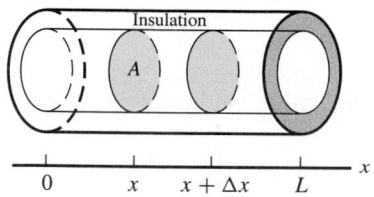

FIGURE 9.20 An insulated rod with cross-sectional area A. The control volume has its left end at x and its right end at $x + \Delta x$.

Away from equilibrium, this conservation law demands that the net change of heat energy within a control volume equal the rate at which heat is added to the control volume less the rate at which heat is lost.

> This conservation law is a precise analog of the law of conservation of population, among others: The amount added less the amount lost equals the net change over a fixed time interval.

Models are derived by accounting for heat energy gained and lost in a control volume such as that in figure 9.20. Since heat is flowing only in the axial direction (horizontally in figure 9.20), we need account only for heat flow through the left and right sides of the control volume. The quantities required by conservation of energy for a change over Δt are

$$\text{change in heat energy over } \Delta t = Q(t + \Delta t) - Q(t),$$

$$\text{heat energy } in \text{ at } x = -kA(x)\frac{\partial T(x, t)}{\partial x}\Delta t,$$

$$\text{heat energy } out \text{ at } x + \Delta x = -kA(x + \Delta x)\frac{\partial T(x + \Delta x, t)}{\partial x}\Delta t.$$

Here $A(x)$ is the area of the cross section of the rod at coordinate x, and $Q(t)$ is the heat energy in the control volume at time t. Having assumed that heat energy is flowing only in the axial direction, we need account only for heat energy flow through the circular sides of the control volume at x and $x + \Delta x$.

> The gradient expression $-k\partial T/\partial x$ gives heat flow *in* on the left side of the control volume and *out* on the right side because x is increasing from left to right and that is the direction in which the gradient is computed. On the left, the direction of the derivative defining the gradient is into the control volume, and on the right it is out.

Stop and Think   **9.45** What expression gives the total heat flow *per unit area* from t to $t + \Delta t$ in through the left end of the control volume? In through the right end?

Conservation of energy requires that the net change over Δt in heat energy within the control volume equals the amount of heat energy that flowed in less the amount that flowed out. In words, conservation of energy demands

Conservation of heat energy

$$\text{change over } \Delta t \text{ of heat energy in control volume}$$
$$= \text{amount gained} - \text{amount lost}.$$

The corresponding mathematical statement is

Conservation of heat energy

$$Q(t + \Delta t) - Q(t)$$
$$= -kA(x)\frac{\partial T(x, t)}{\partial x}\Delta t + kA(x + \Delta x)\frac{\partial T(x + \Delta x, t)}{\partial x}\Delta t.$$

To relate heat energy Q to temperature T, use $Q = cmT$, where c is the specific heat of the material in the rod and m is the mass of the control volume. (See equation (2.14), page 56.) Now the mass of the control volume is just its volume multiplied by the density ρ of the rod material. The volume of the slice of the rod between x and $x + \Delta x$ is approximately

$$A(x)\Delta x,$$

so the mass of that piece is $\rho A(x)\Delta x$. Consequently, the heat energy in the control volume at time t is $c\rho A(x)\Delta x T(x, t)$.

With this substitution, the conservation of energy relation becomes

$$Q(t + \Delta t) - Q(t)$$
$$= c\rho A(x)T(x, t + \Delta t)\Delta x - c\rho A(x)T(x, t)\Delta x$$
$$= -kA(x)\frac{\partial T(x, t)}{\partial x}\Delta t + kA(x + \Delta x)\frac{\partial T(x + \Delta x)}{\Delta x}\Delta t.$$

Stop and Think   **9.46** Explain the connection between each term in this expression and conservation of energy.

Division by both Δt and Δx yields

$$c\rho A(x)\frac{T(x, t + \Delta t) - T(x, t)}{\Delta t}$$
$$= k\frac{A(x + \Delta x)T_x(x + \Delta x, t) - A(x)T_x(x, t)}{\Delta x},$$

where T_x denotes $\partial T/\partial x$. With the limits

$$\lim_{\Delta t \to 0} \frac{T(x, t + \Delta t) - T(x, t)}{\Delta t} = \frac{\partial T(x, t)}{\partial t}$$

and

$$\lim_{\Delta x \to 0} \frac{A(x + \Delta x)T_x(x + \Delta x, t) - A(x)T_x(x, t)}{\Delta x} = \frac{\partial(A(x)T(x, t))}{\partial x},$$

the conservation relation yields the governing equation, the **heat equation**

HEAT EQUATION

$$c\rho A(x)\frac{\partial T(x, t)}{\partial t} = k\frac{\partial(A(x)T(x, t))}{\partial x}.$$

The heat equation is a **partial differential equation** because it involves partial derivatives, in contrast with the ordinary derivatives in the *ordinary differential equations* studied earlier.

Stop and Think   **9.47** Which term in the heat equation comes from the change over time of the heat energy in the control volume? Which term represents diffusion?

If the rod is in equilibrium, then $\partial T/\partial t = 0$ and the heat equation reduces to the diffusion equation considered in section 9.1.

To complete the heat-flow model, we need appropriate auxiliary data. The spatial derivative is of second order $(\partial(AT_x)/\partial x = A\partial^2 T/\partial x^2 + \cdots)$, and it represents the effects of diffusion. Boundary conditions at the left and right ends of the bar might be appropriate; for example, the ends might be kept at zero temperature:

Boundary conditions

$$T(0, t) = 0, \quad T(L, t) = 0, \quad t > 0.$$

The time derivative $\partial T/\partial t$ is of first order and should demand an initial condition. But that initial condition must be specified at each value of x. For example, if the initial temperature distribution in the rod followed the sine curve shown in figure 9.21, the corresponding initial condition would be

Initial condition

$$T(x, 0) = M \sin \pi x/L, \quad 0 < x < L.$$

Using these auxiliary conditions, the complete heat-flow model for the rod is the **initial-boundary-value problem**

MODEL FOR TEMPERATURE IN A THIN ROD

$$c\rho A(x) \frac{\partial T(x, t)}{\partial t} = k \frac{\partial (A(x) T_x(x, t))}{\partial x},$$
$$T(x, 0) = M \sin \pi x/L, \quad 0 < x < L,$$
$$T(0, t) = 0, \quad T(L, t) = 0, \quad t > 0.$$

Stop and Think **9.48** Which equation is the initial condition? Which are the boundary conditions?

$T = 0$ $T = 0$

$T(x, 0)$

M

$M \sin \pi x/L$

0 L x

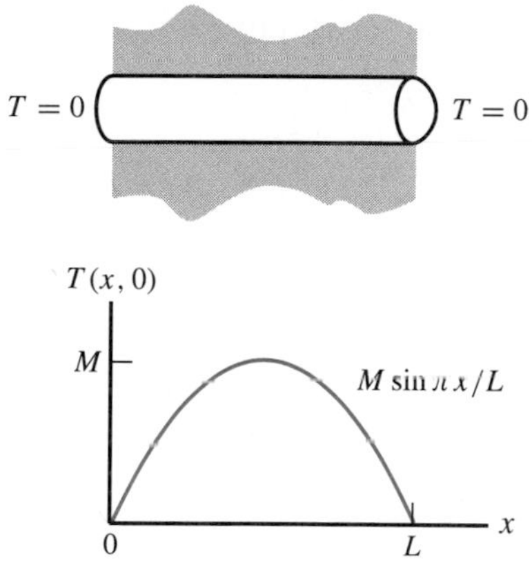

FIGURE 9.21 The ends of a thin rod of length L are maintained at zero temperature. At $t = 0$, the temperature in the rod is given by $M \sin \pi x/L$.

■ **EXAMPLE 15** *Write a model for the temperature distribution in a thin rod of constant cross-sectional area, such as that shown in figure 9.21. Use the initial and boundary conditions shown there.*

If area A is constant, then the heat equation simplifies to

$$\frac{\partial T(x, t)}{\partial t} = \kappa \frac{\partial^2 T(x, t)}{\partial x^2}, \tag{9.17}$$

where $\kappa = k/c\rho$ is the *thermal diffusivity*. The initial and boundary conditions are

$$T(x, 0) = M \sin \pi x/L, \quad 0 < x < L,$$
$$T(0, t) = 0, \quad T(L, t) = 0, \quad t > 0. \ \blacksquare \tag{9.18}$$

The fundamental idea in the derivation of this model is the application of conservation of energy to a control volume. The terms involving the change over time of the net heat energy in the volume lead in the limit to the time derivative T_t. The terms involving the difference in the temperature gradient on either side of the control volume lead to the second-order spatial derivative $(AT_x)_x$.

To emphasize the role of the control volume, the next example examines the case of a wire heated by a current passing through it. The heat generated by the current adds another term to the conservation of energy balance.

■ **EXAMPLE 16**   *Suppose a piece of insulated wire of constant cross-sectional area A and length L is heated by a current passing through it. Assume the ends of the wire are maintained at zero temperature and that the wire is initially at some temperature $f(x)$, $0 \leq x \leq L$. Derive a model for the variation with position and time of the temperature in the wire.*

The situation is essentially that shown in figure 9.21, but the initial temperature is given by the arbitrary function $f(x)$.

To derive the governing partial differential equation, employ a control volume such as that illustrated in figure 9.20. Using the ideas given earlier, write expressions for the heat energy flow through the sides of the control volume and for the net change in heat energy within the control volume. The challenge is incorporating the heat generated by the current in the wire.

Let H denote the *rate* at which the current generates heat energy *per unit volume* of wire. Then, over a time period Δt, the current will add $H \Delta V \Delta t$ units of heat energy (e.g., calories) to a volume ΔV of wire. In the case of the control volume shown in figure 9.20, the volume is $\Delta V = A \Delta x$.

Stop and Think   **9.49** Carefully explain each of the factors in $H \Delta V \Delta t$, the amount of heat energy added to a volume ΔV of wire over time Δt.

Conservation of heat energy demands

change over Δt of heat energy in control volume

CONSERVATION OF ENERGY $=$ heat flow in through left $-$ heat flow out through right

$+$ heat added by current.

The terms involved in this accounting are

change in heat energy over $\Delta t = Q(t + \Delta t) - Q(t)$,

$$\text{heat energy } in \text{ at } x = -kA\frac{\partial T(x, t)}{\partial x}\Delta t,$$

$$\text{heat energy } out \text{ at } x + \Delta x = -kA\frac{\partial T(x + \Delta x, t)}{\partial x}\Delta t,$$

$$\text{heat generated by current} = HA\Delta x \Delta t.$$

Equating the terms as required by conservation of energy leads to

Conservation of energy

$$Q(t + \Delta t) - Q(t)$$
$$= -kA\frac{\partial T(x, t)}{\partial x}\Delta t + kA\frac{\partial T(x + \Delta x, t)}{\partial x}\Delta t + HA\Delta x \Delta t.$$

Once again, we can relate heat energy Q in the control volume to its temperature using $Q = cmT$, where c is the specific heat of the wire and $m = \rho \Delta V = \rho A \Delta x$ is the mass of the control volume expressed in terms of the density ρ of the wire. Making this substitution and dividing out the constant factor A leads to

$$c\rho \frac{T(x, t + \Delta t) - T(x, t)}{\Delta t} = k \frac{T_x(x + \Delta x, t) - T_x(x, t)}{\Delta x} + H.$$

Letting Δt and Δx go to zero yields the governing partial differential equation,

Nonhomogeneous heat equation

$$\frac{\partial T(x, t)}{\partial t} = \kappa \frac{\partial^2 T(x, t)}{\partial x^2} + \frac{H}{c\rho},$$

where $\kappa = k/c\rho$.

Stop and Think   **9.50** Associate each term in this equation with one of the following effects: change in heat energy over time, diffusion of heat energy, heat energy generated by the current.

This is a nonhomogeneous version of the heat equation (9.17). To complete the model, impose both the boundary conditions corresponding to zero end temperature,

Boundary conditions

$$T(0, t) = T(L, t) = 0, \quad t > 0,$$

and the initial condition,

Initial conditions

$$T(x, 0) = f(x), \quad 0 < x < L. \ ■$$

The dependence of the rate of heat flow on the gradient of the temperature can be used to develop alternate boundary conditions.

■ **EXAMPLE 17**   *Suppose a thin insulated rod of constant cross-sectional area has its left end insulated against heat flow while its right end is maintained at temperature T_L, as shown in figure 9.22. Its initial temperature varies linearly from zero at the left end to T_L at the right end. Write a model for the temperature in the rod.*

FIGURE 9.22 The left end of a thin rod is insulated. Its right end is kept at temperature T_L. At $t = 0$, the temperature increases linearly from zero at the left to T_L at the right.

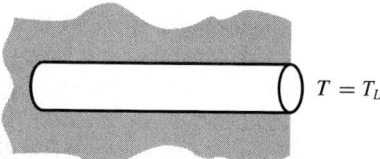

$T = T_L$

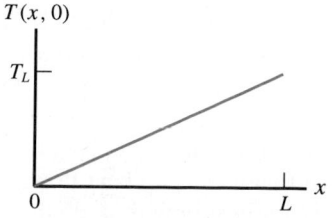

Since cross-sectional area is constant and no heat is generated within the rod, the governing equation is the heat equation (9.17),

Heat equation

$$\frac{\partial T(x,t)}{\partial t} = \kappa \frac{\partial^2 T(x,t)}{\partial x^2}.$$

The initial condition is evidently

Initial condition

$$T(x,0) = \frac{T_L x}{L}, \quad 0 < x < L.$$

Stop and Think   **9.51** Does this initial temperature indeed vary "linearly from zero at the left end to T_L at the right end" of the rod?

The boundary condition at $x = L$ must be

Boundary condition at $x = L$

$$T(L,t) = T_L, \quad t > 0.$$

Stop and Think   **9.52** Why is this the correct boundary condition at $x = L$?

Insulating the left end of the bar means that the rate of heat energy flow through that surface is zero. Since rate of heat flow is proportional to T_x, the insulated boundary condition is

Insulated boundary condition at $x = 0$

$$T_x(0,t) = 0, \quad t > 0. \quad ■$$

Stop and Think   **9.53** By what authority is the "rate of heat flow ... proportional to T_x"?

The insulated boundary condition is another example of a *Neumann* boundary condition.

FIGURE 9.23   A computer chip mounted on a circuit board.

■ **EXAMPLE 18**   *Many of the processor chips in personal computers are narrow, thin rectangular blocks, as illustrated in figure 9.23. They are mounted on circuit boards, which in a poor design could trap heat generated by a power supply positioned below it. Viewing the chip from its narrow end, derive a model that will capture the variation in temperature of the chip from its hot underside to its cool top. Assume the chip starts at temperature zero and that its top remains at that temperature while its bottom is at temperature T_0.*

Since the height of the chip is relatively small compared with its width and depth, simplify the problem by neglecting the variation in temperature from left to right and from front to back in the end view of figure 9.24. That is, assume that heat flows only in the vertical direction from the lower surface $y = 0$ at temperature T_0 to the upper surface $y = h$ at temperature zero.

An appropriate control volume is the box with width Δx, height Δy, and depth Δz shown in that figure. Neglecting temperature variation in all but the vertical direction means no heat flows through the sides of the box. We need account only for heat flow through the top and bottom of the box.

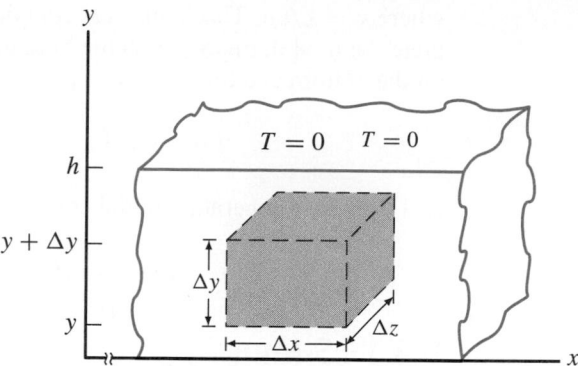

FIGURE 9.24 An end view of the computer chip showing a control volume with height Δy, width Δx, and depth Δz. The lower surface is at temperature T_0, the upper surface at zero temperature.

If the control volume is Δz units deep, then the area of its top and bottom is $\Delta z \Delta x$. The accounting required by conservation of energy is

$$\text{change in heat energy over } \Delta t = Q(t + \Delta t) - Q(t),$$

$$\text{heat energy } in \text{ at } y = -k(\Delta z \Delta x)\frac{\partial T(y, t)}{\partial y}\Delta t,$$

$$\text{heat energy } out \text{ at } y + \Delta y = -k(\Delta z \Delta x)\frac{\partial T(y + \Delta y, t)}{\partial y}\Delta t.$$

Equating these terms as required by conservation of energy leads to

Conservation of energy

$$Q(t + \Delta t) - Q(t)$$
$$= -k(\Delta z \Delta x)\frac{\partial T(y, t)}{\partial y}\Delta t + k(\Delta z \Delta x)\frac{\partial T(y + \Delta y, t)}{\partial y}\Delta t.$$

Once again, we can relate heat energy Q in the control volume to its temperature using $Q = cmT$, where c is the specific heat of the chip and $m = \rho A \Delta x \Delta y \Delta z$ is the mass of the control volume expressed in terms of the density ρ of the chip. Making this substitution and dividing out the common factor $\Delta x \Delta z$ lead to

$$c\rho\frac{T(y, t + \Delta t) - T(y, t)}{\Delta t} = k\frac{T_y(y + \Delta x, t) - T_y(y, t)}{\Delta y}.$$

Letting Δt and Δy go to zero yields the governing partial differential equation,

Heat equation

$$\frac{\partial T(y, t)}{\partial t} = \kappa\frac{\partial^2 T(y, t)}{\partial y^2},$$

where $\kappa = k/c\rho$. This is the heat equation (9.17) with y replacing x. To complete the model, impose both the boundary conditions giving the temperature on the bottom and top of the chip,

Boundary conditions

$$T(0, t) = T_0, \quad T(h, t) = 0, \quad t > 0,$$

and the zero temperature initial condition,

Initial condition

$$T(y, 0) = 0, \quad 0 < y < h. \ \blacksquare$$

9.4.1 Summary

Rate of heat flow through a unit area is proportional to the *negative* temperature gradient perpendicular to the area—e.g., to $-T_x(x, t)$ for the unit area shown in figure 9.19. The proportionality constant is k, the thermal conductivity.

Conservation of heat energy demands that net heat energy flowing into a control volume, such as that in figure 9.20 or 9.24, be balanced by the net increase in energy within the control volume. An example of the appropriate accounting appears on page 467. Relating heat energy Q to temperature T via $Q = cmT$ leads to governing equations such as the heat equation (9.17). A complete model includes boundary and initial conditions, as in (9.18).

9.4.2 Exercises

EXERCISE GUIDE	
To gain experience . . .	**Try exercises**
Finding thermal diffusivity κ	1–2
Defining *homogeneous*, etc. for the heat equation	3
Deriving models of thermal diffusion	4–11
Interpreting terms in the heat equation	12–13

1. According to table 2.5, page 56, the specific heat of copper is 0.09 cal/g·K, and its thermal conductivity is 0.908 cal/s·K·cm²/cm. (The thermal conductivity units include the last "/cm" because the data in table 2.5 are for samples 1 cm thick.) Its density is 8.85 g/cm³. Compute the value (including units) of its thermal diffusivity.

2. Repeat exercise 1 for eastern white pine. Thermal conductivity is 0.0003 cal/s·K·cm, specific heat is 0.42 cal/g·K, and density is 0.373 g/cm³.

3. The text casually refers to $T_t = \kappa T_{xx}$ as the *homogeneous* heat equation and $T_t = \kappa T_{xx} + H/c\rho$ as the *nonhomogeneous* heat equation. Define *homogeneous* and *nonhomogeneous*

in this setting, and verify by substitution that the text's claims are correct.

4. Example 16 derives a model for the temperature in a wire of constant cross-sectional area when it is heated by passing a current through it. Derive a model for the same situation, but allow the cross-sectional area of the wire to vary with position.

5. The sides and both ends of a thin rod of length L are insulated. Initially, its temperature is given by $T_L x/L$, $0 < x < L$. Write an initial-boundary-value problem that models the temperature in the rod.

6. Two copper rods of length L are heated to temperatures T_1 and T_2, respectively. Their ends are quickly joined, the entire assembly is wrapped with insulation, and the outer ends of the double-length rod are maintained at the initial temperature of the corresponding segment. Derive a model for the variation with time and position of the temperature of the complete assembly. Assume heat flows unhindered across the junction of the two rods.

7. Suppose the thermal conductivity k of the rod considered in example 15 varies with position because of manufacturing irregularities. Derive a model for the temperature variations with time in the rod.

8. The top and bottom of a thin circular plate of thickness d and radius L are insulated; see figure 9.9. The boundary of the disk is maintained at the uniform temperature T_L, and its initial temperature is a constant T_i. Since there is no angular variation, we can suppose that heat flows only in the radial (center to circumference) direction.

   (a) Using the doughnut-shaped control volume shown in figure 9.7 derive the following equation for the temperature of the plate as a function of time t and radius x:

   $$x T_t(x, t) = \kappa (x T_x(x, t))_x.$$

   (b) Argue that circular symmetry demands $T_x(0, t) = 0$. Add a boundary condition at $x = L$ and an initial condition to complete the model.

9. Suppose the computer chip considered in example 18 is generating heat internally at a rate of H cal/cm^3·s. Derive a model for the vertical temperature profile of the chip as a function of time. Neglect temperature variations in all but the vertical direction.

10. The model of the temperature profile in the computer chip of example 18 neglected temperature variation in the horizontal direction in the end view shown in figure 9.24. Improve the model by considering temperature variations in both the horizontal and vertical directions as follows. (Let w denote the width of the chip.)

   (a) Consider the control volume in figure 9.24. Argue that the heat flow in through the left side over time Δt is $-k \Delta y \Delta z T_x(x, y, t) \Delta t$. Construct a similar expression for the heat flow out through the right side.

   (b) Conservation of energy demands equality between change in heat energy within the control volume and heat flow through the walls of the control volume. Write out the terms in that energy balance.

   (c) Collect terms and take appropriate limits to obtain

   $$T_t(x, y, t) = \kappa \left(T_{xx}(x, y, t) + T_{yy}(x, y, t) \right).$$

   (d) The bottom of the chip is at temperature T_0, and its sides and top are at temperature zero. Initially, the chip's temperature is zero. Complete the model by using this information to specify boundary conditions on $T(x, y, t)$ at $x = 0, w$ for $0 < y < h$ and at $y = 0, h$ for $0 < x < w$, as well as initial conditions at $t = 0$ for $0 < x < w, 0 < y < h$.

11. How would the boundary conditions in exercise 10 change if the left and right sides of the computer chip were insulated? If the top were insulated instead? (See example 17.)

12. Example 16 showed that $T_t - \kappa T_{xx} = H/c\rho$ governs heat flow in the presence of a heat source $H > 0$. (The heat source in that example is a current passing through the wire.) In the absence of thermal diffusion ($\kappa = 0$), the heat source obviously acts to increase temperature. Using similar logic, argue that the effect of thermal diffusion is to lower temperature; that is, diffusion is dissipative. Is this observation physically reasonable? For example, would a hot spot in the center of an otherwise cool rod diminish by spreading or intensify and become hotter?

13. Examine example 16. What is happening in the rod if $T_t - \kappa T_{xx} = H/c\rho$ is the governing equation and $H < 0$?

9.5 ■ INITIAL-BOUNDARY-VALUE PROBLEMS: FOURIER METHODS

The partial differential equation models of the previous section are solved by combining *separation of variables* for partial differential equations with *eigenfunction expansions*, a technique for representing certain functions as infinite sums of sines and cosines, for example.

The separation method provides a family of solutions to the partial differential equation which also satisfy (homogeneous) boundary conditions. Representing the initial condition in an eigenfunction expansion leads to a superposition of solutions that satisfies the partial differential equation, the boundary conditions, and the initial condition.

9.5.1   Separation of Variables

To illustrate the separation idea, consider the model for heat flow in a thin rod derived in example 15. (See figure 9.21.) The heat equation

$$\frac{\partial T(x,t)}{\partial t} = \kappa \frac{\partial^2 T(x,t)}{\partial x^2}, \tag{9.19}$$

is subject to the initial and boundary conditions

$$
\begin{aligned}
T(x,0) &= M \sin \pi x / L, \quad 0 < x < L, \\
T(0,t) &= 0, \quad T(L,t) = 0, \quad t > 0.
\end{aligned}
\tag{9.20}
$$

Separation of variables begins with the assumption that the unknown temperature can be written as a product of two functions, one dependent only on x, the other only on t:

Assume solution in product form.

$$T(x,t) = X(x)\Theta(t).$$

Substituting this assumed solution form into the heat equation (9.19) yields

$$X(x)\Theta'(t) = \kappa X''(x)\Theta(t).$$

Prime denotes the derivative with respect to the argument shown: $\Theta'(t) = d\Theta(t)/dt$, whereas $X'(x) = dX(x)/dx$.

Collecting all functions of t on one side of the equation and all functions of x on the other leads to

Separate variables.

$$\frac{\Theta'(t)}{\kappa \Theta(t)} = \frac{X''(x)}{X(x)}.$$

Stop and Think   **9.54** Compare this notion of separation of variables with that for a first-order ordinary differential equation.

Now fix t and vary x. Since the quotient on the left is constant as long as t is fixed, the quotient on the right must be constant. Reversing the roles of x and t forces us to conclude that *the two quotients are constant.* That observation is the key to separation of variables.

Denote that common constant by $-\lambda$:

Equate to common constant.

$$\frac{\Theta'(t)}{\kappa \Theta(t)} = \frac{X''(x)}{X(x)} = -\lambda.$$

(The negative sign is for convenience later.) Two *ordinary* differential equations for $\Theta(t)$ and $X(x)$ result:

$$
\begin{aligned}
\Theta'(t) + \lambda \kappa \Theta(t) &= 0, \\
X''(x) + \lambda X(x) &= 0,
\end{aligned}
$$

where λ remains to be determined.

The homogeneous boundary conditions $T(0, t) = X(0)\Theta(t) = 0$ and $T(L, t) = X(L)\Theta(t) = 0$, $t > 0$, lead immediately to boundary conditions on $X(x)$:

Boundary conditions for $X(x)$

$$X(0) = X(L) = 0.$$

To determine the spatial dependence of the temperature (and to evaluate the separation constant λ), we must solve the boundary-value problem

$$X''(x) + \lambda X(x) = 0, \quad X(0) = X(L) = 0. \tag{9.21}$$

For any value of λ, one obvious solution is $X(x) \equiv 0$. Unfortunately, choosing the trivial solution for X leads to $T(x, t) \equiv 0$, a solution of the heat equation that can never satisfy the given initial condition.

Hence, determining X requires finding *nontrivial solutions* of the homogeneous boundary-value problem (9.21). More precisely, we seek those values of λ for which

EIGENVALUE PROBLEM

$$X''(x) + \lambda X(x) = 0, \quad X(0) = X(L) = 0,$$

has *nontrivial solutions*. The values of λ for which there are nontrivial solutions are called **eigenvalues**, and the corresponding nontrivial solutions are **eigenfunctions**. The homogeneous boundary-value problem containing the parameter λ is an **eigenvalue problem**.

> An eigenvalue problem couples a *homogeneous* differential equation with *homogeneous* boundary conditions. That combination *always* has the trivial solution. The trick is finding the values of λ, the eigenvalues, for which the problem has *non* trivial solutions. Compare the matrix eigenvalue problem $\mathbf{Bp} = r\mathbf{p}$, e.g., page 385.

> The separation process must always lead to an eigenvalue problem, a homogeneous differential equation containing a parameter subject to homogeneous initial conditions. If necessary, the original initial-boundary-value problem is transformed before separation of variables to obtain homogeneous boundary conditions and a homogeneous differential equation; see example 22 below.

Find the eigenfunctions and eigenvalues.

To find the eigenvalues and eigenfunctions, we begin with a general solution of $X'' + \lambda X = 0$ and impose the boundary conditions. The characteristic equation of $X'' + \lambda X = 0$ is

$$r^2 + \lambda = 0.$$

Its roots are $r = \pm\sqrt{-\lambda}$. Three cases arise, corresponding to λ negative, zero, and positive. The corresponding solutions are

$$\lambda < 0 : \quad X(x) = C_1 e^{\sqrt{-\lambda}\,x} + C_2 e^{-\sqrt{-\lambda}\,x},$$
$$\lambda = 0 : \quad X(x) = C_1 + C_2 x,$$
$$\lambda > 0 : \quad X(x) = C_1 \cos\sqrt{\lambda}\,x + C_2 \sin\sqrt{\lambda}\,x.$$

When $\lambda < 0$, the boundary conditions $X(0) = X(L) = 0$ require

$$x = 0 : \qquad\qquad C_1 + C_2 = 0,$$
$$x = L : \quad C_1 e^{\sqrt{-\lambda}\,L} + C_2 e^{-\sqrt{-\lambda}\,L} = 0.$$

Using Cramer's rule to solve this pair of equations for C_1, C_2, we find that the determinant of coefficients is

$$\det \begin{pmatrix} 1 & 1 \\ e^{\sqrt{-\lambda}\,L} & e^{-\sqrt{-\lambda}\,L} \end{pmatrix} = e^{\sqrt{-\lambda}\,L} - e^{-\sqrt{-\lambda}\,L} \neq 0.$$

Hence, $C_1 = C_2 = 0$. The eigenvalue problem (9.21) has no negative eigenvalues. A similar analysis, which is left to exercise 5, shows that $\lambda = 0$ is not an eigenvalue.

If $\lambda > 0$, then the boundary condition $X(0) = 0$ requires

$$X(0) = C_1 = 0.$$

The other boundary condition, $X(L) = 0$, then demands

$$X(L) = C_2 \sin \sqrt{\lambda}\, L = 0.$$

Choosing $C_2 = 0$ is possible but undesirable, for we seek nontrivial solutions. Alternatively, we can choose λ to satisfy

$$\sin \sqrt{\lambda}\, L = 0,$$

which requires that

$$\sqrt{\lambda}\, L = 0, \ \pm\pi, \ \pm 2\pi, \ldots$$

or

$$\lambda = \frac{\pi^2}{L^2}, \ \frac{4\pi^2}{L^2}, \ldots.$$

(The value $\lambda = 0$ is dropped because it leads again to the trivial solution.) The constant C_2 remains arbitrary.

Hence, the boundary-value problem

Eigenvalue problem

$$X''(x) + \lambda X(x) = 0, \quad X(0) = X(L) = 0,$$

has the nontrivial solutions, or *eigenfunctions*,

Eigenfunctions $X_n(x)$

$$X_n(x) = C \sin \frac{n\pi x}{L},$$

where C is arbitrary, corresponding to the *eigenvalues*

Eigenvalues λ_n

$$\lambda_n = \frac{n^2\pi^2}{L^2}, \quad n = 1, 2, \ldots.$$

For convenience, let $C = 1$.

For each value of λ, there is a corresponding solution of the Θ equation. A general solution of

$$\Theta'(t) + \lambda_n \kappa \Theta(t) = 0$$

is obviously

$\Theta_n(t)$ corresponding to λ_n

$$\Theta_n(t) = c_n e^{-\lambda_n \kappa t}, \quad n = 1, 2, \dots .$$

For each integer n, the product $T_n(x, t) = X_n(x)\Theta_n(t)$ is a solution of the governing equation (9.19). That is, for $n = 1, 2, \dots$

$T_n = X_n(x)\Theta_n(t)$ solves the partial differential equation . . .

$$T_n(x, t) = c_n e^{-n^2 \pi^2 \kappa t / L^2} \sin \frac{n \pi x}{L}$$

is a solution of

$$\frac{\partial T(x, t)}{\partial t} = \kappa \frac{\partial^2 T(x, t)}{\partial x^2}.$$

Furthermore, for each n, $T_n(x, t)$ satisfies the boundary conditions,

. . . and T_n satisfies the boundary conditions.

$$T_n(0, t) = T_n(L, t) = 0, \quad n = 1, 2, \dots .$$

Satisfying the initial condition demands

Impose initial conditions.

$$T_n(x, 0) = c_n \sin \frac{n \pi x}{L} = M \sin \frac{\pi x}{L}. \tag{9.22}$$

Evidently, $n = 1$ and $c_1 = M$. The solution of the initial-boundary value problem (9.19–9.20) is

Solution of initial-boundary-value problem

$$T(x, t) = M e^{-\pi^2 \kappa t / L^2} \sin \frac{\pi x}{L}. \tag{9.23}$$

Note that $T(x, t) \to 0$ as $t \to \infty$ for each x, $0 \le x \le L$. This solution is decaying to the equilibrium zero temperature we would expect from the zero boundary conditions. That equilibrium temperature is the solution of the governing problem (9.19–9.20) with $\partial T / \partial t = 0$; i.e., $T_{xx} = 0$ subject to $T = 0$ at $x = 0, L$.

The hottest point in the rod is always in its interior, at $x = L/2$ in this case. The rate of decay toward the equilibrium is governed by $\pi^2 \kappa / L^2$. The temperature decays more quickly in a shorter rod (L smaller) or in one that conducts heat better (κ larger).

The steps just executed are:

1. Assume the heat equation (9.19) has a solution in the separated form $T(x, t) = X(x)\Theta(t)$.

2. Substitute $T(x, t) = X(x)\Theta(t)$ into (9.19). Collect functions of t on one side of the equality, functions of x on the other. Equate the two functions of distinct independent variables to a common constant λ and simplify, leading to two *ordinary* differential equations, one for $\Theta(t)$ and one for $X(x)$.

3. Use the boundary conditions on $T(x, t)$ in (9.20) to derive boundary conditions on $X(x)$, leading to an eigenvalue problem in X.

4. Solve the resulting eigenvalue problem to determine the eigenvalues λ_n and the corresponding nontrivial solutions (eigenfunctions) $X_n(x)$. Substitute λ_n into the $\Theta(t)$ equation and solve for $\Theta_n(t)$.

5. Reassemble the product solutions $T_n(x, t) = X_n(x)\Theta_n(t)$ and choose n by matching $T_n(x, 0)$ to the initial condition in (9.20).

9.5.2 Eigenfunction Expansions

What if the initial conditions are other than a sine curve that conveniently matches the spatial dependence of one of the solutions $T_n(x, t)$; that is, what if the initial conditions do not match the solutions of the heat equation as easily as in (9.22)? The following example explores this possibility.

■ **EXAMPLE 19** *Suppose the initial temperature distribution in the rod shown in figure 9.21 is*

$$T(x, 0) = Mx(L - x).$$

Find the resulting temperature in the rod.

In spite of the change in the initial condition, the separation of variables solution process follows the steps illustrated previously:

1. Assume the solution has the product form $T(x, t) = X(x)\Theta(t)$.

2. Substitute in the governing heat equation (9.19), separate variables, and introduce a separation constant $-\lambda$,

$$\frac{\Theta'(t)}{\kappa\Theta(t)} = \frac{X''(x)}{X(x)} = -\lambda.$$

3. Use the boundary conditions $T(0, t) = T(L, t) = 0$ to obtain an eigenvalue problem for $X(x)$,

$$X''(x) + \lambda X(x) = 0, \quad X(0) = X(L) = 0.$$

4. Solve the eigenvalue problem to obtain the eigenfunction-eigenvalue pairs

$$X_n(x) = \sin\frac{n\pi x}{L}, \quad \lambda_n = \frac{n^2\pi^2}{L^2}, \quad n = 1, 2, \ldots,$$

and using the eigenvalues λ_n, solve the equation $\Theta' + \kappa\lambda\Theta = 0$ to find

$$\Theta_n(t) = c_n e^{-\lambda_n \kappa t}, \quad n = 1, 2, \ldots.$$

5. Obtain a family

$$T_n(x, t) = c_n e^{-n^2\pi^2\kappa t/L^2} \sin\frac{n\pi x}{L}$$

of solutions of the heat equation (9.19) that also satisfy the boundary conditions $T(0, t) = T(L, t) = 0$.

Unfortunately, none of the individual solutions $T_n(x, t)$ can satisfy the initial condition

$$T(x, 0) = Mx(L - x),$$

because

$$T_n(x, 0) = c_n \sin \frac{n\pi x}{L}.$$

The individual sine curves of $T_n(x, 0)$ and the parabola can never match exactly.

Since each of the $T_n(x, t)$ satisfies both the *homogeneous*, linear heat equation and the *homogeneous* boundary conditions, any finite sum of these functions is also a solution. Furthermore, the infinite sum

$$T(x, t) = \sum_{n=1}^{\infty} T_n(x, t) = \sum_{n=1}^{\infty} c_n e^{-n^2\pi^2\kappa t/L^2} \sin \frac{n\pi x}{L}$$

is also a solution, if it converges.

> Claiming that a finite sum of solutions of the homogeneous equation is again a solution of the equation is an application of the principle of superposition, a concept that is easily extended from ordinary to partial differential equations. Would a reasonable definition of *linear partial differential equation* include the heat equation?

> Although it is a serious mathematical question, we will take for granted the convergence of the infinite series in order to concentrate on the construction of a solution in this form.

If this infinite sum is to satisfy the initial condition, then we must find constants c_n so that

$$T(x, 0) = \sum_{n=1}^{\infty} c_n \sin \frac{n\pi x}{L} = Mx(L - x), \quad 0 < x < L. \tag{9.24}$$

This request may not be completely preposterous, for one-half cycle of a sine function does resemble the parabolic arc $Mx(L - x), 0 < x < L$. (See figure 9.25.)

The key to finding the constants c_n is the **orthogonality** on $0 \leq x \leq L$ of the family of functions $\sin n\pi x/L$,

ORTHOGONALITY PROPERTY
OF $\sin n\pi x/L$

$$\int_0^L \sin \frac{m\pi x}{L} \sin \frac{n\pi x}{L} \, dx = \begin{cases} 0, & m \neq n \\ L/2, & m = n. \end{cases} \tag{9.25}$$

> The adjective *orthogonal* is used because the integral of the product of the two functions is a natural generalization of the dot product of two vectors when the two sine functions are regarded as members of the vector space of continuous functions. We will see later that this family of functions is orthogonal because it is a family of eigenfunctions.

How does orthogonality help to find c_n? Choose an integer m and multiply both sides of (9.24) by $\sin m\pi x/L$. If integration and the infinite sum can be interchanged, then

$$\sum_{n=1}^{\infty} \int_0^L \sin \frac{m\pi x}{L} \sin \frac{n\pi x}{L} \, dx = \int_0^L \sin \frac{m\pi x}{L} Mx(L - x) \, dx.$$

Because of the orthogonality of the family of sine functions, every integral on the left is zero except for one case, $n = m$, when the integral is $L/2$. The infinite sum collapses to a single term, and we have

$$c_m = \frac{2}{L} \int_0^L \sin \frac{m\pi x}{L} M x(L - x)\, dx.$$

Evaluating the integral yields

$$c_m = \frac{4ML^2}{m^3\pi^3}(1 - (-1)^m),$$

or

$$c_m = \begin{cases} 8ML^2/m^3\pi^3, & m \text{ odd} \\ 0, & m \text{ even}. \end{cases} \tag{9.26}$$

With the choice of c_n given by (9.26),

Eigenfunction expansion of $Mx(L - x)$

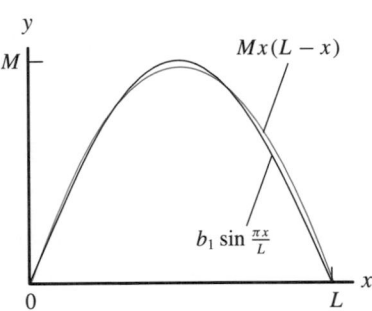

$$\sum_{n=1}^{\infty} c_n \sin \frac{n\pi x}{L} = M x(L - x), \quad 0 \le x \le L; \tag{9.27}$$

the meaning of equality will be made precise in the eigenfunction expansion theorem to be stated shortly. This series is an example of an *eigenfunction expansion* of $Mx(L - x)$ in terms of the eigenfunctions $X_n = \sin n\pi x/L$ of the eigenvalue problem

$$X''(x) + \lambda X(x) = 0, \quad X(0) = X(L) = 0.$$

Figure 9.25 compares the graph of $Mx(L - x)$ with the *first* term $(8ML^2/\pi^3)\sin(\pi x/L)$ in the eigenfunction expansion (9.27). There is a small but pronounced difference between the two. Adding just a few more terms of the expansion (9.27) makes it impossible to distinguish between the expansion and the original function $Mx(L - x)$.

Hence, we have the formal solution

FIGURE 9.25 A comparison of the function $Mx(L - x), 0 \le x \le L$, (blue) and the first term of its eigenfunction expansion, $(8ML^2/\pi^3)\sin(\pi x/L)$ (black).

$$T(x, t) = \sum_{n=1}^{\infty} c_n X_n(x) \Theta_n(t)$$

FORMAL SOLUTION OF INITIAL-BOUNDARY-VALUE PROBLEM

$$= \sum_{n=1,3,\dots}^{\infty} \frac{8ML^2}{m^3\pi^3} e^{-n^2\pi^2\kappa t/L^2} \sin \frac{n\pi x}{L}$$

of the heat equation (9.19) subject to the boundary conditions

$$T(0, t) = T(L, t) = 0, \quad t > 0,$$

and to the initial condition

$$T(x, 0) = Mx(L - x), \quad 0 < x < L.$$

Here c_n is given by (9.26). ■

Odd Functions

Figure 9.25 suggests that the eigenfunction expansion (9.27) is converging rapidly to $Mx(L-x)$ on $0 \le x \le L$. What function does this expansion represent for values of x outside of this interval?

The answer is surprisingly easy. For arbitrary x, let

$$f(x) = \sum_{n=1}^{\infty} c_n \sin \frac{n\pi x}{L},$$

where c_n is given by (9.26). Since $f(x)$ is a sum of sine functions, it must have the same symmetry about the y-axis as the sine function; that is, it must be an **odd function**, one which satisfies

ODD FUNCTION

$$f(x) = -f(-x).$$

An arbitrary odd function is anti-symmetric about the origin, as illustrated in figure 9.26.

Stop and Think

9.55 Sketch a graph of the sine function, and verify that it has the same symmetry properties as the arbitrary odd function sketched in figure 9.26.

9.56 The caption for figure 9.26 says that the graph of an odd function "for $x < 0$ is the reflection of the graph for $x > 0$ across the y-axis and then the x-axis." The relation defining an odd function is $f(x) = -f(-x)$. Which negative signs accounts for reflection across the x-axis? Across the y-axis?

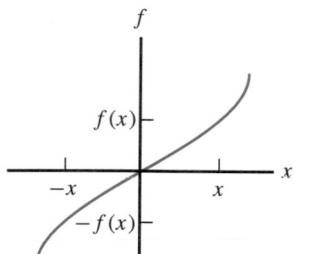

FIGURE 9.26 An odd function, $f(x) = -f(-x)$. Its graph for $x < 0$ is the reflection of the graph for $x > 0$ across the y-axis and then the x-axis.

Since $f(x) = Mx(L-x), 0 \le x \le L$, its definition for $-L \le x < 0$ is immediate from $f(x) = -f(-x) = -M(-x)(L-(-x)) = Mx(L+x)$,

$$f(x) = \begin{cases} Mx(L-x), & 0 \le x \le L \\ Mx(L+x), & -L \le x < 0. \end{cases}$$

The *odd extension* of $Mx(L-x)$ is illustrated in figure 9.27.

To reach beyond the interval $-L \le x \le L$, note that the term of lowest frequency (largest period) in the expansion for f is $\sin \pi x/L$, a function with period $2L$; i.e., $\sin \pi x/L = \sin \pi(x+2L)/L$ for all x. The sum itself—and, consequently, $f(x)$—must have period $2L$. The eigenfunction expansion converges to the *$2L$-periodic extension* of $f(x)$.

Finally, a bit of housecleaning. Since the odd integers $n = 1, 3, 5, \ldots$ can be written as $n = 2p - 1$, $p = 1, 2, 3, \ldots$, the change of index $n = 2p - 1$ transforms the sum in the solution expression to

$$T(x,t) = \sum_{p=1}^{\infty} \frac{8ML^2}{(2p-1)^3\pi^3} e^{-(2p-1)^2\pi^2\kappa t/L^2} \sin \frac{(2p-1)\pi x}{L}.$$

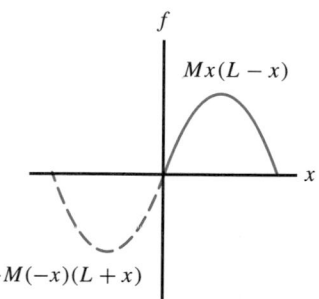

FIGURE 9.27 The odd extension (dashed curve) $-f(-x) = -M(-x)(L+x)$ of the function (solid curve) $f(x) = Mx(L-x)$, $x > 0$.

Eigenfunction Expansion Theorem

The construction illustrated in example 19 is an example of an **eigenfunction expansion**, a series of the form

$$\sum^{\infty} c_n X_n(x),$$

where $\{X_n(x)\}$ are the eigenfunctions of an eigenvalue problem of the form

$$
\begin{aligned}
X''(x)\lambda X(x) &= 0, \\
aX(0) + bX'(0) &= 0, \\
cX(L) + dX'(L) &= 0.
\end{aligned} \tag{9.28}
$$

In the example 19, $a = c = 1$ and $b = d = 0$ and $X_n(x) = \sin n\pi x/L$.

The eigenvalue problem (9.28) is an example of a *self-adjoint* boundary-value problem. See exercise 14 for a more general example.

The important property of eigenfunctions of (9.28) that correspond to distinct eigenvalues is their *orthogonality*,

$$
\int_0^L X_n(x)X_m(x)\,dx = 0, \quad m \neq n,
$$

as in (9.25), for example, where $X_n(x) = \sin n\pi x/L$.

Theorem 1 (Eigenfunction expansion). *Let $\{X_n(x)\}$ be a family of eigenfunctions of the eigenvalue problem (9.28), where neither a and b nor c and d are simultaneously zero.*

i. *Eigenfunctions corresponding to distinct eigenvalues are orthogonal on $0 \leq x \leq L$; i.e.,*

$$
\int_0^L X_n(x)X_m(x)\,dx = 0, \quad m \neq n.
$$

ii. *If $f(x)$ is square integrable on $0 \leq x \leq L$, that is, if*

$$
\int_0^L [f(x)]^2\,dx < \infty,
$$

then its eigenfunction expansion converges in mean square,

Meaning of $\sum\limits^{\infty} c_n X_n(x) = f(x)$

$$
\lim_{p \to \infty} \int_0^L \left(f(x) - \sum^p c_n X_n(x) \right)^2 dx = 0,
$$

where

GENERALIZED FOURIER
COEFFICIENT

$$
c_n = \left(\int_0^L [X_n(x)]^2\,dx \right)^{-1} \int_0^L f(x)X_n(x)\,dx. \tag{9.29}
$$

In example 19 with $X_n(x) = \sin n\pi x/L$, the orthogonality could be verified by direct integration:

$$
\int_0^L \sin\frac{n\pi x}{L} \sin\frac{m\pi x}{L}\,dx = 0, \quad m \neq n.
$$

Since

$$\int_0^L \left(\sin \frac{n\pi x}{L}\right)^2 dx = \frac{L}{2},$$

the (generalized) **Fourier coefficient** c_n given (9.29) in part (ii) of the theorem is just the coefficient c_n we computed in (9.26). The eigenfunction expansion $\sum c_n X_n(x)$ is sometimes called a generalized *Fourier expansion*.

The *mean-square convergence* provided by this theorem permits the infinite sum and $f(x)$ to differ at a large number of points, provided their total "length" on the interval $0 \le x \le L$ is zero. For our purposes, these differences may occur where f suffers a jump discontinuity, and we will consider only functions with a finite number of such jumps.

The proof of the mean-square convergence of the eigenfunction expansion is rather delicate and is omitted. We will illustrate the idea of the proof of orthogonality and the use of the orthogonality relation to compute the generalized Fourier coefficient c_n.

Proof. (Sketch of orthogonality argument and Fourier coefficient formula).

Consider part (i) in the special case of the boundary conditions $X(0) = X(L) = 0$. Suppose $X(x)$ is an eigenfunction of (9.28) corresponding to the eigenvalue λ, and $Y(x)$ is an eigenfunction corresponding to the eigenvalue μ; that is,

$$X''(x) + \lambda X(x) = 0, \quad X(0) = X(L) = 0,$$

and

$$Y''(x) + \mu Y(x) = 0, \quad Y(0) = Y(L) = 0.$$

Further, suppose $\lambda \ne \mu \ne 0$. To establish orthogonality, we must show that

$$\int_0^L X(x)Y(x)\, dx = 0.$$

Substitute $Y - Y''/\mu$ in this integral, integrate by parts using $u = X$, $v = Y'$, and apply the boundary conditions $X(0) = X(L) = 0$:

$$\int_0^L X(x)Y(x)\, dx = -\frac{1}{\mu} \int_0^L X(x)Y''(x)\, dx$$

$$= -\frac{1}{\mu} \left(X(x)Y'(x)\big|_{x=0}^{x=L} - \int_0^L X'(x)Y'(x)\, dx \right)$$

$$= \frac{1}{\mu} \int_0^L X'(x)Y'(x)\, dx.$$

One more integration by parts using $u = X'$, $v = Y'$ and the boundary conditions $Y(0) = Y(L) = 0$ yields

$$\int_0^L X(x)Y(x)\, dx = -\frac{1}{\mu} \int_0^L X''(x)Y(x)\, dx.$$

Now substitute $X'' = -\lambda X$ and move all terms to the left to obtain

$$\left(1 - \frac{\lambda}{\mu}\right) \int_0^L X(x) Y(x)\, dx = 0.$$

Since $\lambda/\mu \neq 1$, the integral must be zero, establishing the orthogonality condition.

The argument for more general boundary conditions is similar; see exercise 22 for the boundary conditions $X'(0) = X(L) = 0$.

To show the plausibility of the formula (9.29) for the generalized Fourier coefficient c_n given in part (ii), suppose that the eigenfunction expansion converges to $f(x)$ at each point of $0 \leq x \leq L$:

Proposed eigenfunction expansion

$$f(x) = \sum_{n=1}^{\infty} c_n X_n(x), \quad 0 \leq x \leq L.$$

Mimicking the calculations in example 19, choose a fixed integer m, multiply through by $X_m(x)$, and assume integration and summation can be interchanged:

Multiply by $X_m(x)$.

$$\int_0^L f(x) X_m(x)\, dx = \sum_{n=1}^{\infty} c_n \int_0^L X_n(x) X_m(x)\, dx.$$

Because of orthogonality, the integral inside the summation is zero except when $n = m$, leaving

Use orthogonality of eigenfunctions.

$$\int_0^L f(x) X_m(x)\, dx = c_m \int_0^L [X_m(x)]^2\, dx,$$

from which the formula for c_m follows immediately. ■

■ **EXAMPLE 20**   *Expand the function*

$$\frac{T_L x}{L}, \quad 0 \leq x \leq L,$$

in the eigenfunctions of the eigenvalue problem

$$X''(x) + \lambda X(x) = 0, \quad X'(0) = X'(L) = 0.$$

We leave to exercise 6 the task of showing that this eigenvalue problem has no nontrivial solutions for $\lambda \leq 0$.

For $\lambda > 0$, the general solution is $X_g = C_1 \cos \sqrt{\lambda}\, x + C_2 \sin \sqrt{\lambda}\, x$. The boundary condition $X'(0) = 0$ immediately forces $C_2 = 0$. The second boundary condition $X'(L) = 0$ then requires

$$-C_1 \sqrt{\lambda} \sin \sqrt{\lambda}\, L = 0.$$

We reject the choice $C_1 = 0$ because we want nontrivial solutions. The alternative is

$$\sin \sqrt{\lambda} L = 0 \implies \sqrt{\lambda} L = n\pi, \quad n = 0, 1, 2, \ldots .$$

That is, the eigenvalues and corresponding eigenfunctions are

Eigenvalues and eigenfunctions

$$\lambda_n = \frac{n^2\pi^2}{L^2}, \quad X_n(x) = \cos \frac{n\pi x}{L}, \quad n = 0, 1, 2, \ldots .$$

To complete the expansion, we need constants c_n such that

$$\sum_{n=0}^{\infty} c_n \cos \frac{n\pi x}{L} = \frac{T_L x}{L}, \quad 0 \le x \le L,$$

where the equality is understood in the sense of mean-square convergence, as in part (ii) of the eigenfunction expansion theorem. To use formula (9.29) for c_n, compute

$$\int_0^L [X_n(x)]^2 \, dx = \int_0^L \cos^2 \frac{n\pi x}{L} \, dx = \begin{cases} L, & n = 0 \\ L/2, & n = 1, 2, \ldots . \end{cases}$$

For $n = 0$, (9.29) yields

$$c_0 = \frac{1}{L} \int_0^L \frac{T_L x}{L} \, dx = \frac{T_L}{2},$$

and for $n = 1, 2, \ldots,$

$$c_n = \frac{2}{L} \int_0^L \frac{T_L x}{L} \cos \frac{n\pi x}{L} \, dx$$

$$= \frac{2T_L}{n^2\pi^2}((-1)^n - 1), \quad n = 1, 2, \ldots,$$

or

$$c_n = \begin{cases} -4T_L/n^2\pi^2, & n = 1, 3, \ldots \\ 0, & n = 2, 4, \ldots . \end{cases}$$

Hence, the eigenfunction expansion is

$$\frac{T_L}{2} - \sum_{n=1,3,\ldots} \frac{4T_L}{n^2\pi^2} \cos \frac{n\pi x}{L} = \frac{T_L x}{L}, \quad 0 \le x \le L, \quad (9.30)$$

where equality is again understood in the sense of mean-square convergence. ■

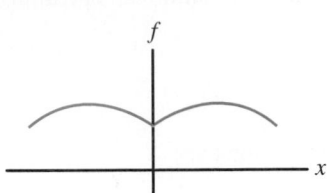

FIGURE 9.28   An even function.

Even Functions

To what function $f(x)$ does the eigenfunction expansion (9.30) converge when x is not in $0 \le x \le L$? Since f is a sum of a constant and cosines, it must have the same symmetry across the y-axis as those two functions; that is, f must be an **even** function,

$$f(x) = f(-x).$$

Figure 9.28 illustrates a generic even function, which is its own reflection across the y-axis.

Stop and Think

9.57 Sketch a graph of the cosine function and verify that it has the same symmetry as the generic even function shown in figure 9.28.

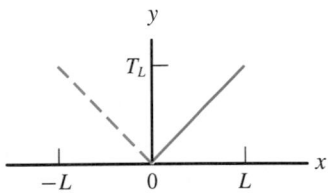

FIGURE 9.29   The function $T_L x/L$, $0 \le x \le L$ (solid line), and its even extension (dashed line) to $-L \le x < 0$.

For $-L \le x < 0$, the eigenfunction expansion above evidently must converge in mean square to the even extension of $T_L x/L$, the dashed line shown in figure 9.29. Since the terms in the eigenfunction expansion each have period $2L$, the expansion converges in mean square to the $2L$-periodic extension of this function for x outside of $-L \le x \le L$.

The rate of convergence of the expansion is illustrated in figure 9.30.

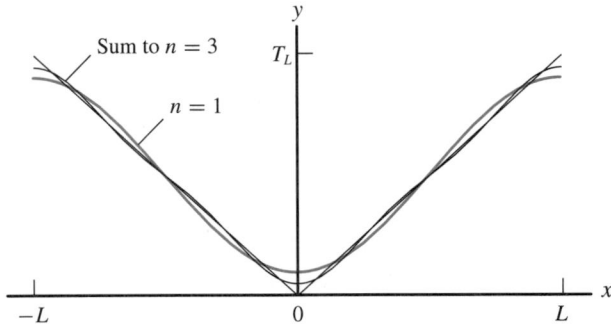

FIGURE 9.30   Convergence of the eigenfunction expansion (9.30) to the even extension of $T_L x/L$ from $0 \le x \le L$.

Solutions of Initial-Boundary-Value Problems

Solving the initial-boundary-value problems of the next examples will put these eigenfunction expansions to use.

■ **EXAMPLE 21**   *Suppose both ends of a thin, insulated rod of length L are insulated and its initial temperature varies linearly from zero at $x = 0$ to T_L at $x = L$. Using the model derived in exercise 5 of section 9.4,*

$$\frac{\partial T(x, t)}{\partial t} = \kappa \frac{\partial^2 T(x, t)}{\partial x^2},$$

$$T_x(0, t) = T_x(L, t) = 0, \quad t > 0,$$

$$T(x, 0) = T_L x/L, \quad 0 < x < L,$$

find the variation with position and time of the temperature of the rod.

Assuming $T(x, t) = X(x)\Theta(t)$, substituting in the partial differential equation, and separating variables leads to

Separate variables.

$$\frac{\Theta'(t)}{\kappa\Theta(t)} = \frac{X''(x)}{X(x)} = -\lambda,$$

where the separation constant λ arises because the variables x and t are independent. The corresponding ordinary differential equations are

$$\Theta'(t) + \kappa\lambda\Theta(t) = 0,$$
$$X''(x) + \lambda X(x) = 0.$$

Applying the boundary condition $T_x(0, t) = X'(0)\Theta(t) = 0$ leads to $X'(0) = 0$. Similarly, $X'(L) = 0$. Coupling these boundary conditions with the differential equation for X, we have the eigenvalue problem

Eigenvalue problem

$$X''(x) + \lambda X(x) = 0, \quad X'(0) = X'(L) = 0.$$

From example 20, its eigenvalues and eigenfunctions are

Eigenvalues and eigenfunctions

$$\lambda_n = \frac{n^2\pi^2}{L^2}, \quad X_n(x) = \cos\frac{n\pi x}{L}, \quad n = 0, 1, 2, \ldots.$$

The corresponding solutions of the Θ equation are

$$\Theta_n(t) = c_n e^{-\kappa n^2\pi^2 t/L^2}.$$

Since every member of the family of functions

$$T_n(x, t) = X_n(x)\Theta_n(t) = c_n e^{-\kappa n^2\pi^2 t/L^2} \cos\frac{n\pi x}{L}, \quad n = 0, 1, 2, \ldots,$$

solves the homogeneous partial differential equation and satisfies the homogeneous boundary conditions, so does the sum

$$T(x, t) = \sum_{n=0}^{\infty} T_n(x, t) = \sum_{n=0}^{\infty} c_n e^{-\kappa n^2\pi^2 t/L^2} \cos\frac{n\pi x}{L},$$

provided the infinite sum converges.

Satisfying the initial condition requires choosing the constants c_n so that

$$\sum_{n=0}^{\infty} c_n \cos\frac{n\pi x}{L} = \frac{T_L x}{L}, \quad 0 < x < L;$$

that is, the constants c_n are the coefficients of the eigenfunction expansion (9.30) of $T_L x/L$ on $0 \le x \le L$. These were found in example 20.

Using the expression for c_n from that example, we have the (formal) solution of the initial-boundary-value problem,

$$T(x, t) = \frac{T_L}{2} - \sum_{n=1,3,\ldots} \frac{4T_L}{n^2\pi^2} e^{-\kappa n^2\pi^2 t/L^2} \cos\frac{n\pi x}{L}.$$

Note that this solution decays to a steady state of $T_L/2$, the average of the initial temperature distribution, as the heat energy trapped in the rod by the insulation redistributes itself uniformly. Chapter exercise 13 asks you to illustrate this property in general. ■

■ **EXAMPLE 22**   *Example 16 of section 9.4 derived a model for an insulated wire being heated by a current passing through it. With the particular choice of initial condition given here, that model is*

$$\frac{\partial T(x,t)}{\partial t} = \kappa \frac{\partial^2 T(x,t)}{\partial x^2} + \frac{H}{c\rho},$$

$$T(0,t) = T(L,t) = 0, \quad t > 0,$$

$$T(x,0) = M + \frac{Hx(L-x)}{2k}, \quad 0 < x < L.$$

The term H arises from the heat generated by the current. Assume H is constant and find a solution of this initial-boundary-value problem.

Separation of variables assumes a solution of the form

$$T(x,t) = X(x)\Theta(t)$$

and substitutes in the differential equation,

$$X(x)\Theta'(t) = \kappa X''(x)\Theta(t) + \frac{H}{c\rho}.$$

But the nonhomogeneous term $H/c\rho$ prevents separation into expressions involving functions of x alone and functions of t alone. We must find a new problem in which the differential equation is *homogeneous*.

The key is the solution $E(x)$ of the *equilibrium problem*,

Equilibrium boundary-value problem

$$0 = \kappa E''(x) + \frac{H}{c\rho}, \quad E(0) = E(L) = 0,$$

the model that governs when temperature is constant in time. We readily find that the equilibrium temperature is

$$E(x) = \frac{Hx(L-x)}{2k c\rho}.$$

Since the equilibrium temperature accounts for the effects of the nonhomogeneous term $H/c\rho$, we expect that the difference

$$u(x,t) = T(x,t) - E(x)$$

should satisfy a *homogeneous* differential equation. Indeed, substituting $T(x,t) = u(x,t) + E(x)$ into $T_t = \kappa T_{xx} + H/c\rho$ yields

HOMOGENEOUS PARTIAL DIFFRENTIAL EQUATION

$$\frac{\partial u(x,t)}{\partial t} = \kappa \frac{\partial^2 u(x,t)}{\partial x^2}, \tag{9.31}$$

a homogeneous heat equation. Furthermore, we find the boundary conditions

HOMOGENEOUS BOUNDARY CONDITIONS

$$u(0, t) = u(L, t) = 0, \quad t > 0, \tag{9.32}$$

and the initial condition

$$u(x, 0) = T(x, 0) - E(x) = M, \quad 0 \le x \le L.$$

The new dependent variable u satisfies the homogeneous partial differential equation (9.31) subject to homogeneous boundary conditions (9.32). To apply separation of variables, assume

$$u(x, t) = X(x)\Theta(t).$$

Then *precisely* the same calculations as in example 19 lead to the eigenvalue problem (9.21). Its eigenfunctions are

$$X_n(x) = \sin \frac{n\pi x}{L}, \quad n = 1, 2, \dots.$$

From the eigenvalues $\lambda_n = n^2\pi^2/L^2$, the corresponding solutions of the $\Theta(t)$ equation are

$$\Theta_n(t) = c_n e^{-n^2\pi^2\kappa t/L^2}, \quad n = 1, 2, \dots.$$

If it converges appropriately, the sum

SOLUTION OF HOMOGENEOUS PROBLEM

$$u(x, t) = \sum_{n=1}^{\infty} c_n e^{-n^2\pi^2\kappa t/L^2} \sin \frac{n\pi x}{L}$$

satisfies the homogeneous differential equation (9.31) and boundary conditions (9.32). The initial condition for u demands

INITIAL CONDITION

$$\sum_{n=1}^{\infty} c_n \sin \frac{n\pi x}{L} = M, \quad 0 < x < L. \tag{9.33}$$

Using the orthogonality relation (9.25) and the formula (9.29) for c_n, we find

FOURIER COEFFICIENT

$$c_n = \frac{2}{L} \int_0^L M \sin \frac{n\pi x}{L} \, dx$$

$$= \frac{2M}{n\pi}(1 - (-1)^n) = \begin{cases} 4M/n\pi, & n = 1, 3, \dots \\ 0, & n = 2, 4, \dots. \end{cases}$$

With this value of c_n, the solution of the original initial-boundary-value problem is

$$T(x, t) = E(x) + u(x, t)$$

SOLUTION OF INITIAL-BOUNDARY-VALUE PROBLEM

$$= \frac{Hx(L - x)}{2k} + \sum_{n=1}^{\infty} c_n e^{-n^2\pi^2\kappa t/L^2} \sin \frac{n\pi x}{L}. \quad ■$$

For $-L \leq x \leq 0$, the series given by (9.33) should converge to the odd extension of $f(x) = M$, $0 \leq x \leq L$. The resulting *square wave* is shown in figure 9.31 along with the first few terms of the corresponding expansion. Note that

- The convergence is slower for this discontinuous function than for either the continuously differentiable function of figure 9.25 or the merely continuous triangular wave of figure 9.30.
- At the jump discontinuity at $x = 0$, the eigenfunction expansion is converging to the *average* of the limiting values from the left and the right.

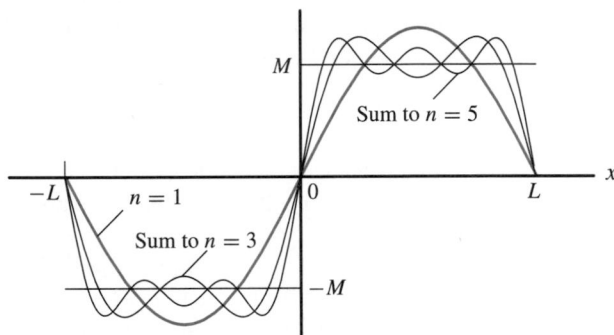

FIGURE 9.31   Convergence of the eigenfunction expansion (9.33) to the odd extension of $f(x) = M$ from $0 \leq x \leq L$.

9.5.3   Fourier Series

The eigenfunction expansion ideas discussed previously also apply to eigenvalue problems with **periodic boundary conditions**

$$X''(x) + \lambda X(x) = 0, \quad -L < x < L,$$
$$X(-L) - X(L) = 0,$$
$$X'(-L) - X'(L) = 0.$$

The eigenfunctions of this problem are 1, $\sin \pi x/L$, $\cos \pi x/L, \ldots,$ $\sin n\pi x/L$, $\cos \pi x/L, \ldots$. (See exercise 26.) The corresponding eigenfunction expansion is a **Fourier series**

FOURIER SERIES

$$a_0 + \sum_{n=1}^{\infty} \left(a_n \cos \frac{n\pi x}{L} + c_n \sin \frac{n\pi x}{L} \right), \tag{9.34}$$

where the **Fourier coefficients** a_n, b_n are defined by

FOURIER COEFFICIENTS

$$a_n = \begin{cases} \dfrac{1}{2L} \displaystyle\int_{-L}^{L} f(x)\, dx, & n = 0, \\[3mm] \dfrac{1}{L} \displaystyle\int_{-L}^{L} f(x) \cos \dfrac{n\pi x}{L}\, dx, & n = 1, 2, \ldots, \end{cases}$$

$$b_n = \frac{1}{L} \int_{-L}^{L} f(x) \sin \frac{n\pi x}{L}\, dx, \qquad n = 1, 2, \ldots.$$

To state (without proof) a stronger convergence result for this particular set of eigenfunctions, we require some terminology. A function is **piecewise continuous** on the interval $-L \leq x \leq L$ if it is continuous at all but a finite number of points, called **jump discontinuities**, where the limits from the left and the right exist but differ in value. (See figure 10.4, page 516, of section 10.1.) The **limit from the left** of f at x is

LIMIT FROM THE LEFT

$$\lim_{h \to 0^+} f(x - h),$$

where $h \to 0^+$ indicates that h approaches zero through positive values. The **limit from the right** is similar with $f(x + h)$ replacing $f(x - h)$.

Theorem 2 (Fourier expansion theorem). *Let f and f' each be piecewise continuous on $-L \leq x \leq L$. Then at every point of $-L \leq x \leq L$, the Fourier expansion (9.34) converges to*

$$\frac{1}{2} \left(\lim_{h \to 0^+} f(x - h) + \lim_{h \to 0^+} f(x + h) \right),$$

where f is defined by its periodic extension $f(x) = f(x \pm 2L)$ for values of x outside of $-L \leq x \leq L$.

In other words, the Fourier expansion converges to the function at points of continuity of f and to the midpoint of the jump at points of discontinuity.

The Fourier coefficient formulas are again consequences of orthogonality. The appropriate relation for the cosine is

ORTHOGONALITY OF
$\cos m\pi x / L$

$$\int_{-L}^{L} \cos \frac{m\pi x}{L} \cos \frac{n\pi x}{L} \, dx = \begin{cases} 0, & m \neq n \\ L, & m = n. \end{cases}$$

There is a corresponding relation with sin replacing cos as well as an orthogonality condition for mixed products,

ORTHOGONALITY OF
$\sin m\pi x / L, \cos n\pi x / L$

$$\int_{-L}^{L} \sin \frac{m\pi x}{L} \cos \frac{n\pi x}{L} \, dx = 0, \quad \text{for all } m, n.$$

See exercise 28.

If the Fourier series (9.34) contains only sine functions (i.e., $a_n \equiv 0$), it is called a *sine series*. As we saw in example 19, an odd function is expanded in a sine series because it has the same symmetry about the origin as the sine function, $f(x) = -f(-x)$, and vice versa.

Similarly, an even function, $f(x) = f(-x)$, is expanded in a series containing only cosines (and the constant term a_0) because it has the same symmetry about the origin as the cosine function. That series is called a *cosine series*. (See example 20 and figure 9.30.)

Since the eigenfunction expansions we derived earlier are either sine or cosine series, the stronger convergence result of the Fourier expansion theorem applies, as shown in figure 9.31 for the square wave function. However, not all eigenfunction expansions are special cases of Fourier series, as the exercises illustrate.

MATLAB

DELAB can calculate Fourier coefficients.   For guidance, select `Help, Textbook`, then go to chapter 9, Fourier coefficient.

9.5.4 Summary

To solve a linear, *homogeneous* partial differential equation by separation of variables:

1. Assume the solution has the product form $T(x, t) = X(x)\Theta(t)$.

2. Substitute in the partial differential equation, separate variables, and introduce a separation constant $-\lambda$.

3. Use the *homogeneous* boundary conditions to obtain an eigenvalue problem for $X(x)$.

4. Solve the eigenvalue problem to obtain eigenfunction, eigenvalue pairs $X_n(x)$, λ_n. Using the eigenvalues λ_n, solve the Θ equation for Θ_n. The family of functions $X_n(x)\Theta_n(t)$ satisfies both the differential equation and the boundary conditions, as does $\sum c_n X_n(x)\Theta(t)$.

5. To satisfy the initial condition $T(x, 0) = f(x)$, require that c_n be the coefficients in an eigenfunction expansion for $f(x)$, $\sum c_n X_n(x) = f(x)$.

6. Compute c_n from (9.29) and write the complete solution of the initial-boundary-value problem as $\sum c_n X_n(x)\Theta_n(t)$.

If the differential equation or the boundary conditions are not homogeneous, subtract the solution of the equilibrium (time-independent) boundary-value problem. All these manipulations assume that differentiation and the infinite sum can be interchanged.

> The eigenfunction expansion theorem stated here is essentially that of Coddington and Levinson [6]. For a more accessible introduction to eigenfunction expansions, see Weinberger's text [27].

9.5.5 Exercises

EXERCISE GUIDE	
To gain experience . . .	**Try exercises**
Separating variables	1–2
Verifying general solutions of eigenvalue problems	4
Finding eigenvalues and eigenfunctions	5–6, 11, 26
Finding eigenfunction expansions	7–12

EXERCISE GUIDE (Continued)	
To gain experience ...	**Try exercises**
With superposition in the heat equation	13
With odd and even extensions	7–10, 14, 29
With orthogonality and generalized Fourier coefficients	7–12, 15, 22, 27–29
Solving initial-boundary-value problems	16–21
With physical interpretations of initial-boundary-value problems	18–21
With equilibrium boundary-value problems	23–25
With nonhomogeneous equations and boundary conditions	25
Analyzing and interpreting solution behavior	3

In exercises 1 and 2, suppose $T(x, t) = X(x)\Theta(t)$, where $T(x, t)$ satisfies the given partial differential equation and boundary conditions.

(i) Find an eigenvalue problem for $X(x)$ and an ordinary differential equation for $\Theta(t)$.

(ii) Solve the eigenvalue problem. Solve the equation defining $\Theta(t)$.

(iii) Write a sum which satisfies both the partial differential equation and the boundary conditions.

1. $T_t = \kappa T_{xx}$, $T_x(0, t) = T(L, t) = 0$

2. $T_t = 4T_{xx}$, $T(0, t) = T_x(L, t) = 0$

3. The thermal diffusivity values for copper and wood are $\kappa = 0.11$ cm^2/s and $\kappa = 0.0019$ cm^2/s, respectively. Compare the rate of temperature decay in identical thin rods with insulated sides made from these two materials; e.g., if the ends of the rods are at temperature zero and the initial temperatures are identical, how do the midpoint temperatures compare as time passes?

4. The text claims that the general solution of $X''(x) + \lambda X(x) = 0$ is

$$\lambda < 0: \quad X(x) = C_1 e^{\sqrt{-\lambda}\, x} + C_2 e^{-\sqrt{-\lambda}\, x},$$
$$\lambda = 0: \quad X(x) = C_1 + C_2 x,$$
$$\lambda > 0: \quad X(x) = C_1 \cos\sqrt{\lambda}\, x + C_2 \sin\sqrt{\lambda}\, x.$$

Verify this claim. Compare your results with those given by DELAB.

5. Show that $\lambda = 0$ is not an eigenvalue of

$$X''(x) + \lambda X(x) = 0, \quad X(0) = X(L) = 0.$$

That is, show that the boundary conditions force $C_1 = C_2 = 0$ in the general solution of this equation when $\lambda = 0$.

6. Show that the eigenvalue problem

$$X''(x) + \lambda X(x) = 0, \quad X'(0) = X'(L) = 0,$$

which arose in example 20, has no nontrivial solutions for $\lambda \leq 0$.

7. Find the expansion of $f(x) = T_L x/L$, $0 \leq x \leq L$, using the eigenfunctions of example 19, $X_n(x) = \sin n\pi x/L$, $n = 1, 2, \ldots$. To what function does the eigenfunction expansion converge for $-L \leq x \leq L$?

8. Repeat exercise 7 for $f(x) = L - x$.

9. Find the expansion of $f(x) = L - x$, $0 \leq x \leq L$, using the eigenfunctions of example 20, $X_n(x) = \cos n\pi x/L$, $n = 0, 1, 2, \ldots$. To what function does the eigenfunction expansion converge for $-L \leq x \leq L$?

10. Repeat exercise 9 for $f(x) = 4$.

11. Solve the eigenvalue problem $X''(x) + \lambda X(x) = 0$, $X'(0) = X(\pi) = 0$. Using those eigenfunctions, expand $f(x) = x/\pi$, $0 \leq x \leq \pi$. To what function does the eigenfunction expansion converge for $-\pi \leq x \leq \pi$?

12. Using the eigenfunctions found in the preceding exercise, expand $f(x) = x^2 - \pi x$, $0 \leq x \leq \pi$. To what function does the eigenfunction expansion converge for $-\pi \leq x \leq \pi$?

13. The text claims that any finite sum of solutions

$$T_n(x, t) = c_n e^{-n^2\pi^2\kappa t/L} \sin \frac{n\pi x}{L}$$

of the heat equation

$$\frac{\partial T(x, t)}{\partial t} = \kappa \frac{\partial^2 T(x, t)}{\partial x^2}$$

subject to the boundary conditions

$$T(0, t) = T(L, t) = 0$$

again satisfies both the partial differential equation and the boundary conditions. Explicitly verify this claim for the sum $T_1(x, t) + T_2(x, t)$ by substituting $T_1 + T_2$ into the heat equation and into the boundary conditions.

(a) Use the explicit formulas for T_1, T_2.

(b) Use only the knowledge that T_1, T_2 individually satisfy the heat equation and the boundary conditions.

14. Sketch and write a formula for the odd extension and the even extension of the following functions.

(a) $\cos \pi x, 0 < x < 1$

(b) $\sin \pi x, 0 < x < 1$

(c) $\sqrt{x}, 0 < x < 4$

(d) $|x|, 0 < x < L$

(e) $\begin{cases} -1, & 0 < x \leq 2 \\ 1, & 2 < x < 4 \end{cases}$

15. Verify the text's calculation in example 19 of the generalized Fourier coefficient c_n for $f(x) = Mx(L - x)$, $0 \leq x < L$.

16. Solve the initial-boundary-value problem

$$T_t(x, t) = \kappa T_{xx}(x, t),$$
$$T(0, t) = T(4, t) = 0, \quad t > 0,$$
$$T(x, 0) = -2, \quad 0 < x < 4.$$

17. Solve the initial-boundary-value problem

$$T_t(x, t) = \kappa T_{xx}(x, t),$$
$$T_x(0, t) = T_x(4, t) = 0, \quad t > 0,$$
$$T(x, 0) = 8 - 2x, \quad 0 < x < 4.$$

18. Solve the initial-boundary-value problem

$$T_t(x, t) = \kappa T_{xx}(x, t),$$
$$T_x(0, t) = T(\pi, t) = 0, \quad t > 0,$$
$$T(x, 0) = x^2 - \pi x, \quad 0 < x < \pi.$$

Describe a physical situation which this problem might model. As time passes, does the solution appear to approach the proper equilibrium solution?

19. Repeat exercise 18 with the initial condition $T(x, 0) = x/\pi$.

20. Solve the initial-boundary-value problem

$$T_t(x, t) = \kappa T_{xx}(x, t),$$
$$T(0, t) = 4, \quad T(L, t) = 2, \quad t > 0,$$
$$T(x, 0) = -2, \quad 0 < x < L.$$

Describe a physical situation which this problem might model. (*Hint:* To solve by separation of variables, find an equilibrium solution $E(x)$ which satisfies the nonhomogeneous boundary conditions. Show that $u(x, t) = T(x, t) - E(x)$ satisfies a problem with homogeneous boundary conditions.)

21. Repeat exercise 20 with the boundary conditions $T_x(0, t) = 0, T(L, t) = 8$.

22. Verify the orthogonality relation $\int_0^L X(x)Y(x) \, dx = 0$ when X, Y are eigenfunctions of

$$X''(x) + \lambda X(x) = 0, \quad X'(0) = X(L) = 0,$$

corresponding to distinct eigenvalues.

23. Solve the equilibrium boundary-value problem

$$0 = \kappa E''(x) + H/c\rho, \quad E(0) = E(L) = 0,$$

encountered in example 22. Compare your solution with that given by DELAB.

24. Example 21 shows that the limit in time of the solution of the initial-boundary-value problem

$$\frac{\partial T(x, t)}{\partial t} = \kappa \frac{\partial^2 T(x, t)}{\partial x^2},$$
$$T_x(0, t) = T_x(L, t) = 0, \quad t > 0,$$
$$T(x, 0) = T_L x/L, \quad 0 < x < L,$$

is $T_L/2$. Obtain this equilibrium solution directly from the model that governs at equilibrium.

25. Verify the computation in example 22: $u(x, t) = T(x, t) - E(x)$ satisfies the homogeneous equation $u_t = \kappa u_{xx}$ if T solves the nonhomogeneous equation $T_t = \kappa T_{xx} + H/c\rho$ and E solves the equilibrium equation $0 = \kappa E''(x) + H/c\rho$. Do you need a formula for $E(x)$? Does the result of this exercise depend on whether H is constant?

26. Verify that $1,\ \sin \pi x/L,\ \cos \pi x/L, \ldots,\ \sin n\pi x/L,$ $\cos \pi x/L, \ldots$ are eigenfunctions of the periodic eigenvalue problem

$$X''(x) + \lambda X(x) = 0,$$
$$X(-L) - X(L) = 0,$$
$$X'(-L) - X'(L) = 0.$$

Find the corresponding eigenvalues.

27. Using formulas involving the sum and difference of angles, evaluate the integrals and verify the following orthogonality relations.

(a) $\displaystyle\int_{-L}^{L} \sin \frac{m\pi x}{L} \sin \frac{n\pi x}{L}\, dx = \begin{cases} 0, & m \neq n \\ L, & m = n \end{cases}$

(b) $\displaystyle\int_{-L}^{L} \cos \frac{m\pi x}{L} \cos \frac{n\pi x}{L}\, dx = \begin{cases} 0, & m \neq n \\ L, & m = n \neq 0 \end{cases}$

(c) $\displaystyle\int_{-L}^{L} \cos \frac{m\pi x}{L} \sin \frac{n\pi x}{L}\, dx = 0$ for all m, n

28. Using the orthogonality relations of exercise 27, derive the Fourier coefficient formulas for $a_n,\ b_n$ on page 486.

Assume that the Fourier expansion converges to $f(x)$ for $-L \leq x \leq L$,

$$f(x) = \frac{a_0}{2} + \sum_{n=1}^{\infty} \left(a_n \cos \frac{n\pi x}{L} + b_n \sin \frac{n\pi x}{L} \right),$$

and that summation and integration can be interchanged.

At a jump discontinuity, the Fourier expansion does not converge to $f(x)$, but there are only a finite number of such points. Hence, they can not affect the values of the integrals in the formulas for the coefficients a_n, b_n.

29. Following example 19, the text constructed a Fourier sine expansion of the odd function

$$f(x) = \begin{cases} Mx(L-x), & 0 \leq x < L \\ Mx(L+x), & -L \leq x < 0. \end{cases}$$

(See figure 9.27.) Using the Fourier coefficient formula on page 486, verify directly that $a_n = 0, n = 0, 1, 2, \ldots.$

9.6 ■ INITIAL-BOUNDARY-VALUE PROBLEMS: NUMERICAL METHODS

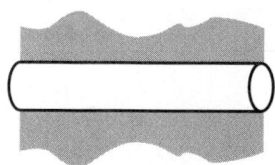

FIGURE 9.32 A thin rod whose sides are insulated.

Models of time-dependent diffusion are initial-boundary-value problems, i.e., partial differential equations in space and time coupled with both boundary conditions and initial conditions. One approach to the numerical solution of such problems approximates the spatial boundary-value problem by finite differences, then analyzes the resulting system of *ordinary* differential equations, perhaps with Runge–Kutta or a similar numerical tool for initial-value problems.

To develop these ideas, consider the model for the temperature in a thin, insulated rod of length L that was developed in example 15, page 463,

$$\frac{\partial T(x,t)}{\partial t} = \kappa \frac{\partial^2 T(x,t)}{\partial x^2}, \qquad (9.35)$$

Initial-boundary-value problem

$$T(x,0) = M \sin \pi x/L, \quad 0 < x < L, \qquad (9.36)$$
$$T(0,t) = 0, \quad T(L,t) = 0, \quad t > 0. \qquad (9.37)$$

See figure 9.32.

Stop and Think **9.58** Which relations are the boundary conditions? What is the temperatures at each end of the rod? Which relation is the initial condition? What is the shape of the initial temperature distribution in the rod?

Divide the rod into n equally spaced subintervals defined by the grid points $x_i = i\Delta x$, where $\Delta x = L/n$. At any one of those fixed locations,

$T(x_i, t)$ is a function of t *only*, and $\partial T(x_i, t)/\partial t$ is just an *ordinary* derivative. The idea is to approximate T_{xx} via finite differences on the spatial grid to obtain a *system of ordinary differential equations* in $T(x_1, t), \ldots, T(x_{n-1}, t)$.

Stop and Think   **9.59** Explain why $T(x_0, t)$ and $T(x_n, t)$ are already known.

To carry out this process, apply the centered-difference second-derivative approximation (9.6),

$$u''(x) \approx \frac{u(x - \Delta x) - 2u(x) + u(x + \Delta x)}{\Delta x^2},$$

to $T_{xx}(x, t)$ at one of the grid points x_i. The result is

Centered-difference
approximation to $T_{xx}(x_i, t)$

$$T_{xx}(x_i, t) \approx \frac{T(x_i - \Delta x, t) - 2T(x_i, t) + T(x_i + \Delta x, t)}{\Delta x^2}$$

$$= \frac{T(x_{i-1}, t) - 2T(x_i, t) + T(x_{i+1}, t)}{\Delta x^2}.$$

At each of the interior grid points $x_1, \ldots, x_{n-1}$, the partial differential equation (9.35) becomes

$$\frac{\partial T(x_i, t)}{\partial t} \approx \kappa \frac{T(x_{i-1}, t) - 2T(x_i, t) + T(x_{i+1}, t)}{\Delta x^2}, \quad i = 1, \ldots, n-1,$$

a system of (approximate) *ordinary* differential equations.

To obtain an approximation $U_i(t)$ to $T(x_{i+1}, t)$, require that $U_i(t)$ satisfy *exactly* these equations,

$$U_1'(t) = (\kappa/\Delta x^2)(U_0(t) - 2U_1(t) + U_2(t))$$

$$\vdots \tag{9.38}$$

$$U_{n-1}'(t) = (\kappa/\Delta x^2)(U_{n-2}(t) - 2U_{n-1}(t) + U_n(t)),$$

a system of $n - 1$ ordinary differential equations in $n + 1$ unknowns.

Just as with finite-difference approximations to boundary-value problems, the boundary conditions (9.37) eliminate the two extra unknowns:

$$T(0, t) = 0 \implies U_0(t) \equiv 0,$$

$$T(L, t) = 0 \implies U_n(t) \equiv 0.$$

Stop and Think   **9.60** Suppose $T(0, t) = \alpha$, $T(L, t) = \beta$. Then $U_0(t) = ?$, $U_n(t) = ?$

The resulting system of ordinary differential equations in the unknowns $U_1(t), \ldots, U_{n-1}(t)$ is

$$U_1'(t) = (\kappa/\Delta x^2)(-2U_1(t) + U_2(t))$$

Approximation to $T_t = \kappa T_{xx}$,
$T(0, t) = 0$, $T(L, t) = 0$

$$\vdots \tag{9.39}$$

$$U_{n-1}'(t) = (\kappa/\Delta x^2)(U_{n-2}(t) - 2U_{n-1}(t)).$$

This system approximates the partial differential equation (9.35) subject to the boundary conditions (9.37).

Stop and Think **9.61** Write the differential equation $U_2'(t) = \cdots$. The equation $U_{n-2}'(t) = \cdots$.

9.62 Is the system of equations (9.39) linear or nonlinear? Homogeneous or non-homogeneous?

The system (9.39) can be written in the matrix form $\mathbf{u}' = \mathbf{Au}$ by defining the vector of unknown functions

$$\mathbf{u}(t) = \begin{pmatrix} U_1(t) \\ \vdots \\ U_{n-1}(t) \end{pmatrix},$$

and the (constant) coefficient matrix

$$\mathbf{A} = \frac{\kappa}{\Delta x^2} \begin{pmatrix} -2 & 1 & 0 & \cdots & 0 & 0 & 0 \\ 1 & -2 & 1 & \cdots & 0 & 0 & 0 \\ & & & \ddots & & & \\ 0 & 0 & 0 & \cdots & 0 & 1 & -2 \end{pmatrix}. \tag{9.40}$$

Stop and Think **9.63** How many rows and columns are there in $\mathbf{A}$? Verify that the second row is correct as shown. Find the next-to-last row.

9.64 Is the system of equations $\mathbf{u}' = \mathbf{Au}$ linear or nonlinear? Homogeneous or nonhomogeneous?

To complete the approximation, add initial conditions to the system $\mathbf{u}' = \mathbf{Au}$. From (9.36), they are

$$T(x, 0) = M \sin \pi x / L, \quad 0 < x < L,$$

$$\Rightarrow \quad U_i(0) = M \sin \pi x_i / L, \quad i = 1, \ldots, n - 1.$$

Stop and Think **9.65** Why are $i = 0$ and $i = n$ omitted?

The complete approximation to the solution of the *partial* differential equation, initial-boundary-value problem (9.35–9.37) is the *ordinary* differential equation, initial-value problem

Approximation to the solution of (9.35–9.37)

$$\mathbf{u}' = \mathbf{Au},$$
$$U_i(0) = M \sin \pi x_i / L, \quad i = 1, \ldots, n - 1,$$

THE METHOD OF LINES

where the $n - 1 \times n - 1$ matrix $\mathbf{A}$ is defined by (9.40), $x_i = i \Delta x / L$, and $\Delta x = L/n$. This approximation is sometimes called the *method of lines* because the functions $U_i(t)$ approximate the solution of (9.35–9.37) along the "lines" $x = x_i$ in the xt-plane. (See Stop and Think 9.75, page 496.)

SUMMARY

To summarize, a method-of-lines approximation to an initial-boundary-value problem is a system of ordinary differential equations in time that are derived by approximating the spatial (or x) derivatives in the partial differential equation by finite differences. The ordinary differential equations approximate the partial differential equation at selected values of x, subject to the given boundary conditions. The initial conditions for the partial differential equation provide the initial conditions for the method-of-lines system of ordinary differential equations.

■ **EXAMPLE 23**   *Using $\Delta x = L/3$, find a method-of-lines approximation to the initial-boundary-value problem (9.35–9.37). Solve the resulting system of ordinary differential equations and comment on the behavior of the solution.*

Choosing $\Delta x = L/3$ corresponds to $n = 3$. There are two points interior to the x-interval, $x = L/3$ and $x_2 = 2L/3$, and the method-of-lines system (9.39) will define $U_1(t)$ and $U_2(t)$, approximations to the time behavior of $T(x_i, t)$ at these two points.

From (9.40), the system of two ordinary differential equations is

$$\begin{pmatrix} U_1(t) \\ U_2(t) \end{pmatrix}' = \frac{\kappa}{\Delta x^2} \begin{pmatrix} -2 & 1 \\ 1 & -2 \end{pmatrix} \begin{pmatrix} U_1(t) \\ U_2(t) \end{pmatrix}.$$

The initial conditions are

$$U_1(0) = M \sin \pi/3 = \frac{M\sqrt{3}}{2},$$

$$U_2(0) = M \sin 2\pi/3 = \frac{M\sqrt{3}}{2}.$$

The solution of this initial-value problem is

Approximate solution with $\Delta x = L/3$ of the initial-boundary-value problem (9.35–9.37)

$$U_1(t) = U_2(t) = \frac{M\sqrt{3}}{2} e^{-9\kappa t/L^2}.$$

Stop and Think   **9.66** At which points of the rod are these functions approximating temperature as a function of time? Is it physically reasonable that the temperature profiles at two different points of the rod could be described by the same function?

The internal temperature of the rod seems to be decaying toward zero, a reasonable outcome given the zero temperature at its ends. The characteristic decay time or time constant is

Approximate time constant for decay

$$\left(\frac{9\kappa}{L^2}\right)^{-1} = \frac{L^2}{9\kappa}.$$

TIME CONSTANT   Recall that the *time constant* for exponential decay is the time required to decay to $1/e \approx 0.37$ of the initial amount.

This approximate behavior is consistent with that predicted by the exact solution given by (9.23) of section 9.5,

Exact solution of (9.35–9.37) $$T(x, t) = Me^{-\pi^2\kappa t/L^2} \sin \pi x/L,$$

because the time constant for the exact decay is comparable with the approximate time constant,

Exact time constant for decay $$\left(\frac{\pi^2\kappa}{L^2}\right)^{-1} = \frac{L^2}{\pi^2\kappa} \approx \frac{L^2}{9.87\kappa} \approx \frac{L^2}{9\kappa}. \quad ■ \qquad (9.41)$$

Stop and Think **9.67** Is the time constant from the method-of-lines approximation an overestimate or an underestimate? Roughly, what is the magnitude of the error?

9.68 What point of the rod is likely to be its hottest point? Does the method-of-lines approximation offer any direct information about the temperature at that point?

■ **EXAMPLE 24**   *Set $L = 5$, $\kappa = 1$, and $M = 2$ in the initial-boundary-value problem (9.35–9.37). Construct a method-of-lines approximation to the temperature at $x = 1, \ldots, 4$ and solve it numerically to describe the behavior of the temperature in the rod.*

To approximate $T(x, t)$ at $x = 1, \ldots, 4$, choose $\Delta x = 1$, corresponding to $n = 5$.

The method-of-lines approximation is $\mathbf{u}' = \mathbf{A}\mathbf{u}$ subject to the initial condition

$$\mathbf{u}(0) = \begin{pmatrix} U_1(0) \\ U_2(0) \\ U_3(0) \\ U_4(0) \end{pmatrix} = \begin{pmatrix} 2\sin \pi/5 \\ 2\sin 2\pi/5 \\ 2\sin 3\pi/5 \\ 2\sin 4\pi/5 \end{pmatrix}. \tag{9.42}$$

Stop and Think **9.69** Justify these initial values.

From 9.40, the coefficient matrix is

$$\mathbf{A} = \begin{pmatrix} -2 & 1 & 0 & 0 \\ 1 & -2 & 1 & 0 \\ 0 & 1 & -2 & 1 \\ 0 & 0 & 1 & -2 \end{pmatrix}.$$

Stop and Think **9.70** Verify $\mathbf{A}$.

To approximate the solution of $\mathbf{u}' = \mathbf{A}\mathbf{u}$ subject to the initial conditions (9.42), use the fourth-order Runge-Kutta method from DELAB with $\Delta t = 0.5$.

Stop and Think **9.71** Experiment with other choices of Δt. Which values seem to be too large? Does $\Delta t = 0.5$ seem small enough to give reasonable accuracy?

9.72 Using the time constant for decay found in example 23, estimate the number of steps of Runge-Kutta using $\Delta t = 0.5$ that would suffice to capture most of the decay of $\mathbf{u}(t)$.

Plots of two components of that numerical approximation are shown in figure 9.33. (The other two are identical.) Both exhibit the expected decay.

Stop and Think **9.73** For which points of the rod are approximate temperature profiles being exhibited in figure 9.33? Which one is closer to the center of the rod? Is the point closer to the center always hotter?

9.74 Which curve in figure 9.33 could also be a plot of $U_3(t)$? Of $U_4(t)$?

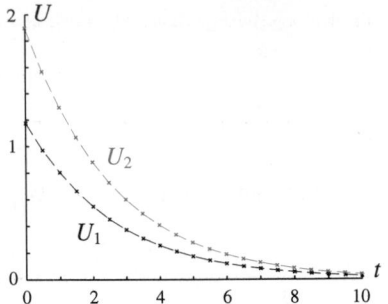

FIGURE 9.33 Two components $U_1(t)$, $U_2(t)$ of a method-of-lines approximation to (9.35–9.37).

These four approximate temperature profiles can be combined into a single surface plot to give a representation of the behavior of the temperature throughout the rod as time passes, as shown in figure 9.34. ■

Stop and Think **9.75** In figure 9.34, identify the lines $x = 1, \ldots, 4$ in the xt-plane that give the method of lines its name.

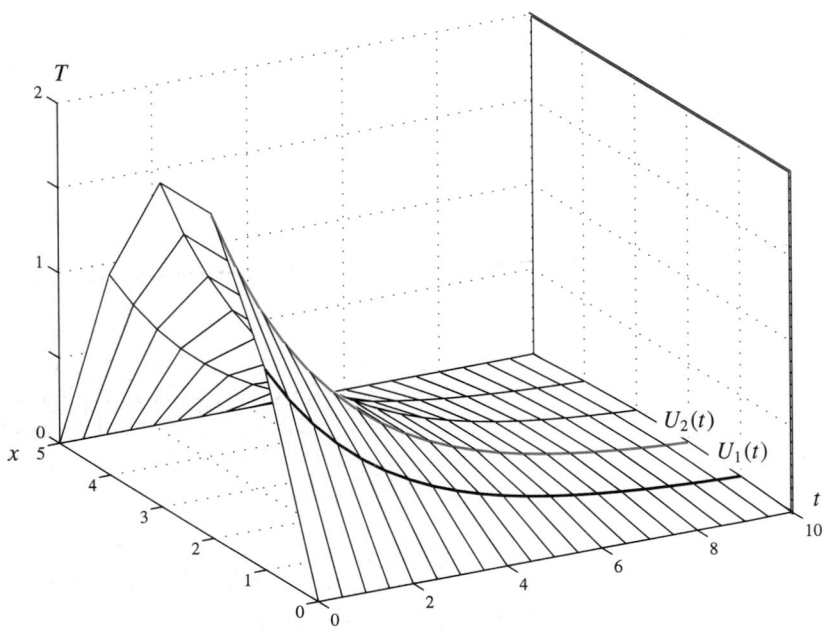

FIGURE 9.34 An approximation to the behavior over time of the temperature throughout the rod of example 24; this surface plot is based on four method-of-lines temperature profiles $U_1(t), \ldots, U_4(t)$ at $x = 1, \ldots, 4$.

MATLAB

DELAB can find numerical approximations to the solutions of the systems of ordinary differential equations produced by the method of lines. For guidance, select **Help, Textbook**, then go to chapter 9, example 24.

■ **EXAMPLE 25** *Using $\Delta x = 0.5$, construct a method-of-lines approximation to the solution of*

$$T_t(x, t) = T_{xx}(x, t),$$
$$T(x, 0) = 1 + 2 \sin \pi x / 4, \quad 0 < x < 2, \qquad (9.43)$$
$$T(0, t) = 1, \quad T(2, t) = 3, \quad t > 0.$$

Use that approximation to explore the steady-state temperature attained by the rod and the rate of decay to that steady state.

Choosing $\Delta x = 0.5$ on an x-interval of length 2 corresponds to $n = 4$.

Using centered differences to approximate T_{xx}, the partial differential equation $T_t = T_{xx}$ is replaced by a system of $4 - 1 = 3$ equations in the $4 + 1 = 5$ unknowns $U_0(t), \ldots, U_4(t)$, as in (9.38):

$$U_1'(t) = 4(U_0(t) - 2U_1(t) + U_2(t)),$$
$$U_2'(t) = 4(U_1(t) - 2U_2(t) + U_3(t)),$$
$$U_3'(t) = 4(U_2(t) - 2U_3(t) + U_4(t)).$$

Stop and Think **9.76** Explain the factor 4 on the right side of each equation.

Since U_0 approximates $T(0, t)$ and U_4 approximates $T(2, t)$, those two unknowns are immediately determined by the boundary conditions

$$T(0, t) = 1 \quad \Rightarrow \quad U_0(t) = 1, \quad t > 0,$$
$$T(2, t) = 3 \quad \Rightarrow \quad U_4(t) = 3, \quad t > 0.$$

The three differential equations now involve just three unknowns, $U_1(t)$, $U_2(t)$, $U_3(t)$,

$$U_1'(t) = 4(1 - 2U_1(t) + U_2(t)),$$
$$U_2'(t) = 4(U_1(t) - 2U_2(t) + U_3(t)),$$
$$U_3'(t) = 4(U_2(t) - 2U_3(t) + 3).$$

This system can be written in the form $\mathbf{u}' = \mathbf{Au} + \mathbf{f}$ by defining

Method-of-lines approximation to (9.43) is $\mathbf{u}' = \mathbf{Au} + \mathbf{f}$ . . .

$$\mathbf{u}(t) = \begin{pmatrix} U_1(t) \\ U_2(t) \\ U_3(t) \end{pmatrix}, \quad \mathbf{A} = 4 \begin{pmatrix} -2 & 1 & 0 \\ 1 & -2 & 1 \\ 0 & 1 & -2 \end{pmatrix}, \quad \mathbf{f} = \begin{pmatrix} 4 \\ 0 \\ 12 \end{pmatrix}.$$

Stop and Think **9.77** Verify the form of $\mathbf{A}$ and $\mathbf{f}$.

Initial conditions for $\mathbf{u}$ come directly from $T(x_i, 0) = 1 + 2 \sin \pi x_i / 4$ evaluated at $x_1 = 0.5$, $x_2 = 1$, and $x_3 = 1.5$:

. . . subject to initial conditions.

$$\mathbf{u}(0) = \begin{pmatrix} 1 + 2 \sin \pi / 8 \\ 1 + 2 \sin \pi / 4 \\ 1 + 2 \sin 3\pi / 8 \end{pmatrix} \approx \begin{pmatrix} 1.7654 \\ 2.4142 \\ 2.8478 \end{pmatrix}. \qquad (9.44)$$

To explore the steady-state temperature predicted by this approximation, solve $0 = \mathbf{A}\mathbf{u}_{ss} + \mathbf{f}$ to find

$$\mathbf{u}_{ss} = \begin{pmatrix} 3/2 \\ 2 \\ 5/2 \end{pmatrix}.$$

Stop and Think   **9.78** Why is the steady-state solution $\mathbf{u}_{ss}$ defined by $0 = \mathbf{A}\mathbf{u}_{ss} + \mathbf{f}$?

The method-of-lines approximation appears to predict a linear temperature profile for the steady state, as shown in figure 9.35.

Stop and Think   **9.79** How were the end values in figure 9.35, the steady-state temperatures at $x = 0$ and $x = 2$, determined?

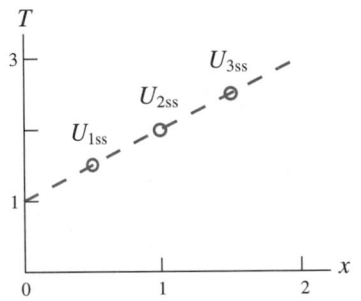

FIGURE 9.35   The steady-state temperature values $U_{1ss} = 3/2$, $U_{2ss} = 2$, $U_{3ss} = 5/2$ predicted by a method-of-lines approximation using $\Delta x = 0.5$ to the solution of (9.43).

What is the rate of decay to the steady state predicted by the method-of-lines approximation? An alternative question is, What is the rate of decay of the general solution of $\mathbf{u}' = \mathbf{A}\mathbf{u} + \mathbf{f}$? An analysis of the general solution will automatically include the special case selected by imposing the initial values (9.44).

Recall from section 8.1 that a general solution of $\mathbf{u}' = \mathbf{A}\mathbf{u} + \mathbf{f}$ has the form

$$\mathbf{u}_g = \mathbf{u}_h + \mathbf{u}_p,$$

where $\mathbf{u}_p$ is a particular solution of the original nonhomogeneous system and $\mathbf{u}_h$ is a general solution of the corresponding homogeneous system $\mathbf{u}' = \mathbf{A}\mathbf{u}$. We already know a particular solution, the steady state: $\mathbf{u}_p = \mathbf{u}_{ss}$. Evidently, *all of the decay behavior occurs in the homogeneous solution* $\mathbf{u}_h$.

Stop and Think   **9.80** This system involves *three* differential equations. Most of the examples in chapter 8 involved only two. Does the notion of general solution of a linear system depend upon the number of equations?

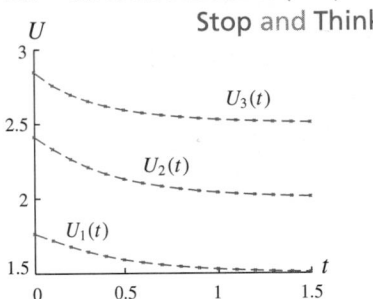

FIGURE 9.36   Fourth-order Runge-Kutta calculation of the temperature decay predicted by a method-of-lines approximation using $\Delta x = 0.5$ to the solution of (9.43).

Time constant for decay given by the method-of-lines approximation

Applying the characteristic equation method (see section 8.2) to a homogeneous, constant-coefficient system like $\mathbf{u}' = \mathbf{A}\mathbf{u}$ shows that the general solution $\mathbf{u}_h$ involves exponentials of the form $e^{r_i t}$, where r_i is an eigenvalue of the coefficient matrix $\mathbf{A}$. A straightforward calculation (or MATLAB) shows that the eigenvalues of $\mathbf{A}$ are

$$r_1 = -8 + 4\sqrt{2}, \quad r_2 = -8, \quad r_3 = -8 - 4\sqrt{2}.$$

Evidently, the least negative of these is $r_1 = -8 + 4\sqrt{2}$. Hence, the slowest decaying term in $\mathbf{u}_h$ is $e^{(-8+4\sqrt{2})t}$. The dominant time constant for decay of the method-of-lines solution is

$$\frac{1}{|-8+4\sqrt{2}|} \approx 0.4268.$$

This decay is exhibited in figure 9.36, which plots a fourth-order Runge-Kutta approximation with $\Delta t = 0.1$ to the solution of the method-of-lines initial-value problem. ■

Stop and Think

9.81 Do the decay curves in figure 9.36 predict the same steady states as are shown in figure 9.35?

9.82 Do the rates of decay exhibited in figure 9.36 appear to be consistent with a decay time constant of 0.4268?

MATLAB

For guidance in using MATLAB and DELAB to solve a system like $0 = \mathbf{A}\mathbf{u}_{ss} + \mathbf{f}$ or to find eigenvalues of a matrix, start DELAB, select `Help, Textbook`, then go to chapter 9, example 25.

9.6.1 Exercises

EXERCISE GUIDE

To gain experience . . .	Try exercises
Formulating method-of-lines approximations	1, 6, 8, 11, 12(b), 15–16, 17(a, c)
Studying method-of-lines systems numerically and analytically	2–4, 8–9, 11(a, b), 12(c–e), 17(b, c)
Comparing exact and approximate solutions	4(b), 8–9
Analyzing models using the method of lines	5, 7, 10–11, 14

1. Formulate method-of-lines approximations to the following initial-boundary-value problems. Write the approximation as the system $\mathbf{u}' = \mathbf{A}\mathbf{u} + \mathbf{f}$, and give the form of $\mathbf{u}$, $\mathbf{A}$, and $\mathbf{f}$. Describe a physical situation of which each problem might be a model.

   (a) $T_t(x, t) = T_{xx}(x, t)$,
   $T(x, 0) = (x - a)(x - b)$, $a < x < b$,
   $T(a, t) = 0$, $T(b, t) = 0$, $t > 0$.

   (b) $T_t(x, t) = T_{xx}(x, t)$,
   $T(x, 0) = F(x)$, $a < x < b$,
   $T(a, t) = \ell(t)$, $T(b, t) = r(t)$, $t > 0$.

2. Compare the coefficient matrix defined by (9.40) with (9.10), page 449, the coefficient matrix for a finite-difference approximation to a boundary-value problem involving a second derivative in x. Explain the differences and the similarities.

3. Example 23 gives an analytic solution to a system $\mathbf{u}' = \mathbf{A}\mathbf{u}$ of two ordinary differential equations with initial conditions that approximate the solution of the initial-boundary-value problem (9.35–9.37).

   (a) Confirm by substitution that the solution given in ex-

ample 23 is correct.

   (b) Find the eigenvalues of $\mathbf{A}$ and use them to derive that analytic solution.

   (c) Compare the results of your work and the claims of example 23 with the solution given by DELAB.

4. Example 23 constructs a method-of-lines approximation of (9.35–9.37) using $\Delta x = L/3$, then finds the analytic solution of the resulting system of ordinary differential equations.

   (i) Duplicate that analysis when the initial condition (9.36) is replaced by each of the following. Comment on the behavior of the analytic solution of the method-of-lines system.

   (ii) Find the exact solution of (9.35–9.37). Compare it with the solution of the approximate system of ordinary differential equations and comment on the relationships you observe.

   (a) $T(x, 0) = M \sin 2\pi x/L, 0 < x < L$

   (b) $T(x, 0) = M \sin 3\pi x/L, 0 < x < L$

   (c) $T(x, 0) = Mx(L - x), 0 < x < L$ (*Hint*: See example 19 of section 9.5.)

5. Example 23 concludes with a discussion of the exact and approximate time constants for temperature decay in the model (9.35–9.37); e.g., see (9.41). How much time must pass before the rod is nearly at zero temperature (in whatever sense you choose to define *nearly*) throughout its length? How could you use the value of the time constant to guide the choice of the stopping time for a numerical solution of a method-of-lines approximation? How does the time constant given by a method-of-lines approximation $\mathbf{u}' = \mathbf{A}\mathbf{u}$ depend upon the initial value $\mathbf{u}(0)$? How does it depend upon the coefficient matrix $\mathbf{A}$? How does the time constant in the exact solution of the initial-boundary-value problem depend upon the initial conditions (9.36)?

6. Carry out the details of the derivation of the method-of-lines approximation developed in example 25. Begin with the centered-difference approximation of T_{xx}. Extend the analysis by giving the form of the approximation for an arbitrary number of subdivisions of $0 \le x \le 2$.

7. Figure 9.33 shows plots of the decay of two components of the method-of-lines approximation to the solution of (9.35–9.37); they were computed in example 24. Estimate the time constant for the decay of each of these solutions. Should the values of the time constants be similar? Why? Are they? How do they compare with the time constant for decay found in the less accurate approximation of example 23? How do they compare with the time constant for decay predicted by the exact solution (9.23), page 473, of (9.35–9.37) ?

8. Suppose the initial condition (9.36) in the initial-boundary-value problem (9.35–9.37) is replaced by $T(x, 0) = 0$, $0 < x < L$. Construct a method-of-lines approximation. Show that it *always* gives the exact solution if (9.35–9.37) has no more than one solution.

9. Repeat the preceding exercise using the initial condition $T(x, 0) = (\beta x + \alpha(L - x))/L$ and the boundary conditions $T(0, t) = \alpha, T(L, t) = \beta$.

10. Find the steady-state temperature attained by the rod modeled by the initial-boundary-value problem (9.43). Is that steady state physically reasonable? Are the predictions of the method-of-lines approximation developed in example 25 consistent with that steady state? Explore the latter question numerically and analytically.

11. Using $\Delta x = 2/3$, construct a method-of-lines approximation to the initial-boundary-value problem (9.43) considered in example 25.

   (a) Determine the time constant for decay predicted by that approximation.

   (b) Show that a numerical solution of the method-of-lines approximation exhibits decay behavior consistent with

that time constant.

   (c) Determine the dominant time constant for decay predicted by the exact solution. Compare that value with the approximate analyses of parts (a) and (b).

12. Consider a variant of the initial-boundary-value problem (9.43) considered in example 25,

$$T_t(x, t) = T_{xx}(x, t),$$
$$T(x, 0) = 1 + 2 \sin \pi x / 4, \quad 0 < x < 2,$$
$$T(0, t) = 1, \quad T_x(2, t) = 0, \quad t > 0.$$

   (a) Describe the physical significance of the altered boundary condition.

   (b) Construct a method-of-lines approximation using $\Delta x = 0.5$, as in example 25, but involving four differential equations for the unknowns $U_1(t), \ldots, U_4(t)$. (*Hint*: Accommodate the boundary condition $T_x(2, t) = 0$ in the same way that example 13, page 453, accommodated $c'(L) = 0$ by introducing a fictitious grid point x_{n+1} outside the original x-interval in order to approximate the derivative boundary condition using a centered difference.)

   (c) Find the steady state of the method-of-lines system. Find the steady state of the original initial-boundary-value problem. Are the two consistent? Is the steady state behavior physically reasonable?

   (d) Use DELAB to conduct a numerical exploration of the behavior of the method-of-lines approximation. Find the exact solution of the original initial-boundary-value problem. Are the decay behaviors described by the two approaches consistent?

   (e) Use the eigenvalues of the coefficient matrix to analyze the decay behavior of the method-of-lines approximation. (You can use DELAB to find the eigenvalues.) Compare that analysis with the results of part (d).

13. To what extent is the following statement correct?

   Suppose $\mathbf{u}' = \mathbf{A}\mathbf{u} + \mathbf{f}$ is a method-of-lines approximation to an initial-boundary-value problem such as (9.43). Then the time constant for decay predicted by that approximation is the least negative eigenvalue of the matrix $\mathbf{A}$.

14. The Runge-Kutta computations used to generate figure 9.36 found that $U_1(1.5) \approx 1.5089$, $U_2(1.5) \approx 2.0126$, $U_3(1.5) \approx 2.5089$; $U_i(t)$, $i = 1, 2, 3$, are the method-of-lines approximations to the solution of the initial-boundary-value problem (9.43) considered in example 25. Are those values of $U_i(1.5)$ consistent with the value of 0.4268 estimated for the decay time constant in that example?

15. Formulate a method-of-lines approximation to the solution of the initial-boundary-value problem

$$y_t(x, t) = Dy_{xx}(x, t) - Vy_x(x, t),$$
$$y(x, 0) = M \sin \pi x / L, \quad 0 < x < L, \qquad (9.45)$$
$$y(0, t) = 0, \quad y(L, t) = 0, \quad t > 0.$$

(*Hint*: Use the centered-difference approximation (9.2), page 445, for the y_x term.) Is the system of ordinary differential equations that results linear or nonlinear? Homogeneous or nonhomogeneous? Describe a physical situation of which this problem might be a model.

16. Write a system of ordinary differential equations in t of the form $\mathbf{u}' = \mathbf{Au} + \mathbf{f}$ that approximates the solution of

$$\frac{\partial T(x, t)}{\partial t} = \kappa \frac{\partial^2 T(x, t)}{\partial x^2}$$
$$T(x, 0) = g(x), \quad 0 < x < L,$$

subject to the given boundary conditions. In each case, describe the physical significance of the boundary conditions and give the form of $\mathbf{u}$, $\mathbf{A}$, and $\mathbf{f}$.

(a) $T(0, t) = \alpha, T(L, t) = \beta, t > 0$

(b) $T(0, t) = \ell(t), T(L, t) = r(t), t > 0$

(c) $T_x(0, t) = 0, T(L, t) = \beta, t > 0$

(d) $T(0, t) = \alpha, T_x(L, t) = 0, t > 0$

(e) $T(0, t) + dT_x(0, t) = 0, T(L, t) = \beta, t > 0$

(f) $T(0, t) = \alpha, T(L, t) + dT_x(L, t) = 0, t > 0$

17. In the initial-boundary-value problem (9.45), set $D = V = L = 1, M = 2$.

(a) Choose $\Delta x = 1/3$ and write the system of ordinary differential equations that approximate its solution.

(b) Using DELAB if you wish, find an analytic solution of this system. What does its behavior reveal about the physical situations of which (9.45) might be a model?

(c) Choose $\Delta x = 1/5$. Use DELAB to approximate numerically the solution of the resulting method-of-lines system. Comment on the solution behavior you observe and on its relation to the $\Delta x = 1/3$ approximation found in previous steps.

9.7 ■ CHAPTER EXERCISES

EXERCISE GUIDE

To gain experience ...	Try exercises
Deriving models and boundary conditions	1–2, 9–11, 20
Solving boundary-value problems	2–8
Separating variables	12
With orthogonality and eigenfunction expansions	14–16
Solving eigenvalue problems	15
Solving initial-boundary-value problems	17–19
Analyzing and interpreting solution behavior	13
Approximating solutions using finite differences and the method of lines	21–23

1. Figure 9.37 shows a cross section of the wall of a house. Suppose that the temperature in the wall varies only from left to right, not vertically, as would happen if the inner and outer temperatures T_{in} and T_{out} were uniform.

(a) Assuming that the thermal conductivity k of the material in the wall is constant, derive the model

$$T''(x) = 0, \quad T(0) = T_{in}, \quad T(L) = T_{out},$$

for the temperature profile $T(x)$ in the wall. Sketch the wall and show the control volume you use.

(b) In fact, the thermal conductivity of the wall of a house is not constant. It varies with the material (wallboard, insulation, sheathing, siding, etc.) one encounters at various positions within the wall. Derive the model

$$(k(x)T'(x))' = 0, \quad T(0) = T_{in}, \quad T(L) = T_{out},$$

for the temperature profile $T(x)$ in the wall. Can you use the same control volume as in part (a)?

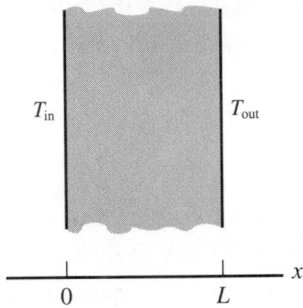

FIGURE 9.37   A cross section of the wall of a house.

2. Is the heat-loss model developed in section 2.3 an equilibrium-temperature model? What is the equilibrium temperature of the house in that model?

In exercises 3–8, solve the given boundary-value problem. Compare your solution with that given by DELAB. Are the two consistent?

3. $y'' - y = 4$, $y'(0) = 4$, $y(2) = -1$

4. $2y'' - 4y' + 10y = 0$, $y(-1) = 1$, $y(1) = 0$

5. $y'' + y = 0$, $y(0) + y'(0) = 0$, $y(\pi/2) - y'(\pi/2) = 0$

6. $y'' + y = 2\cos t$, $y(0) = 0$, $y(\pi/2) = 0$

7. $x^2 y'' - 2xy' + 2y = 0$, $y(1) = 2$, $y(2) + y'(2) = 11$
(Homogeneous solutions are x, x^2.)

8. $x^2 y'' - 2xy' + 2y = 4x^2$, $y(1) = 2$, $y(2) = 4$
(Homogeneous solutions are x, x^2.)

9. Suppose the wall of a house has cooled uniformly to the outside temperature T_{out}. Derive the model

$$T_t = \kappa T_{xx},$$

$$T(0, t) = T_{in}, \quad T(L, t) = T_{out}, \quad t > 0,$$

$$T(x, 0) = T_{out},$$

that describes how the temperature profile through the wall changes when the inner side of the wall is suddenly raised to the temperature T_{in} by turning on the furnace in the house. See figure 9.37 . (Since the wall's height is much greater than its width, the only significant heat flow is from left to right. Construct a control volume within the wall, but consider only heat flow through its left and right boundaries.)

10. The house heat loss model considered in section 2.3 supposed the temperature was uniform throughout the house. This exercise asks you to improve that model by incorporating variations with position.

(a) Sketch the cross section of a flat-roofed (for simplicity) house. Using a rectangular control volume and the law of conservation of energy, derive a governing partial differential equation.

(b) Suppose the outside of the house is at temperature T_{out}, the interior at T_{in}. Complete the model by coupling the differential equation of part (a) with appropriate boundary and initial conditions.

11. Suppose the right end of the rod in figure 9.18, page 460, is covered with a poor insulator having thermal conductivity δ. Argue that the rate of heat flow through this poor insulator is $A(L)\delta(T(L, t) - T_L)$. But conservation of energy requires that the flow of heat energy through the end cap must equal the rate of heat flow through the end of the rod, $-kA(L)T_x(L, t)$. Equate these two expressions to derive a boundary condition of the form

$$T(L, t) + aT_x(L, t) = b, \quad t > 0.$$

Express the constants a, b in terms of k, δ, T_{out}. What would be the form of the boundary condition if the poorly insulated cap were applied at $x = 0$?

This boundary condition is known as a boundary condition of the *third kind*.

12. Suppose $T(x, t) = X(x)\Theta(t)$, where $T(x, t)$ satisfies

$$xT_t = \kappa(xT_x)_x, \quad T_x(0, t) = T(L, t) = 0.$$

(The equation and boundary conditions arise in a model of heat flow in a thin circular disk whose top and bottom are insulated; see exercise 8 of section 9.4.) Find an eigenvalue problem for $X(x)$ and an ordinary differential equation for $\Theta(t)$. Can the equation for $X(x)$ be solved by any methods you have studied?

13. The results of example 21 suggest that the initial heat energy in a rod whose ends are insulated simply redistributes itself uniformly at equilibrium. Verify this observation in general by showing that as $t \to \infty$, the solution of the initial-boundary-value problem

$$\frac{\partial T(x, t)}{\partial t} = \kappa \frac{\partial^2 T(x, t)}{\partial x^2},$$

$$T_x(0, t) = T_x(L, t) = 0, \quad t > 0,$$

$$T(x, 0) = g(x), \quad 0 < x < L,$$

tends to

$$\frac{1}{L} \int_0^L g(x)\, dx.$$

14. The eigenfunction expansion theorem can be extended to the more general *self-adjoint* problem

$$-(p(x)X'(x))' + q(x)X(x) = \lambda\rho(x)X(x),$$
$$aX(0) + bX'(0) = 0,$$
$$cX(L) + dX'(L) = 0,$$

where p is continuously differential and nonzero on $0 \leq x \leq L$ and q, ρ are continuous there. The eigenfunctions are then *orthogonal with weight ρ,*

$$\int_0^L \rho(x)X_n(x)X_m(x)\,dx = 0, \quad m \neq n.$$

(a) Verify the orthogonality relation for the boundary conditions $X(0) = X(L) = 0$.

(b) Verify the orthogonality relation for the boundary conditions $X'(0) = X'(L) = 0$.

(c) Verify the orthogonality relation for the general boundary conditions given in the problem statement.

15. Solve the eigenvalue problem $X''(x) + \lambda X(x) = 0$, $X(0) = X'(L) = 0$. Using those eigenfunctions, expand $f(x) = 2$, $0 \leq x \leq L$. To what function does the eigenfunction expansion converge for $-L \leq x \leq L$?

16. Using the eigenfunctions found in the previous problem, expand $f(x) = x/L$. To what function does the eigenfunction expansion converge for $-L \leq x \leq L$?

17. Solve the initial-boundary-value problem

$$T_t(x, t) = \kappa T_{xx}(x, t),$$
$$T(0, t) = T_x(L, t) = 0, \quad t > 0,$$
$$T(x, 0) = 2, \quad 0 < x < \pi.$$

Describe a physical situation that this problem might model. As time passes, does the solution appear to approach the proper equilibrium solution?

18. Repeat exercise 17 with the initial condition $T(x, 0) = x/L$.

19. Find the vertical temperature profile in the computer chip modeled in example 18, page 466. That is, solve

$$\frac{\partial T(y, t)}{\partial t} = \kappa \frac{\partial^2 T(y, t)}{\partial y^2},$$
$$T(0, t) = T_0, \quad T(h, t) = 0, \quad t > 0,$$
$$T(y, 0) = 0, \quad 0 < y < h.$$

20. A long circular pipe of inner radius r and outer radius R is carrying steam at temperature T_s. Its outer surface is at temperature T_{out}. Show that the equilibrium temperature in the wall of the pipe is modeled by

$$-\kappa(xT'(x))' = 0, \quad T(r) = T_s, \quad T(R) = T_{\text{out}},$$

where x is the radial coordinate. What assumption must you make about heat flow in the axial direction?

21. For each of the following boundary-value problems:
   (i) Write a finite-difference approximation.
   (ii) Set $\Delta x = 0.5$ and write out *all* of the equations in the approximation.
   (iii) Use DELAB to evaluate this approximate solution for $\Delta x = 0.5$.
   (iv) Compare the approximate solution with the exact solution. Explain the accuracy you observe.

   (a) $-c''(x) = 0$, $c(0) = 2$, $c(1) = 4$

   (b) $-c''(x) = 0$, $c(0) = c(1) = 2$

   (c) $-c''(x) = 2$, $c(0) = c(2) = 0$

   (d) $-c''(x) = 2$, $c(0) = 0$, $c'(2) = 0$

22. Formulate a finite-difference approximation to each of the following boundary-value problems. Write the approximation in the form $\mathbf{Ac} = \mathbf{b}$ and define $\mathbf{A}$, $\mathbf{c}$, and $\mathbf{b}$.

   (a) $-y''(x) + P(x)y'(x) + Q(x)y(x) = f(x)$, $y(a) = \alpha$, $y(b) = \beta$

   (b) $-y''(x) + P(x)y'(x) + Q(x)y(x) = f(x)$, $y'(a) = \alpha$, $y(b) = \beta$

   (c) $-y''(x) + P(x)y'(x) + Q(x)y(x) = f(x)$, $y(a) = \alpha$, $y'(b) = \beta$

   (d) $-y''(x) + P(x)y'(x) + Q(x)y(x) = f(x)$, $y(a) + dy'(a) = 0$, $y(b) = \beta$

   (e) $-y''(x) + P(x)y'(x) + Q(x)y(x) = f(x)$, $y(a) = \alpha$, $y(b) + dy'(b) = 0$

23. Formulate a method-of-lines approximation to the solution of each of the following initial-boundary-value problems. Write the approximation in the form $\mathbf{u}' = \mathbf{Au} + \mathbf{f}$ and define $\mathbf{u}$, $\mathbf{A}$, and $\mathbf{f}$. Describe a physical situation of which each might be a model.

   (a) $y_t(x, t) = Dy_{xx}(x, t) - Vy_x(x, t)$,
   $y(x, 0) = M \sin \pi x/L$, $0 < x < L$,
   $y(0, t) = 0$, $y(L, t) = 0$, $t > 0$.

   (b) $y_t(x, t) = Dy_{xx}(x, t) - Vy_x(x, t) + x(L - x)e^{-t}$,
   $y(x, 0) = 4 + \cos \pi x/L$, $0 < x < L$,
   $y(0, t) = 5$, $y(L, t) = 3$, $t > 0$.

   (c) $y_t(x, t) = Dy_{xx}(x, t)$,
   $y(x, 0) = F(x)$, $0 < x < L$,
   $y_x(0, t) = 0$, $y(L, t) = \beta$, $t > 0$.

   (d) $y_t(x, t) = Dy_{xx}(x, t) + y(x, t) + S(x, t)$,
   $y(x, 0) = F(x)$, $0 < x < L$,
   $y(0, t) = 0$, $y(L, t) = 0$, $t > 0$.

9.8 ■ CHAPTER PROJECTS

1. **Shooting method.** Numerical methods for solving initial-value problems, such as the Euler method or Runge-Kutta methods, can be adapted to solve boundary-value problems through the *shooting method*: the solution of a boundary-value problem is the solution of an initial-value problem involving the same differential equation with an appropriate but unknown initial slope. For example, the solution of the boundary-value problem

$$y'' + y = 0, \quad y(0) = 1, \quad y(1) = 2, \tag{9.46}$$

is also a solution of the initial-value problem

$$y'' + y = 0, \quad y(0) = 1, \quad y'(0) = v_i, \tag{9.47}$$

for an appropriate choice of initial slope v_i. The shooting method applies a numerical method to the initial-value problem and varies v_i until the second boundary condition $y(1) = 2$ is satisfied.

Verify the claims of the preceding paragraph by solving the boundary-value problem (9.46), finding the slope of that solution at $x = 0$, solving the initial-value problem (9.47) using the value of v_i just found, and showing that the two solutions are identical.

To implement the shooting method analytically, write the solution of the initial-value problem (9.47) using v_i as a parameter. Require that the solution of the initial-value problem satisfy the second boundary condition $y(1) = 2$ to obtain an equation for the desired value of v_i. Show that when v_i has this value, the solution of the initial-value problem is again the solution of the boundary-value problem.

Implement the shooting method numerically. Use an initial-value solver of your choice from DELAB to compute the numerical solution of several boundary-value problems of your own invention or chosen from the exercises of section 9.2. Vary the initial slope until the second boundary condition is satisfied to sufficient accuracy. How sensitive is the solution of the boundary-value problem to the value of the initial slope? Can you develop an algorithm for varying the initial slope that leads methodically to a solution of the boundary-value problem?

> The bellicose name *shooting* comes from thinking of the graph of the solution of the boundary-value problem as the trajectory of a cannonball fired from a gun located at $x = 0$, $y = 1$. The angle of the cannon must be adjusted until the cannonball hits the point $x = 1$, $y = 2$.

2. **The minimum principle.** The internal temperature of a thin insulated rod is never less than the smaller of its end temperatures; i.e., its minimum temperature occurs at an end. That physical observation is the basis of a theorem known as the *minimum principle* that can be used to establish the uniqueness of solutions of boundary-value problems.

   (a) Suppose a thin insulated rod is heated by a current passing through it while its ends are maintained at zero temperature. Show that its equilibrium temperature is modeled by

$$-\kappa A(x)T''(x) - \kappa A'(x)T'(x) = f(x), \quad T(0) = T(L) = 0,$$

where $A(x)$ is the cross-sectional area of the rod, $\kappa > 0$ is its thermal diffusivity, L is its length, and $f(x) > 0$ is a function proportional to the rate at which heat is generated by the current.

(b) Prove that the minimum of $T(x)$ must occur at $x = 0$ or $x = L$. Argue by contradiction. Suppose a minimum of $T(x)$ occurs at x_0, $0 < x_0 < L$. Then $T'(x_0) = 0$ and $T''(x_0) \geq 0$. How is the differential equation contradicted?

(c) Using these same ideas, prove the following theorem.

Theorem 3 (Minimum principle). *Let* $y(x)$ *be a solution of the boundary-value problem*

$$-y''(x) + P(x)y' + Q(x)y = f(x), \quad y(0) = y(L) = 0,$$

where $f(x) \geq 0$, $Q(x) > 0$, $0 \leq x \leq L$. *Then* $y(x) \geq 0$, $0 \leq x \leq L$.

State and prove the corresponding maximum principle for $f(x) \leq 0$, $0 \leq x \leq L$.

(d) Prove the following theorem.

Theorem 4 (Uniqueness for boundary-value problems). *If* $Q(x) > 0$, $0 \leq x \leq L$, *then the boundary-value problem*

$$-y''(x) + P(x)y' + Q(x)y = f(x), \quad y(0) = a, \quad y(L) = b,$$

has at most one solution.

To begin the proof, suppose the boundary-value problem has two solutions. Of what boundary-value problem is their difference a solution? What do the maximum and minimum principles assert about the solutions of that boundary-value problem?

3. **Temperature distribution in a slab.** Suppose a rectangular slab is resting on the xy-plane. It is thin enough to ignore heat flow in the z direction. However, its boundaries are at different temperatures, and the temperature distribution as a function of x and y is of interest. (For example, the slab might be a cooling fin on a heat sink for a computer chip or on the cylinder of an air-cooled gas engine.)

(a) For convenience, suppose the thin slab covers the square region $0 \leq x \leq 1, 1 \leq y \leq 1$. Considering only heat flow in the x and y directions, derive the following model for the equilibrium temperature of the slab,

$$T_{xx}(x, y) + T_{yy}(x, y) = 0,$$
$$T(0, y) = T_E,$$
$$T(1, y) = T_W,$$
$$T(x, 0) = T_S,$$
$$T(x, 1) = T_N, \quad 0 \leq x \leq 1, \quad 0 \leq y \leq 1.$$

Here $T_E, \ldots, T_N$ are the constant temperatures along the east, west, south, and north edges of the slab. (The partial differential equation is called *Laplace's equation*.)

(b) Divide the unit square covered by the slab into a square grid. Approximate T_{xx} at each point of the grid with a centered difference. Repeat for T_{yy}. Show that an approximation to $T_{xx}(x, y) + T_{yy}(x, y)$ at any point of this grid involves a five-point star or stencil; compare with figure 9.13, page 448.

(c) Limit the grid to four subdivisions in each direction: $\Delta x = \Delta y = 1/4$. Write a finite-difference approximation to the boundary-value problem of part (a). (One convenient way to keep track of the unknown interior grid point temperatures is to number them using so-called typewriter order, from left to right and top to bottom.)

(d) For boundary temperature values of your choice, solve the finite-difference approximation using MATLAB. Be sure to test your approximation using boundary temperature values for which you can guess the exact solution of the boundary-value problem. Is the behavior of the approximate solution consistent with your physical intuition?

(e) Describe how to extend the finite-difference approximation to include boundary temperatures that vary along the edge of the slab.

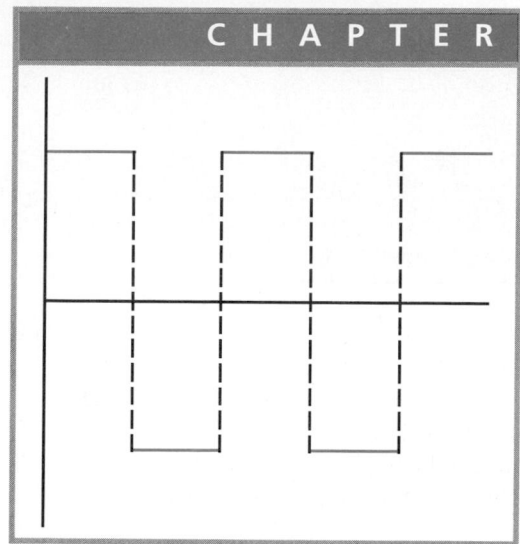

10

The Laplace Transform

The Laplace transform is a useful tool for solving linear constant-coefficient differential equations, among other problems, especially when those equations involve square waves, ramps, and other unusual forcing functions. In addition, Laplace transforms succinctly describe the input-output behavior of systems modeled by such equations. For example, they provide a relation between the forcing frequency applied to a system and the frequencies that appear in the response, or output, of the system.

10.1 ■ THE TRANSFORM IDEA

10.1.1 Motivation

Imagine a problem in chemical testing: Does a given sample of fortified milk contain vitamin D (or iron or some other additive), as required by law? A chemist's job is to devise an assay that will answer that question, and perhaps give the quantity of the particular ingredient as well.

Now imagine a problem in mathematical modeling: Does the solution of a mathematical model of a given system—an initial-value problem—contain exponentials (or sines and cosines or some other function) whose behavior is important? If the solution formula is available, that question is easy to answer by inspection. If there is no formula, then mathematicians need to devise a test that will detect exponentials.

Here is an idea for such a test: Suppose we are testing for exponentials of the form e^{at}. Multiply the function being tested by e^{-st}. If the exponential e^{at} is present, then the product

$$e^{at}e^{-st} = e^{(a-s)t}$$

will be bounded if $s > a$ and constant if $s = a$. Better yet, the *integral*

$$\int_0^\infty e^{(a-s)t}\, dt$$

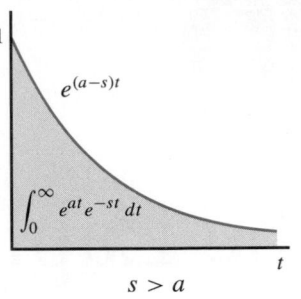
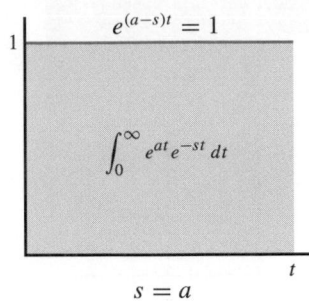

FIGURE 10.1 The test integral $\int_0^\infty e^{(a-s)t}\,dt$—the Laplace transform of e^{at}— is bounded for $s > a$ and unbounded otherwise.

will be bounded if $s > a$ but unbounded if $s = a$, as figure 10.1 illustrates. Roughly, the idea of the test is to multiply by e^{st}, integrate over $0 \le t < \infty$, and monitor the behavior of the test integral as s decreases, watching for the value of s at which the integral becomes unbounded. That test integral is called the **Laplace transform** of e^{at}.

Definition 1. *The **Laplace transform** $\mathcal{L}\{f\}$ of a function $f(t)$ is*

LAPLACE TRANSFORM

$$\mathcal{L}\{f\} = \int_0^\infty f(t)e^{-st}\,dt, \tag{10.1}$$

for those values of s (if any) for which this improper integral exists.

The improper integral in (10.1) is defined by

$$\int_0^\infty f(t)e^{-st}\,dt = \lim_{b\to\infty}\int_0^b f(t)e^{-st}\,dt.$$

Conditions on f that guarantee the existence of this improper integral are given at the end of this section. These conditions require that the integrand $f(t)e^{-st}$ decay to zero rapidly enough to ensure that the integral $\int_0^b f(t)e^{-st}\,dt$ approaches a finite limit for large b.

The Laplace transform can be used to find a solution of a constant-coefficient initial-value problem by turning the differential equation into an algebraic equation that can readily be solved for an alternative representation of the original unknown function. This new form is a language that control engineers, in particular, use to describe the behavior of the systems they study.

The power of the Laplace transform is that this testing process can be applied to solutions of initial-value problems *without knowing a solution formula.* That is, we can test the solution of an initial-value problem for the presence of a term like e^{at} without first finding the solution. The next three examples build up to an illustration of this idea.

The first step is obtaining a precise expression for the behavior of the Laplace transform test integral when it is applied to e^{at}.

■ **EXAMPLE 1**   *Find $\mathcal{L}\{e^{at}\}$; a is a constant.*

Apply the definition (10.1) of Laplace transform with $f(t) = e^{at}$:

$$\mathcal{L}\left\{e^{at}\right\} = \int_0^\infty e^{at} e^{-st}\, dt$$

$$= \int_0^\infty e^{(a-s)t}\, dt$$

$$= \lim_{b\to\infty} \left. \frac{e^{(a-s)t}}{a-s} \right|_{t=0}^{t=b}$$

$$= \lim_{b\to\infty} \frac{e^{(a-s)b}}{a-s} - \frac{1}{a-s}$$

$$= \frac{1}{s-a},$$

provided $s > a$.

Stop and Think   **10.1** Why is the restriction $s > a$ needed? Where is it first required? See exercise 34.

Hence,

Transform of e^{at}

$$\mathcal{L}\left\{e^{at}\right\} = \frac{1}{s-a}, \quad s > a.$$

Loosely, this example says, "If the Laplace transform testing process results in a function containing a term that behaves like $1/(s-a)$, then the exponential e^{at} is present in the original function." Figure 10.2 illustrates the singularity in $\mathcal{L}\left\{e^{at}\right\}$ as s approaches a.

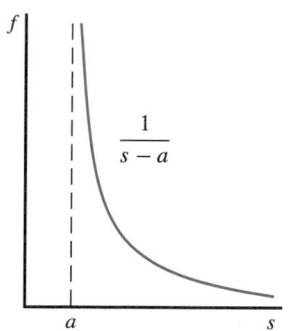

FIGURE 10.2 $F(s) = \mathcal{L}\left\{e^{at}\right\} = 1/(s-a)$ is unbounded as s approaches a, $s > a$.

MATLAB

To find the Laplace transform of a function using DELAB, select **Analytic tools/Laplace** from the menu bar in the main DELAB window. Note that no information is provided about the range of validity of the transform.

Use that selection to confirm $\mathcal{L}\left\{e^{at}\right\} = 1/(s-a)$. (Remember to enter e^{at} as **exp(a*t)**.)

To apply the Laplace transform test to the specific initial-value problem, for example, $y' + ky = e^{-3t}$, $y(0) = 4$, we need an expression for the Laplace transform of the derivative of an arbitrary function like y'.

Stop and Think   **10.2** To get a sense of the exponential terms that might appear in the solution of this initial-value problem, think of applying undetermined coefficients to $y' + ky = e^{-3t}$. What exponentials, if any, would you expect to see in a particular solution of this equation? Think of applying characteristic equations to the homogeneous equation associated with $y' + ky = e^{-3t}$. What exponentials, if any, would you expect to see in a homogeneous solution? In a general solution?

■ **EXAMPLE 2**   *Let f be a differentiable function. Find $\mathcal{L}\{f'\}$ in terms of the transform $F(s) = \mathcal{L}\{f(t)\}$.*

Apply the definition (10.1) of the Laplace transform and integrate by parts using $u = e^{-st}$ and $dv = f'(t)\,dt$:

$$\mathcal{L}\{f'\} = \int_0^\infty f'(t)e^{-st}\,dt$$

$$= \lim_{b \to \infty} f(t)e^{-st}\Big|_{t=0}^{t=b} + s \int_0^\infty f(t)e^{-st}\,dt$$

$$= \lim_{b \to \infty} f(b)e^{-sb} - f(0) + s\mathcal{L}\{f\}.$$

The integral in the middle line is just the definition of $\mathcal{L}\{f\}$; the factor s can be brought outside the integral because it is constant with respect to the variable of integration t. Evaluating the limit for $s > 0$ and writing $F(s) = \mathcal{L}\{f\}$ yields

---

Transform of f'

$$\mathcal{L}\{f'\} = sF(s) - f(0).$$

Now we can use the Laplace transform to test the solution of an initial-value problem for the presence of exponential terms. That calculation will require one other property of the Laplace transform: Because it is defined by an integral, the Laplace transform is a *linear* operator. That is, if $f(t)$, $g(t)$ are functions with Laplace transforms $F(s) = \mathcal{L}\{f\}$, $G(s) = \mathcal{L}\{g\}$, then

THE LAPLACE TRANSFORM IS LINEAR.

$$\mathcal{L}\{cf + dg\} = c\mathcal{L}\{f\} + d\mathcal{L}\{g\} = cF + dG$$

for any constants c, d.

Stop and Think   **10.3** Write out both sides of the preceding equation using the definition of the Laplace transform and verify the claim of linearity.

■ **EXAMPLE 3**   *Find the Laplace transform of the solution of the initial-value problem*

$$y' + ky = e^{-3t}, \quad y(0) = 4.$$

Identify the exponential terms appearing in this solution.

Assume that $y(t)$ does indeed have a Laplace transform and let $Y(s) = \mathcal{L}\{y(t)\}$. Apply the transform operator $\mathcal{L}$ to the entire differential equation, invoke the linearity of the transform, and use the formulas developed in the preceding examples. On the left, we find

Transform left-hand side.

$$\mathcal{L}\{y' + ky\} = \mathcal{L}\{y'\} + \mathcal{L}\{ky\}$$
$$= \mathcal{L}\{y'\} + k\mathcal{L}\{y\}$$
$$= \mathcal{L}\{y'\} + kY(s)$$
$$= sY(s) - y(0) + kY(s).$$

Stop and Think   **10.4** Which step uses the linearity of the Laplace transform?

The last expression came from the previous example, $\mathcal{L}\{f'\} = sF(s) - f(0)$, with y in place of f. Invoking the initial condition $y(0) = 4$ yields

$$\mathcal{L}\{y' + ky\} = (s + k)Y - 4.$$

On the right side of the transformed differential equation, apply example 1 to find

Transform right-hand side.

$$\mathcal{L}\{e^{-3t}\} = \frac{1}{s + 3}.$$

Hence, $\mathcal{L}\{y' + ky\} = \mathcal{L}\{e^{-3t}\}$ becomes

$$(s + k)Y - 4 = \frac{1}{s + 3}.$$

Now solve for Y to find the transform of the solution of the initial-value problem $y' + ky = e^{-3t}$, $y(0) = 4$,

Solve for $Y(s) = \mathcal{L}\{y(t)\}$.

$$Y(s) = \frac{1}{(s + k)(s + 3)} + \frac{4}{s + k}.$$

We can inspect $Y(s)$ in search of terms of the form $1/(s - a)$, a marker of the presence of e^{at} in the solution $y(t)$.

Since the term $4/(s + k)$ is of the form $1/(s - a)$ with $a = -k$, the solution $y(t)$ appears to contain the exponential e^{-kt}. (To be more precise, the exponential term corresponding to $4/(s + k)$ is $4e^{-kt}$.)

The significance of the term $1/(s + k)(s + 3)$ is less clear until we think of it as the result of adding two fractions,

$$\frac{1}{(s + k)(s + 3)} = \frac{A}{s + k} + \frac{B}{s + 3},$$

for some constants A and B. (The coefficients A and B can be found using partial fractions. See appendix A.7.) The second term gives rise to a new exponential, e^{-3t}, for it has the form $1/(s - a)$ with $a = -3$.

Stop and Think   **10.5** Find A and B in the partial fraction expansion of $1/(s + k)(s + 3)$. Identify precisely the exponential terms that give rise to each of these fractions.

We conclude that the solution of $y' + ky = e^{-3t}$, $y(0) = 4$, will contain two different exponentials, e^{-kt} and e^{-3t}. ■

Stop and Think   **10.6** Has the Laplace transform identified all of the exponential terms you would expect to find in the solution of this initial-value problem?

If we could somehow invert the Laplace transform process—that is, if we could find the *inverse Laplace transform*

$$y(t) = \mathcal{L}^{-1} Y(s) = \mathcal{L}^{-1} \left\{ \frac{1}{(s+k)(s+3)} + \frac{4}{s+k} \right\} \qquad (10.2)$$

—then the work of this example would give exactly the solution of the initial-value problem it considered. The inverse transform question is addressed in the next section.

MATLAB

To invert a Laplace transform using DELAB, select **Analytic tools/ Inverse Laplace** from the menu bar in the main DELAB window.
    Use that selection to find the solution $y(t)$ of $y' + ky = e^{-3t}$, $y(0) = 4$, given by (10.2).

The Laplace transform testing idea extends to other functions, such as $\sin at$.

■ **EXAMPLE 4**   *Find the Laplace transform of* $\sin at$, *a constant.*

Assume $s > 0$, apply the Laplace transform definition (1), and integrate by parts twice using $dv = e^{-st}\,dt$:

$$\mathcal{L}\{\sin at\} = \int_0^\infty (\sin at) e^{-st}\,dt$$

$$= -\frac{(\sin at) e^{-st}}{s} \Big|_0^\infty + \frac{a}{s} \int_0^\infty (\cos at) e^{-st}\,dt$$

$$= -\frac{a(\cos at) e^{-st}}{s^2} \Big|_0^\infty - \frac{a^2}{s^2} \int_0^\infty (\sin at) e^{-st}\,dt$$

$$= \frac{a}{s^2} - \frac{a^2}{s^2} \int_0^\infty (\sin at) e^{-st}\,dt.$$

Solve for the integral $\int_0^\infty (\sin at)e^{-st}\, dt$ defining the transform to obtain

$$\mathcal{L}\{\sin at\} = \int_0^\infty (\sin at)e^{-st}\, dt$$

$$= \left(1 + \frac{a^2}{s^2}\right)^{-1} \frac{a}{s^2}$$

$$= \frac{a}{a^2 + s^2}.$$

Hence, for $s > 0$,

> **Transform of** $\sin at$
>
> $$\mathcal{L}\{\sin at\} = \frac{a}{a^2 + s^2}.$$

Terms of the form $a/(a^2 + s^2)$ in a Laplace transform are markers of the presence of $\sin at$ in the original function.

Applying the definition of the Laplace transform to other functions leads to the list of transforms that is shown in table 10.1; e.g., example 1 derived entry 3, and example 4 derived entry 4. Exercises 18–25 ask for the derivation of the balance of the entries in table 10.1.

TABLE 10.1 Basic Laplace transform pairs

	$f(t)$	$F(s) = \mathcal{L}\{f(t)\}$
1.	1	$\dfrac{1}{s}$
2.	$t^n,\, n = 1, 2, \ldots$	$\dfrac{n!}{s^{n+1}}$
3.	e^{at}	$\dfrac{1}{s-a},\, s > a$
4.	$\sin at$	$\dfrac{a}{s^2 + a^2},\, s > 0$
5.	$\cos at$	$\dfrac{s}{s^2 + a^2},\, s > 0$
6.	$\sinh at$	$\dfrac{a}{s^2 - a^2},\, s > a$
7.	$\cosh at$	$\dfrac{s}{s^2 - a^2},\, s > a$

The derivative expression obtained in example 2 is the first in a list of useful transform properties. Table 10.2 starts such a list, and exercise 29 requests a derivation of the second entry. The list will grow in subsequent sections.

TABLE 10.2   The beginning of a list of useful properties of Laplace transforms

Properties of Laplace transforms

$$F(s) = \mathcal{L}\{f(t)\} = \int_0^\infty e^{-st} f(t)\, dt$$

1. Transform of a derivative

$$\mathcal{L}\{f'(t)\} = sF(s) - f(0)$$

$$\mathcal{L}\{f''(t)\} = s^2 F(s) - sf(0) - f'(0)$$

The next example illustrates how easily the Laplace transform can accommodate forcing functions, such as the ramp shown in figure 10.3, that change character at a given time. An appropriately scaled version of this ramp function might represent, for example, the increase in an emigration rate from zero to a constant level.

Stop and Think   **10.7** Could undetermined coefficients or other analytic solution methods you have studied readily find a solution of the equation $y' + ky = r(t)$, where r is the ramp function of figure 10.3?

■ **EXAMPLE 5**   *Let $r(t)$ be the* ramp function

$$r(t) = \begin{cases} t, & 0 \le t \le 1, \\ 1, & 1 < t < \infty, \end{cases}$$

shown in figure 10.3. *Find the Laplace transform of r.*

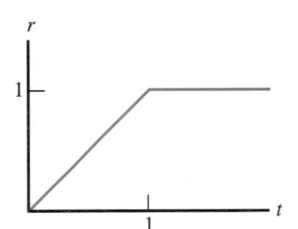

FIGURE 10.3   The ramp function $r(t)$

Apply the Laplace transform definition (10.1) and explicitly evaluate the integrals that result. Note the requirement that s be positive in the evaluation of the integral $\int_1^\infty e^{-st}\, dt$.

$$\mathcal{L}\{r\} = \int_0^\infty r(t) e^{-st}\, dt$$

$$= \int_0^1 t e^{-st}\, dt + \int_1^\infty e^{-st}\, dt$$

$$= \frac{1 - e^{-s}}{s^2} - \frac{e^{-s}}{s} + \frac{e^{-s}}{s}.$$

Hence, the Laplace transform of the ramp function is

Transform of ramp function

$$\mathcal{L}\{r\} = \frac{1 - e^{-s}}{s^2}, \quad s > 0. \quad ■$$

The improper integral $\int_0^\infty f(t)e^{-st}\,dt$ defining the Laplace transform is actually a function of the parameter s. Like the operations of differentiation and antidifferentiation, which transform one function into another, the Laplace transform takes one function, written here as a function of t, into another function, written here as a function of s. We usually write $\mathcal{L}\{f(t)\} = F(s)$, using uppercase letters for Laplace transforms of functions denoted by lowercase letters.

Control engineers often refer to the *time domain*, the place where the original function $f(t)$ is defined, and the *s domain*, the place where its transform $F(s)$ is defined. They might say, "The time-domain function e^{at} behaves like $1/(s-a)$ in the *s* domain."

As another example of a transform of one function into another, consider the *area transform*

$$\mathcal{A}\{f\} = \int_0^s f(t)\,dt.$$

It is defined for all functions f that are nonnegative and integrable for $s \geq 0$. The transformed function $F(s) = \mathcal{A}\{f(t)\}$ is just the area under the graph of f between $t = 0$ and $t = s$; $\mathcal{A}$ transforms a given function into another function that provides the area beneath its graph.

Although the Laplace transform lacks an obvious physical interpretation such as area, like the area transform and like integration, it is linear.

10.1.2   Existence of the Laplace Transform

The Laplace transform is defined by an improper integral,

$$\mathcal{L}\{f\} = \int_0^\infty f(t)e^{-st}\,dt = \lim_{b \to \infty} \int_0^b f(t)e^{-st}\,dt.$$

The existence of the transform thus hinges on two points:

- The existence of the finite integral $\int_0^b f(t)e^{-st}\,dt$ for each $b > 0$.
- The existence of the limit as $b \to \infty$.

To accommodate the former, we require that f be **piecewise continuous**—i.e., that f be continuous on every finite interval $0 \leq t \leq b$ except perhaps at a finite number of *jump discontinuities*, where the limits of f from the left and right both exist but differ in value.

Figure 10.4 illustrates a piecewise-continuous function. The integral over any finite interval of such a function certainly exists; it is just the sum of the integrals of the continuous pieces. The function $f(t) = 1/(1-t)$ is *not* piecewise continuous because neither one-sided limit exists at $t = 1$.

The second condition, that the improper integral $\int_0^\infty f(t)e^{-st}\,dt$ exist, is assured if the integrand $f(t)e^{-st}$ decays to zero rapidly enough to ensure that the integral $\int_0^b f(t)e^{-st}\,dt$ approaches a finite limit for large b. To that end, we will require that f be of *exponential order*: f is of **exponential order** if for some constants $C, M > 0$ and for some $T \geq 0$,

Function of exponential order

$$|f(t)| \leq Ce^{Mt}$$

for all $t \geq T$.

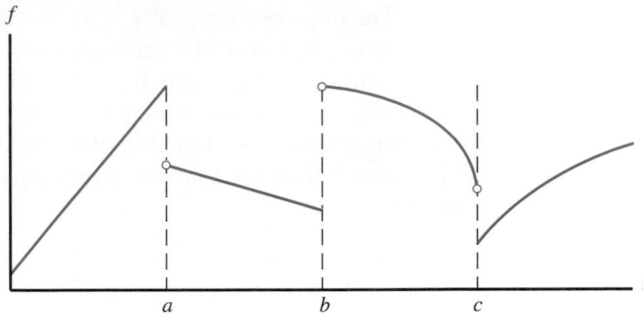

FIGURE 10.4 A piecewise-continuous function with jump discontinuities at $t = a, b, c$.

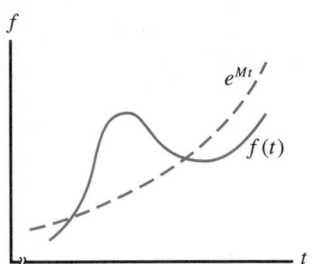

FIGURE 10.5 A function of exponential order

Figure 10.5 illustrates a function of exponential order. It is eventually bounded by e^{Mt}. If f is of exponential order, then the integrand $f(t)e^{-st}$ in the definition of the Laplace transform decays like $e^{-(s-M)t}$, provided $s > M$ and t is large enough. Hence, exponential order is a key to the existence of the Laplace transform.

Bounded functions are certainly of exponential order. On the other hand, e^{t^2} is not of exponential order because $e^{t^2} \leq Ce^{Mt}$ requires $t^2 \leq \ln C + Mt$. But that inequality fails because t^2 grows faster than t.

If f is of exponential order, we will show that the improper integral $\int_T^\infty |f(t)|\, dt$ exists, which guarantees that $\int_T^\infty f(t)\, dt$ converges. (The integral is said to **converge absolutely** in this case, analogous to the absolute convergence of series.)

These ideas lead to a precise statement of conditions for the existence of the Laplace transform.

Theorem 1. *If f is piecewise continuous and of exponential order, then the Laplace transform of f defined by (10.1) exists for s sufficiently large.*

Proof. Since f is of exponential order,

$$|f(t)| < e^{Mt}$$

for some $M > 0$ and $t \geq T$. For large values of b, we have

$$\left| \int_0^b f(t)e^{-st}\, dt \right| \leq \left| \int_0^T f(t)e^{-st}\, dt \right| + \left| \int_T^b f(t)e^{-st}\, dt \right|.$$

The first integral exists because its integrand is piecewise continuous, and its value is independent of b. For the second integral, the exponential bound on *f* gives

$$\left| \int_T^b f(t)e^{-st}\, dt \right| \le \int_T^b \left| f(t)e^{-st} \right|\, dt$$

$$\le \int_T^b \left| e^{-(s-M)t} \right|\, dt$$

$$= \left. \frac{e^{-(s-M)t}}{M-s} \right|_T^b$$

$$= \frac{e^{-(s-M)T} - e^{-(s-M)b}}{s-M}.$$

Requiring $s > M$, we find

$$\left| \int_T^\infty f(t)e^{-st}\, dt \right| = \lim_{b \to \infty} \left| \int_T^b f(t)e^{-st}\, dt \right|$$

$$\le \lim_{b \to \infty} \frac{e^{-(s-M)T} - e^{-(s-M)b}}{s-M}$$

$$= \frac{e^{-(s-M)T}}{s-M}.$$

Hence, if $s > M$, the improper integral defining the Laplace transform converges, and $\mathcal{L}\{f\}$ exists. ■

The Laplace transform is named for Pierre Simon de Laplace (1749-1827), a French mathematician and politician. (*Laplace* is pronounced La*plaz* as in *plaza*, not *place*.) His contributions to mathematics are marred by his failure to give credit to others and by his greed as a politician. Those faults may be balanced by the value of his own work and by the assistance he gave to the young Cauchy, who established calculus much as we study it today.

10.1.3 Exercises

EXERCISE GUIDE	
To gain experience ...	**Try exercises**
Applying the definition of Laplace transform	1–5, 18–23, 26–29, 32–33, 35
With the lincarity of transforms	1–5, 30–31, 36
Finding Laplace transforms of functions	1–17, 24–25, 34
Finding Laplace transforms of solutions of differential equations	14–17, 34
Using entries from table 10.1	1–13
Using entries from table 10.2	24–25

In exercises 1–5,

    (i) Use the definition (1) to find the Laplace transform of each of the following functions. Give the values of s for which each transform is defined.

    (ii) Use the linearity of the Laplace transform and the appropriate entries in table 10.1 to verify that the transform you obtained from the definition is correct. Indicate explicitly those steps which depend upon the linearity of the Laplace transform. Identify by number each entry from table 10.1 that you use.

    (iii) Compare each result with the transform expression given by DELAB.

1. $4 - 9e^{-4t}$

2. $3t^2$ (*Hint*: Integrate by parts.)

3. $7 + \pi$

4. $\cos \pi t$ (*Hint*: Integrate by parts twice as in example 4.)

5. $2t - 5$

In exercises 6–13, use the linearity of the Laplace transform and the appropriate entries in table 10.1 to find the Laplace transform of each of the following functions. Indicate explicitly those steps which depend upon the linearity of the Laplace transform. Identify by number each entry from table 10.1 that you use. Compare each result with the transform expression given by DELAB.

6. 14

7. $2t^3$

8. $14 - 2t^3$

9. $t + 2e^{2t}$

10. $5t^3 + 3 \sin 2t$

11. $A \sin \omega t$, A, ω constants

12. $2 \sin 4t - 9 \cosh 6t$

13. $\sum_{n=1}^{4} nt^n$

In exercises 14–17, assume that the Laplace transform of the solution of each initial-value problem exists.

    (i) Find the transform of the solution of the initial-value problem working directly from the differential equation.

    (ii) Solve each initial-value problem by a method of your choice. Compute the transform of the solution you find and show that it is the same as the one you obtained directly from the differential equation in part (i). Confirm both the solution and the transform using DELAB.

14. $y' - 4y = \sin 2t$, $y(0) = 3$

15. $2y' + 7y = 5t + e^{-2t}$, $y(0) = -1$

16. $x' - 2x = t^2 - 3 \cos \pi t$, $x(0) = 3$

17. $y' - ky = \alpha e^{-t}$, $y(0) = y_i$, k, α constants

In exercises 18–25, use the definition of Laplace transform to derive the indicated entry from table 10.1.

18. Entry 1; of what other entries is this a special case?

19. Entry 2 for $n = 1$

20. Entry 2 for $n = 2$

21. Entry 2 for n an arbitrary nonnegative integer (Use mathematical induction.)

22. Entry 5

23. Entry 6 (*Hint*: Use the definition of the hyperbolic sine function, entry 3 in table 10.1, and the linearity of the Laplace transform.)

24. Entry 7 (See the preceding hint.)

25. Find $\mathcal{L}\{\sinh at + \cosh at\}$ using entry 3.

In exercises 26–27, use the indicated entry from table 10.2 to find the Laplace transform of the given function. If the transform can be evaluated by a second method, do so to verify the accuracy of your answer. If a transform does not exist, explain why.

26. Use $\mathcal{L}\{f'(t)\} = sF(s) - f(0)$ with an appropriate choice of f to find the Laplace transform of

    (a) $4e^{4t}$

    (b) $3t^2$

    (c) $-a \cos at$

27. Use $\mathcal{L}\{f''(t)\} = s^2 F(s) - sf(0) - f'(0)$ with an appropriate choice of f to find the Laplace transform of

    (a) $-a^2 \sin at$

    (b) $a^2 e^{at}$

    (c) $6t$

In exercises 28–29, use the definition of Laplace transform to derive the indicated entry in table 10.2. In each case, $F(s) = \mathcal{L}\{f(t)\}$, and all transforms are assumed to exist.

28. $\mathcal{L}\{f'(t)\} = sF(s) - f(0)$ (*Hint*: Integrate the definition of Laplace transform by parts using $dv = f'(t)\,dt$.)

29. $\mathcal{L}\{f''(t)\} = s^2 F(s) - sf(0) - f'(0)$ (*Hint*: Apply the result of exercise 26 to the function $g'(t)$, where $g(t) = f'(t)$.)

In exercises 30–31, use the definition of Laplace transform to find the transform of the given function. Sketch the graph of each function before you attempt to find its transform.

30. $g(t) = \begin{cases} 4t, & 0 \le t \le 7 \\ 0, & 7 < t < \infty \end{cases}$

31. $g(t) = \begin{cases} 1, & 0 \le t \le 2\pi \\ \cos t, & 2\pi \le t \le 7\pi/2 \\ 0, & 7\pi/2 < t < \infty \end{cases}$

32. The text claims:

Because it is defined by an integral, the Laplace transform is a *linear* operator. That is, if $f(t)$, $g(t)$ are functions with Laplace transforms $F(s) = \mathcal{L}\{f\}$, $G(s) = \mathcal{L}\{g\}$, then

$$\mathcal{L}\{cf + dg\} = c\mathcal{L}\{f\} + d\mathcal{L}\{g\} = cF + dG$$

for any constants c, d.

Verify this claim for $f(t) = r(t)$, the ramp function of example 4, and $g(t) = e^{at}$, a constant. Use the definition of Laplace transform to compute $\mathcal{L}\{cf + dg\}$. Then show that it is identical to $c\mathcal{L}\{f\} + d\mathcal{L}\{g\}$.

33. Verify that the area transform $\mathcal{A}(f) = \int_0^s f(t)\,dt$ is linear. That is, show that if $F(s) = \mathcal{A}(f)$, $G(s) = \mathcal{A}(g)$, then

$$\mathcal{A}(cf + dg) = c\mathcal{A}(f) + d\mathcal{A}(g) = cF + dG$$

for any constants c, d. Does the Laplace transform have this same property?

34. In example 1, the text computes $\mathcal{L}\{e^{at}\}$ from its definition,

$$\mathcal{L}\{e^{at}\} = \int_0^\infty e^{at} e^{-st}\,dt$$

$$= \int_0^\infty e^{(a-s)t}\,dt$$

$$= \lim_{b \to \infty} \frac{e^{(a-s)t}}{a-s}\Big|_{t=0}^{t=b}$$

$$= \lim_{b \to \infty} \frac{e^{(a-s)b}}{a-s} - \frac{1}{a-s}$$

$$= \frac{1}{s-a},$$

but requires $s > a$. Why is this restriction necessary? Where is it first required?

35. Example 2 shows that for any differentiable function f,

$$\mathcal{L}\{f'\} = sF(s) - f(0).$$

Mimic the calculations of that example for $f(t) = e^{at}$, a constant, to show that

$$\mathcal{L}\{ae^{at}\} = s\mathcal{L}\{e^{at}\} - 1.$$

What is the source of the 1 at the end of the last equation? Use the linearity of the Laplace transform to write $\mathcal{L}\{ae^{at}\}$

in terms of $\mathcal{L}\{e^{at}\}$ to confirm that your calculations are correct. (Use

$$\mathcal{L}\{cf + dg\} = c\mathcal{L}\{f\} + d\mathcal{L}\{g\} = cF + dG$$

with $c = a$, $d = 0$. Is it true that you can "bring constants across the transform sign" as you can with integrals?)

36. By working directly with the differential equation itself, we found in example 3 that the Laplace transform $Y(s)$ of the solution of the initial-value problem

$$y' - ky = e^{-3t}, \quad y(0) = 4,$$

is

$$Y(s) = \frac{1}{(s+k)(s+3)} + \frac{4}{s+k}.$$

Find the solution $y(t)$ of this initial-value problem and verify that its Laplace transform is indeed the formula just given.

37. For $s > 0$ and a constant, example 4 finds $\mathcal{L}\{\sin at\}$ as follows:

$$\mathcal{L}\{\sin at\} = \int_0^\infty (\sin at)e^{-st}\,dt$$

$$= -\frac{(\sin at)e^{-st}}{s}\Big|_0^\infty + \frac{a}{s}\int_0^\infty (\cos at)e^{-st}\,dt$$

$$= -\frac{a(\cos at)e^{-st}}{s^2}\Big|_0^\infty - \frac{a^2}{s^2}\int_0^\infty (\sin at)e^{-st}\,dt$$

$$= \frac{a}{s^2} - \frac{a^2}{s^2}\int_0^\infty (\sin at)e^{-st}\,dt.$$

Solve for the integral $\int_0^\infty (\sin at)e^{-st}\,dt$ defining the transform to obtain

$$\mathcal{L}\{\sin at\} = \int_0^\infty (\sin at)e^{-st}\,dt$$

$$= \left(1 + \frac{a^2}{s^2}\right)^{-1} \frac{a}{s^2}$$

$$= \frac{a}{a^2 + s^2}.$$

Fill in the details of this calculation. Indicate clearly those steps that require the condition $s > 0$.

38. For any differentiable function f, define the derivative transform $\mathcal{D}\{f(t)\} = f'(t)$. Verify that this transform is linear. That is, show that

$$\mathcal{D}\{cf + dg\} = c\mathcal{D}\{g\} + d\mathcal{D}\{g\}$$

for any differentiable functions f, g and any constants c, d.

10.2 ■ INVERSE TRANSFORMS AND INITIAL-VALUE PROBLEMS

10.2.1 Inverse of the Laplace Transform

Example 3 of section 11.1 worked directly with the differential equation to find the transform of the solution of the initial-value problem

$$y' + ky = e^{-3t}, \quad y(0) = 4.$$

It assumed that the solution $y(t)$ did indeed have a transform, let $Y(s) = \mathcal{L}\{y(t)\}$, applied the transform operator $\mathcal{L}$ to the entire differential equation, and found that the transform of the solution of this initial-value problem is

$$Y(s) = \frac{1}{(s+k)(s+3)} + \frac{4}{s+k}.$$

We realized that if we could somehow invert the Laplace transform process—that is, if we could find the **inverse Laplace transform**

$$y(t) = \mathcal{L}^{-1}\{Y(s)\} = \mathcal{L}^{-1}\left\{ \frac{1}{(s+k)(s+3)} + \frac{4}{s+k} \right\},$$

—then we would have the solution of the initial-value problem.

> The inverse transform $\mathcal{L}^{-1}\{Y\} = y$ exists only if there is exactly one function $y(t)$ associated with each transform $Y(s)$. To recall an analogy, the function $f(x) = x^2$ has no inverse because for each (positive) value of f, we can choose $x = +\sqrt{f}$ or $x = -\sqrt{f}$. The inverse relation is not single valued, so there is no inverse function. Although we will not prove it here, the Laplace transform is better behaved: Under suitable conditions and in the appropriate setting, the inverse transform relation $Y(s) \to y(t)$ is single valued.

Although we have a straightforward definition for the Laplace transform itself in (10.1), we lack the mathematical tools to state a corresponding definition for the inverse transform. Instead, we must resort to a procedure much like antidifferentiation.

You are accustomed to inverting differentiation by evaluating an expression such as $\int g(t)\, dt$ to answer the question, What is a function whose derivative is g? The actual computations often require some deliberation and judgment, unlike differentiation, which usually entails no more than blind application of a formula.

Inverting Laplace transforms is much the same. We do have a formula (10.1) to compute the transform of a given function f. But evaluating an inverse transform $\mathcal{L}^{-1}\{F\}$ to answer the question, What is a function whose Laplace transform is F? requires careful thought and judgment. Just as a table of derivative formulas provides clues to antiderivatives, a list of simple Laplace transforms such as table 10.1 helps to invert transforms. Learning to invert Laplace transforms sharpens the skills of detection needed to determine the types of time domain behavior that are described by a given transform in the s domain.

At the end of the last section, we found the transform of the solution of the initial-value problem $y' + ky = e^{-3t}$, $y(0) = 4$. We begin with an example which inverts that transform.

■ **EXAMPLE 6**   *Find the inverse transform*

$$y(t) = \mathcal{L}^{-1}\left\{\frac{1}{(s+k)(s+3)} + \frac{4}{s+k}\right\}.$$

Since $\mathcal{L}$ is linear, $\mathcal{L}^{-1}$ is linear as well. The inverse transform can be computed term by term, and constant factors can be brought outside the operator $\mathcal{L}^{-1}$, much like an antiderivative,

$$y(t) = \mathcal{L}^{-1}\left\{\frac{1}{(s+k)(s+3)}\right\} + \mathcal{L}^{-1}\left\{\frac{4}{s+k}\right\}$$

$$= \mathcal{L}^{-1}\left\{\frac{1}{(s+k)(s+3)}\right\} + 4\mathcal{L}^{-1}\left\{\frac{1}{s+k}\right\}.$$

To invert the second expression, search table 10.1 for a transform that looks like $1/(s+k)$. Entry 3 with $-a = k$ is the obvious choice. Since

$$\mathcal{L}\left\{e^{-kt}\right\} = \frac{1}{s-(-k)} = \frac{1}{s+k}$$

we have

$$\mathcal{L}^{-1}\left\{\frac{1}{s+k}\right\} = e^{-kt}.$$

The first term in the inverse is more of a problem, for we find nothing in the right-hand column of table 10.1 of the form

$$\frac{1}{(s+k)(s+3)}.$$

But we could invert the factors $1/(s+k)$ and $1/(s+3)$ if they stood alone. (We just found e^{kt} for the inverse of the first, and the second is the same with $k = 3$.)

An expression such as $1/(s+k)(s+3)$ can be separated into its component fractions using *partial fractions*. (See appendix A.7.) For some unknown constants A and B, write

Partial-fraction decomposition

$$\frac{1}{(s+k)(s+3)} = \frac{A}{s+k} + \frac{B}{s+3}$$

$$= \frac{A(s+3) + B(s+k)}{(s+k)(s+3)}.$$

Since the first and last numerator must be identical, we have

$$1 = A(s + 3) + B(s + k).$$

Equations for A and B result from equating coefficients of like powers of s:

Equate powers of s.

$$s^0 : \quad 3A + kB = 1,$$
$$s^1 : \quad \quad A + B = 0.$$

Solving this pair of simultaneous equations yields

Solve for partial fraction numerators.

$$A = \frac{1}{3 - k}, \quad B = \frac{1}{k - 3}.$$

Using these values of A and B, the inverse transform is

$$\mathcal{L}^{-1}\left\{ \frac{1}{(s + k)(s + 3)} \right\} = A\mathcal{L}^{-1}\left\{ \frac{1}{s + k} \right\} + B\mathcal{L}^{-1}\left\{ \frac{1}{s + 3} \right\}$$

$$= Ae^{-kt} + Be^{-3t}$$

$$= \frac{e^{-kt} - e^{-3t}}{3 - k}. \tag{10.3}$$

Combining the inverse transforms of the two individual terms, we obtain the desired inverse transform. The solution of $y' + ky = e^{-3t}$, $y(0) = 4$, is

Inverse transform, the solution of the initial-value problem

$$y(t) = \mathcal{L}^{-1}\left\{ \frac{1}{(s + k)(s + 3)} \right\} + 4\mathcal{L}^{-1}\left\{ \frac{1}{s + k} \right\}$$

$$= \frac{e^{-kt}}{3 - k} + \frac{e^{-3t}}{k - 3} + 4e^{-kt}$$

$$= \frac{(4k - 13)e^{-kt} + e^{-3t}}{k - 3}, \quad k \neq 3. \quad ■$$

As this example illustrates, we can use Laplace transforms and their inverses to solve linear differential equations with constant coefficients. The steps are as follows:

LAPLACE TRANSFORM SOLUTION PROCESS

1. Transform the entire differential equation, using any initial values that are given. Solve for the transform of the solution.

2. Invert the transform to find the solution of the differential equation or initial-value problem.

Unfortunately, step 2 is easier said than done!

■ **EXAMPLE 7**   *Illustrate these steps by solving the initial-value problem*

$$y' + ky = e^{-3t}, \quad y(0) = 4.$$

The first step was completed in example 3 of the previous section, where from $\mathcal{L}\{y' + ky\} = \mathcal{L}\{e^{-3t}\}$ we found

Transform differential equation and solve for $Y(s)$.

$$Y(s) = \frac{1}{(s+k)(s+3)} + \frac{4}{s+k}.$$

Recall that the initial conditions were incorporated in the transform of the derivative term,

$$\mathcal{L}\{y'\} = sY(s) - y(0) = sY(s) - 4.$$

The second step was completed in the preceding example by inverting the transform to find

Invert $Y(s)$ to find the solution of the initial-value problem.

$$y(t) = \mathcal{L}^{-1}\{Y(s)\} = \mathcal{L}^{-1}\left\{\frac{1}{(s+k)(s+3)} + \frac{4}{s+k}\right\}$$

$$= \frac{(14k-13)e^{-kt} + e^{-3t}}{k-3}, \quad k \neq 0. \ ■$$

■ **EXAMPLE 8**   *Solve the initial-value problem*

$$x''(t) + 4x(t) = 0, \quad x(0) = 2, \quad x'(0) = -1.$$

Let $X(s) = \mathcal{L}\{x(t)\}$. Use entry 1 of table 10.2 and the initial conditions to transform the entire differential equation:

Transform differential equation.

$$\mathcal{L}\{x''(t) + 4x(t)\} = \mathcal{L}\{x''(t)\} + 4\mathcal{L}\{x(t)\}$$

$$= s^2 X(s) - sx(0) - x'(0) + 4X(s)$$

$$= s^2 X - 2s + 1 + 4X$$

$$= (s^2 + 4)X - 2s + 1 = 0.$$

Solve for X:

Solve for $X(s)$.

$$X(s) = \frac{2s-1}{s^2+4}.$$

Now determine the inverse transform,

Invert $X(s)$.

$$x(t) = \mathcal{L}^{-1}\{X(s)\} = \mathcal{L}^{-1}\left\{\frac{2s-1}{s^2+4}\right\}$$

$$= 2\mathcal{L}^{-1}\left\{\frac{s}{s^2+4}\right\} - \mathcal{L}^{-1}\left\{\frac{1}{s^2+4}\right\}.$$

With the aid of entries 5 and 6 in table 10.1, we find

$$\mathcal{L}^{-1}\left\{\frac{s}{s^2+4}\right\} = \cos 2t,$$

$$\mathcal{L}^{-1}\left\{\frac{1}{s^2+4}\right\} = \frac{1}{2}\mathcal{L}^{-1}\left\{\frac{2}{s^2+4}\right\} = \frac{1}{2}\sin 2t.$$

Combining these results gives the solution of the given initial-value problem,

Solution of initial-value problem

$$x(t) = 2\cos 4t - \frac{1}{2}\sin 2t. \quad ■$$

10.2.2 The Convolution Theorem

Some additional Laplace transform properties can ease the inversion in step 2 of the Laplace transform solution process. We develop two of those properties in the next few pages and illustrate their application to the solution of initial-value problems. (These and similar results are summarized in table 10.3.)

TABLE 10.3   **Useful properties of Laplace transforms—first revision.**

Properties of Laplace transforms

$$F(s) = \mathcal{L}\{f(t)\} = \int_0^\infty e^{-st} f(t)\,dt, \quad G(s) = \mathcal{L}\{g(t)\}$$

1. Transform of a derivative:

$$\mathcal{L}\{f'(t)\} = s F(s) - f(0),$$
$$\mathcal{L}\{f''(t)\} = s^2 F(s) - s f(0) - f'(0).$$

2. Convolution theorem:

$$\mathcal{L}\left\{\int_0^t f(r)g(t-r)\,dr\right\} = F(s)G(s).$$

3. Shifted transform:

$$\mathcal{L}\{e^{at} f(t)\} = F(s-a).$$

The motivation for the first property, *the convolution theorem*, is the problem encountered in the first example, finding the inverse transform of the product of two transforms. It involves the **convolution integral**, or simply *convolution*, of two functions f and g,

CONVOLUTION INTEGRAL

$$\int_0^t f(r)g(t-r)\,dr.$$

Theorem 2 (Convolution). *Let f and g be two functions of exponential order. Denote their Laplace transforms by*

$$F(s) = \mathcal{L}\{f(t)\}, \quad G(s) = \mathcal{L}\{g(t)\}.$$

Then the transform of the convolution of f and g is

Transform of convolution

$$\mathcal{L}\left\{\int_0^t f(r)g(t-r)\,dr\right\} = F(s)G(s).$$

Proof. The expression $F(s)G(s)$ is the product of two improper integrals:

$$F(s)G(s) = \left(\int_{r=0}^{\infty} e^{-sr} f(r)\,dr\right)\left(\int_{q=0}^{\infty} e^{-sq} g(q)\,dq\right).$$

(We have used r and q for the variables of integration rather than t to avoid confusion later.) Writing this product as an iterated double integral yields

$$F(s)G(s) = \int_{r=0}^{\infty} \int_{q=0}^{\infty} e^{-s(r+q)} g(q) f(r)\,dq\,dr.$$

Now change variables in the inner integral. Treating r as fixed, let $t = r + q$ to obtain

$$F(s)G(s) = \int_{r=0}^{\infty} \int_{t=r}^{\infty} e^{-st} g(t-r) f(r)\,dt\,dr.$$

This double integral covers the blue triangle in the rt-plane, as illustrated in the left-hand graph in figure 10.6. But that same triangle can be covered equally well by a double integral with the order of integration reversed, as the right-hand graph in the same figure shows:

$$F(s)G(s) = \int_{t=0}^{\infty} \int_{r=0}^{t} e^{-sr} g(t-r) f(r)\,dr\,dt$$

$$= \int_{t=0}^{\infty} e^{-sr} \left(\int_{r=0}^{t} g(t-r) f(r)\,dr\right) dt.$$

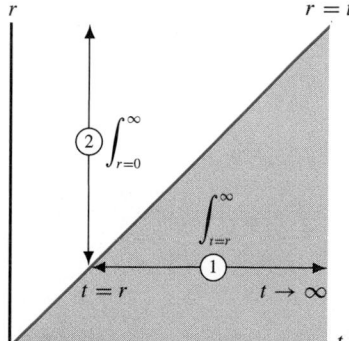 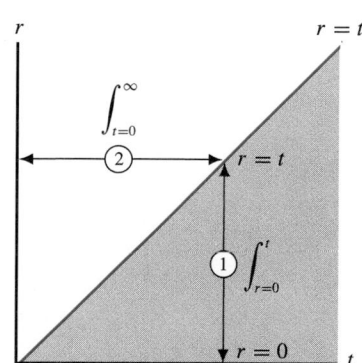

FIGURE 10.6 The double integrals $\int_{r=0}^{\infty}\int_{t=0}^{\infty}\cdots dt\,dr$ and $\int_{t=0}^{\infty}\int_{r=0}^{t}\cdots dr\,dt$ cover the same region in the rt-plane. The inner integral is numbered 1, and the outer integral is numbered 2 in each graph.

The inner integral is the convolution of f and g, and the outer integral is the definition of the Laplace transform. ■

Careful justification of reversing the order of integration of the improper integrals depends on the *absolute convergence* of these integrals, which is guaranteed by the hypothesis that f and g are of exponential order.

■ **EXAMPLE 9**   *Use the convolution theorem 2 to find the inverse transform*

$$\mathcal{L}^{-1}\left\{\frac{1}{(s+k)(s+3)}\right\},$$

which appears in example 6.

The given expression is the product of two transforms F and G:

$$F(s)G(s) = \frac{1}{(s+k)(s+3)}$$

with

$$F(s) = \frac{1}{s+k}, \quad G(s) = \frac{1}{s+3}.$$

From entry 3 of table 10.1, the inverses of these individual transforms are

$$f(t) = \mathcal{L}^{-1}\{F(s)\} = e^{-kt}, \quad g(t) = \mathcal{L}^{-1}\{G(s)\} = e^{-3t}.$$

From the convolution theorem, we have

$$\mathcal{L}^{-1}\{F(s)G(s)\} = \int_0^t f(r)g(t-r)\,dr,$$

or

$$\mathcal{L}^{-1}\left\{\frac{1}{(s+k)(s+3)}\right\} = \int_0^t e^{-kr}e^{-3(t-r)}\,dr$$

$$= e^{-3t}\int_0^t e^{(3-k)r}\,dr$$

$$= \frac{e^{-kt}-e^{-3t}}{3-k},$$

precisely as in equation (10.3) of example 6. ■

10.2.3 The Shifted Transform

Another useful transform property is the *shifted transform theorem*:

Theorem 3 (Shifted transform). *Let $f(t)$ be of exponential order, and let $F(s)$ denote its Laplace transform. Then for any real number a, the transform of $e^{at} f(t)$ exists, and*

Shifted transform
$$\mathcal{L}\left\{e^{at} f(t)\right\} = F(s - a).$$

Proof. If f is of exponential order, the product $e^{at} f(t)$ is, too, with constant $M + a$. Hence, $\mathcal{L}\left\{e^{at} f(t)\right\}$ exists.

Evaluating the transform requires only the identification of the transform variable in the definition of Laplace transform:

$$\mathcal{L}\left\{e^{at} f(t)\right\} = \int_0^\infty e^{-st} e^{at} f(t)\, dt$$

$$= \int_0^\infty e^{-(s-a)t} f(t)\, dt = F(s - a). \quad \blacksquare$$

■ **EXAMPLE 10**   *Use the shifted transform theorem to evaluate*

$$\mathcal{L}\left\{e^{-2t} \cos 6t\right\}.$$

To use the shifted transform theorem with $f(t) = \cos 6t$ and $a = -2$, note that $F(s) = \mathcal{L}\{\cos 6t\} = s/(s^2 + 36)$. Then

$$\mathcal{L}\left\{e^{6t} \cos 2t\right\} = F(s + 2) = \frac{s + 2}{(s + 2)^2 + 36} = \frac{s + 2}{s^2 + 4s + 40}. \qquad (10.4)$$

For future reference, note that we could complete the square in the final denominator $s^2 + 4s + 40$ to recover $(s + 2)^2 + 36$, from which we could identify $a = -2$ and the denominator $s^2 + 36$ of $F(s)$. ■

■ **EXAMPLE 11**   *Solve the initial-value problem*

$$x'' + 4x' + 40x = 0, \quad x(0) = 3, \quad x'(0) = 12.$$

Let $X(s) = \mathcal{L}\{x(t)\}$. Transform the differential equation,

Transform the differential equation.

$$\mathcal{L}\left\{x'' + 4x' + 40x\right\} = s^2 X - sx(0) - x'(0) + 4(sX - x(0)) + 40X$$
$$= (s^2 + 4s + 40)X - 3s - 24 = 0,$$

and solve for X,

Solve for $X(s)$.

$$X(s) = \frac{3s + 24}{s^2 + 4s + 40}.$$

Now seek the inverse transform

Invert $X(s)$.

$$x(t) = \mathcal{L}^{-1}\{X(s)\} = \mathcal{L}^{-1}\left\{\frac{3s + 24}{s^2 + 4s + 40}\right\}$$

$$= 3\mathcal{L}^{-1}\left\{\frac{s}{s^2 + 4s + 40}\right\} + 24\mathcal{L}^{-1}\left\{\frac{1}{s^2 + 4s + 40}\right\}. \tag{10.5}$$

The denominator here is the same as that of (10.4), a transform that was obtained with the shifted transform theorem.

To work backward to find the form of $F(s - a)$, rewrite the denominator by completing the square,

$$s^2 + 4s + 40 = \left(s^2 + 4s + \left(\frac{4}{2}\right)^2\right) - \left(\frac{4}{2}\right)^2 + 40$$

$$= (s^2 + 4s + 4) + 36$$

$$= (s + 2)^2 + 36.$$

The second term in (10.5), for example, is now,

$$\mathcal{L}^{-1}\left\{\frac{1}{s^2 + 4s + 40}\right\} = \mathcal{L}^{-1}\left\{\frac{1}{(s + 2)^2 + 36}\right\},$$

an expression dangerously close to a shifted sine transform, entry 4 in table 10.1, with $a = 6$. A simple manipulation even puts the correct number in the numerator:

$$\mathcal{L}^{-1}\left\{\frac{1}{(s + 2)^2 + 36}\right\} = \frac{1}{6}\mathcal{L}^{-1}\left\{\frac{6}{(s + 2)^2 + 36}\right\} = \frac{1}{6}F(s + 2),$$

with

$$F(s) = \frac{6}{s^2 + 36}.$$

Hence, we can use the shifted transform theorem with

$$f(t) = \mathcal{L}^{-1}\{F(s)\} = \sin 6t$$

to write

$$\mathcal{L}^{-1}\left\{\frac{1}{s^2 + 4s + 40}\right\} = \frac{1}{6}\mathcal{L}^{-1}\left\{\frac{6}{(s + 2)^2 + 36}\right\}$$

$$= \frac{1}{6}\mathcal{L}^{-1}\{F(s + 2)\}$$

$$= \frac{1}{6}e^{-2t}f(t) = \frac{1}{6}e^{-2t}\sin 6t. \tag{10.6}$$

The same logic can be applied to the first term in (10.5). We find

$$\mathcal{L}^{-1}\left\{\frac{s}{s^2 + 4s + 40}\right\} = \mathcal{L}^{-1}\left\{\frac{s}{(s + 2)^2 + 36}\right\}.$$

But to use the shifted transform in entry 3 of table 10.3, *all* appearances of the argument s must be shifted, *including the one in the numerator.* Adding and subtracting 2 in the numerator yields

$$\mathcal{L}^{-1}\left\{\frac{s}{s^2 + 4s + 40}\right\} = \mathcal{L}^{-1}\left\{\frac{s}{(s + 2)^2 + 36}\right\}$$

$$= \mathcal{L}^{-1}\left\{\frac{s + 2 - 2}{(s + 2)^2 + 36}\right\}$$

$$= \mathcal{L}^{-1}\left\{\frac{s + 2}{(s + 2)^2 + 36}\right\} - 2\mathcal{L}^{-1}\left\{\frac{1}{(x + 2)^2 + 36}\right\}$$

$$= \mathcal{L}^{-1}\left\{F(s + 2)\right\} - \frac{1}{3}e^{-2t}\sin 6t,$$

where

$$F(s) = \frac{s}{s^2 + 36}, \quad f(t) = \mathcal{L}^{-1}\left\{F(s)\right\} = \cos 6t.$$

(The $\sin 6t$ term comes from (10.6).)

Entry 3 of table 10.3 with $a = -2$ yields

$$\mathcal{L}^{-1}\left\{F(s + 2)\right\} = e^{-2t}f(t) = e^{-2t}\cos 6t.$$

Combining the expressions we have obtained for the two inverse transforms in (10.5) finally gives the solution

Solution of initial-value problem

$$x(t) = \mathcal{L}^{-1}\left\{X(s)\right\}$$

$$= 3\mathcal{L}^{-1}\left\{\frac{s}{s^2 + 4s + 40}\right\} + 24\mathcal{L}^{-1}\left\{\frac{1}{s^2 + 4s + 40}\right\}$$

$$= 3e^{-2t}\cos 6t + 3e^{-2t}\sin 6t. \quad ■$$

The examples in this section have decomposed quadratic denominators of transforms by partial fractions and by completing the square. Which technique is appropriate when?

USE PARTIAL FRACTIONS OR
COMPLETE THE SQUARE?

- Use partial fractions when the quadratic has real roots; the individual terms lead to exponentials in time. (Alternatively, apply the convolution theorem 2 to the product of the factors.)

- Complete the square when the quadratic has complex roots; the corresponding shifted transform leads to the product of an exponential and a sine or cosine.

Exercises 26 and 27 ask you to prove these claims.

More generally, if the denominator of a transform contains factors that can be identified, use partial fractions to decompose it into simpler expressions. The next example illustrates these ideas.

■ **EXAMPLE 12**   *Invert the transform*

$$Y(s) = \frac{s^3 - 33s^2 + 90s - 11}{(s^2 - 6s + 13)(s^2 + 4s - 12)}.$$

Since factors of the denominator are given, begin with a partial-fraction decomposition,

$$Y(s) = \frac{s^3 - 33s^2 + 90s - 11}{(s^2 - 6s + 13)(s^2 + 4s - 12)}$$

$$= \frac{As + B}{s^2 - 6s + 13} + \frac{Cs + D}{s^2 + 4s - 12},$$

where A, B, C, and D are constants to be determined. Finding a common denominator for the sum and equating coefficients of powers of s in the resulting numerator and the original numerator lead to

$$
\begin{array}{rl}
s^0: & -12B + 13D = -11 \\
s^1: & -12A + 4B + 13C - 6D = 90 \\
s^2: & 4A + B - 6C + D = -33 \\
s^3: & A + C = 1.
\end{array}
$$

Use the first and fourth equations to eliminate A and B from the middle two equations, and then solve that pair of equations for C and D. We obtain

$$A = -3, \quad B = 2, \quad C = 4, \quad D = 1.$$

Hence,

$$Y(s) = \frac{2 - 3s}{s^2 - 6s + 13} + \frac{4s + 1}{s^2 - 4s - 12},$$

and we must find two inverse transforms,

$$y_1(t) = \mathcal{L}^{-1}\left\{\frac{2 - 3s}{s^2 - 6s + 13}\right\}, \tag{10.7}$$

$$y_2(t) = \mathcal{L}^{-1}\left\{\frac{4s + 1}{s^2 - 4s - 12}\right\}. \tag{10.8}$$

The denominator $s^2 - 6s + 13$ of the first transform (10.7) has complex roots because the discriminant $6^2 - 4 \cdot 1 \cdot 13$ is negative. Consequently, we complete the square rather than factor:

$$s^2 - 6s + 13 = (s - 3)^2 + 4.$$

We have a transform $F(s - a)$ with $a = 3$, and we should prepare to use the shifted transform theorem, $\mathcal{L}\left\{e^{at} f(t)\right\} = F(s - a)$.

Write $F(s - 3)$ in terms of $s - 3$ in order to identify $F(s)$. We find

$$F(s - 3) = \frac{2 - 3s}{(s - 3)^2 + 4}$$

$$= \frac{-7 - 3(s - 3)}{(s - 3)^2 + 4}$$

$$= -\frac{7}{2} \frac{2}{(s - 3)^2 + 4} - 3 \frac{(s - 3)}{(s - 3)^2 + 4},$$

and

$$F(s) = -\frac{7}{2} \frac{2}{s^2 + 4} - 3 \frac{s}{s^2 + 4}.$$

Hence, the function f in $\mathcal{L}\left\{e^{at} f(t)\right\} = F(s - a)$ is

$$f(t) = \mathcal{L}^{-1}\{F(s)\} = -\frac{7}{2} \sin 2t - 3 \cos 2t.$$

Finally, the shifted transform theorem 3 yields

$$y_1(t) = e^{3t}\left(-\frac{7}{2} \sin 2t - 3 \cos 2t\right).$$

Since the denominator of the second transform (10.8) can be factored,

$$\frac{4s + 1}{s^2 + 4s - 12} = \frac{4s + 1}{(s - 2)(s + 6)},$$

we can either find a partial-fraction decomposition or use the convolution theorem 2. We choose partial fractions and seek constants A, B satisfying

$$\frac{4s + 1}{(s - 2)(s + 6)} = \frac{A}{s - 2} + \frac{B}{s + 6}.$$

Finding a common denominator and equating coefficients of powers of s lead to

$$\begin{aligned} s^0 &: \quad 6A - 2B = 1 \\ s^1 &: \quad A + B = 4. \end{aligned}$$

We have $A = 9/8$, $B = 23/8$, and

$$y_2(t) = \mathcal{L}^{-1}\left\{\frac{4s + 1}{(s - 2)(s + 6)}\right\}$$

$$= \mathcal{L}^{-1}\left\{\frac{9}{8}\frac{1}{s - 2} + \frac{23}{8}\frac{1}{s + 6}\right\}$$

$$= \frac{9}{8}e^{2t} + \frac{23}{8}e^{-6t}.$$

Combining the results for (10.7) and (10.8) yields the desired inverse transform,

$$\mathcal{L}^{-1}\{Y(s)\} = y_1(t) + y_2(t)$$

$$= e^{3t}\left(-\frac{7}{2}\sin 2t - 3\cos 2t\right) + \frac{9}{8}e^{2t} + \frac{23}{8}e^{-6t}. ■$$

10.2.4 Summary

To use Laplace transforms to solve a constant-coefficient, linear differential equation:

LAPLACE TRANSFORM
SOLUTION PROCESS

1. Transform the entire differential equation, using any initial values that are given. Solve for the transform of the solution.

2. Invert the transform to find the solution of the differential equation or the initial-value problem.

Two useful tools for finding inverse transforms are:

> **Convolution theorem**
>
> $$\mathcal{L}\left\{\int_0^t f(r)g(t-r)\,dr\right\} = F(s)G(s).$$

> **Shifted transform**
>
> $$\mathcal{L}\left\{e^{at}f(t)\right\} = F(s-a).$$

From the point of view of identifying time-domain behavior from a given Laplace transform, note that

- *products of transforms* are markers of *convolutions* of time-domain functions,
- *shifted transforms* are markers of *exponential factors* in the time domain.

10.2.5 Exercises

EXERCISE GUIDE	
To gain experience . . .	**Try exercises**
Using the convolution theorem	1–6, 25, 27
Using partial fractions	1–6, 27
With shifted transforms	8, 10–11, 13, 26, 28
Finding inverse transforms	1–12

EXERCISE GUIDE (Continued)	
To gain experience . . .	**Try exercises**
Finding Laplace transforms of functions	13, 23–24(c)
Solving initial-value problems	14–24
With the fundamental ideas of inverse transforms	26, 27

In exercises 1–6,

    (i) Write the given transform as an appropriate product and use the convolution theorem to find its inverse.

    (ii) Use partial fractions to write the given transform as an appropriate sum and find its inverse.

    (iii) Confirm the inverse result using DELAB.

1. $\dfrac{4}{s(s+4)}$

2. $\dfrac{3}{(s-2)(s+3)}$

3. $\dfrac{1}{s^2-16}$

4. $\dfrac{s}{(s+1)(s^2+4)}$

5. $\dfrac{s}{(s-1)(s^2-4)}$

6. $\dfrac{1}{s^2+9}$ (*Hint:* Use $s^2+9 = (s+3i)(s-3i)$ and $e^{i3t} = \cos 3t + i\sin 3t$.)

In exercises 7–12, find the inverse of the given transform. Confirm the inverse using DELAB.

7. $\dfrac{6}{s^2-5s+4}$

8. $\dfrac{3s}{2s^2-3s+5}$

9. $\dfrac{s-1}{3s^2-15s+12}$

10. $\dfrac{2s+3}{s^2-2s+2}$

11. $\dfrac{3-2s}{s^2-2s+5}$

12. $\dfrac{2+s}{s^2+4s+1}$

13. Use $\mathcal{L}\left\{e^{at}f(t)\right\} = F(s-a)$ with an appropriate choice of f to find the Laplace transform of each of the following.

    (a) $e^{3t}t$

    (b) $e^{-4t}\cos \pi t$

    (c) $e^{-3t/2}(t^2+4)$

In exercises 14–21,

    (i) Working directly from the differential equation, find the transform of the solution of the given initial-value problem.

    (ii) Invert that transform to find the solution of the initial-value problem.

    (iii) Find the solution of the initial-value problem by another method and verify that the solution obtained by Laplace transforms is correct.

    (iv) Use DELAB to confirm both the transform and the solution result.

14. $y' - 6y = 3$, $y(0) = 2$

15. $y' - 6y = \sin 3t$, $y(0) = 5$

16. $y' + 4y = \cos \pi t$, $y(0) = 0$

17. $2y' + 8y = 6e^{-3t}$, $y(0) = -2$

18. $3x''/16 + 1.41x' + 2x = 0$, $x(0) = 0.75$, $x'(0) = 0$ (See example 22 of section 6.4.)

19. $x'' + 4x = \cos \pi t$, $x(0) = 1$, $x'(0) = 4$

20. $x'' + 4x = \sin 2t$, $x(0) = 1$, $x'(0) = 4$

21. $x'' - 4x' + 4x = 0$, $x(0) = 2$, $x'(0) = 2$

22. Example 7 uses Laplace transforms to solve the initial-value problem

$$y' + ky = e^{-3t}, \quad y(0) = 4.$$

It finds

$$y(t) = \frac{(14k-13)e^{-kt} + e^{-3t}}{k-3}.$$

    (a) Verify by direct substitution that this solution is correct.

(b) Solve this initial-value problem by another method and show that you obtain the same solution.

(c) Obviously, this solution is valid only for $k \neq 3$. Other than avoiding dividing by zero in this formula, why must this particular value of k be singled out?

(d) Use Laplace transforms to solve this initial-value problem when $k = 3$.

23. Example 8 uses Laplace transforms to find the solution

$$x(t) = 2 \cos 4t - \tfrac{1}{2} \sin 2t$$

of the initial-value problem

$$x''(t) + 4x(t) = 0, \quad x(0) = 2, \quad x'(0) = -1.$$

(a) Verify by direct substitution that this solution is correct.

(b) Solve this initial-value problem by another method and show that you obtain the same solution.

(c) Find the Laplace transform of the solution you found in part (b). Show that it agrees with the transform of the solution obtained in example 8 directly from the initial-value problem itself,

$$X(s) = \frac{2s - 1}{s^2 + 4}.$$

24. For any differentiable function f, define the derivative transform $\mathcal{D}\{f(t)\} = f'(t)$. Show that this transform does *not* have an inverse. That is, show that

$$\mathcal{D}\{f\} = \mathcal{D}\{g\}$$

for two differentiable functions f, g with $f \neq g$.

25. Example 11 obtains the solution

$$x(t) = 3e^{-2t} \cos 6t + 4e^{-2t} \sin 6t$$

of the initial-value problem

$$x'' + 4x' + 40x = 0, \quad x(0) = 3, \quad x'(0) = 12.$$

(a) Verify by direct substitution that this solution is correct.

(b) Solve this initial-value problem by another method and show that you obtain the same solution.

(c) Find the Laplace transform of the solution you found in part (b). Show that it agrees with the transform of the solution obtained in example 11 directly from the initial-value problem itself,

$$X(s) = \frac{3s + 24}{s^2 + 4s + 40}.$$

26. Verify the following claim made in the text:

Complete the square when the quadratic has complex roots; the corresponding shifted transform leads to the product of an exponential and a sine or cosine.

(a) Consider

$$Y(s) = \frac{1}{s^2 + c_1 s + c_0}.$$

What conditions on c_1, c_0 guarantee that the quadratic denominator has complex roots? Denote those roots by $r_1, r_2 = \alpha \pm i\beta$.

(b) Show that completing the square leads to

$$s^2 + c_1 s + c_0 = (s - \alpha)^2 + \beta^2.$$

(c) Apply the shifted transform theorem to

$$Y(s) = \frac{1}{(s - \alpha)^2 + \beta^2}$$

to find $\mathcal{L}^{-1}\{Y(s)\}$.

27. Verify the following claim made in the text:

Use partial fractions when a quadratic denominator has real roots; the individual terms lead to exponentials in time. (Alternatively, apply the convolution theorem 2 to the product of the factors.)

(a) Consider

$$Y(s) = \frac{1}{s^2 + c_1 s + c_0}.$$

What conditions on c_1, c_0 guarantee that the quadratic denominator has real roots? Denote the real roots by r_1, r_2: $s^2 + c_1 s + c_0 = (s - r_1)(s - r_2)$.

(b) Use a partial-fraction expansion

$$\frac{1}{s^2 + c_1 s + c_0} = \frac{A}{s - r_1} + \frac{B}{s - r_2}$$

to find $\mathcal{L}^{-1}\{Y(s)\}$.

(c) Apply the convolution theorem to

$$Y(s) = \left(\frac{1}{s - r_1}\right)\left(\frac{1}{s - r_2}\right)$$

to find $\mathcal{L}^{-1}\{Y(s)\}$.

10.3 ■ OTHER PROPERTIES OF LAPLACE TRANSFORMS

This section lists some additional properties of Laplace transforms, supplementing those in table 10.3, page 524. The indicated exercises guide your derivation of each of the formulas given.

10.3.1 Transform of a Periodic Function

Exercise 23 asks you to prove that if $f(t)$ is a *periodic function with period* T, then

---

Transform of a periodic function

$$\mathcal{L}\{f\} = \frac{1}{1 - e^{-st}} \int_0^T e^{-st} f(t)\, dt.$$

■ **EXAMPLE 13**   *Find the Laplace transform of the square wave of period* T,

$$(t) = \begin{cases} a, & 0 \le t < T/2, \\ -a, & T/2 \le t < T, \end{cases}$$

shown in figure 10.7.

Using the definition of this function, we have

$$\int_0^T e^{-st} f(t)\, dt = \int_0^{T/2} ae^{-st}\, dt - \int_{T/2}^T ae^{-st}\, dt$$

$$= -\frac{a}{s} e^{-st}\Big|_{t=0}^{t=T/2} + \frac{a}{s} e^{-st}\Big|_{t=T/2}^{t=T}$$

$$= \frac{a}{s}\left(e^{-sT} - 2e^{-sT/2} + 1\right)$$

$$= \frac{a}{s}\left(e^{-sT/2} - 1\right)^2.$$

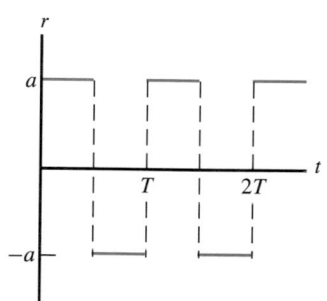

FIGURE 10.7   A square wave with amplitude a and period T.

Since $1 - e^{-sT} = (1 - e^{-sT/2})(1 + e^{-sT/2})$, the transform of the square wave is

$$\mathcal{L}\{f(t)\} = \frac{a(1 - e^{-sT/2})}{s(1 + e^{-sT/2})}. \quad ■$$

■ **EXAMPLE 14**   *Find the Laplace transform of the solution of the initial-value problem*

$$x'' + 4\pi^2 x = f(t), \quad x(0) = 0, \quad x'(0) = -2,$$

when f is a square wave of period 1,

$$f(t) = \begin{cases} a, & 0 \le t < \frac{1}{2}, \\ -a, & \frac{1}{2} \le t < 1. \end{cases} \tag{10.9}$$

From the results of the preceding example, the transform of the forcing term is

Transform of forcing term

$$F(s) = \mathcal{L}\{f(t)\} = \frac{a(1 - e^{-s/2})}{s(1 + e^{-s/2})}.$$

Letting $X(s) = \mathcal{L}\{x(t)\}$ and transforming the differential equation in the usual way yield

Transform the differential equation.

$$s^2 \left[s^2 X(s) - sx(0) - x'(0) \right] + 4\pi^2 X = F(s).$$

Using the initial conditions and solving for X, we find the transform of the solution of this initial-value problem,

Solve for $X(s)$.

$$X(s) = \frac{-2}{s^2 + 4\pi^2} + \frac{a(1 - e^{-s/2})}{s(1 + e^{-s/2})(s^2 + 4\pi^2)}. \quad ■$$

■ **EXAMPLE 15**   *Invert the transform of the previous example and find the solution of the initial-value problem given there.*

Write that transform as $X(s) = X_1(s) + X_2(s)$, where

$$X_1(s) = \frac{-2}{s^2 + 4\pi^2}, \quad X_2(s) = \frac{a(1 - e^{-s/2})}{s(1 + e^{-s/2})(s^2 + 4\pi^2)}.$$

The first term is not difficult to invert:

$$\mathcal{L}^{-1}\{X_1(s)\} = \mathcal{L}^{-1}\left\{ \frac{-2}{s^2 + 4\pi^2} \right\} = -\frac{2}{2\pi} \mathcal{L}^{-1}\left\{ \frac{2\pi}{s^2 + 4\pi^2} \right\} = -\frac{1}{\pi} \sin 2\pi t.$$

We recognize that the second term is the product of two terms whose inverse transforms we know,

$$X_2(s) = \left(\frac{a(1 - e^{-s/2})}{s(1 + e^{-s/2})} \right) \left(\frac{1}{2\pi} \frac{2\pi}{s^2 + 4\pi^2} \right),$$

an observation that suggests using the convolution theorem, entry 2 of table 10.3. Since the first factor is the transform of the square wave $f(t)$ defined in (10.9) and the second is the transform of $(\sin 2\pi t)/2\pi$, we can write $X_2(s)$ as the convolution integral

$$X_2(s) = \mathcal{L}\left\{ \frac{1}{2\pi} \int_0^t f(r) \sin 2\pi (t - r) \, dr \right\}.$$

To evaluate this convolution integral, suppose $n \le t < n + 1$. Then we can write the integral $\int_0^t \cdots dt$ as a sum of integrals over the common period of f and the sine function:

$$\int_0^t f(r) \sin 2\pi(t - r)\, dr$$

$$= \int_0^1 f(r) \sin 2\pi(t - r)\, dr + \int_1^2 f(r) \sin 2\pi(t - r)\, dr + \cdots$$

$$+ \int_{n-1}^n f(r) \sin 2\pi(t - r)\, dr + \int_n^t f(r) \sin 2\pi(t - r)\, dr. \quad (10.10)$$

Because $f(r)$ and $\sin 2\pi(t - r)$ are both periodic in r with period 1, the integrals with integer limits are all equal. The sum collapses:

$$\int_0^t f(r) \sin 2\pi(t - r)\, dr$$

$$= n \int_0^1 f(r) \sin 2\pi(t - r)\, dr + \int_n^t f(r) \sin 2\pi(t - r)\, dr.$$

Now expand $\sin 2\pi(t - r)$ using the formula for the sine of a difference of angles and recall the definition (10.9) of $f(t)$:

$$\int_0^1 f(r) \sin 2\pi(t - r)\, dr$$

$$= \sin 2\pi t \int_0^1 f(r) \cos 2\pi r\, dr - \cos 2\pi t \int_0^1 f(r) \sin 2\pi r\, dr$$

$$= a \sin 2\pi t \left(\int_0^{1/2} \cos 2\pi r\, dr - \int_{1/2}^1 \cos 2\pi r\, dr \right)$$

$$- a \cos 2\pi t \left(\int_0^{1/2} \sin 2\pi r\, dr - \int_{1/2}^1 \sin 2\pi r\, dr \right)$$

$$= \frac{a \sin 2\pi t}{2\pi} \left(\sin 2\pi r \big|_0^{1/2} - \sin 2\pi r \big|_{1/2}^1 \right)$$

$$+ \frac{a \cos 2\pi t}{2\pi} \left(\cos 2\pi r \big|_0^{1/2} - \cos 2\pi r \big|_{1/2}^1 \right)$$

$$= -\frac{2a \cos 2\pi t}{\pi}.$$

The last integral in (10.10) can be treated similarly. If $n \le t < n + \frac{1}{2}$, then

$$\int_n^t f(r) \sin 2\pi (t - r) \, dr$$

$$= \sin 2\pi t \int_n^t f(r) \cos 2\pi r \, dr - \cos 2\pi t \int_n^t f(r) \sin 2\pi r \, dr$$

$$= a \sin 2\pi t \int_n^t \cos 2\pi r \, dr - a \cos 2\pi t \int_n^t \sin 2\pi r \, dr$$

$$= \frac{a \sin 2\pi t}{2\pi} \left(\sin 2\pi r \big|_n^t \right) + \frac{a \cos 2\pi t}{2\pi} \left(\cos 2\pi r \big|_n^t \right)$$

$$= \frac{a}{2\pi} \left(\sin^2 2\pi t + \cos^2 2\pi t - \cos 2\pi t \right)$$

$$= \frac{a(1 - \cos 2\pi t)}{2\pi}.$$

Likewise, if $n + \frac{1}{2} \le t < n + 1$, then

$$\int_n^t f(r) \sin 2\pi (t - r) \, dr$$

$$= \sin 2\pi t \int_n^t f(r) \cos 2\pi r \, dr - \cos 2\pi t \int_n^t f(r) \sin 2\pi r \, dr$$

$$= a \sin 2\pi t \left(\int_n^{n+1/2} \cos 2\pi r \, dr - \int_{n+1/2}^t \cos 2\pi r \, dr \right)$$

$$\quad - a \cos 2\pi t \left(\int_n^{n+1/2} \sin 2\pi r \, dr - \int_{n+1/2}^t \sin 2\pi r \, dr \right)$$

$$= \frac{a \sin 2\pi t}{2\pi} \left(\sin 2\pi r \big|_n^{n+1/2} - \sin 2\pi r \big|_{n+1/2}^t \right)$$

$$\quad + \frac{a \cos 2\pi t}{2\pi} \left(\cos 2\pi r \big|_n^{n+1/2} - \cos 2\pi r \big|_{n+1/2}^t \right)$$

$$= -\frac{a}{2\pi} \left(\sin^2 2\pi t + \cos^2 2\pi t + 3 \cos 2\pi t \right)$$

$$= -\frac{a(1 + 3 \cos 2\pi t)}{2\pi}.$$

Collecting the results of this analysis gives the solution of the initial-value problem:

$$x(t) = -\frac{\sin 2\pi t + 2na \cos 2\pi t}{\pi} + \frac{a(1 - \cos 2\pi t}{2\pi}$$

for $n \le t < n + \frac{1}{2}$, and

$$x(t) = -\frac{\sin 2\pi t + 2na \cos 2\pi t}{\pi} - \frac{a(1 + 3 \cos 2\pi t)}{2\pi}$$

for $n + \frac{1}{2} \le t < n + 1$. Note the presence of the term proportional to n; the amplitude of this solution is growing with time. Could this system be in resonance? ■

10.3.2 Derivative of a Transform

Exercises 24 and 25 ask you to prove the general result

Derivative of a transform

$$\mathcal{L}\left\{(-t)^n f(t)\right\} = F^{(n)}(s), \quad n = 1, 2, \ldots .$$

---

and the special case $n = 1$,

First derivative of a transform

$$\mathcal{L}\{-tf(t)\} = F'(s).$$

■ **EXAMPLE 16** *Use the derivative of a transform formula to find $\mathcal{L}\left\{te^{at}\right\}$.*

The expression te^{at} has the form $-tf(t)$ with $f(t) = -e^{at}$. Using $f(t) = -e^{at}$, we find

$$F(s) = \mathcal{L}\left\{-e^{at}\right\} = -\mathcal{L}\left\{e^{at}\right\} = \frac{-1}{s-a}, \quad s > a.$$

To apply $\mathcal{L}\{-tf(t)\} = F'(s)$, we compute

$$F'(s) = \frac{d}{ds}\left(\frac{-1}{s-a}\right) = \frac{1}{(s-a)^2}.$$

Hence, $\mathcal{L}\left\{te^{at}\right\} = 1/(s-a)^2, s > a$. ■

Alternatively, we could use the shifted transform theorem, entry 3 of table 10.3,

$$\mathcal{L}\left\{e^{at}t\right\} = F(s-a),$$

with $f(t) = t$. From

$$F(s) = \mathcal{L}\{t\} = \frac{1}{s^2},$$

we find

$$\mathcal{L}\left\{e^{at}t\right\} = F(s-a) = \frac{1}{(s-a)^2}.$$

■ **EXAMPLE 17**   *Use the derivative of a transform formula to argue that the undamped system,*

$$x'' + \omega^2 x = A \cos \omega t, \quad x(0) = x_i, \quad x'(0) = v_i,$$

is in resonance; i.e., show that the solution of this system contains at least one term whose amplitude grows with t.

The transform of the solution of this initial-value problem is

$$X(s) = \frac{s x_i + v_i}{s^2 + \omega^2} + \frac{As}{(s^2 + \omega^2)^2}.$$

Stop and Think   **10.8** Derive $X(s)$.

The second term can be written as the derivative of a known transform:

$$\frac{As}{(s^2 + \omega^2)^2} = -\frac{A}{2\omega} \frac{d}{ds} \left(\frac{\omega}{s^2 + \omega^2} \right) = -\frac{A}{2\omega} \mathcal{L}\{-t \sin \omega t\}.$$

Hence, the solution contains a term whose amplitude grows with time; the system is in resonance. ■

10.3.3 Transform of an Integral

Exercise 26 asks you to prove

Transform of an integral

$$\mathcal{L}\left\{ \int_0^t f(r)\, dr \right\} = \frac{F(s)}{s}.$$

■ **EXAMPLE 18**   *Use the transform of an integral formula to invert*

$$Y(s) = \frac{1}{s(s^2 + 4)}.$$

We recognize that we can write Y as a known transform divided by s:

$$Y(s) = \frac{F(s)}{s},$$

where

$$F(s) = \frac{1}{s^2 + 4} = \frac{1}{2} \mathcal{L}\{\sin 2t\}.$$

With $f(t) = (\sin 2t)/2$, the transform of an integral formula yields

$$Y(s) = \mathcal{L}\left\{ \frac{1}{2} \int_0^t \sin 2t\, dt \right\} = \mathcal{L}\left\{ \frac{1}{4}(\cos 2t - 1) \right\}. \quad ■$$

■ **EXAMPLE 19**   *Given that* $\mathcal{L}\left\{t^k\right\} = k!/s^{k+1}$, $s > 0$, *for some positive integer k, show that*

$$\mathcal{L}\left\{t^{k+1}\right\} = \frac{(k+1)!}{s^{k+2}}.$$

(This is the key step in a proof by mathematical induction of entry 2 of table 10.1.)

Apply the transform of an integral formula with $f(t) = t^k$. Since

$$\int_0^t t^k \, dt = \frac{t^{k+1}}{k+1}$$

and

$$F(s) = \mathcal{L}\left\{t^k\right\} = \frac{k!}{s^{k+1}},$$

we have

$$\mathcal{L}\left\{\frac{t^{k+1}}{k+1}\right\} = \mathcal{L}\left\{\int_0^t t^k \, dt\right\} = \frac{F(s)}{s} = \frac{k!}{s^{k+2}}.$$

Multiplying by $k + 1$ completes the argument. ■

10.3.4 Integral of a Transform

Exercise 27 asks you to prove that if $f(t)/t$ is defined at $t = 0$, then

---

Integral of a transform

$$\mathcal{L}\left\{\frac{f(t)}{t}\right\} = \int_s^\infty F(r) \, dr.$$

■ **EXAMPLE 20**   *Argue that the function*

$$g(t) = \frac{\sin at}{t}$$

can be defined at $t = 0$, and use the integral of a transform formula to find its Laplace transform.

Define $g(0)$ by its limiting value:

$$g(0) \equiv \lim_{t \to 0} \frac{\sin at}{t} = a.$$

Then g is continuous at $t = 0$ and, hence, integrable for $0 \le t < \infty$. Certainly, the bounded function g is of exponential order.

To compute the transform whose existence we have just established, use the integral of a transform formula with $f(t) = \sin at$. Since $F(s) = \mathcal{L}\{\sin at\} = a/(s^2 + a^2)$, we have

$$\mathcal{L}\{g(t)\} = \mathcal{L}\left\{\frac{\sin at}{t}\right\}$$

$$= \int_s^\infty \frac{a}{r^2 + a^2}\, dr$$

$$= \tan^{-1}\frac{r}{a}\Big|_{r=s}^{r=\infty}$$

$$= \frac{\pi}{a} - \tan^{-1}\left(\frac{s}{a}\right),$$

the desired transform. ■

■ **EXAMPLE 21**   *Find*

$$\mathcal{L}\left\{\frac{4\cos at}{t}\right\}.$$

The expression $4\cos at/t$ has the form $f(t)/t$ with $f(t) = 4\cos at$, and

$$F(s) = \mathcal{L}\{4\cos at\} = \frac{4s}{s^2 + a^2}, \quad s > 0.$$

To use the integral of a transform formula above, compute

$$\int_s^\infty F(s)\, ds = \int_s^\infty \frac{4s}{s^2 + a^2}\, ds = 2\ln\left(s^2 + a^2\right)\Big|_s^\infty.$$

But when we evaluate this last expression, we find that it diverges to infinity as $s \to \infty$. The Laplace transform of $4\cos at/t$ does not exist!

To explain this difficulty, note that the integrand

$$\frac{4\cos at}{t}e^{-st}$$

appearing in the definition of this Laplace transform is not defined at the lower limit of integration $t = 0$. Indeed, near $t = 0$ this expression is behaving like $1/t$, and $1/t$ can not be integrated at $t = 0$. This Laplace transform does not exist because the integral (10.1) required for its definition does not exist, even for finite values of the upper limit b. ■

Table 10.4 provides an enlarged list of useful properties of the Laplace transform.

TABLE 10.4   **Useful properties of Laplace transforms–second revision.**

Properties of Laplace transforms

$$F(s) = \mathcal{L}\{f(t)\} = \int_0^\infty e^{-st} f(t)\, dt, \quad G(s) = \mathcal{L}\{g(t)\}$$

1. Transform of a derivative:

$$\mathcal{L}\{f'(t)\} = s F(s) - f(0),$$

$$\mathcal{L}\{f''(t)\} = s^2 F(s) - s f(0) - f'(0).$$

2. Convolution theorem:

$$\mathcal{L}\left\{ \int_0^t f(r) g(t - r)\, dr \right\} = F(s) G(s).$$

3. Shifted transform:

$$\mathcal{L}\{e^{at} f(t)\} = F(s - a).$$

4. Transform of a periodic function of period T:

$$\mathcal{L}\{f\} = \frac{1}{1 - e^{-sT}} \int_0^T e^{-st} f(t)\, dt.$$

5. Derivative of a transform:

$$\mathcal{L}\{-t f(t)\} = F'(s),$$

$$\mathcal{L}\{(-t)^n f(t)\} = F^{(n)}(s), \quad n = 1, 2, \ldots.$$

6. Transform of an integral:

$$\mathcal{L}\left\{ \int_0^t f(r)\, dr \right\} = \frac{F(s)}{s}.$$

7. Integral of a transform:

$$\mathcal{L}\left\{ \frac{f(t)}{t} \right\} = \int_s^\infty F(r)\, dr.$$

10.3.5 Exercises

EXERCISE GUIDE	
To gain experience . . .	**Try exercises**
Finding Laplace transforms	1–5, 9–15, 21
Finding inverse transforms	6–8
With transforms of periodic functions	1, 9–10, 19, 23
With derivatives of transforms	2–3, 6, 11–12, 14, 22, 24–25
With transforms of integrals	4, 7, 26
With integrals of transforms	5, 8, 18, 27
Using the convolution theorem	22

In exercises 1–5, use the indicated transform property to evaluate the Laplace transform of the given function. If the transform can be evaluated by a second method, do so to verify your answer. If a transform does not exist, explain why. When possible, compare your result with that of DELAB.

1. Use

$$\mathcal{L}\{f\} = \frac{1}{1 - e^{-sT}} \int_0^T e^{-st} f(t)\, dt,$$

where f has period T, to find the Laplace transform of

(a) $\sin \pi t$

(b) The square wave of period 4, which is defined for $0 \le t \le 4$ by

$$g(t) = \begin{cases} 1, & 0 \le t \le 2, \\ -1, & 2 < t \le 4. \end{cases}$$

2. Use $\mathcal{L}\{-tf(t)\} = F'(s)$ to find the Laplace transform of

(a) t^2 with $f(t) = -t$

(b) t^3 with $f(t) = -t^2$

(c) $t \cos 2t$

3. Use $\mathcal{L}\{(-t)^n f(t)\} = F^{(n)}(s)$ with an appropriate choice of f to find the Laplace transform of

(a) $t^2 e^{at}$

(b) $2t^3 \cos \pi t$

(c) $5t^2 \sinh at$

4. Use $\mathcal{L}\left\{ \int_0^t f(r)\, dr \right\} = F(s)/s$ to find the Laplace transform of

(a) $\dfrac{\sin \pi t}{\pi}$ with $f(t) = \cos \pi t$

(b) $\dfrac{e^{at}}{a}$ with $f(t) = e^{at}$

(c) $\dfrac{t^5}{5}$ with $f(t) = t^4$

5. Use $\mathcal{L}\{f(t)/t\} = \int_s^\infty F(s)\, ds$ with an appropriate choice of f to find the Laplace transform of

(a) $t^{-1} \sinh 2t$

(b) $t^{-1} \sin 2t$

(c) $\dfrac{1 - \cos t}{t}$

In exercises 6–8, use the indicated transform property (and other properties as needed) to find the inverse of the given transform. When possible, compare your result with that of DELAB.

6. Use $\mathcal{L}\{-tf(t)\} = F'(s)$ to invert the Laplace transforms

(a) $\dfrac{4}{(s - 1)^2}$

(b) $\dfrac{2s}{(s^2 + 16)^2}$

(c) $\dfrac{3s}{(1 + s^2)^3}$

7. Use $\mathcal{L}\left\{ \int_0^t f(r)\, dr \right\} = F(s)/s$ to invert the Laplace transforms

(a) $\dfrac{1}{s(s - 1)}$

(b) $\dfrac{6(s - 1)}{s^2(s + 2)}$

8. Use $\mathcal{L}\{f(t)/t\} = \int_s^\infty F(s)\, ds$ with an appropriate choice of f to invert the Laplace transforms

(a) $\dfrac{2}{s - 2}$

(b) $\dfrac{4s}{(s^2 + a^2)^2}$

9. Sketch a graph of the nonsymmetric square wave of period T

$$f(t) = \begin{cases} a, & 0 \le t < b, \\ -a, & b \le t < T, \end{cases}$$

and find its Laplace transform.

10. Sketch a graph of the sawtooth wave of period T

$$f(t) = -a + \frac{2at}{T}, \quad 0 \le t < T,$$

and find its Laplace transform.

In exercises 11–15, find the indicated transform; a and ω are constants. When possible, compare your result with that of DELAB.

11. $\mathcal{L}\{te^{at}\}$

12. $\mathcal{L}\{t^n e^{at}\}$

13. $\mathcal{L}\{e^{at} \sin \omega t\}$

14. $\mathcal{L}\{t \sin \omega t\}$

15. $\mathcal{L}\{\sin \omega t - \omega t \cos \omega t\}$

16. Example 20 begins with the claim

$$\lim_{t \to 0} \frac{\sin at}{t} = a.$$

Use l'Hôpital's rule or some other technique to evaluate this limit.

17. Verify the claim made in example 20 that the function

$$f(t) = \frac{\sin at}{t}, \quad t > 0,$$

is of exponential order.

18. Supply the details of the evaluation of the integral

$$\int_s^\infty \frac{a}{r^2 + a^2} \, dr$$

omitted from example 20.

19. Work through example 15, which solved the initial-value problem forced by a square wave. Show all of the details which were omitted from the text.

20. Example 15 solves the initial-value problem

$$x'' + 4\pi^2 x = f(t), \quad x(0) = 0, \quad x'(0) = -2,$$

where f is a square wave of period one,

$$f(t) = \begin{cases} a, & 0 \le t < \frac{1}{2}, \\ -a, & \frac{1}{2} \le t < 1. \end{cases}$$

It concludes that the solution is growing with time and asks whether this system could be in resonance. Is it? What is this system's natural frequency? What is the forcing frequency?

21. Use Laplace transforms to find the complete solution of the initial-value problem

$$x'' + \omega^2 x = A \cos \omega t, \quad x(0) = x_i, \quad x'(0) = v_i,$$

considered in example 17.

22. Example 17 inverts $As/(s^2 + \omega^2)^2$ using the derivative of a transform formula:

$$\frac{As}{(s^2 + \omega^2)^2} = -\frac{A}{2\omega} \frac{d}{ds}\left(\frac{\omega}{s^2 + \omega^2}\right)$$

$$= -\frac{A}{2\omega}\mathcal{L}\{-t \sin \omega t\}.$$

Obtain the same result using the convolution theorem.

In exercises 23–27, use the definition of Laplace transform to derive the indicated transform property. In every case, $F(s) = \mathcal{L}\{f(t)\}$ and all transforms are assumed to exist.

23. If f has period T, then

$$\mathcal{L}\{f\} = \frac{1}{1 - e^{-sT}} \int_0^T e^{-st} f(t) \, dt.$$

(*Hint*: Argue that the definition of the Laplace transform can be written

$$\mathcal{L}\{f\} = \sum_{n=1}^\infty \int_{(n-1)T}^{nT} f(t)e^{-st} \, dt.$$

The change of variables $r = (n-1)T + t$ and the periodicity of f then reduces the sum to one involving the integral $\int_0^T e^{-sr} f(r) \, dr$ in every term. When this integral is factored out of the sum, all that remains is a geometric series in e^{-sT}. Its sum is $1/(1 - e^{-sT})$.)

24. $\mathcal{L}\{-tf(t)\} = F'(s)$. (*Hint*: Assuming the derivative d/ds can be interchanged with the improper integral in the definition of Laplace transform, compute $dF(s)/ds$.)

25. $\mathcal{L}\{(-t)^n f(t)\} = F^{(n)}(s)$ for $n = 2, 3$. (*Hint*: For $n = 2$, apply the result of exercise 24 to the function $g(t) = tf(t)$. Use a similar idea for $n = 3, \dots$.)

26. $\mathcal{L}\{\int_0^t f(r) \, dr\} = F(s)/s$. (*Hint*: Integrate the definition of Laplace transform by parts using $dv = e^{-st} \, dt$ and $u = \int_0^t f(r) \, dr$. The statement of the fundamental theorem of calculus in appendix section A.2, page 624, will help you compute du/dt.)

27. $\mathcal{L}\{f(t)/t\} = \int_s^\infty F(r) \, dr$. (*Hint*: To evaluate the integral that appears on the right, assume that it can be interchanged with the improper integral in the definition of the Laplace transform.)

10.4 ■ RAMPS AND JUMPS

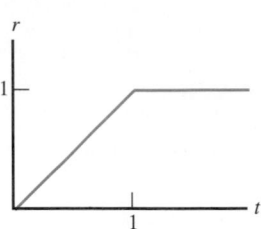

FIGURE 10.8   A ramp function.

Many models have forcing terms that increase steadily for a time and then level off at a fixed value. A good example is the heat output rate of a wood stove when it is first lit; its temperature-versus-time graph could have the ramplike profile of figure 10.8.

Other forcing terms can be modeled by expressions that jump from one value to another. For example, chapter 2 modeled emigration caused by the Irish potato famine as continuing at a constant rate for all time,

$$P' - 0.015P = -E(t), \quad P(0) = 8,$$

where

$$E(t) = 0.209 \text{ million people/year.}$$

But a better model might end emigration after 4 years, when the famine had subsided, by replacing the constant emigration rate with one that jumps to zero after 4 years have passed:

$$E(t) = \begin{cases} 0.209, & 0 \leq t < 4, \\ 0, & 4 \leq t < \infty. \end{cases}$$

The Laplace transform can solve initial-value problems having ramps, jumps, or similar functions in their forcing terms. This section develops the necessary transform tools.

> If the forcing term jumps, then the highest derivative in the differential equation must have a similar jump. For example, the emigration equation above has the form $P' + \cdots = -E(t)$. If $E(t)$ jumps suddenly from 0.209 to 0 at $t = 4$, then P' must exhibit a similar jump. Consequently, this first-order differential equation can not have a solution whose first derivative is continuous. We must relax our definition of *solution* to require only that the *highest-order derivative be piecewise continuous* and that the solution *satisfy the equation on the intervals where the highest-order derivative is continuous*.

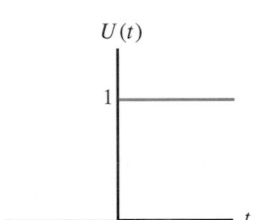

FIGURE 10.9   The unit step function $\mathcal{U}$ is "off" (value 0) for negative values of its argument and "on" (value 1) for positive values.

UNIT STEP FUNCTION

10.4.1 The Unit Step Function

While transforms of ramps or jumps, such as the emigration function above, can be computed directly from the definition of Laplace transform, we can also construct them conveniently using the **unit step function**

$$\mathcal{U}(t) = \begin{cases} 0, & t < 0, \\ 1, & 0 \leq t, \end{cases}$$

shown in figure 10.9. The unit step function $\mathcal{U}$ is "off" for negative values of its argument and "on" for positive values.

> The unit step function is sometimes called the **Heaviside function**, after the British physicist who discovered its utility.

The key transform result, which you are asked to derive in exercise 16, is

---

Transform of a shifted function

$$\mathcal{L}\{f(t-a)\mathcal{U}(t-a)\} = e^{-as}F(s),$$

---

where $F(s) = \mathcal{L}\{f(t)\}$ and $a > 0$. Before computing transforms involving the unit step function, we illustrate how it can be used to construct functions with switch points.

Figure 10.10 illustrates how the expression $f(t-a)\mathcal{U}(t-a)$ "turns on" at $t = a$. The curve on the left shows $\mathcal{U}(t-a)$, an expression that switches from 0 to 1 when t reaches the value a. The middle curve in figure 10.10 shows the graph of an arbitrary function $f(t-a)$; it is the graph of $f(t)$ shifted from the origin to the right by a time units. The graph on the right illustrates the product $f(t-a)\mathcal{U}(t-a)$ turning on at $t = a$. Because of the factor $\mathcal{U}$, the product is zero before $t = a$. After $t = a$, the product simply has the value $f(t-a)$.

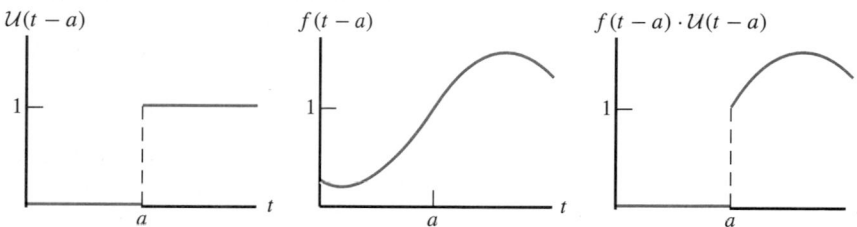

FIGURE 10.10 The "turn-on" property of the unit step function.

Stop and Think **10.9** Why is the graph of $f(t-a)$ the graph of $f(t)$ shifted to the *right* by a units? What function would represent the graph shift to the *left* by a units?

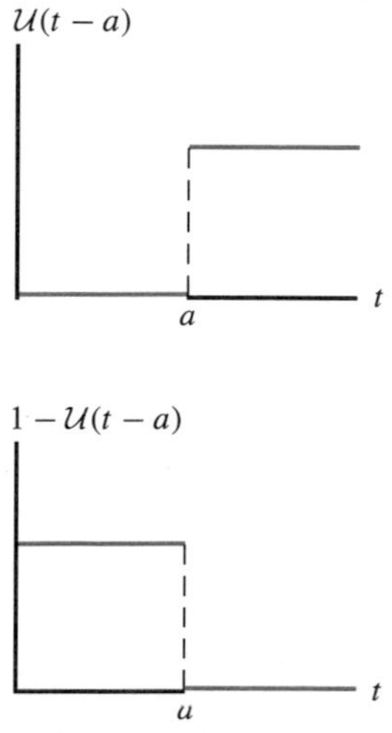

Figure 10.11 illustrates two simple expressions using the unit step function. The function $\mathcal{U}(t-a)$ jumps from off to on at $t = a$,

$$\mathcal{U}(t-a) = \begin{cases} 0, & 0 \le t < a, \\ 1, & a \le t < \infty, \end{cases}$$

and $1 - \mathcal{U}(t-1)$ is its complement, jumping from on to off at $t = a$,

$$1 - \mathcal{U}(t-a) = \begin{cases} 1, & 0 \le t < a, \\ 0, & a \le t < \infty. \end{cases}$$

A difference of unit step functions provides the off/on/off behavior of the **isolated pulse**

$$\mathcal{U}(t-a) - \mathcal{U}(t-b) = \begin{cases} 0, & 0 \le t < a, \\ 1, & a \le t < b, \\ 0, & b \le t < \infty, \end{cases}$$

FIGURE 10.11 $\mathcal{U}(t-a)$ switches on at $t = a$; $1 - \mathcal{U}(t-a)$ switches off at $t = a$.

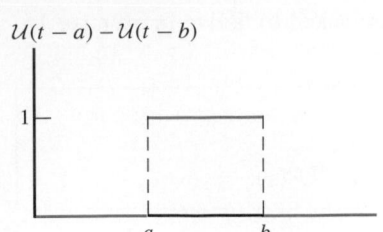

$\mathcal{U}(t-a) - \mathcal{U}(t-b)$

FIGURE 10.12 An isolated pulse.

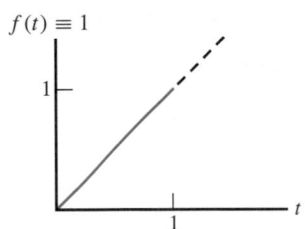

$f(t) \equiv 1$

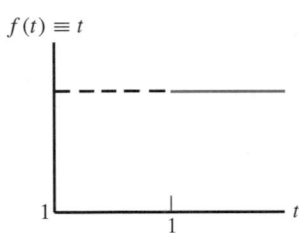

$f(t) \equiv t$

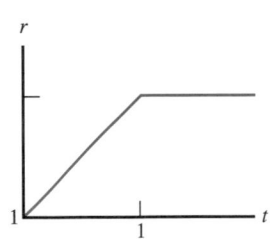

r

FIGURE 10.13 The ramp function, shown in the lower graph, is composed of a linear function that stops at $t = 1$ and a constant function that starts at $t = 1$.

shown in figure 10.12. The term $\mathcal{U}(t - a)$ turns on at $t = a$. When $\mathcal{U}(t - b)$ turns on at a later time $t = b$, it subtracts 1 from $\mathcal{U}(t - a)$ to restore the net value of the expression to zero.

If b grows infinitely large, then the isolated pulse (figure 10.12) collapses to the off/on behavior of $\mathcal{U}(t - a)$, because $t - b$ is always negative and $\mathcal{U}(t - b) = 0$ for any finite value of t in the limit as b grows. If $a = 0$, then $\mathcal{U}(t - 0) = \mathcal{U}(t) = 1$ for $t \geq 0$, and the isolated pulse has the on/off behavior of $1 - \mathcal{U}(t - b)$.

■ **EXAMPLE 22**   *Write the ramp function*

$$r(t) = \begin{cases} t, & 0 \leq t \leq 1, \\ 1, & 1 < t < \infty, \end{cases}$$

shown in figure 10.8 *using the unit step function* $\mathcal{U}$.

Think of this ramp function as composed of two parts, the rising section $r(t) = t$, which appears only for $0 \leq t < 1$, and the flat section $r(t) = 1$, which appears for $1 \leq t < \infty$. This decomposition is illustrated in figure 10.13.

Roughly, we could write

$$r(t) = t \times (\text{function that is on for } 0 \leq t < 1)$$
$$+ 1 \times (\text{function that is on for } 1 \leq t < \infty).$$

The two functions that are "on" for fixed intervals are just isolated pulses. The function that is on for $0 \leq t < 1$ is

$$\mathcal{U}(t - 0) - \mathcal{U}(t - 1) = \begin{cases} 1, & 0 \leq t < 1, \\ 0, & 1 \leq t < \infty. \end{cases}$$

Its counterpart for $1 \leq t < \infty$ is

$$\mathcal{U}(t - 1) - \mathcal{U}(t - \infty) = \begin{cases} 0, & 0 \leq t < 1, \\ 1, & 1 \leq t < \infty. \end{cases}$$

We have no business writing an expression such as $t - \infty$, because ∞ is a symbol used to denote unbounded limits, not a number with which we can perform arithmetic. Nonetheless, this abuse of notation is too suggestive to resist.

Using these expressions, the ramp function is

$$r(t) = t \times [\mathcal{U}(t - 0) - \mathcal{U}(t - 1)]$$
$$+ 1 \times [\mathcal{U}(t - 1) - \mathcal{U}(t - \infty)].$$

Since we are considering only $t \geq 0$, $\mathcal{U}(t - 0) = \mathcal{U}(t) = 1$. Likewise, since $t < \infty$, we have $\mathcal{U}(t - \infty) = 0$. The final form for the ramp function is

$$r(t) = t\,[1 - \mathcal{U}(t - 1)] + \mathcal{U}(t - 1).$$

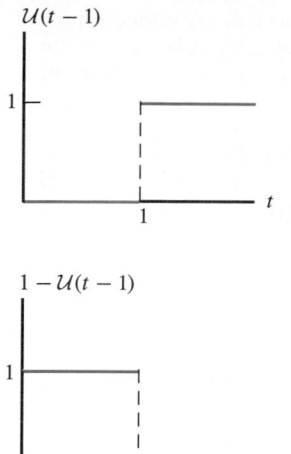

$\mathcal{U}(t-1)$

$1-\mathcal{U}(t-1)$

FIGURE 10.14 The off/on unit step function $\mathcal{U}(t-1)$ and its on/off complement $1-\mathcal{U}(t-1)$ used to write the ramp function.

The first factor turns on the linear function t for $0 \le t \le 1$. The second term turns on the constant 1, the flat part of the ramp function, when $t > 1$. This construction is also consistent with the on/off behavior of $1 - \mathcal{U}(t - 1)$ and the off/on behavior of $\mathcal{U}(t - 1)$ illustrated in figure 10.14 with $a = 1$. ■

■ **EXAMPLE 23**   *Write the function*

$$g(t) = \begin{cases} 0, & 0 \le t \le 2, \\ \sin t, & 2 < t < \infty, \end{cases}$$

using the unit step function.

Using the logic of example 22, think of this function as composed of two sections, one for the interval $0 \le t < 2$ and the other for $2 \le t < \infty$. Write g as a sum of products, each product consisting of the desired value for the function multiplying a term that is unity when t is in the appropriate interval and zero otherwise:

$$g(t) = 0\,[\mathcal{U}(t-0) - \mathcal{U}(t-2)] + (\sin t)\,[\mathcal{U}(t-2) - \mathcal{U}(t-\infty)].$$

The expression $\mathcal{U}(t - \infty)$ is again poor notation, but it reduces to zero since t is finite. Hence, the final expression for g is

$$g(t) = (\sin t)\mathcal{U}(t - 2),$$

consistent with the off/on behavior of $\mathcal{U}(t - 2)$ illustrated in figures 10.10 and 10.11 with $a = 2$. ■

■ **EXAMPLE 24**   *Find the Laplace transform of the function*

$$g(t) = \begin{cases} 0, & 0 \le t \le 2, \\ \sin t, & 2 < t < \infty, \end{cases}$$

considered in the last example.

From example 23, we have

$$g(t) = (\sin t)\,\mathcal{U}(t - 2).$$

To apply the shifted function transform

Identify a, $f(t - a)$.

$$\mathcal{L}\{f(t - a)\mathcal{U}(t - a)\} = e^{-as}F(s),$$

we identify $a = 2$ and $f(t - a) = \sin t$. We must find a formula for $f(t)$, the function without its argument shifted, so that we can compute $F(s) = \mathcal{L}\{f(t)\}$.

Find $f(t)$.

A simple change of variables suffices. Let $r = t - 2$ in $f(t-2) = \sin t$. Since $t = r + 2$, we have $f(r) = \sin(r + 2)$, or, if we replace the place holder r with t,

$$f(t) = \sin(t + 2).$$

Equivalently, we could argue that $f(t-2) = \sin t$ means that $\sin t$ is the formula for f evaluated at t shifted back by two units. To obtain a formula for f at the present time, we must shift t forward by two units, replacing every occurrence of t by $t+2$. Hence, $f(t) = \sin(t+2)$.

Now we can compute the transform $F(s) = \mathcal{L}\{f(t)\}$:

Transform $f(t)$.

$$F(s) = \mathcal{L}\{f(t)\} = \mathcal{L}\{\sin(t+2)\}$$
$$= \mathcal{L}\{\sin t \cos 2 + \cos t \sin 2\}$$
$$= \frac{\cos 2}{s^2+1} + \frac{s \sin 2}{s^2+1}, \quad s > 0.$$

With $F(s)$ in hand, we can use the shifted function transform to write

Apply shifted transform.

$$\mathcal{L}\{g(t)\} = \mathcal{L}\{f(t-a)\mathcal{U}(t-a)\}$$

$$= e^{-2s}\frac{\cos 2 + s \sin 2}{s^2+1}, \quad s > 0. \ ■$$

■ **EXAMPLE 25**   *Find the Laplace transform of the ramp function*

$$r(t) = \begin{cases} t, & 0 \le t \le 1, \\ 1, & 1 < t < \infty, \end{cases}$$

shown in figure 10.8.

Example 22 wrote this ramp function as

Write function in terms of $\mathcal{U}$.

$$r(t) = t\,[1 - \mathcal{U}(t-1)] + \mathcal{U}(t-1).$$

From the linearity of the Laplace transform, we have

Transform term by term.

$$\mathcal{L}\{r(t)\} = \mathcal{L}\{t\} - \mathcal{L}\{t\mathcal{U}(t-1)\} + \mathcal{L}\{\mathcal{U}(t-1)\}.$$

We find in turn each of the transforms in this last expression.
Entry 2 of table 10.1 immediately yields

Transform of first term

$$\mathcal{L}\{t\} = \frac{1}{s^2}.$$

The shifted function formula,

$$\mathcal{L}\{f(t-a)\mathcal{U}(t-a)\} = e^{-as}F(s),$$

will provide the other two transforms involving the unit step function, $\mathcal{L}\{t\mathcal{U}(t-1)\}$ and $\mathcal{L}\{\mathcal{U}(t-1)\}$.
To treat $\mathcal{L}\{t\mathcal{U}(t-1)\}$, we identify $a = 1$ and

In $\mathcal{L}\{t\mathcal{U}(t-1)\}$, identify $f(t-a)$.

$$f(t-1) = t.$$

The change of variables $r = t - 1$ converts $f(t - 1) = t$ to $f(r) = r + 1$, or, when we replace r with t,

Find $f(t)$.

$$f(t) = t + 1.$$

Entries 1 and 2 of table 10.1 immediately yield

Transform $f(t)$.

$$F(s) = \mathcal{L}\{t + 1\} = \mathcal{L}\{t\} + \mathcal{L}\{1\} = \frac{1}{s^2} + \frac{1}{s} = \frac{1+s}{s^2}.$$

Coupling this expression with $\mathcal{L}\{f(t - a)\mathcal{U}(t - a)\} = e^{-as}F(s)$, we find

Transform of second term

$$\mathcal{L}\{t\mathcal{U}(t - 1)\} = e^{-s}\frac{1+s}{s^2}.$$

The last transform is $\mathcal{L}\{\mathcal{U}(t - 1)\}$. Now we use the shifted function transform with $a = 1$ and $f(t - 1) = 1$. Obviously, $f(t) = 1$ and $F(s) = \mathcal{L}\{f(t)\} = 1/s$. Hence,

Transform of last term

$$\mathcal{L}\{\mathcal{U}(t - 1)\} = \frac{e^{-s}}{s}.$$

Combining these three expressions, which are transforms of t, $t\mathcal{U}(t-1)$, and $\mathcal{U}(t - 1)$, we find the transform of the ramp function,

$$\mathcal{L}\{r(t)\} = \mathcal{L}\{t\} - \mathcal{L}\{t\mathcal{U}(t - 1)\} + \mathcal{L}\{\mathcal{U}(t - 1)\}$$

Combine terms to find transform of ramp function.

$$= \frac{1}{s^2} - e^{-s}\frac{1+s}{s^2} + \frac{e^{-s}}{s}$$

$$= \frac{1 - e^{-s}}{s^2}, \quad s > 0,$$

the same result as in example 5, page 514, which used the definition of Laplace transform.

More elegantly, we might have noticed that $r(t) = [1 - \mathcal{U}(t - 1)]t + \mathcal{U}(t - 1)$ can be written

$$r(t) = t - (t - 1)\mathcal{U}(t - 1).$$

The shifted function result $\mathcal{L}\{f(t - a)\mathcal{U}(t - a)\} = e^{-as}F(s)$ can then be applied immediately with $f(t - 1) = t - 1$ and $f(t) = t$. ■

■ **EXAMPLE 26**　*Solve the initial-value problem*

$$y' + 2y = r(t), \quad y(0) = 4,$$

where $r(t)$ is the ramp function shown in figure 10.8.

Let $Y(s) = \mathcal{L}\{y(t)\}$. Transforming the entire differential equation yields

Transform the differential equation.

$$(s + 2)Y(s) - y(0) = \mathcal{L}\{r(t)\}.$$

Using $y(0) = 4$ and the expression for $\mathcal{L}\{r(t)\}$ from example 25, we have

Solve for $Y(s)$.

$$Y(s) = \frac{4}{s+2} + \frac{1}{s^2(s+2)} - e^{-s}\frac{1}{s^2(s+2)}. \qquad (10.11)$$

The first term comes from table 10.1:

Invert $Y(s)$, first term from transform table.

$$\mathcal{L}\{4e^{-2t}\} = \frac{4}{s+2}.$$

The second term in (10.11) can be evaluated using either the convolution theorem or partial fractions. The latter yields

Invert $Y(s)$, second term via partial fractions.

$$\frac{1}{s^2(s+2)} = \frac{1}{4}\left(\frac{2}{s^2} - \frac{1}{s} + \frac{1}{s+2}\right).$$

Hence,

$$\frac{1}{s^2(s+2)} = \frac{1}{4}\mathcal{L}\{2t - 1 + e^{-2t}\}.$$

The third term in (10.11) contains an exponential in the transform variable s, a clue that we must use the shifted function transform,

Invert $Y(s)$, third term via shifted function transform.

$$\mathcal{L}\{f(t-a)\mathcal{U}(t-a)\} = e^{-as}F(s).$$

Comparing $e^{-as}F(s)$ with the given transform,

$$e^{-s}\frac{1}{s^2(s+2)},$$

we identify $a = 1$ and

$$F(s) = \frac{1}{s^2(s+2)}.$$

But from our work in the preceding paragraph, we know

$$F(s) = \mathcal{L}\{f(t)\} = \frac{1}{4}\mathcal{L}\{2t - 1 + e^{-2t}\}.$$

If $f(t) = \frac{1}{4}(2t - 1 + e^{2t})$, then

$$f(t-1) = \frac{1}{4}\left(2(t-1) - 1 + e^{-2(t-1)}\right) = \frac{1}{4}\left(2t - 3 + e^{-2(t-1)}\right).$$

We conclude that the third term in (10.11) must be the transform of

$$f(t-1)\mathcal{U}(t-1);$$

that is, the transform of

$$\tfrac{1}{4}\left(2t - 3 + e^{-2(t-1)}\right)\mathcal{U}(t-1).$$

The solution of the given initial-value problem is the sum of the inverse transforms of the three terms in (10.11),

Combine inverse transforms to find solution of initial-value problem.

$$y(t) = 4e^{-2t} + \tfrac{1}{4}\left(2t - 1 + e^{-2t}\right)$$
$$- \tfrac{1}{4}\left(2t - 3 + e^{-2(t-1)}\right)\mathcal{U}(t-1). \quad ■$$

10.4.2 Summary

The unit step function $\mathcal{U}$ has the value one when its argument is positive and the value zero when its argument is negative. See figure 10.9, page 546.

Functions such as ramps, that switch between several different formulas at various values of t, can be written as sums of isolated pulses. An isolated pulse of amplitude one is

Isolated pulse

$$\mathcal{U}(t-a) - \mathcal{U}(t-b).$$

It is unity when $a \le t < b$ and zero otherwise. See figure 10.12, page 548.

Transforms involving the unit step function are computed using the shifted function transform

---

Transform of a shifted function

$$\mathcal{L}\{f(t-a)\mathcal{U}(t-a)\} = e^{-as}F(s),$$

---

where $F(s) = \mathcal{L}\{f(t)\}$. An exponential in the s domain is a clue to the presence of a shift in the t domain.

A complete list of Laplace transform properties, including this latest formula, appears in table 10.5.

TABLE 10.5 **Properties of Laplace transforms**

Properties of Laplace Transform

$$F(s) = \mathcal{L}\{f(t)\} = \int_0^\infty e^{-st} f(t)\, dt, \quad G(s) = \mathcal{L}\{g(t)\}$$

1. Transform of a derivative:

$$\mathcal{L}\{f'(t)\} = s F(s) - f(0),$$
$$\mathcal{L}\{f''(t)\} = s^2 F(s) - s f(0) - f'(0).$$

2. Convolution theorem:

$$\mathcal{L}\left\{\int_0^t f(r)g(t-r)\, dr\right\} = F(s)G(s).$$

3. Shifted transform:

$$\mathcal{L}\{e^{at} f(t)\} = F(s-a).$$

4. Transform of a periodic function of period T:

$$\mathcal{L}\{f\} = \frac{1}{1 - e^{-sT}} \int_0^T e^{-st} f(t)\, dt.$$

5. Derivative of a transform:

$$\mathcal{L}\{-t f(t)\} = F'(s),$$
$$\mathcal{L}\{(-t)^n f(t)\} = F^{(n)}(s), \quad n = 1, 2, \ldots.$$

6. Transform of an integral:

$$\mathcal{L}\left\{\int_0^t f(r)\, dr\right\} = \frac{F(s)}{s}.$$

7. Integral of a transform:

$$\mathcal{L}\left\{\frac{f(t)}{t}\right\} = \int_s^\infty F(r)\, dr.$$

8. Transform of a shifted function using the unit step function $\mathcal{U}$:

$$\mathcal{L}\{f(t-a)\mathcal{U}(t-a)\} = e^{-as} F(s), \quad a > 0.$$

10.4.3 Exercises

EXERCISE GUIDE	
To gain experience . . .	**Try exercises**
Finding transforms of shifted functions	1–11
Expressing functions in terms of the unit step function	4–11
With graphs involving the unit step function	8–11, 15

In exercises 1–3, use $\mathcal{L}\{f(t-a)\mathcal{U}(t-a)\} = e^{-as}F(s)$ with an appropriate choice of f to find the Laplace transform of the given function.

1. $e^{4t}\mathcal{U}(t-3)$

2. $(2-3t)\mathcal{U}(t-2)$

3. The isolated pulse $g(t) = \begin{cases} 0, & 0 \le t \le a \\ 1, & a < t \le b \\ 0, & b < t \end{cases}$

In exercises 4–7,

(i) Rewrite the given function in terms of the unit step function $\mathcal{U}$.

(ii) Use that expression and $\mathcal{L}\{f(t-a)\mathcal{U}(t-a)\} = e^{-as}F(s)$ to find the Laplace transform of the given function.

4. $g(t) = \begin{cases} 4t, & 0 \le t \le 7 \\ 0, & 7 < t < \infty \end{cases}$

5. $g(t) = \begin{cases} mt, & 0 \le t \le b \\ 0, & b < t < \infty \end{cases}$

6. $g(t) = \begin{cases} 0, & 0 \le t \le 2 \\ t-2, & 2 \le t \le 4 \\ 6, & 4 < t < \infty \end{cases}$

7. $g(t) = \begin{cases} 1, & 0 \le t \le 2\pi \\ \cos t, & 2\pi \le t \le 7\pi/2 \\ 0, & 7\pi/2 < t < \infty \end{cases}$

In exercises 8–11,

(i) Graph the forcing term and write it in terms of the unit step function $\mathcal{U}$.

(ii) Solve the initial-value problem using Laplace transforms.

(iii) Give the interval(s) within which the highest derivative of the solution appearing in the equation is continuous.

(iv) Verify that your solution satisfies the differential equation within each interval of continuity.

8. $y' - 6y = g(t)$, $y(0) = 4$,

$$g(t) = \begin{cases} 0, & 0 \le t < 2 \\ 2, & 2 \le t < \infty \end{cases}$$

9. $y' + 2y = g(t)$, $y(0) = 0$,

$$g(t) = \begin{cases} 2t - 1, & 0 \le t < 2 \\ 3, & 2 \le t < \infty \end{cases}$$

10. $y'' + 4y = g(t)$, $y(0) = 0$, $y'(0) = -2$,

$$g(t) = \begin{cases} -2, & 0 \le t < 4 \\ 4, & 4 \le t < 8 \\ 0, & 8 \le t < \infty \end{cases}$$

11. $P' - 0.015P = -E(t)$, $P(0) = 8$,

$$E(t) = \begin{cases} 0.209, & 0 \le t \le 4 \\ 0, & 4 < t < \infty \end{cases}$$

(This is the Irish potato-famine emigration model with population P in millions and time t in years measured from 1847. In this form, emigration stops in 1851. See equation (2.10), page 46, of section 2.2.)

In exercises 12–13, assume that the Laplace transform of the solution of each initial-value problem exists. Find the transform of the solution of the initial-value problem working directly from the differential equation.

12. $2y' - 5y = g(t)$, $y(0) = 3$, where $g(t)$ is the isolated pulse of exercise 3,

$$g(t) = \begin{cases} 0, & 0 \le t \le a \\ 1, & a < t \le b \\ 0, & b < t \end{cases}$$

13. $y' - ky = r(t)$, $y(0) = y_i$, where k is a constant and $r(t)$ is the ramp function of example 25,

$$r(t) = \begin{cases} t, & 0 \le t \le 1 \\ 1, & 1 < t < \infty \end{cases}$$

14. Use the convolution theorem to find the inverse transform of

$$\frac{1}{s^2(s+2)}.$$

Compare your answer with that in example 26 which used partial fractions.

15. On one graph, sketch both $\mathcal{U}(t-2)$ and $-\mathcal{U}(t-4)$. Illustrate graphically how they combine to form the isolated pulse

$$\mathcal{U}(t-2) - \mathcal{U}(t-4) = \begin{cases} 0, & 0 \leq t < 2 \\ 1, & 2 \leq t < 4 \\ 0, & 4 \leq t < \infty \end{cases}$$

16. Derive the shifted function transform

$$\mathcal{L}\{f(t-a)\mathcal{U}(t-a)\} = e^{-as}F(s),$$

where $F(s) = \mathcal{L}\{f(t)\}$ and $a > 0$. (*Hint:* Because of its definition, the unit step function $\mathcal{U}(t-a)$ changes to a nonzero value the lower limit of the integral in the definition (10.1) of the Laplace transform. Make a change of variable of integration that converts the lower limit back to zero and leaves the upper limit at infinity.)

10.5 ■ THE UNIT IMPULSE FUNCTION

The upper end of a spring carrying a mass is jerked suddenly upward. A freely swinging pendulum is struck sharply with a hammer. A switch is closed in an electric circuit, causing an abrupt jump in voltage. A oil spill kills thousands of sea birds in the wink of an eye. This section develops the *unit impulse function*, a tool for modeling these kinds of instantaneous changes.

Whenever individuals are removed from a population, whether through the tragedy of an ecological disaster or through a planned harvest in a fish farm, the emigration model

$$P' = kP - E(t), \qquad P(0) = P_i,$$

is appropriate. Recall that $E(t)$ is the *rate* at which individuals are removed from the population; units of trout/minute might be appropriate for a fish farm.

If a finite number R of individuals is removed at time $t = a$, then the emigration function $E(t)$ could have the form

$$E(t) = Rd_h(t-a),$$

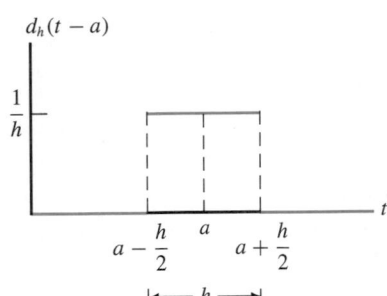

FIGURE 10.15   An isolated pulse of unit area centered at $t = a$

where $d_h(t-a)$ is an isolated pulse of width h and height $1/h$ centered at $t = a$. See figure 10.15.

Because $E(t)$ is a *rate* of population removal, the number removed over all time is the integral

$$\int_{-\infty}^{\infty} E(t)\,dt.$$

But since the isolated pulse $d_h(t-a)$ has unit area, that number removed is

$$\int_{-\infty}^{\infty} E(t)\,dt = R\int_{-\infty}^{\infty} d_h(t)\,dt = R, \qquad (10.12)$$

the desired amount.

In a population model, where the action of removing individuals occurs over a brief but nonzero interval, the simple isolated pulse $d_h(t - a)$ might be adequate. But when the impulse is effectively instantaneous, such as a sudden jerk on the end of a spring, a different tool is needed. That tool is a limit of d_h as the width h of the pulse drops to zero while it retains unit area; figure 10.16 suggests that d_h evolves into a tall, narrow spike as $h \to 0$. Since such a limiting pulse of unit area can not have a finite value at $t = a$, we define it in terms of its action within integrals such as (10.12).

Let $\delta(t - a)$ represent this limiting pulse of zero width and unit area. Define $\delta(t - a)$ by

Definition of $\delta(t - a)$

$$\int_{-\infty}^{\infty} \phi(t)\delta(t - a)\, dt = \lim_{h\to 0} \int_{-\infty}^{\infty} \phi(t)d_h(t - a)\, dt$$

$$= \lim_{h\to 0} \int_{a-h/2}^{a+h/2} \phi(t)d_h(t - a)\, dt,$$

(10.13)

where ϕ is an arbitrary function, continuous for all t and $\phi(t) = 0$ for $|t|$ sufficiently large. This limiting impulse is called the **Dirac delta function** or **unit impulse function**.

The functions ϕ are called **test functions** because they are used to test the action of the limiting impulse $\delta(t - a)$. They are said to have **compact support** since the length of the interval on which they are nonzero is finite.

The delta function is, in fact, not a function but a **distribution**, a concept defined in terms of its action on test functions rather than in terms of the input-output rules of a conventional function. Stakgold [25] provides a clear treatment of distributions.

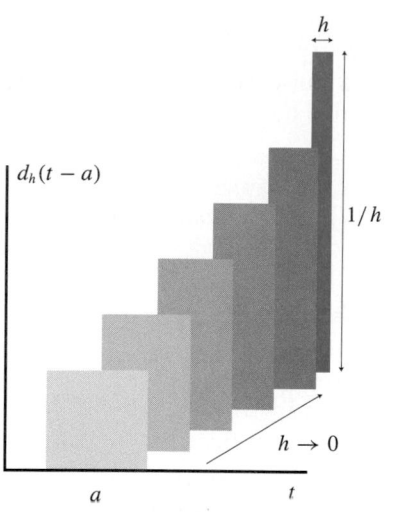

FIGURE 10.16 The isolated pulse of unit area becomes a tall, narrow spike as h is made smaller.

To determine the action of $\delta(t - a)$ on an arbitrary ϕ, use the mean-value theorem for integrals in the integral involving d_h in (10.13):

$$\int_{-\infty}^{\infty} \phi(t)\delta(t - a)\, dt = \lim_{h\to 0} \phi(\tilde{t}) \int_{a-h/2}^{a+h/2} d_h(t - a)\, dt$$

$$= \lim_{h\to 0} \phi(\tilde{t}).$$

Here $a - h/2 \le \tilde{t} \le a + h/2$, and the last equality results because d_h has unit area. Since ϕ is continuous and $\tilde{t}$ approaches a as h shrinks to zero, we conclude that

Action of $\delta(t - a)$

$$\int_{\infty}^{\infty} \phi(t)\delta(t - a)\, dt = \phi(a).$$

Integrating a continuous function with $\delta(t - a)$ picks out the value of the function at $t = a$.

Now choose ϕ to be zero everywhere but in a small neighborhood of some point $t^* \ne a$, where ϕ is positive. Since the integral $\int \phi(t)d_h(t - a)\, dt$ in definition (10.13) is zero once h is small enough, we find

$$\int_{-\infty}^{\infty} \phi(t)\delta(t - a)\, dt = 0.$$

But in a neighborhood of t^*, $\phi(t)$ is positive. We conclude that the integral can be zero only if $\delta(t^* - a) = 0$. That is, $\delta(t - a) = 0, t \neq a$.

The two important properties of the delta function are as follows:

Delta function properties

$$\int_{-\infty}^{\infty} \phi(t)\delta(t - a)\, dt = \phi(a)$$

$$\delta(t - a) = 0, \quad t \neq a.$$

Since $\delta(t - a) = 0$ for $t \neq a$, the function $\phi(t)$ can always be redefined away from $t = a$ without affecting the value of $\int \phi(t)\delta(t - a)\, dt$. Hence, the restriction in (10.13) that $\phi(t)$ be zero outside a finite interval can be dropped.

An immediate consequence of these properties is

$$\int_{-\infty}^{\infty} \delta(t - a)\, dt = 1.$$

The delta function *preserves the unit area property* of the finite pulses $d_h(t - a)$. The delta function somehow represents a pulse of zero width, infinite height, and unit area, requirements that can be met by no ordinary function. It is indeed a *unit impulse*.

■ **EXAMPLE 27**   *Use these properties to find the Laplace transform of* $\delta(t - a), a > 0.$

Formally, the desired transform is defined by

$$\mathcal{L}\{\delta(t - a)\} = \int_0^{\infty} e^{-st}\delta(t - a)\, dt.$$

If $a > 0$, we can choose $\phi(t)$ in (10.13) so that $\phi(t) = e^{-st}$ near $t = a$ and $\phi(t)$ tapers quickly to zero elsewhere. Then the first property of the delta function immediately yields

$$\int_0^{\infty} e^{-st}\delta(t - a)\, dt = e^{-st}\big|_{t=a} = e^{-as}.$$

Hence,

Delta function transform

$$\mathcal{L}\{\delta(t - a)\} = e^{-as}, \quad a > 0.$$

A special case of this argument is $a = 0$: $\mathcal{L}\{\delta(t)\} = 1$.

■ **EXAMPLE 28**   *Solve* $y' = \delta(t - a), y(0) = 0,$ *when* $a > 0.$

Since $\delta(t - a)$ is not defined at $t = a$, we must suspend our usual definition of *solution*. Proceeding formally, let $Y(s) = \mathcal{L}\{y(t)\}$ and transform the differential equation to find

Transform differential equation.

$$sY(s) = e^{-as}.$$

Since $\mathcal{L}^{-1}\{1/s\} = 1$, entry 8 of table 10.5 yields

Find $y(t)$.

$$y(t) = \mathcal{L}^{-1}\left\{\frac{1}{s}e^{-as}\right\} = \mathcal{U}(t - a).$$

That is, in a sense we can not properly define here, the derivative of the unit step function $\mathcal{U}(t - a)$ is the delta function $\delta(t - a)$.

Since $\delta(t - a) = 0$ and $\mathcal{U}(t - a)$ is constant for $t \neq a$, the relation certainly seems reasonable. At $t = a$, $\mathcal{U}(t - a)$ is not even continuous, much less differentiable. But if we think of the derivative in some extended sense, the behavior of the derivative at a jump discontinuity is captured by the delta function's impulse of zero width and unit area.

The delta function is the derivative of $\mathcal{U}(t-a)$ in the sense that the action of $\delta(t-a)$ on a test function $\phi(t)$ is the same as the action of the unit step function on the derivative $\phi'(t)$. Since $\phi(t)$ is zero for t sufficiently large, a formal integration by parts relation holds:

$$\int_{-\infty}^{\infty} \phi(t)\delta(t - a)\,dt = \int_{-\infty}^{\infty} \phi'(t)\mathcal{U}(t - a)\,dt = \int_{a}^{\infty} \phi'(t)\,dt = \phi(a).$$

■ **EXAMPLE 29**   *Solve the initial-value problem $P' = kP - R\delta(t - a)$, $P(0) = P_i$, a model of a population in which R individuals are removed suddenly at time $t = a$.*

Let $Q(s) = \mathcal{L}\{P(t)\}$. Formally transforming the equation yields

Transform differential equation; solve for $Q = \mathcal{L}\{P\}$.

$$Q(s) = \frac{P_i}{s - k} - \frac{R}{s - k}e^{-as}.$$

The first term is inverted using entry 3 of table 10.1. The second is inverted using entry 8 of table 10.5. Hence, the solution is

$$P(t) = P_i e^{kt} - Re^{k(t-a)}\mathcal{U}(t - a).$$

The first term represents familiar exponential population growth. The unit step function in the second term introduces a sudden drop in population at $t = a$, but the exponential factor leads immediately to renewed growth, although from the reduced population level. ■

■ **EXAMPLE 30**   *At time $t = a$, the upper end of a vertical spring-mass system is jerked upward suddenly and returned to its original position. Write a model for this situation.*

The forced model (5.13–5.14), page 210, of section 5.1 applies with the position $h(t)$ of the upper end of the spring given by $h(t) = H\delta(t - a) - h(0)$ for some H. The model is

$$mx'' + kx = kH\delta(t - a), \quad x(0) = x_i, \quad x'(0) = v_i. \ ◼$$

To grasp the physical meaning of an equation such as $mx'' + \cdots = kH\delta(t - a)$, informally integrate both sides. Then the relation is $mx' + \cdots = kH$. The quantity on the left, mx', is the change in the momentum of the mass, and kH represents that change. Since mx'' is a force (mass times acceleration), $kH\delta(t - a)$ is an *impulsive force* of brief duration and sufficient magnitude to produce a change in momentum of kH.

Closing a switch at time $t = a$ in a series RLC circuit can cause a step function change in potential, $E(t) = V\mathcal{U}(t - a)$. The current model $Li'' + Ri' + i/C = E'(t)$ becomes

$$Li'' + Ri + \frac{1}{C}i = V\delta(t - a),$$

where the impulse $V\delta(t - a)$ is the consequence of a step increase V in potential at $t = a$. The time integral of $V\delta(t - a)$ is the magnitude V of the voltage change.

10.5.1 Exercises

EXERCISE GUIDE	
To gain experience . . .	**Try exercises**
With transforms involving $\delta(t - a)$	1–3, 5–7, 9–13
Solving initial-value problems	2, 5–7, 6–12
Analyzing models involving impulses	4–6, 9–12
With $\delta(t - a)$ as the derivative of $\mathcal{U}(t - a)$	7–8, 11

1. Find $\mathcal{L}^{-1}\{1\}$.

2. Solve $y' + ky = 4\delta(t - 6)$, $y(0) = 2$.

3. Invert the transform $\dfrac{s}{s - 2}$.

4. Sketch a graph of the solution $P(t) = P_i e^{kt} - Re^{k(t-a)}\mathcal{U}(t-a)$ of the population model $P' = kP - R\delta(t - a)$, $P(0) = P_i$, considered in example 29. Explain the significance of the jumps in the graph.

5. Consider the heat-loss model $T' + (Ak/cm)T = b\delta(t-a)$, $T(0) = 0$. Solve this initial-value problem. Use your so-

lution to explain the significance of the parameter b. Of what physical situation might this initial-value problem be a model?

6. Solve the initial-value problem

$$mx'' + kx = kH\delta(t - a), \quad x(0) = x_i, \quad x'(0) = v_i,$$

developed in example 30. It models a spring-mass system in which the upper end of the spring is jerked suddenly. Consider $x_i = v_i = 0$. Use your solution to explain the significance of the parameter H.

7. Find the current in a series *RLC* circuit if the potential drop across the three circuit elements is raised from $E = 0$ to $E = V$ at time $t = a$. For simplicity, impose zero initial conditions and neglect resistance.

8. Approximate the unit step function $\mathcal{U}(t - a)$ by

$$U_h(t - a) = \begin{cases} 0, & -\infty < t < a - h/2 \\ (t - a + \frac{h}{2})/h, & a - h/2 \le t \le a + h/2 \\ 1, & a + h/2 < t < \infty \end{cases}$$

Sketch a graph of $U_h(t - a)$. Compute $dU_h(t - a)/dt$, and sketch its graph. For what values of t is this derivative not defined? What connections between $\delta(t - a)$ and $\mathcal{U}(t - a)$ does this approximation suggest?

9. Consider the linearized pendulum equation subject to an impulsive forcing term,

$$L\theta'' + g\theta = b\delta(t - a), \quad \theta(0) = \theta'(0) = 0.$$

   (a) Solve this problem and explain the terms in the solution you obtain.

   (b) Use the solution obtained in part (a) to explain the significance of the parameter b.

   (c) If a pendulum is released from rest from the angle θ_i and left to swing undisturbed, its amplitude is θ_i. Choose b so that the pendulum in this model also has amplitude θ_i for $t \ge a$.

10. An undamped spring-mass system subject to an impulsive disturbance at $t = a$ is modeled by $mx'' + kx = b\delta(t - a)$, $x(0) = x'(0) = 0$.

   (a) Describe the motion of the mass for $t < a$. Solve the governing initial-value problem, confirm your description, and explain the terms that are nonzero for $t \ge a$.

   (b) Choose the parameter b so that the mass achieves a prescribed amplitude A for $t \ge a$.

11. The potential drop across the three circuit elements in a series *RLC* circuit is $E(t) = V\mathcal{U}(t - a)$. Argue that the current in the circuit is governed by

$$Li'' + Ri' + \frac{1}{C}i = V\delta(t - a).$$

Solve this equation subject to zero initial conditions; for convenience, set $R = 0$. Determine the value of V that produces the same amplitude as the initial conditions $i(0) = I$, $i'(0) = 0$.

12. The *unit impulse response* of a system is the solution of the governing initial-value problem with zero initial conditions and the forcing function $\delta(t)$. Find the unit impulse response of the current in a series *RLC* circuit. How could the unit impulse response of such a circuit be realized physically?

13. Derive the formula for $\mathcal{L}\{\delta(t - a)\}$ from the definition (10.13)—that is, evaluate the integrals involving $e^{-st}d_h(t - a)$.

10.6 ■ CHAPTER EXERCISES

EXERCISE GUIDE

To gain experience . . .	Try exercises
Using the definition of Laplace transform	1–2
Finding Laplace transforms	3–10
Finding inverse transforms	15–24
Expressing functions in terms of the unit step function	11–13
With graphs involving the unit step function	11–12, 14
Solving initial-value problems	11–13
Analyzing function behavior, given their Laplace transforms	25–28
Solving systems of equations	29–36

In exercises 1–2, use the definition of Laplace transform to find the transform of the given function. Sketch the graph of each function before you attempt to find its transform.

1. $g(t) = \begin{cases} mt, & 0 \le t \le b \\ 0, & b < t < \infty \end{cases}$

2. $g(t) = \begin{cases} 0, & 0 \le t \le 2 \\ t - 2, & 2 \le t \le 4 \\ 6, & 4 < t < \infty \end{cases}$

In exercises 3–10, find the indicated transform; a and ω are constants. Confirm each result with DELAB.

3. $\mathcal{L}\{\cos 2t \sin 4t\}$

4. $\mathcal{L}\{\cos^2 3t\}$

5. $\mathcal{L}\{e^{-4t} \cos 2t \sin 4t\}$

6. $\mathcal{L}\{e^{-2t} \cos^2 3t\}$

7. $\mathcal{L}\{t(1 - t)\}$

8. $\mathcal{L}\{e^{at} \cos \omega t\}$

9. $\mathcal{L}\{t \cos \omega t\}$

10. $\mathcal{L}\{\sin \omega t + \omega t \cos \omega t\}$

In exercises 11–12,

(i) Graph the forcing term and write it in terms of the unit step function $\mathcal{U}$.

(ii) Solve the initial-value problem, using Laplace transforms.

(iii) Give the interval(s) within which the highest derivative of the solution appearing in the equation is continuous.

(iv) Verify that your solution satisfies the differential equation within each interval of continuity.

(v) As appropriate, confirm each transform and solution result with DELAB.

11. $y' + 2y = g(t)$, $y(0) = 5$,

$g(t) = \begin{cases} -2, & 0 \le t < 4 \\ 4, & 4 \le t < 8 \\ 0, & 8 \le t < \infty \end{cases}$

12. $y'' - 9y = g(t)$, $y(0) = 4$, $y'(0) = 0$,

$g(t) = \begin{cases} 0, & 0 \le t < 2 \\ 2, & 2 \le t < \infty \end{cases}$

13. Assuming that it exists, find the Laplace transform of the solution of $x' + x = g(t)$, $x(0) = -3$, where $g(t)$ is the isolated pulse and cosine wave of exercise 7, page 555,

$g(t) = \begin{cases} 1, & 0 \le t \le 2\pi \\ \cos t, & 2\pi \le t \le 7\pi/2 \\ 0, & 7\pi/2 < t < \infty \end{cases}$

Work directly from the differential equation.

14. Draw a graph like that of figure 10.10, page 547, illustrating the "turn-on" property of the unit step function product $f(t - a)\mathcal{U}(t - a)t - a$ for the function appearing in example 24, page 549,

$g(t) = \begin{cases} 0, & 0 \le t \le 2 \\ \sin t, & 2 < t < \infty \end{cases}$

In exercises 15–24, predict the general behavior of the function of t whose transform is given, then find the function and compare it with your prediction. Comment on differences. As appropriate, confirm each result with DELAB.

15. $\dfrac{4}{s^2 + 5s + 6}$

16. $\dfrac{4e^{-2s}}{s^2 + 5s + 6}$

17. $\dfrac{2s}{s^2 - 4}$

18. $\dfrac{2s}{s^2 + 4}$

19. $\dfrac{2se^{-5s}}{s^2 + 4}$

20. $\dfrac{s - 1}{(s + 1)(s^2 - 4)}$

21. $\dfrac{4}{s^2 + 2s + 5}$

22. $\dfrac{3}{s^3 + 3s^2 + 2s}$

23. $\dfrac{2s + 6}{s^2 + 6s + 18}$

24. $\dfrac{3s}{s^2 - 3s + 10}$

In exercises 25–28, the transform of the response of system is given. In each case,

(i) Determine whether the response contains oscillatory terms. If so, give their frequencies.

(ii) Determine whether the response includes exponential decay or growth. If so, give the factor in the exponential (or the corresponding time constant).

(iii) Determine if the response contains any terms that switch on or off. If so, give the switching time(s).

Do *not* invert the transforms. (As appropriate, use DELAB to check your results.)

25. $\dfrac{s^2+1}{s^4-16}$

26. $\dfrac{s^2 e^{-2s}}{(s^2+4)(s^2+4s+3)}$

27. $\dfrac{2-4e^{-3s}}{(s-1)^3}$

28. $\dfrac{3}{(s-3)(s^2+12s+45)}$

Laplace transforms can be used to solve systems of constant-coefficient differential equations. The steps to follow are: Introduce variables for the transforms of the unknowns (e.g., $Y(s) = \mathcal{L}\{y(t)\}$, $Z(s) = \mathcal{L}\{z(t)\}$), transform the system, solve for $Y(s)$, $Z(s)$, and invert. Use this technique to solve exercises 29–36. Confirm solutions with DELAB.

29. $y' = y + z$, $y(0) = 1$
$z' = -y + z$, $z(0) = 2$

30. $y' = z$, $y(0) = 0$
$z' = y$, $z(0) = 1$

31. $y' = -4y + 2z$, $y(0) = 3$
$z' = -3y + 3z$, $z(0) = 4$

32. $y' = -4z$, $y(0) = 0$
$z' = -4y$, $z(0) = 2$

33. $y' = y/2 + \sqrt{3}\,z/2 + 2t$, $y(0) = 6$
$z' = \sqrt{3}\,y/2 - z/2 - 4t^2$, $z(0) = 4$

34. $y' = z + 4\cos t$, $y(0) = 1$
$z' = y$, $z(0) = 0$

35. $y' = y + 3z + e^{2t}$, $y(0) = 1$
$z' = 3y + z - 1$, $z(0) = 0$

36. $y' = y - z - 2$, $y(0) = 1$
$z' = 2y + 4z - 2e^{-2t}$, $z(0) = 0$

Exercises 37–40 ask you to use Laplace transforms to determine the charge on the capacitor in the series *RLC* circuit of figure 6.10, page 285, section 6.3, when the voltage source E varies as described. Point out the terms in your solution that reflect dramatic changes in E.

The parameters of the circuit in figure 7.10 are $R = 200\Omega$, $L = 100\mu\text{H}$, and $C = 100\,\text{pF}$. Time is measured in microseconds and voltage in millivolts. Assume the voltage drop across the capacitor is initially zero and that no current is flowing at $t = 0$.

37. $E(t) = \begin{cases} 10, & 0 \le t < 4 \\ 0, & 4 \le t \end{cases}$

38. $E(t) = \begin{cases} t, & 0 \le t < 6 \\ 6, & 6 \le t \end{cases}$

39. $E(t) = \begin{cases} 0, & 0 \le t < 2 \\ 5, & 2 \le t \end{cases}$

40. E tapers linearly from 5 mV initially to 0 at 4 μs, where it remains thereafter.

41. Consider the series *RLC* circuit model with zero initial conditions: $Lq'' + Rq' + q/C = E(t)$, $q(0) = q'(0) = 0$; see section 5.3, page 226. Show that the solution of this initial-value problem can be written

$$q(t) = \int_0^t E(r)g(t-r)\,dr,$$

where

$$g(t) = \frac{e^{-Rt/2L}\sin\omega t}{\omega L}$$

is the inverse transform of the *transfer function*

$$G(s) = \frac{1}{Ls^2 + Rs + 1/C}.$$

Here $\omega = \sqrt{4L/C - R^2}/2L$. (*Hint:* Write the transform of the solution in terms of $G(s)$ and use the convolution theorem.)

10.7 ■ CHAPTER PROJECTS

1. **General inverse formula.** Applying the Laplace transform to a differential equation usually leads to an expression of the form

$$Y(s) = \frac{N(s)}{D(s)},$$

where N and D are polynomials in s with real coefficients and the degree of N is no greater than the degree k of D. This project asks you to develop a general expression for the solution of such a differential equation by inverting $Y(s)$.

(a) Let $r_1, \ldots, r_k$ denote the roots of D, each repeated according to its multiplicity. Find the numerators A_i in the partial-fraction expansion

$$\frac{N(s)}{D(s)} = \frac{A_1}{s - r_1} + \cdots + \frac{A_k}{s - r_k}$$

in terms of $N(r_i)$, $D'(r_i)$. (Multiply by $s - r_i$ and let $s \to r_i$.) Then write $y(t) = \mathcal{L}^{-1}\{Y(s)\}$ as a sum of exponentials using the A_i as coefficients. (How are these related to those of the characteristic polynomial?)

(b) Since complex roots must occur in conjugate pairs, argue that the corresponding partial-fraction coefficients A_i are also complex conjugate pairs. Further simplify the expression for $y(t)$ from part (a) by replacing complex exponentials with sines and cosines.

2. **Stability of a time-delayed logistic model.** The familiar logistic equation $P' = (a - sP)P$ could be thought of as a simple population equation $P' = kP$ whose growth rate was modified to decrease with increasing population. That is, k is replaced by $a - sP$ to reflect decreasing birth rate as larger populations competed for the same resources.

In fact, there ought to be some delay in this feedback process. Beyond the fixed delay of gestation periods, a range of prior experiences should affect the present birth rate, with the most immediate and the most distant experiences having the least impact on today's birth rate. This argument suggests replacing k by

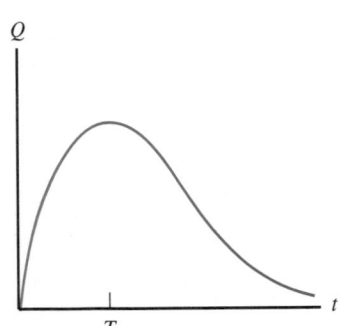

Q

T

FIGURE 10.17 The weighting function Q in the modified logistic equation.

$$a - s \int_0^t P(r)Q(t - r)\, dr,$$

where Q might have the general shape shown in figure 10.17. The experience of large populations T years ago, near the peak in the Q curve, would tend to reduce today's birth rates.

To normalize, assume $\int_0^\infty Q(r)\, dr = 1$. One choice of Q is

$$Q(r) = \frac{r^2}{T^2} e^{-r/T}.$$

In this new model, populations satisfy the *integrodifferential equation*

$$P'(t) = \left(a - s \int_0^t P(r)Q(t - r)\, dr\right) P(t).$$

(a) Derive this equation.

(b) Show that its equilibria are $P = 0$, a/s.

(c) Show that small perturbations $p(t)$ of either of these equilibria satisfy

$$p'(t) = -a \int_0^t p(r)Q(t-r)\,dr.$$

Use Laplace transforms to determine the behavior of p and, hence, the stability of these steady states.

(d) How does the behavior of this model differ from that of the simple logistic model?

3. **Transfer functions.** Think of the forcing term in a differential equation as its *input*. Think of the forced response (the solution subject to *zero* initial conditions) as the *output*. Then the *transfer function* of the system modeled by the equation is the ratio of the Laplace transform of the output to the Laplace transform of the input. Analysis of the transfer function provides a picture of the response of the system.

(a) Show that the transfer function $G(s)$ of the generic second-order, linear, constant-coefficient equation

$$b_2 x'' + b_1 x' + b_0 x = f(t)$$

is

$$G(s) = \frac{X(s)}{F(s)} = \frac{1}{b_2 s^2 + b_1 s + b_0},$$

where $F(s) = \mathcal{L}\{f(t)\}$, $X(s) = \mathcal{L}\{x(t)\}$.

(b) Control engineers say, "In transform space, the output is the product of the transfer function and the input. Multiplication in transform space is convolution in the time domain." Explain these statements using the relation $X(s) = G(s)F(s)$.

(c) The *poles* (or singularities) of the transfer function $G(s)$ are the values of s at which its denominator $F(s) = 0$. What is the connection between the poles of the transfer function and the roots of the characteristic polynomial of the original differential equation?

Justify the following statements.

(i) If the coefficients $b_2 > 0$ and $b_1, b_0 \geq 0$, then the poles of the transfer function lie in the left half of the complex plane; i.e., the poles all have nonnegative real part.

(ii) If the poles of the transfer function are pure imaginary and $b_2 > 0$, $b_1, b_0 \geq 0$, then the system is undamped.

(iii) If the poles of the transfer function are real and $b_2 > 0$, $b_1, b_0 \geq 0$, then the system is overdamped.

(iv) If the poles of the transfer function are complex with negative real part and $b_2 > 0$, $b_1, b_0 \geq 0$, then the system is underdamped.

(v) If a pole of the transfer function has a positive real part, then the response of the system can grow without bound.

(d) Find the transfer function for a *first-order* constant-coefficient system $b_1 x' + b_0 = f(t)$. Use it to justify the following statements about first-order systems:

(i) The transfer function of a time-invariant (or constant-coefficient), first-order system can be written

$$G(s) = \frac{c}{1 + Ts},$$

where c is a constant and T is the time constant of the system. (Recall that the *time constant* of e^{-rt} is $1/r$.)

(ii) At $t = T$ the *unit step response* of the system (the response to the forcing term $f(t) = \mathcal{U}(t)t$ with zero initial condition) is 63.2% of its initial value.

(iii) The initial slope of the unit step response is $1/T$.

4. **Unit step function response.** A prototype control problem is maintaining the relative position of a mass suspended vertically from a spring as the entire assembly is moved up or down. (Automobiles with self-leveling suspension systems and passenger liners with underwater vanes that compensate for rolling in rough seas incorporate more complicated versions of such control systems.) If a sensor detects that the position of the mass has shifted, then the top of the spring is moved up or down by an amount proportional to the error, that is, proportional to the distance between the current position of the mass and the desired position.

Of course, shifting the top of the spring will disturb the mass, and some time must pass before it settles back into position. The time required to return to equilibrium and the maximum displacement before it settles down are important design parameters.

To standardize the discussion of such response characteristics, control engineers analyze the system subject to forcing by a unit step function. For the prototype spring-mass control problem, the governing initial-value problem for a unit step function response analysis would be

$$mx'' + px' + kx = k\mathcal{U}(t), \quad x(0) = x'(0) = 0.$$

The important design characteristics of the response of such a system are:

rise time, the time required to shift from 10% to 90% of the steady-state position of the mass;

peak time, the time required to reach the absolute maximum displacement of the mass;

maximum overshoot, the difference between the maximum displacement of the mass and its steady-state value;

settling time, the time required for the position of the mass to settle to within $\pm 10\%$ of its steady-state value. (This range is called the **tolerance band**.)

(a) Sketch a diagram of such a spring-mass control system. Explain the forcing term $k\mathcal{U}(t)$.

(b) Graph possible responses of the system. Does the level of damping affect the response? Mark the design characteristics (rise time, etc.) on the graphs.

(c) Find expressions for these design characteristics in terms of the system parameters m, p, and k. Are the relationships intuitively reasonable?

(d) Describe the behavior of the system in Laplace transform variables. How do the design characteristics reveal themselves in the s-domain?

(e) Are there design trade-offs among the various response characteristics? Do you have any overall recommendations for the design of a device that attempts to maintain the mass in a fixed position?

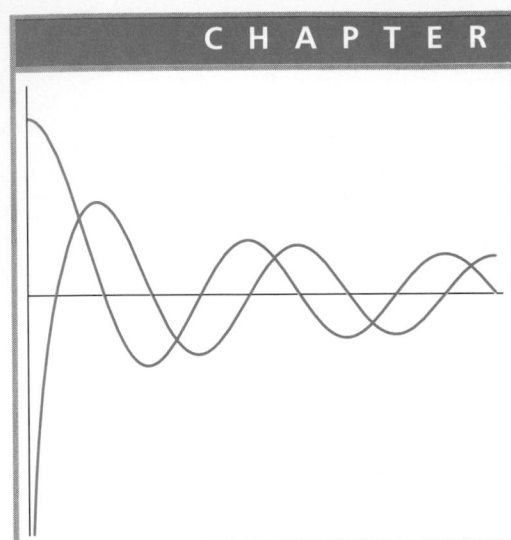

11

More Analytic Tools for Two Dimensions

11.1 ■ REDUCTION OF ORDER

Reduction of order is a technique for finding a second solution of a homogeneous linear equation when we have somehow found a first solution. The steps in the method are straightforward:

1. Given one solution y_1 of a linear, homogeneous, second-order equation, assume there is a second solution in the form $y_2(x) = u(x)y_1(x)$. Substitute this form into the homogeneous equation.
2. Simplify to find a *first-order* equation for u', the derivative of the unknown function. Solve and integrate to obtain u and the second solution $y_2 = uy_1$.

To motivate this technique, suppose we naively applied the characteristic equation method to

$$4y'' - 12y' + 9y = 0,$$

which is solved completely in example 19, page 263, of section 6.3. We would assume $y = e^{rx}$, substitute in the differential equation, and find the characteristic equation

$$4r^2 - 12r + 9 = 4\left(r - \frac{3}{2}\right)^2 = 0.$$

Its single root is $r = \frac{3}{2}$, and one solution of $4y'' - 12y' + 9y = 0$ is $y_1 = e^{3x/2}$.

We know the characteristic equation formalism for repeated real roots: Obtain the second solution as x times the first, $y_2 = xe^{rx}$. Generalizing this idea suggests seeking a second solution in the form

Assume $y_2 = uy_1$.

$$y_2(x) = u(x)y_1(x) = u(x)e^{3x/2},$$

where we must determine the unknown function u. We find that

$$y_2' = u'e^{3x/2} + u\frac{3}{2}e^{3x/2},$$

$$y_2'' = u''e^{3x/2} + 3u'e^{3x/2} + \frac{9u}{4}e^{3x/2}.$$

Substituting into the differential equation yields

Substitute y_2 into the equation.

$$4y_2'' - 12y_2' + 9y_2 = 4u''e^{3x/2} + u'(12e^{3x/2} - 12e^{3x/2}) \tag{11.1}$$

$$+ u\left(4\frac{9}{4}e^{3x/2} - 12\frac{3}{2}e^{3x/2} + 9e^{3x/2}\right). \tag{11.2}$$

The last term in parentheses is zero because $e^{3x/2}$ is a solution of the homogeneous equation. To summarize, we have

$$4y_2'' - 12y_2' + 9y_2 = (4u'')e^{3x/2} = 0,$$

or

$$4u'' = 0.$$

At first glance, the second-order differential equation $4u'' = 0$ for the unknown factor u does not seem to be much of an advance over the original second-order equation (although we could easily integrate this one). A closer look reveals, however, that we actually have a *first-order* equation in the unknown u',

$$4(u')' = 0.$$

To formalize that observation, define $v(x) = u'(x)$ and rewrite $4u'' = 0$ as

First-order equation for $v = u'$

$$4v' = 0,$$

a *first-order* equation. We have reduced the problem of finding another solution of the second-order equation to the problem of solving the first-order equation $4v' = 0$. Hence, the name *reduction of order*.

The first-order equation $4v' = 0$ is solved by a simple antidifferentiation,

Solve for v.

$$v(x) = C,$$

where C is an arbitrary constant. Since $u' = v = C$, another antidifferentiation yields $u(x) = Cx + D$, where D is another arbitrary constant. Because we need only the simplest form of u that meets our needs, not a general family of such functions, we choose $C = 1$ and $D = 0$. From $u(x) = x$, we have the second solution

Use u to write y_2.

$$y_2(x) = xy_1(x) = xe^{3x/2}.$$

To better appreciate why this reduction of order occurs, repeat the calculation that leads from the original second-order equation to the first-order equation $4v' = 0$, this time using the symbol y_1 in place of $e^{3x/2}$. We begin with $y_2 = uy_1$ and find

$$y_2' = u'y_1 + uy_1',$$
$$y_2'' = u''y_1 + 2u'y_1' + uy_1''.$$

Substituting in $4y'' - 12y' + 9y = 0$ yields

$$4y_2'' - 12y_2' + 9y_2 = 4y_1u'' + (8y_1' - 12y_1)u' + u(4y_1'' - 12y_1' + 9y_1).$$

As in equation (11.1), the last expression in parentheses, the coefficient of u, is zero precisely because y_1 is a solution of the original homogeneous differential equation $4y'' - 12y' + 9y = 0$. Hence, we have a homogeneous equation of first order for the unknown u',

$$4y_2'' - 12y_2' + 9y_2 = (4y_1)u'' + (8y_1' - 12y_1)u' = 0.$$

We can argue in general that the substitution $y_2 = uy_1$ reduces *any* homogeneous second-order linear differential equation of which y_1 is a solution to a first-order equation for u', provided the coefficient of y'' is never zero. (See exercise 12.)

■ **EXAMPLE 1**   *One solution of*

$$y'' + y' - 12y = 0$$

is $y_1 = e^{-4x}$. Find a second linearly independent solution.

Assume $y_2 = uy_1$.

We assume the form $y_2 = u(x)e^{-4x}$ for the second solution, substitute in $y'' + y' - 12y = 0$, and obtain

Substitute y_2 into the differential equation.

$$y_2'' + y_2' - 12y_2 = u''e^{-4x} + u'(-8e^{-4x} + e^{-4x})$$
$$+ u(16e^{-4x} - 4e^{-4x} - 12e^{-4x})$$
$$= u''e^{-4x} - 7u'e^{-4x} = 0.$$

As expected, the coefficient of u vanished because y_1 is a solution of the homogeneous equation $y'' + y' - 12y = 0$. We have

$$u'' - 7u' = 0,$$

or, if $v = u'$, the first-order equation

First-order equation for $v = u'$

$$v' - 7v = 0.$$

One solution of the reduced-order equation $v' - 7v = 0$ is $v(x) = e^{7x}$. Since $u = \int v \, dx$,

Solve for v. Find $u = \int v$.

$$u(x) = \int e^{7x} \, dx = \frac{e^{7x}}{7},$$

where we dropped the constant of integration. A second solution of $y'' + y' - 12y = 0$ is

$$uy_1 = \frac{e^{7x}}{7} e^{-4x} = \frac{e^{3x}}{7}.$$

Since a constant multiple of a solution of a linear homogeneous equation is again a solution of that equation, we can ignore the factor $\frac{1}{7}$ and take

Use u to write y_2.

$$y_2(x) = e^{3x}.$$

A straightforward computation shows that $W(e^{-4x}, e^{3x}) \neq 0$ for all x, verifying the linear independence of this second solution. ■

Stop and Think **11.1** Is this the same pair of solutions we would have obtained by the characteristic equation method?

■ **EXAMPLE 2** *One solution of*

$$x^2 y'' - 3xy' + 3y = 0$$

is $y_1(x) = x$. Find a second solution.

Since the given equation is linear and homogeneous, we can use reduction of order to find a second solution in the form $y_2(x) = u(x)x$. Substituting y_2 in $x^2 y'' - 3xy' + 3y = 0$ and simplifying yields

Assume $y_2 = uy_1$.

Substitute y_2 into the differential equation.

$$x^2 y_2'' - 3xy_2' + 3y_2 = x^3 u'' - x^2 u' = 0.$$

The usual reduction of order substitution $v = u'$ reduces the second-order equation $x^3 u'' - x^2 u'$ to the first-order equation

First-order equation for $v = u'$

$$x^3 v' - x^2 v = 0.$$

We solve $x^3 v' - x^2 v = 0$ by separating variables,

$$\frac{dv}{v} = \frac{dx}{x},$$

Solve for v.

and integrating to obtain $v(x) = Cx$. Since we need only one solution of $x^3 v' - x^2 v = 0$, we may choose the arbitrary constant C for convenience; e.g., $C = 1$. Then, from $u' = v = x$, we obtain

Find $u = \int v$.

$$u(x) = \frac{x^2}{2},$$

where we have again chosen an arbitrary constant of integration to give the simplest possible expression.

Hence, a second solution of $x^2 y'' - 3xy' + 3y = 0$ is

Use u to write y_2.

$$uy_1 = \frac{x^2}{2} x = \frac{x^3}{2}.$$

Since we are solving a homogeneous equation, we could multiply this solution by a constant to obtain the simplified second solution

$$y_2 = x^3. \quad \blacksquare$$

11.1.1   Summary

The method of reduction of order obtains a second solution y_2 of a linear, homogeneous, second-order differential equation given a first solution y_1.

REDUCTION OF ORDER

1. Assume $y_2(x) = u(x)y_1(x)$.

2. Substitute y_2 into the given equation. After the simplification that results because y_1 solves the given homogeneous equation, obtain a second-order equation for u.

3. Let $u' = v$ in an equation that contains only u'' and u'. Solve the first-order equation for v, and integrate once to obtain u.

4. Write the second solution as $u(x)y_1(x)$.

11.1.2 Exercises

EXERCISE GUIDE	
To gain experience . . .	**Try exercises**
Using reduction of order	1–11
Solving initial-value problems	8–10
With the foundations of reduction of order	12–14

In exercises 1–7, use the given solution and reduction of order to find a second solution of each of the following differential equations. If you believe that the method is not applicable to a given equation, explain why. As appropriate, confirm your results with DELAB.

1. $x^2 y'' - 3xy' + 3y = 0$, $y_1 = x^3$

2. $y'' - y = 0$, $y_1 = e^{1-x}$

3. $y'' + y' - 12y = 0$, $y_1 = e^{3x}$

4. $y'' + y = \cos \pi x$, $y_1 = (1 - \pi^2)^{-1} \cos \pi x$

5. $x^2 y'' + 2xy' - 6y = 0$, $y_1 = x^{-3}$

6. $x^2 y'' - 3xy' + 4y = 0$, $y_1 = x^2$

7. $3x^2 y'' - 3xy' + 6y = 0$, $y_1 = x \cos(\ln x)$

In exercises 8–10, use the information given to solve the initial-value problem. Confirm your results with DELAB.

8. $t^2 x'' - 3tx' + 4x = 0$, $x(1) = 1$, $x'(1) = 0$. One solution of this equation is t^2.

9. $x'' + x' - 12x = 2t$, $x(0) = 0$, $x'(0) = 4$. One solution of the corresponding homogeneous equation is e^{3t}.

10. $3t^2 w''(t) - 3t w'(t) + 6w(t) = 0$, $w(1) = 2$, $w'(1) = 1$. One solution of this equation is $t \cos(\ln t)$.

11. Example 2 sought a second solution of
$$x^2 y'' - 3xy' + 3y = 0$$
in the form $y_2 = u(x)x$, given the first solution $y_1 = x$.

   (a) Supply the details of the calculations that led from the substitution of y_2 into $x^2 y'' - 3xy' + 3y = 0$ to a second-order equation for u,
   $$x^2 y_2'' - 3xy_2' + 3y_2 = x^3 u'' - x^2 u' = 0.$$

   (b) Make the substitution $u' = v$ in $x^3 u'' - x^2 u' = 0$ and verify by substituting the solution of the resulting first-order equation obtained by example 25 using separation of variables.

   (c) Obtain an expression for $u(x)$. Show that you can ultimately derive the same second solution as in example 2, $y_2 = x^3$.

12. The text claims

   We can argue in general that the substitution $y_2 = uy_1$ reduces *any* homogeneous second-order linear differential equation of which y_1 is a solution to a first-order equation for u', provided the coefficient of y'' is never zero.

   Verify this claim by showing that if y_1 is a solution of
   $$y'' + P(x)y' + Q(x)y = 0,$$
   then the substitution $y_2 = uy_1$ reduces the given equation to
   $$y_1 u'' + (2y_1' + Py_1)u' = 0.$$
   Explain why this demonstration verifies the claim. Make explicit note of the step where you use the fact that y_1 is a solution of the original equation.

13. The preceding problem asks you to show that the substitution $y_2 = uy_1$ reduces the equation
$$y'' + P(x)y' + Q(x)y = 0$$
to
$$y_1 u'' + (2y_1' + Py_1)u' = 0,$$
where y_1 is a solution of the given equation $y'' + P(x)y' + Q(x)y = 0$. Follow the steps given here to show that a second solution of this differential equation is
$$y_2(x) = y_1(x) \int y_1^{-2}(x) \exp\left(-\int^x P(s)\,ds\right)\,dx.$$

   (a) Use the substitution $u' = v$ to rewrite $y_1 u'' + (2y_1' + Py_1)u' = 0$ as a first-order equation.

   (b) Solve the resulting first-order equation. (*Hint*: Use (4.8), page 150.)

   (c) Integrate once to obtain u from the expression for v.

   Compare the complexity of this formula with the relative simplicity of the steps in the method of reduction of order. It is certainly simpler and more reliable to understand the method and apply it to each problem than it is to memorize this formula.

14. Exercise 13 derives a general expression for a second solution y_2 of a linear, second-order, homogeneous equation. Use the linear independence test to show that y_2 is linearly independent of y_1 on any interval on which $y_1 \neq 0$. (*Hint*: Compute the Wronskian, using $y_2 = uy_1$, not the formula obtained here. You will find $W(y_1, y_2) = u'y_1^2$. Can you guarantee $u' \neq 0$? Why would y_1 and y_2 automatically be linearly dependent if $u'(x) = 0$ throughout the interval of interest?) Why is the restriction $y_1 \neq 0$ consistent with the formula obtained in the preceding problem?

11.2 ■ CAUCHY-EULER EQUATIONS

Example 2 of the preceding section uses reduction of order to find that a second solution of
$$x^2 y'' - 3xy' + 3y = 0$$
is x^3, given that a first solution is x. How was that first solution derived?

   One approach is to observe that the power of x in each coefficient is the same as the order of the derivative in that term; x^2 multiplies y'', and so on. Consequently, assuming a solution of the form
$$y(x) = x^r$$

would lead to a kind of characteristic equation for the unknown constant r. Each derivative of y would reduce the exponent by 1, but that reduction would be compensated by the power of x in the coefficient. Every term would still contain x^r, and we could divide out that common factor if $x > 0$.

Explicitly, substituting $y = x^r$ into $x^2 y'' - 3xy' + 3y = 0$ leads to

$$x^2 y'' - 3xy' + 3y = x^2 r(r-1)x^{r-2} - 3xr x^{r-1} + 3x^r$$
$$= r(r-1)x^r - 3rx^r + 3x^r = 0.$$

As long as $x > 0$, we can divide x^r out of the last equation to obtain a quadratic equation for r,

$$r(r-1) - 3r + 3 = r^2 - 4r + 3 = (r-3)(r-1) = 0.$$

The roots of this *indicial equation* are obviously $r = 1, 3$. Hence, solutions of the differential equation $x^2 y'' - 3xy' + 3y = 0$ are $x, x^3, x > 0$.

We have found a pair of linearly independent (verify!) solutions of the given equation by exploiting the observation that the power of x matches the order of the derivative in each term.

Equations of this form are called **Cauchy-Euler equations**. (They are also known as **equidimensional equations**.)

Specifically, Cauchy-Euler equations are linear, second-order, homogeneous equations that can be written in the form

CAUCHY-EULER EQUATION

$$x^2 y'' + b_1 xy' + b_0 y = 0,$$

where b_1, b_0 are constants. To solve such equations,

1. Assume a solution of the form $y = x^r$ and substitute into the differential equation $x^2 y'' + b_1 xy' + b_0 y = 0$.
2. Divide out x^r, $x > 0$ and solve the resulting quadratic equation for r,

INDICIAL EQUATION

$$r(r-1) + b_1 r + b_0 = 0.$$

If the **indicial** (or *auxiliary*) **equation** $r(r-1) + b_1 r + b_0 = 0$ has two distinct real roots r_1, r_2, then the differential equation has the pair of linearly independent solutions

Solutions from distinct real roots

$$x^{r_1}, \quad x^{r_2}.$$

These are shown to be linearly independent for $x > 0$. (See exercise 15.) Later in this section we explore the other possibilities: A single repeated root and complex conjugate roots of the indicial equation.

■ **EXAMPLE 3**   *Find a general solution valid for $x > 0$ of*

$$x^2 y'' + xy' - 4y = 0.$$

This equation is certainly of Cauchy-Euler type, for the power of x in each coefficient matches the order of the derivative of the unknown y. Assuming a solution in the form $y = x^r$ and substituting leads to

Substitute x^r.

$$x^2 y'' + xy' - 4y = r(r-1)x^r + rx^r - 4x^r = 0.$$

Since $x > 0$, we may divide by x^r to obtain

Indicial equation

$$r(r-1) + r - 4 = r^2 - 4 = (r-2)(r+2) = 0.$$

Obviously, $r = \pm 2$, and two solutions of the differential equation are x^2, x^{-2}. We find $W(x^2, x^{-2}) = -4/x$; these solutions are linearly independent for $x > 0$. A general solution of $x^2 y'' + xy' - 4y = 0$ is

General solution

$$y_g(x) = C_1 x^2 + C_2 x^{-2}, \quad x > 0. ■$$

■ **EXAMPLE 4** *Find a general solution valid for $x > 0$ of*

$$x^2 y'' - xy' + y = 0.$$

We seek solutions of this Cauchy-Euler equation in the form $y = x^r$. The usual substitution of x^r in $x^2 y'' - xy' + y = 0$ and division by x^r leads to

Substitute x^r; obtain indicial equation.

$$r(r-1) - r + 1 = r^2 - 2r + 1 = (r-1)^2 = 0.$$

The quadratic has the single repeated root $r = 1$. One solution of $x^2 y'' - xy' + y = 0$ is $y_1 = x$. What is another?

We turn to reduction of order and assume there is a second solution of the form $y_2 = u(x)x$. Substituting this assumed solution in $x^2 y'' - xy' + y = 0$ yields

$$x^2 y_2'' - xy_2' + y_2 = x^2(u''x + 2u') - x(u'x + u) + ux$$
$$= x^3 u'' + x^2 u' = 0.$$

The reduction of order substitution $u' = v$ yields the usual first-order equation

$$x^3 v' + x^2 v = 0,$$

whose solution by separation of variables is

$$\frac{dv}{v} = -\frac{dx}{x},$$

$$v = Ce^{-\ln x} = Ce^{\ln(1/x)} = \frac{C}{x}.$$

We choose $C = 1$ and obtain

$$u(x) = \int v(x)\, dx = \int \frac{1}{x}\, dx = \ln x + D.$$

Choosing the arbitrary constant $D = 0$ finally yields the second solution,

Second solution from reduction of order

$$y_2(x) = x \ln x.$$

In this example with repeated roots of the quadratic equation defining r, we find that the second solution is simply the first multiplied by $\ln x$.

Since $W(x, x \ln x) = x$, the pair of solutions x, $x \ln x$ is linearly independent for $x > 0$. A general solution of $x^2 y'' - xy' + y = 0$ is

General solution

$$y_g(x) = C_1 x + C_2 x \ln x, \quad x > 0. \ \blacksquare$$

The situation observed in this example holds in general. If the quadratic equation $r(r-1) + b_1 r + b_0 = 0$ has the single repeated root r, then two linearly independent solutions of the differential equation $x^2 y'' + b_1 xy' + b_0 y = 0$ are

Solutions from repeated root

$$x^r, \quad x^r \ln x, \quad x > 0.$$

See exercise 16.

■ **EXAMPLE 5**   *Find a general solution of the diffusion equation in a circle,*

$$xc'' + c' = 0,$$

which was derived in example ?, page ?, to model the concentration $c(x)$ of salt in a pond.

To see that this is indeed a Cauchy-Euler equation, multiply the given expression by x: $x^2 c'' + xc' = 0$.

Substitute x^r.

Assuming a solution of the form $c(x) = x^r$ and substituting into $x^2 c'' + xc' = 0$, we find that r must satisfy

Indicial equation

$$r(r - 1) + r = r^2 = 0.$$

The (repeated) roots of this indicial equation are $r = 0$. Linearly independent solutions are $x^0 = 1$ and $x^0 \ln x = \ln x$, $x > 0$. A general solution of $xc'' + c' = 0$ is

General solution

$$c_g(x) = C_1 + C_2 \ln x, \quad x > 0. \ \blacksquare$$

■ **EXAMPLE 6**   *Find two solutions of*

$$x^2 y'' + 5xy' + 13y = 0$$

that are linearly independent for $x > 0$.

Since $x^2 y'' + 5xy' + 13y = 0$ has the form of a Cauchy-Euler equation, we assume $y = x^r$ and substitute. We obtain

Substitute x^r.

$$x^2 y'' + 5xy' + 13y = r(r - 1)x^r + 5rx^r + 13x^r = 0.$$

The unknown constant r evidently must satisfy

Indicial equation

$$r(r-1) + 5r + 13 = r^2 + 4r + 13 = 0.$$

The quadratic formula reveals that the roots of this equation are $r = -2 \pm 3i$. Two solutions of $x^2 y'' + 5xy' + 13y = 0$ are certainly x^{-2+3i}, x^{-2-3i}, but they are not real-valued. Using the identity

$$x = e^{\ln x}$$

and Euler's formula,

$$e^{i\beta} = \cos\beta + i\sin\beta,$$

we can write the first of these two solutions as

Complex-valued solutions

$$x^{-2+3i} = x^{-2} x^{3i} = x^{-2}(e^{\ln x})^{3i}$$
$$= x^{-2} e^{i(3\ln x)} = x^{-2}[\cos(3\ln x) + i\sin(3\ln x)].$$

Similarly, the second may be written

$$x^{-2-3i} = x^{-2}[\cos(3\ln x) - i\sin(3\ln x)].$$

Now the principle of superposition guarantees that linear combinations of these two solutions will again be a solution of the homogeneous linear equation $x^2 y'' + 5xy' + 13y = 0$. We can form linear combinations of x^{-2+3i} and x^{-2-3i} that will give us real-valued solutions. If we take $C_1 = C_2 = \frac{1}{2}$ in the linear combination

Use superposition.

$$C_1 x^{-2+3i} + C_2 x^{-2-3i}$$
$$= x^{-2}[(C_1 + C_2)\cos(3\ln x) + i(C_1 - C_2)\sin(3\ln x)],$$

we obtain

Real-valued solutions

$$y_1(x) = x^{-2}\cos(3\ln x).$$

The choice $C_1 = 1/2i$, $C_2 = -1/2i$ yields

$$y_2(x) = x^{-2}\sin(3\ln x).$$

We leave it to you to verify that y_1, y_2 are indeed linearly independent for $x > 0$. ∎

The arguments of example 6 obviously extend directly to the general situation in which the roots of the quadratic equation $r(r-1) + b_1 r + b_0 = 0$ are the complex conjugate pair $r = \alpha \pm i\beta$. In analogy with the result of the last example, the linearly independent solutions of $x^2 y'' + b_1 xy' + b_0 y = 0$ are

SOLUTIONS FOR COMPLEX ROOTS

$$x^\alpha \cos(\beta \ln x), \quad x^\alpha \sin(\beta \ln x), \quad x > 0.$$

See exercise 17.

■ **EXAMPLE 7**   *Find two linearly independent solutions of*

$$(t-4)^2 y'' - (t-4)y' + 5y = 0.$$

State the values of t for which the solutions you find are linearly independent.

Although this equation is not precisely in the Cauchy-Euler form $x^2 y'' + b_1 x y' + b_0 y = 0$, it obviously could be transformed by the substitution $x = t - 4$. Alternatively, we could seek directly solutions in the form $y(t) = (t-4)^r$.

Substitute $(t-4)^r$.   To pursue the latter approach, substitute $y(t) = (t-4)^r$ in $(t-4)^2 y'' - (t-4)y' + 5y = 0$ and simplify in the usual manner. We obtain

$$(t-4)^2 y'' - (t-4)y' + 5y = r(r-1)(t-4)^r - r(t-4)^r + 5(t-4)^r.$$

If $t > 4$, then r must be a root of

Indicial equation
$$r(r-1) - r + 5 = r^2 - 2r + 5 = 0.$$

The quadratic formula reveals $r = 1 \pm 2i$.

A pair of solutions of $(t-4)^2 y'' - (t-4)y' + 5y = 0$ is evidently

Complex-valued solutions
$$(t-4)^{1+2i}, \quad (t-4)^{1-2i},$$

but they are not real-valued. However, they may rewritten using Euler's formula, just as in example 6. We obtain

$$(t-4)^{1+2i} = (t-4)\,(\cos(2\ln(t-4)) + i\sin(2\ln(t-4)))]$$

and

$$(t-4)^{1-2i} = (t-4)\,(\cos(2\ln(t-4)) - i\sin(2\ln(t-4)))].$$

Mimicking the linear combinations that lead to the real-valued solutions in example 6, we combine the two preceding expressions to obtain the real-valued solutions

Real-valued solutions
$$(t-4)\cos\,[2\ln(t-4)], \quad (t-4)\sin\,[2\ln(t-4)], \quad t > 4.$$

We leave it to you to prove that this pair of solutions is linearly independent for $t > 4$. ■

All the preceding work has treated x^r with $x > 0$. How do we solve

$$x^2 y'' + b_1 x y' + b_0 y = 0$$

when $x < 0$? In chapter project 1, we ask you to show that the appropriate solution form is $y = (-x)^r$ for $x < 0$. The same solution forms appear but with x replaced by $-x$. To encompass both $x > 0$ and $x < 0$ in the following summary, we replace x with $|x|$.

11.2.1 Summary

A Cauchy-Euler equation is a second-order, homogeneous, linear differential equation in which the power of the independent variable in the coefficient matches the order of the derivative; i.e., an equation of the form

Cauchy-Euler equation

$$x^2 y'' + b_1 x y' + b_0 y = 0.$$

To solve, assume a solution of the form $|x|^r$, $x \neq 0$, substitute in the differential equation $x^2 y'' + b_1 x y' + b_0 y = 0$, and find that r must satisfy the quadratic indicial equation

Indicial equation

$$r(r-1) + b_1 r + b_0 = 0.$$

If this quadratic has a pair of distinct real roots r_1, r_2, then

Distinct real roots

$$|x|^{r_1}, \quad |x|^{r_2}$$

are linearly independent solutions of $x^2 y'' + b_1 x y' + b_0 y = 0$ for $x \neq 0$. If $r(r-1) + b_1 r + b_0 = 0$ has the single repeated root r, then

Repeated root

$$|x|^r, \quad |x|^r \ln|x|$$

are linearly independent solutions of $x^2 y'' + b_1 x y' + b_0 y = 0$ for $x \neq 0$. If the quadratic indicial equation has the pair of complex conjugate roots $r = \alpha \pm i\beta$, then

Complex conjugate roots

$$|x|^\alpha \cos(\beta \ln|x|), \quad |x|^\alpha \sin(\beta \ln|x|)$$

are linearly independent solutions of $x^2 y'' + b_1 x y' + b_0 y = 0$ for $x \neq 0$.
These solution forms are summarized in table 11.1.

TABLE 11.1 General solution of the Cauchy-Euler equation $x^2 y'' + b_1 x y' + b_0 y = 0$

$x^2 y'' + b_1 x y' + b_0 y = 0$, $y = |x|^r \Rightarrow r(r-1) + b_1 r + b_0 = 0$	
Type of Root	**General Solution for $x \neq 0$**
Real $r_1 \neq r_2$	$C_1 |x|^{r_1} + C_2 |x|^{r_2}$
Real repeated r	$C_1 |x|^r + C_2 |x|^r \ln|x|$
Complex conjugate pair $\alpha \pm i\beta$	$|x|^\alpha (C_1 \cos(\beta \ln|x|) + C_2 \sin(\beta \ln|x|))$

The pair x^r, $x^r \ln x$ is derived from the single solution x^r by reduction of order. The real-valued pair $x^\alpha \cos(\beta \ln x)$, $x^\alpha \sin(\beta \ln x)$ is obtained by rewriting the pair of complex-valued solutions $x^{\alpha \pm i\beta}$, using Euler's formula, and forming linear combinations.

11.2.2 Exercises

EXERCISE GUIDE	
To gain experience ...	**Try exercises**
Solving Cauchy-Euler equations	1–8, 13, 18, 19–20(a), 21
Solving boundary-value problems	19–20(a), 21
Verifying linear independence	1–8, 15(a), 16(c), 18
Checking solutions	9–12
Finding equations, given solutions	9–12
Obtaining real-valued solutions from complex roots	3, 5, 17–18
With the foundations of the Cauchy-Euler method	13–17
Analyzing and interpreting behavior of solutions	19–20(b), 21

In exercises 1–8 find, if possible, a pair of linearly independent solutions of each of the following equations. Verify linear independence, and state the interval upon which it holds. If you can not find such a pair of solutions, explain why each solution method you tried failed. As appropriate, confirm solutions using DELAB.

1. $x^2 y'' - 2xy' + 2y = 0$

2. $x^2 y'' - 3xy' + 4y = 0$

3. $x^2 y'' - xy' + 5y = 0$

4. $x^2 y'' + 2xy' - 6y = 0$

5. $3x^2 y'' - 3xy' + 6y = 0$

6. $x^2 y'' - 2x^2 y' + 2y = 0$

7. $4(2t - 3)^2 u''(t) - 2(2t - 3)u'(t) + 5u(t) = 0$

8. $(1 - x)^2 y'' + 2(1 - x)y' + 2y = 0$

In exercises 9–12, ...

  (i) Write Cauchy-Euler equations that have the given pair of functions as solutions.

  (ii) Verify by substitution that the equation you found is indeed solved by the pair of functions.

9. $x, 1/x$

10. $(2 + x)^3, 2 + x$

11. $x^2 \cos(4 \ln x), x^2 \sin(4 \ln x)$

12. $x^2 \ln x, x^2$

13. Show by substituting the assumed solution form $y = x^r$ into the general Cauchy-Euler equation,

$$x^2 y'' + b_1 xy' + b_0 y = 0,$$

that the unknown constant r is a solution of the quadratic equation

$$r(r - 1) + b_1 r + b_0 = 0.$$

What restrictions must you place on x?

14. The preceding problem shows that the general Cauchy-Euler equation,

$$x^2 y'' + b_1 xy' + b_0 y = 0,$$

has a solution of the form x^r if r is a root of

$$r(r - 1) + b_1 r + b_0 = 0.$$

Find conditions on the coefficients b_1, b_0 which guarantee that this quadratic equation has:

(a) Distinct real roots.

(b) A single repeated real root.

(c) A pair of complex conjugate roots.

15. (a) Verify the text's claim that x^{r_1}, x^{r_2} are linearly independent on any interval not including $x = 0$ if r_1, r_2 are distinct real numbers. Why must you exclude $x = 0$?

(b) Is the result still true if r_1, r_2 are complex conjugate pairs?

16. The text claims that:

If the quadratic equation

$$r(r-1) + b_1 r + b_0 = 0$$

defining r has the single repeated root r, then two linearly independent solutions of the differential equation

$$x^2 y'' + b_1 x y' + b_0 y = 0$$

for $x \neq 0$ are x^r, $x^r \ln x$.

Verify this claim by completing the following steps.

(a) Show that if $r(r-1) + b_1 r + b_0 = 0$ possesses a single repeated real root, then b_1 can be expressed in terms of b_0. (*Hint*: What is the value of the discriminant?)

(b) Using x^r as the first solution, construct a second solution of the differential equation of the form $u(x)x^r$ via reduction of order. Using part (a), show that $u(x) = \ln x$ by solving the resulting differential equation for u.

(c) Verify that x^r and $x^r \ln x$ are linearly independent for $x \neq 0$.

17. Following example 6, the text claims:

> The arguments of example 6 obviously extend directly to the general situation in which the roots of the quadratic equation $r(r-1) + b_1 r + b_0 = 0$ are the complex conjugate pair $r = \alpha \pm i\beta$ .... The linearly independent solutions of $x^2 y'' + b_1 x y' + b_0 y = 0$ for $x \neq 0$ are
>
> $$x^\alpha \cos(\beta \ln x), \quad x^\alpha \sin(\beta \ln x).$$

Verify this claim by deriving this pair of real-valued solutions from the pair of complex-valued solutions $x^{\alpha+i\beta}$, $x^{\alpha-i\beta}$. Show that these real-valued solutions form a linearly independent pair for $x \neq 0$.

18. Fill in the details in example 7 of the derivation of the solutions

$$(t-4)\cos[2\ln(t-4)], \quad (t-4)\sin[2\ln(t-4)]$$

of

$$(t-4)^2 y'' - (t-4)y' + 5y = 0.$$

Verify that this pair of solutions is linearly independent for $t \neq 4$.

19. The boundary-value problem

$$-(xT'(x))' = 0, \quad T'(0) = 0, \quad T(L) = T_L,$$

models the equilibrium temperature in a thin disk of radius L whose top and bottom are insulated while its circumference is maintained at temperature T_L; see exercise 11, page 437, of section 9.1.

(a) Solve this boundary-value problem.

(b) Explain the physical significance of the solution you obtain.

20. The boundary-value problem

$$-(xT'(x))' = 0, \quad T'(0) = 0, \quad T'(L) = 0,$$

models the equilibrium temperature in a thin disk of radius L whose top, bottom and circumference are insulated; see exercise 12, page 437, of section 9.1.

(a) Solve this boundary-value problem.

(b) Explain the physical significance of the solution you obtain. What additional physical information is required to specify the solution uniquely?

21. Solve the boundary-value problem governing the equilibrium temperature in the wall of a steam pipe,

$$-\kappa(xT'(x))' = 0, \quad T(r) = T_s, \quad T(R) = T_{out}.$$

(See exercise 20, page 503, of the chapter exercises for chapter 9.) Sketch a graph of the resulting temperature profile. Why is the temperature variation through the wall of the pipe different than that through the wall of a house?

11.3 ■ VARIATION OF PARAMETERS: SECOND-ORDER EQUATIONS

Variation of parameters is a method for finding particular solutions of second-order equations without the restrictions of constant coefficients and special forcing terms that the method of undetermined coefficients imposes. In return for increased generality, we are required to know two solutions of the corresponding *homogeneous* equation in order to construct the desired particular solution. It is simply a specialization to second-order equations of the method of the same name introduced earlier for first-order systems.

Recall that in section 8.4 we used variation of parameters to construct a particular solution of the first-order system (written here in scalar form)

$$y' = k_1(t)y + k_2(t)z + F(t),$$
$$z' = \ell_1(t)y + \ell_2(t)z + G(t).$$

We assumed we knew linearly independent solution pairs (y_1, z_1) and (y_2, z_2) of the corresponding homogeneous system. We sought a particular solution (y_p, z_p) of the nonhomogeneous system in the form

$$y_p = u_1(t)y_1 + u_2(t)y_2,$$
$$z_p = u_1(t)z_1 + u_2(t)z_2,$$

where u_1, u_2 are functions to be determined.

Substituting this proposed particular solution into the nonhomogeneous system lead to two simultaneous equations for u_1', u_2',

$$y_1 u_1' + y_2 u_2' = F,$$
$$z_1 u_1' + z_2 u_2' = G.$$

We solved this system (using Cramer's rule, for example) to obtain

$$u_1' = \frac{F z_2 - G y_2}{W}, \quad u_2' = \frac{G y_1 - F z_1}{W},$$

where W is just the Wronskian of the homogeneous solutions,

$$W = y_1 z_2 - z_1 y_2.$$

Stop and Think   **11.2** What problem would arise in solving for u_1', u_2' if the homogeneous solutions were not linearly independent?

To specialize this process to the second-order linear equation

Second-order equation

$$y'' + P(x)y' + Q(x)y = R(x),$$

write it as a first-order system by introducing the variable $z = y'$,

Equivalent first-order system

$$y' = z,$$
$$z' = -Q(x)y - P(x)z + R(x).$$

These equations correspond to the generic linear system with $F = 0$ and $G = R$.

If we suppose we have two linearly independent solutions y_1, y_2 of the homogeneous equation $y'' + P(x)y' + Q(x)y = 0$, then we can seek a particular solution in the form

Variation of parameters solution

$$y_p = u_1(x)y_1 + u_2(x)y_2.$$

The results for the first-order system tell us that u_1', u_2' satisfy the system of equations

$$y_1 u_1' + y_2 u_2' = F = 0,$$
$$y_1' u_1' + y_2' u_2' = G = R.$$

The first of the two equations defining u'_1 and u'_2 can be thought of as a condition imposed to simplify the calculation of y''_p. The second equation is a consequence of substituting y_p into the differential equation.

The solution of this system is

Basic equations of variation of parameters

$$u'_1 = -\frac{R y_2}{W}, \quad u'_2 = \frac{R y_1}{W},$$

where W is the Wronskian $W(y_1, y_2) = y_1 y'_2 - y_2 y'_1$. To write a particular solution of $y'' + P(x)y' + Q(x)y = R(x)$, we need only find u_1, u_2. Unfortunately, that task is often easier said than done.

Stop and Think   **11.3** Compare this process with variation of parameters for a first-order equation, which seeks a particular solution in the form

$$y_p(x) = u(x) y_h(x),$$

where u is an unknown function and y_h is a known solution of the homogeneous equation.

Variation of parameters is sometimes called *variation of constants* because the expression $y_p = u_1(x)y_1 + u_2(x)y_2$ looks like a general solution of the homogeneous equation with C_1, C_2 replaced by the functions u_1, u_2.

■ **EXAMPLE 8**   *Find a particular solution of*

$$y'' + 4y' + 4y = 8e^{-2x}$$

in the form $y_p = u_1 y_1 + u_2 y_2$.

Characteristic equations reveal that linearly independent solutions of the homogeneous equation $y'' + 4y' + 4y = 0$ are

Solutions of homogeneous equation

$$y_1(x) = e^{-2x}, \quad y_2(x) = xe^{-2x}.$$

Now seek a particular solution of $y'' + 4y' + 4y = 8e^{-2x}$ in the form

Variation of parameters solution

$$y_p = u_1(x)y_1(x) + u_2(x)y_2(x) = u_1(x)e^{-2x} + u_2(x)xe^{-2x}.$$

The two equations for the two unknowns u'_1, u'_2 are

$$e^{-2x}u'_1 + xe^{-2x}u'_2 = 0,$$
$$(e^{-2x})'u'_1 + (xe^{-2x})'u'_2 = 8e^{-2x}.$$

Of course, the determinant of coefficients of this system of equations is the Wronskian,

$$W(e^{-2x}, xe^{-2x}) = e^{-2x}(xe^{-2x})' - xe^{-2x}(e^{-2x})'.$$

The solution of the system is

Basic equations

$$u_1' = -\frac{8e^{-2x}xe^{-2x}}{W(e^{-2x}, xe^{-2x})},$$

$$u_2' = \frac{8e^{-2x}e^{-2x}}{W(e^{-2x}, xe^{-2x})}.$$

Since

$$W(e^{-2x}, xe^{-2x}) = e^{-4x},$$

these expressions simplify to

$$u_1' = -8x, \quad u_2' = 8.$$

A single antidifferentiation yields

Integrate to find u_1, u_2.

$$u_1 = -4x^2, \quad u_2 = 8x.$$

The constant of integration was set to zero in both cases because we need *just one* pair of nontrivial functions u_1, u_2 satisfying the equations for u_1', u_2'.

Using these expressions for u_1, u_2 in the solution form $y_p = u_1 y_1 + u_2 y_2$ yields

Particular solution

$$y_p(x) = -4x^2 e^{-2x} + 8x(xe^{-2x}) = 4x^2 e^{-2x}.$$

Compare this particular solution with that obtained by undetermined coefficients in example 37, section 6.6. ■

■ **EXAMPLE 9** *Find a general solution of*

$$x^2 y'' + xy' - 4y = \ln x$$

valid for $x \neq 0$.

Solutions of homogeneous equation

Example 3 of section 11.2 found that solutions of the homogeneous equation $x^2 y'' + xy' - 4y = 0$, a Cauchy-Euler equation, are $y_1 = x^2$, $y_2 = x^{-2}$. These are linearly independent for $x \neq 0$.

To use variation of parameters to find a particular solution of this equation, assume

Variation of parameters solution

$$y_p(x) = u_1(x)x^2 + u_2(x)x^{-2}.$$

For consistency with the present notation, divide the entire equation by x^2 to make the coefficient of y'' be unity; that is, $P(x) = 1/x$, $Q(x) = -4/x^2$, and $R(x) = (\ln x)/x^2$. The functions u_1', u_2' must satisfy

$$u_1' y_1 + u_2' y_2 = x^2 u_1' + x^{-2} u_2' = 0,$$

$$u_1' y_1' + u_2' y_2' = 2x u_1' - 2x^{-3} u_2' = \frac{\ln x}{x^2}.$$

(The first equation simplifies the form of y_p'' and the second is the result of substituting y_p into the differential equation.)

Solving this pair of equations yields

Basic equations

$$u_1' = -\frac{x^2 \ln x}{W(x^2, x^{-2})} = \frac{\ln x}{4x^3},$$

$$u_2' = \frac{x^{-2} \ln x}{W(x^2, x^{-2})} = -\frac{x \ln x}{4},$$

where $W(x^2, x^{-2}) = -4/x$. Integrating these expressions by parts and setting all constants of integration to zero yield

Integrate to find u_1, u_2.

$$u_1(x) = -\frac{2 \ln x + 1}{16x^2},$$

$$u_2(x) = -\frac{x^2(2 \ln x - 1)}{16}.$$

Substituting these expression for u_1, u_2 into $y_p = u_1 y_1 + u_2 y_2$ yields

Particular solution

$$y_p(x) = -\frac{\ln x}{4}.$$

Hence, a general solution of $x^2 y'' + x y' - 4y = \ln x$ is

General solution

$$y_g(x) = -\frac{\ln x}{4} + C_1 x^2 + C_2 x^{-2}, \quad x \neq 0. \ ■$$

■ **EXAMPLE 10** *Find a solution of the initial-value problem*

$$x^2 y'' - x y' + y = \pi x^2 \cos \pi x, \quad y(1) = 0, \quad y'(1) = 3.$$

Solutions of homogeneous equation

Example 4 of section 11.2 found that a pair of linearly independent solutions of the homogeneous equation $x^2 y'' - x y' + y = 0$ is $y_1(x) = x$, $y_2(x) = x \ln x$, $x \neq 0$. Consequently, we assume a particular solution of the form

Variation of parameters solution

$$y_p(x) = u_1(x) y_1(x) + u_2(x) y_2(x) = x u_1(x) + x \ln x \, u_2(x).$$

To put the equation into standard form, divide through by x^2. Then $R(x) = \pi \cos \pi x$.

Proceeding directly to

Basic equations

$$u_1' = -\frac{R y_2}{W}, \quad u_2' = \frac{R y_1}{W},$$

the result of solving the equations for u_1', u_2', we find

$$u_1' = -\frac{(\pi \cos \pi x)(x \ln x)}{W(x, x \ln x)},$$

$$u_2' = \frac{x\pi \cos \pi x}{W(x, x \ln x)},$$

or, since $W(x, x \ln x) = x$,

$$u_1' = -\pi \cos \pi x \, \ln x, \quad u_2' = \pi \cos \pi x.$$

The second expression yields

Integrate to find u_2.
$$u_2(x) = \sin \pi x.$$

Since the first expression lacks an explicit antiderivative, we employ the fundamental theorem of calculus (theorem 2 of the appendix) to write

Integrate to find u_1.
$$u_1(x) = -\pi \int_1^x \cos \pi s \, \ln s \, ds.$$

The lower limit of integration was chosen to make this expression easy to evaluate at the initial point $x = 1$.

Using these expressions for u_1, u_2 in the particular solution form $y_p = u_1 y_1 + u_2 y_2$ yields

Particular solution
$$y_p(x) = -\pi x \int_1^x \cos \pi s \, \ln s \, ds + x \sin \pi x \, \ln x.$$

A general solution of $x^2 y'' - xy' + y = \cos \pi x$ is obviously

General solution
$$y_g(x) = -\pi x \int_1^x \cos \pi s \, \ln s \, ds + x \sin \pi x \, \ln x + C_1 x + C_2 x \ln x.$$

To find C_1, C_2, impose the initial conditions $y(1) = 0$, $y'(1) = 3$, to find

Impose initial conditions.
$$y_g(1) = C_1 = 0,$$
$$y_g'(1) = C_1 + C_2 = 3,$$

and $C_1 = 0$, $C_2 = 3$. The solution of the initial-value problem is

Solution of initial-value problem
$$y(x) = -\pi x \int_1^x \cos \pi s \, \ln s \, ds + (x \sin \pi x + 3x) \ln x. \quad ■$$

11.3.1 Summary

Variation of parameters finds a particular solution of a second-order linear equation

$$y'' + P(x)y' + Q(x)y = R(x).$$

by assuming a particular solution of the form

VARIATION OF PARAMETERS PARTICULAR SOLUTION

$$y_p(x) = u_1(x)y_1(x) + u_2(x)y_2(x),$$

where u_1, u_2 are unknown functions and y_1, y_2 are two *known*, linearly independent solutions of the corresponding homogeneous equation. We find that u_1', u_2' must satisfy the simultaneous equations

Defining the variation of parameters of coefficients

$$u_1'y_1 + u_2'y_2 = 0,$$
$$u_1'y_1' + u_2'y_2' = R(x)$$

whose solution is

Basic equations for the variation of parameters coefficients

$$u_1' = -\frac{Ry_2}{W}, \quad u_2' = \frac{Ry_1}{W}.$$

Here W is the Wronskian $W(y_1, y_2) = y_1y_2' - y_1'y_2$.

These expressions for u_1', u_2' are integrated (not necessarily in closed form) and substituted in $y_p = u_1y_1 + u_2y_2$ to obtain the desired particular solution.

The first of the two equations defining u_1' and u_2' can be thought of as a condition imposed to simplify the calculation of y_p''. The second equation is a consequence of substituting y_p into the differential equation.

11.3.2 Exercises

<table>
<tr><td colspan="2" align="center">**EXERCISE GUIDE**</td></tr>
<tr><td>**To gain experience . . .**</td><td>**Try exercises**</td></tr>
<tr><td>Using variation of parameters</td><td>1–12, 17, 20</td></tr>
<tr><td>Finding and using general solutions</td><td>1–12, 16, 20</td></tr>
<tr><td>With the foundations of variation of parameters</td><td>13–15</td></tr>
<tr><td>Generalizing variation of parameters</td><td>18–19</td></tr>
</table>

Use variation of parameters to find a general solution of the differential equations in exercises 1–12. Compare your solution with that provided by DELAB.

1. $y'' - y = 1$

2. $y'' - y = -2t$

3. $y'' - 9y = e^{-3x}$

4. $y'' - 9y = e^{3x}$

5. $y'' - 9y = \cosh 3x$

6. $y'' + 16y = 2\tan 4x$

7. $w'' + 9w = 2e^{3x}$

8. $y'' + y' - 6y = e^{2t}$

9. $y'' + y = 4t$

10. $t^2 u + tu' - 4u = t\ln t$

11. $t^2 u'' + 2tu' - 6u = 1 + 1/t$

12. $x^2 c'' + 2xc' = Px$, P a constant

13. Why would variation of parameters fail if the two homogeneous solutions y_1 and y_2 were not linearly independent?

14. Find the free and forced response (see exercise 16 of section 6.1) of the initial-value problem considered in example 10,

$$x^2 y'' - xy' + y = \pi x^2 \cos \pi x, \quad y(1) = 0, \ y'(1) = 3.$$

Does this system possess a steady state?

15. Example 8 solves the problem it considers in great detail, while example 9 finds a particular solution of

$$x^2 y'' + xy' - 4y = \ln x$$

but omits many steps. Fill in the details of example 9 in the manner of example 8. In particular, derive the equations that u_1' and u_2' satisfy.

16. The text suggests that the arguments of example 8 can be applied to the general linear equation

$$y'' + P(x)y' + Q(x)y = R(x)$$

to find the equations

$$u_1' = -\frac{Ry_2}{W(y_1, y_2)}, \quad u_2' = \frac{Ry_1}{W(y_1, y_2)}.$$

Supply the details of this argument. In particular, derive the equations that u_1' and u_2' satisfy. (*Hint:* The preceding exercise asks you to perform this same task for a specific equation. Mimic the appropriate steps here.)

17. Suppose y_1, y_2 are linearly independent solutions of

$$a_2(x)y'' + a_1(x)y' + a_0(x)y = r(x)$$

and that $a_2(x) \neq 0$. Find equations for the functions u_1' and u_2' in the particular solution form $y_p = u_1 y_1 + u_2 y_2$.

18. The functions x^2 and x^3 are solutions of the homogeneous form of

$$x^2 y'' - 4xy' + 6y = \frac{1}{x}$$

for $x > 0$. Find a general solution of this equation.

11.4 ■ POWER SERIES METHODS

In this and other chapters, we have followed a pattern of assuming solutions of a particular form and substituting the solution form into the differential equation to determine the unknown constants or functions in that form. We are now about to make what is apparently the most general possible assumption, that the solution can be expanded in a power series whose coefficients we can determine. We proceed formally at first; the last part of this section explores the validity of this bold assumption.

Recall that a **power series about** x_0 is an infinite series of the form

Power series about x_0.

$$\sum_{n=0}^{\infty} c_n (x - x_0)^n,$$

where the c_n are constants. Only integer powers of x appear in a power series. Within the interval of convergence of this series, the function it represents can be integrated or differentiated by integrating or differentiating the series term by term. That is, if

$$f(x) = \sum_{n=0}^{\infty} c_n (x - x_0)^n$$

$$= c_0 + c_1 (x - x_0) + c_2 (x - x_0)^2 + c_3 (x - x_0)^3 + \cdots,$$

then

$$f'(x) = \sum_{n=0}^{\infty} \left(c_n (x - x_0)^n \right)' = \sum_{n=1}^{\infty} c_n n (x - x_0)^{n-1}$$

and

$$\int f(x)\, dx = \sum_{n=0}^{\infty} c_n \int (x - x_0)^n \, dx = C + \sum_{n=0}^{\infty} c_n \frac{(x - x_0)^{n+1}}{n+1}$$

whenever the original series for f converges. (In the series for $\int f$, C is just a constant of integration.)

Stop and Think   **11.4** Why did the starting value of the summation index n in the series for f' change from $n = 0$ to $n = 1$? (See exercise 1.)

For the moment, we will proceed formally with such manipulations. The constants c_n in the power series for f are just the **Taylor coefficients**

$$c_n = \frac{f^{(n)}(x_0)}{n!}.$$

To see this relation for $n = 0$, evaluate $f(x) = \sum_{n=0}^{\infty} c_n (x - x_0)^n$ at $x = x_0$:

$$f(x_0) = c_0.$$

Similarly, evaluating $f'(x) = \sum_{n=1}^{\infty} c_n n (x - x_0)^{n-1}$ at $x = x_0$ yields

$$f'(x_0) = c_1,$$

and so on for $f''(x_0)$, $f^{(3)}(x_0)$, ... Evidently, a formal power series like $f(x) = \sum_{n=0}^{\infty} c_n (x - x_0)^n$ can be constructed for any function that has an infinite number of derivatives about the point x_0.

But if the series does not converge to the value $f(x)$, constructing the series is a waste.

To solve differential equations using power series, we will proceed in the other direction. Rather than construct the series from the known function f, we will attempt to construct as much of the solution function as we can by finding the coefficients in its power series directly from the differential equation. That is, given a linear differential equation and an expansion point x_0, we will attempt to find the coefficients c_n in the power series expansion for the solution about x_0. *We substitute the assumed solution form into the equation and equate coefficients of powers of x to find the unknown constants c_n.*

This method is not as general as you might like to believe, for not every solution of every linear differential equation has a power series expansion about an arbitrary point. For example, the Cauchy-Euler equation

$$x^2 y'' + 7xy' + 9y = 0$$

has the solutions x^{-3}, $x^{-3} \ln x$. Neither of these functions has a power series expansion about $x_0 = 0$.

■ **EXAMPLE 11**   *Find a power series solution about $x_0 = 0$ of*

$$y'' + y = 0.$$

We assume the solution has the form of a power series in x about $x_0 = 0$,

Assume power series

$$y(x) = \sum_{n=0}^{\infty} c_n x^n = c_0 + c_1 x + c_2 x^2 + c_3 x^3 + \cdots .$$

Then we differentiate term by term to compute

$$y'(x) = c_1 + 2c_2 x + 3c_3 x^2 + 4c_4 x^3 + \cdots = \sum_{n=1}^{\infty} n c_n x^{n-1},$$

$$y''(x) = 2c_2 + 6c_3 x + 12c_4 x^2 + 20c_5 x^3 + \cdots = \sum_{n=2}^{\infty} n(n-1) c_n x^{n-2}.$$

Note the change in the starting value of the summation index n each time the series is differentiated.

Substituting the power series expansions for y, y'' in the differential equation $y'' + y = 0$ yields

Substitute.

$$
\begin{aligned}
y'' + y &= 2c_2 + 6c_3 x + 12c_4 x^2 + 20c_5 x^3 + \cdots \\
&\quad + c_0 + c_1 x + c_2 x^2 + c_3 x^3 + \cdots \\
&= (2c_2 + c_0) + (6c_3 + c_1)x + (12c_4 + c_2)x^2 \\
&\quad + (20c_5 + c_3)x^3 + \cdots \\
&= 0.
\end{aligned}
$$

Since the right-hand side is zero, each of the coefficients of a power of x on the left must be zero. Hence, from the successive powers of x, we obtain

Equate coefficients to zero.

$$
\begin{aligned}
x^0 : &\quad 2c_2 + c_0 = 0, \\
x^1 : &\quad 6c_3 + c_1 = 0, \\
x^2 : &\quad 12c_4 + c_2 = 0, \\
x^3 : &\quad 20c_5 + c_3 = 0, \\
&\qquad \vdots
\end{aligned}
$$

We solve each of these equations for the coefficient with the highest index:

Solve for higher index.

$$x^0: \quad c_2 = -\frac{c_0}{2},$$

$$x^1: \quad c_3 = -\frac{c_1}{6},$$

$$x^2: \quad c_4 = -\frac{c_2}{12},$$

$$x^3: \quad c_5 = -\frac{c_3}{20},$$

$$\vdots$$

Evidently, the even indexed coefficients can be expressed in terms of c_0,

Express c_n in terms of c_0 ...

$$c_2 = -\frac{c_0}{2},$$

$$c_4 = -\frac{c_2}{12} = \frac{c_0}{24},$$

$$\vdots$$

while the odd ones can be written in terms of c_1,

... and c_n.

$$c_3 = -\frac{c_1}{6},$$

$$c_5 = -\frac{c_3}{20} = \frac{c_1}{120},$$

$$\vdots$$

Using these values of c_n in the assumed power series solution form $y(x) = \sum_{n=0}^{\infty} c_n x^n$, we obtain

Power series solution

$$y(x) = c_0 \left(1 - \frac{x^2}{2} + \frac{x^4}{24} + \cdots \right) + c_1 \left(-\frac{x^3}{6} + \frac{x^5}{120} + \cdots \right),$$

the power series solution we sought. ■

The power series solution obtained in example 11 contains two apparently arbitrary constants, c_0 and c_1. Are they the two constants we usually encounter in a general solution of a linear, second-order equation?

The strong suggestion that they are comes from the observation that evaluating the power series solution at $x = 0$ gives

$$y(0) = c_0, \quad y'(0) = c_1.$$

For further confirmation, note that $\cos x$ and $\sin x$ are linearly independent solutions of $y'' + y = 0$. Since these functions have the power series expansions

$$\cos x = 1 - \frac{x^2}{2} + \frac{x^4}{24} + \cdots,$$

$$\sin x = x - \frac{x^3}{6} + \frac{x^5}{120} + \cdots,$$

a general solution of $y'' + y = 0$ is

$$y(x) = C_1 \cos x + C_2 \sin x$$

$$= C_1 \left(1 - \frac{x^2}{2} + \frac{x^4}{24} + \cdots \right) + C_2 \left(x - \frac{x^3}{6} + \frac{x^5}{120} + \cdots \right),$$

just a restatement of the power series solution we obtained in example 11 directly from the differential equation.

■ **EXAMPLE 12**   *Find the general expression for the coefficients c_n in a power series about $x_0 = 0$ of a solution of*

$$y'' + y = 0.$$

That is, find a formula for c_n in terms of n.

We begin with the same assumed solution form as in example 11,

Assume power series.

$$y(x) = \sum_{n=0}^{\infty} c_n x^n,$$

but we shall preserve the general index n throughout in order to find a formula for c_n. Taking care to adjust the starting index of the sum to account for terms lost to differentiation, we find

$$y'(x) = \sum_{n=1}^{\infty} n c_n x^{n-1},$$

$$y''(x) = \sum_{n=2}^{\infty} n(n-1) c_n x^{n-2}.$$

Substituting the expressions for y and y'' in the differential equation $y'' + y = 0$ yields

Substitute.

$$y'' + y = \sum_{2}^{\infty} n(n-1) c_n x^{n-2} + \sum_{n=0}^{\infty} c_n x^n = 0.$$

To parallel example 11, we must collect coefficients of like powers of x. However, the two sums in the preceding equation contain different powers of x; the first has x^{n-2} and the second, x^n.

To rewrite the first sum so that the exponent of x is the index itself, as in the second sum, rather than the index less 2, as in the first, we make the change of index $m = n - 2$ in the first sum. Since $n = m + 2$, we obtain

Shift index

$$\sum_{n=2}^{\infty} n(n-1)c_n x^{n-2} = \sum_{m=0}^{\infty} (m+2)(m+1)c_{m+2} x^m.$$

Note that the starting value of the lower index has changed to $m = 0$ because $m = 0$ when $n = 2$, the old starting value.

Using the altered form of the first sum gives us

$$y'' + y = \sum_{m=0}^{\infty} (m+2)(m+1)c_{m+2} x^m + \sum_{n=0}^{\infty} c_n x^n.$$

But the notation m and n for the summation indices in this last expression is entirely arbitrary. We could relabel them both k, for instance. Then the result of substituting in the differential equation would read

$$y'' + y = \sum_{k=0}^{\infty} (k+2)(k+1)c_{k+2} x^k + \sum_{k=0}^{\infty} c_k x^k.$$

Since both summations start at the same index, we may combine them into one. Furthermore, since the power of x in both is the same, we may factor it out of the coefficients of the combined sum to obtain

Collect like powers.

$$y'' + y = \sum_{k=0}^{\infty} \left[(k+2)(k+1)c_{k+2} + c_k \right] x^k = 0.$$

Because this power series sums to zero, each of the coefficients of x^k must be zero. (The power series for the identically zero function is the one that has every coefficient equal to zero.) Hence, we obtain the **recurrence relation**

Recurrence relation from equating coefficients to zero.

$$(k+2)(k+1)c_{k+2} + c_k = 0, \quad k = 0, 1, 2, \ldots,$$

which successively expresses each coefficient in terms of one or more of its predecessors. Solving for the coefficient with the higher index, we find

$$c_{k+2} = -\frac{c_k}{(k+2)(k+1)}.$$

Working our way back toward the coefficient with lower index, we calculate c_k in terms of c_{k-2}, then c_{k-2} in terms of c_{k-4}, and so forth. That is, we use the preceding formula with k replaced by $k - 2$, then with k replaced by $k - 4$, and so on.

$$c_k = -\frac{c_{k-2}}{k(k-1)} = \frac{c_{k-4}}{k(k-1)(k-2)(k-3)} = \cdots.$$

If k is even, then the sequence terminates after $k/2$ steps with c_0,

$$c_k = (-1)^{k/2} \frac{c_0}{k!}, \quad k \text{ even}.$$

If k is odd, then the sequence terminates after $(k-1)/2$ steps with c_1,

$$c_k = (-1)^{(k-1)/2} \frac{c_1}{k!}, \quad k \text{ odd}.$$

To see the pattern of these general calculations, try a few specific cases, such as $k = 6, 7, 8, 9$.

Although we could substitute these expressions for c_k as they stand in the power series solution form, the result has an awkward appearance,

$$y(x) = c_0 \sum_{k=0,2,\ldots} (-1)^{k/2} \frac{x^k}{k!} + c_1 \sum_{k=1,3,\ldots} (-1)^{(k-1)/2} \frac{x^k}{k!}.$$

Alternatively, we can write the even indices as $k = 2n$ and the odd indices as $k = 2n + 1$, where $n = 0, 1, 2, \ldots$. Now we can use the common index n to write this solution in a prettier form,

Power series solution

$$y(x) = c_0 \sum_{n=0}^{\infty} (-1)^n \frac{x^{2n}}{(2n)!} + c_1 \sum_{n=0}^{\infty} (-1)^n \frac{x^{2n+1}}{(2n+1)!}.$$

We immediately identify the two series in this expression as the power series expansions about $x = 0$ of $\cos x$ and $\sin x$. Hence, we have obtained the same solution of $y'' + y = 0$ as the characteristic equation method. The leading coefficients c_0, c_1 are indeed the arbitrary constants in the general solution. ■

■ **EXAMPLE 13**   *Find the first five nonzero terms in a power series solution of the initial-value problem*

$$y'' + (x-1)^2 y' - 2(x-1)^3 y = 0, \quad y(1) = 1, \quad y'(1) = -8.$$

Since the initial conditions are given at $x = 1$ and a power series is most easily evaluated at its expansion point, we assume a power series solution of $y'' + (x-1)^2 y' - 2(x-1)^3 y = 0$ that is expanded about $x_0 = 1$,

Assume power series.

$$
\begin{aligned}
y(x) &= \sum_{0}^{\infty} c_n (x-1)^n \\
&= c_0 + c_1(x-1) + c_2(x-1)^2 \\
&\quad + c_3(x-1)^3 + c_4(x-1)^4 + c_5(x-1)^5 + c_6(x-1)^6 + \cdots.
\end{aligned}
$$

At the beginning of a power series solution, we usually do not know how many terms to keep in order to obtain finally the desired number of nonzero terms. We added two extra to be safe.

We compute

$$y'(x) = c_1 + 2c_2(x-1) + 3c_3(x-1)^2 + 4c_4(x-1)^3$$
$$+ 5c_5(x-1)^4 + 6c_6(x-1)^5 + \cdots,$$

$$y''(x) = 2c_2 + 6c_3(x-1) + 12c_4(x-1)^2$$
$$+ 20c_5(x-1)^3 + 30c_6(x-1)^4 + \cdots.$$

Substituting these expression for y, y', and y'' into the differential equation $y'' + (x-1)^2 y' - 2(x-1)^3 y = 0$ leads to

Substitute.

$$y'' + (x-1)^2 y' - 2(x-1)^3 y$$
$$= 2c_2 + 6c_3(x-1) + 12c_4(x-1)^2$$
$$+ 20c_5(x-1)^3 + 30c_6(x-1)^4 + \cdots$$
$$+ c_1(x-1)^2 + 2c_2(x-1)^3 + 3c_3(x-1)^4 + \cdots$$
$$- 2c_0(x-1)^3 - 2c_1(x-1)^4 + \cdots$$
$$= 0.$$

Each expression has been expanded to the same order (four in this case) as it was substituted in the differential equation. There is no point in keeping a particular power of $x - 1$ in one expression if we have not kept it in another, because we will be required to collect *all* coefficients of a given power of $x - 1$ as we solve for the coefficients c_n.

We collect coefficients of powers of x to obtain

Collect coefficients

$$2c_2 + 6c_3(x-1) + (12c_4 + c_1)(x-1)^2$$
$$+ (20c_5 + 2c_2 - 2c_0)(x-1)^3$$
$$+ (30c_6 + 3c_3 - 2c_1)(x-1)^4 + \cdots = 0.$$

Setting each coefficient to zero yields a series of equations for the c_n,

Equate coefficients to zero.

$$(x-1)^0: \qquad\qquad 2c_2 = 0,$$
$$(x-1)^1: \qquad\qquad 6c_3 = 0,$$
$$(x-1)^2: \qquad 12c_4 + c_1 = 0,$$
$$(x-1)^3: \quad 20c_5 + 2c_2 - 2c_0 = 0,$$
$$(x-1)^4: \quad 30c_6 + 3c_3 - 2c_1 = 0.$$

The solution of these equations is

Recurrence relations

$$c_2 = 0, \quad c_3 = 0, \quad c_4 = -\frac{c_1}{12},$$

$$c_5 = \frac{c_0 - c_2}{10} = \frac{c_0}{10},$$

$$c_6 = \frac{c_1}{15} - \frac{c_3}{10} = \frac{c_1}{15}.$$

Using these values of $c_2, \ldots, c_6$, we find that our solution is

Power series with arbitrary
constants

$$y(x) = c_0 + c_1(x - 1) - c_1 \frac{(x - 1)^4}{12}$$

$$+ c_0 \frac{(x - 1)^5}{10} + c_1 \frac{(x - 1)^6}{15} + \cdots$$

$$= c_0 \left(1 + \frac{(x - 1)^5}{10} + \cdots \right)$$

$$+ c_1 \left((x - 1) - \frac{(x - 1)^4}{12} + \frac{(x - 1)^6}{15} + \cdots \right).$$

To determine the remaining coefficients c_0 and c_1, we turn to the initial conditions, as we have in solving every other initial-value problem. Combining the first initial condition with the solution form $y(x) = \sum_{n=0}^{\infty} c_n (x - 1)^n$, we have

Apply initial conditions.

$$y(1) = c_0 = 1.$$

Similarly, from the second initial condition and

$$y'(x) = c_0 \left(\frac{(x - 1)^4}{2} + \cdots \right) + c_1 \left(1 - \frac{(x - 1)^3}{3} + \frac{2(x - 1)^5}{5} + \cdots \right),$$

we obtain

$$y'(1) = c_1 = -8.$$

Although the coefficients $c_2, c_3, \ldots$ are found in terms of c_0 and c_1 by substituting in the differential equation, the values of c_0 and c_1 depend *only* upon the given initial conditions. We can choose to apply the initial conditions at any convenient point in the solution process. See exercise 18.

When these values of c_0 and c_1 are substituted in the solution $y(x) = \sum_{n=0}^{\infty} c_n (x - 1)^n$, we obtain the solution of the initial-value problem,

Power series solution of
initial-value problem

$$y(x) = 1 - 8(x - 1) + \frac{(x - 1)^5}{10} - \frac{2(x - 1)^4}{3} - \frac{8(x - 1)^6}{15} + \cdots . \blacksquare$$

These ideas apply equally well to nonhomogeneous equations. We only need to expand the forcing term in a series about the same point as the solution.

■ **EXAMPLE 14**   *Find the first five nonzero terms in a series expansion of the solution of*

$$y'' - 2xy = \ln(1 + x), \quad y(0) = y'(0) = 1.$$

We require a series about the initial point $x = 0$. The Taylor series for the forcing term is

Power series for finding forcing term

$$\ln(1 + x) = x - \frac{x^2}{2} + \frac{x^3}{3} + \cdots, \quad |x| < 1.$$

The appropriate form for the solution is

Assume power series for solution.

$$y(x) = c_0 + c_1 x + c_2 x^2 + c_3 x^3 + \cdots.$$

Apply initial conditions.

The initial conditions require $c_0 = c_1 = 1$.

Substituting the proposed solution in the equation and using the series expansion of $\ln(1 + x)$ lead to

Substitute.

$$y'' - 2xy = 2c_2 + 6c_3 x + 12c_4 x^2 + 20c_5 x^3 + \cdots$$
$$- 2x - 2x^2 - 2c_2 x^3 + \cdots$$
$$= x - \frac{x^2}{2} + \frac{x^3}{3} + \cdots.$$

Equating like powers of x yields

Equate coefficients of like powers.

$$
\begin{aligned}
x^0 &: & 2c_2 &= 0, \\
x^1 &: & 6c_3 - 1 &= 1, \\
x^2 &: & 12c_4 - 2 &= -1/2, \\
x^3 &: & 20c_5 - 2c_2 &= 1/3.
\end{aligned}
$$

We conclude that

$$
\begin{aligned}
c_2 &= 0, \\
c_3 &= 1/3, \\
c_4 &= 1/8, \\
c_5 &= 1/60.
\end{aligned}
$$

Hence, the solution we seek is

Power series solution of initial-value problem

$$y(x) = 1 + x + \frac{x^3}{3} + \frac{x^4}{8} + \frac{x^5}{60} + \cdots. \quad ■$$

11.4.1 Existence of a Power Series Solution

We remarked earlier that the Cauchy-Euler equation

$$x^2 y'' + 7xy' + 9y = 0$$

is an example of a linear differential equation that does *not* have power series solutions about $x = 0$. Its linearly independent solutions x^{-3} and $x^{-3} \ln x$ both involve *negative* powers of x; a power series about $x = 0$ could only use nonnegative powers. Here we describe conditions that guarantee the existence of power series solutions.

A function $f(x)$ is **analytic** about the point x_0 if it has a convergent power series representation about x_0; i.e., if for some $\delta > 0$,

$$f(x) = \sum_{n=0}^{\infty} a_n(x - x_0)^n, \quad |x - x_0| < \delta.$$

We are asking the question: Which linear differential equations have a pair of linearly independent analytic solutions?

If a function is analytic at x_0, then the function itself and each of its derivatives must certainly be defined at x_0; we need only evaluate the power series or its term by term derivative at $x = x_0$. (And convergent power series can be differentiated term by term.) That is,

$$f(x_0) = \sum_{n=0}^{\infty} a_n(x_0 - x_0)^n = a_0,$$

$$f'(x_0) = \sum_{n=1}^{\infty} n a_n(x_0 - x_0)^{n-1} = a_1,$$

$$f''(x_0) = \sum_{n=2}^{\infty} n(n - 1)a_n(x_0 - x_0)^{n-2} = 2a_2,$$

$$\vdots$$

Consequently, a function that is not defined at x_0 is not analytic there; e.g.,

$$f(x) = \frac{1}{x^2}$$

is not analytic at $x_0 = 0$. Likewise, a function that is not differentiable at x_0 is not analytic there; e.g.,

$$f(x) = |x|$$

is not analytic at $x_0 = 0$ because $|x|$ has no derivative at $x = 0$. (See figure 11.1.) Likewise, $f(x) = (x + 1)^{3/2}$ is not analytic at $x_0 = -1$ because

$$f'' = \frac{3}{4(x + 1)^{1/2}}$$

is not defined there.

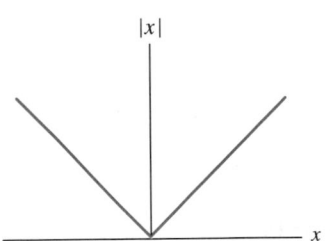

$|x|$

x

FIGURE 11.1 Because of the "corner" at $x = 0$, the function $|x|$ is not differentiable at $x = 0$. Hence, it is not analytic at $x = 0$.

On the other hand, a function which is a polynomial (a finite sum of nonnegative powers) in $x - x_0$ or has known power series is analytic at x_0. For example,

$$f(x) = 4 - (x + 2)^3$$

is analytic at $x_0 = -2$ (or any other point) because it can be written as a finite power series in $x + 2$. The function

$$\sin x = x - \frac{x^3}{3!} + \frac{x^5}{5!} + \cdots$$

is certainly analytic at $x = 0$ because of its well known convergent Taylor series expansion. Other familiar functions that are analytic for all x are $\cos x$ and e^x.

Definition 1. *The point x_0 is an **ordinary point** of*

$$y'' + P(x)y' + Q(x)y = 0$$

*if P and Q are analytic at x_0. Otherwise, x_0 is a **singular point** of the equation.*

■ **EXAMPLE 15**   *The point $x = 0$ is an ordinary point of*

$$y' + (\cos \pi x)y' + \frac{y}{1 - x} = 0.$$

The coefficients $P(x) = \cos \pi x$ and $Q(x) = 1/(1 - x)$ (see example 16) have convergent power series expansions about $x_0 = 0$:

$$\cos \pi x = 1 - \frac{(\pi x)^2}{2!} + \frac{(\pi x)^4}{4!} + \cdots, \quad |x| < \infty,$$

$$\frac{1}{1 - x} = 1 + x + x^2 + \cdots, \quad |x| < 1. \blacksquare$$

■ **EXAMPLE 16**   *The point $x = 1$ is a singular point of*

$$y' + (\cos \pi x)y' + \frac{y}{1 - x} = 0,$$

but the point $x_0 = 0$ is an ordinary point of the same equation.

Although $P(x) = \cos \pi x$ is certainly analytic at $x_0 = 1$, the coefficient $Q(x) = 1/(1 - x)$ is not because it is not even defined there.

On the other hand, at $x_0 = 0$ both $P(x) = \cos \pi x$ and $Q(x) = 1/(1 - x)$ are analytic. The cosine has its familiar Taylor series expansion, and for $1/(1 - x)$ we can construct the Taylor series

$$\frac{1}{1 - x} = 1 + x + x^2 + x^3 + \cdots + x^n + \cdots, \quad |x| < 1,$$

using

$$f(x) = (1-x)^{-1} \quad \Rightarrow \quad f(0) = 1,$$
$$f'(x) = (1-x)^{-2} \quad \Rightarrow \quad f'(0) = 1,$$
$$f''(x) = 2(1-x)^{-3} \quad \Rightarrow \quad f''(0) = 2,$$

$$\vdots$$

$$f^{(n)}(x) = n!(1-x)^{-n-1} \quad \Rightarrow \quad f^{(n)}(0) = n!$$

and the ratio test (or geometric series) to determine the radius of convergence. Hence, $x_0 = 0$ is an ordinary point of this equation. ■

Since both of the coefficients P and Q in the previous example have power series expansions about the ordinary point $x_0 = 0$, we might reasonably expect that solutions of this equation would have convergent power series expansions about $x_0 = 0$ as well. However, it seems equally likely that such series solutions would converge only for $|x| < 1$, the radius of convergence of $Q(x) = 1/(1-x)$, for at $x = 1$ this function is not analytic. The next theorem confirms this suspicion.

Theorem 1 (Existence of analytic solutions). *Let* $P(x)$, $Q(x)$ *be functions that are analytic at* x_0 *with radii of convergence* R_P, R_Q. *Then the differential equation*

$$y'' + P(x)y' + Q(x)y = 0$$

has a pair of linearly independent power series solutions with radius of convergence $R = \min(R_P, R_Q)$.

Rather than prove this theorem, we outline the main steps in the argument:

1. To construct linearly independent solutions, find a pair of solutions that satisfy the two sets of initial conditions $y(x_0) = 1$, $y'(x_0) = 0$ and $y(x_0) = 0$, $y'(x_0) = 1$. If such solutions exist, then at $x = 0$ the Wronskian $W = 1 \neq 0$.

2. Assume a solution of the form

$$y(x) = \sum_{n=0}^{\infty} c_n(x - x_0)^n.$$

   The first set of initial conditions requires $c_0 = 1$, $c_1 = 0$. Substitute into the differential equation to obtain a recursion formula for $c_2, c_3, \ldots$ in terms of the coefficients of the power series for P, Q.

3. Use the convergence of the power series expansions for P, Q to show that the power series for y converges for $|x - x_0| < R$. The convergence of that series then justifies the formal manipulations (interchange of differentiation and summation, etc.) of the previous step.

4. Repeat the preceding two steps for the second set of initial conditions, for which $c_0 = 0$, $c_1 = 1$.

Project 3 asks you to provide a proof of the existence of analytic solutions following this outline.

For example, this theorem tells us that the equation

$$y'' + (\cos \pi x)y + \frac{y}{1-x} = 0$$

will have power series solutions about the ordinary point $x_0 = 0$ whose radius of convergence is the smaller of infinity, the radius of convergence of the series for $\cos \pi x$, and 1, the radius of convergence of the series for $1/(1-x)$.

Full understanding of power series requires the perspective of the complex plane. For example, the power series about $x_0 = 0$ for $1/(1 + x^2)$ converges only for $|x| < 1$, even though this function is continuous and differentiable for all x. The problem is a "hidden" singularity at $x = i$, a point off the real axis, that limits the radius of convergence to $R = 1$.

11.4.2 Summary

The steps in finding a **power series solution** of a linear differential equation follow.

1. Assume a solution of the form

POWER SERIES

$$\sum_{n=0}^{\infty} c_n (x - x_0)^n,$$

where x_0 is an ordinary point of the equation. If initial conditions are given, choose x_0 to be the initial point. Use the initial conditions to evaluate c_0 and c_1.

2. Compute the derivatives of the power series term by term. Substitute in the differential equation and collect coefficients of like powers of x. Take care to account for all terms containing a given power of x. If necessary, adjust summation indices so that all series are written with the same exponent of x.

3. Set the coefficient of each power of x to zero. Solve the resulting equations for the highest numbered coefficient c_n in terms of the lower numbered coefficient(s) in each equation. Working recursively, express each coefficient in terms of c_0 and c_1 and write the series solution using just c_0 and c_1.

4. If initial conditions are given, use them to find c_0 and c_1.

11.4.3 Exercises

EXERCISE GUIDE	
To gain experience . . .	**Try exercises**
Differentiating, integrating, manipulating power series	1–3, 11, 14–15, 16, 19
Locating ordinary points	4–10, 16
Finding general solutions	4–10, 13(a–c)
Solving initial-value problems	6–7, 12, 13(e), 18
Using Taylor series	11, 13(a), 16, 19
With validity of power series solutions	13(d), 16
With recurrence relations	17(a)
With analytic functions	19–20
Finding radius of convergence	2–3(c), 12(b), 17(b)

1. After writing $f(x) = \sum_{n=0}^{\infty} c_n(x - x_0)^n$, the text asks, "Why did the starting value of the summation index n in

$$f'(x) = \sum_{n=0}^{\infty} (c_n(x - x_0)^n)'$$

$$= \sum_{n=1}^{\infty} c_n n(x - x_0)^{n-1}$$

change from $n = 0$ to $n = 1$?" Answer this question.

2. Using the convergent power series expansion

$$e^x = \sum_{n=0}^{\infty} \frac{x^n}{n!},$$

verify the following.

(a) $d(e^x)/dx = e^x$, using term by term differentiation.

(b) $\int_0^t e^x \, dx = e^t - 1$, using term by term integration.

(c) Determine the radius of convergence of this power series.

3. Given the convergent power series expansions

$$\sin x = \sum_{n=0}^{\infty} (-1)^{(n+1)} \frac{x^{2n-1}}{(2n-1)!},$$

$$\cos x = \sum_{n=0}^{\infty} (-1)^n \frac{x^{2n}}{(2n)!},$$

verify the following.

(a) $d(\cos x)/dx = -\sin x$, using term by term differentiation.

(b) $d(\sin x)/dx = \cos x$, using term by term differentiation.

(c) Determine the radius of convergence of each power series.

For exercises 4–10,

(i) Verify that the given expansion point or initial point is an ordinary point of the equation.

(ii) Find the first four nonzero terms in power series solutions of the differential equation or initial-value problem.

4. $y'' + xy = 0$ (about $x = 0$)

5. $y'' + (x + 2)y = 0$ (about $x = -2$)

6. $y'' - y = 0$, $y(0) = 2$, $y'(0) = 2$

7. $y'' - y = 0$, $y(1) = -1$, $y'(1) = 4$

8. $x'' - (t - 2)x' - 3(t - 2)x = 0$ (about $t = 2$)

9. $y' + y = x^2$ (about $x = 0$)

10. $y'' - 3y' + 2x^2 y = e^x$ (about $x = 0$)

11. Formally compute the derivatives f'', $f^{(3)}$, and $f^{(4)}$ of the power series function

$$f(x) = \sum_{n=0}^{\infty} c_n(x - x_0)^n;$$

evaluate each at $x = x_0$; and verify the Taylor coefficient formula

$$c_n = \frac{f^{(n)}(x_0)}{n!}$$

for $n = 2, 3, 4$.

12. In example 11, the text claims that the constants c_0 and c_1 that remained undetermined at the end of the power series solution of

$$y'' + y = 0$$

satisfy

$$y(0) = c_0, \quad y'(0) = c_1.$$

(a) Use the power series solution

$$y(x) = c_0 \left(1 - \frac{x^2}{2} + \frac{x^4}{24} + \cdots \right)$$

$$+ c_1 \left(x - \frac{x^3}{20} + \frac{x^5}{120} + \cdots \right)$$

obtained in example 11 to verify that claim.

(b) Verify this same claim for the alternate form of this solution,

$$y(x) = c_0 \sum_{n=0}^{\infty} (-1)^n \frac{x^{2n}}{(2n)!}$$

$$+ c_1 \sum_{n=0}^{\infty} (-1)^n \frac{x^{2n+1}}{(2n+1)!},$$

which was obtained in example 12, and find its radius of convergence.

(c) Verify in general that if $y(x)$ is defined by the convergent power series

$$y(x) = \sum_{n=0}^{\infty} c_n (x - x_0)^n,$$

then $y(x_0) = c_0, y'(x_0) = c_1$.

13. (a) Use the Taylor coefficient formula

$$c_n = \frac{f^{(n)}(x_0)}{n!}$$

to verify that power series expansions of $\cos x$ and $\sin x$ about $x_0 = 0$ are

$$\cos x = 1 - \frac{x^2}{2} + \frac{x^4}{24} + \cdots,$$

$$\sin x = x - \frac{x^3}{6} + \frac{x^5}{120} + \cdots.$$

(b) Verify by substitution that $\cos x$, $\sin x$ are indeed solutions of the differential equation considered in example 11,

$$y'' + y = 0.$$

(c) Use the results of parts (a) and (b) to show that a general solution of $y'' + y = 0$ may be written in the power series form

$$y(x) = C_1 \left(1 - \frac{x^2}{2} + \frac{x^4}{24} + \cdots \right)$$

$$+ C_2 \left(x - \frac{x^3}{6} + \frac{x^5}{120} + \cdots \right),$$

where C_1, C_2 are the usual arbitrary constants.

(d) Verify by direct substitution in the differential equation $y'' + y = 0$ that the series expression in part (c) is indeed a solution.

(e) Express the arbitrary constants C_1, C_2 appearing in part (c) in terms of the initial values $y(0), y'(0)$.

14. Verify that the first three terms in each of the sums in

$$y(x) = c_0 \sum_{n=0}^{\infty} (-1)^n \frac{x^{2n}}{(2n)!}$$

$$+ c_1 \sum_{n=0}^{\infty} (-1)^n \frac{x^{2n+1}}{(2n+1)!},$$

the solution of $y'' + y = 0$ obtained in example 12, are the same as those in the solution of the same equation found in example 11,

$$y(x) = c_0 \left(1 - \frac{x^2}{2} + \frac{x^4}{24} + \cdots \right)$$

$$+ c_1 \left(x - \frac{x^3}{6} + \frac{x^5}{120} + \cdots \right).$$

15. Examples 11 and 12 both find power series solutions about $x_0 = 0$ of

$$y'' + y = 0.$$

Example 11 works with the first few terms of the proposed series solution, while example 12 uses the summation notation $\sum$ throughout. Write out the two examples side by side, matching the equations in each example that represent identical steps. (For example, match the equations that represent the calculation of y' in each example.) Add any computational details omitted from either example. What are the major differences between the two approaches?

16. Verify that the Cauchy-Euler equation

$$x^2 y'' + 7xy' + 9y = 0$$

has the solutions $x^{-3}, x^{-3} \ln |x|$. Show that neither of these functions has a convergent power series expansion about $x_0 = 0$. (Recall that if a function has a convergent power series expansion, then its coefficients must be defined by the Taylor coefficient formula $c_n = f^{(n)}(x_0)/n!$.) Is $x = 0$ an ordinary point of this equation?

17. In example 12, the text derives from the recurrence relation

$$c_k + (k+2)(k+1)c_{k+2} = 0, \quad k = 0, 1, 2, \ldots,$$

the expressions

$$c_k = (-1)^{k/2} \frac{c_0}{k!}, \quad k \text{ even}$$

and

$$c_k = (-1)^{(k-1)/2} \frac{c_1}{k!}, \quad k \text{ odd}$$

for the coefficients in a power series expansion. Then the text says:

> To get the pattern of these general calculations, try a few specific cases, such as $k = 6, 7, 8, 9$.

(a) Do just that; e.g., write c_6 in terms of c_4, then in terms of c_2, and so on. Do enough concrete cases to convince yourself that the exponent of -1 in the two expressions for c_k is correct.

(b) Find the radius of convergence of this series.

18. Example 13 found the first five nonzero terms in a power series solution of the initial-value problem

$$y'' + (x-1)^2 y' - 2(x-1)^3 y = 0,$$
$$y(1) = 1, \quad y'(1) = -8.$$

It began with the series expansion

$$y(x) = \sum_0^\infty c_n(x-1)^n$$

and found expressions for c_2, c_3, ... in terms of c_0, c_1. Then it used the initial conditions to determine c_0, c_1. Reverse these steps: Find the first five nonzero terms in a power series expansion of the solution of this initial-value problem by *first* evaluating c_0, c_1 in this series expansion, and then substituting in the differential equation to evaluate the remaining coefficients in the series. Show that you obtain the same solution as example 13.

19. Suppose $f(x)$ is analytic at x_0. Verify that

$$f(x_0) = \sum_{n=0}^\infty a_n(x_0 - x_0)^n = a_0,$$

$$f'(x_0) = \sum_{n=1}^\infty na_n(x_0 - x_0)^{n-1} = a_1,$$

$$f''(x_0) = \sum_{n=2}^\infty n(n-1)a_n(x_0 - x_0)^{n-2} = 2a_2,$$

$$\vdots$$

20. Show that a polynomial in x (a finite sum of nonnegative powers of x) is analytic at every finite value of x.

11.5 ▪ REGULAR SINGULAR POINTS

The previous section constructed power series solutions of linear equations

$$y'' + P(x)y' + Q(x)y = 0$$

about *ordinary* points, those where the coefficients P and Q are both analytic. But what happens at a *singular* point? For example, $x = 0$ is a singular point of the equation

$$y'' + \frac{3}{2x}y' - \frac{1}{2x^2}y = 0$$

because neither $P = 3/2x$ nor $Q = -1/2x^2$ is defined, much less analytic, at $x = 0$. What is the behavior at $x = 0$ of the solutions of such an equation? Is the singularity in the coefficients reflected in the solutions as well?

We can solve this particular example. Multiplying by $2x^2$ reduces it to the Cauchy-Euler equation

$$2x^2 y'' + 3xy - y = 0$$

whose indicial equation

$$2r(r-1) + 3r - 1 = 2r^2 + r - 1 = (2r-1)(r+1) = 0$$

has roots $r = -1, \frac{1}{2}$. Linearly independent solutions of this equation are

$$y_1 = x^{-1}, \quad y_2 = x^{1/2}.$$

Neither solution is analytic at $x = 0$.

But these solutions could be thought of as nearly analytic in the sense that they are either a negative or a noninteger power of x multiplying a (trivial) power series

$$x^{-1} = x^{-1}(1 + 0x + 0x^2 + \cdots),$$
$$x^{1/2} = x^{1/2}(1 + 0x + 0x^2 + \cdots).$$

That observation is the key to the technique of this section, the *method of Frobenius*, which seeks solutions of linear equations in the form of an arbitrary power of x multiplying a power series. Identifying the leading powers of x (x^{-1} and $x^{1/2}$ in this example) is important because those leading terms capture the behavior of the solutions near the singular point.

The method of Frobenius will be restricted to linear equations whose coefficients suffer only a mild singularity at the expansion point. In the preceding example, both $xP(x) = \frac{3}{2}$ and $x^2 Q(x) = -\frac{1}{2}$ are analytic at $x = 0$; the singularity in P is no worse than $1/x$, and that in Q is no worse than $1/x^2$. This weak singularity in P and Q is an example of a *regular singular point*.

Definition 2. *The point x_0 is a **regular singular point** of*

$$y'' + P(x)y' + Q(x)y = 0$$

if x_0 is a singular point (P and/or Q is not analytic at x_0) but both

$$(x - x_0)P(x) \quad and \quad (x - x_0)^2 Q(x)$$

*are analytic there. A singular point that is not regular is called **irregular**.*

To construct a series solution of a linear equation in a neighborhood of a *regular singular point* x_0, the **method of Frobenius** seeks a solution in the form

An arbitrary power of x multiplies the power series.

$$y(x) = (x - x_0)^r \sum_{n=0}^{\infty} c_n (x - x_0)^n,$$

where both the exponent r and the power series coefficients c_n are to be determined by substitution in the given differential equation. To avoid ambiguity in determining r, we assume that $c_0 \neq 0$; that is, we have factored the highest possible power of $x - x_0$ from the series and placed it in $(x - x_0)^r$.

We illustrate the method of Frobenius with an example. The techniques are much like those of finding power series expansions with the added wrinkle of determining r as well.

■ **EXAMPLE 17**   *Find two linearly independent series solutions about* $x_0 = 0$ *of*

$$2x^2 y'' + 3x(1-x)y' - y = 0.$$

Written in the usual form, this equation is

$$y'' + \frac{3(1-x)}{2x} y' - \frac{1}{2x^2} y = 0.$$

Evidently, $x_0 = 0$ is a singular point of this equation. But both $xP(x) = 3(1-x)/2$ and $x^2 Q(x) = \frac{1}{2}$ are analytic at the singular point so we seek a solution about a *regular* singular point.

To use the method of Frobenius, we seek a solution in the form

Assume Frobenius series.

$$y(x) = x^r \sum_{n=0}^{\infty} c_n x^n = c_0 x^r + c_1 x^{r+1} + c_2 x^{r+2} + \cdots,$$

where we assume $c_0 \neq 0$. The derivatives we need to substitute into the differential equation are

$$y'(x) = c_0 r x^{r-1} + c_1(r+1)x^r + c_2(r+2)x^{r+1} + \cdots,$$
$$y''(x) = c_0 r(r-1)x^{r-2} + c_1(r+1)rx^{r-1} + c_2(r+2)(r+1)x^r + \cdots.$$

Substituting the expressions for y, y', and y'' into $2x^2 y'' + 3x(1-x)y' - y = 0$ and keeping terms up to x^{r+2} yields

Substitute.

$$
\begin{aligned}
2c_0 r(r-1)x^r &+ 2c_1(r+1)rx^{r+1} + 2c_2(r+2)(r+1)x^{r+2} + \cdots \\
+ 3c_0 r x^r &+ 3c_1(r+1)x^{r+1} + \quad 3c_2(r+2)x^{r+2} \quad + \cdots \\
&- \quad 3c_0 r x^{r+1} - \quad 3c_1(r+1)x^{r+2} \quad - \cdots \\
- c_0 x^r &- \quad c_1 x^{r+1} - \quad c_2 x^{r+2} \quad\quad\; - \cdots = 0.
\end{aligned}
$$

The top line is $x^2 y''$, the next two are $3x(1-x)y'$, and the last is $-y$. To ease the task of collecting coefficients of like powers of x, corresponding powers appear in the same column, x^r, x^{r+1}, and x^{r+2} from left to right. Try a similar technique to organize your own work.

Since the sum is zero, we require that the coefficient of each power of x be zero:

Equate coefficients to zero.

$$
\begin{aligned}
x^r : \quad & c_0[2r(r-1) + 3r - 1] = 0, \\
x^{r+1} : \quad & c_1[2r(r+1) + 3(r+1) - 1] - 3c_0 = 0, \qquad\qquad (11.3)\\
x^{r+2} : \quad & c_2[2(r+2)(r+1) + 3(r+2) - 1] - 3c_1(r+1) = 0,
\end{aligned}
$$

$$\vdots$$

The first equation,

$$c_0[2r(r-1)+3r-1]=0,$$

the one coming from the lowest power of x, requires either

$$c_0 = 0 \quad \text{or} \quad 2r(r-1)+3r-1=0.$$

Since we assumed $c_0 \neq 0$, we must choose r to satisfy the **indicial equation**

Indicial equation
$$2r(r-1)+3r-1=(2r-1)(r+1)=0.$$

Hence,

The indicial equation determines r.
$$r = -1, \frac{1}{2}.$$

To ensure real-valued expressions in $x^{1/2}$ and to avoid the singularity at $x = 0$ in x^{-1}, we restrict consideration to $x > 0$.

Now the other equations in (11.2) provide recurrence relations involving r that give c_n in terms of c_{n-1}, $n = 1, 2, \ldots$:

Recurrence relations
$$c_1 = \frac{3r}{2r(r+1)+3(r+1)-1}c_0,$$

$$c_2 = \frac{3(r+1)}{2(r+2)(r+1)+3(r+2)-1}c_1,$$

$$\vdots \tag{11.4}$$

For each choice of r, we obtain a set of values for the constants c_n in the power series.

$r = -1$
Using $r = -1$ and setting $c_0 = 1$ for convenience, these equations give

$$c_1 = \frac{-3}{-1}c_0 = 3,$$

$$c_2 = \frac{0}{-4}c_1 = 0.$$

Thus, the solution corresponding to $r = -1$ is

Solution with $r = -1$
$$y_1 = x^{-1}(1 + 3x + \cdots), \quad x > 0.$$

$r = \frac{1}{2}$
Repeating these steps with $r = \frac{1}{2}$ and $c_0 = 1$ gives

$$c_1 = \frac{3/2}{10/2}c_0 = \frac{3}{10},$$

$$c_2 = \frac{-9/2}{29/2}c_1 = \frac{-9}{29}.$$

The solution corresponding to $r = 1/2$ is

Solution with $r = \frac{1}{2}$

$$y_2 = x^{1/2} \left(1 + \frac{3x}{10} - \frac{9x^2}{29} + \cdots \right), \quad x > 0.$$

Taking linear independence for granted (because of the different leading powers of x), a general solution of $2x^2 y'' + 3x(1-x)y' - y = 0$ for $x > 0$ is

$$y_g = C_1 y_1 + C_2 y_2.$$

Had we kept the undetermined constant c_0 in either of the solutions for $r = -1$, $1/2$, it would have been absorbed into C_1 or C_2. ■

Much of the value of the solution obtained in the preceding example is its picture of the behavior near the singular point $x = 0$. One solution ($y_2 = x^{1/2}$) is bounded. The other is not, growing like x^{-1}. Neither solution is differentiable at the origin.

This example illustrates the basic idea of the *method of Frobenius*.

Method of Frobenius. If $x = x_0$ is a regular singular point of

$$y'' + P(x)y' + Q(x)y = 0,$$

assume a solution of the form

FROBENIUS SERIES

$$y(x) = (x - x_0)^r \sum_{n=0}^{\infty} c_n (x - x_0)^n,$$

require $c_0 \neq 0$, and substitute into the differential equation. Collect like powers of x. Apply the condition $c_0 \neq 0$ to the coefficient of the lowest power of x to obtain the *indicial equation*, a quadratic equation for r. Find the roots r_1, r_2 of the indicial equation, and determine one set of power series coefficients for each value of r from the recurrence relations giving c_n in terms of c_{n-1}, $n = 1, 2, \ldots$.

When r is other than a nonnegative integer, $(x - x_0)^r$ will either be singular at $x = x_0$ or complex-valued for $x < x_0$. The cases $x < x_0$ and $x > x_0$ can be considered separately and eventually consolidated, using $|x - x_0|$ in place of $x - x_0$. For simplicity, we consider only $x > x_0$.

Unfortunately, several complications lie below the tip of this iceberg.

- If $r_1 = r_2$, there is no second independent Frobenius solution because the two sets of recurrence relations collapse into one.

- If $r_1 - r_2$ is an integer, then the series for one value of r may or may not be a multiple of the other series; the factor $x^{r_1-r_2}$ might be absorbed into the power series part of one solution. A more detailed analysis is needed.

The following examples hint at some of these difficulties.

■ **EXAMPLE 18**   *Find series solutions about $x = 0$ of Bessel's equation of order one-half,*

Bessel's equation, order one-half

$$x^2 y'' + xy' + \left(x^2 - \tfrac{1}{4}\right) y = 0.$$

> Bessel's equation of order p is $x^2 y'' + xy' + (x^2 - p^2)y = 0$. It arises in the course of solving heat-flow problems in circular domains, among many other settings.

We leave it to you to verify that $x = 0$ is a regular singular point. As with the previous example, we assume a solution of the form

Assume Frobenius series.

$$y(x) = x^r \sum_{n=0}^{\infty} c_n x^n = c_0 x^r + c_1 x^{r+1} + c_2 x^{r+2} + \cdots,$$

where sufficient powers of x have been factored from the sum to ensure $c_0 \neq 0$. Following the pattern of example 17, we differentiate, substitute in the differential equation, and collect like powers of x to obtain (through terms in x^{r+2}):

Substitute.

$$
\begin{aligned}
c_0 r(r-1)x^r &+ c_1(r+1)rx^{r+1} + c_2(r+2)(r+1)x^{r+2} + \cdots \\
+ c_0 r x^r \quad &+ \quad c_1(r+1)x^{r+1} + \quad c_2(r+2)x^{r+2} \quad + \cdots \\
- c_0 x^r/4 \quad &- \quad c_1 x^{r+1}/4 \quad - \quad c_2 x^{r+2}/4 \quad + \cdots \\
&+ \quad c_0 x^{r+2} \quad + \cdots = 0.
\end{aligned}
$$

Equating to zero the coefficients of each power of x yields

Equate coefficients to zero.

$$
\begin{aligned}
x^r : \quad & c_0[r^2 - 1/4] = 0, \\
x^{r+1} : \quad & c_1[(r+1)^2 - 1/4] = 0, \\
x^{r+2} : \quad & c_2[(r+2)^2 - 1/4] + c_0 = 0, \\
& \quad \vdots
\end{aligned}
\tag{11.5}
$$

From the coefficient of x^r and the requirement that $c_0 \neq 0$, we have the indicial equation

$$r^2 - \tfrac{1}{4} = 0$$

The indicial equation determines r.

and the roots

$$r = \pm \tfrac{1}{2}.$$

These roots differ by 1.

The coefficients of x^{r+1} and x^{r+2}, as before, provide relations that define coefficients in the power series.

$r = \tfrac{1}{2}$

With $r = \tfrac{1}{2}$, we find from the coefficient of x^{r+1} that

$$c_1 \left[\left(\tfrac{3}{2}\right)^2 - \tfrac{1}{4} \right] = 2c_1 = 0.$$

Hence, $c_1 = 0$ for $r = \frac{1}{2}$. From the coefficient of x^{r+2} with $r = \frac{1}{2}$, we find

$$c_2 = \frac{-c_0}{6}.$$

Setting $c_0 = 1$, we can take one solution of Bessel's equation of order one-half to be

$$y_1 = x^{1/2} \left(1 - \frac{x^2}{6} + \cdots \right).$$

$r = -\frac{1}{2}$ With $r = -\frac{1}{2}$, we find a more surprising result from the coefficient of x^{r+1}:

$$c_1 \left[\left(\tfrac{1}{2} \right)^2 - \tfrac{1}{4} \right] = c_1 \cdot 0 = 0;$$

c_1 is arbitrary! With a bit of hesitation, perhaps, we choose $c_1 = 0$, the same value obtained when $r = \frac{1}{2}$. Then the coefficient of x^{r+2} gives

$$c_2 = \frac{-c_0}{2}.$$

Setting $c_0 = 1$, a second solution of Bessel's equation of order one-half is

$$y_2 = x^{-1/2} \left(1 - \frac{x^2}{2} + \cdots \right),$$

and it appears to be linearly independent of y_1, even though the roots of the indicial equation differed by a positive integer. ■

■ **EXAMPLE 19**   *Attempt to mimic the previous analysis to obtain solutions of Bessel's equation of order two,*

Bessel's equation, order two

$$x^2 y'' + xy' + (x^2 - 4)y = 0,$$

about $x = 0$.

Proceeding exactly as in example 18, start with a Frobenius series solution, substitute in the differential equation, collect powers of x, and find the following relations from the coefficients of the lowest five powers of x:

$$
\begin{aligned}
x^r : \quad & c_0[r^2 - 4] = 0, \\
x^{r+1} : \quad & c_1[(r+1)^2 - 4] = 0, \\
x^{r+2} : \quad & c_2[(r+2)^2 - 4] + c_0 = 0, \\
x^{r+3} : \quad & c_3[(r+3)^2 - 4] + c_1 = 0, \\
x^{r+2} : \quad & c_4[(r+4)^2 - 4] + c_2 = 0,
\end{aligned}
$$

$$\vdots$$

As usual, the first equation and $c_0 \neq 0$ produce the indicial equation

$$r^2 - 4 = 0$$

and the indicial roots

$$r = \pm 2.$$

The second relation forces $c_1 = 0$, and the third requires that

$$c_2[r^2 + 4r] + c_0 = 0.$$

When $r = 2$, we quickly find $c_2 = -c_0/12$. Setting $c_0 = 1$, we have one solution,

$$y_1 = x^2 \left(1 - \frac{x^2}{12} + \cdots \right).$$

But when $r = -2$, trouble arises. The relation $c_2[r^2 + 4r] + c_0 = 0$ reduces to $c_2 = c_0/4$ while $c_4[(r+4)^2 - 4] + c_2 = 0$ becomes

$$c_4 \cdot 0 + c_2 = c_4 \cdot 0 + \frac{c_0}{4} = 0.$$

The recurrence relations collapse at this point because there is no c_4 satisfying this equation as long as $c_0 \neq 0$. The coefficient $[(r+4)^2 - 4]$ with $r = 2$ reduces to the indicial equation with $r = 2$ because the roots differ by 4. ■

When separation of variables is applied to the heat equation in a circular domain, the resulting eigenvalue problem involves Bessel's equation of order zero. Consequently, we explore the solution of that equation in a bit more detail.

■ **EXAMPLE 20**   *Find a solution of Bessel's equation of order zero,*

$$x^2 y'' + x y' + x^2 y = 0,$$

subject to the initial conditions $y(0) = 1$, $y'(0) = 0$. Determine the behavior at $x = 0$ of a second linearly independent solution.

Since $x = 0$ is a regular singular point, we assume a solution in the form of a Frobenius series,

$$y(x) = x^r \sum_{n=0}^{\infty} c_n x^n,$$

where $c_0 \neq 0$. Substituting in the differential equation and collecting similar powers of x lead to

$$c_0[r(r-1) + r]x^r + c_1[(r+1)r + (r+1)]x^{r+1}$$
$$+ (c_2[(r+2)(r+1) + (r+2)] + c_0)\, x^{r+2} + \cdots$$
$$+ (c_n[(r+n)(r+n-1) + (r+n)] + c_{n-1})\, x^{r+n} + \cdots = 0.$$

Bessel's equation, order zero

Assume Frobenius series.

Substitute

The coefficient of x^r and $c_0 \neq 0$ yield the indicial equation,

$$r(r-1) + r = r^2 = 0.$$

The (repeated) indicial root is $r = 0$. Setting $r = 0$, the coefficient of x^{r+1} reduces to

$$c_1 = 0.$$

Setting to zero the coefficient of x^{r+n}, $n = 2, 3, \ldots$ and using $r = 0$ gives the general recurrence relation

Recurrence relation

$$c_n = \frac{-c_{n-2}}{n(n-1)+n} = \frac{-c_{n-2}}{n^2}, \quad n = 2, 3, \ldots .$$

Since $c_1 = 0$, all coefficients with odd indices are zero.

Representing the remaining even indices as $n = 2m$, $m = 1, 2, \ldots$, the recurrence relation becomes

$$c_{2m} = \frac{-c_{2(m-1)}}{4m^2}, \quad m = 1, 2, \ldots .$$

By writing $c_{2(m-1)}$ in terms of $c_{2(m-2)}$, and so on, we find

$$c_{2m} = \frac{(-1)^m c_0}{4^m (m!)^2}, \quad m = 1, 2, \ldots .$$

Since $r = 0$, the solution $y(x) = x^r (c_0 + c_1 x + \cdots)$ can be evaluated at $x = 0$: $y(0) = c_0$. The initial condition $y(0) = 1$ forces $c_0 = 1$. We have found the solution commonly denoted $J_0(x)$, the **Bessel function of first kind of order zero**,

First solution, a Bessel function

$$J_0(x) = \sum_{m=0}^{\infty} \frac{(-1)^m x^{2m}}{4^m (m!)^2}.$$

Since $J_0(x) = 1 - x^2/4 + \cdots$ and $J_0^2 = 1 - x^2/2 + \cdots$, it obviously satisfies the remaining initial condition, $J_0'(0) = 0$.

Since the indicial equation has only one root, we can not find a second linearly independent solution using a Frobenius series. Instead, we turn to reduction of order and assume the second solution has the form

Reduction of order

$$y(x) = J_0(x)u(x),$$

where the function $u(x)$ is to be determined. Substituting into Bessel's equation and letting $v = u'$ in the usual way yield

$$x^2 J_0(x)v'(x) + \left(2x^2 J_0'(x) + x J_0(x)\right) v(x) = 0.$$

With caution because of the singularity at $x = 0$, we use separation of variables to obtain the solution

$$v(x) = \frac{1}{x(J_0(x))^2}.$$

Hence,

$$u(x) = \int v(x)\, dx = \int \frac{dx}{x(J_0(x))^2}.$$

Our only concern is the behavior of this second solution near $x = 0$. Consequently, we estimate the behavior of this integral rather than evaluate it exactly. Since $J_0(x) = 1 - x^2/4 + \cdots$, we have

$$\frac{1}{(J_0(x))^2} = \frac{1}{1 - x^2/2 + \cdots} \approx 1 + \frac{x^2}{2} + \cdots$$

for x small.

The last approximation used $(1 - z)^{-1} \approx 1 + z$, $|z| < 1$, a consequence of a Taylor expansion (or the binomial expansion).

Consequently, when x is small and positive, we find

$$u(x) = \int \frac{dx}{x(J_0(x))^2} \approx \int \frac{1}{x}\left(1 + \frac{x^2}{2} + \cdots\right) dx = \ln x + \cdots.$$

The terms other than $\ln x$ are all bounded near zero since they involve positive powers of x.

Behavior of second solution

We conclude that a second linearly independent solution of Bessel's equation of order zero must behave near $x = 0$ as does $J_0(x) \ln x$. *The second solution of Bessel's equation of order zero has a logarithmic singularity at $x = 0$, as does the second solution of a Cauchy-Euler equation with the repeated indicial root $r = 0$.* ■

A standard form for the second solution of Bessel's equation of order zero is the *Bessel function of second kind of order zero*. It is defined by

$$Y_0(x) = \left(\ln\left(\frac{x}{2}\right) + \gamma\right) J_0(x) - \frac{2}{\pi} \sum_{m=1}^{\infty} \frac{(-1)^m x^{2m}}{4^m (m!)^2} S_m.$$

Here S_m is the partial sum of the harmonic series,

$$S_m = 1 + \frac{1}{2} + \frac{1}{3} + \cdots + \frac{1}{m},$$

and the Euler constant γ is defined by

$$\gamma = \lim_{n \to \infty} (S_n - \ln n).$$

Figure 11.2 shows graphs of J_0 and Y_0. Tables of values of these and other solutions of Bessel's equation appear in [1]. These functions are also well represented in MATLAB.

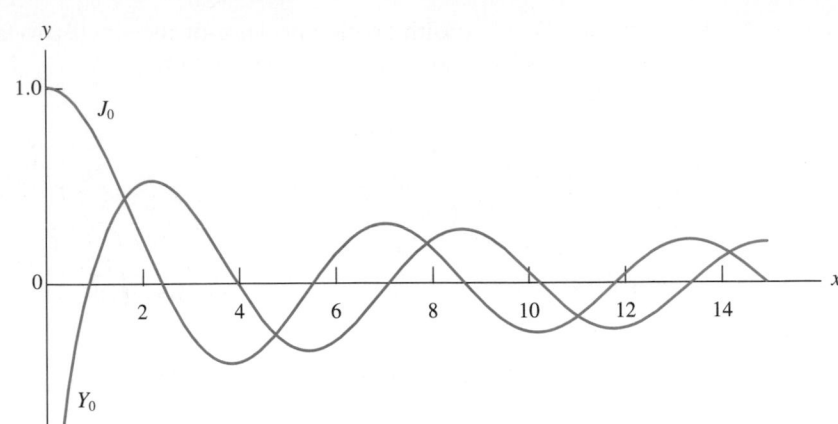

FIGURE 11.2   Graphs of the solutions J_0 and Y_0 of Bessel's equation of order zero

As the last example suggests, the problems that arise when indicial roots differ by zero or an integer can be overcome to produce a second linearly independent solution involving more than just a Frobenius series. Those issues are not pursued here.

When the *roots of the indicial equation do not differ by zero or a positive integer*, the method of Frobenius can be shown to *always yield two linearly independent solutions*.

11.5.1 Exercises

EXERCISE GUIDE	
To gain experience . . .	**Try exercises**
Locating ordinary points, regular singular points	1–6, 9, 13–19
Finding indicial equations, indicial roots	1–6, 10–12, 22(a–b)
Identifying analytic functions	7–8
Using the method of Frobenius	10–11, 13–19, 20(b)
With the limitations of the method of Frobenius	19
With recurrence relations	20(a), 21
Finding radius of convergence	20(c), 23
With the method of Frobenius in a general setting	22

For each of the differential equations in exercise 1–6,

(i) Find all finite ordinary points,

(ii) Find all finite regular singular points,

(iii) Find the indicial equation and its roots at each regular singular point.

1. $x^2 y'' + 4xy' + (x^2 - 4)y = 0$

2. $(x - 4)^2 y'' + 4xy' + (x^2 - 4)y = 0$

3. $(x^2 - 4)y'' + 4xy' + (x^2 + 4)y = 0$

4. $y'' \sin 2x + xy' - y = 0$

5. $x^2 y'' + xy' + \left(x^2 - \frac{1}{16}\right) y = 0$ (Bessel's equation of order one-quarter)

6. $4xy'' + 2y' + y = 0$

7. Verify the claim in the text that neither of the solutions

$$y_1 = x^{-1}, \quad y_2 = x^{1/2}$$

of

$$y'' + \frac{3}{2x} y' - \frac{1}{2x^2} y = 0$$

are analytic at $x = 0$.

8. Example 17 used the method of Frobenius to find the solutions

$$y_1 = x^{-1} (1 - 3x + \cdots)$$

and

$$y_2 = x^{1/2} \left(1 + \frac{3x}{14} - \frac{9x^2}{31} + \cdots\right)$$

of $2x^2 y'' + 3x(1 - x)y' - y = 0$ in a neighborhood of the regular singular point $x = 0$. Explain why neither of these solutions is

(a) analytic at $x = 0$;

(b) a power series about $x = 0$.

9. Verify that $x = 0$ is a regular singular point of Bessel's equation of order p, $x^2 y'' + xy' + (x^2 - p^2)y = 0$. Does this equation have any other finite singular points?

10. Example 18 skips the details of substituting the Frobenius series into Bessel's equation of order one-half, $x^2 y'' + xy' + \left(x^2 - \frac{1}{4}\right) y = 0$. Supply them and verify that the resulting indicial equation and recurrence relations are correct.

11. Example 19 skips the details of substituting the Frobenius series into Bessel's equation of order two, $x^2 y'' + xy' + (x^2 - 4)y = 0$. Supply them and verify that the resulting indicial equation and recurrence relations are correct.

12. Show that the indicial equation for Bessel's equation of order p, $x^2 y'' + xy' + (x^2 - p^2) = 0$, is

$$r^2 - p^2 = 0.$$

What choices of p guarantee that the roots of the indicial equation will differ by *other* than zero or a positive integer?

In exercises 13–18, find the first three nonzero terms in series solutions of the given differential equations about a regular singular point.

13. $4xy'' + 2y' + y = 0$

14. $16x^2 y'' + 16xy' + (16x^2 - 1)y = 0$

15. $x^2 y'' + xy' + \left(x^2 - \frac{1}{16}\right) y = 0$
(Bessel's equation of order one-quarter)

16. $2x^2 y'' + (1 - x)y' + y = 0$

17. $3(x - 2)y'' + y' - y = 0$

18. $6(x + 1)^2 y'' + 5(x + 1)y' + x(x + 2)y = 0$

19. Try to apply the method of Frobenius to the equation $x^3 y'' + y = 0$ at $x = 0$. Is $x = 0$ a regular singular point? Demonstrate how the method of Frobenius fails.

20. Building on the work of example 17,

(a) Show that the terms in the Frobenius series about $x = 0$,

$$x^r \sum_{n=0}^{\infty} c_n x^n,$$

for the differential equation $2x^2 y'' + 3x(1 - x)y' - y = 0$ are determined by the recurrence relation

$$c_{k+1} = \frac{3(r + k)}{2(r + k)(r + k + 1) + 3(r + k + 1) + 1} c_k.$$

(b) Write the general terms in a pair of linearly independent solutions of this equation.

(c) Determine the radius of convergence of the *power series* portion of these solutions.

21. Show that the general recurrence relation term for Bessel's equation of order p, $x^2 y'' + xy' + (x^2 - p^2) = 0$, is

$$c_n[(r + n)(r + n - 1) + r + n - p^2] + c_{n-2} = 0,$$

$n = 2, 3, 4, \ldots$. Using the indicial equation $r^2 - p^2 = 0$, deduce that this relation reduces to

$$c_n[2nr + n^2] + c_{n-2} = 0.$$

What problem occurs whenever one of the roots of the indicial equation is a negative, even integer?

22. Suppose $x = 0$ is a regular singular point of the equation

$$y'' + P(x)y' + Q(x)y = 0.$$

Write the power series expansions for the analytic functions $xP(x)$ and $x^2 Q(x)$ as

$$xP(x) = \sum_{n=0}^{\infty} p_n x^n, \quad x^2 Q(x) = \sum_{n=0}^{\infty} q_n x^n.$$

(a) Show that the indicial equation for $y'' + P(x)y' + Q(x)y = 0$ is

$$r(r-1) + p_0 r + q_0 = 0;$$

i.e., it is the indicial equation for the Cauchy-Euler equation

$$x^2 y'' + p_0 xy' + q_0 y = 0$$

whose coefficients are the leading terms in the power series expansions for xP and $x^2 Q$.

(b) Use this result to recover the indicial equation for

$$2x^2 y'' + 3x(1-x)y' - y = 0,$$

the equation considered in example 17.

(c) State and derive the corresponding result for an arbitrary regular singular point $x = x_0$.

23. Find the radius of convergence of the series defining the Bessel function of first kind of order zero,

$$J_0(x) = \sum_{m=0}^{\infty} \frac{(-1)^m x^{2m}}{4^m (m!)^2}.$$

11.6 ■ SOLUTION METHOD SUMMARY

The only general solution methods for arbitrary *nonlinear* second-order equations are numerical: Euler, Heun, Runge-Kutta, etc.

For *linear* equations, the options are as follows.

1. Homogeneous equation:
   (a) Constant-coefficient equation:
       i. Characteristic equations: $y = e^{rx}$
   (b) Variable-coefficient equation:
       i. Cauchy-Euler equation: $y = x^r$
       ii. Reduction of order: $y_2 = u(x)y_1$
       iii. Power series: $y = c_0 + c_1(x - x_0) + \cdots$, x_0 an ordinary point
       iv. Frobenius: $y = (x - x_0)^r (c_0 + c_1(x - x_0) + \cdots)$, x_0 a regular singular point

2. Nonhomogeneous equation:
   (a) Constant-coefficient equation:
       i. Undetermined coefficients: y_p similar to forcing terms of sine, cosine, exponential, polynomial
       ii. Variation of parameters: $y_p = u_1(x)y_1 + u_2(x)y_2$
   (b) Variable-coefficient equation:
       i. Variation of parameters: $y_p = u_1(x)y_1 + u_2(x)y_2$
       ii. Power series: $y = c_0 + c_1(x - x_0) + \cdots$, x_0 an ordinary point, and forcing term expanded about x_0 as well.

Figure 11.3 provides a decision tree for choosing among these possible solution methods. Note that all linear equations, regardless of order, can be written as systems so that the methods of chapter 8 can be applied.

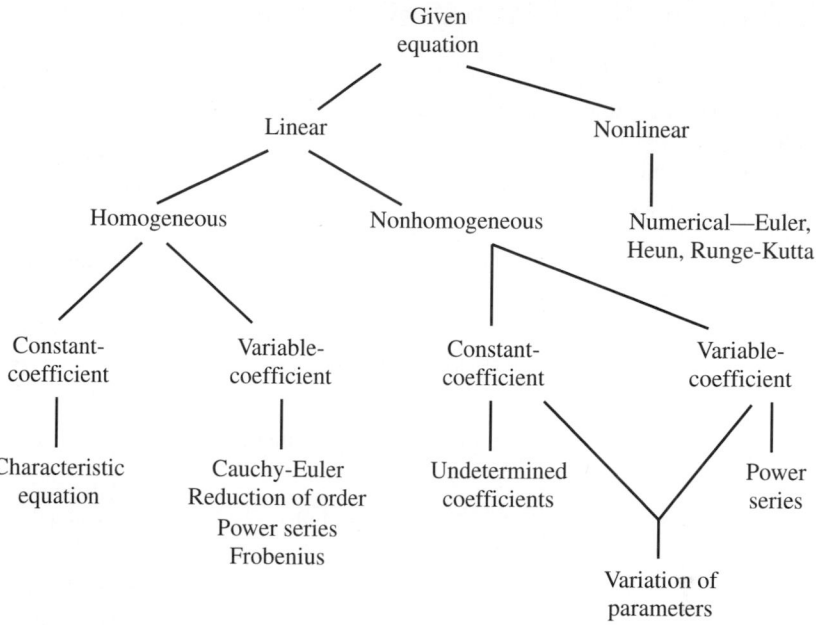

FIGURE 11.3 A decision tree for choosing solution methods for second-order equations.

11.7 ■ CHAPTER EXERCISES

EXERCISE GUIDE	
To gain experience ...	**Try exercises**
Finding general solutions	1–10, 22–23
Solving initial-value problems	7–8
Locating ordinary points, regular singular points	11–21
Using reduction of order	24
Solving Cauchy-Euler equations	1–2, 5–6, 9
Using variation of parameters	2–6
Using series solution methods	7–8, 10, 22–23
Solving eigenvalue problems	25

In exercises 1–10, either solve the given initial-value problem or find a general solution of the differential equation. If you use a series method, find at least the first three nonzero terms in the series.

1. $4(x + 1)^2 y'' - 12(x + 1)y' + 16y = 0$

2. $x^2 y'' + 2xy' + 3y = \tan x$

3. $y'' + y = \sin t$

4. $y'' + y = \sin \pi t$

5. $t^2 u'' + 2tu' - 6u = 1$

6. $t^2 u'' + 2tu' - 6u = 1/t$

7. $xy'' + y = 0,\ y(1) = 2,\ y'(1) = 0$

8. $x^2 y'' + xy' + y = 0,\ y(1) = y'(1) = 1$

9. $(x-1)^2 y'' - 2(x-1)y' + 2y = 4(x-2)^2$

10. $t^2 z'' - 2tz' + 2z = (t-1)^2$

For each of the differential equations in exercises 11–21, find all finite

   (i) ordinary points,

   (ii) regular singular points.

11. $x^2 y'' + xy' + x^2 y = 0$ (Bessel's equation of order zero)

12. $x^2 y'' + xy' + (x^2 - 1)y = 0$ (Bessel's equation of order one)

13. $x^2 y'' + xy' + (x^2 - p^2)y = 0$ (Bessel's equation of order p)

14. $x(x-1)y'' - y' - y = 0$ (a form of the hypergeometric equation)

15. $(1 - x^2)y'' - xy' + p^2 y = 0$ (Chebyshev's equation)

16. $xy'' + (1-x)y' - y = 0$ (another form of the hypergeometric equation)

17. $xy'' + (1-x)y' + ny = 0$, n a nonnegative integer (Laguerre's equation of order n)

18. $y'' + (3 - x^2)y = 0$ (a form of Hermite's equation)

19. $y'' + 4\pi^2 m(2E - kx^2)y/h^2 = 0$, m, h, E, k all constants (a form of the Schrödinger wave equation; m is mass, h is Planck's constant, E is total energy, and k is a kind of spring constant).

20. $(1 - x^2)y'' - 2xy' + n(n+1)y = 0$, n a nonnegative integer (Legendre's equation)

21. $y'' - xy = 0$ (Airy's equation)

22. Find the first three nonzero terms in a pair of linearly independent series solutions about $x = 0$ of Hermite's equation,

$$y'' - 2xy' + 2py = 0,$$

where p is a constant.

23. Find the first three nonzero terms in a pair of linearly independent series solutions about $x = 0$ of Airy's equation,

$$y'' + xy = 0.$$

24. The purpose of this problem is to show via reduction of order that every constant-coefficient, linear, homogeneous, second-order equation

$$y'' + b_1 y' + b_0 y = 0$$

possessing the single, repeated, real characteristic root r has the pair of linearly independent solutions e^{rx}, xe^{rx}.

  (a) Show that if the characteristic equation of $y'' + b_1 y' + b_0 y = 0$ possesses a single repeated real root, then b_1 can be expressed in terms of b_0. (*Hint*: What is the value of the discriminant in this case?)

  (b) Using e^{rx} as the first solution, construct a second solution of $y'' + b_1 y' + b_0 y = 0$ of the form $u(x)e^{rx}$ via reduction of order. Using part (a), show that $u(x) = x$ by deriving the equation $u'' = 0$.

  (c) Verify that e^{rx} and xe^{rx} are indeed linearly independent.

25. The eigenvalue problem

$$xX''(x) + X'(x) + x\lambda X(x) = 0, \quad X'(0) = X(L) = 0,$$

arises when separation of variables is applied to a model of heat flow in a thin circular disk of radius L; see exercise 12, page 502, of the chapter exercises for chapter 9.

  (a) Show that solutions of the differential equation and the boundary condition $X'(0) = 0$ are $X(x) = J_0(\sqrt{\lambda}x)$, $\lambda > 0$. Recall that $J_0(x)$, the Bessel function of first kind of order zero, is a solution of Bessel's equation of order zero, $x^2 y''(x) + xy'(x) + x^2 y(x) = 0$.

  (b) Argue that the eigenvalues are defined by $J_0(\sqrt{\lambda}L) = 0$. Use the graph of J_0 in figure 11.2, page 614, to write the first few eigenvalues in terms of L. How do those compare with the corresponding eigenvalues of $X''(x) + \lambda X(x) = 0$ subject to the same boundary conditions?

11.8 ■ CHAPTER PROJECTS

1. Cauchy-Euler equations for $x < 0$. The text treats the Cauchy-Euler equation

$$x^2 y'' + b_1 xy' + b_0 y = 0$$

for $x > 0$. This project asks you to extend that work to the case $x < 0$, thereby verifying the solution forms given in table 11.1, page 579.

Begin by assuming a solution of the form $y = (-x)^r$, $x < 0$. Using the chain rule *carefully*, compute dy/dx and d^2y/dx^2. Rewrite the coefficients x^2 and x in the differential equation in terms of $-x$. Substitute the assumed solution form, and show that you obtain the same indicial equation as when $x > 0$. Write the appropriate forms of the solution when the indicial equation has distinct real roots, repeated real roots, and complex conjugate roots. Finally, use the definition of absolute value ($|x| = x$ for $x \geq 0$, $|x| = -x$ for $x < 0$) to obtain the entries in table 11.1. State your results in the form of a theorem.

2. **Change of variables in Cauchy-Euler equations.** Complete steps (a–d) to show that the change of variables $x = e^t$ reduces the Cauchy-Euler equation

$$x^2 y''(x) + b_1 xy'(x) + b_0 y(x) = 0$$

to the constant-coefficient equation

$$\frac{d^2Y}{dt^2} + (b_1 - 1)\frac{dY}{dt} + b_0 Y = 0,$$

which then may be solved by the characteristic equation method. What restrictions, if any, must be placed on x? On t?

(a) Define a new function $Y(t)$ by $Y(t) = y(e^t)$. Use the chain rule to show that

$$\frac{dY}{dt} = \frac{dy}{dx}\frac{dx}{dt} = xy'(x).$$

Derive a similar expression for d^2Y/dt^2, beginning with

$$\frac{d^2Y}{dt^2} = \frac{d(xy'(x))}{dt} = \frac{d(xy'(x))}{dx}\frac{dx}{dt}.$$

(b) Substitute the expressions obtained in part (a) into $x^2 y''(x) + b_1 xy'(x) + b_0 y(x) = 0$ to obtain the differential equation for Y.

(c) If the differential equation for Y has a solution $Y(t) = e^{rt}$, where r is a root of the characteristic equation

$$r^2(b_1 - 1)r + b_0 = 0,$$

argue that x^r is a solution of $x^2 y''(x) + b_1 xy'(x) + b_0 y(x) = 0$.

(d) If r is a repeated characteristic root of the equation for Y, argue that the corresponding solution te^{rt} transforms into the solution $x^r \ln x$ of $x^2 y''(x) + b_1 xy'(x) + b_0 y(x) = 0$.

(e) Use this transformation technique to solve three Cauchy-Euler equations, one whose indicial equation has distinct real roots, one with repeated real roots, and one with complex conjugate roots.

3. **Proof of existence of analytic solutions.** Following the steps outlined after the statement of theorem 1, page 600, prove that a second-order linear equation has a pair of linearly independent analytic solutions in a neighborhood of an ordinary point. In accordance with usual professional practice, give credit for any ideas you obtain from other sources.

Appendix

This appendix briefly reviews some of the ideas from calculus and algebra that are frequently used in studying differential equations.

A.1 ■ DERIVATIVES

The derivative $f'(t)$ of the function $f(t)$ is defined by

$$f'(t) = \lim_{\Delta t \to 0} \frac{f(t + \Delta t) - f(t)}{\Delta t},$$

if the limit exists.

The derivative is often interpreted as a rate of change. If the position s of an object is given as a function of time t, then the derivative function ds/dt gives velocity as a function of time. When $ds/dt > 0$, then velocity is positive, and position s is increasing. If $v = ds/dt$ defines the velocity function, then $dv/dt = d^2s/dt^2$ is acceleration. When $d^2s/dt^2 = d(ds/dt)/dt > 0$, then ds/dt is increasing; i.e., velocity v is increasing.

In a graphical setting, the value of the derivative $f'(x)$ of a function $f(x)$ is the slope of the line tangent to the graph of the function at x. The value of the second derivative $f''(x)$ gives the rate of change of the slope of the tangent line at x.

If $f'(x) > 0$, then the function is *increasing* at x; the slope of its graph is positive there. If $f''(x) = d(f'(x))/dx > 0$, then the rate of change of the slope is positive at x; the slope is increasing. If $f''(x) > 0$ over an interval, then the slope of the graph is increasing over that interval; the graph is *concave up* there.

Derivatives with respect to different variables are related by the *chain rule for derivatives*:

Theorem 1 (Chain rule for derivatives). *If s is a function of t and t is a function T and both are differentiable, then*

$$\frac{ds}{dT} = \frac{ds}{dt} \frac{dt}{dT}.$$

For example, suppose s represents position measured in meters, t is time measured in seconds, and T is time measured in minutes. Then ds/dt is velocity in meters/second, and ds/dT is velocity in meters/minute.

The chain rule tells us how these two derivatives of the same function with respect to different independent variables are related. The functional relationship between t and T is

$$t = 60T.$$

Hence, $dt/dT = 60$ seconds/minute , and the chain rule yields

$$\frac{ds}{dT} \text{ meters/minute} = 60\frac{ds}{dt} \text{ meters/second.}$$

More frequently, you have used the chain rule to compute derivatives of composite functions, as in the following example.

■ **EXAMPLE 1**   *Find* dy/dx *if* $y = e^{\cos x}$.

To see clearly that this is a composite function, write it as $y = e^{(\cos x)}$. The "outside" function is the exponential, and the "inside" function is the cosine. If you were evaluating this function on a calculator, you would enter the value of x and then push the function keys *cos* and *exp* in that order, from inside out.

The chain rule can be stated informally as "derivative of the inside times derivative of the outside." To relate that phrase to the formal statement of the chain rule, let $u = \cos x$. Then we can write the given function in two parts as

$$y = e^u, \quad \text{and} \quad u = \cos x.$$

The chain rule yields

$$\frac{dy}{dx} = \frac{dy}{du}\frac{du}{dx} = (e^u)(-\sin x) = -e^{\cos x}\sin x.$$

The term $e^u = e^{\cos x}$ is the derivative of the "outside" function, and $-\sin x$ is the derivative of the "inside" function. ■

If necessary, review the section of your calculus text that discusses the chain rule as well as the derivative formulas for polynomials, trigonometric functions, the logarithm, and the exponential.

For functions of more than one variable, the rate of change with respect to one of those variables, independent of the others, is the partial derivative. For example, if t and y are independent variables, then we can define the partial derivatives

$$\frac{\partial f(t, y)}{\partial t} = \lim_{\Delta t \to 0} \frac{f(t + \Delta t, y) - f(t, y)}{\Delta t}$$

and

$$\frac{\partial f(t, y)}{\partial y} = \lim_{\Delta y \to 0} \frac{f(t + \Delta y, y) - f(t, y)}{\Delta y},$$

provided the limits in question exist.

Now suppose that y is not an independent variable but a function of t. Then the expression $z(t) = f(t, y(t))$ is a function of t alone. Another form of the chain rule provides a formula for the derivative $z'(t)$,

$$z'(t) = \frac{\partial f(t, y(t))}{\partial t} + \frac{\partial f(t, y(t))}{\partial y}\frac{dy(t)}{dt}.$$

A.2 ■ ANTIDERIVATIVES

The antiderivative notation

$$\int f(x)\,dx = ?$$

asks the question,

What is a function whose derivative with respect to x is f?

The answer always includes an additive constant (the famous "plus C") because adding a constant to a function does not change its derivative. More formally, if F is a function whose derivative with respect to x is f (i.e., if $dF/dx = f$), then the derivative of $F + C$ is also f:

$$\frac{d(F + C)}{dx} = \frac{dF}{dx} + \frac{dC}{dx} = f + 0 = f.$$

We can assign a specific value to C only if we have some added information, such as the value of the antiderivative at a specific point.

■ **EXAMPLE 2** *Find a function whose derivative with respect to x is $5x^2$ and that assumes the value 6 at $x = 2$.*

To find the family of functions whose derivative with respect to x is $5x^2$, we evaluate

$$\int 5x^2\,dx = \frac{5x^3}{3} + C.$$

(In the notation of the paragraph preceding the example, $f(x) = 5x^2$ and $F(x) = 5x^3/3$.)

We must choose C so that $5x^3/3 + C$ has the value 6 when $x = 2$; i.e., so that $5 \cdot \frac{8}{3} + C = 6$. Hence, $C = -\frac{22}{3}$, and the desired function is

$$\frac{5x^3}{3} - \frac{22}{3}. ■$$

A common error is writing an expression such as

$$\int t^2\,dT = \frac{t^3}{3} + C.$$

This expression would be correct if the antiderivative on the left were $\int t^2\,dt$. Since t and T are different variables, the expression $\int t^2\,dT$ is a mixture of "apples and oranges." It is asking for a function whose derivative with respect to T is t^2.

We can salvage this expression only when we know the functional relationship between t and T, as we did, for example, in the discussion in section

A.1, where we had the relation $t = 60T$ between time t in seconds and time T in minutes.

In many cases, we can not write a closed-form expression for an antiderivative, e.g., $\int e^{x^2} dx$. Nonetheless, we often need notation for that one particular antiderivative which assumes a specified value at a given point. We make use of one part of the *fundamental theorem of calculus*:

Theorem 2 (Fundamental theorem of calculus). *If f is an integrable function and a is a constant, then*

$$\frac{d}{dx} \int_a^x f(s) \, ds = f(x).$$

In other words, $\int_a^x f(s) \, ds$ is an antiderivative of f. (The dummy variable s is used within the definite integral to avoid confusion with the independent variable x.)

If f is integrable, then $\int_a^x f(s) \, ds$ certainly defines a function of x. If f is positive-valued and $a \leq x$, the value of the function for a given x is the area under the graph of f between a and x. In particular, the value of this function at $x = a$ is zero. For a given f, you could even evaluate the function $\int_a^x f(s) \, ds$ with a single keystroke on a calculator that includes a definite integral key.

■ **EXAMPLE 3** *Find the function whose derivative with respect to x is $5x^2$ and that assumes the value 6 at $x = 2$.*

Using the fundamental theorem of calculus, we can immediately write one function whose derivative is $5x^2$ as

$$\int_a^x 5s^2 \, ds,$$

where a is a constant. That is,

$$\int 5x^2 \, dx = \int_a^x 5s^2 \, ds + C.$$

(Note that the $\int$ symbol on the left is an *antiderivative*, whereas the $\int_a^x$ on the right is a *definite integral*.)

We must determine C to give our particular antiderivative the value 6 at $x = 2$. That is, C must satisfy

$$\int_a^2 5s^2 \, ds + C = 6.$$

To avoid evaluating the integral, choose $a = 2$. Then we immediately have $C = 6$. The desired antiderivative is

$$F(x) = \int_2^x 5s^2 \, ds + 6.$$

Is this function the same as the one we obtained in example 2, which posed the same problem? Evaluating the definite integral defining $F(x)$ yields

$$F(x) = \int_2^x 5s^2 \, ds + 6 = \frac{5s^3}{3} \Big|_2^x + 6$$

$$= \frac{5x^3}{3} - \frac{40}{3} + 6 = \frac{5x^3}{3} - \frac{22}{3},$$

which agrees with example 2. ■

■ **EXAMPLE 4**   *Find the function whose derivative is e^{x^2} and that has the value 2.6 at $x = -\pi$.*

We use the fundamental theorem. For ease of evaluation, choose $a = -\pi$. Add the constant 2.6 to give the antiderivative the proper value at $x = -\pi$. The desired function is

$$\int_{-\pi}^x e^{s^2} \, ds + 2.6. \quad ■$$

A.3 ■ TAYLOR'S THEOREM

Complex functional behavior is often approximated by something simpler, such as a straight line. ("If I keep traveling at this rate, I'll be home by ...," and so on.) Taylor's theorem extends that approximation idea to polynomials of arbitrary order, and it gives the error in such an approximation.

Theorem 3 (Taylor's theorem). *Let $y(x)$ have n continuous derivatives on the interval $x_0 \le x \le x_1$. Then*

$$y(x) = y(x_0) + y'(x_0)(x - x_0) + \frac{y''(x_0)(x - x_0)^2}{2}$$

$$+ \cdots + \frac{y^{(n-1)}(x_0)(x - x_0)^{n-1}}{(n-1)!} + R_n,$$

where the remainder term is

$$R_n = \frac{y^{(n)}(\xi)(x - x_0)^n}{n!} \tag{A.1}$$

for some ξ, $x_0 \le \xi \le x$.

The remainder term R_n gives the error in approximating $y(x)$ by a polynomial in x of degree $n - 1$.

To formally validate the terms in the Taylor polynomial, evaluate both sides at $x = x_0$: $y(x_0) = y(x_0)$. Differentiating both sides and evaluating at $x = x_0$ again yields an identity, $y'(x_0) = y'(x_0)$, and so on.

■ **EXAMPLE 5**   *Find the first two nonzero terms in the Taylor polynomial for* $\cos 4x$ *about* $x_0 = 0$. *Estimate the error in approximating this function by this Taylor polynomial.*

With $y(x) = \cos 4x$ and $x_0 = 0$, we find

$$y(x_0) = \cos 0 = 1,$$
$$y'(x_0) = -4 \sin 0 = 0,$$
$$y''(x_0) = -16 \cos 0 = -16.$$

In addition, for the remainder term we will need

$$y^{(3)}(x) = 64 \sin 4x.$$

A total of three terms ($n = 3$) in the Taylor polynomial leaves us with two nonzero terms, as required. The Taylor polynomial is

$$\cos 4x = 1 - \frac{16}{2!}(x - 0)^2 + R_3 = 1 - 8x^2 + R_3,$$

where the remainder uses the third derivative of y,

$$R_3 = \frac{64 \sin 4\xi}{3!}(x - 0)^3 = \frac{32 \sin 4\xi}{3}x^3.$$

How good is the approximation

$$\cos 4x \approx 1 - 8x^2$$

that results from keeping only the first three terms of the Taylor polynomial (of which only two are nonzero)? The answer is in the size of the remainder term R_3, and it depends on the range of x values.

For example, if $-0.5 \leq x \leq 0.5$, then

$$\left|\cos 4x - (1 - 8x^2)\right| = |R_3|$$

$$= \left|\frac{32 \sin 4\xi}{3}x^3\right|$$

$$\leq \frac{32 \cdot 1}{3}(0.5)^3 \approx 1.33.$$

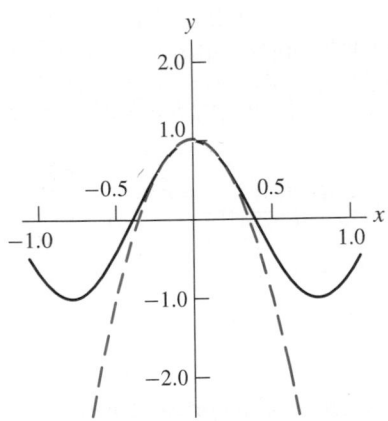

FIGURE A.1   The second-order Taylor polynomial approximation $1 - 8x^2$ (blue curve) to $\cos 4x$ (black curve)

This bound is rather conservative, as the graphs in figure A.1 of $\cos 4x$ and the Taylor polynomial $1 - 8x^2$ reveal.

The real power of the remainder term is its ability to tell how the accuracy of the Taylor approximation varies with the size of the x interval; e.g., for the smaller interval $-0.1 \leq x \leq 0.1$, the approximation $\cos 4x \approx 1 - 8x^2$ has a much smaller error,

$$|\cos 4x - (1 - 8x^2)| = |R_3| \leq \frac{32 \cdot 1}{3}(0.1)^3 \approx 0.011. \ ■$$

To approximate a function of two variables $f(t, y)$ in a neighborhood of (t_0, y_0), the formal analog of the Taylor expansion is

$$f(t, y) = f(t_0, y_0) + f_t(t_0, y_0)(t - t_0) + f_y(t_0, y_0)(y - y_0)$$
$$+ f_{tt}(t_0, y_0)(t - t_0)^2 + 2f_{ty}(t_0, y_0)(t - t_0)(y - y_0)$$
$$+ f_{yy}(t_0, y_0)(y - y_0)^2 + \cdots.$$

Subscripts denote partial derivatives (e.g., $f_{ty} = \partial^2 f / \partial t \partial y$), which are all assumed to exist.

A.4 ■ POWER SERIES

A **power series about** x_0 is an infinite sum of the form

$$\sum_{n=0}^{\infty} c_n(x - x_0)^n = c_0 + c_1(x - x_0) + c_2(x - x_0)^2 + \cdots.$$

Only nonnegative, integer powers of x may appear.

A power series **converges at** x if

$$\lim_{k \to \infty} \sum_{n=0}^{k} c_n(x - x_0)^n$$

exists; that is, the sequence of partial sums, each of which adds one more term to the sum, must have a limit. A power series has **radius of convergence** R if it converges for each x in $|x - x_0| < R$—i.e., for $x_0 - R < x < x_0 + R$. Within its radius of convergence, a power series may be *differentiated and integrated term by term*.

> The ability to interchange summation and differentiation or integration is a property of *uniformly convergent series*. Roughly, we can get a finite sum within a specified tolerance of the limiting value, using a designated number of terms that is independent of x. Power series converge uniformly if they converge.

The **ratio test** provides the standard method for determining the radius of convergence of a power series.

Theorem 4 (Ratio test). *The radius of convergence of the power series*

$$\sum_{n=0}^{\infty} c_n(x - x_0)^n$$

is

$$R = \lim_{n \to \infty} \frac{c_n}{c_{n+1}}$$

if this limit exists. The series diverges for $|x - x_0| > R$. If $c_n/c_{n+1} \to \infty$, then the power series converges for all x.

Convergence at the end points $x = x_0 \pm R$ of the interval of convergence must be examined separately.

Suppose a power series has radius of convergence $R > 0$. Its limit is a function of x:

$$f(x) = \sum_{n=0}^{\infty} c_n(x - x_0)^n, \quad |x - x_0| < R.$$

Since we can differentiate a convergent power series term by term, we easily recover the **Taylor coefficients** ,

$$c_0 = f(x_0),$$
$$c_1 = f'(x_0),$$
$$c_2 = \frac{f''(x_0)}{2!},$$
$$\vdots$$
$$c_n = \frac{f(n)(x_0)}{n!},$$
$$\vdots$$

If $f(x)$ has derivatives of all order for $|x - x_0| < R$ for some $R > 0$, then we can formally construct the **Taylor series**,

$$f(x) = \sum_{n=0}^{\infty} \frac{f^{(n)}(x_0)}{n!}(x - x_0)^n.$$

To determine if this series converges to f (as it often does), we ask whether the remainder term R_n in the Taylor polynomial of theorem 3 decays to zero as n grows.

■ **EXAMPLE 6**   *Construct the Taylor series expansion of e^x about $x_0 = 0$, and determine its radius of convergence.*

Since e^x and all its derivatives are unity at $x_0 = 0$, the Taylor coefficients of this function are just

$$c_n = \frac{1}{n!}.$$

The formal Taylor series for e^x is

$$\sum_{n=0}^{\infty} \frac{x^n}{n!}.$$

The usual convention is $0! = 1$.

From the ratio test, its radius of convergence is

$$R = \lim_{n\to\infty} \frac{1/n!}{1/(n+1)!} = \lim_{n\to\infty} (n+1).$$

Hence, this series converges for all x.

To determine if this series actually converges to e^x, examine the remainder term (A.1) in the Taylor polynomial. It is

$$R_n = \frac{e^\xi}{n!}.$$

Suppose $|x| < S$ for any fixed $S > 0$. Then as $n \to \infty$,

$$|R_n| < \frac{e^S}{n!} \to 0.$$

On any bounded interval, the Taylor expansion converges to e^x. Hence, it converges to e^x for all x and we can write

$$e^x = \sum_{n=0}^{\infty} \frac{x^n}{n!}, \quad |x| < \infty. \quad ■$$

A.5 ■ MATRICES AND DETERMINANTS

An $m \times n$ **matrix** is a rectangular array of numbers, called *elements* or *entries*, arranged in m rows and n columns,

$m \times n$ matrix

$$A = \begin{pmatrix} a_{11} & a_{12} & \cdots & a_{1n} \\ a_{21} & a_{22} & \cdots & a_{2n} \\ \vdots & \vdots & & \vdots \\ a_{m1} & a_{m2} & \cdots & a_{mn} \end{pmatrix}.$$

The arbitrary entry a_{ij} appears in the ith row and the jth column. *The first subscript denotes row; the second subscript denotes column.*

Transpose of a matrix

The **transpose** A^T is the $n \times m$ matrix whose jth row is the jth column of A. For example, if

$$A = \begin{pmatrix} 11 & 12 & 13 \\ 21 & 22 & 23 \end{pmatrix}, \tag{A.2}$$

then

$$A^T = \begin{pmatrix} 11 & 21 \\ 12 & 22 \\ 13 & 23 \end{pmatrix}.$$

Square matrix
Row vector
Column vector

A **square** matrix has an equal number of rows and columns. A matrix that consists of a single row (or a single column) is often called a **row vector** (or a **column vector**). The transpose of a row vector is a column vector and vice versa.

The transpose of a matrix provides a compact way to write column vectors; e.g. the column vector

Example of a column vector

$$\mathbf{x} = \begin{pmatrix} x \\ y \\ z \end{pmatrix} \tag{A.3}$$

can be written on one line as $\mathbf{x} = \begin{pmatrix} x & y & z \end{pmatrix}^T$, saving considerable space on the page.

Matrices of the same size are added or subtracted by adding or subtracting corresponding elements; e.g.,

MATRIX ADDITION

$$\begin{pmatrix} a & b \\ c & d \end{pmatrix} + \begin{pmatrix} 1 & 2 \\ 3 & 4 \end{pmatrix} = \begin{pmatrix} a+1 & b+2 \\ c+3 & d+4 \end{pmatrix}.$$

MATRIX EQUALITY

Two matrices of the same size are *equal* if corresponding *elements are equal.*

SCALAR MULTIPLICATION

Scalar multiplication is *multiplication* of a matrix *by a single number.* Each entry of the matrix is multiplied by that number. For example, using the matrix $\mathbf{A}$ defined by (A.2),

$$-2\mathbf{A} = \begin{pmatrix} -22 & -24 & -26 \\ -42 & -44 & -46 \end{pmatrix}.$$

Definition 1. *Suppose $\mathbf{A}$ is an $m \times n$ matrix and $\mathbf{B}$ is an $n \times p$ matrix. Then the $m \times p$ product matrix $\mathbf{C} = \mathbf{AB}$ is defined by*

MATRIX MULTIPLICATION

$$c_{ij} = \sum_{k=1}^{n} a_{ik}b_{kj}, \quad i = 1, \ldots, m, \quad j = 1, \ldots, p.$$

Here a_{ik}, b_{kj}, and c_{ij} are entries in $\mathbf{A}$, $\mathbf{B}$, and $\mathbf{C}$, respectively.

Loosely, matrix multiplication is *row times column*: The entry in the ith row, jth column of the product $\mathbf{AB}$ is the sum of the product of the corresponding entries from the ith row of $\mathbf{A}$ and the jth column of $\mathbf{B}$.

■ **EXAMPLE 7**   *Using the 2×3 matrix $\mathbf{A}$ defined by (A.2) and the 3×1 matrix (or column vector) $\mathbf{x}$ defined by (A.3), compute $\mathbf{Ax}$.*

Because $\mathbf{A}$ has two rows, the product $\mathbf{Ax}$ will have two rows. Because $\mathbf{x}$ has one column, the product $\mathbf{Ax}$ will have one column: "2×3 times 3×1 gives 2×1." The "row times column" rule computes the quantity that appears in each position:

$$\mathbf{Ax} = \begin{pmatrix} 11 & 12 & 13 \\ 21 & 22 & 23 \end{pmatrix} \begin{pmatrix} x \\ y \\ z \end{pmatrix} = \begin{pmatrix} 11x + 12y + 13z \\ 21x + 22y + 23z \end{pmatrix}. \quad ■$$

■ **EXAMPLE 8** *Write the system of simultaneous equations*

$$11x + 12y + 13z = 100,$$
$$21x + 22y + 23z = 200 \tag{A.4}$$

in the compact matrix form $\mathbf{Ax} = \mathbf{b}$. *Use the matrices* $\mathbf{A}$ *and* $\mathbf{x}$ *defined by* (A.2) *and* (A.3), *respectively, and define the matrix* $\mathbf{b}$ *appropriately.*

The previous example obtained the left side of the system (A.4) via the matrix multiplication $\mathbf{Ax}$. If $\mathbf{b}$ is defined by

$$\mathbf{b} = \begin{pmatrix} 100 \\ 200 \end{pmatrix},$$

then the matrix equation $\mathbf{Ax} = \mathbf{b}$ is equivalent to the system of equations (A.4) because matrix equality requires that corresponding entries be equal. That is, equality of the first entry in $\mathbf{Ax}$ with the first entry in $\mathbf{b}$ gives the first equation in (A.4), and so on:

$$\mathbf{Ax} = \begin{pmatrix} 11 & 12 & 13 \\ 21 & 22 & 23 \end{pmatrix} \begin{pmatrix} x \\ y \\ z \end{pmatrix}$$

$$= \begin{pmatrix} 11x + 12y + 13z \\ 21x + 22y + 23z \end{pmatrix} = \begin{pmatrix} 100 \\ 200 \end{pmatrix} = \mathbf{b}. \ \blacksquare$$

■ **EXAMPLE 9** *Using the matrix* $\mathbf{A}$ *defined by* (A.2) *and*

$$\mathbf{B} = \begin{pmatrix} 1 & 2 & 3 \\ 3 & 4 & 5 \\ -1 & -2 & -3 \end{pmatrix},$$

find $\mathbf{AB}$. *What can you say about* $\mathbf{BA}$?

Since the number of columns of $\mathbf{A}$ matches the number of rows of $\mathbf{B}$ (three), the product $\mathbf{AB}$ is defined. Using row times column, we find

$$\mathbf{AB} = \begin{pmatrix} 11 & 12 & 13 \\ 21 & 22 & 23 \end{pmatrix} \begin{pmatrix} 1 & 2 & 3 \\ 3 & 4 & 5 \\ -1 & -2 & -3 \end{pmatrix}$$

$$= \begin{pmatrix} 11 \cdot 1 + 12 \cdot 3 + 13 \cdot -1 & 11 \cdot 2 + 12 \cdot 4 + 13 \cdot -2 & \cdots \\ 21 \cdot 1 + 22 \cdot 3 + 23 \cdot -1 & 21 \cdot 2 + 22 \cdot 4 + 23 \cdot -2 & \cdots \end{pmatrix}$$

$$= \begin{pmatrix} 34 & 44 & 54 \\ 64 & 84 & 104 \end{pmatrix}.$$

The size of the product is correct: A 2×3 matrix times a 3×3 matrix yields a 2×3 matrix.

The product **BA** is not defined; **B** has three columns, but **A** has only two rows. ■

Matrix multiplication is *not commutative*: **AB** $\neq$ **BA**. Indeed, in the previous example, **BA** is not even defined.

The **identity** matrix, denoted by **I**, is a square matrix with 1's on the diagonal and 0's elsewhere; e.g., the 2×2 identity is

Identity matrix

2 × 2 identity matrix

$$\mathbf{I} = \begin{pmatrix} 1 & 0 \\ 0 & 1 \end{pmatrix}.$$

Multiplication on the left or on the right by the identity matrix of the proper size leaves the original unchanged.

Stop and Think

A.1 Verify that $\mathbf{I}_2\mathbf{A} = \mathbf{A}\mathbf{I}_3 = \mathbf{A}$, where **A** is the 2×3 matrix defined by (A.2) and $\mathbf{I}_2$ and $\mathbf{I}_3$ are the 2×2 and 3×3 identity matrices.

If it exists, the **inverse** of a square matrix **A** is the matrix $\mathbf{A}^{-1}$ with the property that

$$\mathbf{A}^{-1}\mathbf{A} = \mathbf{A}\mathbf{A}^{-1} = \mathbf{I}.$$

A matrix that possesses an inverse is called **invertible**, or **nonsingular**.

If the square matrix **A** is invertible, then one way to represent the solution of the system of linear equations $\mathbf{Ax} = \mathbf{b}$ is $\mathbf{x} = \mathbf{A}^{-1}\mathbf{b}$.

■ **EXAMPLE 10**   *Verify that the inverse of the matrix*

$$\mathbf{A} = \begin{pmatrix} 1 & 1 \\ 2 & 4 \end{pmatrix}$$

is

$$\mathbf{A}^{-1} = \begin{pmatrix} 2 & -1/2 \\ -1 & 1/2 \end{pmatrix}.$$

In addition, verify that $x = 1$, $y = -1$ is a solution of the system of simultaneous equations

$$\begin{array}{rcr} x + y &=& 0 \\ 2x + 4y &=& -2, \end{array} \tag{A.5}$$

verify that the system (A.5) is equivalent to $\mathbf{Ax} = \mathbf{b}$ with

$$\mathbf{x} = \begin{pmatrix} x \\ y \end{pmatrix}, \quad \mathbf{b} = \begin{pmatrix} 0 \\ -2 \end{pmatrix},$$

and verify that the solution of (A.5) can be written $\mathbf{x} = \mathbf{A}^{-1}\mathbf{b}$.

Verify that $\mathbf{A}^{-1}\mathbf{A} = \mathbf{I}$ by carrying out the matrix multiplication:

$$\mathbf{A}^{-1}\mathbf{A} = \begin{pmatrix} 2 & -1/2 \\ -1 & 1/2 \end{pmatrix} \begin{pmatrix} 1 & 1 \\ 2 & 4 \end{pmatrix}$$

$$= \begin{pmatrix} 2(1) + (-1/2)2 & 2(1) + (-1/2)4 \\ -1(1) + (1/2)2 & -1(1) + (1/2)4 \end{pmatrix}$$

$$= \begin{pmatrix} 1 & 0 \\ 0 & 1 \end{pmatrix} = \mathbf{I}.$$

Stop and Think   **A.2** Use similar steps to demonstrate that $\mathbf{A}\mathbf{A}^{-1} = \mathbf{I}$.

Substitute into (A.5) to verify that $x = 1$, $y = -1$ is a solution:

$$1 + (-1) = 0,$$
$$2(1) + 4(-1) = -2.$$

Note that this last calculation is equivalent to the matrix multiplication $\mathbf{A}\mathbf{x} = \mathbf{b}$ that verifies the matrix form of (A.5).

Finally, a matrix multiplication confirms that the inverse of $\mathbf{A}$ finds the solution of $\mathbf{A}\mathbf{x} = \mathbf{b}$:

$$\mathbf{A}^{-1}\mathbf{b} = \begin{pmatrix} 2 & -1/2 \\ -1 & 1/2 \end{pmatrix} \begin{pmatrix} 0 \\ -2 \end{pmatrix}$$

$$= \begin{pmatrix} 2(0) + (-1/2)(-2) \\ (-1)(0) + (1/2)(-2) \end{pmatrix} = \begin{pmatrix} 1 \\ -1 \end{pmatrix} = \mathbf{x}. \ ■$$

Determinant

The **determinant** of a 2×2 matrix $\mathbf{A}$ is the number $\det \mathbf{A}$ defined by

2 × 2 DETERMINANT

$$\det \mathbf{A} = \det \begin{pmatrix} a & b \\ c & d \end{pmatrix} = ad - bc.$$

Determinants of larger square matrices are computed recursively using the *minors* M_{ij} of $\mathbf{A}$; the **minor** M_{ij} of the $n \times n$ matrix $\mathbf{A}$ is the $n - 1 \times n - 1$ matrix obtained by deleting row i and column j from $\mathbf{A}$.

To compute the determinant of the $n \times n$ matrix $\mathbf{A}$, using its minors, chose a row (or column) k. Then

$$\det \mathbf{A} = \sum_{j=1}^{n} (-1)^{k+j} \det M_{kj} = \sum_{i=1}^{n} (-1)^{i+k} \det M_{ik}.$$

The first sum is the *expansion in minors along the kth row*; the second is the expansion along the kth column.

> A proof is required to show that the value of the determinant is independent of the choice of row or column along which expansion occurs.

A matrix $\mathbf{A}$ is nonsingular (invertible) if and only if $\det \mathbf{A} \neq 0$. If the entries in $\mathbf{A}^{-1}$ are denoted by c_{ij}, then

$$c_{ij} = \frac{(-1)^{i+j} M_{ji}}{\det \mathbf{A}}.$$

If $\det \mathbf{A} \neq 0$, then the equation $\mathbf{A}\mathbf{x} = \mathbf{0}$ has only one solution, $\mathbf{x} = \mathbf{A}^{-1}\mathbf{0} = \mathbf{0}$. (In two dimensions, the two lines defined by $\mathbf{A}\mathbf{x} = \mathbf{0}$ intersect at a single point, the origin.) Conversely, if $\det \mathbf{A} = 0$, then $\mathbf{A}\mathbf{x} = \mathbf{0}$ has many nontrivial solutions. (In two dimensions, the two lines defined by $\mathbf{A}\mathbf{x} = \mathbf{0}$ are in fact identical. They intersect at the origin and along their entire length.)

> The determinant of a matrix is a powerful theoretical tool, and it is useful for hand calculations with tiny matrices. But the number of multiplications required to evaluate a determinant grows like $n!$, while matrices can be inverted with computational effort that grows only like n^3. To appreciate the difference, compare those two numbers when $n = 100$.
>
> Large systems of equations $\mathbf{A}\mathbf{x} = \mathbf{b}$ are solved numerically by techniques based on row elimination—eliminating unknowns below the diagonal by adding and subtracting rows, as illustrated in section 9.3. These techniques are much more efficient than computing matrix inverses or determinants.

MATLAB

MATLAB was designed to manipulate matrices easily, accurately, and efficiently. For guidance in using some of its matrix tools, start DELAB , select `Help, Textbook`, then go to chapter A, Matrices.

If the entries in a matrix are functions rather than numbers, all of these ideas still apply. In addition, we can integrate or differentiate a matrix function by applying the appropriate operation to each of its elements.

A.6 ■ CRAMER'S RULE

Cramer's rule uses determinants to solve systems of linear equations. To review determinants, read the concluding paragraphs of the previous section. We demonstrate Cramer's rule with a simple example.

■ **EXAMPLE 11** *Solve*

$$2x + 3y = -4$$
$$6x + 4y = -2.$$

Cramer's rule gives the values of x and y as quotients. The denominator is the determinant of coefficients

$$D = \det \begin{pmatrix} 2 & 3 \\ 6 & 4 \end{pmatrix} = (2 \cdot 4) - (3 \cdot 6) = -10.$$

The numerator is the determinant formed by replacing the column corresponding to the desired unknown with the constants from the right-hand side of the system of equations. The numerator for the unknown x is

$$N_x = \det \begin{pmatrix} -4 & 3 \\ -2 & 4 \end{pmatrix} = (-4 \cdot 4) - (3 \cdot -2) = -10,$$

and Cramer's rule gives the value of x as

x via Cramer's rule

$$x = \frac{N_x}{D} = \frac{-10}{-10} = 1.$$

The numerator for the unknown y is

$$N_y = \det \begin{pmatrix} 2 & -4 \\ 6 & -2 \end{pmatrix} = (2 \cdot -2) - (-4 \cdot 6) = 20,$$

and Cramer's rule gives the value of y as

y via Cramer's rule

$$y = \frac{N_y}{D} = \frac{20}{-10} = -2.$$

Substituting $x = 1$, $y = -2$ into the original system confirms the accuracy of this answer. ■

$\det \mathbf{A} \neq 0 \Rightarrow \mathbf{Ax} = \mathbf{b}$ has exactly one solution.

If the determinant of coefficients is *not* zero, then Cramer's rule provides the *unique* solution of the system of equations. In this case, *homogeneous* systems of linear equations, those with only zeros on the right-hand side, *can have only zero solutions*. The numerator for each unknown is zero because it is a determinant containing a column of zeros.

$\det \mathbf{A} = 0 \Rightarrow \mathbf{Ax} = \mathbf{0}$ has many solutions.

If the determinant of coefficients is zero, then Cramer's rule can not provide numerical values for the unknowns, for division by zero is either indeterminate or undefined. For homogeneous systems, Cramer's rule would give the unknowns the indeterminate values $0/0$, so a homogeneous system whose determinant of coefficients is zero has *many* solutions. A *nonhomogeneous* system may have *many* solutions if the numerator for each unknown happens to be zero, or it may have *no* solutions if one of the numerators is not zero.

Cramer's rule is useful theoretically; it provides an easy test for determining when small systems of simultaneous equations have a unique solution. But it is too inefficient to be a practical computation tool.

A.7 ■ PARTIAL FRACTIONS

Partial-fraction expansions reverse the common denominator process, separating a fraction with a complicated polynomial denominator into a sum of fractions with simpler denominators. Finding those simpler fractions can ease both integration and the inversion of Laplace transforms.

For example, we can readily add two fractions by finding a common denominator:

$$\frac{2}{s+1} + \frac{3}{s-2} = \frac{2(s-2)+3(s+1)}{(s+1)(s-2)} = \frac{5s-1}{s^2-s-2}.$$

A partial-fraction expansion guides us in reversing that process, in going from the fraction on the right back to the simpler terms of which it is the sum.

The steps in the process are:

PARTIAL FRACTION
EXPANSION

1. If necessary, divide the denominator into the numerator to obtain a **proper fraction**, a quotient of polynomials with the degree of the numerator *smaller* than the degree of the denominator.

2. Factor the denominator completely.

3. Write the original fraction as a sum of fractions whose denominators are the factors of the original denominator. Provide each simpler (proper) fraction with a numerator containing appropriate unknown coefficients.

   The numerator of a linear factor is a constant. The numerator of a quadratic factor with no real roots is linear, and so on.

4. Find a common denominator and sum the simpler fractions. To find the unknown coefficients, equate the numerator of the sum to the original numerator.

■ **EXAMPLE 12**　　*Find a partial-fraction expansion for*

$$\frac{5s-1}{s^2-s-2}.$$

Since the degree of the numerator is 1 (from $5s$) and the degree of the denominator is 2 (from s^2), the given fraction is proper.

Factor the denominator:

$$s^2 - s - 2 = (s+1)(s-2).$$

Write the original fraction as a sum involving the factors of the denominator:

$$\frac{5s-1}{s^2-s-2} = \frac{A}{s+1} + \frac{B}{s-2}.$$

The coefficients A, B are to be determined. Note that each of the fractions on the right is proper. Because the highest power in the denominator is 1, the highest power in the numerator is 0.

Find a common denominator and sum:

$$\frac{5s - 1}{s^2 - s - 2} = \frac{A}{s + 1} + \frac{B}{s - 2}$$

$$= \frac{A(s - 2) + B(s + 1)}{(s + 1)(s - 2)}$$

$$= \frac{(A + B)s + B - 2A}{s^2 - s - 2}.$$

The first and last fractions can be equal only if their numerators are equal,

$$(A + B)s + B - 2A = 5s - 1.$$

Equate like powers of s to find A, B:

$$s^0 : \quad -2A + B = -1,$$
$$s^1 : \quad A + B = 5.$$

Solving this pair of equations by Cramer's rule or other methods yields $A = 2$, $B = 3$.

Hence, the desired partial-fraction expansion is

$$\frac{5s - 1}{s^2 - s - 2} = \frac{2}{s + 1} + \frac{3}{s - 2}. \quad ■$$

If the denominator contains a repeated factor, e.g., $(s - 2)^3$, then the fraction could have been formed by combining simple terms with the denominators $(s-2)^i$, $i = 1, 2, 3$, each with a constant numerator. This idea applies to repeated quadratic factors as well, but the numerators are linear. For example, we would construct the expansion

$$\frac{4 - 2s}{(s^2 + 1)(s - 1)^2} = \frac{As + B}{s^2 + 1} + \frac{C}{s - 1} + \frac{D}{(s - 1)^2}.$$

Then A, B, C, D would be determined by finding a common denominator and equating like powers of s in the numerators.

A.8 ■ HOW TO READ A PROBLEM

Few of the problems you will encounter in your career will come complete with instructions for their solution. Indeed, life may be just an ambiguous word problem. This section outlines some strategies for solving the sorts of technical problems you will encounter in this text and later in your academic and professional career.

Problem *Complete the following steps to derive a model describing the height above the surface of the earth of a rocket launched vertically upward. Use Netwon's law ($F = ma$) and the following experimental facts:*

Fact 1: The force of gravity on the rocket is proportional to its mass and *inversely* proportional to the square of its distance from the *center* of the earth.

Fact 2: A known function $T(t)$ gives the thrust force of the rocket at time t measured from the instant of ignition.

1. Introduce a coordinate system and write a formula for the force F_g of gravity given by the first experimental fact in terms of the height $y(t)$ of the rocket above the *surface* of the earth at time t.

2. Use Newton's law to obtain a differential equation relating the height $y(t)$, one or more of its derivatives, the gravity expression from part 1, and the thrust $T(t)$ of the rocket to one another.

3. Complete the model by adding appropriate initial conditions. (Is the rocket moving at the instant its motor is ignited?)

Begin your attack on the problem by deciding what the problem is about:

*Follow the steps given below to derive a **model** describing the height above the surface of the earth **of a rocket launched vertically upwards**.*

As you study a problem, underline important words and phrases.

The words that describe the heart of the problem are shown here in bold type. What information have we been given?

... Use Netwon's law ($F = ma$) and the following experimental facts:

Fact 1: The **force of gravity** on the rocket is proportional to its mass and *inversely* proportional to the square of its distance from the *center* of the earth.

Fact 2: A known function $T(t)$ gives the **thrust force of the rocket** at time t measured from the instant of ignition.

We see that we are given $F = ma$, information about the force of gravity on the rocket, and the name of a function describing the thrust force of the rocket.

What specifically does the problem ask of us? What constitutes a solution? The first part directs:

1. **Introduce a coordinate system and write a formula** for the force F_g of gravity given by the first experimental fact in terms of the height $y(t)$ of the rocket above the surface of the earth at time t.

We will have completed this part of the problem when we have defined a coordinate system and when we have written a formula relating F_g from fact 1 to $y(t)$. (This notation is defined in the problem statement.)

A natural coordinate system is one that points vertically upward and places its origin at the surface of the earth. Such a vertical coordinate is just the height y, as illustrated in figure A.2.

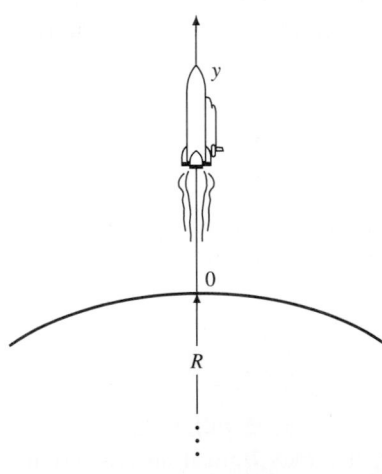

FIGURE A.2 A vertical coordinate system whose origin is at the surface of the earth

To write a formula for F_g, the force of gravity on the rocket, re-read Fact 1. It provides the recipe for the rest of the answer to part 1:

> The **force of gravity** on the rocket is **proportional to its mass** and **inversely proportional to the square of its distance from the center of the earth**.

To write this functional relation, we need notation for a proportionality constant, for the mass of the rocket, and for the distance of the rocket from the center of the earth.

Figure A.2 uses R for the radius of the earth. Then the distance from the rocket to the center is $R + y$. A natural choice for the mass of the rocket is m. How about G for the proportionality constant? We will leave to you the job of using this notation to turn the recipe in fact 1 into a formula for F_g.

Having settled part 1 (if you did your share by writing a formula for F_g), we turn to part 2. It directs:

> **2.** Use Newton's law to **obtain a differential equation** relating the height $y(t)$, one or more of its derivatives, the gravity expression from part 1, and the thrust $T(t)$ of the rocket to one another.

The problem says "… obtain a differential equation …" But how? Look again:

> **2.** Use Newton's law …

Newton's law is $F = ma$; F is the total force on the rocket, m is its mass, and a is its acceleration.

What forces sum to give F? The problem statement lists

> … the gravity expression from part 1, and the thrust $T(t)$ of the rocket ….

In part 1, you have written a formula for F_g, the gravity force. You are handed the name $T(t)$ of the formula for the thrust force. Hence, $F = F_g + T(t) = ma$.

> You do not know the *rule* for the thrust force, just the name of the function. But you only need the name. If the actual formula appears later, you can always substitute it into your results.

The acceleration a of the rocket is the second derivative of displacement. That observation relates a to y. Now you can turn $F = ma$ into the required differential equation.

> … relating the height $y(t)$, one or more of its derivatives, the gravity expression from part 1, and the thrust $T(t)$ of the rocket to one another.

Part 2 is finished when you have written this differential equation.

Part 3 gives one more instruction, and it asks a parting question.

> **3. Complete the model** by adding appropriate initial conditions. (**Is the rocket moving** at the instant its motor is ignited?)

"Complete the model ..." How? " ... by adding appropriate initial conditions." When you complete the details of part 2, you will find that the differential equation contains the second derivative d^2y/dt^2. Hence, two initial conditions are required. Natural choices are initial position and initial velocity. What is the initial position in the coordinate system of figure A.2?

"Is the rocket moving at the instant its motor is ignited?" Who knows? But it is natural to suppose that the rocket was fixed on its launch pad and that its initial velocity $y'(0)$ is zero. The question suggests a specific numerical value for the second initial condition.

We have stepped through this problem item by item, emphasizing the need to obtain from the problem statement the information you need to decide what constitutes a solution. We have not completed all the details. They are left to exercise 17 of the chapter 1 exercises.

We close by summarizing this approach. By underlining appropriate phrases, by making a separate list, or by some other means of your choice, identify

- What constitutes a solution of the problem you face.
- What information you are given.
- What directions, if any, you are given for obtaining a solution.

Use these three categories of information to formulate a reasonable solution.

A.9 ■ APPENDIX EXERCISES

1. Use the fundamental theorem of calculus to find df/dx:

(a) $f(x) = \exp\left(\int_0^x \cos \pi s \, ds\right)$

(b) $f(x) = \exp\left(\int_{-1}^x (s+1)^3 \, ds\right)$

2. Evaluate the definite integral in each part of exercise 1 and recompute the derivative directly. Show that each answer agrees with that obtained by using the fundamental theorem of calculus.

3. Use partial fractions to evaluate

$$\int_1^x \frac{dr}{ar - sr^2},$$

where a and s are constants, $0 < s < a$. Is there an upper limit on x?

4. Use the fundamental theorem of calculus,

$$\frac{d}{dx} \int_a^x f(s) \, ds = f(x),$$

to write solutions of the following initial-value problems.

(a) $y'(x) = e^{x^2}$, $y(0) = 2$

(b) $y'(x) = \cos(\ln x)$, $y(1) = -3$

5. (a) Find the first three nonzero terms in a Taylor expansion about $x = \pi/2$ of $\sin x$.

(b) Use the remainder term in Taylor's theorem to estimate the error in this Taylor approximation to $\sin x$ on two intervals, $|x - \pi/2| \le 0.5$ and $|x - \pi/2| \le 0.25$. Reducing the size of the x interval by a factor of 2 will reduce the error by what factor?

(c) Improve the approximation of example 5 by extending it to three *nonzero* terms. Estimate the error in this approximation on the intervals $-0.1 \le x \le 0.1$ and $-0.5 \le x \le 0.5$.

6. Use Taylor's theorem to estimate the error in the approximation $\sqrt{1-x} \approx 1 - x/2$, $0 \le x \le 1/2$.

7. Consider the system

$$3x - 2y = 13,$$
$$4x + 2y = \ 8.$$

(a) Use Cramer's rule to solve this system.

(b) Let $\mathbf{x} = \begin{pmatrix} x & y \end{pmatrix}^T$. Define matrices $\mathbf{A}$ and $\mathbf{b}$ so that this system can be written in the form $\mathbf{Ax} = \mathbf{b}$.

(c) Using the values of x and y from part (a) in $\mathbf{x}$ and the matrices $\mathbf{A}$ and $\mathbf{b}$ defined in part (b), carry out the matrix multiplication to verify that $\mathbf{Ax} = \mathbf{b}$.

(d) Solve the system $\mathbf{Ax}_1 = \begin{pmatrix} 1 & 0 \end{pmatrix}^T$ for the column vector $\mathbf{x}_1$. Solve the system $\mathbf{Ax}_2 = \begin{pmatrix} 0 & 1 \end{pmatrix}^T$ for the column vector $\mathbf{x}_2$. (Note that the right-hand sides of these systems are the columns of the 2×2 identity matrix $\mathbf{I}_2$.) Let $\mathbf{B}$ be the 2×2 matrix whose columns are $\mathbf{x}_1$ and $\mathbf{x}_2$, respectively. Show that $\mathbf{B}$ is the inverse of $\mathbf{A}$ by verifying that $\mathbf{BA} = \mathbf{AB} = \mathbf{I}_2$.

(e) Use the process described in part (d) to find the inverse of the matrix considered in example 10,

$$\mathbf{A} = \begin{pmatrix} 1 & 1 \\ 2 & 4 \end{pmatrix}$$

Verify your results by computing $\mathbf{A}^{-1}\mathbf{A}$ and $\mathbf{AA}^{-1}$.

(f) Use the inverse just computed to find the solution of $\mathbf{Ax} = \begin{pmatrix} -1 & -8 \end{pmatrix}^T$. Verify the solution by substituting into the equivalent system of simultaneous equations and by finding it via Cramer's rule.

8. Find the determinant and the inverse of the matrix function

$$\mathbf{A}(x) = \begin{pmatrix} 1 & 1 \\ e^x & e^{-x} \end{pmatrix}.$$

Confirm that $\mathbf{A}^{-1}\mathbf{A} = \mathbf{AA}^{-1} = \mathbf{I}$.

9. Let $W(y_1, y_2)$ be the determinant

$$W(y_1, y_2) = \det \begin{pmatrix} y_1(t) & y_2(t) \\ y_1'(t) & y_2'(t) \end{pmatrix}$$

for some differentiable functions y_1, y_2. (W is called the *Wronskian*; see section 6.2 for its use in determining the linear independence of two functions.) Verify each of the following.

(a) Let $y_1(t) = e^{r_1 t}$, $y_2(t) = e^{r_2 t}$ for some constants r_1, r_2. Then $W \neq 0$ for all t if $r_1 \neq r_2$ and $W \equiv 0$ if $r_1 = r_2$.

(b) Let $y_1(t) = e^{rt}$, $y_2(t) = te^{rt}$ for some constant $r \neq 0$. Then $W \neq 0$ for all t.

(c) Let $y_1(t) = \sin \beta t$, $y_2(t) = \cos \beta t$ for some constant $\beta \neq 0$. Then $W \neq 0$ for all t.

(d) Let $y_1(t) = e^{\alpha t} \sin \beta t$, $y_2(t) = e^{\alpha t} \cos \beta t$ for some constants $\alpha, \beta \neq 0$. Then $W \neq 0$ for all t.

10. Let

$$\mathbf{A} = \begin{pmatrix} a & b \\ c & d \end{pmatrix}, \quad \mathbf{x} = \begin{pmatrix} x \\ y \end{pmatrix}.$$

Show that the matrix equation $\mathbf{Ax} = \mathbf{0}$ defines two lines that intersect at the origin. Find the slope of each line in terms of a, b, c, d. Show that the two lines are parallel if and only if $\det \mathbf{A} = 0$. Argue geometrically that $\mathbf{Ax} = \mathbf{0}$ has exactly one solution if and only if $\det \mathbf{A} \neq 0$.

11. Use Cramer's rule to determine conditions on the constant a that guarantee that the system

$$\begin{aligned} C_1 - aC_2 &= 0, \\ C_1 + C_2 &= 0 \end{aligned}$$

has

(a) Exactly one solution (C_1, C_2).

(b) Infinitely many solutions.

12. Find a partial-fraction expansion for $\dfrac{4 - 2s}{(s^2 + 1)(s - 1)^2}$.

13. Find a partial-fraction expansion for $\dfrac{s^3 - s - 2s^2 - 2}{s^4 - 1}$.

Bibliography

[1]   M. Abramowitz and I. A. Stegun (eds.), *Handbook of Mathematical Functions*, Dover Publishing Co., New York, 1965

[2]   G. L. Baker and J. P. Golub, *Chaotic Dynamics*: *An Introduction*, Cambridge University Press, Cambridge, 1990

[3]   *The Boston Globe*, June 22, 1997, p. E8

[4]   Bureau of the Census, *Statistical Abstract of the United States*, 109th ed., Washington, D. C. 1989

[5]   F. Cavallini, Fitting a logistic curve to data, *College Math. J.*, **24**(3) May 1993, 247–253

[6]   E. A. Coddington and N. Levinson, *Theory of Ordinary Differential Equations*, McGraw-Hill, New York, 1955

[7]   Department of International Economic and Social Affairs, *Demographic Yearbook 1986*, United Nations, New York, 1988

[8]   L. Edelstein-Keshet, *Mathematical Models in Biology*, Random House, Inc., New York, 1988

[9]   J. Frauenthal, *Introduction to Population Modeling*, UMAP Monographs, Birkhauser, Boston, 1979

[10] J. Gleick, *Chaos*: *Making a New Science*, Viking Penguin, Inc., New York, 1987

[11] *Handbook of Chemistry and Physics* (various editions), The Chemical Rubber Publishing Co., Cleveland

[12] F. C. Hoppensteadt and C. S. Peskin, *Mathematics in Medicine and the Life Sciences*, Springer-Verlag, New York, 1992

[13] W. Hurewicz, *Lectures on Ordinary Differential Equations*, The MIT Press, Cambridge, 1958

[14] *Information Please Almanac 1997*, 50th ed., Houghton Mifflin Company, New York 1997

[15] G. T. Kurian, *Datapedia of the United States 1790–2000*: *America Year by Year*, Bernan Press, Lanham, MD 1992

[16] A. C. Lazer and P. J. McKenna, Large-Amplitude Periodic Oscillations in Suspension Bridges: Some New Connections with Nonlinear Analysis, *SIAM Review*, **32**(4), 537–578

[17] C. C. Lin and L. A. Segel, *Mathematics Applied to Deterministic Problems in the Natural Sciences*, Macmillan Publishing Co., New York, 1974; corrected reprinting by the Society for Industrial and Applied Mathematics, Philadelphia, 1988

[18] J. D. Murray, *Mathematical Biology*, Springer-Verlag, New York, 1989

[19] J. M. Ortega and W. G. Poole, Jr., *An Introduction to Numerical Methods for Differential Equations*, Pitman Publishing, Inc., Marshfield, Massachusetts, 1981

[20] D. A. Sánchez, *Ordinary Differential Equations and Stability Theory*: *An Introduction*, W. H. Freeman and Co., San Francisco, 1968

[21] J. Schnakenberg, Simple chemical reaction systems with limit cycle behavior, *J. Theor. Bio.* **81** (1979), 389–400

[22] G. Schwartz and P. W. Bishop, (eds.), *Moments of Discovery*, Basic Books, Inc., New York, 1958

[23] G. F. Simmons, *Differential Equations with Applications and Historical Notes*, 2d ed., McGraw-Hill, Inc., New York, 1991

[24] D. A. Smith, Human population growth: stability or extinction?, *Math. Magazine* **50** (1977), 186-197

[25] I. Stakgold, *Green's Functions and Boundary Value Problems*, John Wiley & Sons, New York, 1979

[26] B. van der Pol, On "Relaxation-Oscillations," *Phil. Mag.*, **2**(7) (1927), 978–992

[27] H. F. Weinberger, *A First Course in Partial Differential Equations with Complex Variables and Transform Methods*, Blaisdell Publishing Co., Waltham, Masschusetts, 1965

Solutions to Selected Exercises

CHAPTER 1

Chapter Exercises

1. Use $v(t_p) = 0$ to find $t_p = v_i/g$; t_p is the t-intercept in figure 1.3.

3. $v(0) = v_i \Rightarrow v_1(t) = -gt + v_i$;
$v(t_p) = 0 \Rightarrow v_2(t) = g(t_p - t)$; $v_1(t) \equiv v_2(t)$ only if $t_p = v_i/g$. The conditions in this problem specify the slope ($v' = -g$) and *two* points $(0, v_i)$ and $(t_p, 0)$ of the line in figure 1.3.

5. (a) Use $y' = v$; $y'' = -g \Rightarrow y(t) = -gt^2/2 + C_1 t + C_2$. Could determine C_1, C_2 from $y(0) = y_i$, $y'(0) = v_i$.

(b) If t_r is roundtrip time, then
$y(t_r) = y_i \Rightarrow t_r = 0, 2v_i/g$. Clearly, $t_r = 2v_i/g$;
$y'(t_r) = -v_i$. Time from peak to release is $t_r/2$.
A parabola has constant second derivative; three items of data (e.g., curvature y'', y-intercept y_i, and initial slope v_i) define it.

(c) Using $y'(t_p) = 0$ yields $t_p = v_i/g$. Then maximum height is $y(t_p) = v_i^2/2g$; e.g., doubling the initial velocity increases the maximum height by 4. Note $t_p = t_r/2$.

(d) $y(t) = -gt^2/2 + v_i t + y_i$.

7. (a) $y = \sin t$.

(b) $w = 4t^{5/2}/15 - 2t + 2$.

(c) $y = 4e^x - 4e^2 + \pi$.

9. (a) Use a vertical coordinate system with up as positive; $F_g = -mg$. Since air resistance opposes velocity, $F_a = kv^2$ (acts upward against falling pellet).

(b) $F_t = F_g + F_a \Rightarrow mv' = -mg + kv^2$. Since shot is dropped, $v(0) = 0$.

(c) $v' = -g + kv^2/m > -g \Rightarrow$ slope of the v versus t curve for this model is always greater than that of model 3. Also, as v grows in magnitude, slope increases (decreases in magnitude) toward zero. See figure S.1.

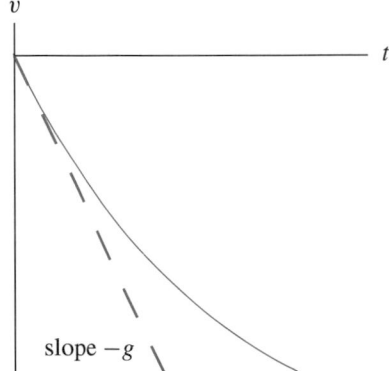

FIGURE S.1   Chapter 1, exercise 9(c).

(d) $v' = 0 \Rightarrow -g + kv^2/m = 0 \Rightarrow v = \pm\sqrt{mg/k}$. Terminal velocity is $v_{\text{ter}} = -\sqrt{mg/k}$.

11. Draw a careful version of figure 1.3.

13. (a) $y = Ce^{-8t}$, $-\infty < t < \infty$.

(b) $P = Ce^{0.0102t}$, $-\infty < t < \infty$.

(c) $P = Ce^{kt}$, $-\infty < t < \infty$.

(d) $P = Ce^{0.015t} + 0.209/0.015$, $-\infty < t < \infty$.

(e) Use partial fractions (appendix section A.7) to find

$$\frac{1}{y^3 - y} = \frac{-1}{y} + \frac{1/2}{y - 1} + \frac{1/2}{y + 1}.$$

Then evaluate $\int dy/(y^3 - y) = -\int dy/y + \cdots = \int dt$ and solve for y to find $y = \pm(1 - Ce^{2t})^{-1/2}$, valid for t such that $1 - Ce^{2t} > 0$.

(f) $T = Ce^{-0.0002t} + 5$, $-\infty < t < \infty$.

(g) $T = e^{-Akt/cm} + T_{\text{out}}$, $-\infty < t < \infty$.

(h) $y = e^{2x}/2 + C$, $-\infty < x < \infty$.

(i) $y = \ln(1/\sqrt{C - 2x})$, $-\infty < x < C/2$.

(j) $u = y^3/3 - y + C$, $-\infty < y < \infty$.

15. Equations (a–c), and (e) have the trivial solution; the others do not.

17. See appendix section A.8, page 637.

19. (a) $y = y(0)e^{kt}$; clearly increasing if $y(0) > 0$ and $k > 0$. The exponential with a positive argument is increasing.

(b) The exponential with a negative argument is decreasing.

(c) If $k > 0$, then $y' = ky > 0$ whenever $y > 0$. If $y(0) > 0$, then y starts positive and remains positive as it increases.

(d) Reverse the preceding argument.

21. For 21(a–d), see figure S.2 on next page.

(a) Increasing for $y < 0$.

(b) Increasing for $P > 0$.

(c) Increasing for $P > 0.209/0.015 \approx 14$.

(d) Increasing for $-1 < y < 0$, $1 < y$.

(e) Increasing for $T < 5$.

(f) Increasing for all y.

(g) Increasing for all y.

(h) Increasing for $u > 1$, $u < -1$.

23. $\Delta t = 0.5$

23. (a) $y(1) \approx y_2 = 18$; initially decreasing

(b) $P(1) \approx P_2 = 8.0152$; initially decreasing.

(c) $T(1) \approx T_2 = 4.0199$; initially increasing.

(d) $u(1) \approx u_2 = -0.81699$; initially decreasing.

25. Substitute to verify both solutions. DELAB finds y_1, not y_2. Formal separation of variables also finds only y_1 but $1/\sqrt{1 - y^2}$ is undefined at $y = 1$, warning of possible complications.

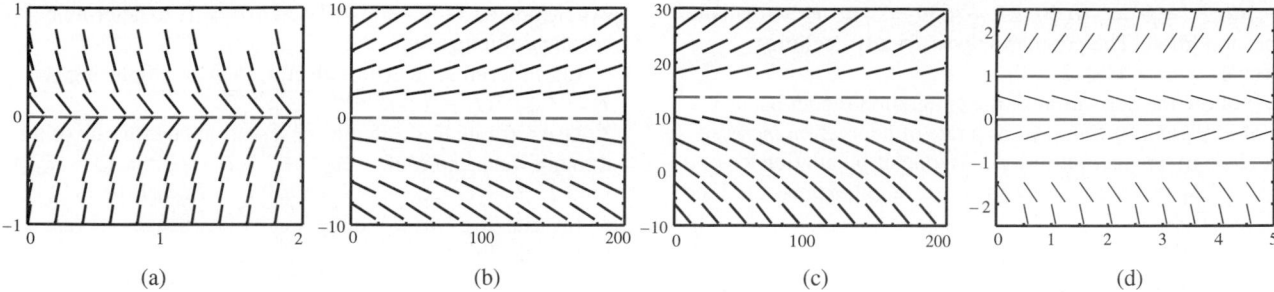

FIGURE S.2 Direction field graphs for chapter 1 exercises 21(a–d).

27. $\dfrac{dg}{dy}\dfrac{dy}{dt} = \dfrac{1}{Y(y)}\dfrac{dy}{dt} = \dfrac{dh}{dt} = T(t).$

CHAPTER 2

Section 2.1

1. **(a)** Substitute in differential equation to verify solution; substitute $t = 1970$ to verify initial condition.

(b) $P(1988) = 246.37$ versus 246.11 in table 2.2

3 **(a)** **(i)** $P' = 0.0183P$, $P(1978) = 4{,}258$;

(ii) $P(t) = 4{,}258e^{0.0183(t-1978)}$;

(iii) $P(1985) = 4{,}840$ versus $4{,}837$ in table 2.1;

$P(1997) = 6{,}029$ vs $5{,}840$ in table 2.1.

5. $P' = -0.0012P$, $P(1985) = 16.64 \Rightarrow P(t) = 16.64e^{-0.0012(t-1985)}$; $P(t_{1/2}) = 8.32 \Rightarrow t_{1/2} - 1985 = (\ln 2)/0.0012 \Rightarrow t_{1/2} = 2563$. Using the 1978 data with $k = -0.0007$, $t_{1/2} = 1978 + (\ln 2)/0.0007 = 2968$.

(c) $P(t_2) = P_i e^{kt_2} = 2P_i \Rightarrow t_2 = (\ln 2)/k.$

9. **(a)** $N' = -sN$ and $s, N > 0 \Rightarrow N' < 0.$

(b) $N = N_i e^{-st}$ is decaying.

(c) See 7(c) with $1/2$ replacing 2; $t_{1/2} = (\ln 2)/s.$

11. **(a)** $N' = iN$ and $i, N > 0 \Rightarrow N' > 0.$

(b) $N = N_i e^{it}$ is increasing.

(c) See 7(c); $t_2 = (\ln 2)/i.$

13. **(a)** $k \approx (5.30 - 3.93)/3.93 = 0.35$; a model is $P' = 0.35P$, $P(1790) = 3.93.$

(b) $P(t) = 3.93e^{0.35(t-1790)} \Rightarrow P(1970) = 9.01 \times 10^{27}$, a little higher than the recorded 204.88 million! Addition of land area increased population in 1790–1800, a factor omitted from the model being fit to this data.

(c) $k \approx (31.51 - 23.26)/23.26 = 0.35$; a model is $P' = 0.35P$, $P(1850) = 23.26$. Then $P(t) = 23.26e^{0.35(t-1850)} \Rightarrow P(1970) = 4.05 \times 10^{19}$, still much too high. Perhaps net birth rate is decreasing as population increases.

15. $P_1 = P_0 + kP_0 = (1 + k)P_0$,
$P_2 = P_1 + kP_1 = (1 + k)P_1 = (1 + k)^2 P_0, \ldots ,$
$P_n = (1 + k)^n P_0$. Use $n = 10$, $k = 0.01$ or $k = 0.03$.

17. Use the argument of exercise 15 with $P_0 = P(1979)$, $P_1 = P(1980)$, etc.; e.g., the first line has slope $0.0282P_0$. See section 3.1.

19. For India: choose $P(1997) = 952.11$, $k = 0.0203$. Using Euler with $\Delta t = 1$ yr, $P(2100) \approx P_{103} = 7{,}545.$

21. With $P_1 = 2P_0$, Euler is $P_1 = 2P_0 = P_0 + (0.0102P_0)\Delta t$. Solve $2 = 0.0102\Delta t$ for $\Delta t = 2/0.0102 \approx 196$ yr. Exact doubling time is $(\ln 2)/0.0102 \approx 68$ yr. (See 23.) Euler overestimates because it ignores increase in rate of population growth with population.

23. $P' = kP$, $P(0) = P_i \Rightarrow P(t) = P_i e^{kt}$. Solve $P(t_2) = 2P_i = P_i e^{kt_2}$ to obtain $t_2 = (\ln 2)/k.$

25. Compare example 4. Solution of $P' = kP$, $P(t_0) = P_0$ is $P(t) = P_0 e^{k(t-t_0)}$. Insert $P(t_1) = P_1$ and solve for k to find $k = (\ln(P_1/P_0)/(t_1 - t_0)).$

Section 2.2

1. **(a)** Left side is $\ln|0.015P - .209|/0.015$, right is $t + C.$

(b) Multiply by 0.015 before exponentiating; use $C = e^{0.015c}.$

(c) Substitute $t = 1847$, $P = 8.$

3. If population is constant, then $dP/dt = 0$. From $P' = kP - E$, obtain $kP - E = 0$. The emigration rate to maintain a constant population level is $E = kP$. Using the data of example 5, $E = 0.015 \cdot 8 = 0.12$ million people per year, or 120,000 people per year.

5. **(a)** Follow the derivation of the emigration model with $E = 12{,}000$ people per day $= 4.39$ million people per year; $k = k_b - k_d = -0.0012$ million per year (using the only available value, that for 1985). Find $P' = -0.0012P - 4.39$, $P(1989) = 16.64$ (again using a 1985 value).

(b) Write $P(t) = (16.64 + 4.39/.0012)e^{.0012(t-1989)} - 4.39/.0012 = 3{,}674.97\, e^{-.0012(t-1989)} - 3{,}658.33$, then solve $P(t) = 0$: $t = 1989 - [\ln(3{,}658.33/3{,}674.97)]/.0012 \approx 1{,}992.78$. The country would have been deserted late in the year 1992.

(c) $P' = 0 \Rightarrow -0.0012P - 4.39 = 0$; no $P > 0$ satisfies this condition. The population was already declining because $k = -0.0012 < 0$.

7. Follow the derivation of the emigration model but use $I(t)$, the *immigration* rate, as a rate of population *increase* in place of *emigration* rate E as a rate of population loss; i.e., $+I\Delta t$ replaces $-E\Delta t$ in conservation expression. Find $P' = kP + I$, $P(t_0) = P_i$.

9. (a) $bA\Delta t$.

(b) $dA^2\Delta t$.

(c) $r(A - A_{\text{out}})\Delta t$.

(d) Net change equals number added less number lost $\Rightarrow A' = bA + r(A - A_{\text{out}}) - dA^2$, $A(0) = A_i$.

11. Even if interest rate I is 0, deposits increase at rate m. Reject (b), (d) because $m' = -D < 0$ when $I = 0$. Analogously, reject (a) (and (d)) because $m' = -Im < 0$ when $D = 0$, $m > 0$. Hence, accept model (c) as plausible.

13. $m(t) = (S + D/I)e^{It} - D/I$.

15. $P' = 0 \Rightarrow kP - E = 0 \Rightarrow P_{\text{crit}} = E/k$; $P_i > E/k \Rightarrow P'(0) > 0 \Rightarrow P(t) > E/k \Rightarrow P'(t) > 0$ for all t. Reverse inequalities for $P_i < E/k$. Alternatively, the maximum emigration rate satisfies $kP_i - E_{\text{max}} = 0$, or $E_{\text{max}} = kP_i$.

17. See exercise 15. In addition, E constant in $P' = kP - E \Rightarrow P'' = kP'$.
Since P'' is defined for all t and never zero unless $P' = 0$, there are no local maxima or minima. The case $P' = 0$ is the steady-state solution $P \equiv E/k$. If $P_i < E/k$, the solution function can decrease to negative values.

19. Suppose birth rate constant is k. Then over time Δt, $P(t + \Delta t) - P(t) = kP(t)\Delta t - [rP(t) + s(P(t))^2]\Delta t$. The usual manipulations yield $dP/dt = (k - r)P - sP^2$, the logistic equation with $a = k - r$, a net (constant) birth rate parameter.

21. $aP - sP^2 = P(a - sP) = 0 \Rightarrow P_{ss} = 0, a/s$. If $0 < P < a/s$, then $P' = P(a - sP) = + \cdot + > 0$; P is increasing. If $a/s < P$, then $P' = P(a - sP) = + \cdot - < 0$; P is decreasing. Since the increase or decrease of the continuous function P is bounded by a/s and $P' = 0$ if $P \neq a/s$, $\lim_{t \to \infty} P(t) = a/s$.
$P_{ss} = a/s$ is the *carrying capacity* because all populations, regardless of their initial size, tend toward it in time. Substitute $K = a/s$ to verify that $P' = aP - sP^2 = aP(1 - P/K)$.

Section 2.3

1. Direction field of each equation is identical in form to figure 2.12 but magnitude of slope varies depending upon the size of the factor in front of $(T - T_{\text{out}})$.

(a) $T = T_{\text{out}} + (293 - T_{\text{out}})e^{-0.012t}$.

(b) Separate variables, $dT/(T - T_{\text{out}}) = -(Ak/cm)\,dt$, integrate to obtain $\ln|T - T_{\text{out}}| = -(Ak/cm)t + C$, and

exponentiate to solve for T. Use $T(0) = T_i$ to determine constant of integration; $T = T_{\text{out}} + (T_i - T_{\text{out}})e^{-Akt/cm}$.

(c) Proceed as in 1(b) with $A_r k_r + A_w k_w$ replacing Ak; $T = T_{\text{out}} + (T_i - T_{\text{out}})e^{-(A_r k_r + A_w k_w)t/cm}$.

3. Use separate terms in the conservation law for rate of loss through walls and roof: $-A_w k_w(T - T_{\text{out}}) - A_r k_r(T - T_{\text{out}})$. Find $T' = -[(A_w k_w + A_r k_r)/cm](T - T_{\text{out}})$, $T(0) = T_i$.

5. (a) Halving k by doubling insulation changes Ak/cm from 0.012 to 0.006.

(b) With $Ak/cm = 0.006$ min^{-1} and $T_i = 293°$K, evaluate $T(60) = T_{\text{out}} + (293 - T_{\text{out}})e^{-0.006 \cdot 60}$. Hence, $T(60) = 293 \cdot 0.6977 + T_{\text{out}} \cdot 0.3023 = 204 + 0.3023T_{\text{out}}$.

(c) If $T_{\text{out}} = 0°C = 273°$K, then the better insulated house of part (b) is at temperature $287°$K or $14°$C one hour after the furnace fails. The house of example 4 is at $282°$K or $9°$C, significantly colder.

7. To estimate m, first estimate volume of wood, plaster, and other building materials; then use density values from a source like [11]. Add an estimate of mass of interior furnishings.

9. Use solution of 1(b) with $T_i = 25 + 273 = 298°$K, $T_{\text{out}} = 271°$K, and $T(45) = 292°$K. Use logarithms to solve $T(45) = 271 + (298 - 271)e^{-45(Ak/cm)} = 292$ for Ak/cm: $Ak/cm = -(\ln 21/27)/45 = 0.0056$ min^{-1}.

11. (a) On pages 57–58, replace $Ak(T - T_{\text{out}})$ with $A\kappa(T^4 - T_{\text{out}}^4)$.

(b) The sign of T' in the new model equation is still determined by the relative sizes of T and T_{out}.

(c) The rates of temperature change in the two models are in the ratio $(T - T_{\text{out}})/(T^4 - T_{\text{out}}^4) = T^3 + T_{\text{out}}T^2 + T_{\text{out}}^2 T + T_{\text{out}}^3$.

(d) Separation of variables would lead to an integral of the form $\int dT/(T^4 - T_{\text{out}}^4)$. Factor the denominator as $(T - T_{\text{out}})(T + T_{\text{out}})(T^2 + T_{\text{out}}^2)$ and use partial fractions. The first two factors would give rise to logarithms, the third to an inverse tangent. Obtaining an explicit analytic expression for $T(t)$ is not easy. DELAB has nothing to offer.

(e) For the familiar model, Euler with $\Delta t = 0.05$ predicts $T(1) \approx 273.99°$K. For the radiation model, the result is $T(1) \approx 273°$K. The rate of decrease of temperature via radiation is much greater than by conduction, even with $\kappa \ll k$.

13. Conservation of energy: $Q(t + \Delta t) - Q(t) = S\Delta t - Ak(T - T_{\text{out}})$. Divide by Δt, use $Q = cmT$, etc. to obtain $T' = -(Ak/cm)(T - T_{\text{out}}) + S/cm$.

15. From $T(t) = T_{\text{out}} + (293 - T_{\text{out}})e^{-0.012t}$, compute $\partial T(t)/\partial T_{\text{out}} = 1 - e^{-0.012t}$. Given a change ΔT_{out} in the outside temperature, the corresponding change in the temperature of the house is $\Delta T(t) \approx (\partial T(t)/\partial T_{\text{out}}) \Delta T_{\text{out}} = (1 - e^{-0.012t})\Delta T_{\text{out}}$. At $t = 0$, $1 - e^{-0.012t} = 0$, and $\Delta T(t) = 0$; the temperature

of the house is independent of the outside temperature at the instant the furnace stops. At $t = 60$ min, $1 - e^{-0.012t} \approx 0.5$, and the variation in the house temperature is about half that of T_{out}. After 5 hours ($t = 300$), $1 - e^{-0.012t} \approx 0.97$ and the two changes are essentially equal.

17. The furnace turns on when $T = T_{set}$. If the heat output rate of the furnace exactly balances the rate of loss through the walls and roof, the temperature of the house will be constant at T_{set}. That is, at the critical value of T_{out}, $T' = 0 = -(Ak/cm)(T_{set} - T_{out}^{crit}) + R/cm$. Hence, the outside temperature can not be lower than $T_{out}^{crit} = T_{set} - R/Ak$.

Section 2.4

1. In the conservation-of-rabbits expression, add the loss term $-d_R R \Delta t$. Replace b_R with the *net* birth rate $b_R - d_R$. The final form of the rabbit equation is $R' = (b_R - d_R - \beta F)R$.

3. The steady state is interior to the closed curve in the phase plane that corresponds to the periodic oscillations in the fox and rabbit populations.

5. (a) Use conservation of moose:
$(S_{MR}R)M\Delta t M(t + \Delta t) - M(t) =$
$a_M M\Delta t - (s_M M)M\Delta t - (s_{MR}R)M\Delta t$.

(b) Conservation of rabbits:
$(S_{RM}M)R\Delta t R(t + \Delta t) - R(t) =$
$a_R R\Delta t - (s_R R)R\Delta t - (s_{RM}M)R\Delta t$.

(c) Initial conditions: $M(0) = M_i$, $R(0) = R_i$.

7. (a) Set $R = 0$. Find the corresponding steady state: $M' = 0 = M(a_M - s_M M) \Rightarrow M_{ss} = 0, a_M/s_M$; use a conventional logistic equation analysis.

(b) $R_{ss} = 0, a_R/s_R$.

9. The y and z variables can be interchanged without changing the equations. Hence, the initial-value problem in the example is the same as that in the exercise. They will have the same solution (with the roles of y and z reversed) so long as the initial-value problem has only one solution.

11. A fixed fraction of the recovereds become susceptible again. Conservation of susceptibles now has an additional gain term, $S(t + \Delta t) - S(t) = \cdots + gR\Delta t$. Conservation of recovereds has an additional loss term, $R(t + \Delta t) - R(t) = \cdots - gR\Delta t$.

13. $I' = 0 \Rightarrow I = 0$ or $S = r/b$. Use $I_{ss} = 0$ and $S' = 0$ in first equation to find $S_{ss} = P$; $I_{ss} = 0, S_{ss} = P$ corresponds to no infectives but everyone is susceptible.
Use $S_{ss} = r/b$ and $S' = 0$ in first equation to find $I_{ss} = g(P - r/b)/(r + g)$, a physically reasonable steady state so long as $r/b < P$; $S_{ss} = r/b$,
$I_{ss} = g(P - r/b)/(r + g) = g(P - S_{ss})/(r + g)$ defines the state in which rate of infection just balances rate of recovery.

15. Because of the y-z symmetry in this particular pair of equations, the time history of the extinction of y is exactly the same as that shown in the text in figure 2.16 with the y

and z labels reversed: y declines steadily to $y = 0$ while z first declines, then increases to $z = 1$.

Chapter exercises

1. (a) Conservation of mass with zero rate of adding isotope: $y(t + \Delta t) - y(t) = 0 - ky\Delta t$. Divide by Δt, take limit as $\Delta t \to 0$ to find $y' = -ky, y(0) = y_i$.

(b) $y_i > 0 \Rightarrow y'(0) < 0 \Rightarrow y$ is decaying, as expected with radioactive decay.

(c) $y = y_i e^{-kt} \to 0$ as t grows, as expected.

(d) Use $y(t_{1/2}) = y_i/2$ in solution found in previous part to find $t_{1/2} = (\ln 2)/k$.

(e) y might have units of mass, k units of inverse time.

3. (a) $v_R = Rdq/dt, v_C = q/C$.

(b) $v_R + v_C = 0$.

(c) Substitute 3(a) into 3(b) and add initial condition: $Rdq/dt + q/C = 0, q(0) = q_i$.

5. (a) $q' = -q/RC \Rightarrow q > 0$ decays; negative charge grows. In $q' = -q/RC + E/R$ growth or decay depends on the sign of $-q/RC + E/R$.

(b) $q' = -q/RC, q(0) = q_i \Rightarrow q(t) = q_i e^{-t/RC}$; all solutions decay to zero. With a constant imposed voltage E, $q' = -q/RC + E/R, q(0) = q_i$
$\Rightarrow q(t) = (q_i - EC)e^{-t/RC} + EC$; all solutions decay to EC.

(c) Steady state is $q = EC$.

7. Nothing in the application of the conservation law hinges on the relative values of T and T_{out}; the derivation of the model is virtually identical to that in the text.

9. (a) Rate of addition of mass of chemical per unit volume due to reaction is kc; rate of loss due to dilution is d. Conservation of mass requires $c(t + \Delta t)$
$- c(t) = kc(t)\Delta t - d\Delta t$. Usual arguments yield 8(c), $c' - kc = -d, c(0) = c_i$. (Analogous to emigration model.)

(b) $c(t) = (c_i - d/k)e^{kt} + d/k$. Growth or decay of c over time depends on sign of $c_i - d/k$; i.e., on whether initial concentration is sufficiently large so that growth from reaction exceeds loss due to dilution.

(c) Initial-value problems (a) and (b) of exercise 8 have decaying solutions $c(t) = (c_i \mp d/k)e^{-kt} \pm d/k$; solution of (d), $c(t) = (c_i + d/k)e^{kt} - d/k$, *grows* if $c_i = 0$.

11. Reject (b), (d) because $c' \cdots = -i < 0$ means *decay* due to neutron beam; reject (c), (d) because $c' = kc > 0$ means *growth* in the absence of the neutron beam ($i = 0$). Hence, (a) is the only reasonable choice.

13. (a) Let c denote BOD concentration, k the decay constant, d the dumping rate. Conservation of BOD: $c(t + \Delta t) - c(t) = d\Delta t - kc(t)\Delta t$. Usual manipulations yield $c' = -kc + d, c(0) = c_i$.

(b) $c(t) = (c_i - d/k)e^{-kt} + d/k \to d/k$; BOD level in the lake approaches an equilibrium as rate of natural decay balances rate of addition of pollution.

(c) With $d = 0$, $c \to 0$; lake cleanses itself with time constant $1/k$.

15. Need $y' - gy = \cdots$ for growth so reject (b), (d). Need $y' \cdots = -d$ for loss due to insecticide so reject (a), (b). Accept (c). (Analogous to emigration model.)

17. k is growth (positive) or decay (negative) constant; e.g., k in population models, $k = -Ak/cm$ in heat-loss model; f is source (positive) or sink (negative); e.g., $f = -E$ in emigration model, $f = AkT_{out}/cm$ in heat-loss model.

19. $F' = kF + A - R$, $F(0) = F_i$.

21. (a) Heat loss $S\Delta t$, heat gain $-Ak(T - T_{out})$.

(b) Heat loss through walls to cold water.

(c) Heat loss through walls to cold outside, heat gain from stove.

23. The F equation has little in common with either y or z equation; F exhibits population-dependent decay $(-d_F F)$ and interaction-dependent growth $(\alpha R F)$; y and z exhibit population-dependent growth (y or rz), logistic-like decay $(-y^2$ or $-rz^2)$, and interaction-dependent decay $(-eyz$ or $-fzy)$.

The R equation has in common with both y and z equations population-dependent growth $(b_R R)$ and interaction-dependent decay $(-\beta F R)$.

25. $y_{ss} = a$ is stable for $k < 0$ because $\lim_{t\to\infty} y(t) = 0$ for any solution $y(t) = (y_i - a)e^{kt} + a$ of $y' = k(y - a)$, $y(0) = y_i$. This steady state is unstable for $k \geq 0$. For $k > 0$, this is the emigration equation, for $k < 0$, the heat-loss equation.

27. (a) Balance units of quantity Q (mass, population, etc.) and time T: $dy/dt = ky$ requires $Q/T = kQ$. Hence, k has units of inverse time, $1/T$.

(b) From the chain rule (appendix section A.1, page 622), with $s = kt$ and $Y(s) = y(s/k)$,

$$\frac{dY}{ds} = \frac{dy}{dt}\frac{dt}{ds} = \frac{1}{k}\frac{dy}{dt}.$$

Substitute $dy/dt = kdY/ds$, $y = Y$ into $y' = ky$.

(c) For $P' = kP$, let $s = kt$, $Q(s) = P(s/k)$. Then $kdQ/ds = kQ - E$ or $dQ/ds = Q - E/k$. To write $T' = -(Ak/cm)(T - T_{out})$ in dimensionless form, repeat these steps with Ak/cm in place of k, $(Ak/cm)T_{out}$ in place of E to obtain $dQ/ds = -(Q - T_{out})$, where $Q(s) = T(cmt/Ak)$.

CHAPTER 3

Section 3.1

1. (a) 0.03076.

(b) The error is -0.09595.

3. (a) 1.1875.

(b) Error $= .2185$.

5. $y = -1.0252$.

7. (a) To estimate the time for the population to double its initial size, solve $2P_i = P_i + k\Delta t P_i$ for Δt to find $t_{approx} = 0 + \Delta t = 1/k$.

(b) To find the exact time for the population to double, solve $2P_i = P_i e^{kt}$ to find $t_{exact} = \ln 2/k$. The error $(\ln 2 - 1)/k \approx -0.31/k$.

9. (a) To estimate the time for the temperature to decrease to the point where $T - T_{out} = (T_i - T_{out})/2$ (i.e., to the point where $T = T_{out} + (T_i - T_{out})/2 = (T_i + T_{out})/2$, solve $(T_i + T_{out})/2 = T_i - (Ak/cm)T_i\Delta t$ for Δt to find $t_{approx} = 0 + \Delta t = cm/Ak$.

(b) To find the exact time for $T - T_{out}$ to decrease to half its initial size, solve $(T_i + T_{out})/2 = T_{out} + (T_i - T_{out})e^{-Akt/cm}$ to find $t = cm\ln 2/Ak$. The error is $(\ln 2 - 1)cm/Ak$.

11. The exact solution is $P(t) = 115.40\,e^{0.0282(t-1978)}$. See table S.1.

TABLE S.1   Section 3.1, exercise 11

n	t	P_n	Error
1	1988	147.9428	0.05052×10^2
2	1998	189.6627	0.13175×10^2
3	2008	243.1475	0.25770×10^2
4	2018	311.7151	0.44810×10^2
5	2028	399.6188	0.73054×10^2
6	2038	512.3113	1.14349×10^2
7	2048	656.7831	1.74030×10^2

---

13. (a) Carry out the indicated calculations.

(b) The efficient form saves one evaluation of $f(t, y)$ at each step; here, $f = 0.0282P$, saving one multiplication per step. If evaluating f requires 100 multiplications, then 100 multiplications are saved per step.

15. (a) (i) 161.8616.

(iii) $P(t) = 115.40\,e^{0.0282(t-1978)} \Rightarrow P(1990) = 161.8718$.

(iv) 0.0102.

(b) (i) 10.0374.

(iii) $T(t) = 18e^{-0.72t} + 10 \Rightarrow T(10) = 10.0134$.

(iv) 0.0240.

(c) (i) 7.7269.

(iii) $P(t) = -5.93e^{0.015(t-1847)} + 13.93$

$\Rightarrow P(1850) = 7.7269$.

(iv) 0.0000.

17. See figure 2.9, p. 50 of the text, which was computed with $a = 0.12$, $s = 0.024$, and the initial values $P(0) = 0.25, 3.0, 9.5$ mm^2.

19. See figure S.3; Heun with $\Delta t = 0.1$ and initial values $\pm 1.05, \pm 0.99, 0.3$. Evidently, $y_{ss} = 0$ is a stable steady state; $y_{ss} = \pm 1$ are unstable.

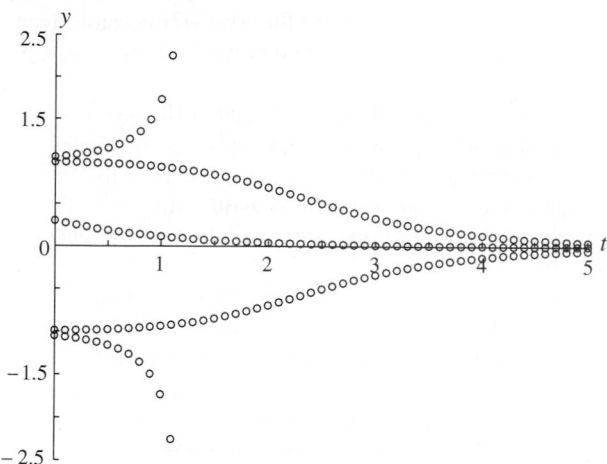

FIGURE S.3   Section 3.1, exercise 19.

21. Heun with $\Delta t = 0.1$, $y(0) = 0$, selects figure 3.21(d), page 117, which is also consistent with $y' = 1 - y^2 > 0$ and $y'' = -2yy' < 0$ for $0 < y < 1$.

25. (b) Use $\Delta t = 5$. Since $1978 + 6 \cdot 5 = 2008$, $n = 6$. The error decreased by a factor of approximately 4.

 (c) The error in Euler is proportional to Δt, that in Heun proportional to Δt^2. Decreasing Δt by a factor of 2 decreases the error in the Heun method by a factor of $2^2 = 4$ and in the Euler method by a factor of $2^1 = 2$.

26. (a) Carry out the indicated calculation; with $\Delta t = 10$, $268.9174 = P(2008) \approx P_3 = 243.1475$. The error is 25.7699.

 (b) With $\Delta t = 5$, $P(2008) \approx P_6 = 254.6339$. The error is 14.2815.

 (c) The error has decreased by a factor of 1.8. The Heun method error improves by a factor of 4 when Δt is halved.

27. (a) Using the Euler method with $f = kP$ leads to $P(t_{n+1}) = (1 + k\Delta t)P_n = (1 + k\Delta t)^2 P_{n-1} = \cdots = (1 + k\Delta t)^{n+1} P_0$.

 (b) A Taylor expansion of the exact solution yields $P(t_{n+1}) = P_0 e^{k(n+1)\Delta t} = P_0 \left[1 + k(n + 1)\Delta t + [k(n + 1)\Delta t]^2/2 + \cdots \right]$. A binomial expansion of the Euler approximation yields $P_{n+1} = (1 + k\Delta t)^{n+1} P_0 = P_0(1 + k(n + 1)\Delta t + n(n + 1)(k\Delta t)^2 + \cdots$. One could also evaluate the limit of P_{n+1} as $n \to \infty$ with $n\Delta t$ fixed.

 (c) At the end of each time step, the population is increased by the factor $1 + k\Delta t$, rather than the "continuous compounding" of the actual model.

29. Since the logistic equation is nonlinear, we can not obtain a simple recurrence relation $P(t_{n+1}) = (1 + k\Delta t)P_n$ as in exercise 27.

31. Obtain a figure similar to figure 3.7 in the text.. The error grows at each step because the Heun approximation

uses an average slope that does not account for the continuously increasing slope of the exact solution.

33. $P_{n+1} = P_n(1 + k\Delta t + (k\Delta t)^2/2)$ (Note the resemblance to the Taylor polynomial for $P_n e^{k\Delta t}$.)

35. Euler with $\Delta t = 20$ min produces believable results with temperature minima at $t \approx 700$ and $2,250$ min (about 11.7 and 37.5 hours past noon of the first day) and temperature maxima at $t \approx 1,500$ and $3,000$ min. (about 1 and 26 hours past noon of the second day). These times are close to intuitive predictions.

37. Runge-Kutta produces good results with $\Delta t = 120$ min and plausible results with $\Delta t = 180$ min, consistent with its relatively high accuracy.

38. See table S.2. The exact solution is $y = 4 - 2e^{-x^2/2}$.

TABLE S.2   **Section 3.1, exercise 38**

n	Δt	Error $\times 10^4$	Error $/\Delta t$
20	0.2	6.6519	3.3260×10^{-3}
40	0.1	5.3620	5.3620×10^{-3}
60	0.0667	4.1890	6.2832×10^{-3}
80	0.05	3.3951	6.7902×10^{-3}

39. See table S.3. The exact solution is $y = 4 - 2e^{-x^2/2}$.

TABLE S.3   **Section 3.1, exercise 39**

n	Δt	Error $\times 10^5$	Error $/\Delta t^2$
20	0.2	63.088	1.5772×10^{-2}
40	0.1	9.6375	9.6875×10^{-3}
60	0.0667	3.8275	8.6110×10^{-3}
80	0.05	2.0475	8.1900×10^{-3}

41. See exercise 38, which considers $y' + xy = 4$, $y(0) = 2$. For that example, the error in the Euler method is about $0.006\Delta t$.

43. The global error in the Euler method is proportional to Δt, that of the Heun method to Δt^2. To reduce each by 1/4 requires reducing Δt by 1/4 for the Euler method and 1/2 for the Heun method; that is, the improvement in accuracy costs four times as many steps with Euler but only twice as many steps with Heun. Euler requires one function evaluation per step and Heun two. Hence, the computational cost as measured by number of evaluations of $f(t, y)$ will increase by a factor of four using either method. In general, an increase in accuracy of $1/n$ multiplies the number of Euler function evaluations by n and the number of Heun function evaluations by $2\sqrt{n}$. So Heun is more efficient for $n > 4$.

45. (a) The midpoint method is defined by
$y_{n+1} = y_n + f(t_n + \Delta t/2, y_n + \Delta t/2)\Delta t$. The method uses
a step whose slope is that at the midpoint of the interval
$[t_n, t_{n+1}]$.

(b) Expand $f(t_0 + \Delta t/2, y_0 + \Delta t/2)$ in a Taylor series
about (t_0, y_0) to obtain

$$y_1 = y_0 + y_0'\Delta t + y_0''\Delta t^2/2 + \mathcal{O}(\Delta t^3).$$

(Recall that $y'' = f_t + f_y y'$.) Since the terms through Δt^2
match the Taylor expansion of the exact solution $y(t)$ about
$t = 0$, $E_L = \mathcal{O}(\Delta t^3)$.

(c) The data in table S.4 confirms that the error in the
mid-point method is proportional to $-0.06\Delta t^2$. The exact
solution is $y = e^{-t}$.

TABLE S.4   Section 3.1, exercise 45(c)

n	Δt	**Error** $\times 10^4$	**Error** $/\Delta t^2$
10	0.1	-6.6159	-6.62×10^{-2}
20	0.05	-1.5918	-6.37×10^{-2}
40	0.025	-0.3905	-6.25×10^{-2}

---

47. To obtain the Euler method from (3.12), choose $a = 1$,
$b = 0$; α and β are arbitrary. Because $b = 0$, this choice
fails to satisfy the second and third equation of (3.16).
49. In the second-order Runge-Kutta method, the slopes
$k_1/\Delta t, k_2/\Delta t$ are averaged with weights a, b, respectively.
51. Evaluating the next term in the expansion (3.20) yields
$b(f_{tt}\alpha^2 + 2f_{ty}\alpha\beta + f_{yy}\beta^2)$. Comparing this with the Taylor
expansion for $y(t)$ yields the local error term
$E_L = y'''(\xi)/6 - b(f_{tt}\alpha^2 + 2f_{ty}\alpha\beta + f_{yy}\beta^2)$. Without some
simplifying assumptions on the behavior of y''' and the
derivatives of f, we can not minimize this error term.
53. In general, $y_n = (1 + k\Delta t)^n y_0$. With $y_0 = 1$ and
$k = -30$, $y_n = (1 - 30 \cdot 0.05)^n = (-0.5)^n$. Hence,
$y_1 = -0.5$, $y_2 = 0.25$, etc. as shown in the figure. With
$k = -40$, $y_n = (1 - 40 \cdot 0.05)^n = (-1)^n$. Hence, $y_1 = -1$,
$y_2 = 1, \ldots$; $y_n = \pm 1$ according to whether n is even or
odd. If $k = -41$, then $y_n = (1 - 41 \cdot 0.05)^n = (-1.05)^n$,
and y_n oscillates in sign as it grows without bound.

Section 3.2

1. (a) All arrows slope downward with constant slope $-g$.
See figure 1.2 of the text.

(b) The arrows would have a shallower negative slope.

(c) All arrows point to the left (horizontal phase line) or
down (vertical phase line).

3. (a) See example 10 of section 2.2, page 49.

(b) The graph of P versus t is increasing and concave up
for $0 < P < a/2s$, increasing and concave down for
$a/2s < P < a/s$, and decreasing and concave up for

$a/s < P$. No curve crosses the line $P = a/s$ because slope
is zero when $P = a/s$. The solution having $P(t_0) = a/s$ is
just the line $P(t) \equiv a/s$.
See the direction field diagram of figure 3.18, page 113.

(c) Since all population curves, regardless of starting
value, tend toward the value a/s, it is reasonable to
designate this limiting value the *carrying capacity* of the
environment. It is the population level the environment can
sustain.

5. If $0 < u(0) < 1$, then $u'(0) < 0$ and the solution
decreases. If $1 < u(0)$, then $u'(0) > 0$ and the solution
increases. If $u(0) = 1$, then $u(t) \equiv 1$, a steady state.

7. Graphs (b) and (c) can be eliminated since $y' > 0$ for
$y < 1$; (a) can be eliminated since $y'' = -2yy' < 0$. Graph
(d) is consistent with both the slope and concavity data
presented here.

9. Graphs (c) and (d) can be eliminated since $y' > 0$ for all
y. Graph (a) can be eliminated since $y'' = 2yy' > 0$ for
$y > 0$. The curve in (b) is the only possible choice, and its
concavity is consistent with $y'' > 0$ for $y > 0$.

11. Use the right-hand side of the differential equation to
determine the sign of the first derivative. Compute the
second derivative using the chain rule (e.g.,
$u' = u^2 - u \Rightarrow u'' = 2uu' - u'$) and find its sign.

Section 3.3

1. The solution formulas shown in the text remain valid,
regardless of the relative sizes of T_i and T_{out}. When
$T_{out} > T_i$, then this model might be called more properly a
heat-gain model. The solution curve now increases to the
steady state $T = T_{out}$.

3. Intuitively, adding insulation increases decay time.
Hence, the time constant cm/Ak should increase, as it does
since added insulation decreases the thermal conductivity k.

(b) Define the perturbation q by $v = -mg/k + q$.
Substitute in $v' = -g - kv/m$ to obtain $q' + kq/m = 0$.
Since solutions of this equation are decreasing,
$v_{ss} = -mg/k$ is stable.

5. (a) Since $v_{ss}' = 0$, $kv_{ss}|v_{ss}| = -mg$. Hence,
$v_{ss} = -\sqrt{mg/k}$. Because the body is falling downward,
$v_{ss} < 0$.

(b) Define the perturbation q by $v = v_{ss} + q$. Substitute
in $v' = -g - kv|v|/m$ to obtain

$$q' = -g - \frac{k}{m}(v_{ss} + q)|v_{ss} + q|$$

$$= -g + \frac{k}{m}(v_{ss} + q)^2.$$

The second equality follows because $v_{ss} + q < 0$ when q is
small; hence, $|v_{ss} + q| = -(v_{ss} + q)$. Now expand the

square and use $v_{ss} = -\sqrt{mg/k}$ to obtain

$$q' = 2v_{ss}q + q^2$$
$$q' \approx 2v_{ss}q.$$

Since $v_{ss} < 0$, the solutions of the linearized perturbation equation $q' = 2v_{ss}q$ are decreasing, and v_{ss} is stable. As in the linear model of exercise 4, the falling body remains at its terminal velocity once it reaches it.

6. (a) If there is a steady-state (constant) yeast population y_{ss}, it must satisfy the differential equation $y' - Ry = -H$. That is, since $y'_{ss} = 0$, $-Ry_{ss} = -H$. The steady-state yeast population is H/R.

(b) Define the perturbation q by $y = H/R + q$. Substitute in $y' - Ry = -H$ to obtain $q' - Rq = 0$. Since solutions of this equation are increasing, $y_{ss} = H/R$ is unstable.

Figure 3.26 is a direction field diagram for this equation if H replaces E and R replaces k. Indeed, harvesting is precisely analogous to emigration.

7. If $q' = -aq$, then $q = Ce^{-at}$; all nontrivial solutions decay to zero.

9. (a–b) See exercises 6(a–b) with c replacing y, k replacing R, and d replacing H.

(c) The equilibrium $c_{ss} = d/k$ represents the (unstable) balance between loss due to dilution and gain from the reaction.

10. (a) $c_{ss} = i/k$.

(b) Solve the initial-value problem and show that $c(t) \to c_{ss}$ as $t \to \infty$. Alternatively, follow exercise 4(b).

(c) The rate of isotope creation by the neutron beam balances the natural decay of the isotope. (The situation is complementary to that described in exercise 9.)

11. (a–b) See exercises 10(a–b) with q replacing c, $1/RC$ replacing k, and E/R replacing i; $q_{ss} = EC$.

(c) Decay of charge on the capacitor through the circuit is balanced by the charge forced onto the capacitor by the emf E. Compare with exercise 10(c).

13. (a) Define the perturbation q by $P = 0 + q$. Substitute into $P' = aP - sP^2$ to obtain (trivially) $q' = aq - sq^2$. Since the linearized equation $q' = aq$ has growing solutions, $P_{ss} = 0$ is unstable.

(b) The carrying capacity $P = a/s$ is well named because initial populations tend toward it in the limit of large time. Since $P_{ss} = 0$ is unstable, any population, no matter how small, will grow toward the carrying capacity a/s.

15. From $y'_{ss} + ky_{ss} = f(x)$ and y_{ss} constant conclude that f must be constant to achieve a steady state. Under that condition, substitute $y(t) = y_{ss} + q(t)$ to conclude that $q' + kq = 0$. Hence, $q = Ce^{-kt}$ decays if $k > 0$, and $y' + ky = f(x)$ has a stable steady state if f is constant and $k > 0$.

17. The transient arises from the decaying term $Ce^{-t/RC}$. Hence, there is always a decaying transient. But there will be no steady-state charge if $E(t)$ varies with time.

19. If $g \neq 0$, a linear differential equation can have at most one steady state. If a steady state exists, it is defined by $y_{ss} = f(x)/g(x)$, $g \neq 0$. If $g \equiv 0$ and $f \equiv 0$, then $y' = 0$ has the multiplicity of steady states $y_{ss} = c$, c an arbitrary constant.

21. Let p denote the perturbation of the steady state $y_{ss} = 0$: $y(t) = y_{ss} + p(t)$. Then $p' = (y_{ss} + p)^3 = p^3 \approx 0$ if p is small. However, if $y(0) > 0$, then $y'(0) = y^3(0) > 0$ and y is increasing. Similarly, if $y(0) < 0$, then y is decreasing. Hence, $y_{ss} = 0$ is unstable. Similar arguments show that the steady state $y_{ss} = 0$ of $y' = -y^3$ is stable.

Chapter exercises

1. $y_{tran} = 3e^{-x^2}/2$; $y_{ss} = 1/2$.

3. (a) Separation of variables or DELAB yields $s = 50 + (s_i - 50)e^{-0.1t}$.

(b) Since the exponential term decays to zero as time grows, the transient is $s_{trans} = (s_i - 50)e^{-0.1t}$. The remaining constant term is the steady state $s_{ss} = 50$, representing the equilibrium between the addition of 5 pounds of salt per minute and the removal each minute of 10% of a total weight of 50 pounds. The transient represents the transition from the initial weight of salt in the tank to this equilibrium weight of 50 pounds.

(c) If no salt is added to the tank, that is, if the forcing term 5 is replaced by 0, then $s' = -0.1s < 0$ means dilution is indeed lowering the amount of salt in the tank. If dilution is eliminated, that is, if 0.1 in the differential equation is replaced by 0, then $s' = 5 > 0$, consistent with addition of salt without dilution.

5. Neither solution exhibits a steady state. The transient part of each solution comprises the terms with the factor $e^{-Akt/cm}$. In the following, let $\alpha = Ak/cm$.

(i)
$$T(t) = T_{out} + \frac{H}{cm}\left(\frac{\alpha \sin \omega t - \omega \cos \omega t}{2(\alpha^2 + \omega^2)}\right) + \frac{H}{2Ak} + Ce^{-\alpha t}$$
where $C = H\dfrac{\omega^2 - \alpha\omega + \alpha^2}{2Ak(\alpha^2 + \omega^2)} + T_i - T_{out}$.

(ii) $T(t) = \dfrac{H}{4Ak(4\alpha^2 + \omega^2)}(\alpha^2 + 4\omega^2 - 2\alpha\omega \sin 2\omega t -$

$\alpha^2 \cos 2\omega t) + Ce^{-\alpha t}$, where $C = \dfrac{H\omega^2}{Ak(4\alpha^2 + \omega^2)} + T_i - T_{out}$.

7. This initial-value problem could model a house with an air conditioner: $-F/cm$ is a heat sink, reducing T in the absence of other effects. From $T'_{ss} = 0$, find $T_{ss} = T_{out} - F/Ak$. Stability is unaffected by the sign of F; the direction diagram is that of figure 3.25 with the steady state shifted.

8. (a) If b is constant, then $a' \pm 0.000493a = \pm b$ has equilibrium $a_{ss} = b/.000493$; $a' \pm 0.000493a = \mp b$ has equilibrium $a_{ss} = -b/0.000493$. The equilibrium is stable for $a' + 0.00493a = \pm b$, unstable otherwise.

9. (a) Follow the derivation of the simple population model $P' = kP$ in section 2.1 with m in place of P, I in place of k.

(b) $m_n = m_i(1 + I)^n$, the formula for the bank balance when interest is compounded daily at the rate I.

11. (a) Compare with 8(a).

(b) We expect the same behavior as in the emigration model; the insecticide acts like emigration to remove individuals from the population. Hence, choose iv.

(c) All the equations have steady states; (i) and (iii) are stable.

13. (a) $y_{ss} = -2$ is unstable, $y_{ss} = 2$ is stable.

(b) $y_{ss} = 0$ is stable, $y_{ss} = 1$ is unstable.

(c) $y_{ss} = 0$ is unstable, $y_{ss} = 3$ is stable.

(d) y_{ss} solves $\sin y_{ss} = -1$; hence,
$y_{ss} = -\pi + 2n\pi = (2n - 1)\pi$, $n = 0, \pm 1, \pm 2, \ldots$.
DELAB shows $y_{ss} = \pi$ is stable.

CHAPTER 4
Section 4.1

1. Linear, nonhomogeneous; $f(x, y) = -9.8$; $\mathcal{M}y = y'$, $r(x) = -9.8$

3. Nonlinear, homogeneous; $f(x, y) = y^2 \cos \pi x$; $\mathcal{M}y = y' - y^2 \cos \pi x, r(x) = 0$.

5. Linear, nonhomogeneous; $f(x, y) = 4e^x$; $\mathcal{M}y = y'$, $r(x) = 4e^x$.

7. Nonlinear, nonhomogeneous;
$f(x, y) = -g - (k/m)y|y|, \mathcal{M}y = y' + (k/m)y|y|$,
$r(x) = -g$.

9. Linear, nonhomogeneous; $f(x, y) = -0.0002(y - 5)$; $\mathcal{M}y = y' + 0.0002y, r(x) = 0.0002 \cdot 5$.

11. Linear, nonhomogeneous; $f(x, y) = x^2 - 1$; $\mathcal{M}y = y'$, $r = x^2 - 1$.

13. Linear, homogeneous; $f(x, y) = ky, \mathcal{M}y = y' - ky$, $r(x) = 0$.

15. The equation $T' = -(Ak/cm)(T - T_{out})$ is linear $(a_1 = 1, a_0 = Ak/cm, r = AkT_{out}/cm)$ and nonhomogeneous (unless $T_{out} = 0$);
$f(x, y) = -(Ak/cm)(y - T_{out}); \mathcal{M}y = y' + (Ak/cm)y$,
$r(x) = (Ak/cm)T_{out}$.

17. Linear, nonhomogeneous $(a_1 = 1, a_0 = 3, r = -x)$;
$f(x, y) = -3y - 3x, \mathcal{M}y = y' + 3y, r(x) = -3x$.

19. Linear, homogeneous $(a_1 = 1, a_0 = x + 3, r = 0)$;
$f(x, y) = -xy - 3y, \mathcal{M}y = y' + (x + 3)y, r(x) = 0$.

21. Nonlinear, homogeneous; $f(x, y) = y^2 - y$;
$\mathcal{M}y = y' - y^2 + y, r(x) = 0$.

23. Linear, nonhomogeneous; $f(x, y) = (\sin \pi x)/x^3$,
$\mathcal{M} = x^3 y' - \sin \pi y, r(x) = 0$.

25. $dT/dt + (Ak/cm)T = (Ak/cm)T_{out}$. Replace $(Ak/cm)T_{out}$ with 0 to obtain homogeneous form.

27. $T_{out} = 0, k = 0, A = 0$, no values of c and m.

29. (a) $a_1 \leftrightarrow 1, a_o \leftrightarrow (Ak/cm), r \leftrightarrow (Ak/cm)T_{out}$,
$y \leftrightarrow T, x \leftrightarrow t$.

(b) Same as part (a) with $r = 0$.

(c) $T = 0 \Rightarrow 0 = (Ak/cm)T_{out} \Rightarrow$ nonhomogeneous (See Exercise 27).

(d) Substitute $T = 0$ to obtain $0 = 0$; homogeneous.

(e) Substitute $T = T_{out}$ into (4.4).

(f) Substitute $T = e^{-Akt/cm}$ into (4.5).

31. 0.

33. (a) $y' + y = 0$, linear, homogeneous; $y' + y = x$, linear, nonhomogeneous.

(b) Substitute $y = 0$, obtain $0 = r(x)$; have trivial solution only if $r(x) \equiv 0$.

35. (a) $P = Ce^{kt} = $ (constant) $\times$ (nontrivial solution of homogeneous).

(b) $P = P_i e^{k(t - t_0)}$.

37. (a) $v = -gt + C$; $v_p = -gt, v_h = 1$.

(b) $v = -gt + v_i$.

39. $dP_h/dt = kP_h; dP_p/dt = kP_p - E$.

41. (a) $(E/k)' = k(E/k) - E$.

(b) $P' = kP; (e^{kt})' = ke^{kt}$.

(c) $(E/k + Ce^{kt})' = k(E/k + Ce^{kt}) - E$.

43. (b) $dT_{out}/dt = -(Ak/cm)(T_{out} - T_{out}) \Rightarrow 0 = -(Ak/cm)(0)$.

(c) $T_h = Ce^{-Akt/cm}$.

(d) Substitute:
$(Ce^{-Akt/cm})' + (Ak/cm)(Ce^{-Akt/cm}) = 0$. Reduces to $0 = 0$.

(e) $T_g = Ce^{-Akt/cm} + T_{out}$.

45. (a) *General solution* is defined only for linear equations.

(b) $y = Ce^x + \sqrt{x}$.

(c) $y = Ch(x) + p(x)$.

(d) Insufficient information—no homogeneous solution is given. (But separation of variables quickly yields $y_h = e^x$. Then a general solution is $y_g = (x + C)e^x$.)

47. Use
$y'(x) = C \exp\left(-\int_0^x (a_0(s)/a_1(s))\, ds\right)(-a_0(x)/a_1(x))$.

49. Let $y' = f(x, y)$ be a homogeneous differential equation. Then $0 = f(x, 0)$. Now suppose $f(x, 0) = 0$. Substitute the trivial solution $y \equiv 0$:
$y' = 0 = f(x, 0) = f(x, y)$. Hence, $y' = f(x, y)$ is homogeneous because it has the trivial solution.

Section 4.2

1. $y = -1/(\sin x + C), \sin x + C \neq 0$.

3. $P = P_i e^{k(t - t_0)}$ for all t.

5. $T = (T_i - T_{out})e^{-Akt/cm} + T_{out}$ for all t.

7. $P = (8 - .209/0.015)e^{0.015(t - 1847)} + .209/0.015$ for all t.

9. $y = Ce^{2x} - 1/2$ for all x.

11. $\sqrt{2} \arctan(\sqrt{2}y) = 2x + C$.

15. (a) $\mu = Ce^{x^2/2}$.

    (b) $\mu = Ce^{\sin x}$.

    (c) $\mu = x^2$.

    (d) $\mu = Ce^{Akt/cm}$.

    (e) $\mu = e^{\int g(x)dx}$.

17. (a) Substitute given expression for y.

    (b) $3\cos x + 8 \neq 0, \infty \Rightarrow y \neq 0$.

    (c) Find $y = (3\cos x + 1/8)^{-1/3}$ — undefined when $\cos x = -1/24$.

19. Use partial fractions for integral;

$$\frac{1}{aP - sP^2} = \frac{A}{P} + \frac{B}{a - sP}, \quad A = 1/a,\ B = s/a.\ \text{See}$$

exercise 3 of the appendix.

21. (a) $P_g = Ce^{kt} + E/k$.

    (b) $P_g = (P_i - E/k)e^{k(t-t_0)} + E/k$.

    (c) $P_i > E/k \Rightarrow P' = kP - E > 0 \Rightarrow P$ increasing.
$P_i < E/k \Rightarrow P' < 0$. Hence $P_{\text{crit}} = E/k$.

23. (a) Use $y' = C\exp\left(-\int_0^x (a_0/a_1)\,ds\right)(-a_0(x)/a_1(x))$.

    (b) From (a), $\exp\left(-\int_0^x (a_0/a_1)\,ds\right)$ is a nontrivial solution to homogeneous equation.

    (c) Choose $C = y_i$.

25. (a) If f is constant, separate as $\dfrac{a_1}{f - a_0 y}dy = dx$.

    (b) $y = Ce^{-a_0 x/a_1} + f/a_0$.

    (c) Choose $f = ka_0(x), a_0 \neq 0$.

Section 4.3

1. $r - 6 = 0, r = 6, y = Ce^{6x}$.

3. $2r + 7 = 0, r = -7/2, u_g = Ce^{-7t/2}, u = 6e^{-7(x+2)/2}$.

5. $r = 1, w = Ce^x$.

7. Nonhomogeneous.

9. $r - 3 = 0, r = 3, y_g = Ce^{3x}$. Then $y(2) = 4e^6$ forces $C = 4e^{-6}$ and $y = 4e^{3(x-2)}$.

11. Nonhomogeneous.

13. (a) Assume $y = e^{rx}$. Then
$b_1 y' + b_0 y = b_1 re^{rx} + b_0 e^{rx} = e^{rx}(b_1 r + b_0) = 0$ and
$b_1 r + b_0 = 0$.

    (b) If $y = Ce^{-b_0 x/b_1}$, then $y' = (-b_0/b_1)Ce^{-b_0 x/b_1}$, so
$b_1 y' + b_0 y = -b_0 Ce^{-b_0 x/b_1} + b_0 Ce^{-b_0 x/b_1} = 0$.

15. (a) Identify $a_0(x) = b_0, a_1(x) = b_1$.

    (b) $\mathcal{L}(C_1 y + C_2 z) =$
$b_1(C_1 y + C_2 z)' + b_0(C_1 y + C_2 z) =$
$C_1(b_1 y' + b_0 y) + C_2(b_1 z' + b_0 z) = C_1 \mathcal{L}y + C_2 \mathcal{L}z$.

17. (a) From solving $2P_i = P_i e^{kt}$ for t, the time to double is $(\ln 2)/k$; the time constant is $1/k$. The time constant for the world's population is about 61 years; the time to double is about 42 years. The corresponding values for Indonesia are 46 and 32 years.

    (b) The time constant is cm/Ak; e.g., increasing area A decreases the time constant, which reflects the decrease in the time required for the house to cool in the absence of a furnace. Increasing insulation decreases thermal conductivity k and increases the time constant.

19. Assume $y = e^{rx}$, where r is a constant. Substitute and attempt to cancel e^{rx}: $r - 2 = x^3 e^{-rx}$; r is not constant, contrary to assumption.

21. Assume $y = e^{rx}$, where r is a constant. Substitute and attempt to cancel e^{rx}: $r + 4e^{-rx}\sin e^{rx} = 0$; r is not constant, contrary to assumption. The differential equation is homogeneous because it has the trivial solution: $0' + 4\sin 0 = 0$.

Section 4.4

1. $4e^{\pi x}/(\pi - 4)$.

3. Undetermined coefficients is not appropriate for a tangent forcing term.

5. $-(4x + 1)/4$.

7. Nonconstant coefficients.

9. Homogeneous.

11. $-3(2x + 1)/4$.

13. Nonlinear.

15. Nonlinear.

17. $t - 3$.

19. (a) (1) $u_h = Ce^{4x}$;
(5) $u_h = Ce^{4x}$;
(11) $y_h = Ce^{2x}$;
(17) $x_h = Ce^t$.

    (b) (1) $4e^{\pi x}/(\pi - 4) + Ce^{4x}$;
(3) $e^{4x}\left(4\int_0^x e^{-4s}\tan(\pi s)\,ds + C\right)$;
(5) $Ce^{4x} + x + 1/4$; (11) $Ce^{2x} - 3x/2 - 3/4$;
(17) $Ce^t + t - 3$.

    (c) (1) $C = 1 - 4/(\pi - 4)$;
(3) $C = 1$;
(5) $C = 5/4$;
(11) $C = 7/4$;
(17) $C = 4$.

21. (a) $Ax^4 + Bx^3 + Cx^2 + Dx + E + Fe^{3x} + G\cos 3x + H\sin 3x$.

    (b) $Ax^4 + Bx^3 + Cx^2 + Dx + E + Fxe^{-3x} + G\cos 3x + H\sin 3x$. (Same as (a).)

    (c) $Ax^4 + Bx^3 + Cx^2 + Dx + E + Fxe^{-3x} + G\cos 3x + H\sin 3x$. (Same as (a).)

    (d) $Ax^2 + Bx + C + D\sin 2x + E\cos 2x + F\sin(\pi x/4) + G\cos(\pi x/4)$.

    (e) $Ax^3 + Bx^2 + Cx + D + Exe^{-3x}$.

    (f) $Be^{kt} + C\sin kt + D\cos kt$.

23. Since $b_1 y' + b_0 y = 0$, the characteristic equation is $b_1 r + b_0 = 0$. Hence, $r = -b_0/b_1$ and $y = Ce^{-b_0 x/b_1}$ if $b_0 \neq 0$ and $y = C$ if $b_0 = 0$.

25. Assume $y_p = Ae^{4x}$ and substitute:
$4Axe^{4x} + Ae^{4x} = e^{4x}$. Collecting terms yields
$e^{4x}(4Ax + A) = e^{4x}$. Equating coefficients of like terms gives two equations for A, $4A = 0$ (from coefficients of xe^{4x}), $A = 1$ (from coefficients of e^{4x}). No value of A satisfies these two equations simultaneously.

27 (a) (i) Guess $y_p = (Ax + B)e^x$;

where $a = RC\omega$ and D is an arbitrary constant. (Compare exercise 21.)

   (c) Choose $D = q_i + ACa^2/(1 + a^2)$.

   (d) The charge oscillates with frequency ω.

38. (a) Use undetermined coefficients, for example: $q_p = Fe^{-\alpha t}$, where $F = AC/(1 - \alpha RC)$.

39. The period of T_{out} is $2\pi/(\pi/12) = 24$ h;

$$T(t) = \frac{Ak}{cmD}\left(L\pi \sin\frac{\pi t}{12} + \frac{12Ak}{cm}\cos\frac{\pi t}{12}\right)$$
$$+ \ell + L/2 + Ce^{-Akt/cm},$$

where $D = 24(Ak/cm)^2 + \pi^2/6$ and

$$C = T_i - \ell - \frac{288(Ak/cm)^2 + \pi^2}{288(Ak/cm)^2 + 2\pi^2}L.$$

41. Decay is obvious: $y' = -ky$ forces y' to have the sign opposite y. There is a steady state $y_{ss} = 0$. Suppose some solution $Y(t)$ with $Y(0) \neq 0$ reaches $Y = 0$ at some time t_0. Then y_{ss} and $Y(t)$ are two distinct solutions satisfying the same initial condition, $y(t_0) = 0$. Uniqueness theorems 3 or 5 are violated. (Theorem 5 obviously applies because $f(t, y) = -ky$ and $\partial f/\partial y = -k$ is defined for all y.)

CHAPTER 5

Section 5.1

1. (a) $2x'' + 122.5x = 0$, $x(0) = -0.04$, $x'(0) = 0$.

   (b) $7x'' + 0.9x' + 122.5x = 0$, $x(0) = 0$, $x'(0) = 0.2$.

   (c) $2x'' + 122.5x = 9.335 \sin(3\pi t/2)$, $x(0) = -0.04$, $x'(0) = 0$. (Use (5.5) with $k = 122.5$ N/m, $h = 3\sin(3\pi t/2)$ in $= 0.0762\sin(3\pi t/2)$ m.)

3. $x = \sin(\sqrt{k/m}\, t)$, $\cos(\sqrt{k/m}\, t)$, $r = \sqrt{k/m}$.

5. (a–b) Graph shows one period of sine function.

   (c) Since displacement is a solution of this differential equation, both the function and the mass should have same period, $2\pi\sqrt{m/k}$.

   (d) Mass oscillates faster with stiffer spring, faster with smaller mass. Replace springs of rapidly bouncing car with weaker springs or fill the trunk with rocks to increase period. (But *amplitude* of oscillations may change, leading to other problems.)

7. (a) $mx''(t) + px'(t) + kx(t) = k(h(t) - h(0))$, $x(0) = x_i$, $x'(0) = v_i$.

   (b) $F_g = -mg$, $F_s = kz = k(h(t) - h(0) - x(t) + mg/k)$, $F_d = -px'(t)$; total force $F_t = F_g + F_s + F_d \Rightarrow mx''(t) + px'(t) + kx(t) = k(h(t) - h(0))$, $x(0) = x_i$, $x'(0) = v_i$.

9. $F_s = -kx$.

11. Use $h(t) = h_0 + (1/4)\sin(3\pi t/2)$ ft.

13. (a) 0.5 inch above the fixed position.

   (b) $2x'' + 8gx/3 = (4g\sin\omega t)/3 - 2g$, $x(0) = 0$ in, $x'(0) = 0$ in/s; $x = 0$ is the rest position without quarters.

15. Verify directly that $d(mv^2(t)/2) = mv(t)v'(t)$; similarly for $kx^2(t)/2$. At $t = 0$, find $C = mv^2(0)/2 + kx^2(0)/2$. Since the sum of kinetic and potential energy is constant over time, energy is conserved.

17. Assuming $x = e^{rt}$, r constant, and substituting in $3x''/16 + 1.41x' + 2x = 0$ yields $3r^2/16 + 1.41r + 2 = 0$. The roots of this quadratic are $r = -1.8744, -5.6456$ so the differential equation has the decaying exponential solutions $e^{-1.8744t}$ and $e^{-5.6456t}$. We have not *proven* that every solution is a decaying exponential. Indeed, this homogeneous equation has the trivial solution $x \equiv 0$, which is not a decaying exponential.

19. The amplitude grows as the forcing frequency ω approaches the so-called *natural frequency*, $\sqrt{2/(1/8)} = 4$ rad/s, of the solutions $\sin 4t$ and $\cos 4t$ of $x''/8 + 2x = 0$, the equation that governs the unforced system. (Guess these solutions as in section 5.1.5, page 212. Confirm them by substituting into the differential equation.)

Section 5.2

1. $L \to m$, $g \to k$, $\theta \to x$, $\theta' \to x'$, $\theta'' \to x''$, $\theta_i \to x_i$, $\omega_i \to v_i$; for example, initial linear velocity v_i and initial angular velocity ω_i are analogous.

3. Substitute to verify solution claims.

5. (a) Accurate to within 0.001.

   (b) 1.752 s.

7. Measuring angle ϕ in degrees, the displacement of the mass along the arc of its path is $s = L\pi\phi/180$. Then follow the derivation in the text to obtain $L\phi'' + (g180/\pi)\sin(\pi\phi/180) = 0$.

9. $L\theta'' + p\theta' + g\theta = 0$, $\theta(0) = \theta_i$, $\theta'(0) = \omega_i$.

Section 5.3

1. $L \to m$, $R \to p$, $C \to 1/k$, $E \to k(h(t) - h(0))$, $q \to x$, $q' \to x'$; for example, resistance is like damping.

3. (a) Substitute $i = \sin(t/\sqrt{LC})$ and $i = \cos(t/\sqrt{LC})$ into the equation.

   (b) Substitute $i = C_1 \sin(t/\sqrt{LC}) + C_2 \cos(t/\sqrt{LC})$ into the equation.

   (c) $C_1 = v_L(0)\sqrt{C/L}$ and $C_2 = i_i$.

5. Kirchhoff's voltage law $\Rightarrow V_R + V_C + E = 0$. Substituting the experimental fact yields $R(dq/dt) + q/C = -E$.

6. Following the solution of the previous problem, add the imposed emf E to Kirchhoff's voltage law to obtain $V_R + V_C - E = 0$. Substituting the experimental fact as before yields $R(dq/dt) + q/C = -E$ with initial condition $q(0) = q_i$.

7. See exercise 5 of chapter exercises in chapter 2.

8. Treat as series RC circuit as in figure 5.23 with potential E between kite and ground, resistance R in kite string, capacitance C of Leyden jar. Model is that of exercise 6. Steady-state solution is $q = EC$. Measure charge in Leyden jar to estimate $E = q/C$.

Chapter exercises

1. (a) If we chose up as the positive direction for this coordinate system and place its origin at the location of the end of the spring *before* the mass is suspended from it, then the extension z appearing in Hooke's law is just $-y$. Hence, the force exerted by the spring is $F_s = -ky$. The force of gravity on the mass remains $F_g = -mg$. Using Newton's law and $a = y''$, we then employ the usual logic: $F_t = F_s + F_g$, or $ma = -ky - mg$, or $my'' + ky = -mg$. The two initial conditions $y(0) = y_i$, $y'(0) = v_i$ complete this second-order model.

(b) Use $y(t) = x(t) - x_u$. Then substitute $y' = x'$, etc. into the differential equation for y to obtain the equation for x.

2. Same as forced spring mass system with forcing $h(t) = h_0 + b \sin(St/L)$. (Car traveling at velocity S crosses S/L periods of the bumps in the road per unit time; i.e., the frequency of the forcing term is S/L.)

3. Use Kirchhoff's current law to write $i_R + i_C + i_L = 0$. Differentiate once to obtain equation.

5. The term $g \sin \theta$ must represent the spring force for a displacement of θ. But $\sin \theta$ varies periodically with θ, while spring force varies linearly with extension. In particular, spring force does not decrease and then change sign with increasing displacement.

CHAPTER 6

Section 6.1

1. Linear, homogeneous, constant coefficients.
3. Linear, nonhomogeneous, constant coefficients.
5. Linear, homogeneous, constant coefficients.
7. Nonlinear, homogeneous.
9. Linear, nonhomogeneous, variable coefficients.
11. Both equations have constant coefficients.
13. Substitute given solution.
15. (a) y_{ss} satisfies only the nonhomogeneous equation. It is a particular solution. The steady state satisfies $y(0) = -1.96$, $y'(0) = 0$.

(b) y_{tran} satisfies the homogeneous equation.

(c) Use y_{tran} as a homogeneous solution and y_{ss} as a particular solution to construct a general solution.

(d) $\lim_{t \to \infty} y(t) = -1.96$ for all values of C_1, C_2.

17. Mass of 1 kg suspended vertically from a spring in a damping medium; spring constant is 5 N/m; damping coefficient is 2 N·s/m; origin of coordinates is at the end of the spring after the mass is suspended from it. The mass is raised 9 m above its rest position and released with

downward velocity 47 m/s. The upper end of the spring is being displaced $h \sin 4t$ m about its original position, where $h = 185/k = 185/5 = 37$ m(!).

19. Saying applies to second derivatives as well. Roughly, superposition says, "a linear differential expression acting on a sum is the sum of the differential expressions."

Section 6.2

1. (a) $W = \omega \neq 0 \Rightarrow$ linearly independent for all x.

(b) $W = \sin x (1 + 2 \cos^2 x) = 0$ at $x = 0, \pm \pi, \ldots$. Hence, Wronskian guarantees linear independence only for $x \neq \pm n\pi$, n an integer. Apply Wronskian test in any open interval between these points to evaluate $C_1 = C_2 = 0$ and conclude linear independence for all x.

(c) Linearly dependent since $W(\cos 2x, \sin(2x - \pi/2)) = 0$. (In fact, $\sin(2x - \pi/2) = \cos 2x$.)

(d) $W = (s - r)e^{(r+s)x} \neq 0 \Rightarrow$ linearly independent for all x.

(e) $W = e^{2rx} \neq 0 \Rightarrow$ linearly independent for all x.

(f) Linearly dependent since $W(e^{rx}, e^{r(x-1)}) = 0$. (In fact, $e^{r(x-1)} = e^{-1} e^{rx}$.)

3. (a) (i) $W = -2 \neq 0 \Rightarrow$ linearly independent for all x; (iii) $y_g = C_1 e^x + C_2 e^{-x}$.

(b) (i) $W = -2e \neq 0 \Rightarrow$ linearly independent; $e^{1-x} = e \cdot e^{-x}$;

(ii) Members of this pair of functions are constant multiples of those in part (a); (iii) $y_g = C_1 e^x + C_2 e^{1-x}$.

(c) (i) $W = 0 \Rightarrow$ linearly dependent;

(ii) $e^{1-x} = e \cdot e^{-x}$, so these functions are constant multiples of one another and, hence, linearly dependent.

5. $x_g = C_1 x^2 + C_2 x^{-2}$, $x \neq 0$; $x_g = 1 + C_1 x^2 + C_2 x^{-2}$, $x \neq 0$.

7. (a) Substitute, using fundamental theorem of calculus to evaluate $W'(x)$.

(b) Use separation of variables.

9. Show that the exponential expression for W is either identically zero or never zero.

11. Theorem 3 asserts that the Wronskian of smooth solutions is either never zero or always zero. Theorems 2 and 3 together assert linear independence if $W \neq 0$ at one point. Theorem 4 asserts linear dependence of solutions if $W = 0$ at one point.

Section 6.3

1. $y = -e^{2x} + 2e^{-x}$.
3. $y = (1/2 - \sqrt{2}/4)e^{(\sqrt{2}-1)x} + (1/2 + \sqrt{2}/4)e^{(-\sqrt{2}-1)x}$.
5. $w = 2e^{x/2}$.
7. $y = (7x - 3)e^x$.
9. $u = (-4/11)e^{(x-1)/4} + (15/11)e^{2(1-x)/3}$.
11. $y'' - y = 0$.
13. $y'' - 2y' + y = 0$.
15. $y'' - 3y' + 2y = 0$.

17. $y'' - y' = 0$; choose second solution as a constant.

19. $W = (r_2 - r_1)e^{(r_1 + r_2)x} \neq 0$ if and only if $r_1 \neq r_2$, whether r_i are real or complex.

21. Use $b_1^2 - 4b_2b_0 = 0$ to write $r = -b_1/2b_2$.

23. (a) $b_1^2 - 4b_0b_2 \neq 0 \Rightarrow$ linearly independent solutions are $\exp\left(-b_1 + \sqrt{b_1^2 - 4b_2b_0})x/2b_2\right)$,

$\exp\left(-b_1 - \sqrt{b_1^2 - 4b_2b_0})x/2b_2\right)$.

   (b) $b_1^2 - 4b_0b_2 = 0 \Rightarrow$ solutions are $\exp(-b_1x/2b_2)$, $x \exp(-b_1x/2b_2)$.

25. $u'' = 0 \Rightarrow u = C_1x + C_2$; choose $C_1 = 1, C_2 = 0$.

Section 6.4

1. (i) $r^2 - 9 = 0$;
(ii) e^{3x}, e^{-3x};
(iv) $y = C_1e^{3x} + C_2e^{-3x}$.

3. (i) $r^2 + 9 = 0$;
(ii) $\sin 3x, \cos 3x$;
(iv) $w = C_1 \sin 3x + C_2 \cos 3x$.

5. (i) $r^2 - 4r + 40 = 0$;
(ii) $e^{2x} \cos 6x, e^{2x} \sin 6x$;
(iv) $y = e^{2x}(C_1 \cos 6x + C_2 \sin 6x)$.

7. (i) $r^2 + r - 6 = 0$;
(ii) e^{-3t}, e^{2t};
(iv) $x = C_1e^{-3t} + C_2e^{2t}$.

9. Nonlinear.

11. Nonhomogeneous.

13. Nonlinear.

15. (i) $mr^2 + pr + k = 0$;
(ii) $C_1e^{-pt/2m}, C_2te^{-pt/2m}$;
(iv) $x = C_1e^{-pt/2m} + C_2te^{-pt/2m}$.

17. (i) $u_g = e^{2t}(C_1 \cos 6t + C_2 \sin 6t)$;
(ii) $u_g = e^{2t}(-3 \cos 6t + (1 - e^{\pi/3}) \sin 6t)/3e^{\pi/3}$.

19. (i) $y_g = e^{-x}(C_1 \sin 2x + C_2 \cos 2x)$;
(ii) $y = -2e^{\pi - x} \sin 2x$.

21. (i) $x_g = C_1 \sin 3t + C_2 \cos 3t$;
(ii) $x = 3 \sin 3t - 3 \cos 3t$.

23. (i) $x_g = C_1e^{3t} + C_2e^{-3t}$;
(ii) $x = 12e^{3t} + 6e^{-3t}$.

25. (a) (i) If $p^2 - 4mk > 0$,

$$x_g = C_1 \exp\left(\left(-p + \sqrt{p^2 - 4mk}\right)t/2m\right) +$$
$$C_2 \exp\left(\left(-p - \sqrt{p^2 - 4mk}\right)t/2m\right).$$

(ii) If $p^2 - 4mk = 0$, $x_g = C_1e^{-pt/2m} + C_2te^{-pt/2m}$.
(iii) If $p^2 - 4mk < 0$, $x_g = e^{-pt/2m}(C_1 \cos \beta t + C_2 \sin \beta t)$, where $\beta = \sqrt{4mk - p^2}/2m$.

   (b) The factor e^{-rt}, $r > 0$, appears in every case, forcing decay to zero as t grows.

   (c) It oscillates for $p^2 - 4mk < 0$. Frequency $\beta = \sqrt{4mk - p^2}/2m$ is different than that of the undamped model.

(d) (i) If $p^2 - 4mk > 0$,
$C_1 = x_i(1 + (p - B)/2B) + mv_i/B$,
$C_2 = -mv_i/B + x_i(-p + B)/2B$, where $B = \sqrt{p^2 - 4mk}$.
(ii) If $p^2 - 4mk = 0$, $C_1 = x_i - (v_i + \alpha x_i)/(\alpha + 1)$,
$C_2 = (v_i + \alpha x_i)/(\alpha + 1)$, where $\alpha = p/2m$.
(iii) If $p^2 - 4mk < 0$, $C_1 = 2m(v_i + px_i)/\beta$, $C_2 = x_i$,
where $\beta = \sqrt{4mk - p^2}$.

27. (a) $C_1e^{3.27it} + C_2e^{-3.27it}$.
   (b) $C_1 = -0.076i, C_2 = 0.076i$.
   (c) Use values of C_1, C_2 above and
$e^{\pm 3.27it} = \cos 3.27t \pm i \sin 3.27t$.

29. Use $C_1 = C_2 = 1/2$ and Euler's formula to obtain $e^{\alpha x} \cos \beta x$ term; $C_1 = -C_2 = 1/2i$ for sine term.

31. Homogeneous equation has nonzero steady state if its characteristic polynomial has root $r = 0$; always has zero steady state (trivial solution); stable if all roots have negative real part.

33. $b_1/b_2 > 0$.

Section 6.5

1. (a) The spring constant is $k = 0.1 \cdot 9.8/0.04$ $= 24.5$ N/m. The governing initial-value problem is $0.6x'' + 24.5x = 0, x(0) = 0.06, x'(0) = 0$. Its solution is $x(t) = 0.06 \sin(6.39t + \pi/2)$.

   (b) The governing initial-value problem is $2x'' + 24.5x = 0, x(0) = 0.07, x'(0) = 0$. Its solution is $x(t) = 0.07 \sin(3.5t + \pi/2)$.

3. (a) No. The period of motion is independent of initial velocity.

   (b) In magnitude, yes. Amplitude depends upon both initial velocity *and* initial position, which is specified as $x(0) = 0$ (released from rest):
$A^2 = 14.1^2 = x_i^2 + (v_i/\omega_n)^2 = (v_i/\omega_n)^2$. Then $v_i = \pm|A\omega_n|$.
For the given data, compute $\omega_n = \sqrt{k/m} = \sqrt{24.5/2} = 3.5$ rad/s. Then $v_i = \pm|A\omega_n| = \pm 50.75$ cm/s.

4. $4 \exp(-pt/2m) = \exp(-4 \cdot 60t/2 \cdot 2) = 5.4/10 \Rightarrow p = 0.01$ N-m/s

5. (a) Since the general solution is $x_g = C_1 \cos \omega_n t + C_2 \sin \omega_n t$, the solution will be the sum of two simple periodic functions, regardless of the values of the arbitrary constants C_1, C_2, which are determined by the initial conditions.

   (b) The period of the sine and cosine appearing in this general solution is $T = 2\pi/\omega_n = 2\pi/\sqrt{k/m}$.

   (c) When $x_i = v_i = 0$, we obtain only the trivial solution $x \equiv 0$.

7. (a) The system will oscillate slower, provided it is underdamped.

   (b) $k \leq 4$.

9. (a) The roots of the characteristic equation are real and distinct. Verify that the system is overdamped by showing $p^2 - 4mk > 0$.

(b) No. Since x starts positive and decays to zero, it can become negative only if it has a (negative) local minimum. But letting

$x'(t) = -1.13 \cdot 1.90e^{-1.90t} + 0.38 \cdot 2.83e^{-2.83t} = 0$ forces

$$\frac{1.13 \cdot 1.90}{0.38 \cdot 2.83} = 2.00 = e^{-0.93t},$$

a relation that holds for no nonnegative value of t.

(c) Yes; use an analysis similar to part (b).

(d) For example, $x_i = -0.75$, $v_i = 0$, or $x_i = 0$, $v_i = -0.25$.

(e) For $x_i > 0$, negative displacement is impossible; for $x_i < 0$, any v_i would yield some negative displacement; for $x_i = 0$, v_i must be negative.

11. Since $r < 0$, both e^{rt} and te^{rt} decay to zero. (For the second exponential, apply l'Hôpital's rule to the ratio t/e^{-rt}.)

13. Since $A \sin(\omega t + \phi) = A \sin \omega t \cos \phi + A \cos \omega t \sin \phi$, we find $C_1 = A \sin \phi$, $C_2 = A \cos \phi$. To obtain just $\cos \omega t$, choose $\phi = \pm\pi/2, \ldots$ to force $C_2 = 0$, etc.

15. Critical points of $x(t) = C_1 \cos \omega t + C_2 \sin \omega t$ occur when $x'(t) = \omega(-C_1 \sin \omega t + C_2 \sin \omega t) = 0$, or when $\tan \omega t = C_1/C_2$. One such value of ωt corresponds to $\cos \omega t = C_1/\sqrt{C_1^2 + C_2^2}$, $\sin \omega t = C_2/\sqrt{C_1^2 + C_2^2}$. Hence, the maximum, or amplitude, of $x(t)$ is $\sqrt{C_1^2 + C_2^2}$.

17. A shift of ϕ rad corresponds to a time shift of $t = \phi T/2\pi$, where T is the period.

19. $3.76 = p/2m = (1.41)16/6$, $2.68 = \omega = \sqrt{4mk - p^2}/2m = \sqrt{3 \cdot 16/16 - (1.41)^2}/(2 \cdot 3/16)$. Use

the formula for the sine of a sum of angles to expand $\sin(2.68t + \phi)$ to show the equivalence of the two solutions; $A = 1.3$, $\phi = 35.3° = 0.62$ rad.

21. $A = \sqrt{C_1^2 + C_2^2}$, $\tan \phi = C_1/C_2$. (Compare with exercise 19.)

23. (a) Period $T = 2\pi/\sqrt{g/L} \Rightarrow L = g(T/2\pi)^2 = 9.8 \cdot (2/2\pi)^2 = 0.99295\cdots$ m.

25 (i) Underdamped: $R^2 - 4L/C < 0$.
(ii) Critically damped: $R^2 - 4L/C = 0$.
(iii) Overdamped: $R^2 - 4L/C > 0$.

27. See table S.5.

29. See table S.6.

31 (i) Underdamped: $1/R^2 - 4C/L < 0$.
(ii) Critically damped: $1/R^2 - 4C/L = 0$.
(iii) Overdamped: $1/R^2 - 4C/L > 0$.

Section 6.6

1. (i) $y_p = Ax + B$.
(ii) $y_p = (2x - 14)/9$.
(iii) $y_g = C_1 e^{3x} + C_2 e^{-3x} + y_p$.
3. (i) $y_p = Ae^{-2x}$.
(ii) $y_p = -e^{-2x}/5$.
(iii) $y_g = C_1 e^{3x} + C_2 e^{-3x} + y_p$.
5. (i) $y_p = A + Bxe^{-3x}$.
(ii) $y_p = -4/9 + xe^{-3x}/6$.
(iii) $y_g = C_1 e^{3x} + C_2 e^{-3x} + y_p$.
7. (i) $y_p = Axe^{3x} + Bxe^{-3x} + Cx + D$.
(ii) $y_p = xe^{3x}/3 + xe^{-3x}/6 + 5x/9$.
(iii) $y_g = C_1 e^{3x} + C_2 e^{-3x} + y_p$.
9. (i) $x_p = A \sin(t/2) + B \cos(t/2)$.

TABLE S.5 **Section 6.5, exercise 27**

Physical analogies: spring-mass and parallel *RLC* circuit	
Spring-mass System	**Parallel *RLC* Circuit**
Inertia mx''	Rate of change Cv'' of capacitor current
Damping force px'	Rate of change v'/R of resistor current
Restoring force kx	Rate of change v/L of inductor current

TABLE S.6 **Section 6.5, exercise 29**

Physical analogies: series *RLC* circuit and parallel *RLC* circuit	
Series *RLC* Circuit	**Parallel *RLC* Circuit**
Inductive potential $Lq'' = Li'$	Rate of change Cv'' of capacitor current
Voltage drop $Rq' = Ri$ across resistor	Rate of change v'/R of resistor current
Capacitive potential q/C	Rate of change v/L of inductor current

(ii) $x_p = 3(\sin(t/2) - \cos(t/2))/4$.

(iii) $x_g = e^{-t}(C_1 \cos(t/2) + C_2 \sin(t/2)) + x_p$.

11. (i) The equation $w'' + 9w = x \sin 3x + 2\cos 4x$ has the homogeneous solutions $\cos 3x$, $\sin 3x$. Since the forcing term contains a polynomial of degree one multiplying a sine function as well as a cosine of different frequency, the first form for the particular solution is

$$x_p = (A_1 x + A_0)\cos 3x + (B_1 x + B_0)\sin 3x$$
$$+ D\cos 4x + E \sin 4x,$$

where A_1, A_0, B_1, B_0, D, E are coefficients to be determined. However, the terms $A_0 \cos 3x$, $B_0 \sin 3x$ are solutions of the homogeneous equation; they arise from the forcing term $x \sin 3x$. Hence, the part of the proposed particular solution which is due to that forcing term is multiplied by x. The final form of the particular solution is

$$x_p = x(A_1 x + A_0)\cos 3x + x(B_1 x + B_0)\sin 3x$$
$$+ D\cos 4x + E\sin 4x.$$

This expression contains no solutions of the homogeneous equation.

(ii) $w_p = (x \sin 3x)/36 - (x^2 \cos 3x)/12 - (2\cos 4x)/7$.

(iii) $w_g = C_1 \cos 3x + C_2 \sin 3x + w_p$.

13. (i) $w_p = (Ax + B)e^{-x}$.

(ii) $w_p = (5x + 1)e^{-x}/50$.

(iii) $w_g = C_1 \cos 3x + C_2 \sin 3x + w_p$.

15. (i) $w_p = A + x(B \sin 3x + C \cos 3x)$.

(ii) $w_p = 2/9 - (x \sin 3x)/6$.

(iii) $w_g = C_1 \cos 3x + C_2 \sin 3x + w_p$.

17. (i) $z_p = Ate^{3t}$.

(ii) $z_p = 6te^{3t}/5$,.

(iii) $z_g = C_1 e^{3t} + C_2 e^{t/2} + z_p$.

19. (i) $z_p = e^{t/2}(A \sin \pi t + B \cos \pi t) + Cte^{3t} + D$.

(ii) $z_p = e^{t/2}(-(2/R) \sin \pi t + (5/\pi R) \cos \pi t) + 6te^{3t}/5 - 1/3$, where $R = 25 + 4\pi^2$.

(iii) $z_g = C_1 e^{3t} + C_2 e^{t/2} + z_p$.

21. (i) $y_p = (Ax + B)e^{-2x}$.

(ii) $y_p = xe^{-2x}/36$.

(iii) $y_g = e^{-2x}(C_1 \cos 6x + C_2 \sin 6x)$.

23. (i) $x_p = At + B$.

(ii) $x_p = t/2 + 1/12$.

(iii) $x_g = C_1 e^{2t} + C_2 e^{-3t} + x_p$.

25. (i) $y_p = (Ax + B)e^{-2x/3}$.

(ii) $y_p = (-2x - 3)e^{-2x/3}/32$.

(iii) $y_g = C_1 e^{(-2+4\sqrt{2}/3)t} + C_2 e^{(-2-4\sqrt{2}/3)t} + y_p$.

27. (i) Homogeneous solutions are e^{-3x}, e^{2x}. An appropriate form for a particular solution is $w_p = xAe^{2x} + xBe^{-3x}$.

(ii) $w_p = (xe^{2x} + xe^{-3x})/5$.

(iii) $w_g = C_1 e^{2x} + C_2 e^{-3x} + w_p$.

29. Show that $-gt^2/2$ is a particular solution and that t and 1 are solutions of the homogeneous equation.

31. Show that $x^3 e^{-2x}$ is a particular solution and that e^{-2x}, xe^{-2x} are solutions of the homogeneous equation.

33. (a) $x_g = x_h + \dfrac{A(k - m\omega^2)}{(p\omega)^2 + (m\omega^2 - k)^2} \sin \omega t - \dfrac{Ap\omega}{(p\omega)^2 + (m\omega^2 - k)^2} \cos \omega t$, where:

(i) $x_h = C_1 \exp\left(-p + \sqrt{p^2 - 4mk}\,)t/2m\right) + C_2 \exp\left(-p - \sqrt{p^2 - 4mk}\,)t/2m\right)$ if $p^2 > 4mk$,

(ii) $x_h = (C_1 t + C_2)e^{-pt/2m}$ if $p^2 = 4mk$,

(iii) $x_h = e^{-pt/2m}(C_1 \cos \beta t + C_2 \sin \beta t)$ if $p^2 < 4mk$, where $\beta = \sqrt{4mk - p^2}/2m$. Determine C_1, C_2 from $x_g(0) = x_i, x_g'(0) = v_i$.

(b) The solution is not affected by whether or not $\omega = \sqrt{k/m}$. The solution is bounded for all time.

35. $y_{p1} = (3x/52 - 159/1352)\cos 2x - (15x/32 + 3/169)\sin 2x$,

$y_{p2} = (25/78)\cos 3x + (5/78)\sin 3x$,

$y_{p3} = (-x^4/20 - x^3/25 - 3x^2/125 - 6x/625)e^{-2x}$,

$y_{p4} = (-x^3/4 - 9x^2/16 - 39x/32 - 153/128)e^{2x}$,

$y_{p5} = x^4/6 - x^3/9 + 7x^2/18 - 13x/54 - \pi/6 + 55/324$,

$y_p = y_{p1} + y_{p2} + y_{p3} + y_{p4} + y_{p5}$.

37. Direct substitution of y_{p1}, y_{p2} makes the left-hand side zero. Direct substitution shows y_{p3} does not solve the homogeneous equation; y_{p1} and y_{p2} are unsatisfactory candidates for particular solutions because they are, in fact, solutions of the homogeneous equation.

39. $y_g = (C_1 x + C_2)e^{-2x} + y_p$.

41. (a) $A \cosh px + B \sinh px$.

(b) $(A_n x^n + \cdots + A_1 x + A_0) \cosh px + (B_n x^n + \cdots + B_1 x + B_0) \sinh px$.

(c) $Ae^{qx} \cosh px + Be^{qx} \sinh px$.

None of these entries add any truly new information because the hyperbolic sine and cosine are just linear combinations of exponentials.

Section 6.7

1. (a) A spring-mass system with mass of 2 gm (say), a damping coefficient of 3 dyne-s/cm and spring constant of 1 dyne/cm; it starts 1 cm above the equilibrium point with initial velocity of 2 cm/s downwards.

(b) A spring-mass system, similar to part (a), with the top of the spring oscillating with frequency ω and amplitude 5 cm.

2. (a) $k = mg/z = 0.1 \cdot 9.8/0.04 = 24.5$ N/m;

$\omega_r = \omega_n = \sqrt{k/m} = \sqrt{24.5/0.6} = 6.39$ rad/s.

(b) $\omega_r = \omega_n = \sqrt{24.5/2} = 3.50$ rad/s.

3. (a) ω_r does not depend on x_i.

(b) ω_r does not depend on v_i.

5. From exercise 4 of section 6.5, the damping coefficient is $p = 0.01$ N-m/s. Using the expression on page 280, the resonant frequency of this underdamped system is

$$\omega_r = \sqrt{\frac{k}{m} - \frac{p^2}{2m^2}} = \sqrt{\frac{24.5}{2} - \frac{(0.01)^2}{2 \cdot 2^2}} = 3.50 \text{ rad/s}.$$

Compare with exercise 2(b); the slight damping has negligible effect on the resonant frequency.

7. The form of the solution does not change at resonance because the system is damped. A particular solution is

$$x_p(t) = -\frac{F(p\omega \cos \omega t + (m\omega^2 - k) \sin \omega t)}{m^2 \omega^4 + (p^2 - 2km)\omega^2 + k^2}.$$

9. Supply the algebra omitted from example 43.

11. $x(t) = -\dfrac{2F\omega}{16 - \omega^2} \sin 4t + \dfrac{8F}{16 - \omega^2} \sin \omega t.$

13. (a) $k > (1.41)^2 \cdot 16/4 \cdot 3 \approx 8/3.$

   (b) Let $D = \sqrt{(1.41\omega)^2 + (k - 3\omega^2/16)^2}$. Then $A = F/D$.

   (c) Set $dA/d\omega = 0$ to find the forcing frequency which produces maximum amplitude. Then show that $d^2 A/d\omega^2 < 0$ at this maximal value of ω for $k > 8/3$. The resonant frequency is $\omega_r = 8\sqrt{3k/4 - 1.41^2/3}$. The system is overdamped for $k \leq 8/3$.

15. A particular solution of this equation is $\dfrac{8 \sin \omega t}{16 - \omega^2}$. To eliminate (in theory) the homogeneous solution from numerical solutions, use the initial conditions $x(0) = 0$, $x'(0) = \dfrac{8\omega}{16 - \omega^2}$. The resonant frequency is $\omega_r = \omega_n = 4$ rad/s. See table S.7 showing forcing frequencies and amplitudes.

TABLE S.7   Section 6.7, exercise 15

Forcing frequency ω	Amplitude of particular solution
$0.1\omega_r$	0.51
$0.5\omega_r$	0.67
$0.9\omega_r$	2.63
$1.1\omega_r$	2.38
$2\omega_r$	0.17
$4\omega_r$	0.03

A plot of amplitude vs. forcing frequency has the qualitative shape of figure 6.18. Theoretically, forcing this system at its resonant frequency would cause the amplitude of the particular solution to grow without bound. In practice, the device would disintegrate.

17. An overdamped system does not oscillate.

19. $\tan \phi = \dfrac{-1/R}{1/L - C\omega^2}.$

21. (a) Solve $e^{-Rt/2L} = 0.01$ for $t = 4.6$ μs.

   (b) Solve $e^{-Rt/2L} = \delta$ for $t = \ln(1/\delta)$ μs.

   (c) Solve $e^{-Rt/2L} = \delta$ for $t = (2L \ln(1/\delta))/R$.

23. Carry out the missing algebra or mimic example 44.

25. Yes; $317°$ and $-43°$ represent the same angle. The response lags the forcing by $43°$. For example, in figure 6.21 a response minimum occurs at $t = 6$ μs (lower graph) shortly *after* a minimum in the forcing (upper graph).

27. $A = 7.6 \times 10^{-3}$ V; it agrees with the lower graph of figure 6.20 after the transient has decayed; R, L, C, ω, and V affect the amplitude. Doubling V doubles the amplitude.

Section 6.8

1. The pendulum equation is nonlinear because of the $\sin \theta$ term. The approximate equation is linear with $b_2 = L$, $b_1 = 0$, and $b_0 = g$. They are both homogeneous.

3. $\theta(t) = \theta_i \cos(\sqrt{g/L} \, t) + \omega_i \sqrt{g/L} \sin(\sqrt{g/L} \, t).$ Frequency is the magnitude of the imaginary part of the solution of the characteristic equation, and it is independent of the initial conditions.

5. The "characteristic equation" $r^2 Le^{rt} + g \sin(e^{rt}) = 0$ can not be solved for a constant value of r. The method applies only to linear (constant-coefficient) differential equations; this equation is nonlinear.

7. (a) (i) The natural frequency $\omega_n = \sqrt{g/L}$ is independent of initial position θ_i and initial angular velocity ω_i.

(ii) Working from the solution formula found in exercise 3, the amplitude of the response, $A = \sqrt{\theta_i^2 + \omega_i^2 L/g}$, depends on θ_i, ω_i, L, and g.

   (b) The initial angular velocity ω_i is not necessarily zero; the pendulum is not simply released from rest.

9. Period $T = 2\pi \sqrt{L/g}$ increases on the moon. Amplitude $A = \sqrt{\theta_i^2 + \omega_i^2 L/g}$ also increases. For large oscillations, a nonlinear model must be used, but we expect the trend of increasing period to continue.

11. (a) $T = 2\pi \sqrt{L/g} = 2$ s agrees with the period values read from the graph.

   (b) This is a pendulum of length of 0.9930 m that is located on the surface of the earth. The linearized pendulum model is reasonable only if θ_i is small.

13. (a) When $\theta = 0.15$ rad, $\sin \theta = 0.1494$; $\sin \theta \approx \theta$ is in error by about 0.4%.

   (b) There is no apparent difference in the periods of the solutions of the linear and nonlinear problems.

   (c) The lack of apparent difference in periods is consistent with the accuracy of $\sin \theta \approx \theta$.

15. Define $\theta(t) = -\pi + q(t)$ and follow example 47 with $-\pi$ replacing π to find, after linearization, that q again satisfies $Lq'' - gq = 0$. Hence, $\theta_{ss} = -\pi$ is unstable. The straight-up position is unstable, regardless of the direction the pendulum is moved to reach it. (Look ahead to figure 7.5, page 337.)

17. Substitute $\theta(t) = \theta_{ss} + q(t)$ to find that a perturbation $q(t)$ of the single steady state $\theta_{ss} = 0$ satisfies $Lq'' + pq' + gq = 0$. Since all nontrivial solutions of this damped equation decay, $\theta_{ss} = 0$ is a stable steady state.

19. The damping term $p\theta'$ could represent the force of friction in the pivot or air resistance. Since $F = ma = mL\theta''$, in the form $L\theta'' + p\theta' + \cdots$, the term $p\theta'$ would have units of force per unit mass, such as N/kg or m/s^2. The coefficient p would involve the inverse of the mass of the pendulum bob.

21. (a) $q_g = C_1 \cos(\sqrt{g/L}\, t) + C_2 \sin(\sqrt{g/L}\, t)$; $q_g = C_1 e^{\sqrt{g/L}\, t} + C_2 e^{-\sqrt{g/L}\, t}$.

(b) Solutions of the first equation are always bounded because sine and cosine are bounded.

(c) The term $C_1 e^{\sqrt{g/L}\, t}$ guarantees exponential growth unless $C_1 = 0$, which happens only if initial conditions are such that $q'(0) = -\sqrt{g/L}\, p(0)$.

23. The steady states are $x_{ss} = 0, 1$. Perturbations defined by $x(t) = x_{ss} + q(t)$ satisfy $0 = q'' + q^2 - q \approx q'' - q$ (for $x_{ss} = 0$) and $0 = q'' + q^2 + q \approx q'' + q$ (for $x_{ss} = 1$). Since solutions of $q'' - q = 0$ include e^t, $x_{ss} = 0$ is an unstable steady state. Since solutions of $q'' + q = 0$ are $\sin t$, $\cos t$, $x_{ss} = 1$ is neutrally stable. (See the text of exercise 14, page 328, for a description of *neutral stability*.)

Chapter exercises

1. A particular solution is any function that satisfies the differential equation, whether it is periodic or otherwise.

3. (a) $W = x^2 \Rightarrow$ linearly independent for $x \neq 0$.

(b) $W = -2 \Rightarrow$ linearly independent for all x.

(c) Linearly dependent since $W(\ln|x|, \ln(x^2)) = 0$.

5. (a) $W = x^3 \Rightarrow$ linearly independent for $x \leq -2$.

(b) $W = 2x^3 \Rightarrow$ linearly independent for $x \leq -2$; $x(1-x)(1+x) = x - x^3$ so this pair is a linear combination of pair in part (a).

(c) Same as 5(b); $y_g = C_1 x + C_2 x(4 + x^2)$.

7. $y = C_1 e^{3x} + C_2 e^{-3x} - xe^{-3x}/6$.

9. $y = e^{2x}(C_1 \sin 6x + C_2 \cos 6x) - e^{-2x}/52 + e^{2x}/36$.

11. $z = (3 + \pi/36) \sin 3t + (\pi^2/12 - 3) \cos 3t + (t \sin 3t)/36 - (t^2 \cos 3t)/12$.

13. $w = C_1 e^{3t} + C_2 e^{t/2} - 1/3$.

15. (a) Need pure imaginary characteristic roots: $\alpha = 0$, $\beta > 0$.

(b) Need characteristic roots with negative real part: either $\alpha > 0$, $\alpha^2 - 4\beta < 0$ (complex roots) or $\alpha > 0$, $\alpha^2/4 \geq \beta > 0$ (negative real roots).

(c) Need real characteristic roots: $\alpha^2/4 - \beta \geq 0$.

17. (a) Need pure imaginary characteristic roots: $a^2 - 4b < 0$.

(b) Forcing term has period π, so need characteristic roots with negative real part; see 15(b).

(c) Need homogeneous solution with frequency $2\pi/4 = \pi/2$, so require pure imaginary characteristic roots with imaginary part $\pi/2$: $a = 0$, $b = (\pi/2)^2$.

(d) Need undamped system in resonance with forcing frequency 2, so pure imaginary characteristic roots with imaginary part 2: $a = 0$, $b = 4$.

19. $y_g = C_1 + C_2 e^{2x} + C_3 e^{-2x}$

21. $y_g = C_1 e^{3x} + C_2 \cos 2x + C_3 \sin 3x$

23. $y_g = e^{4t} + C_1 + C_2 e^{2t} + C_3 e^{-2t}$

25. $t = 0.46$ s.

27. Physically, the system is damped. Mathematically, the characteristic roots $r = (-p \pm \sqrt{p^2 - 4mk})/2m$ always have negative real part, so long as $m, p, k > 0$, forcing the solution to decay to zero.

29. Neglect damping and use $A = \sqrt{x_i^2 + (v_i/\omega_n)^2} = 0.13$ m, $x_i = 0$, $k = mg/z = 0.1 \cdot 9.8/0.04 = 24.5$ N/m, and $\omega_n = \sqrt{k/m} = \sqrt{24.5/0.6} = 6.39$ rad/s to find $x_i' = 0.83$ m/s. Had the blow been directed downward, the mass would have moved just as far.

31. $x'(0) = 0$.

33. The amplitude of an undamped system forced at its resonant frequency grows linearly with time until the spring breaks. The amplitude of a damped system attains a finite maximum amplitude at its resonant frequency.

35. (a) $x_p = \dfrac{F}{D^2}\left[(k - m\omega^2)\sin \omega t - p\omega \cos \omega t\right]$. Use the formula for the sine of a sum of angles to show $A = F/D$; see, for example, exercise 15 of section 6.5.

(b) Since $A = F/D$, A is maximum (at resonance) when D is minimum. Solve $d(D^2)/d\omega = 0$ to find the resonant frequency, the value of ω which minimizes D. (Note: $D_{min} = \sqrt{kp^2/m - p^4/4m^2}$.)

(c) As $p \to 0$, $\omega_r \to \omega_n = \sqrt{k/m}$; $\omega_r < \omega_n$.

(d) $\omega_r = 3.77$ rad/s.

37. See exercise 35(b): $\omega_r = \sqrt{1/LC - R^2/L}$.

39. (a) Use $\sin(\pm 2n\pi + p) = \sin p \approx p$ to show that perturbations of $\theta_{ss} = 0, \pm 2\pi, \pm 4\pi, \ldots$ are always governed by the linearized equation $p'' + p = 0$, whose bounded solutions imply neutral stability (defined in text of exercise 14, page 328).

(b) Use $\sin(\pm(2n-1)\pi + p) = -\sin p \approx -p$, $n = 1, 2, \ldots$ to show that perturbations of $\theta_{ss} = \pm\pi, \pm 3\pi, \ldots$ are always governed by the linearized equation $p'' - p = 0$, whose growing exponential solution implies instability.

41. $p^2 - 4mk = (d^2 - 1)/4mk$.

CHAPTER 7

Section 7.1

1. Find the characteristic roots of the corresponding linearized second-order equation, $L\theta'' + p\theta' + g\theta = 0$.

Conclude that the pendulum is underdamped if $p > 0$ and $p^2 - 4Lg \leq 0$ (guaranteeing complex characteristic roots), overdamped if $p > 0$ and $p^2 - 4Lg > 0$ (guaranteeing negative real characteristic roots).

3. (a) $(\theta_{ss}, \omega_{ss}) = ((0, \pm\pi, \pm 2\pi, ...), 0)$.

(b) $(\theta_{ss}, \omega_{ss}) = ((0, \pm\pi, \pm 2\pi, ...), 0)$.

(c) $(\theta_{ss}, \omega_{ss}) = (0, 0)$.

(d) $(x_{ss}, y_{ss}) = (0, 0)$.

(e) $(F_{ss}, R_{ss}) = (0, 0)$, $(b_R/\beta, d_F/\alpha)$.

(f) S_{ss} arbitrary, $I_{ss} = 0$

(g) $y_{ss}^1 = z_{ss}^1 = 0$; if $1 - ef \neq 0$, $y_{ss}^2 = (1 - e)/(1 - ef)$, $z_{ss}^2 = (1 - f)/(1 - ef)$.

4. (d) The maximum value of θ occurs at the 3 o'clock position. Since initial velocity is zero, the magnitude of the initial displacement is the amplitude.

5. (a) $\theta = \theta_i \cos(\sqrt{g/L}\, t)$, $\omega = -\sqrt{g/L}\, \theta_i \sin(\sqrt{g/L}\, t)$. Plotting these parametric equations yields a closed curve similar to that of figure 7.2.

(b) Time t increases clockwise.

(c) From exercise 5(a),

$$\left(\frac{\theta}{\theta_i}\right)^2 + \left(\frac{\omega}{\theta_i\sqrt{g/L}}\right)^2 = 1,$$

the equation of an ellipse in the (θ, ω) plane.

(d) See exercise 4(d).

(e) The trajectory drawn in this exercise is similar to that of figure 7.2.

7. (a) $\theta(t) = \theta_i \cos\sqrt{g/L}\, t + \dfrac{\omega_i}{\sqrt{g/L}} \sin\sqrt{g/l}\, t$,

$\omega(t) = -\theta_i\sqrt{g/L} \sin\sqrt{g/L}\, t + \omega_i \cos\sqrt{g/l}\, t$. Plotting these parametric equations yields a closed curve similar to that of figure 7.2.

(b) Time t increases clockwise.

(c) Treat the $\theta(t)$ and $\omega(t)$ equations in 7(a) as a system of equations for $\cos\sqrt{g/L}\, t$ and $\sin\sqrt{g/L}\, t$. Use Cramer's rule to solve for the sine and cosine. Set the sum of their squares equal to one to obtain the equation of an ellipse,

$$\theta^2 + \left(\frac{\omega}{\sqrt{g/L}}\right)^2 = A^2,$$

where the squared amplitude is $A^2 = \theta_i^2 + (\omega_i/\sqrt{g/L})^2$.

(d) The amplitude is the point at the 3 o'clock position where the curve crosses the positive θ-axis. To achieve amplitude A, choose θ_i, ω_i so that

$$A = \sqrt{\theta_i^2 + (\omega_i/\sqrt{g/L})^2}.$$

(e) The trajectory drawn in this exercise is similar to that of figure 7.2.

9. (a) When $\theta_i < 0$, then $\omega' = -(g/L) \sin\theta$ is positive near $t = 0$. The trajectory moves from the 9 o'clock position

up into the second quadrant as ω increases from its initial value of zero. The same trajectory is traced clockwise.

(b) The initial velocity is positive; the pendulum will swing to the right. Since θ must grow larger, the trajectory is traced in the clockwise direction.

(c) The initial velocity is negative; the pendulum will swing to the left. Since θ must grow smaller, the trajectory is traced in the clockwise direction.

(c) The velocity is negative, the pendulum will be swinging to the left. The trajectory is the same with a clockwise direction.

11. (a) $x' = v$, $mv' = -kx$.

(b) The solution trajectory is always an ellipse, as in figure 7.2, representing a periodic solution. The only stationary point is the origin. If the initial values are large, the spring breaks.

13. Clockwise; see the argument of exercise 9(a).

15. The trajectory is an ellipse around -2π traveled in a clockwise direction. Use an argument like that of exercise 9(a).

17. The trajectory is an ellipse around 2π traveled in a clockwise direction. Use an argument like that of exercise 9(a).

19. Clockwise. Use an argument like that of exercise 9(a).

21. Trajectories decay toward the $\omega = 0$ axis until they are close enough to begin a spiral around a stable equilibrium point. That equilibrium point is determined by the number of revolutions the pendulum makes before it stops rotating and begins swinging to and fro.

Section 7.2

1. Nullclines are $\omega_{\theta'=0} = 0$, $\theta_{\omega'=0} = 0$, i.e., the coordinate axes. As $\tilde{p} \to 0$, the $\omega' = 0$ nullcline in figure 7.9, page 344 (the dashed line in figure 7.9(a)) rotates clockwise until it is the vertical axis in the limit. The suggestion of closed trajectories is confirmed by setting $\tilde{p} = 0$ in (7.4).

3. Mimic exactly the steps between equations (7.3), page 343, and (7.4) with x in place of θ, v in place of ω.

5. (a) In the first quadrant, $\theta > 0$, $\omega > 0$. Hence, $\theta' = \omega > 0$ (direction field points to the right because horizontal component is positive), $\omega = -\theta < 0$ (direction field points down).

(b) The argument for the third quadrant mimics the first with signs reversed: direction arrows point left and up. Second and fourth quadrants are divided into two regions by $\omega' = 0$ nullcline. According to figure 7.9(b), $\omega' = -\theta - \tilde{p}\omega < 0$ above the nullcline $\omega = -(1/\tilde{p})\theta$, consistent with the second differential equation: $\omega > -(1/\tilde{p})\theta$ (a point that is above the $\omega' = 0$ nullcline—equivalent to $-\tilde{p}\omega < \theta$) forces $\omega' = -\theta - \tilde{p}\omega < -\theta + \theta = 0$. Hence, above the $\omega' = 0$ nullcline, direction arrows point down, below this nullcline they point up. In the second quadrant, $\theta' = \omega > 0$; all arrows point right. In the fourth, $\theta' = \omega < 0$; all arrows point left.

7. (a) Mimic the linearization of example 48, p. 324, with ϕ in place of q.

(b) From $\sin(\phi + 2\pi) \approx \phi$, obtain the damped system (7.3) with ϕ in place of θ. Nullcline analysis that follows (7.3) and leads to figures 7.9 and 7.11 now applies. To the extent that (7.3) is an accurate linear representation of the nonlinear system (7.5) near the steady state $\theta_{ss} = 2\pi$, $\omega_{ss} = 0$, figure 7.11 is accurate and this steady state is stable.

9. (a) Substitute $S = 0$, P.

(b) Via the quotient rule or MATLAB 's symbolic toolbox: $dI_{S'=0}(S)/dS = -g(g + bP)/(bS + g)^2 < 0$.

(c) Via the power rule or MATLAB 's symbolic toolbox: $d^2 I_{S'=0}(S)/dS^2 = 2bg(g + bP)/(bS + g)^2 > 0$.

10. If $I_i = 0$, then $I' = I(bS - r)$ forces $I'(t) \equiv 0$. Then $S' = -bIS + g(P - S - I)$ reduces to $S' = g(P - S)$, a first-order constant-coefficient equation. Characteristic equation yields the homogeneous solution $S_h = Ce^{-gt}$, and undetermined coefficients yield the particular solution $S_p = P$. Applying the initial condition $S(0) = S_i$ gives $S(t) = (S_i - P)e^{-gt} + P$.

18. (a) The natural frequency of the undamped pendulum is $\omega_n = \sqrt{g/L}$ rad/s; $\tau = \omega_n t$. One unit of τ is the time required for the pendulum to swing through an arc of 1 rad. The pendulum always has period 2π τ units.

(b) Substitute $\theta' = \sqrt{g/L}\,\dot{\Theta}$, $\omega' = \sqrt{g/L}\,\dot{\Omega}$.

(c) Substitute Ω from the first equation into the second; $P = p\sqrt{L/g}$.

(d) Identify $\theta \leftarrow \Theta$, $\omega \leftarrow \Omega$, $\tilde{p} \leftarrow p/L\omega_n$.

19. See exercise 18.

21. Use definition of s and differential equations (7.5): $s' = 2\theta\theta' + 2\omega\omega' = 2\theta\omega + 2\omega(-\sin\theta - \tilde{p}\omega) \approx 2\theta\omega + 2\omega(-\theta - \tilde{p}\omega) = -2\tilde{p}\omega \le 0$. Compare (7.4).

Section 7.3

1. (a) $i' = 0$ nullcline: $0 = -\epsilon(i^2/3 - 1)i - q$ defines a cubic in i that crosses the i axis at $i = 0, \pm\sqrt{3}$. Direction arrows in figure 7.15(a) are on $i = 0$, the $q' = 0$ nullcline. Hence, they are horizontal. If $q > -\epsilon(i^2/3 - 1)i$ (above $i' = 0$ nullcline), then $i' = -\epsilon(i^2/3 - 1)i - q < 0$ and arrows point left, increasing in length with distance from $i' = 0$ nullcline.

(b) $q' = 0$ nullcline: $i = 0$ (vertical axis) defines $q' = 0 = i$ nullcline. Arrows are on $i' = 0$ nullcline and hence must be vertical. Sign of $q' = i$ is positive (arrows up) on right of $i = 0$, negative (arrows down) on left.

(c) Combine preceding results to determine signs of i' (horizontal direction), q' (vertical direction) in regions bounded by nullclines; e.g., in first quadrant far from origin, $i' < 0$ (figure 7.15(a)) and $q' > 0$ (figure 7.15(b)) so arrows are left and up.

3. (a) $mx'' + \epsilon(x^2 - 1)x' + (1/C)x = 0$. Spring-force term is familiar; spring constant is $1/C$. Friction-force term is not: friction coefficient $\epsilon(x^2 - 1)$ varies with x, negative for $|x| < 1$ so it is aiding the motion, not resisting it.

(b) "Friction" term can supply energy when $|x| < 1$ to make up for what is lost when $|x| > 1$.

5. (a) p small, $i(t) = i_{ss} + p(t) = 0 + p(t) \Rightarrow \epsilon(i^2 - 1)i' = \epsilon(p^2 - 1)p = \epsilon(p^3 - p) \approx -\epsilon p$.

(b) Characteristic roots of $p'' - \epsilon p' + p = 0$ are $r_{\pm} = (\epsilon \pm \sqrt{\epsilon^2 - 4})/2$. Complex roots ($\epsilon < 2$) have positive real part $\epsilon/2$; real roots ($\epsilon \ge 2$) are always positive.

(c) Linear analysis of stability of $\theta_{ss} = \pi$ for nonlinear pendulum leads from $\theta(t) = \theta_{ss} + q(t)$ to $q'' + \tilde{p}q' - q = 0$; characteristic roots are $(-\tilde{p} \pm \sqrt{\tilde{p}^2 + 4})/2$, real and of opposite sign. Linear analysis of van der Pol system never exhibits negative roots.

7. (a) Compare exercise 3(a).

(b) Mimic solution of exercise 1.

(c) $\theta_{ss} = 0$, $\omega_{ss} = 0$.

Chapter exercises

1. (a) Phase portraits near stable equilibria ($\theta_{ss} = 0, \pm 2\pi, \ldots, \omega_{ss} = 0$) change with $\tilde{p}$: spirals for small $\tilde{p}$ (underdamped), none for large $\tilde{p}$ (overdamped). Define over- and underdamping in terms of linearized approximate pendulum equation, as in exercise 1, section 7.1.

3. System $\begin{cases} Li' &= -Ri - q/C \\ q' &= i \end{cases}$ has steady-state $i_{ss} = 0$, $q_{ss} = 0$. Analogous to many mechanical systems; e.g., damped linear pendulum with $q \leftarrow \theta$, $i \leftarrow \omega$, inductance $L \leftarrow$ length L, $R \leftarrow p$, $1/C \leftarrow g$.

5. (a) $y' = 0$ nullclines defined by $f(y_{y'=0}, z_{y'=0}) = 0$.

(b) By definition, $y' = 0$ on this nullcline. On a yz phase plane, direction arrows on $y' = 0$ nullcline will have zero horizontal component; i.e., they will be vertical unless $z' = 0$ as well (an equilibrium point).

(c) $z' = 0$ nullclines defined by $g(y_{z'=0}, z_{z'=0}) = 0$.

(d) Phase curves are horizontal when they cross $z' = 0$ nullclines; compare 5(b).

(e) Equilibria (y_{ss}, z_{ss}) simultaneously satisfy $f(y_{ss}, z_{ss}) = 0$ and $g(y_{ss}, z_{ss}) = 0$.

(f) Equilibrium points lie at the intersection of $y' = 0$ and $z' = 0$ nullclines.

7. (a) Choose $u = i$, $v = q$, $g(u) = \epsilon(u^2/3 - 1)u$.

(b) Use chain rule in first equation: $u'' = -g'(u)u' - v' = -g'(u)u' - u$; nonlinear unless g' is constant.

(c) Substitute $u_{ss} = 0$, $v_{ss} = 0$.

(d) Study fate of perturbation p defined by $u(t) = u_{ss} + p(t) = 0 + p(t)$: $p'' + g'(p)p' + p = 0$. If p is small ($p \approx 0$), then the nonlinear term $g'(p)p' \approx g'(0)p'$, a linear term. Linear stability of $u_{ss} = 0$ is determined by

$p'' + g'(0)p' + p = 0$. Characteristic roots are $r = \left(-g'(0) \pm \sqrt{g'(0)^2 - 4}\right)/2$. Roots are negative or have negative real part and steady state is stable if $g'(0) > 0$ (compare with a damping coefficient); $u_{ss} = 0$ is unstable if $g'(0) < 0$. Key information needed is sign of $g'(0)$.

(e) See solution to exercise 1, section 7.3. Compare figure 7.15 with $a = -\sqrt{3}, b = \sqrt{3}$.

(f) $g'(0) = -ab$; $u_{ss} = 0$ is unstable.

(g) Compare figure 7.15 with $a = -\sqrt{3}, b = \sqrt{3}$.

(h) Adapt figure 7.16 with vertical sides at $u = 2a$, $u = 2b$, say.

(j) Choose g to mimic behavior of cubic defined in part (e); i.e., $g(u) = 0$ at $u = a, 0, b$ with $a < 0 < b$, $g'(0) < 0$ for $u < a, \dots$.

9. Define perturbations ϕ, ρ by $F(t) = b_r/\beta + \phi(t)$, $R(t) = d_F/\alpha + \rho(t)$. Substitute and linearize by neglecting products of small perturbations to obtain

$$\phi' = (1/m)\rho$$
$$\rho' = -(1/n)\phi,$$

where $m = \beta/(\alpha b_R), n = \alpha/(\beta d_F)$. Then $m\phi\phi' + n\rho\rho' = 0$. Integrate to obtain $m\phi^2 + n\rho^2 =$ constant, an ellipse.

10. (a) The discriminant of the characteristic roots of $\theta'' + \tilde{p}\theta' + \theta = 0$ is $\tilde{p}^2 - 4$; the pendulum is underdamped if $0 < \tilde{p} < 2$.

(b) Use $\phi = \arctan(\omega/\theta)$ to compute

$$\phi' = \frac{1}{1 + (\omega/\theta)^2} \left(\frac{\omega}{\theta}\right)'.$$

Continue with the indicated differentiation via the quotient rule, substitute $\theta' = \omega, \omega' = -\theta - \tilde{p}\omega$, and simplify.

(c) Compute $\partial f/\partial \theta = \omega(\omega^2 - \theta^2)/r^4$ with an analogous expression for $\partial f/\partial \omega$. Hence, critical points are $\theta = \pm\omega$, where $f = \pm 1/2$. For large values of (θ, ω), consider $\theta = a\omega$ for some constant a. Then $|f| = |a/(1 + a^2)| < 1/2$ if $|a| \neq 1$. Hence, $|f| \leq 1/2$. From $f \geq -1/2$, conclude that $\phi' \leq -1 + \tilde{p}/2$.

(d) If $0 < \tilde{p} < 2$ (underdamped system), then $\phi' \leq -1 + \tilde{p}/2 < 0$ and ϕ decreases for all time. (Indeed, $\phi(t) < (-1 + \tilde{p}/2)t$.) As the trajectory winds around the origin with each 2π increase in ϕ, θ passes through zero twice; i.e., the pendulum swings twice through vertical, once in each direction. Since ϕ is decreasing, the trajectory is winding in the clockwise (negative angle) direction.

11. (a) Compute r', θ' from $r^2 = y^2 + z^2$, $\tan\theta = z/y$; e.g., $d(z/y)/dt = d(\tan\theta)/dt = \sec^2(\theta)d\theta/dt$. Write $\sec^2\theta = y/r$, compute $d(z/y)/dt$ using the quotient rule, and simplify using the differential equations.

(b) Solve the equations in the uncoupled system (7.16) by separation of variables (first equation) and integration (second) or verify by substitution.

(c) $r \to 1$, a circle (from inside if $k < 0$, from outside if $k > 0$) that constitutes the limit cycle; $\theta \to -\infty$, corresponding to trajectories perpetually winding clockwise around the origin. The amplitude is 1, the period 2π.

(d) Define perturbation $p(t)$ of $r_{ss} = 0$ by $r(t) = r_{ss} + p(t) = 0 + p(t)$. Linearize: $p' = p(1 - p^2) \approx p$. Since p grows exponentially, linear analysis predicts $r_{ss} = 0$ is unstable. Other steady states are $r_{ss} = \pm 1$ (the limit cycle); $r(t) = r_{ss} + p(t)$ leads to the linearized equation $p' = -3p$, predicting decay of p and stability for $r_{ss} = \pm 1$.

(e) $r \geq 1 + \delta \Rightarrow r' = r(1 - r^2) \leq -\delta(1 + \delta)(2 + \delta) < 0$

CHAPTER 8

Section 8.1

1. (a) $y' = w, 4w' = 16w - 2y$.

(b) $y' = w, 4w' = 16w - 2y + e^{2t}$; $y(0) = 1$, $w(0) = -2$.

(c) $y' = w, 4w' = w(16y^2 - 15)$.

(d) $x' = v, mv' = -pv - kx$; $x(0) = x_i, v(0) = v_i$.

(e) $x' = v, mv' = -kx + A\sin\omega t$.

(f) $q' = i, Ri' = -Ri - q/C + E(t)$.

(g) $\theta' = \omega, L\omega' = -g\theta$; $\theta(0) = \theta_i, \omega(0) = \omega_i$.

(h) $\theta' = \omega, L\omega' = -p\omega - g\sin\theta$.

(i) $y' = z, a_2z' = -a_0y - a_1z + h(t)$; linear, homogeneous, constant coefficients.

3. $y' = z, z' = f(t, y, z)$.

5. (a) Linear, homogeneous, constant coefficients.

(b) Linear, nonhomogeneous, constant coefficients.

(c) Nonlinear, homogeneous.

(d) Linear, homogeneous, constant coefficients.

(e) Linear, nonhomogeneous, constant coefficients.

(f) Linear, nonhomogeneous, constant coefficients.

(g) Nonlinear, homogeneous.

(h) Nonlinear, homogeneous.

(i) Nonlinear, nonhomogeneous.

(j) Linear, nonhomogeneous, constant coefficients.

(k) Linear, homogeneous, constant coefficients.

7. Substitute $\mathbf{y}_1$ and $\mathbf{y}_2$ into the system and show each satisfies the equations; i.e., substitute $y = e^{4t}, z = 4e^{4t}$ and similarly for $\mathbf{y}_2$. Since $y'' - 16y = 0$ is equivalent to the given first-order system, the first components of $\mathbf{y}_1$ and $\mathbf{y}_2$ are solutions of this scalar equation.

9. Assume the system is homogeneous. Then, by definition $\mathbf{0} = (0 \quad 0)^T$ is a solution. Substituting $\mathbf{0}$ yields $F = 0$, $G = 0$. Now assume $F = G = 0$. The system becomes $y' = k_1y + k_2z$ and $z' = \ell_1y + \ell_2z$. Obviously $\mathbf{0}$ is a solution.

11. (i) Substitute each of $\mathbf{y}_1, \mathbf{y}_2$ and $\mathbf{y}_3$ into the system and show they satisfy both equations.

(ii) Pairs ((a), (c)) and ((b), (c)) are linearly independent because their Wronskians are nonzero. Solutions (a) and (b) are negatives of one another.

(iii) $\mathbf{y}_{1g} = C_1\mathbf{y}_1 + C_2\mathbf{y}_3$; $\mathbf{y}_{2g} = C_3\mathbf{y}_2 + C_4\mathbf{y}_3$.

(iv) $\mathbf{Y}_1 = \begin{pmatrix} y_1 & y_3 \\ z_1 & z_3 \end{pmatrix}$, etc.

12. Let $\mathbf{y}_j = (\, y_j \;\; z_j \,)^T$, $j = 1, 2$. Substituting in the first equation yields

$$\begin{aligned} y_h' &= C_1 y_1' + C_2 y_2' \\ &= C_1[k_1 y_1 + k_2 z_1] + C_2[k_2 y_1 + k_2 z_2] \\ &= k_1 y_h + k_2 z_h, \end{aligned}$$

verifying that $C_1\mathbf{y}_1 + C_2\mathbf{y}_2$ solves the first equation. Verification of the second is similar. Linear independence is irrelevant.

13. (a) Let $\mathbf{y}_i' = \mathbf{A}\mathbf{y}_i$, $i = 1, 2$ and substitute: $\mathbf{y}' = C_1\mathbf{y}_1' + C_2\mathbf{y}_2' = C_1\mathbf{A}\mathbf{y}_1 + C_2\mathbf{A}\mathbf{y}_2 = \mathbf{A}(C_1\mathbf{y}_1 + C_2\mathbf{y}_2) = \mathbf{A}\mathbf{y}$.

(b) Yes, but exercise 12 was limited to two equations. first equation yields

$$\begin{aligned} \mathbf{y}_g' &= \mathbf{y}_p' + \mathbf{y}_h' \\ &= k_1 y_p + k_2 z_p + k_1 y_h + k_2 z_h \\ &= F + 0 = F, \end{aligned}$$

verifying that the first equation is satisfied. Verification of the second is similar. If $\mathbf{y}_h$ is the trivial solution, then $\mathbf{y}_g$ is still a solution, just not a general solution.

17. (a) $x_1 = \cos 2t$.

(b) $x' = y$, $y' = -4x$; $\mathbf{x}_1 = (\, \cos 2t \;\; -2\sin 2t \,)^T$.

(c) $x_g = C_1 \cos 2t + C_2 \sin 2t$.

(d) Using $x(t) = \sin 2t$ from part (c), construct $\mathbf{x}_2 = (\, \sin 2t \;\; 2\cos 2t \,)^T$; $\mathbf{x}_g = C_1\mathbf{x}_1 + C_2\mathbf{x}_2$.

(e) $W(t) = 2 \neq 0$.

(f) $x_p = 2t$.

(g) $x' = y$, $y' = -4x + 8t$; $\mathbf{x}_p = (\, 2t \;\; 2 \,)^T$.

(h) $\mathbf{x}_g = \mathbf{x}_p + C_1\mathbf{x}_1 + C_2\mathbf{x}_2$.

(i) $\mathbf{X} = \begin{pmatrix} \cos 2t & \sin 2t \\ -2\sin 2t & 2\cos 2t \end{pmatrix}$; $\mathbf{x}_g = \mathbf{x}_p + \mathbf{X}\mathbf{c}$.

19. $C_1 y_1 + C_2 y_2 = 0$, $C_1 z_1 + C_2 z_2 = 0$. Cramer $\Rightarrow C_1 = 0/W(t) = 0$ if $W \neq 0$. Same for C_2.

21. (a) Using the initial conditions, $C_1 y_1 + C_2 y_2 = 0 \Rightarrow C_1 = 0$ and $C_1 z_1 + C_2 z_2 = 0 \Rightarrow C_2 = 0$. Therefore the two solutions are linearly independent by definition.

(b) Use determinant : $y_1(0)z_2(0) - y_2(0)z_1(0) \neq 0$.

23. In matrix form, this linear, constant coefficient homogeneous system is $\mathbf{y}'(t) = \mathbf{B}\mathbf{y}(t)$, where

$$\mathbf{y}(t) = \begin{pmatrix} y(t) \\ z(t) \end{pmatrix}, \quad \mathbf{B} = \begin{pmatrix} -1/2 & 3/2 \\ 3/2 & -1/2 \end{pmatrix}.$$

In vector notation, the given solutions are

$$\mathbf{y}_1(t) = \begin{pmatrix} e^t \\ e^t \end{pmatrix}, \quad \mathbf{y}_2(t) = \begin{pmatrix} e^{-2t} \\ -e^{-2t} \end{pmatrix}.$$

They are linearly independent for all t because

$$\begin{aligned} W(\mathbf{y}_1(t), \mathbf{y}_2(t)) &= \det(\, \mathbf{y}_1(t) \;\; \mathbf{y}_2(t) \,) \\ &= \det \begin{pmatrix} e^t & e^{-2t} \\ e^t & -e^{-2t} \end{pmatrix} \\ &= -2e^{-t} \neq 0. \end{aligned}$$

A fundamental matrix is

$$\mathbf{Y}(t) = (\, \mathbf{y}_1(t) \;\; \mathbf{y}_2(t) \,) = \begin{pmatrix} e^t & e^{-2t} \\ e^t & -e^{-2t} \end{pmatrix}.$$

A general solution is

$$\mathbf{y}_g(t) = \mathbf{Y}(t)\mathbf{c} = C_1 \begin{pmatrix} e^t \\ e^t \end{pmatrix} + C_2 \begin{pmatrix} e^{-2t} \\ -e^{-2t} \end{pmatrix}.$$

25. (a) Define $y_1 = u$, $y_2 = u'$, $y_3 = u''$, $y_4 = u'''$:

$$\begin{aligned} y_1' &= y_2 \\ y_2' &= y_3 \\ y_3' &= y_4 \\ y_4' &= \frac{1}{b_4}(-b_3 y_4 - b_2 y_3 - b_1 y_2 - b_0 y_1 + h(t)) \end{aligned}$$

(b) $\begin{cases} y_1' = y_2 \\ y_2' = y_3 \\ y_3' = y_4 \\ y_4' = f(t, y_1, y_2, y_3, y_4) \end{cases}$

(c) Part (a) is a special case of part (b); choose

$$f(t, x, y, z, w) = \frac{-b_3 w - b_2 z - b_1 y - b_0 x + h(t)}{b_4}.$$

Section 8.2

1. (i) $(y_g, z_g) = C_1(e^{2t} \cos 4t, -e^{2t} \sin 4t) + C_2(e^{2t} \sin 4t, e^{2t} \cos 4t)$;

(ii) $C_1 = 2$, $C_2 = -2$.

3. (i) $(y_g, z_g) = C_1(e^{2\sqrt{7}t}, (\sqrt{7}/2 - 1)e^{2\sqrt{7}t}) + C_2(e^{-2\sqrt{7}t}, (-\sqrt{7}/2 - 1)e^{-2\sqrt{7}t})$;

(ii) $C_1 = 2 + 5\sqrt{7}/7$, $C_2 = 2 - 5\sqrt{7}/7$.

5. (i) $(y_g, z_g) = C_1(e^{-3t} \cos \omega t, (-2/5)e^{-3t} \cos \omega t + (\sqrt{6}/5)e^{-3t} \sin \omega t) + C_2(e^{-3t} \sin \omega t, (-\sqrt{6}/5)e^{-3t} \cos \omega t - (2/5)e^{-3t} \sin \omega t)$, where $\omega = \sqrt{6}/2$;

(ii) $(y, z) = (0, 0)$.

7. (i) $(y_g, z_g) = C_1(e^t, \sqrt{3}e^t/3) + C_2(e^{-t}, -\sqrt{3}e^{-t}/3)$;

(ii) $C_1 = 3 + 2\sqrt{3}$, $C_2 = 3 - 2\sqrt{3}$.

9. (i) $(x_g, v_g) = C_1(\cos \sqrt{k/m}\, t, -\sqrt{k/m} \sin \sqrt{k/m}\, t) + C_2(\sin \sqrt{k/m}\, t, \sqrt{k/m} \cos \sqrt{k/m}\, t)$;

(ii) $C_1 = x_i$, $C_2 = 0$.

11. (1) If $p^2 > 4mk$,

(i) $(x_g, v_g) = C_1(e^{Qt}, Qe^{Qt}) + C_2(e^{Rt}, Re^{Rt})$, where
$Q = (-p + \sqrt{p^2 - 4mk})/2m$ and
$R = (-p - \sqrt{p^2 - 4mk})/2m$;

(ii) $C_1 = (1/2 - p/2\sqrt{p^2 - 4mk})x_i$,
$C_2 = (1/2 + p/2\sqrt{p^2 - 4mk})x_i$.

(2) If $p^2 = 4mk$,

(i) $(x_g, v_g) = C_1(e^{-pt/2m}, (-p/2m)e^{-pt/2m}) + C_2(te^{-pt/2m}, (-p/2m)te^{-pt/2m} + e^{-pt/2m})$;

(ii) $C_1 = x_i$, $C_2 = (p/2m)x_i$.

(3) If $p^2 < 4mk$,

(i) $(x_g, v_g) =$
$C_1(e^{-dt}\cos\omega t, -de^{-dt}\cos\omega t - \omega e^{-dt}\sin\omega t) +$
$C_2(e^{-dt}\sin\omega t, -de^{-dt}\sin\omega t + \omega e^{-dt}\cos\omega t)$;

(ii) $C_1 = x_i$, $C_2 = dx_i/\omega$, where $\omega = \sqrt{4mk - p^2}/2m$ and $d = p/2m$.

12. (i) $(\theta_g, \omega_g) = C_1\left(\cos\sqrt{g/L}\,t, -\sqrt{g/L}\sin\sqrt{g/L}\,t\right)$
$+ C_2\left(\sin\sqrt{g/L}\,t, \sqrt{g/L}\cos\sqrt{g/L}\,t\right)$.

(ii) $C_1 = \theta_i$, $C_2 = 0$.

13. $r_1 = 3$, $\mathbf{p}_1 = (\,0\quad 1\,)^T$; $r_2 = 1$, $\mathbf{p}_2 = (\,-1\quad 2\,)^T$
$$\mathbf{Y}(t) = \begin{pmatrix} 0 & -e^{-t} \\ e^{3t} & 2e^t \end{pmatrix},$$
$\mathbf{y}_g = \mathbf{Y}\mathbf{c}$; $\mathbf{c} = (\,4\quad -4\,)^T$.

15. $r_1 = 12$, $\mathbf{p}_1 = (\,2\quad 1\,)^T$; $r_2 = 9$, $\mathbf{p}_2 = (\,1\quad -1\,)^T$
$$\mathbf{Y}(t) = \begin{pmatrix} 2e^{12t} & e^{9t} \\ e^{12t} & -e^{9t} \end{pmatrix}.$$
$\mathbf{y}_g = \mathbf{Y}\mathbf{c}$; $\mathbf{c} = (\,0\quad 4\,)^T$.

17. $r_i = 4 \pm 3i$, $\mathbf{p}_i = (\,1\quad \pm i\,)^T$
$$\mathbf{Y}(t) = \begin{pmatrix} e^{4t}\cos 3t & e^{4t}\sin 3t \\ -e^{4t}\sin 3t & e^{4t}\cos 3t \end{pmatrix}.$$
$\mathbf{y}_g = \mathbf{Y}\mathbf{c}$; $\mathbf{c} = (\,4\quad -4\,)^T$.

19. An eigenvector for $r_1 = 1$ is $\mathbf{p}_1 = (\,1\quad 1\quad 1\,)^T$. For $r_2 = (-1 + i\sqrt{3})/2$, use $\mathbf{p}_2 = (\,r_2\quad \bar{r}_2\quad 1\,)^T$. For $r_3 = \bar{r}_2$, use $\mathbf{p}_3 = \bar{\mathbf{p}}_2$. (Recall complex conjugate: $\bar{r}_2 = (-1 - i\sqrt{3})/2$.) $\mathbf{p}_1 e^{r_1 t}$ is real valued. Write complex-valued solution in terms of real and imaginary parts: $\mathbf{p}_2 e^{r_2 t} = \mathbf{R}(t) + i\mathbf{I}(t)$ where

$$\mathbf{R}(t) = e^{-t/2}\begin{pmatrix} -(\cos\beta t)/2 - \beta\sin\beta t \\ -(\cos\beta t)/2 + \beta\sin\beta t \\ \cos\beta t \end{pmatrix},$$

$$\mathbf{I}(t) = e^{-t/2}\begin{pmatrix} -(\sin\beta t)/2 + \beta\cos\beta t \\ -(\sin\beta t)/2 - \beta\cos\beta t \\ \sin\beta t \end{pmatrix}$$

and $\beta = \sqrt{3}/2$. A fundamental matrix is
$\mathbf{Y}(t) = (\,\mathbf{p}_1 e^t\quad \mathbf{R}(t)\quad \mathbf{I}(t)\,)^T$; a general solution,
$\mathbf{y}_g = \mathbf{Y}(t)\mathbf{c}$.

21. (a) $r_1 = -8$, $r_{2,3} = -8 \pm 4\sqrt{2}$;

$$\mathbf{p}_1 = \begin{pmatrix} -1 \\ 0 \\ 1 \end{pmatrix}, \qquad \mathbf{p}_{2,3} = \begin{pmatrix} 1 \\ \pm\sqrt{2} \\ 1 \end{pmatrix}.$$

(b) $\mathbf{y}_h(t) = \mathbf{Y}(t)\mathbf{c}$,
$$\mathbf{Y}(t) = \left(\, \mathbf{p}_1 e^{-8t}\quad \mathbf{p}_2 e^{(-8+4\sqrt{2})t}\quad \mathbf{p}_3 e^{(-8-4\sqrt{2})t}\,\right)^T.$$

(c) $\mathbf{u}'_{ss} = 0 \Rightarrow \mathbf{A}\mathbf{u}_{ss} = -\mathbf{f}$. Solve these simultaneous equations to find $\mathbf{u}_{ss} = (\,3/2\quad 2\quad 5/2\,)^T$.

(d) Use $\mathbf{u}_{ss}$ as a particular solution: $\mathbf{y}_g = \mathbf{Y}(t)\mathbf{c} + \mathbf{u}_{ss}$.

(e) Since $r_i < 0$, $i = 1, 2, 3$, $\lim_{t\to\infty} \mathbf{y}_h(t) = 0$. The steady state $\mathbf{u}_{ss}$ is stable, representing a linear temperature profile.

23. (a) The determinant of coefficients is $(a - r)(d - r) - bc$. According to Cramer's rule, a system of homogeneous linear equations has a nontrivial solution only if that determinant is zero.

(b) Carry out the indicated calculations. Then $p = 0/[(a - r)(d - r) - bc]$ and $p = 0$ unless $(a - r)(d - r) - bc = 0$, in which case $0 \cdot p = 0$, an indeterminate form that permits $p \neq 0$. To eliminate p, multiply the first equation by c, the second by $a - r$ and subtract. The indeterminacy argument is identical.

25. (a) $W(t) = (p_1 q_2 - p_2 q_1)e^{(r_1+r_2)t}$. The solutions are linearly independent if $p_1 q_2 - p_2 q_1 \neq 0$. Use $p_i/q_i = b/(r_i - a)$, $r_1 \neq r_2$, to show that this condition is satisfied.

(b) Re $(y, z) = (e^{\alpha t}(P_1\cos\beta t - P_2\sin\beta t)$, $e^{\alpha t}(Q_1\cos\beta t - Q_2\sin\beta t)$; Im $(y, z) = (e^{\alpha t}(P_2\cos\beta t + P_1\sin\beta t)$, $e^{\alpha t}(Q_2\cos\beta t + Q_1\sin\beta t)$. The solutions are linearly independent if $W(t) = P_1 Q_2 - P_2 Q_1 \neq 0$. To show that this condition is satisfied, use the same argument as in part (a).

27. $z' = -2z \Rightarrow z = Ce^{-2t} \Rightarrow y' - y = 3z = 3Ce^{-2t} \Rightarrow y = Ce^{-2t} + De^t$, where C, D are arbitrary constants.

29. Carry out the multiplication $\mathbf{B}\mathbf{p}_2$.

31. $\det(\mathbf{B} - r\mathbf{I}) = (r - a)^2 \Rightarrow r = a$. Substituting $r = a$ in $(\mathbf{B} - r\mathbf{I})\mathbf{p} = 0$ yields $0 = 0$; p, q are arbitrary. Choose $p = 1, q = 0$ for one eigenvector and $p = 0, q = 1$ for the other.

33. Compute $\mathbf{Y}'$ entry by entry.

35. $r_1 = -3$, $\mathbf{p}_1 = (\,2\quad 1\,)^T$, $r_2 = 2$, $\mathbf{p}_2 = (\,1\quad 3\,)^T$. (Eigenvectors and, hence, columns of two fundamental matrices can differ by a constant multiple; order of columns is arbitrary.) Solution of initial-value problem is $\mathbf{y} = \mathbf{Y}(\,1\quad -1\,)^T$.

37. If r is an eigenvalue of $\mathbf{B}$, then the determinant of coefficients of the system of equations for the components of the eigenvector is $\det(\mathbf{B} - r\mathbf{I}) = 0$. Hence, Cramer's rule cannot be used to solve for the components of $\mathbf{p}$ (but

Cramer's rule does tell us that this homogeneous system can have nontrivial solutions).

Section 8.3

1. Positive distinct eigenvalues $r = 1, 3$: unstable node at the origin.

3. Positive distinct eigenvalues $r = 9, 12$: unstable node at the origin.

5. Complex conjugate eigenvalues $r = 4 \pm 3i$ with positive real part: unstable spiral point at the origin.

7. By definition, a homogeneous system has the trivial solution $\mathbf{y} \equiv \mathbf{0}$ and, hence, the steady state $\mathbf{y}_{ss} = \mathbf{0}$.

9. (a) $\mathbf{J} = \begin{pmatrix} 0 & 1 \\ -k/m & 0 \end{pmatrix}$, $\mathbf{y}_{ss} = \mathbf{0}$, $r = \pm i \sqrt{k/m}$,

neutrally stable center (see exercise 14, page 328).

(b) $\mathbf{J}$ is equation (8.23) with $\cos \theta_{ss} = 1$;
$r = (-p \pm \sqrt{p^2 - 4Lg})/2$, stable spiral if $p^2 < 4Lg$ (underdamped, complex conjugate eigenvalues with negative real part) or stable node if $p^2 > 4Lg$ (overdamped, negative real eigenvalues).

(c) See 9(d).

(d) $\mathbf{J} = \begin{pmatrix} \alpha R - d_f & \alpha F \\ -\beta R & b_R - \beta F \end{pmatrix}$, $\mathbf{y}_{ss}^1 = \mathbf{0}$, second

steady state is $F_{ss} = b_R/\beta$, $R_{ss} = d_F/\alpha$.
At first equilibrium, $\mathbf{J} = \text{diag}(-d_F, b_R)$, $r = -d_F, b_R$; saddle point. Compare lower time plots in figure 2.14, page 69.

At second, $\mathbf{J} = \begin{pmatrix} 0 & \alpha b_R/\beta \\ -\beta d_F/\alpha & 0 \end{pmatrix}$, $r = \pm i\sqrt{b_R/d_F}$,

neutrally stable center (see exercise 14, page 328). Compare upper time plots in figure 2.14, page 69.

(e) $\mathbf{J} = \begin{pmatrix} 1 - 3y^2 - z^2 & 1 - 2yx \\ -1 - 2yz & 1 - 3z^2 - y^2 \end{pmatrix}$, $\mathbf{y}_{ss} = \mathbf{0}$,

$r = 1 \pm i$, unstable spiral.

(f) $\mathbf{J} = \begin{pmatrix} -2I - 1 & -2S - 1 \\ 2I & 2S - 1 \end{pmatrix}$; $I_{ss}^1 = 3/4$, $S_{ss}^1 = 1/2$,

and $I_{ss}^2 = 0$, $S_{ss}^2 = 2$. Compare section 7.2.3, page 349.
At first equilibrium, eigenvalues of $\mathbf{J}$ are
$r = (-5 \pm i\sqrt{23})/4$, a stable spiral. At second, $r = -1, 3$, a saddle point. Compare figure 7.13(c), page 350.

(g–j) $\mathbf{J} = \begin{pmatrix} 1 - 2y - ez & -ey \\ -fz & 1 - 2z - fy \end{pmatrix}$, $\mathbf{y}_{ss}^1 = \mathbf{0}$,

$\mathbf{y}_{ss}^2 = (0 \ 1)^T$, $\mathbf{y}_{ss}^3 = (1 \ 0)^T$; if $1 - ef \neq 0$, additional steady state is $y_{ss}^4 = (1 - e)/(1 - ef)$,
$z_{ss}^4 = (1 - f)/(1 - ef)$.

(k) $\mathbf{J} = \begin{pmatrix} -\epsilon(i^2 - 1) & -1 \\ 1 & 0 \end{pmatrix}$, $\mathbf{y}_{ss} = \mathbf{0}$,

$r = (\epsilon \pm \sqrt{\epsilon^2 - 4})/2$, unstable spiral ($\epsilon < 2$) or saddle ($\epsilon > 2$).

11. $r = \pm 4$, saddle point; compare figure 8.6, page 408.

13. $r = -2, 1$, saddle point; compare figure 8.6, page 408.

15. Real eigenvalues of opposite sign; (unstable) saddle point. The stable axis is the direction $\mathbf{p}_2$.

17. Real, positive eigenvalues; unstable node.

19. Mimic the Taylor expansion argument at the bottom of page 403 with g replacing f.

21. Solve initial-value problem; e.g., trajectory through $(x_i, 0)$ is given by $(x_i \cos \sqrt{k/m}\, t, -\beta x_i \sin \sqrt{k/m}\, t)$. Derive ellipse $(k/m)x(t)^2 + v(t)^2 = $ constant as in exercise 2(b), section 7.2: $(k/m)xx' + vv' = 0$, etc. The period, the time for one circuit of the ellipse, is $2\pi/\sqrt{k/m}$.

23. Associated eigenvectors are $\mathbf{p}_+ = (1 \ -1)^T$, $\mathbf{p}_- = (1 \ 1)^T$. General solutions of this system have the same form as those of example 20 except that increasing and decreasing directions have been reversed. The phase diagram of this system is figure 8.6 with all arrows reversed.

25. From (8.25), page 407, solutions approach the origin only if $C_+ = 0$, $C_- \neq 0$ arbitrary, i.e., only if the initial point lies on $\mathbf{p}_-$, or $z_i = -y_i$.

27. (a) Verify $\mathbf{Bp} = r\mathbf{p}$ for each eigenvalue-eigenvector pair.

(b) $\mathbf{y}_g = C_1\mathbf{p}_2 e^{2t} + C_2\mathbf{p}_8 e^{8t}$.

(c–e) Replace t with $-t$ throughout example 22. If $\mathbf{y}(t)$ solves example 22, $\mathbf{y}' = \mathbf{By}$, then $\mathbf{z}(t) = \mathbf{y}(-t)$ satisfies the system of this exercise, $\mathbf{z}' = \mathbf{B}_1\mathbf{z}$, because
$\mathbf{z}'(t) = d\mathbf{z}(t)/dt = d\mathbf{y}(-t)/dt = -d\mathbf{y}(t)/dt = -\mathbf{y}'(t)$ and $\mathbf{B}_1 = -\mathbf{B}$.

29. (a) $\mathbf{y}_g = C_1\mathbf{p}_1 e^{rt} + C_2\mathbf{p}_2 e^{rt} \to \mathbf{0}$ since $r < 0$.

(b) $y = y_i e^{-t}$, $z = z_i e^{-t}$ decay to zero along the line with slope z_i/y_i.

(c) $y(t)$, $z(t)$ decay along the line with slope
$\dfrac{z(t)}{y(t)} = \dfrac{C_1 p_1 + C_2 p_2}{C_1 q_1 + C_2 q_2} = \dfrac{z_i}{y_i}$. (The common term e^{rt} cancels; C_i are determined by the initial conditions.)

31. The characteristic equation is $r^2 + g/L = 0$. The eigenvalues of the coefficient matrix are pure imaginary, $r = \pm i\sqrt{g/L}$.

33. See the solution to exercise 9(d).

35. (a) $(a - r)(d - r) - bc = r^2 - Tr + D = 0 \Rightarrow r_\pm = (T \pm \sqrt{T^2 - D})/2$. Eigenvalues are real and distinct if $T^2 > D$, complex conjugate if $T^2 < D$. By direct calculation, $r_+r_- = D$: real eigenvalues have the same sign if $D > 0$, opposite sign if $D < 0$.

(b) Saddle if real eigenvalues of opposite sign, i.e., if $T^2 > D$, $D < 0$.

(c) Node if eigenvalues are real and of same sign, i.e., if $T^2 > D$ and $D > 0$; stable node if eigenvalues are negative ($T < 0$), unstable if positive ($T > 0$).

(d) Spiral point if complex conjugate eigenvalues ($T^2 < D$), stable if real part is negative ($T < 0$), unstable if positive ($T > 0$).

Section 8.4

1. (i) $(e^{2t}, 3e^{2t})$, $(e^{-3t}, e^{-3t}/2)$.
(ii) $u_1 = (3t + 19/4)e^{-2t}$, $u_2 = (6t + 2/3)e^{3t}$.
(iii) $(y_p, z_p) = ((90t + 65)/12, (90t + 175)/12)$.
(iv) $(y_g, z_g) = C_1(e^{2t}, 3e^{2t}) + C_2(e^{-3t}, e^{-3t}/2) + (y_p, z_p)$.
(v) $C_1 = -15/4$, $C_2 = 4/3$.

3. (i) $(e^{2t} \cos 4t, -e^{2t} \sin 4t)$, $(e^{2t} \sin 4t, e^{2t} \cos 4t)$.
(ii) $u_1 = e^{-2t} \cos 4t - e^{2t}$, $u_2 = e^{-2t} \sin 4t$.
(iii) $(y_p, z_p) = (1, 0)$.
(iv) $(y_g, z_g) = C_1(e^{2t} \cos 4t, -e^{2t} \sin 4t) + C_2(e^{2t} \sin 4t, e^{2t} \cos 4t) + (1, 0)$.
(v) $C_1 = 1$, $C_2 = -2$.

5. (i) $(e^{4t}, \sqrt{3}e^{4t}/3)$, $(e^{-2t}, -\sqrt{3}e^{-2t})$.
(ii) $u_1 = (-24 + 3\sqrt{3} + 12\sqrt{3}\,t)e^{-4t}/32$,
$u_2 = (4 - \sqrt{3} + 6\sqrt{3}\,t)e^{-4t}/8$.
(iii) $(y_p, z_p) = ((36\sqrt{3}t - 90\sqrt{3} - 8)/32,$
$-(60t + 24\sqrt{3} - 39)/32)$.
(iv) $(y_g, z_g) = C_1(e^{4t}, \sqrt{3}e^{4t}/3) +$
$C_2(e^{-2t}, -\sqrt{3}e^{-2t}) + (y_p, z_p)$.
(v) $C_1 = 21/4 - 3\sqrt{3}/32$, $C_2 = 3\sqrt{3}/8 + 1$.

6. (i) See exercise 5 of section 8.2. (ii) Let $\omega = \sqrt{6}/2$;

$$u_1 = \frac{e^{3t}}{21}\left(4\sqrt{6}\sin \omega t + 3 \cos \omega t\right),$$

$$u_2 = \frac{e^{3t}}{21}\left(3\sin \omega t + 4\sqrt{6}\cos \omega t\right).$$

(iii) $(y_p, z_p) = (3/7, -2/7)$.

7. (i)–(iv) See exercise 6.
(v) $C_1 = -3/7$, $C_2 = \sqrt{6}/7$.

8. (i) See exercise 9 of section 8.2.
(ii) $u_1 =$

$$-F\sqrt{\frac{m}{k}}\left(\frac{\sin(\omega - \sqrt{k/m})t}{2(\omega - \sqrt{k/m})} - \frac{\sin(\omega + \sqrt{k/m})t}{2(\omega + \sqrt{k/m})}\right),$$

$u_2 =$

$$F\sqrt{\frac{m}{k}}\left(-\frac{\cos(\omega + \sqrt{k/m})t}{2(\omega + \sqrt{k/m})} - \frac{\cos(\omega - \sqrt{k/m})t}{2(\omega - \sqrt{k/m})}\right).$$

(iii) $(x_p, v_p) =$
$(u_1 \cos \sqrt{k/m}\,t + u_2 \sin \sqrt{k/m}\,t, -\sqrt{k/m}\,u_1 \sin \sqrt{k/m}\,t + \sqrt{k/m}\,u_2 \cos \sqrt{k/m}\,t)$.
(iv) $(x_g, v_g) = C_1(\cos \sqrt{k/m}\,t, -\sqrt{k/m} \sin \sqrt{k/m}\,t) + C_2(\sin \sqrt{k/m}\,t, \sqrt{k/m} \cos \sqrt{k/m}\,t) + (x_p, v_p)$.
(v) $C_1 = x_i$, $C_2 = \sqrt{\dfrac{m}{k}}\dfrac{F m \omega}{m \omega^2 - k}$.

9. (i)–(iv) See exercise 8.
(v) $C_1 = 0$, $C_2 = \sqrt{\dfrac{m}{k}}\left(v - \dfrac{F m \omega}{m \omega^2 - k}\right)$.

11. (i) See exercise 12 of section 8.2

(ii) $u_1 = -2\sqrt{\dfrac{L}{g}}\left(-\dfrac{\cos(\sqrt{g/L} + \pi)t}{2(\sqrt{g/L} + \pi)}\right.$

$$\left. - \dfrac{\cos(\sqrt{g/L} - \pi)t}{2(\sqrt{g/L} - \pi)}\right),$$

$$u_2 = 2\sqrt{\dfrac{L}{g}}\left(\dfrac{\sin(\sqrt{g/L} + \pi)t}{2(\sqrt{g/L} + \pi)}\right.$$

$$\left. + \dfrac{\sin(\sqrt{g/L} - \pi)t}{2(\sqrt{g/L} - \pi)}\right).$$

(iii) $(\theta_p, \omega_p) = (u_1 \cos \sqrt{g/L}\,t + u_2 \sin \sqrt{g/L}\,t, -u_1\sqrt{g/L} \sin \sqrt{g/L}\,t + u_2 \cos \sqrt{g/L}\,t)$.
(iv) $(\theta_g, \omega_g) = C_1(\cos \sqrt{g/L}\,t, -\sqrt{g/L} \sin \sqrt{g/L}\,t) + C_2(\sin \sqrt{g/L}\,t, \sqrt{g/L} \cos \sqrt{g/L}\,t) + (\theta_p, \omega_p)$.
(v) $C_1 = \theta_i - \dfrac{2}{g/L - \pi^2}$, $C_2 = 0$.

13. (a) Substitute (y_g, z_g) into the system.
(b) $(y, z) =$
$\left(\dfrac{85}{18} - \dfrac{5t}{3} + 2e^{-3t} + e^{2t}, \dfrac{85}{18} - \dfrac{10t}{3} + e^{-3t} + 3e^{2t}\right)$.

14. Keeping the constants of integration C_1, C_2 found when integrating u_1, u_2 adds to the variation of parameters particular solution of example 25 the term $C_1(2e^{-3t}, e^{-3t}) + C_2(e^{2t}, 3e^{2t})$, a linear combination of independent solutions of the homogeneous system. Hence, the resulting solution is a general solution.

15. $W(t) = 0$, and the only solution is the trivial solution.

17. $\mathbf{y}_p = \begin{pmatrix} -2e^{2t}/3 - 1/4 \\ e^{2t}/3 + t \end{pmatrix}$.

19. Using $\mathbf{Y}(t)$ from exercise 13, section 8.2,

$$\mathbf{y}_g = \mathbf{Y}(t)\mathbf{c} + \begin{pmatrix} -4e^{-2t}/3 \\ -6e^{2t} - 8e^{t}/3 + 16e^{-2t}/15 \end{pmatrix}.$$

21. Using $\mathbf{Y}(t)$ from exercise 15, section 8.2

$$\mathbf{y}_g = \mathbf{Y}(t)\mathbf{c} + \begin{pmatrix} -2t - 277/18 \\ 11t + 95/36 \end{pmatrix}.$$

23. $\mathbf{y}(t) = \mathbf{Y}(t)\int_0^t \mathbf{Y}(s)^{-1}\mathbf{f}(s)\,ds + \mathbf{y}_i$. If $\mathbf{Y}(0) = \mathbf{I}$ (and we choose $a = 0$), then $\mathbf{c} = \mathbf{y}_i$.

Chapter exercises

1. (i) $(y_g, z_g) = C_1(e^t, e^t) + C_2(e^{-t}, -e^{-t})$.
(ii) $C_1 = 1/2$, $C_2 = -1/2$.
3. (i) $(y_g, z_g) = C_1(-e^{-4t}, e^{4t}) + C_2(e^{-4t}, e^{-4t})$.
(ii) $C_1 = C_2 = 1$
5. (i) $(e^t, \sqrt{3}e^t/3)$, $(e^{-t}, -\sqrt{3}e^{-t}/3)$.
(ii) $u_1 = (2\sqrt{3}t^2 + (4\sqrt{3} - 1)t + 4\sqrt{3} - 1)e^{-t}$,
$u_2 = (2\sqrt{3}t^2 + (1 - 4\sqrt{3})t + 4\sqrt{3} - 1)e^{t}$.
(iii) $(y_p, z_p) = (4\sqrt{3}t^2 + 8\sqrt{3} - 2, (8 - 2\sqrt{3}/3)t)$.

(iv) $(y_g, z_g) =$
$C_1(e^t, \sqrt{3}e^t/3) + C_2(e^{-t}, -\sqrt{3}e^{-t}/3) + (y_p, z_p)$.
(v) $C_1 = 4 - 2\sqrt{3}, C_2 = 4 - 6\sqrt{3}$.
7. (i) $(e^{-2t}, -e^{-2t}), (e^{4t}, e^{4t})$.
(v) $(5e^{4t} - e^{2t} + e^{-2t} + 3, 5e^{4t} - 3e^{2t} - e^{-2t} - 1)/8$
9. See section 11.3, page 581.

11. $y_g = \begin{pmatrix} -e^{-t} & e^{-4t} \\ 2e^{-t} & -e^{-4t} \end{pmatrix} c + \begin{pmatrix} -3t + 12 \\ 9t - 27 \end{pmatrix}$.

13. $y_g = \begin{pmatrix} -e^{-t} & -2e^t \\ 2e^{-t} & 5e^t \end{pmatrix} c + \begin{pmatrix} 2te^t + 5te^{-t} \\ -5te^t - 10te^{-t} \end{pmatrix}$.

15. $y_g = \begin{pmatrix} -\cos 4t - \sin 4t & -\sin 4t + \cos 4t \\ 2\cos 4t & 2\sin 4t \end{pmatrix} c$.

17. $y_g =$
$\begin{pmatrix} e^{-2t}(5\cos 3t - \sin 3t) & e^{-2t}(\sin 3t - 5\cos 3t) \\ 6e^{-2t}\cos 3t & 6e^{-2t}\sin 3t \end{pmatrix} c$.

19. $y_g =$
$\begin{pmatrix} e^{-t}(-\cos 2t - \sin 2t) & e^{-t}(\sin 2t + \cos 2t) \\ 2e^{-t}\cos 2t & 2e^{-t}\sin 2t \end{pmatrix} c$.

21. $y_g = \begin{pmatrix} 2e^{3t} & -2e^t & e^{-4t} \\ e^{3t} & e^t & -3e^{-4t} \\ -2e^{3t} & 4e^t & 13e^{-4t} \end{pmatrix} c + \begin{pmatrix} -1/12 \\ 1/12 \\ 1/12 \end{pmatrix}$.

23. $y_g = \begin{pmatrix} e^t & e^{3t} & e^{-4t} \\ e^t & 3e^{3t} & -4e^{-4t} \\ e^t & 9e^{3t} & 16e^{-4t} \end{pmatrix} c$.

25. $y_g = \begin{pmatrix} 2e^{3t} & e^{-2t} & 2e^t \\ e^{3t} & e^{-2t} & -e^t \\ -e^{3t} & -e^{-2t} & -e^t \end{pmatrix} c$.

27. See exercise 24 of section 8.4.
29. The $u' = 0$ nullcline is the horizontal axis, $v = 0$; flow is vertical across it. The $v' = 0$ nullcline is the line $v = u/2$; flow is downward ($v' < 0$) above it, upward below it.
31. The $u' = 0$ nullcline is the horizontal axis, $v = 0$; flow is vertical across it. The $v' = 0$ nullcline is the line $v = -u/4$; flow is downward ($v' < 0$) above it, upward below it.
33. Exercise 15: $r = \pm 4i$, origin is a center. Exercise 16: $r = \pm i$, origin is a center. Exercise 17: $r = -2 \pm 3i$, origin is a stable spiral. Exercise 18: $r = -2 \pm 3i$, origin is a stable spiral. Exercise 19: $r = -1 \pm 2i$, origin is a stable spiral.

CHAPTER 9

Section 9.1

1. Repeat the derivation in the text.
3. The derivation is the same with $D(x)$ replacing D, but do not factor $D(x)$ out of the derivative expressions.
5. This diffusion equation is linear and nonhomogeneous. It would have constant coefficients if A, V were independent of x.
7. The rate of flow in through the inner boundary is $-A(x)Dc'(x)$, while the rate of flow out through the outer

boundary is $-A(x + \Delta x)Dc'(x + \Delta x)$ with $A(x) = 2\pi x d$. (The area through which the flow occurs is the wall of a cylinder with diameter $2x$ and height d.) Complete the derivation by applying conservation of mass as in the text.
9. Repeat the preceding arguments. At the insulated end of the bar, no heat flows. Hence, $-kAT'(L) = 0$, or $T'(L) = 0$. Using the solution $T(x) = C_1 x + C_2$ and the boundary conditions $T(0) = T_0$ ($\Rightarrow C_2 = T_0$), $T'(0) = 0$ ($\Rightarrow C_1 = 0$), the equilibrium temperature distribution is $T(x) = T_0$. At equilibrium, heat energy is uniformly distributed in the rod when one end is insulated.
11. Use the same technique as exercise 7 but replace the diffusion coefficient D with thermal conductivity k and concentration $c(x)$ with temperature $T(x)$. Since no heat flows at the origin because of symmetry, $T'(0) = 0$, as in exercise 9.
13. Apply conservation of energy to the control volume of figure 9.8, page 436, but the "rate in" term includes the addition of energy at the rate $-H(x)A\Delta x$ (H is the rate per unit volume, and $A\Delta x$ is the volume of the "slice" in figure 9.8): $-kAT'(x) + H(x)A\Delta x + kAT'(x + \Delta x) = 0$. Divide by Δx and let $\Delta x \to 0$ to obtain $-kAT''(x) = H(x)$. Obtain the boundary conditions as usual.

Section 9.2

1. (i) $y_g = C_1 e^{-x} + C_2 e^x$.
(ii) $C_1 = -\dfrac{e^4}{1 - e^4}, C_2 = \dfrac{1}{1 - e^4}$.
3. (i) $y_g = C_1 e^{-x} + C_2 e^x - \dfrac{\sin \pi x}{1 + \pi^2}$.
(ii) $C_1 = C_2 = 0$.
5. (i) $y_g = C_1 e^{-x} + C_2 e^x - \dfrac{\sin \pi x}{1 + \pi^2}$.
(ii) $C_1 = \dfrac{e^2((e^2 - 4)(\pi^2 + 1) - \pi)}{(e^4 + 1)(\pi^2 + 1)}$,
$C_2 = \dfrac{e^2(\pi^2 + \pi + 4) + \pi^2}{(e^4 + 1)(\pi^2 + 1)}$.
7. (i) $y_g = C_1 e^t + C_2 t e^t + t^2 e^t / 2$.
(ii) $C_1 = 0, C_2 = 1/e - 1/2$.
9. (i) See exercise 9, section 9.1: $T = T_0$.
(ii) Insulation at one end of the bar makes the temperature uniform throughout.
11. (i) $T = C_1$, C_1 arbitrary.
(ii) We need additional information to find the (uniform) temperature of this perfectly insulated body.
12. (i) A general solution is

$$T_g(x) = C_1 x + C_2 + \frac{x^4 - 2Lx^3}{12kA}.$$

The solution of the boundary-value problem is

$$T(x) = \frac{x^4 - 2Lx^3 + L^3 x}{12kA} + \frac{T_0(L - x) + T_L x}{L}.$$

(ii) The second term is the solution of the homogeneous version of this problem. The current acts as a source of heat within the rod, raising its temperature. (The first term is indeed nonnegative. Its second derivative is $12x(x - L) \leq 0, 0 \leq x \leq L$. Since it is zero at $x = 0, L$ and since it is concave down, this term is never negative on $0 \leq x \leq L$.)

13. (i) See exercise 12; $T(x) = \dfrac{x^4 - 2Lx^3 + L^4}{12kA} + T_L$.

(ii) As in exercise 12, the heat source has increased the temperature of the rod above the uniform temperature $T(x) = T_L$ obtained without it.

15. The error in this approximation is $e^{VL/D} - \left(e^{VL/D} - 1\right) = 1$. The error is less than 1% when $1/e^{VL/D} < .01$, or (using logarithms and $-\ln .01 = \ln 100$) when $VL/D > \ln 100 \approx 4.61$, a condition certainly satisfied by the value in example 6, $VL/D \approx 1.02 \times 10^{12}$.

17. The correct concentration graph is the horizontal line $c \equiv c_L$.

19. (a) $c(x) = P(x^4 - 2Lx^3 + L^3 x)/12$. In this case, $D = 1$, $V = 0$ (no flow in channel), toxin source is described by $Px(L - x)$ (zero at ends of channel, maximum in the center of the channel), zero toxin concentration at either end of the channel.

(b) $D = 1$, flow velocity $V = 2$ (positive is flow to the right), toxin source is described by $4P$ (constant throughout the channel), zero toxin concentration at either end of the channel;

$$c(x) = P\left(\frac{xL}{16} - \frac{x}{32} + \frac{Lx^2}{8} - \frac{x^2}{16} - \frac{x^3}{12}\right)$$

$$+ \frac{\left(\dfrac{L}{32} - \dfrac{L^3}{24}\right)(1 - e^{4x})}{1 - e^{4L}}.$$

(c) $c(x) = P(x^4 - 2Lx^3 + 2L^3 x)/12$. In this case, $D = 1$, $V = 0$ (no flow in channel), toxin source is described by $Px(L - x)$ (zero at ends of channel, maximum in the center of the channel), zero toxin concentration at left end of the channel, no diffusion of toxin through right end of channel (perhaps because of a dam there).

Section 9.3

1. (a–b) Forward-, backward-, and centered-difference formulas are exact for lines; e.g., $(u(x+\Delta x) - u(x))/\Delta x = (x + \Delta x - x)/\Delta x = 1 = u'(x)$.

(c–d) Since $u'''(x) \equiv 0$, the centered difference is exact; see E_{centered}, page 446. Since u'' is constant, the error in the forward and backward differences is exactly a constant multiple of Δx; see E_{forward}, page 445.

3. Mimic the analysis on page 445 with $-\Delta x$ in place of Δx.

5. Use $u(x \pm \Delta x) = u(x) \pm u'(x)\Delta x + u''(x)\Delta x^2/2 \pm u'''(x)\Delta x^3/6 + u^{(4)}(x)\Delta x^4/24 + \cdots$ to show that $u^{(4)}(x)/12$ is the leading term in the error expression for the centered-difference approximation to $u''(x)$. Centered-difference approximations to second derivatives are exact for functions with $u^{(4)} \equiv 0$.

7. (a) $c(x) = (4/\pi)^2 \sin(\pi x/4)$

9. Common points are $x = 1, 2, 3$. At $x = 1$, the ratio of absolutes errors ($\Delta x = 1$ error to $\Delta x = 0.5$) is $0.06080/0.0149 = 4.0856$; $x = 3$ is the same. At $x = 2$, the ratio is $0.0860/0.0210 = 4.1028$. Reducing Δx by 2 has reduced error by about 4; this finite difference approximation appears to be of second-order.

11. (a) See 11(b) with $f(x) = 2x$.

(b) $-\dfrac{c_{i-1} - 2c_i + c_{i+1}}{\Delta x^2} = f(x_i), i = 1, \ldots, n - 1$.

(c) $-\dfrac{c_{i-1} - 2c_i + c_{i+1}}{\Delta x^2} + 3\dfrac{c_{i+1} - c_{i-1}}{2\Delta x} = f(x_i)$, $i = 1, \ldots, n - 1$.

(d) $-\dfrac{c_{i-1} - 2c_i + c_{i+1}}{\Delta x^2} + 3\dfrac{c_{i+1} - c_{i-1}}{2\Delta x} - 2c_i = f(x_i)$, $i = 1, \ldots, n - 1$.

13. (a) Approximate $c'(L) = 0$ by $c_n - c_{n-1} = 0$. Then the approximation is (9.14), page 454, with the last equation replaced by $-c_{n-2} + c_{n-1} = \Delta x^2(x_{n-1}(2L - x_{n-1}) + 2)$.

15. (a) Forward-difference error : $e(x) = u''(\xi)\Delta x^2/2$ for some $x < \xi < x + \Delta x$.

(b) Exact for linear functions: $u'' \equiv 0$.

(c) Centered-difference error: $e(x) = u'''(\xi)\Delta x^2/3$ for some $x < \xi < x + \Delta x$.

(d) Exact for quadratic functions: $u''' \equiv 0$.

Section 9.4

1. $\kappa = \dfrac{0.908}{0.09 \cdot 8.85} = 1.14$ cm²/s.

3. A partial differential equation is *homogeneous* if it has the trivial solution. Otherwise, it is *nonhomogeneous*; $0_t = \kappa 0_{xx}$ proves $T_t = \kappa T_{xx}$ is homogeneous; $T_t = \kappa T_{xx} + H/c\rho$ is nonhomogeneous because $0 \neq H/c\rho$.

5. $T_t = \kappa T_{xx}$, $T_x(0, t) = T_x(L, t) = 0$, $T(x, 0) = T_L x/L$.

7. Assume $A(x)$ is constant. Then the governing equation is $c\rho T_t = (k(x)T_x)_x$ with the same boundary and initial conditions as in example 15.

9. Incorporate the heat generation term H as in example 16 to derive the governing equation $T_t = \kappa T_{yy} + H/c\rho$ with the same boundary and initial conditions as in example 18.

11. If the sides of the chip were insulated, then the second and third boundary conditions in exercise 10(d) would be replaced by $T_x(0, y, t) = T_x(w, y, t) = 0$. If the top were insulated, the fourth boundary condition would be replaced by $T_x(x, h, t) = 0$.

13. If $H < 0$, then the rod is absorbing heat; e.g., cooling fluid is being pumped through its core. If $\kappa = 0$, then

$T_t = H < 0$, and the temperature is decreasing with the passage of time.

Section 9.5

1. (i) $X'' + \lambda X = 0$, $X'(0) = X(L) = 0$; $\Theta' + \kappa\lambda\Theta = 0$.
(ii) $X_n = \cos n\pi x/2L$, $\lambda_n = n^2\pi^2/4L^2$, n an odd integer; $\Theta_n = \exp(-\kappa\lambda_n t)$.
(iii) $\sum\limits_{n \text{ odd}} c_n X_n(x)\Theta_n(t)$.

3. Temperature decay is dominated by the $n = 1$ term, $\exp(-\kappa\pi^2 t/L^2)$. The ratio of temperatures copper:wood is $\exp(\kappa_w - \kappa_c)\pi^2 t/L^2) = \exp(-1.138\pi^2 t/L^2)$. Both midpoint temperatures decrease with t, but that of the copper rod decreases more rapidly.

5. $\lambda = 0 \Rightarrow X_g = C_1 + C_2 x$; $X(0) = 0 \Rightarrow C_1 = 0$; $X(L) = 0 \Rightarrow C_2 L = 0 \Rightarrow C_2 = 0$. Hence, $X(x) \equiv 0$. But the trivial solution can not be an eigenfunction.

7. $c_n = \dfrac{(-1)^{n-1}}{n\pi}$, $n = 1, 2, \ldots$. The series $\sum c_n X_n(x)$ converges in the mean to the $2L$-periodic extension of $T_L x/L$, $-L \le x < L$.

9. $c_0 = L/2$, $c_n = 2L/n^2\pi^2$, n odd; $c_n = 0$, n even. For $-L \le x \le 0$, the series $\sum c_n X_n(x)$ converges in the mean to the even extension of $L - x$ from $0 \le x \le L$,

$$\begin{cases} L + x, & -L \le x < 0 \\ L - x, & 0 \le x \le L. \end{cases}$$

For other x, the series converges in the mean to the $2L$-periodic extension of this function.

11. $X_n(x) = \cos nx/2$, $\lambda = n^2/4$, n an odd integer, or $X_m(x) = \cos(2m-1)x/2$, $m = 1, 2, \ldots$;

$$c_m = \frac{8}{(2m-1)^2\pi^2}\left(\frac{(-1)^{m-1}(2m-1)\pi}{2} - 1\right).$$

The series $\sum c_m X_m(x)$ converges in the mean to the even, 2π-periodic extension of x/π from $0 \le x \le \pi$; i.e., to the 2π-periodic extension of

$$\begin{cases} -x/\pi, & -\pi < x < 0 \\ x/\pi, & 0 \le x < \pi. \end{cases}$$

13. (a) Substitute the formulas for T_1, T_2 into the partial differential equation and into the boundary conditions.
(b) Use $T_n = X_n(x)\Theta_n(t)$, where $X_n'' + (n\pi/L)^2 X_n = 0$, $X_n(0) = X_n(L) = 0$, and $X_n(x) = \sin n\pi x/L$.

15. Carry out the indicated integration.

17. The eigenvalue problem is $X'' + \lambda X = 0$, $X'(0) = X'(4) = 0$. The eigenvalues and eigenfunctions are $X_n(x) = \cos n\pi x/4$, $\lambda_n = n^2\pi^2/16$, $n = 0, 1, 2, \ldots$; see exercise 9 with $L = 4$.

$$T(x, t) = 4 + \sum_{n=1}^{\infty} c_n e^{-\kappa n^2\pi^2 t/16}\cos\frac{n\pi x}{4},$$

where $c_n = \dfrac{8}{n^2\pi^2}$, n odd, and $c_n = 0$, n even.

19. The eigenvalues, eigenfunctions, and expansion are given in exercise 11. The solution $T(x, t)$ has the same form as in exercise 18 with c_m from exercise 11.
The physical situation is the same as exercise 18. Only the initial condition has been changed.

21. Solve $E''(x) = 0$, $E(0) = 0$, $E(L) = 8$, to obtain $E(x) = 8x/L$. Then solve the homogeneous initial-boundary-value problem $u_t = \kappa u_{xx}$, $u(0, t) = u(L, t) = 0$, $u(x, 0) = -2 - 8x/L$, where $T(x, t) = u(x, t) + E(x)$. The solution is $T(x, t) = E(x) + u(x, t)$, where

$$u(x, t) = \sum_{n=1}^{\infty} c_n e^{-\kappa n^2\pi^2/L^2}\sin\frac{n\pi x}{L},$$

$c_n = -\dfrac{12}{n\pi}$, n odd, and $c_n = \dfrac{8}{n\pi}$, n even. (Compare with exercise 7.)

23. The general solution is

$$E_g = -\frac{Hx^2}{2\kappa c\rho} + C_1 + C_2 x.$$

$E(0) = 0 \Rightarrow C_1 = 0$; $E(L) = 0 \Rightarrow C_2 = \dfrac{HL}{2\kappa c\rho}$.

Hence, $E(x) = \dfrac{Hx(L - x)}{2\kappa c\rho}$.

25. Substitute, using the differential equations defining E, T. Solution formulas are not required, and H need not be constant.

27. Use $\int \sin x \cos x\, dx = (\sin^2 x)/2$, etc.

29. Argue that $a_n = \dfrac{1}{2L}\displaystyle\int_{-L}^{L} f(x)\cos\frac{n\pi x}{L}\, dx = 0$, either by direct evaluation or because the integrand is an odd function.

Section 9.6

1. (a) $\mathbf{u}' = \mathbf{Au}$, $\mathbf{A}$ as in (9.40), page 493, with $\kappa = 1$, $\mathbf{u}(0) = ((x_1 - a)(x_1 - b) \cdots (x_{n-1} - a)(x_{n-1} - b))^T$, $x_i = a + i\Delta x$, $i = 1, \ldots, n-1$, $\Delta x = (b - a)/n$.
(b) $\mathbf{u}' = \mathbf{Au} + \mathbf{f}$, $\mathbf{A}$ as in (9.40), page 493, with $\kappa = 1$, $\mathbf{f}(t) = \Delta x^{-2}(\ell(t)\ 0 \cdots 0\ r(t))^T$, $\mathbf{u}(0) = (F(x_1) \cdots F(x_{n-1}))^T$, $x_i = a + i\Delta x$, $i = 1, \ldots, n-1$, $\Delta x = (b - a)/n$.

3. (a) Substitute.
(b) $r_1 = -a$, $\mathbf{p}_1 = (1\ \ 1)^T$, $r_2 = -3a$, $\mathbf{p}_2 = (-1\ \ 1)^T$, $a = 9\kappa/L$. From $\mathbf{u}_g = C_1\mathbf{p}_1 e^{-at} + C_2\mathbf{p}_2 e^{-3at}$ and $\mathbf{u}(0) = (M\sqrt{3}/2)\mathbf{p}_1$, find $C_1 = M\sqrt{3}/2$, $C_2 = 0$.

5. Exact time constant is $\tau = \pi^2 \kappa / L$; the temperature reduction factor $e^{-t/\tau} = \epsilon$ when $t = -\tau \ln \epsilon$, $0 < \epsilon < 1$. The method-of-lines decay time constant is determined by the eigenvalues of $\mathbf{A}$, which are independent of initial conditions. Likewise, the Fourier expansion time constants are independent of initial conditions.

7. From exercise 5 with $\epsilon = 0.5$: $\tau = -t_{1/2}/\ln \epsilon \approx 1.4\,t_{1/2}$, where $t_{1/2}$ is time to decay to one-half of initial value. For U_1 and U_2 in figure 9.33, we observe $t_{1/2} \approx 2$ and $\tau \approx 2.8$. (Exact value from initial-boundary-value problem is $L^2/(\pi^2 \kappa) \approx 2.5$.) Time constant τ from the method-of-lines approximation is determined by the least negative eigenvalue of $\mathbf{A}$.

9. The method-of-lines approximation is $\mathbf{u}' = \mathbf{Au} + \mathbf{f}$ with $\mathbf{A}$ defined by (9.40) and $\mathbf{f} = (\kappa/\Delta x^2)(\,\alpha\ \ 0\ \cdots\ 0\ \beta\,)^T$ (from boundary conditions for T). The components of the initial vector $\mathbf{u}$ are defined by $U_i(0) = (\beta x + \alpha(L - x))/L$, $i = 1, \ldots, n - 1$ (from initial conditions for T).

But the (linear in x) initial condition on T solves $T_t = \kappa T_{xx}$ exactly, and it satisfies the boundary conditions: the initial condition is the equilibrium state. The second-order centered-difference approximation for T_{xx} is certainly exact for linear functions (see section 9.3, exercise 5). Since $\mathbf{u}' = \mathbf{Au} + \mathbf{f}$ is that approximation, $\mathbf{Au}(0) + \mathbf{f} = 0$; that is, $\mathbf{u}(0)$ is the steady-state solution of the method-of-lines approximation and it is equal to the exact solution of this version of (9.35–9.37) at $x = x_i$, $i = 1, \ldots, n - 1$.

11. (a–c) $\Delta x = 2/3$, $L = 2 \Rightarrow n = 3$; $\kappa/\Delta x^2 = 9/4$. Method-of-lines approximation to $T(x, t)$ at $x = 2/3, 4/3$ is $\mathbf{u}' = \mathbf{Au} + \mathbf{f}$, where

$$\mathbf{A} - \frac{9}{4}\begin{pmatrix} -2 & 1 \\ 1 & -2 \end{pmatrix}, \quad \mathbf{f} = \frac{9}{4}\begin{pmatrix} 1 \\ 3 \end{pmatrix}.$$

(Compare example 23, page 494, with $L = 2$.)

Eigenvalues of $\mathbf{A}$ are $-27/4, -9/4$; the dominant time constant for decay predicted by $\mathbf{u}' = \mathbf{Au} + \mathbf{f}$ is $1/(9/4) = 4/9 \approx 0.4444$. From (9.41), the exact time constant is $L^2/(\pi^2 \kappa) = 4/\pi^2 \approx 4/9.87 \approx 0.4053$.

13. The statement is correct: the time constant *predicted* by the *approximation* is that of the slowest decaying term in the general solution of $\mathbf{u}' = \mathbf{Au}$; e.g., see (8.24), page 406. The slowest rate of decay corresponds to the least negative eigenvalue of $\mathbf{A}$. Of course, decay behavior in the finite-difference approximation arises from the decay inherent in the heat equation (9.35).

15. Use (9.2), page 445: $y_x \approx (y_{i+1} - y_{i-1})/2\Delta x$. Then the differential equation and the (zero) boundary conditions are approximated by $\mathbf{y}' = (\mathbf{A} - \mathbf{B})\mathbf{y}$, where $\Delta x = L/n$, $\mathbf{A}$ is

defined by (9.40), page 493, with D in place of κ and

$$\mathbf{B} = \frac{V}{2\Delta x}\begin{pmatrix} 0 & -1 & 0 & \cdots & 0 & 0 & 0 \\ 1 & 0 & -1 & \cdots & 0 & 0 & 0 \\ & & & \ddots & & & \\ 0 & 0 & 0 & \cdots & 1 & 0 & -1 \\ 0 & 0 & 0 & \cdots & 0 & 1 & 0 \end{pmatrix}.$$

Initial conditions are $y_i(0) = M \sin \pi x_i / L$, $i = 1, \ldots, n - 1$. This is a linear, homogeneous, constant-coefficient system. It could model time-dependent diffusion in a river flowing with velocity V.

17. See exercise 15.

Chapter exercises

1. (a) In figure 9.37, page 502, construct a control volume using two vertical lines at $x, x + \Delta x$. The rate of heat flow in at x per unit area is $-kT'(x)$; the rate per unit area of heat flow out at $x + \Delta x$ is $-kT'(x + \Delta x)$. Conservation of energy yields $-kT'(x) + kT'(x + \Delta x) = 0$. Divide by Δx and let $\Delta x \to 0$ to obtain $T'' = 0$. The boundary conditions prescribe the temperature at the left and right walls.

(b) Repeat exercise 1(a), but keep the variable $k(x)$ inside the derivative.

3. $y_g = C_1 e^{-x} + C_2 e^x - 4$; $C_1 = \dfrac{3e^2 - 4e^4}{e^4 + 1}$, $C_2 = \dfrac{4 + 3e^2}{e^4 + 1}$.

5. $y_g = C_1 \sin x + C_2 \cos x$; $y \equiv 0$.

7. $y_g = C_1 x^2 + C_2 x$; $y = x + x^2$.

9. Construct a control volume 1 unit deep, Δx wide, Δy high. Put its lower left corner at (x, y). If the temperature of the wall is uniform in the vertical direction, then heat flows only through the sides of the control volume and not through its top or bottom. The balance of the derivation is then the same as that for an insulated bar with $A(x) = 1 \cdot \Delta y$.

11. The temperature on the left side of the end cap is $T(L, t)$; the temperature on the right is T_L. Hence, as with the house heat-loss model of section 2.3, the rate of heat energy flow through a unit area of this "wall" is $\delta(T(L, t) - T_L)$. Since this heat loss balances the rate of energy flow through the end of the cylinder, we find $-kT_x(L, t) = \delta(T(L, t) - T_L)$. The final boundary condition is $T(L, t) + aT_x(L, t) = b$, with $a = k/\delta$, $b = T_L/\delta$.

13. The eigenfunctions and eigenvalues of this problem are those of exercise 17, section 9.5; $\Theta_n(t) = e^{-\kappa \lambda_n t} \to 0$, $n = 1, 2, \ldots$. Hence,

$$T(x, t) = \sum_{n=0}^{\infty} c_n X_n(x)\Theta_n(t)$$

$$\to \quad c_0 X_0(x)\Theta_0(t) = c_0 = \frac{1}{L}\int_0^L f(x)\,dx.$$

15. The eigenfunctions and eigenvalues are

$$X_m(x) = \sin\frac{(2m-1)\pi x}{2L}, \ \lambda = \left(\frac{(2m-1)\pi}{L}\right)^2,$$

$m = 1, 2, \ldots$. The expansion $\sum_{m=1}^{\infty} c_m X_m(x)$ with

$$c_m = \frac{8}{(2m-1)\pi} \text{ converges in the mean to the odd,}$$

$2L$-periodic extension of $f(x) = 2$ from $0 \leq x \leq L$; i.e., on $-L \leq x \leq L$, the series converges to

$$\begin{cases} 0, & x = -L \\ -2, & -L < x < 0 \\ 0, & x = 0 \\ 2, & 0 < x < L \\ 0, & x = L \end{cases}$$

17. Using the eigenfunctions and eigenvalues of exercise 15, we find $T(x,t) = \sum_{m=1}^{\infty} c_m X_m(x) e^{-\kappa \lambda_m t}$, where c_m is given in the solution to exercise 15. This problem could model the temperature of a thin bar with insulated sides whose left end is at zero temperature and whose right end is insulated.

19. To obtain homogeneous boundary conditions, introduce the equilibrium temperature $E(y)$ defined by $0 = \kappa E''(y)$, $E(0) = T_0$, $E(h) = 0$. We find $E(y) = T_0(h-y)/h$. Then $u(y,t)$ defined by $T(y,t) = u(y,t) + E(y)$ solves the homogeneous problem $u_t = \kappa u_{yy}$, $u(0,t) = u(h,t) = 0$, $u(y,0) = -T_0(h-y)/h$. Mimicking example 19, page 474, (with u replacing T, y replacing x, h replacing L, and the initial condition altered), we find

$$u(y,t) = \sum_{n=1}^{\infty} c_n e^{-\kappa n^2 \pi^2 t/h^2} \sin\frac{n\pi y}{h}, \ c_n = -\frac{2T_0}{n\pi}.$$

(a–c) Let $\Delta x = 1/n$, $x_i = i\Delta x$, $i = 0, \ldots, n$. In (a–c), define the $n-1 \times n-1$ matrix $\mathbf{A}$ by (9.10), page 449, and write the finite-difference approximation to $c(x_1), \ldots, c(x_{n-1})$ as $\mathbf{A}\mathbf{c} = \mathbf{b}$ with

(a) $\mathbf{b} = (\ 2 \ 0 \ \cdots \ 0 \ 4\ )^T$.

(b) $\mathbf{b} = (\ 2 \ 0 \ \cdots \ 0 \ 2\ )^T$.

(c) $\mathbf{b} = \Delta x^2(\ 2 \ 2 \ \cdots \ 2 \ 2\ )^T$.

(d) Approximate $c(x_1), \ldots, c(x_n)$ by $\mathbf{A}\mathbf{c} = \mathbf{b}$ with the $n \times n$ matrix $\mathbf{A}$ as on the top of page 456 (for $n = 4$), $\mathbf{b} = \Delta x^2(2 \ 2 \ \cdots \ 2 \ 2\ )^T$.

23. **(a)** See exercise 15, section 9.6.

(b) The differential equation and boundary conditions are approximated by $\mathbf{y}' = (\mathbf{A} - \mathbf{B})\mathbf{y} + \mathbf{f}(t)$, where $\Delta x = L/n$, $\mathbf{A}$ is defined by (9.40), page 493, with D in place of κ, $\mathbf{B}$ is defined in the solution of exercise 15,

section 9.6, and

$$\mathbf{f}(t) = \begin{pmatrix} x_1(L-x_1)e^{-t} + 5\left(\dfrac{D}{\Delta x^2} - \dfrac{V}{2\Delta x}\right) \\ x_2(L-x_2)e^{-t} \\ \vdots \\ x_{n-2}(L-x_{n-2})e^{-t} \\ x_{n-1}(L-x_{n-1})e^{-t} + 3\left(\dfrac{D}{\Delta x^2} - \dfrac{V}{2\Delta x}\right) \end{pmatrix}$$

Initial conditions: $y_i(0) = M \sin\pi x_i/L$, $i = 1, \ldots, n-1$. Could model diffusion with convection and a source.

(c) The differential equation and boundary conditions are approximated by $\mathbf{y}' = \mathbf{A}\mathbf{y} + \mathbf{f}$, where $\Delta x = L/n$, $\mathbf{y}(t)$ approximates $y(x_i, t)$, $i = 0, \ldots, n-1$,

$$\mathbf{A} = \frac{D}{\Delta x^2}\begin{pmatrix} 2 & -2 & 0 & \cdots & 0 & 0 & 0 \\ -1 & 2 & -1 & \cdots & 0 & 0 & 0 \\ & & & \ddots & & & \\ 0 & 0 & 0 & \cdots & -1 & 2 & -1 \\ 0 & 0 & 0 & \cdots & 0 & -1 & 2 \end{pmatrix}$$

and $\mathbf{f} = (\ 0 \ \cdots \ 0 \ \beta D/\Delta x^2\ )^T$. Initial conditions: $y_i(0) = F(x_i)$, $i = 0, \ldots, n-1$. Could model heat flow in a rod with left end insulated.

(d) The differential equation and boundary conditions are approximated by $\mathbf{y}' = (\mathbf{A} + \mathbf{I}_{n-1})\mathbf{y} + \mathbf{f}(t)$, where $\Delta x = L/n$, $\mathbf{A}$ is defined by (9.40), page 493, with D in place of κ, $\mathbf{I}_{n-1}$ is the $n-1 \times n-1$ identity matrix, and

$$\mathbf{f}(t) = \begin{pmatrix} S(x_1, t) \\ \vdots \\ S(x_{n-1}, t) \end{pmatrix}.$$

Initial conditions: $y_i(0) = F(x_i)$, $i = 1, \ldots, n-1$.

CHAPTER 10

Section 10.1

1. $4/s - 9/(s+4)$.

3. $(\pi + 7)/s$.

5. $2/s^2 - 5/s$.

7. $12/s^4$.

9. $1/s^2 + 2/(s-2)$.

11. $A\omega/(s^2 + \omega^2)$.

13. $1/s^2 + 4/s^3 + 18/s^4 + 96/s^5$.

15. $Y(s) = \dfrac{-2s^3 - 3s^2 + 5s + 10}{s^2(s+2)(2s+7)}$.

17. $Y(s) = \dfrac{\alpha}{(s+1)(s-k)} + \dfrac{y_i}{s-k}$.

19. Integrate $\int_0^\infty te^{-st}\,dt$ by parts using $u = t, dv = e^{-st}\,dt$.

21. Use induction and integration by parts; see 19.

23. $\mathcal{L}\{\sinh t\} = (\mathcal{L}\{e^t\} - \mathcal{L}\{e^{-t}\})/2.$

25. $\mathcal{L}\{\sinh at + \cosh at\} = \mathcal{L}\{e^{at}\} = 1/(s-a)$

27. (a) $-a^3/(s^2 + a^2).$

(b) $a^2/(s-a).$

(c) $6/s^2.$

29. Let $g(t) = f'(t).$ Then
$\mathcal{L}\{g'(t)\} = sG(s) - g(0) = s\mathcal{L}\{f'(t)\} - f'(0) =$
$s(sF(s) - f(0)) - f'(0) = s^2 F(s) - sf(0) - f'(0).$

31. $\dfrac{1}{s} - \dfrac{e^{-2\pi s}}{s} + \dfrac{se^{-2\pi s}}{s^2 + 1} + \dfrac{e^{-7\pi s/2}}{s^2 + 1}.$

33. $\mathcal{A}(cf + dg) = \int_0^s (cf(t) + dg(t))dt =$
$c\int_0^s f(t)dt + d\int_0^s g(t)dt = c\mathcal{A}(f) + d\mathcal{A}(g);$ the Laplace operator is linear.

35. $\mathcal{L}\{ae^{at}\} = \int_0^\infty ae^{at}e^{-st}dt =$
$\lim_{b\to\infty} e^{at}e^{-st}|_{t=0}^{t=b} + s\int_0^\infty e^{at}e^{-st}dt =$
$\lim_{b\to\infty} e^{ab}e^{-sb} - 1 + s\mathcal{L}\{e^{at}\} = s\mathcal{L}\{e^{at}\} - 1.$ Since a is constant, $\mathcal{L}\{ae^{at}\} = a\mathcal{L}\{e^{at}\}$ for all a.

Section 10.2

1. $1 - e^{-4t}.$

3. $e^{4t}/8 - e^{-4t}/8.$

5. $(e^{2t} + 2e^{-t} - 3e^{-2t})/6.$

7. $2e^{4t} - 2e^t.$

9. $e^{4t}/3.$

11. $e^t(\sin 2t - 4\cos 2t)/2.$

13. (a) $1/(s-3)^2.$

(b) $(s+4)/(s^2 + 8s + \pi^2 + 16).$

(c) $4/(s+3/2) + 2/(s+3/2)^3.$

15. $(76/15)e^{6t} - (2\sin 3t + \cos 3t)/15.$

17. $(3e^t - 5)e^{-4t}.$

19. $\cos \pi t/(\pi^2 - 4) + 2\sin 2t + (\pi^2 - 5)\cos 2t/(\pi^2 - 4).$

21. $(2 - 2t)e^{2t}.$

(b) Use the characteristic equation $r^2 + 4r + 40 = 0;$
$r = -2 \pm 6i.$

27. (a) $C_1^2 - 4C_0 \geq 0.$

(b) $A = -1/(r_2 - r_1), B = 1/(r_2 - r_1).$

(c) $\dfrac{e^{r_2 t} - e^{r_1 t}}{r_2 - r_1}.$

Section 10.3

1. (a) $\dfrac{\pi}{s^2 + \pi^2}.$

(b) $\dfrac{1}{1 - e^{-4s}}\left(\dfrac{1 + e^{-4s}}{s} - \dfrac{2e^{-2s}}{s}\right) = \dfrac{1 - e^{-2s}}{s(1 + e^{-2s})}.$

3. (a) $2/(s-a)^3.$

(b) $12\left[\dfrac{1}{(s^2 + \pi^2)^2} - \dfrac{8s^2}{(s^2 + \pi^2)^3} + \dfrac{8s^4}{(s^2 + \pi^2)^4}\right].$

(c) $10a\left(\dfrac{4s^2}{(s^2 - a^2)^3} - \dfrac{1}{(s^2 - a^2)^2}\right).$

5. (a) $\dfrac{1}{2}\ln(\dfrac{s+2}{s-2}).$

(b) $\pi/2 - \arctan(s/2).$

(c) $(1/2)\ln(1 + 1/s^2).$

7. (a) $e^t - 1.$

(b) $-9e^{-2t}/2 - 3t + 9/2.$

9. $\dfrac{a(e^{-sT} - 2e^{-bs} + 1)}{s(1 - e^{-sT})}.$

11. $1/(s-a)^2.$

13. $w/(s^2 - 2as + w^2 + a^2).$

15. $2\omega/(s^2 + \omega^2) - 2\omega s^2/(s^2 + \omega^2)^2.$

17. $(\sin at)/t \leq ae^{at}.$

21. $(A\omega t \sin \omega t + A\cos \omega t)/2\omega^2 + (v_i/\omega - At/2\omega)\sin \omega t + (x_i - 9/\omega^2)\cos \omega t.$

23. $\mathcal{L}\{f\} = \sum_{n=1}^\infty e^{-(n-1)sT}\int_0^T e^{-sr}f(r)\,dr.$

25. $\mathcal{L}\{(-t)^2 f(t)\} = \mathcal{L}\{(-t)(-tf(t))\} =$
$d\mathcal{L}\{-tf(t)\}/ds = F''(s),$ etc.

27. $\int_s^\infty F(r)\,dr = \int_0^\infty \int_s^\infty f(t)e^{-rt}\,dr\,dt =$
$\int_0^\infty (-f(t)e^{-rt}/t)\,|_{r=s}^\infty\,dt = \int_0^\infty f(t)e^{-st}/t\,dt$

Section 10.4

1. $e^{(12-3s)}/(s-4).$

3. $g(t) = \mathcal{U}(t-a) - \mathcal{U}(t-b); G(s) = e^{-as}/s - e^{-bs}/s.$

5. $g(t) = mt - m(t-b)\,\mathcal{U}(t-b) - mb\,\mathcal{U}(t-b);$
$G(s) = m/s^2 - e^{-bs}(m/s^2 + mb/s).$

7. $g(t) = 1 - \mathcal{U}(t-2\pi) + \cos t\,\mathcal{U}(t-2\pi) - \cos t\,\mathcal{U}(t-7\pi/2);$
$G(s) = \dfrac{1}{s} - \dfrac{e^{-2\pi s}}{s} + \dfrac{se^{-2\pi s}}{s^2 + 1} - \dfrac{e^{-7\pi s/2}}{s^2 + 1}.$

9. $g(t) = (2t - 1)(\mathcal{U}(t-0) - \mathcal{U}(t-2)) + 3\,\mathcal{U}(t-2);$
$y = 1 + e^{-2t} + \mathcal{U}(t-2)/2 - e^{-2(t-2)}/2.$

11. $-E(t) = -0.209 + 0.209\,\mathcal{U}(t-4);$
$y = 13.93 - 5.93\,e^{0.015t} + 13.93(e^{0.015(t-4)} - 1)\mathcal{U}(t-4).$

13. $Y(s) = \dfrac{1 - e^{-s}}{s^2(s-k)} + \dfrac{y_i}{s-k}.$

Section 10.5

1. $\delta(t).$

3. $\delta(t) + 2e^{2t}.$

5. $T(t) = be^{-Ak(t-a)/cm};$ b represents the magnitude of the change in T_{out} due to an impulse of heat energy.

7. $i(t) = (V/L)\sin(t - a)/\sqrt{LC}.$

9. (a) $\theta(t) = (b/L)\sin[\sqrt{g/L}\,(t-a)].$

(b) Since b/L is an angle, b is a change in displacement around the circumference of the circle described by the pendulum.

(c) $b = L\theta_i.$

11. $E'(t) = V\delta(t-a);$ see exercise 7 for the solution formula. Choose $V = LI$ to achieve amplitude I.

13. Use $\int_{-\infty}^\infty e^{-st}d_h(t-a)\,dt = \int_{a-h/2}^{a+h/2} e^{-st}d_h(t-a)\,dt,$ etc.

Chapter exercises

1. $\dfrac{m}{s^2} - e^{-bs}(\dfrac{m}{s^2} + \dfrac{mb}{s})$.

3. $(4s^2 + 48)/(s^4 + 40s^2 + 144)$.

5. $\dfrac{4s^2 + 32s + 112}{s^4 + 16s^3 + 136s^2 + 576s + 1040}$.

7. $s^{-2} - 2s^{-3}$.

9. $2s^2/(s^2 + \omega^2)^2 - 1/(s^2 + \omega^2)$.

11. $y = 6e^{-2t} - 1 - 3(e^{-2(t-4)} - 1)\mathcal{U}(t - 4) + 2(1 - e^{-2(t-8)})\,\mathcal{U}(t - 8)$.

13. $X(s) = \dfrac{1}{s(s + 1)} - \dfrac{e^{-2\pi s}}{s(s + 1)} + \dfrac{se^{-2\pi s}}{(s + 1)(s^2 + 1)} - \dfrac{e^{-7\pi s/2}}{(s + 1)(s^2 + 1)} - \dfrac{3}{s + 1}$.

15. $4e^{-2t} - 4e^{-3t}$.

17. $e^{2t} + e^{-2t}$.

19. $2\cos(2(t - 5))\,\mathcal{U}(t - 5)$.

21. $2e^{-t} \sin 2t$.

23. $2e^{-3t} \cos 3t$.

25. $s^2 + 4 \Rightarrow$ sine and/or cosine of frequency 2; $(s - 2)(s + 2) \Rightarrow$ exponentials with factor ± 2; no on-off behavior.

27. No oscillations; $s - 1 \Rightarrow$ exponential with factor 1; $e^{-3s} \Rightarrow$ on-off behavior at $t = 3$.

29. $(e^t \cos t + 2e^t \sin t, -e^t \sin t + 2e^t \cos t)$.

30. $(e^t - e^{-t}, e^t + e^{-t})/2$.

31. $(e^{2t} + 2e^{-3t}, 3e^{2t} + e^{-3t})$.

32. $(-e^{4t} + e^{-4t}, e^{4t} + e^{-4t})$.

33. $(4 - 2\sqrt{3})(e^t, \sqrt{3}e^t/3) + (4 - 6\sqrt{3})(e^{-t}, -\sqrt{3}e^{-t}/3) + (4\sqrt{3}t^2 + 8\sqrt{3} - 2, (8 - 2\sqrt{3}/3)t)$.

35. $(y, z) = (5e^{4t} - e^{2t} + e^{-2t} + 3, 5e^{4t} - 3e^{2t} - e^{-2t} - 1)/8$.

37. $q(t) = 10[h(t) - h(t - 4)\mathcal{U}(t - 4)]$, where $h(t) = 1/L(\omega^2 + a^2) - e^{-at}(a \sin \omega t + \omega \cos \omega t)/\omega L(\omega^2 + a^2)$, $a = R/2L$, $\omega = \sqrt{4L/C - R^2}/2L \approx 9.9499$ rad/μs. The $t - 4$ term reflects the change in $E(t)$ at $t = 4$.

39. $q(t) = 5h(t - 2)\mathcal{U}(t - 2)$, where $h(t)$, a, and ω are as in exercise 37. The $t - 2$ term reflects the change in $E(t)$ at $t = 2$.

41. Since $\mathcal{L}\{q(t)\} = G(s)\mathcal{L}\{E(t)\}$, the convolution theorem yields the desired result. To obtain $g(t)$, note that the roots of the denominator of $G(s)$ are $r = -R/2L \pm i\omega$ and use partial fractions (or complete the square).

CHAPTER 11

Section 11.1

1. x.

3. e^{-4x}.

5. x^2.

7. $x \sin(\ln x)$.

9. $x = (38/63)e^{3t} - (33/56)e^{-4t} - (12t + 1)/72$.

11. Carry out the details as directed.

13. Use (4.8), page 150, and $\int 2y_1'/y_1\,dx = 2 \ln y_1 = \ln y_1^2$.

Section 11.2

1. $W = -x^2 \Rightarrow x^2, x$ are linearly independent for $x \neq 0$.

3. $W = -2x \Rightarrow x \sin(2 \ln |x|), x \cos(2 \ln |x|)$ are linearly independent for $x \neq 0$.

5. $W = -x \Rightarrow x \sin(\ln x), x \cos(\ln x)$ are linearly independent for $x \neq 0$.

7. $W = (5 + \sqrt{5})(t - 3/2)^{(1+\sqrt{5})/4}/8 - (5 - \sqrt{5})(t - 3/2)^{(1-\sqrt{5})/4}/8 \mid_{t=5/2} = \sqrt{5}/4 \neq 0 \Rightarrow (t - 3/2)^{(5+\sqrt{5})/8}$, $(t - 3/2)^{(5-\sqrt{5})/8}$ are linearly independent for $t \neq 3/2$.

9. $x^2 y'' + xy' - y = 0$.

11. $x^2 y'' - 3xy' + 20y = 0$

13. $x \neq 0$.

15. $W(x^{r_1}, x^{r_2}) = (r_2 - r_1)x^{r_1+r_2-1} \neq 0 \Rightarrow$ linearly independent for $x \neq 0, r_1 \neq r_2, r_i$ real or complex.

17. $x > 0 \Rightarrow x^{\alpha+i\beta} = (e^{\ln x})^{\alpha+i\beta} = e^{\alpha x}[\cos(\beta \ln x) + i \sin(\beta \ln x)]$. Find a similar expression for the other root. Form linear combinations to isolate the cosine and sine factors. Use the Wronskian to verify linear independence for $x \neq 0$.

19. (a) See example 5: $x^2 T'' + xT' = 0$, $T = x^r \Rightarrow r^2 = 0 \Rightarrow T_g = C_1 + C_2 \ln x$; $T'(0) = 0 \Rightarrow C_2 = 0$; $T(L) = T_L \Rightarrow C_1 = T_L \Rightarrow T(x) = T_L$.

(b) Circular disk has uniform equilibrium temperature determined by (uniform) temperature at boundary.

21. See exercise 19(a) or example 5: $T_g = C_1 + C_2 \ln x$. Boundary conditions require $C_1 + C_2 \ln r = T_s$, $C_1 + C_2 \ln R = T_{out}$. Solve these equations, for example by Cramer's rule, to obtain $T(x) = (T_s \ln R - T_{out} \ln r)/\ln(R/r) + [(T_{out} - T_S)/\ln(R/r)] \ln x$.

Section 11.3

1. $C_1 e^x + C_2 e^{-x} - 1$.

3. $C_1 e^{3x} + C_2 e^{-3x} - xe^{-3x}/6$.

5. $C_1 e^{3x} + C_2 e^{-3x} + xe^{3x}/12 - xe^{-3x}/12$.

7. $C_1 \sin 3x + C_2 \cos 3x + e^{3x}/9$.

9. $C_1 \sin t + C_2 \cos t + 4t$.

11. $C_1 t^2 + C_2 t^{-3} - (t + 1)/6t$.

13. Could not uniquely solve linear equations defining u_1', u_2'.

15. See example 8.

17. $u_1' = \dfrac{-y_2 r(x)}{a_2(x)W}$; $u_2' = \dfrac{y_1 r(x)}{a_2(x)W}$, where $W = y_1 y_2' - y_1' y_2$.

Section 11.4

1. $\sum_{n=0}^{\infty} c_n n(x - x_0)^{n-1} = \sum_{n=1}^{\infty} c_n n(x - x_0)$.

5. (a) $x + 2$ is analytic at $x = -2$.

(b) $y(x) = c_0 + c_1(x + 2) - c_0(x + 2)^3/6 - c_1(x + 2)^4/12$.

7. (a) Constant coefficients are analytic everywhere.

(b) $y = -1 + 4(x - 1) - (x - 1)^2/2 + 2(x - 1)^3/3$.

9. (a) x^2 is analytic at $x = 0$.

(b) $y = c_0 - c_0 x + (1/2)c_0 x^2 - (1/6)c_0 x^3$.

11. $f''(x_0) = \sum_{n=2}^{\infty} c_n n(n - 1)(x - x_0)^{n-2}\big|_{x=x_0} = 2c_2$, etc.

13. (a) Let $\cos x = f(x)$ and $\sin x = f(x)$. Substitute into $c_n = f^{(n)}(x_0)/n!$ for $x_0 = 0$.

(b) Let $y_1 = \cos x$ and $y_2 = \sin x$.

(c) $y_g = C_1 y_1 + C_2 y_2$.

(d) Compute $(1 - x^2/2 + \cdots)'' = -1$, etc. and substitute into differential equation.

(e) $y(0) = C_1$, $y'(0) = C_2$.

15. Using full series permits determination of radius of convergence.

19. $f(x) = a_0 + a_1(x - x_0) + a_2(x - x_0)^2 + \cdots \Rightarrow$ $f'(x_0) = a_1 + 2a_2(x - x_0) + \cdots$.

Section 11.5

1. (i-ii) All real x, $x \neq 0$, are finite ordinary points; $x = 0$ is a finite regular singular point;

(iii) $r^2 + 3r - 4 = 0 \Rightarrow r_1 = 1, r_2 = -4$.

3. (i-ii) $x = \pm 2$ are finite regular singular points; all other real x are ordinary points;

(iii) For $x = \pm 2$, $r^2 + r = 0 \Rightarrow r_1 = 0, r_2 = -1$.

5. (i-ii) $x = 0$ is a regular singular point; all other real x are ordinary points;

(iii) $r^2 - 1/16 = 0 \Rightarrow r_1 = 1/4, r_2 = -1/4$.

7. $y_1 = x^{-1}$ and $y_2' = x^{-1/2}$ do not exist at $x = 0$.

9. $P = 1/x$, $Q = 1 - p^2/x^2$ do not exist at $x = 0$, but $xP = 1$ and $x^2 Q = x^2 - p^2$ are analytic at $x = 0$.

13. $y_1 = 1 - x/2 + x^2/24 + \cdots$, $y_2 = x^{1/2}(1 + x/20 + x^2/42 + \cdots)$.

15. $y_1 = x^{1/4}(1 - x/5 + x^2/90)$, $y_2 = x^{-1/4}(1 - x/5 + x^2/90)$.

17. $y_1 = (x - 2)^{2/3}(1 + (x - 2)/5 + (x - 2)^2/40 + \cdots)$, $y_2 = x - 2 + (x - 2)^2/8 + \cdots$.

19. $x = 0$ is not a regular singular point since $x^2 Q = 1/x$ is not analytic at $x = 0$; $c_0, c_1, c_2, \ldots = 0$.

21. When $n = -2r$, recurrence relation becomes $c_n \cdot 0 + c_{n-2} = 0$, contradicting $c_{n-2} \neq 0$. A different solution form is required.

23. $4^{m+1}[(m + 1)!]^{m+1}/4^m (m!)^m = 4(m + 1)!(m + 1)^m \to \infty \Rightarrow J_0$ series converges for all x.

Chapter exercises

1. $(x + 1)^2(C_2 \ln(x + 1) + C_1)$.

3. $C_1 \sin t + C_2 \cos t - (t \cos t)/2$.

5. $C_1 t^2 + C_2 t^{-3} - 1/6$.

7. $y = 2 - (x - 1)^2 + (x - 1)^3/3 - (x - 1)^4/12 + \cdots$.

9. $y = (4x^2 - 4) \ln(x - 1) + 2 + C_1(x - 1)^2 + C_2(x - 1)$.

11. $x = 0$ is a regular singular point; all other x are ordinary points.

13. $x = 0$ is a regular singular point; all other x are ordinary points.

15. $x = \pm 1$ are regular singular points; all other x are ordinary points.

17. $x = 0$ is a regular singular point; all other x are ordinary points.

19. All x are ordinary points.

21. All x are ordinary points.

23. $y = c_0 \left(1 - x^3/6 + x^6/180 + \cdots\right) + c_1 \left(x - x^4/12 + x^7/504 + \cdots\right)$.

25. (a) Using chain rule, substitute $X(x) = J_0(\sqrt{\lambda}x)$ into differential equation.

(b) The boundary condition $X(L) = J_0(\sqrt{\lambda}L) = 0$ forces $\sqrt{\lambda}L = 2.40, 5.52, 8.65, 11.79, \ldots$, the zeros of J_0. Hence, $\lambda = (2.40/L)^2, \ldots$ $= 5.76/L^2$, $30.47/L^2$, $74.82/L^2$, $139.00/L^2, \ldots$. Eigenvalues of $X'' + \lambda X = 0$ with the same boundary conditions are defined by $\cos \sqrt{\lambda}L = 0$; $\lambda = n^2\pi^2/L^2 = 9.87/L^2$, $39.48/L^2$, $88.83/L^2$, $157.91/L^2, \ldots$.

APPENDIX

1. (a) $\cos \pi x \exp\left(\int_0^x \cos \pi s \, ds\right)$.

(b) $(x + 1)^3 \exp\left(\int_{-1}^x (s + 1)^3 \, ds\right)$.

3. Let $\dfrac{1}{ar - sr^2} = \dfrac{A}{r} + \dfrac{B}{a - sr}$. Then $A = 1/a$, $B = s/a$. Using logarithms, integral is $\dfrac{1}{a} \ln\left(\dfrac{(a - s)x}{a - sx}\right)$.

5. (a) $\sin x = 1 - \dfrac{(x - \pi/2)^2}{2!} + \dfrac{(x - \pi/2)^4}{4!} + R_5(\xi)$.

(b) $R_5(\xi) = \dfrac{(\cos \xi)(x - \pi/2)^5}{5!}$; $|R_5| \leq |x - \pi/2|^5/120$. Maximum approximation error is $2.6 \cdot 10^{-4}$ if $|x - \pi/2| < 0.5$, $8.1 \cdot 10^{-6}$ if $|x - \pi/2| < 0.25$. Reduction factor is $2^5 = 32$.

(c) $y^{(4)}(x) = 4^4 \cos 4x$, $y^{(5)}(x) = -4^5 \sin 4x$,

$$\cos 4x = 1 - \frac{16}{2!}x^2 + \frac{256}{4!}x^4 + R_5(\xi)$$

$$= 1 - 8x^2 + \frac{32}{3}x^4 + R_5(\xi),$$

$$R_5(\xi) = -\frac{4^5 \sin \xi}{5!}x^5 = -\frac{128 \sin \xi}{15}x^5.$$

$|R_5| \leq 128|x|^5/15 \leq 8.5 \cdot 10^{-5}$ if $|x| \leq 0.1$, $|R_5| \leq .27$ if $|x| \leq 0.5$.

7. (a) From $\det \begin{pmatrix} 3 & -2 \\ 4 & 2 \end{pmatrix} = 3 \cdot 2 - 4 \cdot (-2) = 14$,

Cramer yields

$$x = \frac{1}{14} \det \begin{pmatrix} 13 & -2 \\ 8 & 2 \end{pmatrix} = 3,$$

$$y = \frac{1}{14} \det \begin{pmatrix} 3 & 13 \\ 4 & 8 \end{pmatrix} = -2.$$

(b) $\mathbf{A} = \begin{pmatrix} 3 & -2 \\ 4 & 2 \end{pmatrix}, \quad \mathbf{b} = \begin{pmatrix} 13 \\ 8 \end{pmatrix}.$

(c) Multiply $\mathbf{A}(\, 3 \quad -2\,)^T$.

(d) $\mathbf{B} = \mathbf{A}^{-1} = \begin{pmatrix} 1/7 & 1/7 \\ -2/7 & 3/14 \end{pmatrix}.$

(e) See example 10, page 632, for $\mathbf{A}^{-1}$.

(f) $\mathbf{A}^{-1}\mathbf{b} = \begin{pmatrix} 2 & -1/2 \\ -1 & 1/2 \end{pmatrix} \begin{pmatrix} -1 \\ -8 \end{pmatrix} = \begin{pmatrix} 2 \\ -3 \end{pmatrix} = \mathbf{x}.$

9. (a) $W = (r_2 - r_1)e^{(r_1 + r_2)t}$; $W \neq 0$ if and only if $r_1 \neq r_2$.

(b) $W = 2re^{rt} \neq 0$.

(c) $W = -\beta(\sin^2 \beta t + \cos^2 \beta t) = -\beta \neq 0$.

(d) $W = -\beta e^{2\alpha t}(\sin^2 \beta t + \cos^2 \beta t) = -\beta e^{2\alpha t} \neq 0$.

11. (a) To have exactly one solution, the determinant of coefficients $1 + a^2$ must be nonzero. (That solution is $C_1 = C_2 = 0$.)

(b) If $1 + a^2 = 0$, then there are infinitely many solutions.

12. $\dfrac{2s + 1}{s^2 + 1} - \dfrac{2}{s - 1} + \dfrac{1}{(s - 1)^2}.$

13. $\dfrac{s}{s^2 + 1} + \dfrac{1}{s + 1} - \dfrac{1}{s - 1}.$

Index

Fourier or eigenfunction expansion for IBVP: Assume $T(x, t) = X(x)\Theta(t)$ and substitute in $T_t(x, t) = T_{xx}(x, t)$, for example. Separate into $\Theta' = -\lambda\Theta(t)$ and $X'' + \lambda X = 0$ plus boundary conditions (e.g., p. 470-471). Find eigenfunctions X_n, eigenvalues λ_n (p. 472). Solve for $\Theta_n(t)$ (p. 473). Choose c_n in $T(x, t) = \sum_n c_n\Theta_n(t)X_n(x)$ to satisfy initial conditions (p. 473, 478).

Laplace transform: $\mathcal{L}\{f\} = \int_0^\infty f(t)e^{-st}\, dt$. Transform differential equation, solve for transform of unknown (e.g., example 3, p. 510), identify solution behavior or find solution formula using transform pairs and properties (e.g., section 10.2.1, p. 520-524). See accompanying tables.

Laplace transform pairs

$f(t)$	$F(s) = \mathcal{L}\{f(t)\}$
1	$\dfrac{1}{s}$
$t^n, n = 1, 2, \ldots$	$\dfrac{n!}{s^{n+1}}$
e^{at}	$\dfrac{1}{s - a}, s > a$
$\sin at$	$\dfrac{a}{s^2 + a^2}, s > 0$
$\cos at$	$\dfrac{s}{s^2 + a^2}, s > 0$
$\sinh at$	$\dfrac{a}{s^2 - a^2}, s > a$
$\cosh at$	$\dfrac{s}{s^2 - a^2}, s > a$
$\delta(t - a)$	$e^{-as}, a > 0$

Properties of Laplace Transforms

$$F(s) = \mathcal{L}f(t), \quad G(s) = \mathcal{L}g(t)$$

Transform of a derivative:

$$\mathcal{L}\{f'(t)\} = sF(s) - f(0)$$
$$\mathcal{L}\{f''(t)\} = s^2F(s) - sf(0) - f'(0)$$

Convolution theorem:

$$\mathcal{L}\left\{\int_0^t f(r)g(t - r)\, dr\right\} = F(s)G(s)$$

Shifted transform:

$$\mathcal{L}\{e^{at} f(t)\} = F(s - a)$$

Transform of a periodic function of period T:

$$\mathcal{L}\{f\} = \frac{1}{1 - e^{-sT}} \int_0^T e^{-st} f(t)\, dt$$

Derivative of a transform:

$$\mathcal{L}\{-tf(t)\} = F'(s)$$
$$\mathcal{L}\{(-t)^n f(t)\} = F^{(n)}(s), \quad n = 1, 2, \ldots$$

Transform of an integral:

$$\mathcal{L}\left\{\int_0^t f(r)\, dr\right\} = \frac{F(s)}{s}$$

Integral of a transform:

$$\mathcal{L}\left\{\frac{f(t)}{t}\right\} = \int_s^\infty F(r)\, dr$$

Transform of a shifted function using unit step function $\mathcal{U}$:

$$\mathcal{L}\{f(t - a)\mathcal{U}(t - a)\} = e^{-as} F(s), \quad a > 0$$

Analytic tools: page references

	1st-order DE	2nd-order DE	1st-order system
Separation of variables	150	—	—
Characteristic equation	158	273	393
Undetermined coefficients	167	293–294	421
Variation of parameters	180	587	419–420
Reduction of order	—	572	—
Cauchy-Euler	—	579	—
Laplace	510	523	563
Series	—	601, 608	—
Fourier	Partial DE: p. 473, 488		
Linear stability analysis	123	246–247, 326	414–415